Carpentry

This book is dedicated to my wife, Lorraine,
for her help, encouragement, and understanding.

Carpentry

Gaspar Lewis

Building for Success
& Section Reviews
Contributed by
Larry Kness

Delmar Publishers

I(T)P An International Thomson Publishing Company

Albany • Bonn • Boston • Cincinnati • Detroit • London • Madrid • Melbourne
Mexico City • New York • Pacific Grove • Paris • San Francisco • Singapore • Tokyo
Toronto • Washington

NOTICE TO THE READER

Cover Photos: Randall Perry Photography

Delmar Staff:
Publisher: Susan Simpfenderfer
Administrative Editor: Mark Huth
Developmental Editor: Jeanne Mesick
Project Editor: Patricia Konczeski
Art/Design Coordinator: Cheri Plasse
Production Coordinator: Dianne Jensis
Marketing Manager: Lisa Reale

COPYRIGHT © 1995
By Delmar Publishers
a division of International Thomson Publishing Inc.

The ITP logo is a trademark under license

Printed in the United States of America

Delmar Publishers' Online Services
To access Delmar on the World Wide Web, point your browser to:
http://www.delmar.com/delmar.html
To access through Gopher: gopher://gopher.delmar.com
(Delmar Online is part of "thomson.com", an Internet site with information on more than 30 publishers of the International Thomson Publishing organization.)
For information on our products and services:
email: info@delmar.com
or call 800-347-7707

For more information, contact:

Delmar Publishers
3 Columbia Circle, Box 15015
Albany, New York 12212-5015

International Thomson Publishing Europe
Berkshire House 168 - 173
High Holborn
London WC1V7AA
England

Thomas Nelson Australia
102 Dodds Street
South Melbourne, 3205
Victoria, Australia

Nelson Canada
1120 Birchmount Road
Scarborough, Ontario
Canada M1K5G4

International Thomson Editores
Campos Eliseos 385, Piso 7
Col Polanco
11560 Mexico D F Mexico

International Thomson Publishing GmbH
Königswinterer Strasse 418
53227 Bonn
Germany

International Thomson Publishing Asia
221 Henderson Road
#05 - 10 Henderson Building
Singapore 0315

International Thomson Publishing - Japan
Hirakawacho Kyowa Building, 3F
2-2-1 Hirakawacho
Chiyoda-ku, Tokyo 102
Japan

5 6 7 8 9 10 X X X 01 00 99 98

Library of Congress Cataloging-in-Publication Data

Lewis, Gaspar J.
    Carpentry / Gaspar Lewis. — 2nd ed.
      p.   cm
    Includes index.
    ISBN 0-8273-5979-9
    1. Carpentry. I. Title.
TH56.6.L48   1994
    694--dc20
                                    94-8421
                                CIP

CONTENTS

Preface

This textbook is designed, primarily, for use by carpentry students in secondary and postsecondary vocational programs, and by on-the-job carpentry apprentices. Carpentry is also a great resource for anyone interested in learning about the different aspects of carpentry as a trade. It is written by an experienced carpenter, who has kept abreast of new materials, tools and techniques by working at the trade, and has incorporated the best of both the old and the new in the text. It is the aim of the author that the book be used as an aid to help students reach a level of competency in duties up to, and including that of foreman, in the field of carpentry.

Chapters on Engineered Panels and Engineered Lumber, building materials of the future, manufactured with little or no waste, are in this edition. Also, because of their growing popularity, description of the use of cordless tools, including hoseless nailers and staplers are included, as are other tools relatively new to the industry, such as the compound miter saw and the laser level. Concrete forming systems of aluminum, wood, and steel are discussed, and construction methods to control termites, fungi, and other wood destroying pests are expanded upon. Also added, roof framing, a chapter on building porches, decks, and fences, and describes the installation of new millwork products such as circular, elliptical, arched, and other geometrically shaped windows.

Throughout the book, in appropriate locations pertinent to the text, illustrated "Helpful Hints" appear, which are techniques that the author has acquired from almost fifty years of experience to help make the job at hand easier, faster, finer, and more accurate.

The book is divided into four sections. Section One discusses basic and new tools and materials, including engineered panels and engineered lumber. Section Two deals with rough carpentry consisting of building layout, concrete forms, scaffolding, insulation, and floor, wall, stair, and roof framing. Section Three explains how to apply exterior finish such as cornices, roofing, doors, windows, siding, and how to construct porches, decks, and fences. Section Four presents information on the application of interior finish such as drywall, paneling, ceiling tile and suspended ceilings, window and door trim, baseboard, stair finish, ceramic tile, floor tile, wood finish floors, and kitchen and bath cabinets.

Each unit in the book begins with a list of objectives that explain exactly what the student is expected to accomplish. Material is presented through the use of information, illustrations, and step-by-step procedures. This three-fold approach gives the student many opportunities to grasp the material. After reading the explanations, students gain practical, hands-on experience by following the highly illustrated step-by-step procedures. A "Building for Success" section occurs at the end of each unit. This feature is intended to stimulate thought-provoking discussion or to present issues that face today's carpenter. Readers will become familiar with issues, methods, and procedures necessary for success in the carpentry vocation.

The "Student Activity" feature helps students apply their knowledge. These activities encourage students to conduct research, present new information, and solve a variety of problems in a practive environment either individually, in small groups, or in a class setting. These activities will help students develop their construction-related skills.

A sixteen-page color insert details the stages of a residential construction project and outlines career opportunities associated with each step. This photo section focuses on the necessary processes and procedures every carpenter needs to know.

Terms appear in italics and are defined both on first use in the text and in the Glossary at the end of the book. A multiple choice test at the end of each unit measures the student's knowledge of the material.

Additional study questions and suggested activities are provided in a Workbook. The suggested activities may be used to supplement the text material by giving students extra work to generalize their knowledge in the same area, or may be used to complement the material presented in the unit. The workbook correlates directly with the text. Exercises and activities focus on critical thinking skills, planning, preparation, safety, cutting, assembly, tool use, plan reading, communications, calculations, and site cleanup.

The Instructor's Resource Guide suggests further methods of teaching carpentry. A correlation guide will help instructors make the switch from the text they are currently using to this one. It offers a significant number of resource materials that will help facilitate classroom instruction—instructor activities, lesson plans, student activities and projects, suggested readings, worksheets, and sources of literature and visual aids. Transparency masters taken from art in the text will facilitate presentation. A separate solutions manual to all the questions in the text and workbook is also available.

Acknowledgments

The author and Delmar Publishers, gratefully acknowledge the contributions of the following focus group participants and text reviewers:

Ronald Norrie, Mid-Plains Community College, North Platte, NE
Bruce Purdy, Grove City HS, Grove City, OH
Terry L. Smith, Dorchester HS, Dorchester, NE
Robert A. Berry, Hartnell College, Salinas, CA
Andrew Wermes, United Brotherhood of Carpenters and Joiners of America, Washington, D.C.
Larry Kness, Southeast Community College, Milford, NE
Timothy W. Lockley, George Jr Republic Vo Tech, Grove City, PA
Michael C. Davis, Henry Ford Community College, Dearborn Heights, MI
Kathy Suthphin, Bates Technical College, Tacoma, WA
Robert Richardson, Franklin County Technical School, Turners Falls, MA
Jeffry Lew, Purdue University, West Lafayette, IN
John Weeks, Eastern Michigan Univ, Ypsilanti, MI

Several contributing authors also need to be recognized for their critical roles in this project. Our appreciation to Larry Kness for contributing the Building For Success features and the end of section activities, Kathy Sutphin who created the Instructor's Resource Kit and Tim Lockley who wrote the workbook.

Special thanks to Randall Perry and John Garrett of Randall Perry Photography for providing many of the new, outstanding photos throughout the text and all the photos in the color insert. Their patience and skills in coordinating the numerous photo shoots to complete this project are sincerely appreciated. We would also like to thank the businesses who were so accommodating in allowing us in to their work places to complete the photography:

Home Quarters, Latham, NY
The Michaels Group, Latham, NY
Charlew Construction, Schenectady, NY
Morizio Builders, Schaghticoke, NY
Marini Builders, Albany, NY
Masullo Brothers Builders Inc, Schenectady, NY
Moak Builders Inc, Albany, NY
Sheltertherm Industries, Cohoes, NY
Town of Colonie Building Department, Latham, NY

Tri City Saw & Tool Co, Schenectady, NY
Daigle Drywall & Enterprises Inc, Latham, NY
Charles E. Hartnett Land Surveyor & Planning, Melrose, NY
Dave Valcik Construction, Clifton Park, NY
Harbrook Windows, Albany, NY
C. T. Male Associates, Latham, NY

Delmar Publishers would also like to acknowledge Carl Simpfenderfer for his photo contributions and his consulting services for the art program. And thanks to the many businesses and organizations who provided technical information, photographs and illustrations.

Introduction

The history of carpentry goes back to 8000 B.C., when primitive people used stone axes to shape wood to build shelters. Stone Age Europeans built rectangular timber houses more than 100 feet long—proving the existence of carpentry even at this early date. The Egyptians used copper woodworking tools as early as 4000 B.C. By 2000 B.C. they had developed bronze tools and were proficient in the drilling, dovetailing, mitering, and mortising of wood.

In the Roman Empire, two-wheeled chariots, called *carpentum* in Latin, were made of wood. A person who built such chariots was called a *carpentarius*, from which the English word *carpenter* is derived. Roman carpenters handled iron adzes, saws, rasps, awls, gouges, and planes.

During the Middle Ages, most carpenters were found in larger towns where work was plentiful. They would also travel with their tools to outlying villages or wherever there was a major construction project in progress. By this time, they had many efficient, steel-edged hand tools. During this period skillful carpentry was required for the building of timber churches and castles.

In the twelfth century, carpenters banded together to form *guilds*. The members of the guild were divided into *masters, journeymen*, and *apprentices*. The master was a carpenter with much experience and knowledge who trained apprentices. The apprentice lived with the master and was given food, clothing, and shelter and worked without pay. After a period of five to nine years, the apprentice became a journeyman who could work for wages. Eventually, a journeyman could become a master. Guilds were the forerunners of the modern labor unions and associations.

Starting in the fifteenth century, carpenters used great skill in constructing the splendid buildings of the Renaissance period and afterward. With the introduction of the balloon frame in the early nineteenth century, more modern construction began to replace the slower mortise-and-tenon frame. In 1873 electric power was used for the first time to drive machine tools. The first electric hand drill was developed in 1917, and in 1925 electric portable saws were being used.

At present, many power tools are available to the carpenter to speed up the work. Although the volume of the carpenter's work has been reduced by the use of manufactured parts, some of the same skills carpenters used in years past are still needed for the intricate interior finish work in buildings.

Carpenters construct and repair structures and their parts using wood, plywood, and other building materials. They lay out, cut, fit, and fasten the materials to erect the framework and apply the finish. They build houses, factories, banks, schools, hospitals, churches, bridges, dams, and other structures. In addition to new construction, a large part of the industry is engaged in remodeling and repair of existing buildings.

The majority of workers in the construction industry are carpenters (Fig. I-1). They are the first trade workers on the job, laying out excavation and building lines. They take part in every phase of the construction, working below the ground, at ground level, or at great heights. They are the last to leave the job when they put the key in the lock.

Specialization

In large cities, where there is a great volume of construction, carpenters tend to specialize in one area of the trade. They may be specialists in rough carpentry and are called rough carpenters or framers (Fig. I-2). Rough carpentry does not mean that the workmanship is crude. Just as much care is taken in the rough work as in any other work. Rough carpentry will be covered eventually by the finish work or dismantled, as in the case of concrete form construction.

Finish carpenters specialize in applying exterior and interior finish, sometimes called trim (Fig. I-3). Other specialties are constructing concrete forms (Fig. I-4), laying finish flooring, building stairs, applying gypsum board (Fig. I-5), roofing, insulating, and installing suspended ceilings.

Fig. I-1 Carpenters make up the majority of workers in the construction industry.

Fig. I-2 Some carpenters specialize in framing.

Fig. I-4 Erecting concrete formwork may be a speciality.

Fig. I-3 Carpenters may be specialists in applying interior trim. (*Courtesy of Senco Products, Inc.*)

Fig. I-5 Some carpenters choose to do only drywall application.

In smaller communities, where the volume of construction is lighter, carpenters perform tasks in all areas of the trade from the rough to the finish. The general carpenter needs a more complete knowledge of the trade than the specialist does.

Requirements

Carpenters need to know how to use and maintain hand and power tools. They need to know the kinds, grades, and characteristics of the materials with which they work— how each can be cut, shaped, and most satisfactorily joined. Carpenters must be familiar with the many different fasteners available and choose the most appropriate for each task.

Carpenters must know how to lay out and frame floors, walls, stairs, and roofs. They must know how to install windows and doors, and how to apply numerous kinds of exterior and interior finish. They must use good judgment to decide on proper procedures to do the job at hand in the most efficient and safest manner.

Carpenters must be in good physical condition because much of the work is done by hand and sometimes requires great exertion. They must lift large sheets of plywood, heavy wood timbers, and bundles of roof shingles; they also have to climb ladders and scaffolds.

An attitude of care and concern for the job and for other workers is vital. The carpenter must feel that nothing less than a first-class job is acceptable.

Training

Vocational training in carpentry is offered in many high schools for those who become seriously interested at an early age. For those who have completed high school, carpentry training programs are offered at many postsecondary vocational schools and community colleges. Most of these programs train students for duties up to and including that of foreman. Because there is so much more to learn, vocational school training should then be followed by an apprenticeship.

Some schools participate in programs in which students go to school part-time and work as on-the-job apprentices part-time. Upon graduation, they usually continue with the same employer as full-time apprentices.

Apprenticeship training programs (usually four years in length) are offered by the United Brotherhood of Joiners and Carpenters of America, a carpenters' union, and by contractors' associations such as the National Association of Home Builders of the United States (NAHB), and the Associated General Contractors of America (AGC) in cooperation with the Bureau of Apprenticeship and Training of the U. S. Department of Labor (Fig. I-6). Usually these organizations give apprenticeship credit for completion of previous vocational school training in carpentry, resulting in a shortened apprenticeship and a higher starting wage.

The apprentice (Fig. I-7), must be at least sixteen years of age and must learn the trade while working on the job with experienced journeymen. Under the apprenticeship agreement, basic standards provide, among other things, that there can be only a certain number of apprentices hired by a particular contractor in relation to the number of journeymen. This is to insure that apprentices receive proper supervision and training on the job.

Basic standards also provide that there is a progressively increasing schedule of wages. Starting pay for the apprentice carpenter is usually about 50 percent of the journeyman carpenter's rate. Due to periodic increases about every six months, the apprentice should receive 95 percent of the journeyman's wage during the last six months of the apprenticeship.

APPRENTICESHIP AND TRAINING SYSTEM OF THE CARPENTRY TRADE

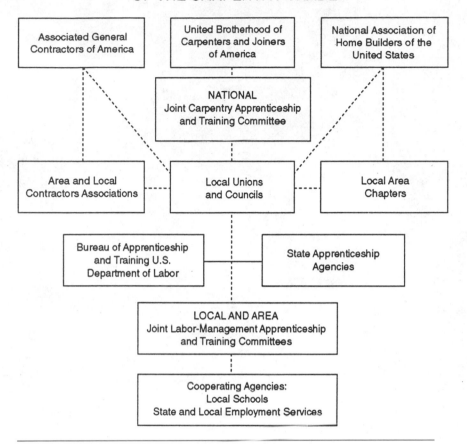

Fig. I-6 Industry, labor, and government work together in carpentry apprenticeship programs.

Under the terms of agreement, the apprentice is required to attend classes for a certain number of hours each year. A minimum of 144 hours per year is normally considered necessary. These classes are usually held in the evening at local schools, twice a week for about thirty-six weeks during the school year, for the length of the apprenticeship. The apprentice becomes accepted as a journeyman carpenter when the training is completed and is awarded a Certificate of Completion of Apprenticeship. The newly graduated apprentice is now expected, within reason, to do jobs required of the journeyman.

Although it is in the best interests of the apprentice to be indentured, that is, to have a written contract with an organization, with conditions of the apprenticeship agreed upon, it is possible to learn the trade as a helper until enough skills have been acquired to demand the recognition and rewards of a journeyman carpenter. However, self-discipline is required to gain knowledge of both the practical and theoretical aspects of the trade. There may be time on the job to explain to a helper or apprentice how to do a certain task, but there usually is not time to explain concepts.

Fig. I-7 Many opportunities lie ahead for the apprentice carpenter.

Fig. I-8 Construction superintendents usually rise from the ranks of the carpenters. (*Courtesy of Brotherhood of Carpenters and Joiners of America*)

Many opportunities exist for the journeyman carpenter. Advancement depends on dependability, skill, productivity, and ingenuity, among other characteristics. Carpentry foremen, construction superintendents, and general contractors usually rise from the ranks of the journeymen carpenters (Fig. I-8). Many who start as apprentice carpenters eventually operate their own construction firms. A survey revealed that 90 percent of the top officials (presidents, vice presidents, owners, and partners) of construction companies who replied began their careers as apprentices. Many of the project managers, superintendents, and craft supervisors employed by these companies also began as apprentices.

Summary

Carpentry is a trade in which there is a great deal of self-satisfaction, pride, and dignity associated with the work. It is an ancient trade and the largest of all trades in the building industry.

Skilled carpenters who have labored to the best of their ability can take pride in their workmanship whether the job was a rough concrete form or the finest finish in an elaborate staircase (Fig. I-9). At the end of each working day, carpenters can stand back and actually see the results of their labor. As the years roll by, the buildings that carpenters' hands had a part in creating still can be viewed with pride in the community.

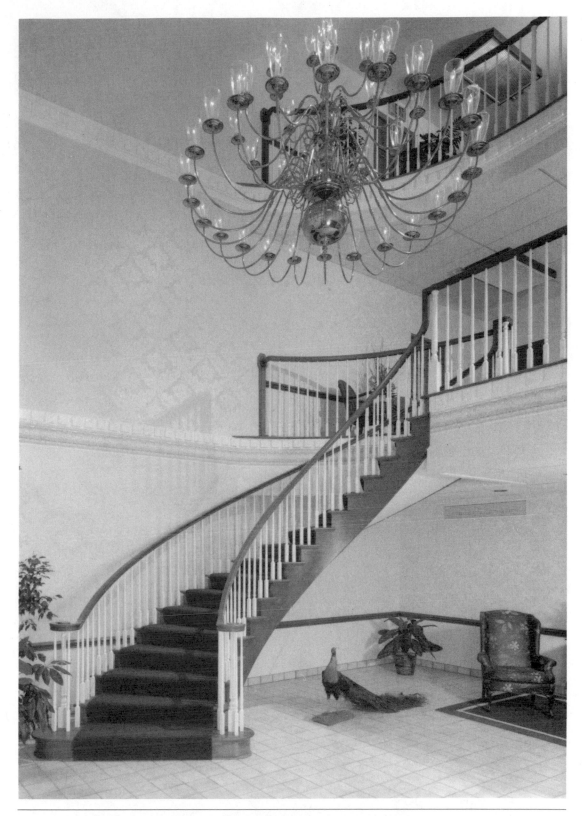

Fig. I-9 After completion of a complicated piece of work, such as this intricate staircase, the carpenter can take pride in and view his or her accomplishment. (*Courtesy of L. J. Smith, Inc.*)

Tools and Materials

UNIT

Wood and Lumber

OBJECTIVES

After completing this unit, the student will be able to:

- name the parts of a tree trunk and state their function.

- describe methods of cutting the log into lumber.

- define hardwood and softwood, give examples of some common kinds, and list their characteristics.

- explain moisture content at various stages of seasoning, tell how wood shrinks, and describe some common lumber defects.

- state the grades and sizes of lumber and compute board measure.

UNIT CONTENTS

An understanding of the nature of wood and the characteristics of lumber is necessary for all those who work with it. With this knowledge, the carpenter is able to protect lumber from decay, select it for appropriate use, work it with proper tools, and join and fasten it to the best advantage.

CHAPTER 1 WOOD

The carpenter works with wood more than any other material and must understand its characteristics in order to use it intelligently. Wood is a remarkable substance. It can be cut, shaped, or bent into just about any form. It is an efficient insulating material. It takes almost 6 inches of brick, 14 inches of concrete, or over 1,700 inches of aluminum to equal the insulating value of only one inch of wood.

There are many kinds of wood that vary in strength, workability, elasticity, color, grain, and texture. It is important to keep these qualities in mind when selecting wood. For instance, baseball bats, diving boards, and tool handles are made from hickory and ash because of their greater ability to bend without breaking (elasticity). Oak and maple are used for floors because of their hardness and durability. Redwood, cedar, cypress, and teak are used in exterior situations because of their resistance to decay. Cherry, mahogany, and walnut are typically chosen for their beauty.

With proper care, wood will last indefinitely. It is a material with beauty and warmth that has thousands of uses. Wood is one of our greatest natural resources. With wise conservation practices, wood will always be in abundant supply. It is fortunate that we have perpetually producing forests that supply this major building material to construct homes and other structures that last for hundreds of years. When those structures have served their purpose and are torn down, the wood used in their construction can be salvaged and used again (recycled) in new building, remodeling, or repair. Wood is biodegradable and when it is considered not feasible for reuse, it is readily absorbed back into the earth with no environmental harm.

Structure and Growth

Wood is made up of many hollow cells held together by a natural substance called *lignin*. The size, shape, and arrangement of these cells determine the strength, weight, and other properties of wood. Tree growth takes place in the *cambium layer*, which is just inside the protective shield of the tree called the *bark*. The tree's roots absorb water that passes upward through the *sapwood* to the leaves, where it is combined with carbon dioxide from the air. Sunlight causes these materials to change into food, which is then carried down and distributed toward the center of the trunk through the *medullary rays*.

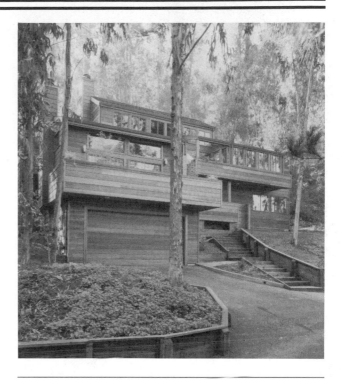

Fig. 1-1 Redwood is often used for exterior siding. (*Courtesy of California Redwood Association*)

As the tree grows outward from the *pith* (center), the inner cells become inactive and turn into *heartwood*. Heartwood is the central part of the tree and is usually darker in color and more durable than sapwood. The heartwood of cedar, cypress, and redwood, for instance, is extremely resistant to decay and is extensively used for outdoor furniture, patios, and exterior siding (Fig. 1-1). Used for the same purposes, sapwood decays more quickly.

Each spring and summer, a tree adds new layers to its trunk. Wood grows rapidly in the spring; it is rather porous and light in color. In summer, tree growth is slower; the wood is denser and darker, forming distinct rings. Because these rings are formed each year, they are called *annular rings* (Fig. 1-2). By counting the dark rings, the age of a tree can be determined. By studying the width of the rings, periods of abundant rainfall and sunshine or periods of slow growth can be discerned. Some trees, like the Douglas fir, grow rapidly to great heights and have very wide and pronounced annular rings. Mahogany, which grows in a tropical climate where the weather is more constant, has

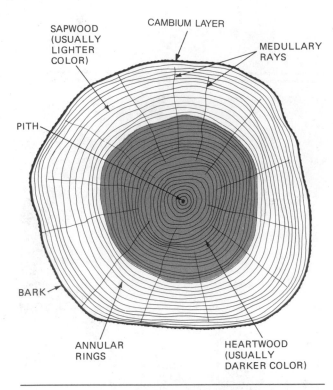

Fig. 1-2 A cross-section of a tree showing its structure. (*Courtesy of Western Wood Products Association*)

annular rings that are not so contrasting and sometimes are hardly visible.

Hardwoods and Softwoods

Woods are classified as either hardwood or softwood. There are different methods of classifying these woods. The most common method of classifying wood is by its source. Hardwood comes from *deciduous* trees that shed their leaves each year. Softwood is cut from *coniferous*, or cone-bearing, trees, commonly known as evergreens (Fig. 1-3). In this method of classifying wood, some of the softwoods may actually be harder than the hardwoods. For instance, fir, a softwood, is harder and stronger than basswood, a hardwood. There are other methods of classifying hardwoods and softwoods, but this method is the one most widely used.

Some common hardwoods are ash, birch, cherry, hickory, maple, mahogany, oak, and walnut. Some common softwoods are pine, fir, hemlock, spruce, cedar, cypress, and redwood.

Wood may also be divided into two groups according to cell structure. *Open-grained* wood has large cells that show tiny openings or pores in the surface. In order to obtain a smooth finish these pores must be filled with a specially prepared paste wood filler. Examples of open-grained wood are oak, mahogany, and walnut. All softwoods are *close-grained*. Some close-grained hardwoods are birch, cherry, maple, and poplar. (See Fig. 1–4 for common kinds of softwoods and their characteristics and Fig. 1–5 for hardwoods.)

Identification of Wood

Identifying different kinds of wood can be very difficult because some closely resemble each other. For instance, ash and white oak are hard to distinguish from each other, as are some pine, hemlock, and spruce. Not only are they the same color, but the

Fig. 1-3 Hardwood is from broad-leaf trees, softwood from cone-bearing trees.

grain pattern and weight are about the same. Only the most experienced workers are able to tell the difference.

It is possible to get some clues to identifying wood by studying literature, but the best way to learn the different kinds of wood is by working with them. Each time you handle a piece of wood, examine it. Look at the color and the grain; feel if it is heavy or light, if it is soft or hard; and smell it for a characteristic odor. Aromatic cedar, for instance, can always be identified by its pleasing moth-repelling odor, if for no other reason. After studying the characteristics of the wood, ask or otherwise find out the kind of wood you are holding, and remember it. In this manner, after a period of time, those kinds of wood that are used regularly on the job can be easily identified. Identification of kinds of wood that are seldom worked with can be accomplished in the same manner, but, of course, the process will take a little longer.

SOFTWOODS

KIND	COLOR	GRAIN	HARDNESS	STRENGTH	WORK-ABILITY	ELASTICITY	DECAY RESISTANCE	USES	OTHER
Red Cedar	Dark Reddish Brown	Close Medium	Soft	Low	Easy	Poor	Very High	Exterior	Cedar Odor
Cypress	Orange Tan	Close Medium	Soft to Medium	Medium	Medium	Medium	Very High	Exterior	
Fir	Yellow to Orange Brown	Close Coarse	Medium to Hard	High	Hard	Medium	Medium	Framing Millwork Plywood	
Ponderosa Pine	White with Brown Grain	Close Coarse	Medium	Medium	Medium	Poor	Low	Millwork Trim	Pine Odor
Sugar Pine	Creamy White	Close Fine	Soft	Low	Easy	Poor	Low	Patternmaking Millwork	Large Clear Pieces
Western White Pine	Brownish White	Close Medium	Soft to Medium	Low	Medium	Poor	Low	Millwork Trim	
Southern Yellow Pine	Yellow Brown	Close Coarse	Soft to Hard	High	Hard	Medium	Medium	Framing Plywood	Much Pitch
Redwood	Reddish Brown	Close Medium	Soft	Low	Easy	Poor	Very High	Exterior	Light Sapwood
Spruce	Cream to Tan	Close Medium	Medium	Medium	Medium	Poor	Low	Siding Subflooring	Spruce Odor

Fig. 1-4 Common kinds of softwood and their characteristics.

HARDWOODS

KIND	COLOR	GRAIN	HARDNESS	STRENGTH	WORK-ABILITY	ELASTICITY	DECAY RESISTANCE	USES	OTHER
Ash	Light Tan	Open Coarse	Hard	High	Hard	Very High	Low	Tool Handles Oars Baseball Bats	
Basswood	Creamy White	Close Fine	Soft	Low	Easy	Low	Low	Drawing Bds Veneer Core	Imparts No Taste Or Odor
Beech	Light Brown	Close Medium	Hard	High	Medium	Medium	Low	Food Containers Furniture	
Birch	Light Brown	Close Fine	Hard	High	Medium	Medium	Low	Furniture Veneers	
Cherry	Lt. Reddish Brown	Close Fine	Medium	High	Medium	High	Medium	Furniture	
Hickory	Light Tan	Open Medium	Hard	High	Hard	Very High	Low	Tool Handles Diving Boards	
Lauan	Lt. Reddish Brown	Open Medium	Soft	Low	Easy	Low	Low	Veneers Paneling	
Mahogany	Russet Brown	Open Fine	Medium	Medium	Excellent	Medium	High	Quality Furniture	
Maple	Light Tan	Close Medium	Hard	High	Hard	Medium	Low	Furniture Flooring	
Oak	Light Brown	Open Coarse	Hard	High	Hard	Very High	Medium	Flooring Boats	
Poplar	Greenish Yellow	Close Fine	Medium Soft	Medium Low	Easy	Low	Low	Furniture Veneer Core	
Teak	Honey	Open Medium	Medium	High	Excellent	High	Very High	Furniture Boat Trim	Heavy Oily
Walnut	Dark Brown	Open Fine	Medium	High	Excellent	High	High	High Quality Furniture	

Fig. 1-5 Common kinds of hardwood and their characteristics.

CHAPTER 2   LUMBER

Manufacture of Lumber

When logs arrive at the sawmill, the bark is removed first. Then a huge bandsaw slices the log into large planks, which are passed through a series of saws. The saws slice, edge, and trim them into various dimensions, and the pieces become *lumber*.

Once trimmed of all uneven edges, the lumber is stacked according to size and grade and taken outdoors where *sticking* takes place. Sticking is the process of restacking the lumber on small cross-sticks that allow air to circulate between the pieces. This air-seasoning process may take six months to two years due to the large amount of water found in lumber.

Following air-drying, the lumber is then dried in huge ovens. Once dry, the rough lumber is surfaced to standard sizes and shipped (Fig. 2-1).

The long, narrow surface of a piece of lumber is called its edge, the long, wide surface is termed its side, and its extremities are called ends. The distance across the edge is called its thickness, across its side is called its width, and from end to end is called its length (Fig. 2-2). The best-appearing side is called the face side and the best-appearing edge is called the face edge.

In certain cases, the surfaces of lumber may acquire different names. For instance, the distance from top to bottom of a beam is called its depth, and the distance across its top or bottom may be called its width or its thickness. The length of posts or columns, when installed, may be called their height.

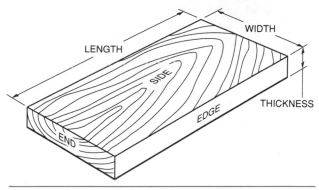

Fig. 2-2 Lumber surfaces are distinguished by specific names.

Plain-Sawed Lumber

A common way of cutting lumber is called the *plain-sawed* method, in which the log is cut tangent to the annular rings. This method produces a distinctive grain pattern on the wide surface (Fig. 2-3). This method of sawing is the least expensive and produces greater widths. However, plain-sawed lumber shrinks more during drying and it warps easily. Plain-sawed lumber is sometimes called slash-sawed lumber.

Quarter-Sawed Lumber

Another method of cutting the log, called *quarter-sawing*, produces pieces in which the annular rings are at or almost at right angles to the wide surface. Quarter-sawed lumber has less tendency to warp

Fig. 2-1 Manufacture of lumber from forest to finished product. (*Courtesy of California Redwood Association*)

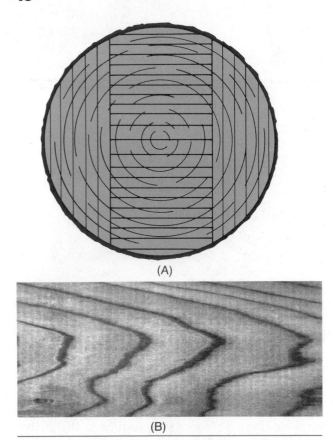

(A)

(B)

Fig. 2-3 a. Plain-sawed lumber; b. surface of plain-sawed lumber. (*Courtesy of California Redwood Association*)

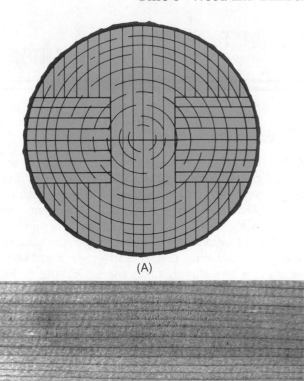

(A)

(B)

Fig. 2-4 a. Quarter-sawed lumber; b. surface of quarter-sawed lumber. (*Courtesy of California Redwood Association*)

and shrinks less and more evenly when dried. This type of lumber is durable because the wear is on the end of the annular rings. Quarter-sawed lumber is frequently used for flooring.

A distinctive and desirable grain pattern is produced in some wood, such as oak, because the lumber is sawed along the length of the medullary rays. Quarter-sawed lumber is sometimes called vertical-grain or edge-grain (Fig. 2-4).

Combination Sawing

Most logs are cut into a combination of plain-sawed and quarter-sawed lumber. The *sawyer* must use great skill in determining how to cut the log with as little waste as possible in the shortest amount of time to get the desired amount and kinds of lumber (Fig. 2-5).

Moisture Content and Shrinkage

When a tree is first cut down, it contains a great amount of water. Lumber, when first cut from the log, is called *green lumber* and is very heavy because most of its weight is water. A piece 2 inches thick, 6 inches wide, and 10 feet long may contain as much as 4 1/4 gallons of water weighing about 35 pounds (Fig. 2-6).

Fig. 2-5 Combination-sawed lumber. (*Courtesy of Western Wood Products Association*)

Green lumber cannot be used in construction because it will eventually dry to the same moisture content as the surrounding air. As green lumber

Fig. 2-6 Green lumber contains a large amount of water.

dries, it shrinks considerably and unequally because of the large amount of water that leaves it. When it shrinks, it usually warps depending on the way it was cut from the log. The use of green lumber in construction results in cracked ceilings and walls, squeaking floors, sticking doors, and many other problems caused by shrinking and warping of the lumber as it dries. Therefore, lumber must be dried to a suitable degree before it can be surfaced and used.

Green lumber is also subject to decay. Decay is caused by *fungi*, low forms of plant life that feed on wood. This decay is commonly known as dry rot because it is usually not discovered until the lumber has dried. Decay will not occur unless wood moisture content is in excess of 19 percent. Wood construction maintained at moisture content of less than 20 percent will not decay. It is important that lumber with an excess amount of moisture be exposed to an environment that will allow the moisture to evaporate. Seasoned lumber must be protected to prevent the entrance of moisture that allows the growth of fungi and decay of the wood. (The subject of fungi and wood decay is discussed in detail in Chapter 33.)

Moisture Content

The *moisture content (M.C.)* of lumber is expressed as a percentage of its total weight. Therefore, a moisture content of 30 percent indicates that about 1/3 the weight of the wood is water. Lumber used for framing and exterior finish should have an M.C. that does not exceed 19 percent, preferably 15 percent. For interior finish, an M.C. of 10 to 12 percent is recommended.

Green lumber has water in the hollow part of the wood cells and also in the cell walls. When wood starts to dry, the water in the cell cavities, called *free water*, is first removed. When all of the free water is gone, the wood has reached the *fiber-*

saturation point; about 1/3 of the remaining weight is water. No noticeable shrinkage of wood takes place up to this point.

As wood continues to dry, the water in the walls of the cells is removed and the wood starts to shrink. It shrinks considerably from the size at its fiber-saturation point of approximately 30 percent M.C. to the desired percent M.C. suitable for construction. Lumber at this stage is called *dry* or *seasoned*, and now must be prevented from getting wet.

It is important to understand not only that wood shrinks as it dries, but also how it shrinks. So little shrinkage occurs along the length of lumber that it is not considered. Most shrinkage takes place in the direction of the annular rings, with more shrinkage taking place on the longer rings. When viewing plain-sawed lumber in cross-section (looking at the end), it can be seen that the piece warps as it shrinks because of the unequal length of the annular rings. A cross-section of quarter-sawed lumber shows annular rings of equal length. Therefore, although the piece shrinks somewhat, it shrinks evenly with no warp (Fig. 2-7). Wood warps as it dries according to the way it was cut from the tree. Cross-section views of the annular rings are different along the length of a piece of lumber; therefore, various kinds of warp result when lumber dries and shrinks.

When the moisture content of lumber reaches that of the surrounding air (about 10 to 12 percent M.C.) it is at *equilibrium moisture content*. At this point, lumber shrinks or swells only slightly with changes of moisture in the air. Realizing that lum-

Fig. 2-7 Lumber shrinks in the direction of the annular rings.

SEAL ALL JOINTS IN FINISH WORK.

PRIME ALL EXTERIOR TRIM AS SOON AS POSSIBLE AFTER INSTALLATION.

BACK PRIME EXTERIOR FINISH.

SEAL BOTTOM ENDS OF DOORS BEFORE HANGING.

PRIME ALL BOTTOM EDGES AND ENDS.

MAKE JOINTS THAT SHED WATER.

MAKE JOINTS THAT STOP ENTRANCE OF WATER.

DETAIL A   DETAIL B   DETAIL B

Fig. 2-8

ber undergoes certain changes when moisture is absorbed or lost, the experienced carpenter uses techniques to deal with this characteristic of wood (Fig. 2-8).

Drying Lumber

Lumber is either *air-dried* or *kiln-dried*, or a combination of both. In air-drying the lumber is stacked in piles with spacers, which are called stick-ers, placed between each layer to permit air to circulate through the pile (Fig. 2-9). Kiln-dried lumber is stacked in the same manner, but is dried in buildings, called kilns, which are like huge ovens (Fig. 2-10).

Fig. 2-9 Drying lumber in the air. (*Courtesy of Western Wood Products Association*)

Fig. 2-10 Drying lumber in a kiln. (*Courtesy of Western Wood Products Association*)

Under carefully controlled temperatures, humidity, and air circulation, the removal of moisture takes place. First, the humidity level is raised and the temperature is increased; the humidity is then gradually decreased. Kiln-drying has the advantage of drying lumber in a shorter period of time, but is more expensive than air-drying.

The recommended moisture content for lumber to be used for exterior finish at the time of installation is 12 percent, except in very dry climates, where 9 percent is recommended. Lumber with low moisture content (8 to 10 percent) is necessary for interior trim and cabinet work.

The moisture content of lumber is determined by the use of a *moisture meter* (Fig. 2-11). Points on the ends of the wires of the meter are driven into the wood, and the moisture content is read off the meter.

The experienced worker knows when lumber is green (because it is much heavier than dry lumber), and he or she can estimate fairly accurately the moisture content of lumber simply by lifting it.

When lumber is sufficiently dry, it is brought to the planer mill where it is straightened, smoothed, and uniformly sized. Most construction lumber is surfaced on four sides (S4S) to standard thicknesses and widths. Some may be surfaced on only two sides (S2S) to required thicknesses.

Lumber Storage

Lumber should be delivered to the job site so that materials are accessible in the proper sequence; that is, those that are to be used first are on the top and those to be used last are on the bottom.

Lumber stored at the job site should be adequately protected from moisture and other hazards. A common practice that must be avoided is placing unprotected lumber directly on the ground. Use short lengths of lumber running at right angles to the length of the pile and spaced close enough to keep the pile from sagging and coming into contact with the ground. The base on which the lumber is to be placed should be fairly level to keep the pile from falling over.

Protect the lumber with a tarp or other type of cover. Leave enough room at the bottom of the pile for circulation of air. A cover that reaches to the ground will act like a greenhouse, trapping ground moisture within the stack.

Keep the piles in good order. Lumber spread out in a disorderly fashion can cause accidents as well as subject the lumber to stresses that may cause warping.

Lumber Defects

A defect in lumber is any fault that detracts from its appearance, function, or strength. One type of defect is called a *warp*. Warps are caused by, among other things, drying lumber too fast, careless handling and storage, or surfacing the lumber before it is thoroughly dry. Warps are classified as *crooks, bows, cups,* and *twists* (Fig. 2-12).

Splits in the end of lumber running lengthwise and across the annular rings are called *checks* (Fig. 2-13). Checks are caused by faster drying of the end than of the rest of the stock. Checks can be prevented to a degree by sealing the ends of lumber with paint, wax, or other material during the drying period. Cracks that run parallel to and between the

Fig. 2-11 Moisture meter. (*Courtesy of Moisture Register Company*)

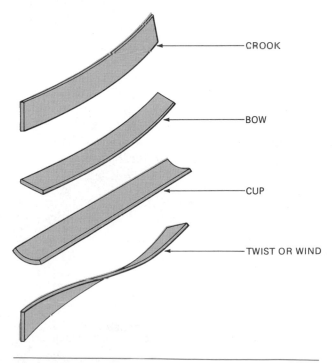

CROOK

BOW

CUP

TWIST OR WIND

Fig. 2-12 Kinds of warp.

Fig. 2-13 A severe check in the end of a piece of lumber.

annular rings are called *shakes* and may be caused by weather or other damage to the tree.

The *pith* is the spongy center of the tree. If possible it should not be used as finish if it is on, or close to, the surface. *Knots* are cross-sections of branches in the trunk of the tree. Knots are not necessarily defects unless they are loose or weaken the piece. *Pitch pockets* are small cavities that hold pitch, which sometimes runs out. A *wane* is bark on the edge of lumber or the surface from which the bark has fallen. *Pecky* wood has small grooves or channels running with the grain. This is common in cypress. Pecky cypress is often used as an interior wall paneling when that effect is desired. Some other defects are *stains*, *decay*, and *wormholes*.

Lumber Grades and Sizes

Lumber grades and sizes are established by wood products associations of which many wood mills are members. Member wood mills are closely supervised by the associations to assure that standards are maintained. The grade stamp of the association is assurance that lumber grade standards have been met.

Member mills use the association grade stamp to indicate strict quality control. A typical grade stamp is shown in Figure 2-14 and shows the association trademark, the mill number, the lumber grade, the species of wood, and the moisture content.

Softwood Grades

The largest softwood association of lumber manufacturers is the Western Wood Products Association (WWPA), which grades lumber in three categories: *boards* (under 2 inches thick), *dimension* (2 to 4 inches thick), and *timbers* (5 inches and thicker). The board group is divided into boards, sheathing, and form lumber. The dimension group is divided into light framing, studs, structural light framing, and structural joists and planks. Timbers are divided into beams and stringers. The three main categories are further classified according to strength and appearance as shown in Figure 2-15.

Hardwood Grades

Hardwood grades are established by the National Hardwood Lumber Association. *Firsts and seconds (FAS)* is the best grade of hardwood and must yield about 85 percent clear cutting. Each piece must be at least 6 inches wide and 8 feet long. The next best grade is called *select*. For this the minimum width is 4 inches, and the minimum length is 6 feet. *No. 1 common* allows even narrower widths and shorter lengths, with about 65 percent clear cutting.

Lumber Sizes

Rough lumber that comes directly from the sawmill is close in size to what it is called (nominal size).

b. MILL NUMBER

c. GRADE OF LUMBER (IN THIS CASE IT IS STANDARD)

d. KIND OF WOOD (DOUGLAS FIR)

a. ASSOCIATION'S TRADEMARK (WESTERN WOOD PRODUCTS ASSN)

e. MOISTURE CONTENT (S-DRY STANDS FOR 19 PERCENT M.C.)

Fig. 2-14 Typical softwood lumber grade stamp. (*Courtesy of Western Wood Products Association*)

Grade Selector Charts
Boards

APPEARANCE GRADES			
SELECTS	B & BETTER C SELECT D SELECT	(IWP—SUPREME)* (IWP—CHOICE) (IWP—QUALITY)	
FINISH	SUPERIOR PRIME E		
PANELING	CLEAR (ANY SELECT OR FINISH GRADE) NO. 2 COMMON SELECTED FOR KNOTTY PANELING NO. 3 COMMON SELECTED FOR KNOTTY PANELING		
SIDING **(BEVEL,** **BUNGALOW)**	SUPERIOR PRIME		
BOARDS **SHEATHING** **& FORM** **LUMBER**	NO. 1 COMMON (IWP—COLONIAL) NO. 2 COMMON (IWP—STERLING) NO. 3 COMMON (IWP—STANDARD) NO. 4 COMMON (IWP—UTILITY) NO. 5 COMMON (IWP—INDUSTRIAL)		
	ALTERNATE BOARD GRADES SELECT MERCHANTABLE CONSTRUCTION STANDARD UTILITY ECONOMY		

SPECIFICATION CHECK LIST
- [ ] Grades listed in order of quality.
- [ ] Include all species suited to project.
- [ ] Specify lowest grade that will satisfy job requirement.
- [ ] Specify surface texture desired.
- [ ] Specify moisture content suited to project.
- [ ] Specify (WP) grade stamp. For finish and exposed pieces, specify stamp on back or ends.

Western Red Cedar

FINISH PANELING AND CEILING	CLEAR HEART A B
BEVEL SIDING	CLEAR — V.G. HEART A — BEVEL SIDING B — BEVEL SIDING C — BEVEL SIDING

*Idaho White Pine carries its own comparable grade designations.

Dimension/All Species 2″ to 4″ thick (also applies to finger-jointed stock)

LIGHT FRAMING 2″ to 4″ Thick 2″ to 4″ Wide	CONSTRUCTION STANDARD UTILITY	This category for use where high strength values are **NOT** required; such as studs, plates, sills, cripples, blocking, etc.
STUDS 2″ to 4″ Thick 2″ and Wider	STUD	An optional all-purpose grade. Characteristics affecting strength and stiffness values are limited so that the "Stud" grade is suitable for all uses, including load bearing walls.
STRUCTURAL LIGHT FRAMING 2″ to 4″ Thick 2″ to 4″ Wide	SELECT STRUCTURAL #1$_T$Btr.* NO. 1 NO. 2 NO. 3	These grades are designed to fit those engineering applications where higher bending strength ratios are needed in light framing sizes. Typical uses would be for trusses, concrete pier wall forms, etc.
STRUCTURAL JOISTS & PLANKS 2″ to 4″ Thick 5″ and Wider	SELECT STRUCTURAL #1$_T$Btr.* NO. 1 NO. 2 NO. 3	These grades are designed especially to fit in engineering applications for lumber five inches and wider, such as joists, rafters and general framing uses.

*Douglas fir/Larch$_T$ Hem-Fir only (DF–L)

Timbers 5″ and thicker

BEAMS & STRINGERS 5″ and thicker Width more than 2″ greater than thickness	SELECT STRUCTURAL NO 1 NO. 2** NO. 3**		**POSTS & TIMBERS** 5″ x 5″ and larger Width not more than 2″ greater than thickness	SELECT STRUCTURAL NO. 1 NO. 2** NO. 3**

**Design values are not assigned.

Fig. 2-15 Softwood lumber grades. (*Courtesy of Western Wood Products Association*)

There are slight variations to nominal size because of the heavy machinery used to cut the log into lumber. When rough lumber is planed, it is reduced in thickness and width to standard and uniform sizes. Its nominal size does not change even though the actual size does. Therefore, when *dressed* (surfaced), although a piece may be called a 2×4, its actual size is 1 1/2 inches by 3 1/2 inches. The same applies to all surfaced lumber; the nominal size (what it is called) and the actual size are not the same. Figure 2-16 shows the standard nominal and dressed sizes of softwood lumber, based on WWPA rules. Hardwood lumber is usually purchased in the rough and straightened, smoothed, and sized as needed.

Standard Lumber Sizes / Nominal, Dressed, Based on WWPA Rules

Product	Description	Nominal Size		Dressed Dimensions		
		Thickness In.	Width In.	Thicknesses and Widths In. Surfaced Dry	Surfaced Unseasoned	Lengths Ft.
DIMENSION	S4S .... Other surface combinations are available. See "Abbreviations" below.	2 3 4	2 3 4 5 6 8 10 12 Over 12	1-½ 2-½ 3-½ 4-½ 5-½ 7-½ 9-½ 11-½ ¾ off nominal	1-⁹⁄₁₆ 2-⁹⁄₁₆ 3-⁹⁄₁₆ 4-⅝ 5-⅝ 7-½ 9-½ 11-½ Off ½	6' and longer, generally shipped in multiples of 2'
SCAFFOLD PLANK	Rough Full Sawn or S4S .... (Usually shipped unseasoned)	1¼ & Thicker	8 and Wider	If Dressed refer to 'DIMENSION" sizes		6' and longer, generally shipped in multiples of 2'
TIMBERS	Rough or S4S .... (Shipped unseasoned)	5 and Larger		½" off nominal (S4S) See 3.20 of WWPA Grading Rules for Rough		6' and longer, generally shipped in multiples of 2'
		Nominal Size		Dressed Dimensions		
		Thickness In.	Width In.	Thickness In.	Width In.	Lengths Ft.
DECKING	2" Single T&G	2	5 6 8 10 12	1½	4 5 6¾ 8¾ 10¾	6' and longer, generally shipped in multiples of 2'
	3" and 4" Double T&G	3 4	6	2½ 3½	5¼	
FLOORING	(D & M), (S2S & CM)............	⅜ ½ ⅝ 1 1¼ 1½	2 3 4 5 6	⁵⁄₁₆ ⁷⁄₁₆ ⁹⁄₁₆ ¾ 1 1¼	1⅛ 2⅛ 3⅛ 4⅛ 5⅛	4' and longer, generally shipped in multiples of 2'
CEILING AND PARTITION	(S2S & CM)	⅜ ½ ⅝ ¾	3 4 5 6	⁵⁄₁₆ ⁷⁄₁₆ ⁹⁄₁₆ ¹¹⁄₁₆	2⅛ 3⅛ 4⅛ 5⅛	4' and longer, generally shipped in multiples of 2'
FACTORY AND SHOP LUMBER	S2S	1 (4/4) 1¼ (5/4) 1½ (6/4) 1¾ (7/4) 2 (8/4) 2½ (10/4) 3 (12/4) 4 (16/4)	5" and wider (except 4" and wider in 4/4 No. 1 Shop and 4/4 No. 2 Shop, and 2" and wider in 5/4 & Thicker No. 3 Shop)	¾ (4/4) 1³⁄₃₂ (5/4) 1¹³⁄₃₂ (6/4) 1¹³⁄₃₂ (7/4) 1¹³⁄₁₆ (8/4) 2⅜ (10/4) 2¾ (12/4) 3¾ (16/4)	Usually sold random width	6' and longer, generally shipped in multiples of 2'

ABBREVIATIONS
Abbreviated descriptions appearing in the size table are explained below.
S1S — Surfaced one side.
S2S — Surfaced two sides.

S4S — Surfaced four sides.
S1S1E — Surfaced one side, one edge.
S1S2E — Surfaced one side, two edges.
CM — Center matched.

D & M — Dressed and matched.
T & G — Tongue and grooved.
Rough Full Sawn — Unsurfaced green lumber cut to full specified size.

Product Classification

	thickness in.	width in.		thickness in.	width in.
board lumber	1"	2" or more	beams & stringers	5" and thicker	more than 2" greater than thickness
light framing	2" to 4"	2" to 4"	posts & timbers	5" x 5" and larger	not more than 2" greater than thickness
studs	2" to 4"	2" to 6" 10' and shorter	decking	2" to 4"	4" to 12" wide
structural light framing	2" to 4"	2" to 4"	siding		thickness expressed by dimension of butt edge
structural joists & planks	2" to 4"	5" and wider	mouldings		size at thickest and widest points

Lengths of lumber generally are 6 feet and longer in multiples of 2'

Nailing Diagram

BOARD ON BOARD — Over 8" wide, use 2 nails 3-4" apart

TONGUE AND GROOVE — Over 8" wide, use 2 nails 3-4" apart

BOARD AND BATTEN — Over 8" Wide use 2 nails 3-4" apart in center

CHANNEL RUSTIC — Over 8" wide, use 2 nails 3-4" apart at exposed edge

Fig. 2-16 Softwood lumber sizes. (*Courtesy of Western Wood Products Association*)

Board Measure

Softwood lumber is usually purchased by specifying the thickness, width, number of pieces, and length—in addition to the grade. Often, when no particular lengths are required, the thickness, width, and total number of linear feet (length in feet) are ordered. The length of the pieces then may vary and are called *random* lengths. Another method of purchasing softwood lumber is by specifying the thickness, width, and total number of *board feet*. Lumber purchased in this manner may also contain random lengths.

Hardwood lumber is purchased by specifying the grade, thickness, and total number of board feet. Large quantities of both softwood and hardwood lumber are priced and sold by the board foot.

A board foot is a measure of lumber. It is equivalent to a piece one inch thick, 12 inches wide, and one foot long. A piece of lumber one inch thick and 6 inches wide must be 2 feet long to equal one board foot. A piece 2 inches thick has twice as many board feet as a piece one inch thick of the same width and length (Fig. 2-17).

To calculate the number of board feet, use the formula: number of pieces × thickness in inches × width in inches × length in feet divided by 12 = number of board feet. For example: 16 pieces × 2 inches × 4 inches × 8 feet = 85 1/3 board feet.

$$\frac{\overset{4}{\cancel{16}} \times 2 \times 4 \times 8}{\underset{3}{\cancel{12}}} = \frac{256}{3} = 85\ 1/3\ \text{bd ft}$$

Fig. 2-17 Each piece contains one board foot.

Review Questions

Select the most appropriate answer.

1. The center of a tree is called the
    a. heartwood.   c. pith.
    b. lignin.   d. sapwood.

2. New wood cells of a tree are formed in the
    a. heartwood.   c. medullary rays.
    b. bark.   d. cambium layer.

3. Tree growth is faster in the
    a. spring.   c. fall.
    b. summer.   d. winter.

4. Quarter-sawed lumber is sometimes called
    a. tangent-grained.   c. vertical-grained.
    b. slash-sawed.   d. plain-sawed.

5. Lumber is called "green" when
    a. it is stained by fungi.
    b. the tree is still standing.
    c. it is first cut from the log.
    d. it has reached equilibrium moisture content.

6. Wood will not decay unless its moisture content is in excess of
    a. 15 percent.   c. 25 percent.
    b. 19 percent.   d. 30 percent.

7. When all of the free water in the cell cavities of wood is removed and before water is removed from cell walls, lumber is at what is called
    a. fiber-saturation point.
    b. 30 percent moisture content.
    c. equilibrium moisture content.
    d. shrinkage commencement.

8. Air-dried lumber cannot be dried less than
    a. 8 to 10 percent moisture content.
    b. equilibrium moisture content.
    c. 25 to 30 percent moisture content.
    d. its fiber-saturation point.

9. A commonly used and abundant softwood is
    a. ash.   c. basswood.
    b. fir.   d. birch.

10. One of the woods extremely resistant to decay is
    a. pine.   c. cypress.
    b. spruce.   d. hemlock

■ B U I L D I N G F O R S U C C E S S ■

Why Construction Education?

You have probably made a decision to pursue a career in the carpentry field or at least have sensed the need to become better informed about a particular carpentry process. Regardless of the reason, you have taken it upon yourself to become better educated. This is a process that will never stop during your many years in the construction industry. As you set goals for your career, the process of obtaining those goals will undoubtedly involve some form of education.

Assuming that your interests have led you in the direction of a carpentry career, how have you decided to pursue that career? Planning for your career can begin at any time during your school years. Today many people change careers at some point in their lives. Some job changes may be made due to economic influences, new interests, or other opportunities where job prospects appear to be very good. Many times job changes are intentionally planned in order to gain the variety of work experiences necessary to reach an ultimate career goal.

There are a few very good publications available to help you undertake an intensive study of the specific jobs available in the construction arena. A full array of job descriptions can be found in resources such as the occupational outlooks and guidebooks at your school or local library. These guidebooks are quite complete and will provide you with the initial information to begin a more detailed research of any occupation. Your instructor or librarian can assist you in locating these guidebooks. The federal and state governments publish literature that addresses the many state or local job opportunities for each area of the country. Some of these informative publications are available free of charge; you may have to pay a small fee for others.

The planning process must begin with your sincere desire to investigate in depth the general career area. You must then begin to eliminate jobs that do not seem appealing. Once you have decided on one or two careers in which you have the most interest, you may intensify and focus your research. By looking at the exact job tasks that are conducted regularly in that line of work, you will be able to make the best decisions for your education.

Today's construction careers require more education and training than in past years. The competition for quality construction jobs is increasing rapidly. The economy, both nationwide and global, will have a tremendous impact on the progress and growth of every industry. Since the construction industry is the pacesetter for much of the nation during both prosperous economic times and recession, we can use it as one of the gauges for the country's economic wellness.

There are various ways to gain the education necessary to enter the career of your choice. Perhaps by using this text, you have entered one of those avenues. The specific job you have selected will reveal the type and amount of informal and formal education required for an entry-level job. If you are pursuing a professional career such as architect, engineer, or construction manager, a four-year degree may be the best avenue. If your concentration is more on the technical side, several options for entry-level jobs exist. One popular way to gain the technical skills required of carpenters, for example, is by way of the local carpenters' union or a professional construction organization. The Associated General Contractors, Associated Builders and Contractors, and the National Association of Home Builders are examples of associations that have excellent construction education opportunities. All of these professional associations have state or local chapters listed in most metropolitan area phone directories.

Other educational programs are offered through various postsecondary institutions such as trade schools, junior colleges, and community colleges. In each state there are institutions that offer a vast array of programs that prepare technicians for construction careers. In most cases these schools prepare their graduates for an entry-level job in two years or less. This option allows the graduates a very quick return on their education investment. Another advantage of institutional construction training programs across the country is that many offer some form of continuing education. There will always be a need to continue gaining new skills and knowledge for almost every trade or occupation where high technology has an impact.

Quality construction education programs must be interwoven with the basic academic curriculum. There can be no division between academics and the vocational/technical realm. The tie has become so necessary that it cannot be neglected in a quality construction curriculum. The basic skills that business and industry require are dependent on the proper blend of academic and specific technical courses. From now on carpenters and other skilled workers will be expected to perform not only technical tasks on the job, but leadership, business, and interpersonal communication tasks as well.

With the fast pace of computer technology, and the introduction of new products and assembly systems, the need for updating will continue. The time is past when only basic technical training and preparation for work will be adequate for a lifetime. Our global society has dictated that we should be fully committed to ongoing education and training programs not only to advance in our employment, but simply to maintain our jobs.

The need for continued education may be a good topic for discussion during the interview process. If the employer sees that continuing education is important for quality growth, he or she will be more willing to assist people in finding the time and financial resources to pursue additional training once they are employed.

As you select a career area and begin the necessary preparation for becoming gainfully employed, keep in mind the fact that the education process has only started and that it will continue at least until retirement. Now is the best time to enter into or continue your education for your desired career. Very seldom do the opportunities become more appropriate, convenient, or economical than at the present. Once you establish a family and make financial commitments, you will need to become the best trained or prepared worker possible.

The competition will continue to increase for better-paying jobs. There will be no reduction of necessary requirements such as technical skills, special abilities, knowledge, or other unique qualifications for the jobs of the twenty-first century. "This is the first day of the rest of your life." What decisions do you have to make to properly pursue a technical career area? Have you approached knowledgeable people for answers to your questions? Have you arrived at the best educational avenue for gaining the necessary entry-level job skills? By answering these questions, you will be well on your way to a more productive and meaningful career.

FOCUS QUESTIONS: For individual or group discussion.

1. Do you know what your career goals are? Can you identify how you are going to reach those goals?

2. What might be some of the considerations that must be part of making a career decision?

3. How does education play a part in helping you attain your career goals?

4. Who do you think could be potential resources to work with as you plan your career? Are you acquainted with knowledgeable people that represent your area of interest?

Engineered Panels

Growing concern over efficient use of forest resources led to the development of reconstituted wood in the form of large sheets, commonly called panels or boards. Since the advent of plywood, the first of these products, many more are now available for use in construction. The carpenter must know the different kinds, sizes, and recommended applications to use them correctly and to the best advantage.

OBJECTIVES

After completing this unit, the student will be able to describe the composition, kinds, sizes, grades, and several uses of:

- plywood
- oriented strand board
- composite panels
- particleboard
- hardboard
- medium-density fiberboard
- softboard

UNIT CONTENTS

CHAPTER 3

RATED PLYWOOD AND PANELS

The term *engineered panels* refers to man-made products in the form of large reconstituted wood sheets, sometimes called *panels* or *boards*. In some cases, the tree has been taken apart and its contents have been redistributed into sheet or panel form. The panels are widely used in the construction industry. They are also used in the aircraft, automobile, and boat-building industries, as well as in the making of road signs, furniture, and cabinets.

With the use of engineered panels, construction progresses at a faster rate because a greater area is covered in a shorter period of time (Fig. 3-1). These panels, in certain cases, present a more attractive appearance and give more protection to a surface than does solid lumber. It is important to know the kinds and uses of various engineered panels in order to use them to the best advantage.

APA-Rated Panels

Many mills belong to associations that inspect, test, and allow mills to stamp the product to certify that it conforms to government and industrial standards. The grade stamp assures the consumer that the product has met the rigid quality and performance requirements of the association.

The trademarks of the largest association of this type, the American Plywood Association (APA),
appear only on products manufactured by APA member mills (Fig. 3-2). This association is concerned not only with quality supervision and testing of *plywood* (cross-laminated wood veneer), but also of *composites* (veneer faces bonded to reconstituted wood cores) and nonveneered panels, commonly called *oriented strand board* (Fig. 3-3).

Plywood

One of the most extensively used engineered panels is plywood. Plywood is a sandwich of wood. Most plywood panels are made up of sheets of veneer

Fig. 3-1 Sheet material covers a greater area in a shorter period of time. (*Courtesy of American Plywood Association*)

Fig. 3-2 The grade stamp is assurance of a high-quality performance-rated panel. (*Courtesy of American Plywood Association*)

COMPOSITE PANEL-LAYERS OF WOOD FIBER ARE BONDED BETWEEN WOOD VENEER. FIVE LAYER PANEL HAS A WOOD VENEER IN THE CENTER.

PLYWOOD-ORIGINAL STRUCTURAL WOOD PANEL. COMPOSED OF THIN SHEETS OF VENEER CALLED PLIES, TO FORM A PANEL.

ORIENTED STRAND BOARD-COMPOSED OF COMPRESSED WOOD STRANDS ARRANGED IN LAYERS, AT RIGHT ANGLES TO ONE ANOTHER.

Fig. 3-3 APA performance-rated panels. (*Courtesy of American Plywood Association*)

(thin pieces) called *plies*. These plies, arranged in layers, are bonded under pressure with glue to form a very strong panel. The plies are glued together so that the grain of each layer is at right angles to the next one. This cross-graining results in a sheet that is as strong as or stronger than the wood it is made from. Plywood always contains an odd number of layers so that the face grain on both sides of the sheet runs in the direction of the long dimension of the panel (Fig. 3-4). Softwood plywood is commonly made with three, five, or seven layers. Because of its construction, plywood is more

ARROWS SHOW DIRECTION OF GRAIN IN EACH LAYER

Fig. 3-4 In plywood construction, the grains of veneer plies are at right angles to each other.

stable with changes of humidity and is resistant to shrinking and swelling.

Manufacture of Veneer Core Plywood. Specially selected "peeler logs" are mounted on a huge lathe in which the log is rotated against a sharp knife. As the log turns, a thin layer is peeled off like paper unwinding from a roll (Fig. 3-5). The entire log is used. The small remaining spindles are utilized for making other wood products.

The long ribbon of veneer is then cut into desired widths, sorted, and dried to a moisture content of 5 percent. After drying, the veneers are fed through glue spreaders that coat them with a uniform thickness. The veneers are then assembled to make panels (Fig. 3-6). Large presses bond the

Fig. 3-5 The veneer is peeled from the log like paper unwinding from a roll. (*Courtesy of American Plywood Association*)

Fig. 3-6 Gluing and assembling plywood veneers into panels. (*Courtesy of American Plywood Association*)

assembly under controlled heat and pressure. From the presses, the panels are either left unsanded, touch-sanded or smooth-sanded, cut to size, inspected, and stamped.

Veneer Grades. In declining order, the letters *N, A, B, C plugged, C,* and *D* are used to indicate the appearance quality of panel veneers. Two letters are found in the grade stamp of veneered panels. One letter indicates the quality of one face, while the other letter indicates the quality of the opposite face. The exact description of these letter grades is shown in Figure 3-7. Panels with B-grade or better veneer faces are always sanded smooth. Some panels, such as APA-Rated Sheathing, are unsanded because their intended use does not require sanding. Other panels used for such purposes as subflooring and underlayment require only a touch-sanding to make the panel thickness more uniform.

Strength Grades. Softwood veneers are made of many different kinds of wood. These woods are classified in groups according to their strength (Fig. 3-8). Group 1 is the strongest. Douglas fir and southern pine are in Group 1 and used to make

most of the softwood plywood. The group number is also shown in the grade stamp.

Oriented Strand Board

Oriented strand board (OSB) is a nonveneered performance-rated structural panel composed of small oriented (lined up) strand-like wood pieces arranged in three to five layers with each layer at right angles to the other (Fig. 3-9). The cross-lamination of the layers achieves the same advantages of strength and stability as in plywood.

Manufacture of OSB. Logs from specially selected species are debarked and sliced into strands that are between 25/1000 and 30/1000 of an inch thick, 3/4 and one inch wide, and between 2 1/2 and 4 1/2 inches long. The strands are dried, loaded into a blender, coated with liquid resins, formed into a mat consisting of three or more layers of systematically oriented wood fibers, and fed into a press where, under high temperatures and pressure, they form a dense panel. As a safety measure, the panels have one side textured to help prevent slippage.

Nonveneered panels have previously been sold with such names as waferboard, structural particleboard and others. At present almost all panels manufactured with oriented strands or wafers are called oriented strand board. Various manufacturers have their own particular brand names for OSB, such as Oxboard, Aspenite, and many others.

Composite Panels

Composite panels are manufactured by bonding veneers of wood to both sides of reconstituted wood panels. More efficient use of wood is thus allowed with this product while retaining the wood grain appearance on both faces of the panel.

Composite panels rated by the American Plywood Association are called *COM-PLY* and are manufactured in three or five layers. A three-layer panel has a reconstituted wood core with wood veneers on both sides. A five-layer panel has a wood veneer in the center as well as on both sides.

Performance Ratings

A Performance Rated Panel meets the requirements of the panel's end use. The three end uses for which panels are rated are single-layer flooring, exterior siding, and sheathing for roofs, floors, and walls. Names given to designate end uses are *APA-Rated Sheathing, Structural I, APA-Rated Sturdi-I-Floor,* and *APA-Rated Siding* (Fig. 3-10). Panels are tested to meet standards in areas of resistance to moisture, strength, and stability.

Veneer Grades

A	Smooth, paintable. Not more than 18 neatly made repairs, boat, sled, or router type, and parallel to grain, permitted. Wood or synthetic repairs permitted. May be used for natural finish in less demanding applications.
B	Solid surface. Shims, sled or router repairs, and tight knots to 1 inch across grain permitted. Wood or synthetic repairs permitted. Some minor splits permitted.
C Plugged	Improved C veneer with splits limited to 1/8-inch width and knotholes or other open defects limited to 1/4 x 1/2 inch. Wood or synthetic repairs permitted. Admits some broken grain.
C	Tight knots to 1 - 1/2 inch. Knotholes to 1 inch across grain and some to 1 - 1/2 inch if total width of knots and knotholes is within specified limits. Synthetic or wood repairs. discoloration and sanding defects that do not impair strength permitted. Limited splits allowed. stiching permitted.
D	Knots and knotholes to 2 - 1/2 inch width across grain and 1/2 inch larger within specified limits. Limited splits are permitted. Stiching permitted. Limited to Exposure 1 or interior panels.

Fig. 3-7 Veneer letter grades define veneer appearance. (*Courtesy of American Plywood Association.*)

CLASSIFICATION OF SPECIES				
Group 1	Group 2	Group 3	Group 4	Group 5

Apitong
Beech,
  American
Birch
  Sweet
  Yellow
Douglas
  Fir 1[a]
Kapur
Keruing
Larch,
  Western
Maple, Sugar
Pine
  Caribbean
  Ocote
Pine, South
  Loblolly
  Longleaf
  Shortleaf
  Slash
Tanoak
	Cedar, Port
	  Orford
	Cypress
	Douglas
	  Fir 2[a]
	Fir
	  Balsam
	  California
	  Red
	  Grand
	  Noble
	  Pacific
	  Silver
	  White
	Hemlock,
	  Western
	Lauan
	  Almon
	  Bagtikan
	  Mayapis
	  Red
	  Tangile
	  White
	Maple, Black
	Mengkulang
	Meranti,
	  Red[b]
	Mersawa
	Pine
	  Pond
	  Red
	  Virginia
	  Western
	  White
	Spruce
	  Black
	  Red
	  Sitka
	Sweetgum
	Tamarack
	Yellow-
	  Poplar
		Alder, Red
		Birch, Paper
		Cedar, Alaska
		Fir,
		  Subalpine
		Hemlock,
		  Eastern
		Maple
		  Bigleaf
		Pine
		  Jack
		  Lodgepole
		  Ponderosa
		  Spruce
		Redwood
		Spruce
		  Engelmann
		  White
			Aspen
			  Bigtooth
			  Quaking
			Cativo
			Cedar
			  Incense
			  Western
			  Red
			Cottonwood
			  Eastern
			  Black
			  (Western
			  Poplar)
			Pine
			  Eastern
			  White
			  Sugar
				Basswood
				Poplar,
				  Balsam

(a) Douglas Fir from trees grown in the states of Washington, Oregon, California, Idaho, Montana, Wyoming, and the Canadian Provinces of Alberta and British Columbia shall be classed as Douglas Fir No. 1. Douglas Fir from trees grown in the states of Nevada, Utah, Colorado, Arizona and New Mexico shall be classed as Douglas Fir No. 2.

(b) Red Meranti shall be limited to species having a specific gravity of 0.41 or more based on green volume and oven dry weight.

Fig. 3-8 Plywood is classified in groups according to strength and stiffness. (*Courtesy of American Plywood Association*)

Fig. 3-9 Oriented strand board being used for wall sheathing.

Exposure Durability

APA performance-rated panels are manufactured in three exposure durability classifications: *Exterior, Exposure 1,* and *Exposure 2.* Panels marked Exterior are designed for permanent exposure to the weather or moisture. Panels marked Exposure 1 are intended for use where long delays in construction may cause the panels to be exposed to the weather before being protected. Panels marked Exposure 2 are designed for use when only moderate delays in providing protection from the weather are expected. The exposure durability of a panel may be found in the grade stamp.

Span Ratings

The span rating in the grade stamp on APA-Rated Sheathing appears as two numbers separated by a

APA Rated Sheathing Typical Trademark	—APA— RATED SHEATHING 24/16 7/16 INCH SIZED FOR SPACING **EXPOSURE 1** —— 000 —— NER-QA397 PRP-108 HUD-UM-40C	Specially designed for subflooring and wall and roof sheathing. Also good for a broad range of other construction and industrial applications. Can be manufactured as plywood, as a composite, or as OSB. EXPOSURE DURABILITY CLASSIFICATIONS: Exterior, Exposure 1, Exposure 2. COMMON THICKNESSES: 5/16, 3/8, 7/16, 15/32, 1/2, 19/32, 5/8, 23/32, 3/4.
APA Structural I Rated Sheathing(c) Typical Trademark	—APA— RATED SHEATHING STRUCTURAL I 32/16 15/32 INCH SIZED FOR SPACING **EXPOSURE 1** —— 000 —— PS 1-83 C-D NER-QA397 PRP-108 —APA— RATED SHEATHING 32/16 15/32 INCH SIZED FOR SPACING **EXPOSURE 1** —— 000 —— STRUCTURAL I RATED DIAPHRAGMS-SHEAR WALLS PANELIZED ROOFS NER-QA397 PRP-108	Unsanded grade for use where shear and cross-panel strength properties are of maximum importance, such as panelixed roofs and diaphragms. Can be manufactured as plywood, as a composite, or as OSB. EXPOSURE DURABILITY CLASSIFICATIONS: Exterior, Exposrue 1. COMMON THICKNESSES: 5/16, 3/8, 7/16, 15/32, 1/2, 19/32, 5/8, 23/32, 3/4.
APA Rated Sturd-I-Floor Typical Trademark	—APA— RATED STURD-I-FLOOR 20 oc 19/32 INCH SIZED FOR SPACING T&G NET WIDTH 47-1/2 **EXPOSURE 1** —— 000 —— NER-QA397 PRP-108 HUD-UM-40C	Specially designed as combination subfloor-underlayment. Provides smooth surface for application of carpet and pad and possesses high concentrated and impact load resistance. Can be manufactured as plywood, as composite, or as OSB. available square edge or tonge-and-groove. EXPOSURE DURABILITY CLASSIFICATIONS: Exterior, Exposure 1, Exposure 2. COMMON THICKNESSES: 19/32, 5/8, 23/32, 3/4, 1, 1-1/8.
APA Rated Siding TYPICAL TRADEMARK	—APA— RATED SIDING 24 oc 19/32 INCH SIZED FOR SPACING **EXTERIOR** —— 000 —— NER-QA397 PRP-108 HUD-UM-40C —APA— RATED SIDING 303-18-S/W 16 oc 11/32 INCH GROUP 1 SIZED FOR SPACING **EXTERIOR** —— 000 —— PS 1-83 FHA-UM-64 NER-QA397 PRP-108	For exterior siding, fencing, etc. Can be manugactured as plywood, as a composite or as an overlaid OSB. Both panel and lap siding available. Special surface treatment such as V-groove, channel groove, deep groove (such as APA Texture 1-11), brushed, rough sawn and overlaid (MDO) with smooth- or texture-embossed face. Span Rating (stud spacing for siding qualified for APA Sturd-I-Wall applications) and face grade classification (for veneer-faced siding) indicated in trademark. EXPOSURE DURABILITY CLASSIFICATION: Exterior. COMMON THICKNESSES: 11/32, 3/8, 7/16, 15/32, 1/2, 19/32, 5/8.

(a) Specific grades, thicknesses and exposure durability classifications may be in limited supply in some areas. Check with your supplier before specifying.

(b) Specify Performance Rated Panels by thickness and Span Rating. Span Ratings are based on panel strength and stiffness. Since these properties are a function of panel composition and configuration as well as thickness, the same Span Rating may appear on panels of different thickness. Conversely, panels of the same thickness may be marked with different Span Ratings.

(c) All plies in Structural I plywood panels are special improved grades and panels marked PS 1 and limited to Group 1 species. Other panels marked Structural I rated qualify through special performance testing.

Structural II plywood panels are also provided for, but rarely manufactured. Application recommendations for Structural II plywood are identical to those for APA RATED SHEATHING plywood.

Fig. 3-10 Guide to APA performance-rated panels. (*Courtesy of American Plywood Association*)

slash, such as 32/16 or 48/24. The left-hand number denotes the maximum recommended spacing of supports when the panel is used for roof sheathing. The right hand number indicates the maximum recommended spacing of supports when the panel is used for subflooring. In both cases, the long dimension of the panel must be placed across three or more supports. A panel marked 32/16, for example, may be used for roof sheathing over rafters not more than 32 inches on center or for subflooring over joists not more than 16 inches on center.

The span ratings on APA-Rated Sturd-I-Floor and APA-Rated Siding appear as a single number. APA Rated Sturd-I-Floor panels are designed specifically for combined subflooring-underlayment applications and are manufactured with span ratings of 16, 20, 24, and 48 inches.

APA-Rated Siding is produced with span ratings of 16 and 24 inches. The rating applies to vertical installation of the panel. All siding panels may be applied horizontally direct to studs 16 to 24 inches on center provided horizontal joints are blocked.

CHAPTER 4

NONSTRUCTURAL PANELS

Plywood

All the rated products discussed in the previous chapter may be used for nonstructural applications. In addition, other plywood products, grade-stamped by the American Plywood Association, are available for nonstructural use. They include sanded and touch-sanded plywood panels (Fig. 4-1) and specialty plywood panels (Fig. 4-2).

APA A-A Typical Trademark `A-A•G-1•EXPOSURE 1-APA•000•PS 1-83`	Use where appearance of both sides is important for interior applications such as built-ins, cabinets, furniture, partitions; and exterior applications such as fences, signs, boats, shipping containers, tanks, ducts, etc. Smooth surfaces suitable for painting. EXPOSURE DURABILITY CLASSIFICATIONS: Interior, Exposure 1, Exterior. COMMON THICKNESSES: 1/4, 11/32, 3/8, 15/32, 1/2, 19/32, 5/8. 23/32, 3/4.
APA A-B Typical Trademark `A-B•G-1•EXPOSURE 1-APA•000•PS 1-83`	For use where appearance of one side is less important but where two solid srurfaces are necessary. EXPOSURE DURIBILITY CLASSIFICATIONS: Interior, Exopsure 1, Exterior. COMMON THICKNESS: 1/4, 11/32, 3/8, 15/32. 1/2. 19/32, 5/8, 23/32, 3/4.
APA A-C Typical Trademark —APA— A-C GROUP 1 EXTERIOR — 000 — PS 1-83	For use where apperrance of only one side is important in exterior applications, such as soffits, fences, farm buildings, ect.[f]EXPOSURE DURABILITY CLASSSIFICATION: Exterior. COMMON THICKNESSES: 1/4. 11/32. 3/8. 15/32, 1/2, 19/32, 5/8, 23/32, 3/4.
APA A-D Typical Trademark —APA— A-D GROUP 1 EXPOSURE 1 — 000 — PS 1-83	For use where apperarnce of only one side is important in interior applications, such as paneling, built-ins, shelving, partitions, flow racks, ect.[f]EXPOSURE DURABILITY CLASSIFICATIONS: Interior, Exposure 1. COMMON THICKNESS: 1/4, 11/32, 3/8, 15/32, 5/8, 23/32, 3/4.
APA B-B Typical Trademark `B-B•G-2•EXPOSURE 1-APA•000•PS 1-83`	Utility panels with two solid sides. EXPOSURE DURABILITY CLASSIFICATIONS: Interior, Exposure 1, Exterior. COMMON THICKNESSES: 1/4, 11/32, 3/8, 15/32, 1/2, 19/32, 5/8, 23/32, 3/4.
APA B-C Typical Trademark —APA— B-C GROUP 1 EXTERIOR — 000 — PS 1-83	Utility panel for farm service and work buildings, boxcar and truck linings, containers, tanks, agricultural equipment, as a base for exterior coatings and other exterior uses or applications subject to high or continuous moisture.[f]EXPOSURE DURABILITY CLASSIFICATION: Exterior. COMMON THICKNESSES: 1/4, 11/32, 3/8, 15/32, 1/2, 19/32, 5/8, 23/32, 3/4.

Fig. 4-1 Guide to APA sanded and touched-sanded plywood panels. (*Courtesy of American Plywood Association*)

APA B-D Typical Trademark	—APA— **B-D** GROUP 2 EXPOSURE 1 —— 000 —— PS 1-83	Utility panel for backing, sides of built-ins, industry shelving, slip sheets, separator boards, bins and other interior or protected applications.[f] EXPOSURE DURABILITY CLASSIFICATIONS: Interior, Exposure 1. COMMON THICKNESSES: 1/4, 11/32, 3/8, 15/32, 1/2, 19/32, 5/8, 23/32, 3/4.
APA Underlayment Typical Trademark	—APA— UNDERLAYMENT GROUP I EXPOSURE 1 —— 000 —— PS 1-83	For application over structural subfloor. Provides smooth surface for application of carpet and pad and possesses high concentrated and impact load resistance. For areas to be covered with resilient flooring, specify panels with "sanded face."[e] EXPOSURE DURABILITY CLASSIFICATIONS: Interior, Exposure 1. COMMON THICKNESSES[d]: 1/4, 11/32, 3/8. 15/32, 1/2, 19/32, 5/8, 23/32, 3/4.
APA C-C Plugged[g] Typical Trademark	—APA— C-C PLUGGED GROUP 2 EXTERIOR —— 000 —— PS 1-83	For use as an underlayment over structrual subfloor, refrigerated or controlled atmosphere storage rooms, pallet fruit bins, tanks, boxcar and truck floors and linings, open soffits, and other similar applications where continuous or severe moisture may be present. Provides smooth surface for application of carpet and pad and possesses high concentrated and impact load resistance. For areas to be covered with resilient flooring, specify panels with "sanded face."[e]EXPOSURE DURABILITY CLASSIFICATION: Exterior. COMMON THICKNESSES:[d] 11/32, 3/8, 15/32, 1/2, 19/32, 5/8, 23/32, 3/4.
APA C-D Plugged Typical Trademark	—APA— C-D PLUGGED GROUP 2 EXPOSURE1 —— 000 —— PS 1-83	For open soffits, built-ins, cable reels, separator boards and other interior or protected applications. Not substitute for Underlayment or APA Rated Sturd-I- Floor as it lacks their puncture resistance. EXPOSURE DURABILITY CLASSIFICATIONS: Interior, Exposure 1. COMMON THICKNESSES: 3/8, 15/32, 1/2, 19/32, 5/8, 23/32, 3/4.

(a) Specific plywood grades, thicknesses and exposure durability classifications may be in limited supply in some areas. Check with your supplier before specifying.

(b) Sanded Exterior plywood panels, C-C Plugged, C-D Plugged and Underlayment grades can also be manufactured in Structural I (allplies limited to Group 1 species)

(c) Some manufactures also produce plywood panels with premium N-grade veneer on one or both faces. Available only by special order. Check with the manufacturer.

(d) Panels 1/2 inch and thicker are Span Rated and do not contain species group number in trademark.

(e) Also available in Underlayment A-C or Underlayment B-C grades, marked either "touch sanded" or "sanded face."

(f) For nonstructural floor underlayment, or other applications requiring improved inner ply construction, specify panels marked either "plugged inner plies" (may also be designated plugged crossbands under face or plugged crossbands or core); or "meets underlayment requirements."

(g) Also may be designated APA Underlayment C-C Plugged.

Fig. 4-1 (Continued)

Hardwood Plywood

Plywood is available with hardwood face veneers, of which the most popular are birch, oak, and lauan. Beautifully grained hardwoods are sometimes matched in a number of ways to produce interesting face designs. Hardwood plywood is used in the interior of buildings for such things as wall paneling, built-in cabinets, and fixtures. Much of this kind of plywood is manufactured with a lumber core (Fig. 4-3), although some are manufactured with a reconstituted wood core.

Particleboard

Particleboard is a reconstituted wood panel made of wood flakes, chips, sawdust, and planer shavings (Fig. 4-4). These wood particles are mixed with an adhesive, formed into a mat, and pressed into sheet form. The kind, size, and arrangement of the wood particles determine the quality of the board.

The highest-quality particleboard is made up of large wood flakes in the center. The flakes become gradually smaller toward the surfaces where the finer particles are found. This type of construction results in an extremely hard board with a very smooth surface. Softer and lower-quality boards contain the same size particles throughout. These boards usually have a rougher surface texture. In addition to the size, kind, and arrangement of the particles, the quality of the board is determined by the method of manufacture.

APA DECORATIVE Typical Trademark	─APA─ DECORATIVE GROUP 2 INTERIOR 000 PS 1-83	Rough-sawn, brushed, grooved, or striated faces. For paneling, interior accent walls, built-ins, counter facing, exhibit displays. Can also be made by some manufactures in Exterior for exterior siding, gable ends, fences and other exterior applications. Use recommendations for Exterior panels vary with the particular product. Check with the manufacutrer. EXPOSURE DURABILITY CLASSIFICATONS: Interior, Exposure1, Exterior. COMMON THICKNESSES: 5/16, 3/8, 1/2, 5/8.
APA HIGH DENSITY OVERLAY (HDO) (b) Typical Trademark HDO•A-A•G-1•EXT-APA•000•PS 1-83		Has a hard semi-opaque resin-fiber overlay on both faces. Abrasion resistant. For concrete forms, cabinents. countertops, signs, tanks. Also available with skid-resistant screen-grid surface. EXPOSURE DURABILITY CLASSIFICATION: exterior. COMMON THICKNESS: 3/8, 1/2, 5/8, 3/4.
APA MEDIUM DENSITY OVERLAY (MDO) (b) Typical Trademark	─APA─ M. D. OVERLAY GROUP 1 EXTERIOR 000 PS 1-83	Smooth, opaque, resin-fiber overlay on one or both faces. Ideal base for paint, both indoors and outdoors. For exterior siding, paneling, shelving, exhibit displays, cabinets, signs. EXPOSURE DURABILITY CLASSIFICATION: Exterior. COMMON THICKNESS: 11/32, 3/8, 15/32, 1/2, 19/32, 5/8, 23/32, 3/4.
APA MARINE TYPICAL TRADEMARK MARINE•A-A•EXT-APA•PS 1-83		Ideal for boat hulls. Made only with Douglas-fir or western larch. Subject to special limitations on core gaps and face repairs. Also available with HDO faces. EXPOSURE DURABILITY CLASSIFICATION: Exterior. COMMON THICKNESSES: 1/4, 3/8, 1/2, 5/8, 3/4.
APA B-B PLYFORM CLASS I (b) Typical Trademark	─APA─ PLYFORM B - B CLASS1 EXTERIOR 000 PS 1-77	Concrete form grades with high reues factor. Sanded both faces and mill-oiled unless otherwise specified. Special restrictions on species. Also available in HDO for very smooth concrete finish, and with special overlays. EXPOSURE DURABILITY CLASSIFICATION: Exterior. COMMON THICKNESSES: 19/32, 5/8, 23/32, 3/4,
APA PLYRON Typical Trademark PLYRON•EXPOSURE 1-APA•000		Hardboard face on both sides. Faces tempered, untempered, smooth or screened. For conuntertops, shelving, cabinet doors, flooring. EXPOSURE DURABILITY CLASSIFICATIONS: Interior, Exposure 1, Exterior. COMMON THICKNESSES: 1/2, 5/8, 3/4.

(a) Specific plywood grades, thicknesses and exposure durability classifications may be in limited supply in some areas. Check with your supplier before specifying.

(b) Can also be manufactured in Structural 1 (all plies limited to Group 1 species).

Fig. 4-2 Guide to APA specialty plywood panels. (*Courtesy of American Plywood Association*)

Particleboard Grades

The quality of particleboard is indicated by its density (hardness), which ranges from 28 to 55 pounds per cubic foot. Nonstructural particleboard is used in the construction industry as an underlayment for finish floors (Fig. 4-5), for the construction of kitchen cabinets and countertops, and for the core of veneer doors and similar panels. *Duraflake, Novoply,* and *Tuf-Flake* are some brand names of particleboard.

Fiberboards

Fiberboards are manufactured as *high-density, medium-density,* and *low-density* boards.

Hardboards

High density fiberboards are called *hardboards* and are commonly known by the trademark, *Masonite,* regardless of the manufacturer. The hardboard industry makes almost complete use of the great nat-

FACE VENEERS

SOLID LUMBER CORE

CROSSBAND VENEERS

Fig. 4-3 Construction of lumber-core plywood.

Fig. 4-4 Particleboard is made from wood flakes, shavings, resins, and waxes. (*Courtesy of Duraflake Division, Williamette Industries, Inc.*)

Fig. 4-5 Particleboard is used extensively as underlayment for finish floors.

ural resource of wood by utilizing the wood chips and board trimmings, which were once considered waste.

Wood chips are reduced to fibers and water is added to make a soupy pulp. The pulp flows onto a traveling mesh screen where water is drawn off to form a mat. The mat is then pressed under heat to weld the wood fibers back together by utilizing lignin, the natural adhesive in wood.

Some panels are *tempered* (coated with oil and baked to increase hardness, strength, and water resistance). Carbide-tipped saws trim the panels to standard sizes.

Sizes of Hardboard. The most popular thicknesses of hardboard range from 1/8 to 3/8 inch. The most popular sheet size is 4 feet by 8 feet, although sheets may be ordered in practically any size.

Classes and Kinds of Hardboard. Hardboard is available in three different classes: tempered, standard, and service-tempered (Fig. 4-6). It may be obtained smooth-one-side (S1S) or smooth-two-sides (S2S). Hardboard is available in many forms, such as perforated, grooved, and striated.

Uses of Hardboard. Hardboard may be used inside or outside. It is widely used for exterior siding and interior wall paneling. It is also used extensively for cabinet backs and drawer bottoms. It can be used wherever a dense, hard panel is required (Fig. 4-7).

Because it is a wood-base product, hardboard can be sawed, routed, shaped, and drilled with standard woodworking tools. It can be securely fastened with glue, screws, staples, or nails.

CLASS	SURFACE	NOMINAL THICKNESS
1 Tempered	S1S and S2S	1/8 1/4
2 Standard	S1S and S2S	1/8 1/4 3/8
3 Service-tempered	S1S and S2S	1/8 1/4 3/8

Fig. 4-6 Kinds and thicknesses of hardboard.

Fig. 4-7 Hardboard is widely used for exterior siding and interior paneling.

Hardboard products are manufactured to comply with U.S. Department of Commerce Voluntary Product Standards. Most hardboard producers belong to the American Hardboard Association.

Medium-density Fiberboard

Commonly called *medium-density fiberboard (MDF)* is known by such names as *Medite, Baraboard* and *Fibrepine,* among others. It is manufactured in a manner similar to that used to make hardboard except that the fibers are not pressed as tightly together. The refined fiber produces a fine-textured, homogeneous board with an exceptionally smooth surface. Densities range from 28 to 65 pounds. It is available in thicknesses ranging from 3/16 to 1 1/2 inches and comes in widths of 4 feet and 5 feet. Lengths run from 6 to 18 feet.

MDF may be used for case goods, drawer parts, kitchen cabinets, cabinet doors, signs, and some interior wall finish.

Softboard

Low-density fiberboard is called *softboard.* Common brand names include *Temlok, Celotex,* and *Baracore,* Softboard is very light and contains many tiny air spaces because the particles are not compressed tightly.

The most common thicknesses range from 1/2 to one inch. The most common sheet size is 4 feet by 8 feet, although many sizes are available.

Uses of Softboard. Because of their lightness, softboard panels are used primarily for insulating or sound control purposes. They are used extensively as decorative panels in suspended ceilings and as ceiling tiles (Fig. 4-8). Much use is made of softboard for exterior wall sheathing. This type may be coated or impregnated with asphalt to protect it from moisture during construction.

Softboard panels can be easily cut with a knife, hand saw, or power saw. They cannot be hand-planed with any satisfactory results. Wide-headed nails, staples, or adhesive are used to fasten softboards in place, depending on the type and use.

Other

The preceding chapters have been limited to engineered wood panels and boards. There are many more products used in the construction industry other than those already mentioned. It is recommended that the student study *Sweet's Architectural File* to become better acquainted with the thousands of building material products on the market. This reference is well-known by architects, contractors, and builders, and is revised and published annually. Two publications, *Products for General Building and Renovations* and *Products for Homebuilding and Remodeling,* may be available at the school or city library.

Fig. 4-8 Softboards are used extensively for decorative ceiling panels. (*Courtesy of Armstrong World Industries, Inc.*)

Review Questions

Select the most appropriate answer.

1. Plywood always contains
   a. three layers.
   b. four layers.
   c. an even number of layers.
   d. an odd number of layers.

2. Most softwood plywood is made of
   a. cedar.   c. western pine.
   b. fir.   d. spruce.

3. The best-appearing face veneer of a softwood plywood panel is indicated by the letter
   a. A.   c. E.
   b. B.   d. N.

4. Which is a good selection of plywood for exterior wall sheathing?
   a. APA Structural Rated Sheathing, Exposure 1
   b. APA A-C, Exterior
   c. APA-Rated Sheathing, Exposure 2
   d. CD, Plugged, Exterior

5. A number "0" on the right of the slash in the Identification Index of Plywood means
   a. there are no knots in the inner plies.
   b. it should not be used for subflooring.
   c. it should not be used for roof sheathing.
   d. it should not be used for any structural purpose.

6. Particleboard not rated as structural may be used for
   a. underlayment.   c. wall sheathing.
   b. subflooring.   d. roof sheathing.

7. Hardboard may be used in
   a. interior applications only.
   b. exterior and interior applications.
   c. applications protected from moisture.
   d. cabinet and furniture work only.

8. Much of the softboard is used in the construction industry for
   a. underlayment for wall-to-wall rugs.
   b. roof covering.
   c. decorative ceiling panels.
   d. interior wall finish.

9. For use as an underlayment over structural subflooring where continuous or severe moisture may be present, select
   a. A-C, Group 1, Exterior plywood.
   b. B-D, Group 2, Exposure 1 plywood.
   c. Underlayment, Group 1, Exposure 1 plywood.
   d. C-C Plugged, Group 2, Exterior plywood.

10. The recommended selection of plywood where the appearance of one side is important for interior applications such as built-ins and cabinets is
    a. A-A, Group 1, Exposure 1 plywood.
    b. M.D. Overlay, Group 1, Exterior plywood.
    c. Decorative, Group 2, Interior plywood.
    d. A-D, Group 2, Exposure 2 plywood.

■ BUILDING FOR SUCCESS ■

Listening Skills

In order to be effective, the carpenter has to be receptive to information, instruction, and details. These are criteria for quality work. Not only what we listen to, but *how* we listen, forms the basis for our gaining knowledge and understanding.

Many of us have played the game in which a group of people sit in a circle and someone begins to relay a story or situation to the person at his or her side. As this message is being whispered into a person's ear, many times it becomes garbled or distorted as it is passed around the circle. Each person interprets what is heard and passes it on until the message comes back to the original communicator. The final report of what was received by the person who first stated the story usually has changed quite a bit. A variety of versions can be shared as to what people thought they heard. This exercise can help us understand the need for clear communication.

The world of construction is similar to other industries across the nation, and calls for good listening skills. These skills must be utilized and nurtured daily in order for communication to be productive. Listening skills must be developed.

Just like the variety of readers that range from rapid scanners to detail seekers, people tend to listen in different ways. Usually the contractor has a daily orientation session with superintendents, supervisors, and foremen in order to provide clear understandings about the day's activities. This is where information is shared about the goals, work assignments, and schedules the company has planned. There may be discussion about the previous day's accomplishments and problems before the new day is brought into focus. Many times this information exchange occurs before the normal work day begins.

As these initial meetings end, the supervisors then relate the new information to each work crew. The leader relates what he or she heard and briefs the workers. It is imperative that the information does not change.

Listening at the construction site may be difficult due to noise created by construction equipment and assembly processes. Hearing protection may need to be worn in order to prevent hearing loss. Today much of the communication linking is conducted by mobile phones and radios in the field. These higher-technology forms of audio communication are almost a necessity and require constant access by each worker.

Assuming that the conditions are good for hearing, how does the carpenter develop good listening skills?

First, all workers must be open to listening. Being receptive to instruction and information is a necessity for learning to take place. A closed mind may be demonstrated when instruction is not being followed.

Second, we must not be selective listeners. Selective listening may influence whom we are willing to listen to and what portion of the total information we actually hear. Appropriate listening will include all information. As each person studies for construction work, he or she must develop the ability to listen to everything that is said.

Third, we must focus our attention on those who know the industry and serve as our advisors. Usually the builder or other leader will work directly with us to provide an immediate communication link. The best skills will be developed by listening to those with more experience. This textbook has been developed by an experienced professional who has a vast array of information that needs to be absorbed by each student. Some of the content of the text will be received through lectures or discussions led by the course instructor. Your full attention will be required in order for learning to take place.

The instruction being presented by each educational outreach using this text is also being reinforced with personal construction work experience. Most of the information in those work experiences will be communicated orally.

One example of good listening skills is demonstrated by a custom home builder who was "all ears" as a student. During the two years of preparation at Southeast Community College, Sam Manzitto, a custom home builder in Lincoln, Nebraska, was continually absorbing information that would be necessary in the home building industry. He had a desire to gain all of the technical data and information possible for the years ahead. He and his brother Mike now are recognized as excellent home builders in the Lincoln community. Last January, Manzitto was recognized by the National Association of Home Builders as one of "America's Best Builders." His company was one of eleven to receive this prestigious award. If Manzitto's listening skills had been poor, he would not have even been in a position to be considered as a candidate for this honor.

How can we develop good listening skills or refine those we currently possess?

1. We should try to determine the speaker's motive, purpose, and perspective. Premature evaluation of what is being said could lead to the wrong assumption. This abruptness will be the breeding grounds for frustration, ill feelings, and errors.

2. Listen for what the speaker actually means. Just as we must sometimes read between the lines, we must try to concentrate on what is meant by the statements.

3. Listen for the speaker's outlook or attitude. Motives shape attitudes and attitudes are visible in behaviors. This may

help us understand the speaker and see where the motivation is based.

4. Don't give negative feedback. Both verbal and nonverbal messages can be sent by the listener. This could have an effect on the speaker's delivery. For example, crossed arms and a stare at the ceiling will undoubtedly tell the speaker you are disagreeing even before hearing the entire message.

5. Record and use what has been said to help you make personal decisions. Most of what is said has value. The value may simply be new information, reinforcement, or verification for your decisions. Somehow put it to use.

A key to good listening will continue to be how receptive the listener is to all information. A closed mind receives very little that is meaningful. You have the responsibility to sort through what is being said and apply the real substance to your life.

FOCUS QUESTIONS: For individual or group discussion.

1. What might be some good reasons for developing good listening skills?

2. Do you personally feel you listen well? Why?

3. In the past, have you communicated with someone who did not even hear you out? What was his or her reaction? What was yours?

4. How do you feel you can improve your own listening skills? Will those skills reflect your attitude toward the issue and person talking?

Engineered Lumber Products

The modern lumber industry is faced with many problems. Insuring the survival of endangered wildlife, making wise use of forest resources, and harvesting logs are some of the problems. Although the supply of *old-growth* trees available for harvesting is steadily diminishing, enough lumber must still be manufactured to meet increasing demands.

OBJECTIVES

After completing this unit, the student will be able to describe the manufacture, composition, uses, and sizes of:

- laminated veneer lumber

- parallel strand lumber and laminated strand lumber

- wood I-beams

- glue-laminated beams

UNIT CONTENTS

Old-growth trees are large, tall, tight-grain trees that take over two hundred years to mature. Most of the large dimension, structural, solid-sawn lumber comes from old-growth trees because not enough can be cut from smaller second- and third-growth trees. Yet it is not economical to plant slow-growing trees and wait for them to mature. Although the supply of old-growth trees is decreasing, the supply of smaller second- and third-growth trees is abundant and actually increasing as the result of replanting and reforestation.

An inevitable result of the decreasing supply of large old-growth trees and the abundance of smaller trees is the development and use of *engineered lumber products*.

Engineered lumber products, commonly called *ELPs*, are reconstituted wood products and assemblies designed to replace traditional structural lumber. Engineered lumber products consume less wood and can be made from smaller trees. Traditional lumber processes typically convert 40 percent of a log to structural solid lumber. Engineered lumber processes convert up to 75 percent of a log into structural lumber. In addition, the manufacturing processes of engineered lumber consume far less energy than those of solid lumber. Also, some ELPs make use of abundant, fast-growing species not currently harvested for solid lumber.

The final engineered lumber product has greater strength, and consequently can span greater distances. It is predicted that engineered lumber will be used more than solid lumber in the near future. It is important that present and future builders be thoroughly informed about engineered lumber products.

This unit describes how engineered lumber products are made, where they are used, and what sizes are available. Construction details for ELPs are shown in following units on floor, wall, and roof framing.

CHAPTER 5

LAMINATED VENEER LUMBER

Laminated Veneer Lumber

Laminated veneer lumber, commonly called *LVL*, is one of several types of engineered lumber products (Fig. 5-1). It was first used to make airplane propellers during World War II. The world's first commercially produced LVL for building construction was patented as MICRO-LAM laminated veneer lumber in 1970. Since then, because of increased demand for the product, several manufacturers are marketing it under similar names, such as GANG-LAM (Fig. 5-2), STRUCLAM, and VERSA-LAM. LVL is now widely used in wood frame construction.

Manufacture of LVL

Like plywood, LVL is a wood veneer product. The grain in each layer of veneer in LVL runs in the same direction, parallel to its length (Fig. 5-3). This is unlike plywood, in which each layer of veneer is laid with the grain at right angles to each other.

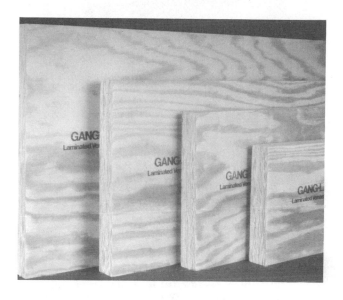

Fig. 5-2 Laminated veneer lumber is marketed under several brand names. (*Courtesy of Louisiana-Pacific Corporation*)

PARALLAM POSTS

WOOD I-BEAMS

LVL BEAMS

PARALLAM BEAMS

Fig. 5-1 There are several types of engineered lumber. (*Courtesy of Trus Joist MacMillan, Boise, Idaho*)

Fig. 5-3 In LVL, the grain in each layer of veneer runs in the same direction.

Laminated veneer lumber is made from sheets of veneer peeled from logs, similar to the first step in the manufacture of plywood (see Fig. 3–5). Douglas fir or southern pine is used because of their strength. The veneer is peeled in widths of 27 inches or 54 inches and from 1/10 to 3/16 inch in thickness. It is then dried, cut into sheets, "ultrasonically" graded for strength, and sorted.

The veneer sheets are laid up in a staggered pattern so that the ends overlap. They are then permanently bonded together with an exterior type adhesive in a continuous press under precisely controlled heat and pressure. Unlike plywood, the

LVL veneers are *densified*, that is, the thickness is compressed and made more compact; fifteen to twenty layers of veneer make up a typical 1 3/4 inch thick beam. The edges of the bonded veneers are then edge trimmed to specified widths and end-cut to specified lengths (Fig. 5-4).

LVL Sizes

Laminated veneer lumber is manufactured up to 3 1/2 inches thick, 18 inches wide, and 80 feet long. The usual thickness are 1 1/2 inches and 1 3/4 inches. The 1 1/2 inch thickness is the same as nominal 2 inch thick framing lumber, while doubling the 1 3/4 inch thickness equals the width of nominal 4 inch framing. LVL widths are usually 9 1/4, 11 1/4, 11 7/8, 14, 16, and 18 inches. LVL beams may be fastened together to make a thicker and stronger beam (Fig. 5-5).

Uses of LVL

Laminated veneer lumber is intended for use as high-strength, load-carrying beams to support the weight of construction over window and door openings, and in floor and roof systems of residential and light commercial wood frame construction (Fig. 5-6). Its relatively light weight makes it easy to handle and suitable for use also in concrete forming, as scaffold planks, in manufactured hous-

Fig. 5-4 LVL manufacturing process, simplified. (*Courtesy of Trus Joist MacMillan*)

TWO ROWS FOR BEAMS UP TO 12" DEEP
THREE ROWS FOR BEAMS OVER 12" DEEP

16D NAILS - 1-3/4" LVL
12D NAILS - 1-1/2" LVL

12" O.C. ON
EACH FACE

NOTES:
THESE FASTENING RECOMMENDATIONS APPLY:
• TO 2-PLY AND 3-PLY BEAMS ONLY.
• WHEN BEAM IS LOADED UNIFORMLY ON THE
  TOP EDGE OF EACH PLY.

Fig. 5-5 Recommended nailing pattern for fastening LVL beams together. (*Courtesy of Louisiana-Pacific Corporation*)

ing, and in many other specialties where a light-weight, strong beam or plank is required. It can be cut with regular tools and requires no special fasteners.

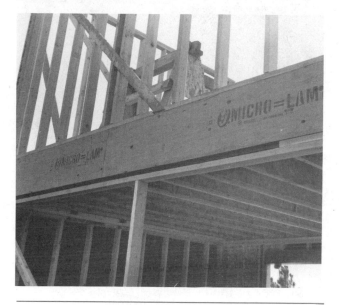

Fig. 5-6 LVL is designed to be used as load-carrying beams. (*Courtesy Trus Joist MacMillan*)

CHAPTER 6

PARALLEL STRAND LUMBER AND LAMINATED STRAND LUMBER

Parallel Strand Lumber

Parallel strand lumber (PSL) commonly known by its brand name, Parallam (Fig. 6-1), was developed, like all engineered lumber products, to meet the need of the building industry. PSL provides large dimension lumber (beams, planks, and posts). PSL also utilizes small diameter, second-growth trees, thus protecting the diminishing supply of old-growth trees.

Manufacture of PSL

Parallam PSL is manufactured by peeling veneer of Douglas fir and southern pine from logs in much the same manner as for plywood and LVL. The veneer is then dried and clipped into *strands* (narrow strips) of up to 8 feet in length and 1/8 or 1/10 inch in thickness. Small defects are removed and

the strands are then coated with a waterproof adhesive. The oriented strands are fed into a rotary belt press and bonded using a patented microwave pressing process. The result is a continuous timber, up to 11 inches thick by 17 inches wide, which can then be factory-ripped into widths and thicknesses to fit builders' needs (Fig. 6-2). The four surfaces are sanded smooth before the product is shipped.

PSL Sizes

PSL comes in many thicknesses and widths and is manufactured up to 66 feet long. PSL is available in square and rectangular shapes for use as posts and beams. Beams are sold in convenient 1 3/4 thickness for installation of single and multiple laminations. Solid 3 1/2 inch thicknesses are compatible with 2×4 wall framing. A 2 11/16 inch thickness is also available. A list of beam and post sizes is

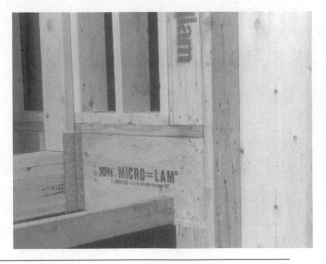

Fig. 6-1 Parallel strand lumber is commonly called Parallam and is used as beams and posts to carry heavy loads. (*Courtesy of Trus Joist MacMillan*)

Fig. 6-2 The manufacture of parallel strand lumber. (*Courtesy of Trus Joist MacMillan*)

shown in Figure 6-3. Also available is Parallam 269, which measures 2 11/16 inches thick.

Uses of PSL

Parallel strand lumber can be used wherever there is a need for a large beam or post. The differences between PSL and solid lumber are many. Solid lumber beams may have defects, like knots, checks, and shakes, which weaken them, while PSL is consistent in strength throughout its length. PSL is readily available in longer lengths and its surfaces are sanded smooth, eliminating the need to cover them by boxing the beams.

Laminated Strand Lumber

The registered brand name of *laminated strand lumber (LSL)* is *TimberStrand* (Fig. 6-4). While LVL and PSL are made from Douglas fir and southern pine,

THICKNESS	WIDTH
3 1/2"	3 1/2"
3 1/2"	5 1/4"
3 1/2"	7"
5 1/4"	5 1/4"
5 1/4"	7"
7"	7"

PARALLAM COLUMN & POST SIZES

THICKNESS	DEPTHS
1 3/4"	7 1/4"
3 1/2"	9 1/4"
5 1/4"	11 1/4"
7"	11 1/2"
	11 7/8"
	12"
	12 1/2"
	14"
	16"
	18"

PARALLAM BEAM SIZES

Fig. 6-3 Post and beam sizes of parallel strand lumber.

Fig. 6-4 Laminated strand lumber is commonly called TimberStrand. (*Courtesy of Trus Joist MacMillan*)

LSL can be made from very small logs of practically any species of wood; its strands are much shorter than those of parallel strand lumber. At present, LSL is being manufactured from surplus, overmature aspen trees that are usually are not large, strong, or straight enough to produce ordinary wood products.

Fig. 6-5 Laminated strand lumber can be manufactured from practically any species of wood. (*Courtesy of Trus Joist MacMillan*)

Manufacture of LSL

The TimberStrand LSL manufacturing process begins by cleaning and debarking 8-foot aspen logs. The wood is then cut into strands up to 12 inches long, dried, and treated with a resin. The treated strands are aligned parallel to each other to take advantage of the wood's natural strength. The strands are pressed into solid *billets* (large blocks) up to 5 1/2 inches thick, 8 feet wide, and 35 feet long (Fig. 6-5). Scraps from the process fuel the furnace that provides heat and steam for the plant.

Sizes and Uses of LSL

The long billet is resawn and sanded to sizes as required by customers. It is used for a wide range of *millwork*, such as doors, windows, and virtually any product that requires high-grade lumber. It is also used for truck decks, manufactured housing, and some structural lumber, such as window and door headers.

One of the differences between PSL and LSL is the kind of wood from which they are manufactured. Because PSL is made from stronger wood and with longer strands, it is designed to carry heavy loads, while LSL is not.

CHAPTER 7 WOOD I-BEAMS

Wood I-Beams

There are several manufacturers of *wood I-beams* and each has its own registered trademark or brand name, such as BCI joist, GNI joist, StrucJoist, TJI joist, and Wood I-Beam. The TJI joist was invented in 1969 as a substitute for solid lumber *joists* (structural members of a floor frame).

Wood I-beams are engineered wood assemblies that utilize an efficient "I" shape, common in steel beams, which gives them tremendous strength in relation to their size and weight (Fig. 7-1). Consequently, they are able to carry heavy loads over long distances while using considerably less wood than solid lumber of a size necessary to carry the same load over the same span.

The flanges of the beam may be made of laminated veneer lumber or specially selected *finger-jointed* solid wood lumber (Fig. 7-2). The web of the beam may be made of plywood, laminated veneer lumber, or oriented strand board.

The Manufacture of Wood I-Beams

Regardless of who makes them, the manufacturing process of wood I-beams consists of gluing top and

OSB WEB

LVL FLANGE

Fig. 7-1 Wood I-beams are available in many sizes. (*Courtesy of Louisiana-Pacific Corporation*)

Fig. 7-2 Finger-joints are used to join the ends of short pieces of lumber to make a longer piece.

Fig. 7-3 Manufacture of wood I-beams. (*Courtesy of Trus Joist MacMillan*)

Fig. 7-4 Wood I-beam sizes. Sizes vary with the manufacturer.

bottom flanges to a connecting center web. First the web material is ripped to specified width and then the edges and ends are shaped for joining to flanges and adjacent web sections. The ends of the flange material are finger-jointed for gluing end to end. One side of the flange material is grooved to receive the beam's web. Flanges and webs are then assembled with waterproof glue by pressure-fitting the web into the flanges. The wood I-beam is end-trimmed and the adhesive cured in an oven or at room temperature (Fig. 7-3). As with most engineered wood products, wood I-beams are produced to approximate equilibrium moisture content.

Wood I-Beam Sizes

Wood I-beams may have webs of various thicknesses, and flanges may vary in thickness and width, depending on intended end use and the manufacturer. Beam depths are available from 9 1/4 to 30 inches (Fig. 7-4). Beams with larger webs and flanges are designed to carry heavier loads. Wood I-beams are available up to 80 feet long.

Uses of Wood I-Beams

Wood I-beams are intended for use in residential and commercial construction as floor joists, roof rafters, and *headers* for window, entrance door, and garage door openings (Fig. 7-5). Window and door headers are beams that support the load above wall openings.

Fig. 7-5 Wood I-beams are used as roof rafters as well as for floor joists and window and door headers. (*Courtesy of Louisiana-Pacific Corporation*)

CHAPTER 8

GLUE-LAMINATED LUMBER

Glue-Laminated Lumber

Glue-laminated lumber, commonly called *glulam*, is constructed of solid lumber glued together, side against side, to make beams and joists of large dimension that are stronger than natural wood of the same size (Fig. 8-1). Even if it were possible, it would not be practical to make solid wood beams as large as most glulams. They are used for structural purposes, but are decorative as well and, in most cases, their surfaces are left exposed to show the natural wood grain.

Fig. 8-1 Glue-laminated lumber is commonly called glulams. (*Courtesy of American Plywood Association*)

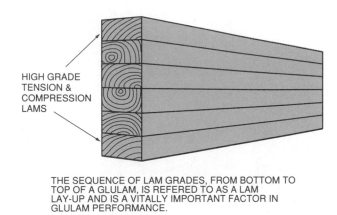

HIGH GRADE
TENSION &
COMPRESSION
LAMS

THE SEQUENCE OF LAM GRADES, FROM BOTTOM TO TOP OF A GLULAM, IS REFERED TO AS A LAM LAY-UP AND IS A VITALLY IMPORTANT FACTOR IN GLULAM PERFORMANCE.

Fig. 8-2 Glulams are made by gluing stacks of solid lumber.

Manufacture of Glulams

Glulam beams are made by gluing stacks of *lam* stock into large beams, headers, and columns (Fig. 8-2). Lams are the individual pieces in the glued-up stack. The lams, which are at approximate equilibrium moisture content, are glued together with exterior adhesives. Thus glulams are rated in terms of weather resistance.

Lam Layup. The lams are arranged in a certain way for maximum strength and minimum shrinkage. Different grade lams are placed where they will do the most good under load conditions. High-grade *tension* lams are used in tension faces, and high-grade *compression* lams are used in compression faces.

Tension is a force applied to a member that tends to increase along its length, while compression is a force tending to decrease along the length of the lam (Fig. 8-3). When a load is imposed on a glulam beam that is supported on both ends, the topmost lams are in compression while those at the bottom are in tension.

More economical lams are used in the lower-stressed middle sections of the glulam. The sequence of lam grades, from bottom to top of a glulam, is referred to as *lam layup*, and it is a vitally important factor in glulam performance. Because of this lam layup, glulam beams come with one edge stamped "TOP." *Always remember to install glulam beams with the "TOP" stamp facing toward the sky* (Fig. 8-4).

Glulam Grades, Sizes, and Uses

Grades. The American Plywood Association-Engineered Wood Systems trademark, APA-EWS (Fig. 8-5), appears on all beams manufactured by American Wood Systems member mills. The AWS is a related corporation of the APA. The trademark guarantees that the glulams meet all the requirements of the American National Standards Institute (ANSI).

Glulams are manufactured in three grades for appearance. The *industrial appearance grade* is used in warehouses, garages, and other structures in which appearance is not of primary importance or

Fig. 8-3 The load on a beam places lams in tension and compression. (*Courtesy of Bohemia, Inc.*)

Fig. 8-4 Glulam beams must be installed with the edge stamped "TOP" pointed toward the sky.

when the beams are not exposed. *Architectural appearance grade* is for projects where appearance is important, and the *premium appearance grade* is used when appearance is critical (Fig. 8-6). There is no difference in strength among the different appearance grades. All glulams with the same design values are rated for the same loadings, regardless of appearance grade.

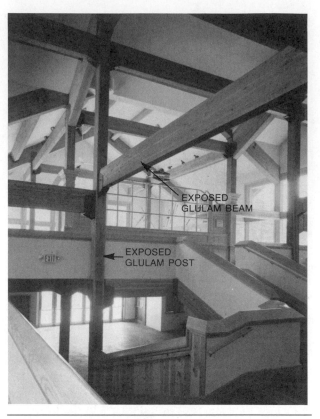

Fig 8–6 The appearance of an exposed beam is important. Some glulam beams are manufactured for appearance. (*Courtesy of Bohemia, Inc.*)

Sizes. The dimensions of glulam beams are indicated by width, depth, and length (Fig. 8-7). Widths range from 2 1/2 to 8 3/4 inches, depths from 6 to

(1) Indicates structural use:
    B - Simple span bending member.
    C - Compression member.
    T - Tension member.
    CB - Continuous or cantilevered span bending member.
(2) Mill number.
(3) Identification of ANSI Standard A190.1, Structural Glued Laminated Timber.
(4) Code recognition of *American Wood Systems* as a quality assurance agency for glued structural members.
(5) Applicable laminating specification.
(6) Applicable combination number.
(7) Species of lumber used.
(8) Designates appearance grade, INDUSTRIAL, ARCHITECTURAL, PREMIUM.

Fig. 8-5 The grade stamp assures that the beam has met all the necessary requirements. (*Courtesy of American Plywood Association*)

Fig. 8-7 The dimensions of glulam beam are width, depth, and length.

28 1/2 inches, and lengths are generally available from 10 to 40 feet in 2-foot increments (Fig. 8-8).

Uses. Various wood species are used to produce straight, curved, arched, and special shapes for all structures—from elegant homes and churches to large malls, warehouses, and civic centers. In all, the American Institute of Timber Construction (AITC) recognizes over one hundred glulam beam

Combination	Species	Use
24F-V4	Df/Df	Simple Span
24F-V8	Df/Df	Multi Span & Cantilever
24F-V5	Df/Hf	Simple Span
24F-V10	Df/Hf	Multi Span & Cantilever
Combination #1	Df/Df	Columns

Combination Number Table Notes
- The 24F means the beam's tension face is rated at 2,400 psi bending stress.
- V stands for visually graded laminations, and the 4, 5, 8, or 10 refer to the lamination lay up.
- The V4 and V8 combinations use only Douglas Fir. V5 and V10 combine Douglas Fir with Hemfir core lams.
- Combination #1 is for columns and uses only Douglas Fir L3 laminates.

Fig. 8-9 Only five glulam combinations, out of over a hundred, are commonly used. (*Courtesy of Bohemia, Inc.*)

combinations, most of which are for specialized applications. However, five glulam beam combinations are the most commonly used and meet most consumer needs (Fig. 8-9).

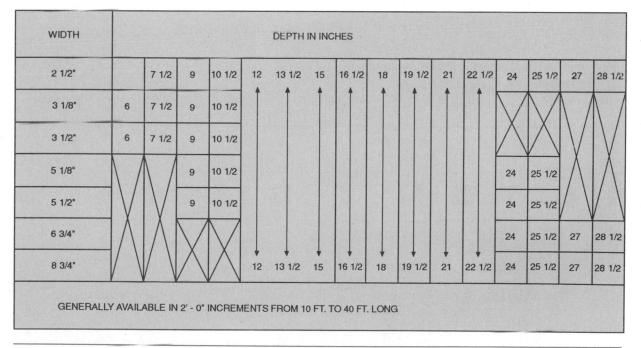

Fig. 8-8 Glulam beam sizes.

Review Questions

Select the most appropriate answer.

1. Engineered lumber products are designed as replacements or substitutes for
   a. sawn solid lumber.
   b. second-growth lumber.
   c. steel framing.
   d. structural lumber.

2. Unlike plywood, the veneers of LVL are
   a. deciduous.   c. diversified.
   b. densified.   d. double-faced.

3. Laminated veneer lumber is manufactured in lengths up to
   a. 40 feet.   c. 80 feet.
   b. 50 feet.   d. 100 feet.

4. Parallel strand lumber is made from
   a. Alaskan cedar and California redwood.
   b. Douglas fir and southern pine.
   c. Idaho pine and eastern hemlock.
   d. Englemann spruce and western pine.

5. Parallel strand lumber is manufactured in lengths up to
   a. 36 feet.   c. 66 feet.
   b. 50 feet.   d. 80 feet.

6. The web of wood I-beams may be made of
   a. hardboard.   c. solid lumber.
   b. particleboard.   d. strand board.

7. The flanges of wood I-beams are generally made from
   a. glue-laminated lumber.
   b. laminated veneer lumber.
   c. parallel strand lumber.
   d. laminated strand lumber.

8. The lams used in glue-laminated lumber are at approximately
   a. the fiber-saturation point.
   b. 19 percent moisture content.
   c. equilibrium moisture content.
   d. 30 percent moisture content.

9. In glulam beams, specially selected tension lams are placed
   a. at the top of the beam.
   b. at the bottom of the beam.
   c. in the center of the beam.
   d. throughout the beam.

10. Although over one hundred different glulam combinations are recognized for use, the number of combinations commonly used are
    a. five.   c. twelve.
    b. ten.   d. eighteen.

■ B U I L D I N G F O R S U C C E S S ■

Characteristics of Construction Materials

Carpenters and other construction workers need to know many things about the materials they use. The most successful builders understand the properties, characteristics, uses, and limitations of the materials they install. For example, a basic requirement for selecting the correct species of wood for a framing job is to know how the wood will perform under loads and the span capability for that species.

Each construction job will vary and call for specific materials. It is true that the architect, engineer, and designer will possess a good understanding of material requirements for a job. The builder must also have some of the same information. The product manufacturer, lumber producer, and materials supplier will have access to that information, but it must be passed on to the builder. For the builder, this becomes a continual learning process and reinforces the need for maintaining a personal construction materials and products file.

High technology continues to create new products from existing or new materials at an accelerated pace. Economic and environmental factors have forced the manufacturing and fabricating industries to look at new ways of producing construction materials. This in turn obligates everyone to look at new or alternate methods of construction. The lumber industry has been drastically affected by environmental issues that show little or no chance for reversal. This turn of events has prompted new thinking and new products.

New or improved construction products such as structural and nonstructural wood panels, laminated veneer lumber, parallel and laminated strand lumber, wood I-beams, glue-laminated beams, and other preengineered products are now the norm in many areas of the country. These innovative products are allowing for the consumption of smaller trees with minimal waste. The use of traditional dimensional lumber will probably continue to decline in many areas. Environmental factors will continue to affect the products used in the conventional platform frame system. We must continually learn as much as possible about these new products as they become a part of our materials and products inventory. The carpenter of the twenty-first century will be expected to know more about construction materials and products in order to be effective.

With the information about construction materials presented in this text, manufacturers' data sheets, and other technical publications, construction students and workers will be able to assemble quality products. The selection and installation of construction products become less complex with a better understanding of each material's characteristics. For example, wood technology must be studied to gain a good working knowledge of defects and how those defects affect the appearance or performance of dimensional lumber and boards. Knowing how various species will perform and why will help us select which wood to use in every case. Softwoods and hardwoods possess many different characteristics that make one or the other more suitable for a certain job. The carpenter must know the machineability, strength, color, texture, decay resistance, and so forth of the woods he or she uses.

A carpenter can read a grade stamp on lumber or sheet goods to determine if code requirements have been met. This in turn helps assure quality. Using inferior materials for a project will cause problems even before the project is completed.

More than one builder has had to correct errors after inappropriate lumber was installed in a frame system. In one incident a carpenter installed floor joists with a number 3 grade when the specs called for a number 2 grade. The problem surfaced when a building inspector noticed the grade stamp, after the subflooring was in place and the walls had been erected. The most economical solution was to place an additional number 2 grade joist alongside the original, thus doubling the cost of the floor joists. By using lumber with the appropriate grade stamp, the code requirements were met. With a knowledgeable and alert carpenter that problem could have been avoided.

The utilization of building materials will be greatly enhanced by their proper placement and storage at the construction site. As the lumber, sheet goods, and preengineered products arrive at the building site, they must be properly stored until used. A proper base system such as pallets or support runners must be used to keep the materials off the ground and free of moisture. The location should be relatively flat and offer easy access for a forklift or loader. Additionally the material should be within easy reach of construction workers but not block vehicles from unloading other materials.

Polyethylene film or other impermeable barriers should be used to cover the materials and protect them from excess moisture. Air circulation is necessary to prevent condensation from forming on the underside of the barrier during long periods of storage.

Many construction sites need security to prevent pilferage and vandalism. Some estimates have placed theft and vandalism damage in the billions of dollars across the country. All of this adds to the cost of construction, which eventually is paid by the owners and their customers.

There will continue to be a need for carpenters to maintain a good working knowledge of the materials and products they install. By knowing the basic composition of a product, its structural rating, and its appropriate uses, the carpenter can help guarantee quality. Quality never seems to go out of style. People are continually looking for quality and expect it in the construction of their homes and places of business. A carpenter's problem-solving skills will be greatly enhanced by knowing as much as possible about building materials. The end result will normally reflect greater safety, quality construction, and profit for the builder.

FOCUS QUESTIONS: For individual or group discussion.

1. What are some reasons for a carpenter to understand the characteristics of the materials he or she installs?

2. How can being able to fully read and interpret a lumber grade stamp help avoid potential problems while framing?

3. Assuming that someday you will be involved in the design and selection of construction materials, how do you think your decisions will be made? Will a good knowledge of material characteristics play a part? How?

Fasteners

Many kinds of fasteners are used in the construction industry. It is important to know the best kind and size to use for each job.

OBJECTIVES

After completing this unit, the student will be able to name and describe the following commonly used fasteners and select them for appropriate use:

- nails
- screws and lag screws
- bolts
- solid and hollow wall anchors
- adhesives

CONTENTS

CHAPTER 9

NAILS, SCREWS, AND BOLTS

ails, screws, and bolts are the most widely used of all fasteners. They come in many styles and sizes. The carpenter must know what is available and wisely select those most appropriate for fastening various materials under different conditions.

Nails

There are hundreds of kinds of nails manufactured for just about any kind of fastening job. They differ according to purpose, shape, material, coating, and in other ways. Nails are made of aluminum, brass, copper, steel, and other metals. Some nails are hardened so that they can be driven into masonry without bending. Only the most commonly used nails are described in this chapter (Fig. 9-1).

Uncoated steel nails are called *bright* nails. Various coatings may be applied to reduce corrosion, increase holding power, and enhance appearance. To prevent rusting, steel nails are coated with *zinc*. These nails are called *galvanized* nails. They may be coated by being dipped in molten zinc (*hot-dipped galvanized* nails), or they may be *electroplated* with corrosion-resistant metal (*plated* nails). Hot-dipped nails have a heavier coating than plated nails and are more resistant to rusting. Many manufacturers specify that their products be fastened with hot-dipped nails because of the heavier coating.

When fastening metal that is going to be exposed to the weather, use nails of the same material if possible. For example, when fastening aluminum, copper, or galvanized iron, use nails made of the same metal. Otherwise, a reaction with moisture and the two different metals, called *electrolysis,* will cause the softer metal to disintegrate over time.

When fastening some woods, such as cedar, redwood, and oak, exposed to the weather, use hot-dipped or stainless steel nails. Otherwise, a reaction between the acid in the wood and bright nails causes dark, ugly stains to appear around the fasteners.

Nail Sizes

The sizes of some nails are designated by the *penny* system. The origin of this system of nail measurement is not clear. Although many people think it should be discarded, it is still in use. Some believe it originated years ago when nails cost a certain number of pennies for a hundred of a specific length. Of course, the larger nails cost more per hundred than smaller ones, so nails that cost 8 pennies were larger

Fig. 9-1 Kinds of commonly used nails.

than those that cost 4 pennies. The symbol for penny is *d*; perhaps it is the abbreviation for *denarius*, an ancient Roman coin.

In the penny system the shortest nail is 2d, and is one inch long. The longest nail is 60d, and is 6 inches long (Fig. 9-2). A sixpenny nail is written as 6d, and is 2 inches long. Eventually, a carpenter can determine the penny size of nails just by looking at them.

The thickness or diameter of the nail is called its gauge. In the penny system, gauge depends on the kind and length of the nail. The gauge increases with the length of the nail. Long nails, 16d and over, are called *spikes.* The length of nails, not included in the penny system, is designated in inches and fractions of an inch, and the gauge may be specified by a number. As a general rule, select a nail that is three times longer than the material being fastened.

Kinds of Nails

Most nails, cut from long rolls of metal wire, are called *wire nails. Cut nails,* used only occasionally, are wedge-shaped pieces stamped from thin sheets of metal. The most widely used wire nails are the common, box, and finish nails (Fig. 9-3).

Common Nails. *Common nails* are made of wire, are of heavy gauge, and have a medium-sized head. They have a pointed end and a smooth shank. A barbed section just under the head increases the holding power of common nails.

PENNY (d) SIZES

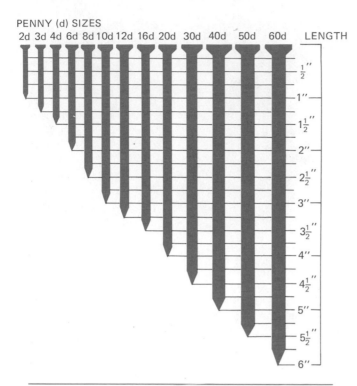

Fig. 9-2 Some nails are sized according to the penny system.

Fig. 9-3 Most widely used nails are the common, box, and finish.

Box Nails. *Box nails* are similar to common nails, except they are thinner. Because of their small gauge, they can be used close to edges and ends with less danger of splitting the wood. Many box nails are coated with cement to increase their holding power.

Finish Nails. *Finish nails* are of light gauge with a very small head. They are used mostly to fasten interior trim. The small head is sunk into the wood with a nail set and covered with a filler. The small head of the finish nail does not detract from the appearance of a job as much as a nail with a larger head.

Casing Nails. *Casing nails* are similar to finish nails. Many carpenters prefer them to fasten exterior finish. The head is cone-shaped and slightly larger than that of the finish nail, but smaller than that of the common nail. The shank is the same gauge as that of the common nail.

Duplex Nails. On temporary structures, such as wood scaffolding and concrete forms, the *duplex nail*, sometimes called the *staging nail*, is often used. The lower head ensures that the piece is fastened tightly. The projecting upper head makes it easy to pry the nail out when the structure is dismantled (Fig. 9-4).

Brads. *Brads* are small finishing nails (Fig. 9-5). They are sized according to length in inches and

Fig. 9-4 Duplex nails are used on temporary structures.

Fig. 9-5 Brads are small finishing nails.

gauge. Usual lengths are from 1/2 inch to 1 1/2 inches, and gauges run from #14 to #20. The higher the gauge number, the thinner the brad. Brads are used for fastening thin material, such as small molding.

Roofing Nails. *Roofing nails* are short nails of fairly heavy gauge with wide, round heads. They are used for such purposes as fastening roofing material and softboard wall sheathing. The large head holds thin or soft material more securely.

Some roofing nails are coated to prevent rusting. Others are made from noncorrosive metals such as aluminum and copper. The shank is usually barbed to increase holding power. Usual sizes run from 3/4 inch to 2 inches. The gauge is not specified when ordering.

Masonry Nails. *Masonry nails* may be cut nails or wire nails (Fig. 9-6). These nails are made from hardened steel to prevent them from bending when being driven into concrete or other masonry. The cut nail has a blunt point that tends to prevent splitting when it is driven into hardwood. Some masonry and flooring nails have round shanks of various designs for better holding power.

Miscellaneous Fasteners. *Tacks* are small pointed nails with a head (Fig. 9-7) used to fasten thin material such as wire-mesh screen. They are available in sizes 1 to 24, with the larger number designating the longer tack.

Staples are U-shaped fasteners (Fig. 9-8). They come in a number of sizes, which are designated by length. Two most popular sizes are 3/4 inch and 1 1/2 inches, with the gauge of the staple increasing with the length. They are usually galvanized and are frequently used to fasten wire fencing and heavy wire mesh. They can be driven individually with a hammer. Widely used in the industry are

Fig. 9-6 Masonry nails are made of hardened steel.

Fig. 9-7 Tacks are sometimes used to fasten screen mesh or other thin material.

Fig. 9-8 Staples are U-shaped nails.

Fig. 9-9 Staples may come glued together in strips for use in power staplers.

staples that come glued together in rows to be driven by hand (Fig. 9-9) or power staplers (Fig. 9-10).

Corrugated fasteners are thin, wrinkled metal pieces used to fasten joints (Fig. 9-11). They are used on light or rough work and on finish work where the fastener is not exposed. They are available in individual pieces to be driven with a hammer or in glued strips for use in power drivers. The usual widths are 1/2 inch and 5/8 inch.

Fig. 9-10 Using a power stapler to fasten plywood roof sheathing (*Courtesy of American Plywood Association*).

Fig. 9-11 Corrugated fasteners are used to fasten joints (*Courtesy of American Plywood Association*).

Screws

Wood screws are used when greater holding power is needed, and when the work being fastened must at times be removed. For example, door hinges must be applied with screws because nails would pull loose after a while, and the hinges may, at times, need to be removed. Screws cost more than nails and require more time to drive. When ordering screws, specify the length, gauge, type of head, coating, kind of metal, and screwdriver slot.

Kinds of Screws

A wood screw is identified by the shape of the screwhead and screwdriver slot. For instance, a screw may be called a *flat head Phillips* or a *round head common* screw. Three of the most common shapes of screwheads are the *flat head*, *round head*, and *oval head*. Other shapes include the *truss head* and *fillister head*.

The pointed end of a screw is called the *gimlet point*. The threaded section is called the *thread*. The smooth section between the head and thread is called the *shank*. Screw lengths are measured from the point to that part of the head that sets flush with the wood when fastened (Fig. 9-12).

Screw Slots. Screwheads made with a straight, single slot are called *common screws*. *Phillips head screws* have a crossed slot. There are many other types of screwdriver slots, each with a different name (Fig. 9-13).

Sheet Metal Screws. Wood screws are threaded, in most cases, only partway to the head.

Fig. 9-12 Common kinds of screws and screw terms.

Fig. 9-13 Common kinds of screw slots.

Fig. 9-14 The point of a self-drilling screw has cutting edges that drill a hole as the screw is driven.

The threads of sheet metal screws extend for the full length of the screw and are much deeper. Sheet metal screws are used for fastening thin metal. They are also recommended for fastening hardboard and particleboard because their deeper thread grabs better in softer and fiberless material.

Another type of screw, used with power screwdrivers, is the *self-drilling screw,* which is used extensively to fasten metal framing. This screw has a cutting edge on its point to eliminate predrilling a hole (Fig. 9-14). It is important that the drilling process be completed before the threading process begins. Drill points are available in various lengths and must be equal to the thickness of the metal being fastened.

Many other screws are available that are designed for special purposes. Like nails, screws come in a variety of metals and coatings. Steel screws with no coating are called *bright* screws.

Screw Sizes

Screws are made in many different sizes. Usual lengths range from 1/4 inch to 4 inches. Gauges run from 0 to 24 (Fig. 9-15). Unlike brads and some nails, the higher the gauge number, the greater the diameter of the screw. Screw lengths are not available in every gauge. The lower gauge numbers are for shorter, thinner screws. Higher gauge numbers are for longer screws.

Lag Screws

Lag screws (Fig. 9-16) are similar to wood screws except that they are larger and have a square or hex head designed to be turned with a wrench instead of a screwdriver. This fastener is used when great holding power is needed to join heavy parts and where a bolt cannot be used.

WOOD SCREWS

LENGTH	GAUGE NUMBERS												
1/4 INCH	0	1	2	3									
3/8 INCH			2	3	4	5	6	7					
1/2 INCH			2	3	4	5	6	7	8				
5/8 INCH				3	4	5	6	7	8	9	10		
3/4 INCH					4	5	6	7	8	9	10	11	
7/8 INCH							6	7	8	9	10	11	12
1 INCH							6	7	8	9	10	11	12	14
1 1/4 INCH								7	8	9	10	11	12	14	16
1 1/2 INCH							6	7	8	9	10	11	12	14	16	18
1 3/4 INCH									8	9	10	11	12	14	16	18	20
2 INCH									8	9	10	11	12	14	16	18	20
2 1/4 INCH										9	10	11	12	14	16	18	20
2 1/2 INCH													12	14	16	18	20
2 3/4 INCH														14	16	18	20
3 INCH															16	18	20
3 1/2 INCH																18	20	24
4 INCH																18	20	24

WHEN YOU BUY SCREWS, SPECIFY (1) LENGTH, (2) GAUGE NUMBER, (3) TYPE OF HEAD — FLAT, ROUND OR OVAL, (4) MATERIAL — STEEL, BRASS, BRONZE, ETC. (5) FINISH — BRIGHT, STEEL, CADMIUM, NICKEL OR CHROMIUM PLATED.

Fig. 9-15 Wood screw sizes.

Fig. 9-16 Lag screws are large screws with a square or hex head.

Fig. 9-17 Kinds of commonly used bolts.

Lag screws are sized by diameter and length. Diameters range from 1/4 inch to one inch, with lengths from one inch to 12 inches and up. Shank and pilot holes to receive lag screws are drilled in the same manner as for wood screws. (See Chapter 13.) Place a flat washer under the head to prevent the head from digging into the wood as the lag screw is tightened down. Apply a little wax to the threads to allow the screw to turn more easily and to prevent the head from twisting off.

Bolts

Most bolts are made of steel. To retard rusting, galvanized bolts are used. As with nails and screws, they are available in different kinds of metal and coatings. Many kinds are used for special purposes, but only a few are generally used.

Kinds of Bolts

Commonly used bolts are the carriage, machine, and stove bolts (Fig. 9-17).

Carriage Bolts. The *carriage bolt* has a square section under its oval head. The square section is embedded in wood and prevents the bolt from turning as the nut is tightened.

Machine Bolts. The *machine bolt* has a square or hex head. This is held with a wrench to keep the bolt from turning as the nut is tightened.

Stove Bolts. *Stove bolts* have either round or flat heads with a screwdriver slot. They are usually threaded all the way up to the head. *Machine screws* are very similar to stove bolts.

Bolt Sizes

Bolt sizes are specified by diameter and length. Carriage and machine bolts range from 3/4 inch to 20 inches in length and from 3/16 to 3/4 inch in diameter. Stove bolts are small in comparison to other bolts. They commonly come in lengths from 3/8 inch to 6 inches and from 1/8 to 3/8 inch in diameter.

Drill holes for bolts the same diameter as the bolt. Use flat washers under the head (except for carriage bolts), and under the nut to prevent the nut from cutting into the wood and to distribute the pressure over a wider area. Apply a little light lubricating oil to the threads before the nut is turned. Use wrenches of the correct size to tighten the bolt.

CHAPTER 10 — ANCHORS AND ADHESIVES

Anchors

Special kinds of fasteners used to attach parts to solid masonry and hollow walls and ceilings are called *anchors*. There are hundreds of types available. Those most commonly used are described in this chapter.

Solid Wall Anchors

Solid wall anchors may be classified as heavy-, medium-, or light-duty. Heavy-duty anchors are used to install such things as machinery, hand rails, dock bumpers, and storage racks. Medium-duty anchors may be used for hanging pipe and ductwork, securing window and door frames, and installing cabinets. Light-duty anchors are used for fastening such things as junction boxes, bathroom fixtures, closet organizers, small appliances, smoke detectors, and other lightweight objects.

Heavy-Duty Anchors. The *wedge anchor* is used when high resistance to pullout is required. The anchor and hole diameter are the same, simplifying installation (Fig. 10-1). The hole depth is not critical as long as the minimum is drilled. Proper installation requires cleaning out the hole.

The *self-drilling anchor* drills its own hole. After using the anchor to drill the hole, remove the anchor. Clean both the hole and the anchor. Insert the external tapered plug into the toothed end of the anchor. Place the assembly, plug down, into the hole until it bottoms out. Drive the anchor over the plug and shear off the top (Fig. 10-2).

The *stud anchor* and hole diameter are the same. This feature eliminates hole layout time because holes are drilled in masonry through the mounting holes in the fixture being installed. However, this anchor requires precise hole depth for proper expansion (Fig. 10-3).

The *sleeve anchor* and its hole size are also the same, but the hole depth need not be exact. After inserting the anchor in the hole, it is expanded by tightening the nut (Fig. 10-4.)

The *drop-in anchor* consists of an expansion shield and a cone-shaped, internal expander plug (Fig. 10-5). The hole must be drilled at least equal to the length of the anchor. A setting tool, supplied with the anchors, must be used to drive and expand the anchor. This anchor takes a machine screw or bolt.

Medium-Duty Anchors. *Split fast anchors* are one-piece steel with two sheared expanded halves

(A)

DRILL - SIMPLY DRILL A HOLE THE SAME DIAMETER AS THE ANCHOR. DO NOT WORRY ABOUT DRILLING TO DEEP SINCE THE ANCHOR WORKS IN A "BOTTOMLESS-HOLE". YOU CAN DRILL INTO THE CONCRETE WITH THE LOAD POSITIONED IN-PLACE; SIMPLY DRILL THROUGH THE PRE-DRILLED MOUNTING HOLES.

INSERT - DRIVE THE ANCHOR FAR ENOUGH INTO THE HOLE SO THAT AT LEAST SIX THREADS ARE BELOW THE TOP SURFACE OF THE FIXTURE.

ANCHOR - MERELY TIGHTEN THE NUT. RESISTANCE WILL INCREASE RAPIDLY AFTER THE THIRD OR FOURTH COMPLETE TURN.

(B)

Fig. 10-1 The wedge anchor has high resistance to pullout. (Courtesy of U.S. Anchor, Pompano Beach, Florida)

Insert anchor into chuck adapter and drill hole in concrete.

Remove anchor from hole and clean hole. Insert tapered plug in tooth end of anchor. Place anchor, plug down, in hole.

Allow assembly to bottom out. Then hammer drive anchor over expander plug. Remove anchor from hammer and shear off flush with surface.

Fasten fixture to anchor with screw or bolt.

Fig. 10-2 The self-drilling anchor requires no predrilled hole. (Courtesy of U.S. Anchor)

STUD BOLT ANCHOR

FIXTURE

EXPANDER PLUG

DRILL HOLE IN MASONRY, SAME DIAMETER AS ANCHOR, TO PRECISE DEPTH. PLACE ANCHOR IN HOLE, EXPANDER END DOWN.

PARTIALLY INSTALL NUT OR, PREFERABLY, USE SETTING TOOL AND DRIVE ANCHOR OVER EXPANDER PLUG. ANCHORING IS COMPLETE WHEN IT CANNOT BE DRIVEN ANY FURTHER.

INSTALL WASHER AND TIGHTEN NUT TO COMPLETE INSTALLATION.

Fig. 10-3 The stud anchor requires precise hole depth.

Fig. 10-4 Sleeve anchors eliminate the problem of exact hole depth requirements. (Courtesy of U.S. Anchor)

at the base. When driven, these halves are compressed and exert immense outward force on the inner walls of the hole as they try to regain their original shape (Fig. 10-6). They come in both flat and round head styles.

Single and *double expansion anchors* are used with machine screws or bolts. Drill a hole of recommended diameter to a depth equal to the length of the anchor. Place the anchor into the hole, flush with or slightly below the surface. Position the object to be fastened and bolt into place (Fig. 10-7). Once fastened, the object may be unbolted, removed, and refastened, if desired.

The *lag shield* is used with a lag screw. The shield is a split sleeve of soft metal, usually a zinc alloy. It is inserted into a hole of recommended diameter and a depth equal to the length of the shield plus 1/2 inch or more. The lag screw length is determined by adding the length of the shield, the thickness of the material to be fastened, plus 1/4 inch. The tip of the lag screw must protrude from the bottom of the anchor to ensure proper expan-

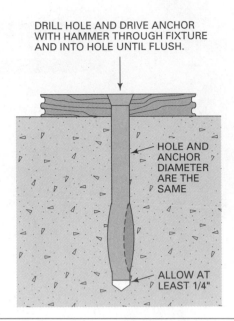

DRILL HOLE AND DRIVE ANCHOR WITH HAMMER THROUGH FIXTURE AND INTO HOLE UNTIL FLUSH.

HOLE AND ANCHOR DIAMETER ARE THE SAME

ALLOW AT LEAST 1/4"

Fig. 10-6 The split fast is a one-piece, all-steel anchor for hard masonry.

sion. As the fastener is threaded in, the shield expands tightly and securely in the drilled hole (Fig. 10-8).

The *concrete screw* utilizes specially fashioned high and low threads that cut into a properly sized hole in concrete (Fig. 10-9). Screws come in 3/16 and 1/4 inch diameters and up to 6 inches in length. The hole diameter is important to the per-

USE HAMMER TO DRIVE SETTING TOOL

SETTING TOOL

FIXTURE

BOLT

DROP-IN ANCHOR

DRILL PROPER DIAMETER HOLE TO A DEPTH AT LEAST EQUAL TO THE LENGTH OF THE ANCHOR. CLEAN HOLE AND INSERT ANCHOR.

USE SETTING TOOL TO DRIVE EXPANDER PLUG AND SET ANCHOR. SETTING TOOL PREVENTS DAMAGE TO INTERNAL THREADS OF ANCHOR.

POSITION FIXTURE OVER HOLE AND FASTEN BY INSERTING SCREW OR BOLT AND TIGHTENING.

Fig. 10-5 The drop-in anchor is expanded with a setting tool. (Courtesy of U.S. Anchor)

Fig. 10-7 Two opposing wedges of the double expansion anchor pull toward each other, expanding the full length of the anchor body. (Courtesy of U.S. Anchor)

Fig. 10-8 Lag shields are designed for light- to medium-duty fastening in masonry.

formance of the screw. It is recommended that a minimum of one inch and a maximum of 1 3/4 inch embedment be used to determine the fastener length. The concrete screw system eliminates the need for plastic or lead anchors.

The *coil anchor* is a lag type screw that uses a wire coil around its threads. The hole is drilled the same diameter as the anchor. Anchor, washer, and coil are assembled, inserted, and tightened (Fig. 10-10). The hole depth is determined by the anchor length minus the fixture thickness plus 1/4 inch minimum.

The *machine screw anchor* consists of two parts. A lead sleeve slides over a threaded, cone-shaped piece. Using a special setting punch that comes with the anchors, the lead sleeve is driven over the cone-shaped piece to expand the sleeve and hold it securely in the hole (Fig. 10-11). A hole of recommended diameter is drilled to a depth equal to the length of the anchor.

Fig. 10-9 The concrete screw system eliminates the need for an anchor when fastening into concrete.

INSTALLING RECOIL ANCHOR

DRILL HOLE SAME DIAMETER AS ANCHOR | ASSEMBLE WASHER AND COIL | INSERT COIL ANCHOR TO MARK | TIGHTEN

Fig. 10-10 The coil anchor utilizes a small metal coil around the threads of the screw. (Courtesy of U.S. Anchor)

Light-Duty Anchors. Three kinds of *drive* anchors are commonly used for quick and easy fastening in solid masonry. The *hammer drive anchor* (Fig. 10-12) has a body of zinc alloy containing a steel expander pin. In the *aluminum drive anchor,* both the body and the pin are aluminum to avoid the corroding action of electrolysis. The *nylon nail anchor* uti-

Fig. 10-11 Machine screw anchors are commonly called tamp-ins.

lizes a nylon body and a threaded steel expander pin. All are installed in a similar manner.

Lead and *plastic anchors,* also called *inserts,* are commonly used for fastening lightweight fixtures to masonry walls. These anchors have an unthreaded hole into which a wood or sheet metal screw is driven. The anchor is placed into a hole of recommended diameter, and 1/4 inch or more deeper than the length of the anchor. As the screw is turned, the threads of the screw cut into the soft material of the insert. This causes the insert to expand and tighten in the drilled hole (Fig. 10-13). Ribs on the sides of the anchors prevent them from turning as the screw is driven.

Fig. 10-12 Hammer drive anchors come assembled for quick and easy fastening. (Courtesy of U.S. Anchor)

← FIBER INSERT

← PLASTIC INSERT

← LEAD INSERT

Fig. 10-13 Lead and plastic anchors or inserts are used for light-duty fastening.

Chemical Anchoring Systems. Threaded studs, bolts, and re-bar (concrete reinforcing rod) may be anchored in solid masonry with a chemical bond using an *epoxy resin compound*. Two types of systems commonly used are the *epoxy injection* and *chemical capsule*.

In the injection system, a dual cartridge is inserted into a tool similar to a caulking gun. The chemical is automatically mixed as it is dispensed. Small cartridges are available to accurately dispense epoxy from an ordinary caulking gun (Fig. 10-14).

Chemical capsules contain the exact amount of all chemicals needed for one installation. Each capsule is marked with the appropriate hole size to be used. Drill holes according to diameter and depth indicated on each capsule. It is important to thoroughly clean and clear the hole of all concrete dust before inserting the capsule (Fig. 10-15).

With all solid masonry anchors, follow the specifications in regard to hole diameter and depth, minimum embedment, maximum fixture thickness, and allowable load on anchor.

MIXING TUBE

ULTRABOND

DUAL CARTRIDGE

**ADHESIVE ANCHOR
TYPE II (MIXTURE)**

MIX THE TWO-COMPONENT ADHESIVE SYSTEM AND PLACE IN HOLE. PUSH THE ANCHOR ROD INTO THE HOLE AND ROTATE SLIGHTLY TO COAT WITH ADHESIVE. ALLOW TO CURE.

Fig. 10-14 The epoxy injection system is designed for high-strength anchoring. (Courtesy of U.S. Anchor)

**ADHESIVE ANCHOR
TYPE I (CAPSULE ANCHOR)**

ATTACH THE ANCHOR ROD TO A ROTARY HAMMER ADAPTER. THE COMBINATION OF THE ROTATION AND HAMMERING ACTION MIXES THE CAPSULE CONTENTS TOGEATHER. ALLOW TO CURE.

Fig. 10-15 The chemical capsule anchoring system provides for easy, premeasured application.

Hollow Wall Fasteners

Toggle Bolts. *Toggle bolts* (Fig. 10-16) may have a wing or a tumble toggle. The wing toggle is fitted with springs, which cause it to open. The tumble toggle falls into a vertical position when passed through a drilled hole in the wall. The hole must be drilled large enough for the toggle of the bolt to slip through. A disadvantage of using toggle bolts is that, if removed, the toggle falls off inside the wall.

Plastic Toggles. The *plastic toggle* consists of four legs attached to a body that has a hole through the center and fins on its side to prevent turning during installation. The legs collapse to allow insertion into the hole. As sheet metal screws are turned through the body, they draw in and expand the legs against the inner surface of the wall (Fig. 10-17).

Expansion Anchors. Hollow wall *expansion anchors* are commonly called *molly screws* (Fig. 10-18). The anchor consists of an expandable sleeve, a machine screw, and a fiber washer. The collar on the outer end of the sleeve has two sharp prongs, that grip into the surface of the wall material. This prevents the sleeve from turning when the screw is tightened to expand the anchor. After expanding the sleeve, the screw is then removed, inserted through the part to be attached, and then screwed back into the anchor. Some types require that a hole be drilled, while other types have pointed ends that may be driven through the wall material.

Installed fixtures may be removed and refastened or replaced by removing the anchor screw without disturbing the anchor. Anchors are manufactured for various wall thicknesses. Make sure to use the right size anchor for the wall thickness in which the anchor is being installed.

Fig. 10-16 Toggle bolts are used for fastening in hollow walls.

Fig. 10-18 Hollow wall expansion anchors are commonly called molly screws.

SQUEEZE TOGGLE WINGS FLAT AND PUSH INTO HOLE DRILLED IN WALL.

TAP ANCHOR IN AND FLUSH WITH WALL.

FIXTURE

PLACE FIXTURE OVER HOLE, INSERT SHEET METAL SCREW AND TIGHTEN.

Fig. 10-17 The plastic toggle is a unique removable and reusable hollow wall anchor. (Courtesy of U.S. Anchor)

Conical Screws. The deep threads of the *conical screw* anchor resist stripping out when screwed into gypsum board, strand board, and similar material. After the plug is seated flush with the wall, the fixture is placed over the hole and fastened by driving a screw through the center of the plug (Fig. 10-19).

Universal Plugs. The *universal plug* is made of nylon and is used for a number of hollow wall and some solid wall applications. A hole of proper diameter is drilled. The plug is inserted, and the screw is driven to draw or expand the plug (Fig. 10-20).

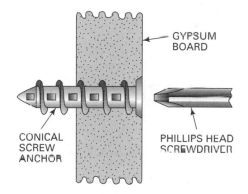

HOLLOW BRICK   PLASTER BOARD

HIGH VALUES IN PLASTER   AERATED CONCRETE

Fig. 10-20 The universal plug is used for many types of hollow wall fastening. (Courtesy of U.S. Anchor)

GYPSUM BOARD

CONICAL SCREW ANCHOR

PHILLIPS HEAD SCREWDRIVER

DRIVE ANCHOR IN WALL BY TURNING WITH SCREWDRIVER UNTIL HEAD IS FLUSH WITH SURFACE.

SHEET METAL SCREW

FIXTURE

PLACE FIXTURE OVER HOLE IN ANCHOR AND FASTEN WITH PROPER SIZE SHEET METAL SCREW.

Fig. 10-19 The conical screw anchor is a self-drilling, hollow wall anchor for lightweight fastenings. (Courtesy of U.S. Anchor)

Fig. 10-21 Framing ties and anchors are manufactured in many unique shapes. (Courtesy of Simpson Strong-Tie Co.)

Connectors

Widely used in the construction industry are devices called *connectors*. They are metal pieces formed into various shapes to join wood to wood, or wood to concrete or other masonry. They are called specific names depending on their function.

Wood-to-Wood. *Framing anchors* and *seismic* and *hurricane ties* are used to join parts of a wood frame (Fig. 10-21). *Post* and *column caps* and *bases* are used at the top and bottom of those members (Fig. 10-22). *Joist hangers* and *beam hangers* are available in many sizes and styles (Fig. 10-23).

Wood-to-Concrete. Some wood-to-concrete connectors are *sill anchors, anchor bolts,* and *holdowns* (Fig. 10-24). A *girder hanger* and a *beam seat* make beam-to-foundation wall connections (Fig. 10-25). *Post bases* come in various styles. They are used to anchor posts to concrete floors or footings.

Many other specialized connectors are used in frame construction. Some are described in the framing sections of this book.

Fig. 10-23 Hangers are used to support joists and beams. (Courtesy of Simpson Strong-Tie Co.)

Fig. 10-22 Caps and bases help fasten tops and bottoms of posts and columns. (Courtesy of Simpson Strong-Tie Co.)

Fig. 10-24 Sill anchors, anchor bolts, and holdowns connect frame members to concrete. (Courtesy of Simpson Strong-Tie Co.)

and are excellent for joining wood parts not subjected to moisture.

Urea Resin Glue. When a moisture-resistant glue is needed, *urea resin*, commonly called *plastic resin*, glue is used. This type comes in powder form and is mixed with water to the consistency of heavy cream. The glue sets slowly, and the joint must remain under pressure for eight hours or more, depending on the temperature.

Resorcinol Resin Glue. For a completely waterproof joint, *resorcinol resin* glue is used. A dry powder is mixed with a liquid part to make this glue. Pieces glued with this substance must be clamped under pressure for twelve to twenty-four hours.

If the work being glued is to receive a penetrating stain, it is extremely important that all excess glue be removed from the surface to be stained. A clean cloth, dampened with water, is recommended for this purpose. If excess glue is allowed to dry on the surface of the wood, it seals the surface. This prevents the stain from penetrating, and results in a blotchy stain job. Not so much care is needed to remove excess glue if the work is eventually to be painted or given a clear finish.

Contact Cement. *Contact cement* is so named because pieces coated with it bond on contact and need not be clamped under pressure. It is extremely important that pieces are positioned accurately before contact is made. Contact cement is widely used to apply plastic laminates for kitchen countertops. It is also used to bond other thin or flexible material that otherwise might require elaborate clamping devices.

Fig. 10-25 Girder and beam seats provide support from concrete walls. (Courtesy of Simpson Strong-Tie Co.)

Adhesives

The carpenter seldom uses any glue in the frame or exterior finish. Glue is used on some joints and other parts of the interior finish work. A number of *mastics* (heavy, paste-like adhesives) are used throughout the construction process.

Glue

White and Yellow Glue. Most of the glue used by the carpenter is the so-called white glue or yellow glue. The white glue is *polyvinyl acetate;* the yellow glue is *aliphatic resin.* Neither type is resistant to moisture. Both are fast-setting, so joints should be made quickly after applying the glue. They are available under a number of trade names

Mastics

There are several types of mastics used throughout the construction trades. They come in cans or cartridges used in hand or air guns. With these adhesives, the bond is made stronger and fewer, if any, fasteners are needed.

Construction Adhesive. One type of mastic is called *construction adhesive.* It is used in a glued floor system, described in a following unit on floor framing. It can be used in cold weather, even on wet or frozen wood. It is also used on stairs to increase stiffness and eliminate squeaks.

Panel Adhesive. *Panel adhesive* is used to apply such things as wall paneling, foam insulation, gypsum board, and hardboard to wood, metal, and masonry. It is usually dispensed with a caulking gun (Fig. 10-26).

Troweled Mastics. Other types of mastics may be applied by hand for such purposes as installing

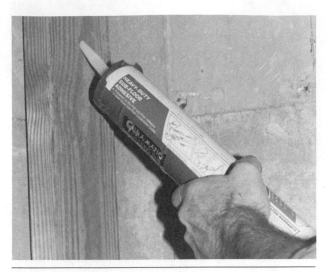

Fig. 10-26 Applying panel adhesive to stud with a caulking gun.

vinyl base, vinyl floor tile, or ceramic wall tile. A notched trowel is usually used to spread the adhesive. The depth and spacing of the notches along the edges of the trowel determine the amount of adhesive left on the surface.

It is important to use a trowel with the correct notch depth and spacing. Failure to follow recommendations will result in serious consequences. Too much adhesive causes the excess to squeeze out onto the finished surface. This leaves no alternative but to remove the applied pieces, clean up, and start over. Too little adhesive may result in loose pieces.

Review Questions

Select the most appropriate answer.

1. The length of an eight penny nail is
 a. 1 1/2 inches.   c. 2 1/2 inches.
 b. 2 inches.   d. 3 inches.

2. Fasteners coated with zinc to retard rusting are called
 a. coated.   c. electroplated.
 b. dipped.   d. galvanized.

3. Brads are
 a. types of screws.
 b. small box nails.
 c. small finishing nails.
 d. kinds of stove bolts.

4. When a moisture-resistant glue is required use
 a. aliphatic resin.
 b. plastic resin.
 c. polyvinyl acetate.
 d. yellow glue.

5. Many carpenters prefer to use casing nails to fasten
 a. interior finish.
 b. exterior finish.
 c. door casings.
 d. roof shingles.

6. The blunt point on the end of a cut nail helps
 a. drive the nail straight.
 b. prevent splitting the wood.
 c. hold the fastened material more securely.
 d. start the nail in the material.

7. On temporary structures, such as wood scaffolding and concrete forms, use
 a. common nails.
 b. duplex nails.
 c. galvanized nails.
 d. spikes.

8. As a general rule, how should the length of a nail compare to the thickness of the material being fastened?
 a. The same.
 b. Twice as long.
 c. 2 1/2 times as long.
 d. 3 times as long.

9. The name of a one-piece, all steel, hammer-driven anchor is
 a. wedge anchor.
 b. drop-in anchor.
 c. split fast anchor.
 d. expansion anchor.

10. For a waterproof bond, use
 a. aliphatic resin glue.
 b. urea resin glue.
 c. resorcinol resin glue.
 d. polyvinyl acetate glue.

■ BUILDING FOR SUCCESS ■

Fasteners: So What's the Connection?

Along with each new engineered construction product, manufacturers have designed sophisticated fasteners. The traditional nail, screw, and bolt have all been improved and complemented in many ways. Holding power for fasteners and corrosion resistance have been increased. These three mainstays are still being used to securely attach components in the assembly process. Mechanical and automation devices such as electric and pneumatic systems have greatly increased the speed of construction during the past decade. A vast array of anchors and adhesives have made the assembling of materials and products easier.

Along with each new fastener has come a need to assure proper installation. Manufacturers' recommendations become the basis for numerous building code requirements. Testing for public safety by both manufacturers and independent testing agencies has helped assure proper and safe installation procedures. With adherence to these recommended fastening procedures, the carpenter will assemble a safe and long-lasting product. In all too many cases, if proper fastening procedures are neglected, a tragedy can occur.

A number of years ago a carpenter was installing the track and other hardware for an overhead garage door. On this type of door, the track and mounting brackets for the door suspension system are lagged into the door trimmer and header. The pilot holes for the lag screws are to be drilled to the root diameter of the lag. In this case, the pilot holes greatly exceeded the root diameter. Some were driven in with a hammer, not torqued in with the required wrench. The lag screws holding the cable connecting S-hook and bracket pulled out under tension when the door was in the closed position. At the time, the carpenter had his arm over the track and received the full impact of the S-hook and mounting bracket as they pulled loose. His upper arm was severely damaged in the process.

A mistake of this magnitude should not have happened. It was costly both physically and financially for the carpenter. An informed and knowledgeable carpenter should always know the capabilities of fasteners and their proper installation requirements. It is the worker's responsibility to select the correct fastener for the job and to install it properly. Fasteners, whether anchors, bolts, nails, screws, mastics, or adhesives, are to be used with care. Directions are to be followed to avoid tragedies.

Students, apprentices, journeymen, and other skilled workers must "make the proper use connection." Safety must connect with common sense each time fasteners are used. Shortcuts may appear to save time, but usually affect quality and worker safety. There can be no substitute for safe and proper installation procedures in today's construction world.

FOCUS QUESTIONS: For individual or group discussion.

1. What might be sound reasons for selecting and installing the proper fastener on the job?

2. What can be learned from the accident described in this study?

3. What are some suggestions for selecting and installing fasteners so they do not cause problems?

4. What are some good ideas to implement if you do not have the proper fastener for the job at hand?

UNIT

5

Hand Tools

OBJECTIVES

After completing this unit, the student will be able to:

- identify and describe the hand tools that are commonly used by the carpenter
- use each of the hand tools in a safe and appropriate manner
- sharpen and maintain hand tools in suitable working condition

CONTENTS

Even though there has been a substantial increase in the use of power tools, there are still many occasions when hand tools must be used. There may be times when it is easier and faster to use hand tools rather than power tools.

Knowing how to choose the proper hand tool, how to use it skillfully, and how to keep it in good working condition is essential to the carpentry trade. A student should not underestimate the importance of hand tools and neglect their proper use and care. Carpenters are expected to have their own hand tools and keep them in working condition.

Buy tools of good quality. Inferior tools sometimes cannot be brought to a sharp edge and dull rapidly. Keep hand tools sharp and in good condition. If they get wet on the job, dry them as soon as possible, and coat them with a little light oil to prevent them from rusting.

Layout Tools

Many layout tools must be used by the carpenter. They are used, among other things, to measure distances, lay out lines and angles, test for depths of cuts, and set vertical and level pieces.

Measuring Tools

The ability to take measurements quickly and accurately must be mastered early in the carpenter's training. Practice reading the rule or tape to gain skill in fast and precise measuring.

Rules and Tapes. *Rules* used in construction are divided into feet, inches, and usually 16ths of an inch. Most rules and tapes used by the carpenter have clearly marked increments of 16 inches to help in laying out spaced framing members.

Some carpenters prefer the *6-foot folding rule* (Fig. 11-1) while others use the *pocket tape* (Fig. 11-2). The folding rule sometimes has a metal extension

Fig. 11-1 Six-foot folding rule. (Courtesy of Stanley Tools)

Fig. 11-2 Pocket tape. (Courtesy of Stanley Tools)

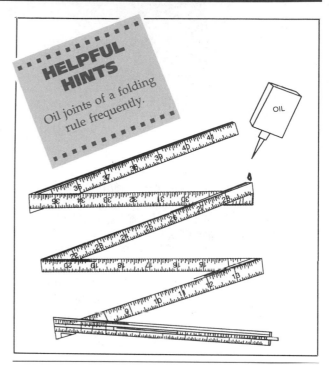

Fig. 11-3

on one end for taking inside measurements. Oil the joints occasionally to prevent breaking the rule when opening and closing it (Fig. 11-3). Pocket tapes are available in 6- to 25-foot lengths. A hook on the end slides back and forth slightly, to compensate for the thickness of its metal, when taking outside and inside measurements.

Steel tapes of 50- and 100-foot lengths are commonly used to lay out longer measurements. The end of the tape has a steel ring with a folding hook attached. The hook may be unfolded to go over the edge of an object. It may also be left in the folded position and the ring placed over a nail when extending the tape. Remember to place the nail so that the *outside* of the ring, which is the actual end of the tape, is to the mark (Fig. 11-4). Rewind the tape when not using it. If the tape is stepped on when extended, it may snap or become kinked. Keep it out of water. If it gets wet, dry it while rewinding.

Slip-sticks. A pair of *slip-sticks* is a very helpful and accurate device for laying out stock that has to be fitted between two surfaces. In some cases, it is better than a folding rule or tape. For instance, the distance from floor to ceiling is too great for a rule, and a tape is not rigid enough to stand straight for that height without support. Slip-sticks can be made from two narrow strips of a convenient length of scrap lumber, more than half the distance

STEEL TAPE

Fig. 11-4 Steel tape. (Courtesy of Stanley Tools)

to be determined. Hold the slip-sticks together. Slide them until the end of one touches a surface while the end of the other touches the opposite surface (Fig. 11-5). Hold the pieces without moving them and lay them on the stock to be marked for length. This is an accurate and convenient method of laying out the length of wall studs, closet shelves, baseboard, and other trim.

Squares

The carpenter has the use of a number of different kinds of squares to lay out for square and other angle cuts.

Combination Squares. The *combination square* (Fig. 11-6) consists of a movable blade, one inch wide and 12 inches long, which slides along the

SLIP-STICKS MAY BE MADE FROM THE EDGES OF TONGUE AND GROOVED LUMBER.

CROSS-SECTION

SLIP-STICKS

Fig. 11-5 Using slip-sticks is a practical and accurate way to lay out inside dimensions.

Fig. 11-6 The combination square is often used on the job.

body of the square. It is used to lay out or test 90 and 45 degree angles. Hold the body of the square against the edge of the stock and mark along the blade. It can function as a depth gauge to lay out or test the depth of *rabbets*, *grooves*, and *dadoes*. It can also be used with a pencil as a marking gauge to draw lines parallel to the edge of a board. Drawing lines in this manner is called *gauging* lines. Lines may also be gauged by holding the pencil and riding the finger along the edge of the board (Fig. 11-7). Finger-gauging takes practice, but once mastered saves a lot of time.

Fig. 11-7 The combination square has many uses.

Speed Squares. Some carpenters prefer to use a triangular shaped square known by the brand name of *Speed Square* (Fig. 11-8). Speed Squares are made of one-piece aluminum alloy and are available in two sizes. They can be used to lay out 90 and 45 degree angles and as guides for portable power saws. A degree scale allows angles to be laid out; other scales may be used to lay out rafters.

Framing Squares. The *framing square*, often called the *steel square* (Fig. 11-9) is an L-shaped tool made of thin steel or aluminum. The longer of the two legs is called the *blade*, and is 2 inches wide and 24 inches long. The shorter leg is called the *tongue*, and is 1 1/2 inches wide and 16 inches long. The outside corner is called the *heel*.

The side that has the manufacturer's name stamped on it is called the *face* side. A number of different tables are stamped on both sides of the square. Most are rarely used except for the *rafter table*, found on the face side of the blade, which is used to find the length of several kinds of rafters.

(How to use this table is explained in the following units on roof framing.) On the same side of the square, on the tongue, can be found the *octagon scale*, which was used to lay out eight-sided timbers from square ones.

On the back side of the square, the *Essex board foot table* was used to calculate the number of board feet in lumber. The *brace table* was used to figure the length of diagonal braces. The *hundredths scale*, consisting of an inch divided into one hundred parts, was used to find 1/100ths of an inch.

On the face side, the edges are divided into inches which are graduated into 1/8ths on the inside and into 1/16ths on the outside. The edges on the back side are divided into inches and 1/12ths on the outside, while one inside edge is graduated into 1/16ths and the other into 1/10ths.

The framing square has been used for centuries. Entire books have been written about it. Based on the use of the right angle triangle, many layout techniques have been devised and used throughout the years. Although those techniques, and the scales, tables, and graduations retained on the square, are now rarely used, the carpenter once depended on them.

At present, the framing square is very useful for laying out angles for roof rafters, bridging, and stair framing. It is also used to lay out 90 and 45 degree angles on stock too wide for smaller squares (Fig. 11-10).

Sliding T-Bevels. The *sliding T-bevel*, sometimes called a *bevel square* or just a *bevel* (Fig. 11-11), consists of a body and a sliding blade that can be turned to any angle and locked in position. It is used to lay out or test angles other than those laid out with squares. The body of the tool is held against the edge of the stock, and the angle is laid out by marking along the blade.

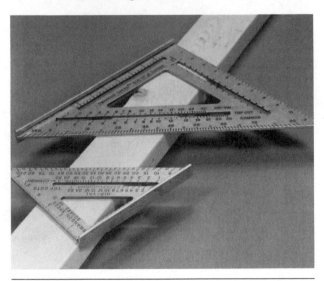

Fig. 11-8 Speed Squares are used by many carpenters.

Fig. 11-9 Framing square. (Courtesy of Stanley Tools)

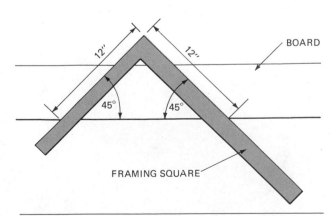

Fig. 11-10 Laying out 45 degree angles with a framing square.

Fig. 11-11 Sliding T-bevel. (Courtesy of Stanley Tools)

Straightedges

A *straightedge* can be made of metal or wood. It can have any thickness, width, or length, as long as the size is convenient for its intended use and it has at least one edge that is absolutely straight from one end to the other. To determine if it is straight, sight along the edge. Another way is to lay the piece on its side and mark along its edge from one end to the

STEP 1 - LAY STRAIGHTEDGE ON MARKING SURFACE AND DRAW LINE ALONG ITS EDGE (POSITION 1)

STEP 2 - TURN STRAIGHTEDGE OVER WITHOUT TURNING END FOR END. LINE UP EDGE WITH PENCIL MARK. IF EDGE COINCIDES WITH LINE FOR TOTAL LENGTH, EDGE IS STRAIGHT

Fig. 11-12

STEP BY STEP
P R O C E D U R E S

1. ALONG EDGE OF STOCK, LAYOUT A CONVENIENT LENGTH. FROM EACH END LAYOUT SAME LENGTH UPWARD AND TOWARD EACH OTHER UNTIL THEY INTERSECT TO FORM TRIANGLE WITH 60° ANGLES.

2. FIND CENTER OF ONE SIDE OF TRIANGLE. DRAW LINE FROM CORNER TO CENTER OF OPPOSITE SIDE TO FORM 30° ANGLE.

3. ALONG SIDES OF 30° ANGLE, LAYOUT EQUAL LENGTHS. CONNECT THE END POINTS. FIND CENTER AND DRAW LINE TO FORM ANGLE OF 15°.

4. IN A MANNER SIMILAR TO (3), LAYOUT EQUAL LENGTHS ON SIDES OF 15° ANGLE. CONNECT ENDS WITH STRAIGHT LINE, FIND CENTER AND DRAW LINE FOR 7-1/2°.

5. LAYOUT ANY COMBINATION TO FORM PRACTICALLY ANY DESIRED ANGLE.

6. LAYOUT A TRIANGLE WITH 3, 4, & 5 OR MULTIPLES THEREOF TO FORM A 90° ANGLE.

Fig. 11-13 Angles of many different sizes can be laid out with only a rule, straightedge, and pencil.

other. Turn the piece over. Hold each end on the line just marked, and mark another line. If both lines coincide, the edge is straight (Fig. 11-12).

Straightedges are useful for many purposes. The framing square, the blade of the combination square, the back of a saw, or a yardstick could be used as a straightedge for drawing short, straight lines. Other larger straightedges may be used for leveling over long distances, or testing the straightness of the sides of door frames and the edges of intersecting wall studs. Although not manufactured and sold as a tool, it is handy to know what a straightedge is, how to make or select one, and how to use it.

Angle and Circle Layout Techniques

Angles. Practically any angle can be laid out accurately using just a straightedge and a rule. Use a scrap piece of plywood or similar material. and lay out any convenient length along its edge (in this case, use 12 inches). Connect the end points with straight lines, forming a triangle with equal sides of 12 inches each. The angle formed by the sides is 60 degrees. This is the angle used to make joints on arrangements of six equal sides called *hexagons*. Fig. 11–13 shows how to make this angle and others.

Arcs and Circles. A pair of tools called *trammel* points may be used to draw circles or parts of circles, called arcs (Fig. 11-14), which may be too large for a compass. They can be clamped to a strip of wood any distance apart according to the desired radius of the circle to be laid out. One trammel point can be set on the center while the other, which may have a pencil attached, is swung to lay out the circle or arc.

In place of trammel points, the same kinds of layouts can be made by using a thin strip of wood with a brad or small finish nail through it for a center point. Measure from the end of the strip a distance equal to the desired radius. Drive the brad through the strip until the point comes through. Set

Fig. 11-14 Trammel points are used to lay out arcs of large diameter.

Fig. 11-15 A thin strip of wood can be used to lay out circles or arcs.

the point of the brad on the center, and hold a pencil against the end while swinging the strip to form the circle or arc (Fig. 11-15). To keep the pencil from slipping, a small V may be cut on the end of the strip or a hole may be drilled near the end to insert the pencil. Make sure measurements are taken from the bottom of the V or the center of the hole.

Levels

In construction, the term *level* is used to indicate that which is *horizontal* and the term *plumb* is used to mean the same as *vertical*. An important point to remember is that level and plumb lines, or objects, must also be straight throughout their length or height. Parts of a structure may have their end points level or plumb with each other. If they are not straight in between, however, they are not level or plumb for their entire length (Fig. 11-16).

Carpenter's Levels The *carpenter's level* (Fig. 11-17) is used to test both level and plumb surfaces. Accurate use of the level depends on accurate reading. The air bubble in the slightly crowned glass tube of the level must be exactly centered between the lines marked on the tube.

The pair of tubes located in the center are used to test for level. The pairs located at each end are used to test for plumb.

Levels are made of wood or metal, usually aluminum. Most carpenters prefer the metal type in lengths of 24 to 28 inches. Lengths of 48, 72, and 78 inches are also available.

Note: Care must be taken not to drop the level because this could break the glass or disturb the accuracy of the level.

Line Levels The *line level* (Fig. 11-18) consists of one glass tube encased in a metal sleeve with hooks on each end. The hooks are attached to a stretched

Fig. 11-16 To be level or plumb for their entire length, pieces must be straight from end to end.

IN HOT WEATHER, OR IF LAYING IN THE SUN, THE LIQUID IN THE TUBE EXPANDS MAKING THE AIR BUBBLE SMALLER.

AT MODERATE TEMP-ERATURE THE AIR BUBBLE SHOULD FIT EXACTLY BETWEEN THE TWO LINES ON THE TUBE.

AT COLDER TEMP-ERATURES THE LIQUID IN THE TUBE CONTRACTS MAKING THE AIR BUBBLE LARGER.

REGARDLESS OF CONDITIONS, THE AIR BUBBLE MUST BE CENTERED BETWEEN THE TWO LINES ON THE TUBE.

Fig. 11-17 Carpenter's level.

line, which is then moved up or down until the bubble is centered. However, this is not an accurate method and gives only an approximate levelness. Care must be taken that the level be attached close to the center of the suspended line because the weight of the level causes the line to sag. If the line level is off center to any great degree, the results are very faulty.

Plumb Bobs. The *plumb bob* (Fig. 11-19) is very accurate and is used frequently for testing and establishing plumb lines. Suspended from a line, the plumb bob hangs absolutely vertical when it stops swinging. However, it is difficult to use outside when the wind is blowing. Plumb bobs come in several different weights. Heavy plumb bobs stop swinging quicker than lighter ones. Some have hollow centers that are filled with heavy metal to increase the weight without enlarging the size.

The plumb bob is very useful for quick and accurate plumbing of posts, studs, door frames, and

Fig. 11-18 Line level.

Fig. 11-19 Plumb bob.

Fig. 11-20 The post is plumb when the distance between it and the plumb line is the same.

other vertical members of a structure (Fig. 11-20). It can be suspended from a great height to establish a point that is plumb with another. Its only limitation is the length of the line.

Chalk Lines

Long straight lines are laid out by using a *chalk line.* A line coated with chalk dust is stretched tightly between two points and snapped against the surface (Fig. 11-21). The chalk dust is dislodged from the line and remains on the surface.

Widely used at present is the *chalk line reel,* also called a *chalk box* (Fig. 11-22). The box is filled with chalk dust that comes in a number of colors. The most popular colors are blue, yellow, maroon, and white. The dust saturates the line, which is on a reel inside the box. The line is ready to be snapped when it is pulled out of the box. After several snaps, the line will need more chalk and must be reeled in to be recoated with chalk. Shaking the box helps recoat the line.

Some carpenters still use the *spool and awl* type of chalk line, which was widely used before the chalk box became popular. The line, which is wound around the spool, has a small awl fastened to its end. A few inches from the awl, a small loop is tied into the line. The spool has a hole through its center for storing the awl when it is not being used. The awl is used to hold the end of the stretched line. Put the point through the loop and then insert the awl into the material, at the mark, on which the line is to be snapped.

Fig. 11-22 Chalk line reel.

Fig. 11-21 Snapping a chalk line.

Fig. 11-23 Attaching and chalking a line with cake chalk.

To chalk the line, a semispherical cake of chalk is rubbed over the line to prevent the line from cutting into the cake. At the same time, the line is unwound by holding the spool ends between the middle finger and the thumb of the other hand (Fig. 11-23). It is not necessary to rewind the line each time it needs more chalk. The cake of chalk can be rubbed over the line by walking from one end to the other. This type of chalk can be used for other purposes. It can coat the edges of electrical outlet boxes to mark their location on wallboard or mark wood shingles to be fitted between others.

Chalk Line Techniques. When unwinding and chalking the line, keep it off the surface until snapped. Otherwise many lines will be made on the surface and this could be confusing. Make sure lines are stretched tight in order to snap a straight and true line. Sight long lines by eye for straightness when snapping lines on vertical surfaces, to make sure there is no sag in the line. If there is a sag, take it out by supporting the line with a nail near the center. Over long distances, stretch and fasten the line at both ends. Press the center of the line to the deck and snap the line on both sides of the center. Keep the line from getting wet. A wet line is practically useless.

Wing Dividers

Wing dividers can be used as a compass to lay out circles and arcs, and as dividers to space off equal distances. However, this tool is used mainly for *scribing* and is often called a *scriber*. Scribing is the technique of laying out stock to fit against an irregular surface (Fig. 11-24). For easier and more accurate

Fig. 11-24 Scribing is laying out a piece to fit against an irregular surface.

scribing, heat and bend the end of the solid metal leg outward (Fig. 11-25). Pencils are usually used in place of the interchangeable steel marking leg. Use pencils with hard lead that keep their points longer.

HELPFUL HINTS

Bend the leg of the dividers as shown for easier and more accurate scribing.

METAL LEG IS INTERCHANGEABLE WITH PENCIL AND IS USED TO SCRIBE LINES ON DARK SURFACES.

HEAT AND BEND THE LEG OUTWARD FOR EASIER AND MORE ACCURATE SCRIBING.

SCRIBING SURFACE

POINT DOES NOT RIDE ON SURFACE.

DIVIDER LEG BEFORE HEATING AND BENDING OUTWARD.

SCRIBING MAY NOT BE ACCURATE BECAUSE POINT IS NOT RIDING ON SURFACE.

SCRIBING SURFACE

POINT RIDES ON SURFACE. FILE TO SHAPE, IF NEEDED.

LEG AFTER BENDING. DO NOT TRY TO BEND WITHOUT HEATING. LEG MAY BREAK.

MORE ROOM HERE TO CLEAR OBSTRUCTIONS.

SCRIBING IS MORE ACCURATE WHEN POINT RIDES ON SURFACE.

Fig. 11-25

Butt Gauges

The *butt gauge* (Fig. 11-26) is used to lay out the gain (cutout) for butt hinges. Butt hinges are commonly used on doors. (The use of the butt gauge is explained in detail in a following unit on doors.)

Butt markers (Fig. 11-27) are available in three sizes. They are often used to mark hinge gains. The marker is laid on the door edge at the hinge location and tapped with a hammer to outline the cutout for the hinge.

Fig. 11-26 Butt gauge.

Fig. 11-27 Butt hinge markers come in three sizes.

CHAPTER 12 BORING AND CUTTING TOOLS

Boring Tools

The carpenter is often required to cut holes in wood and metal. Boring denotes cutting larger holes in wood. Drilling is often thought of as making holes in metal or smaller holes in wood. Boring tools include those that actually do the cutting and those used to turn the cutting tool.

Bit Braces

The *bit brace* (Fig. 12-1) is used to hold and turn auger bits to bore holes in wood. It may also be used to hold screwdriver bits to drive large screws. Its size is determined by its *sweep* (the diameter of the circle made by its handle). Sizes range from 8 to 12 inches. Most bit braces come with a ratchet that

Fig. 12-1 Bit brace. (Courtesy of Stanley Tools)

Fig. 12-3 Expansive bit.

can be used when there is not enough room to make a complete turn of the handle.

Bits

Auger Bits. *Auger bits* (Fig. 12-2) are available with coarse or fine *feed screws*. Bits with coarse feed screws are use for fast boring in rough work. Fine feed bits are used for slower boring in finish work. As the bit is turned, the *spurs* score the circle in advance of the *cutting lips*. The cutting lips lift the chip up and through the twist of the bit.

Auger Bit Sizes. A full set of auger bits ranges in sizes from 1/4 to one inch, graduated in 1/16-inch increments. The bit size is designated by the num-

ber of 1/16-inch increments in its diameter. For instance, a #12 bit will bore a 3/4-inch diameter hole.

Expansive Bits. To bore holes over one inch in diameter, the carpenter usually uses an *expansive bit* (Fig. 12-3). With two interchangeable and adjustable cutters, holes up to 3 inches in diameter may be bored. This tool is handy for boring holes in doors for locksets when power tools are not available. Usually these locksets require a hole of 2 1/8 inches in diameter.

Boring Techniques. To avoid splintering the back side of a piece when boring all the way through, stop when the point comes through. Finish by boring from the back side. This is especially important when both sides are exposed to view as in the case of a door. Care must be taken not to strike any nails or other objects that might cause blunting and shortening of the spurs. If the spurs become too short, the auger bit is ruined.

Sharpening Bits. Auger bits are sharpened with an auger bit file. File across the inside of the spurs only enough to bring them to a sharp edge. Excessive filing of the spurs reduces their height, eventually making the bit useless. Do not file on the outside because this may reduce the diameter of the bit.

File the cutting lips on the top side and on the original bevel only until the edge is sharp (Fig. 12-4).

COARSE AND FINE AUGER BIT FEED SCREWS

Fig. 12-2 A. Auger bit. (Courtesy of Stanley Tools) B. Coarse and fine feed screws.

Fig. 12-4 Sharpening an auger bit.

Fig. 12-5 A. Hand drill. (Courtesy of Stanley Tools)
B. Twist drills.

Drills

Hand Drills. *The hand drill* is used to turn *twist drills* (Fig. 12-5) to make holes in wood or metal that are 1/4 inch or less in diameter. It is a very useful tool when a few holes are to be drilled or when no power is available.

Fig. 12-6 Countersink.

Twist Drills. *Twist drills* range in size from 1/16 to 1/4 inch in increments of 1/64 inch. These drills are particularly useful for drilling holes for screws and are also used in power drills.

Countersinks. The *countersink* (Fig. 12-6) may be turned in a hand drill or power drill. It forms a recess for a flat head screw to set flush with the surface of the material in which it is driven.

Combination Drills. *Combination drills and countersinks* are used to drill shank and pilot holes for screws and countersink in one operation (Fig. 12-7).

Fig. 12-7 Combination drill and countersink

Cutting Tools

The carpenter uses many kinds of cutting tools and must know which to select for each job, and how to use, sharpen, and maintain them.

Edge-cutting Tools

Wood Chisels. The *wood chisel* (Fig. 12-8), is used to cut recesses in wood for such things as door hinges and locksets and to make joints.

Chisels are sized according to the width of the blade and are available in widths of 1/8 inch to 2 inches. Most carpenters can do their work with a set consisting of chisels that are 1/4, 1/2, 3/4, one, and 1 1/2 inches in size.

Firmer chisels have long, thick blades and are used on heavy framing. *Butt* chisels are short, with a thinner blade. They are preferred for finish work.

Fig. 12-8 Wood chisel.

CAUTION Improper use of chisels has caused many accidents. When using chisels, keep both hands back of the cutting edge at all times (Fig. 12-9). When not in use, the cutting edge should be shielded (Fig. 12-10). Never put or carry chisels or other sharp or pointed tools in pockets.

Fig. 12-9 Keep both hands back of the chisel's cutting edge.

Fig. 12-10 Keep the chisel edge shielded when not being used.

Bench Planes. *Bench planes* (Fig. 12-11) come in several sizes. They are used for smoothing rough surfaces and bringing work down to the desired size. Large planes are used on work like door edges to produce a straight surface over a long distance. Long planes will bridge hollows in a surface and cut high spots better than short plane (Fig. 12-12). Small planes are more easily used for shorter work.

Bench planes are given names according to their length. The longest is called the *jointer*. In declining order are the *fore, jack,* and *smooth* planes. It is not necessary to have all the planes. The jack plane is 14 inches long and of all the bench planes is considered the best for all-around work.

Block Planes. *Block planes* are small planes designed to be held in one hand. They are often used to smooth the edges of short pieces and for trimming end grain to make fine joints (Fig. 12-13).

Block planes are designed differently than bench planes. On bench planes, the cutting edge bevel is on the bottom side. On block planes, it is on the top. In addition, the bench plane iron has a plane iron cap attached to it, while the block plane iron has none (Fig. 12-14). Practice assembling and adjusting both types, until proficient, to produce a fine shaving.

Unlike bench planes, block planes are available with their blades set at a high angle or at a low angle. Most carpenters prefer the low-angle block

Fig. 12-11 The jack plane is a general purpose plane.

Fig. 12-12 Longer planes bridge hollows to plane long, straight edges.

Fig. 12-13 Block plane.

plane because it seems to have a smoother cutting action and because it fits into the hand more comfortably.

Using Planes. When planing, have the stock securely held against a stop. Always plane with the grain. When starting, push forward while applying pressure downward on the toe (front). When the heel (back) clears the end, apply pressure downward on both ends while pushing forward. When the opposite end is approached, relax pressure on the toe and continue pressure on the heel until the cut is complete (Fig. 12-15). This method prevents tilting the plane over the ends of the stock and helps ensure a straight, smooth edge.

Sharpening Chisels and Plane Irons. The bevel of the wood chisel or plane iron is shaped by a grinding wheel. The bevel should have a concave surface, which is called a *hollow grind*. In order to obtain a hollow grind, a grinding attachment may be used.

CAUTION Use safety goggles or otherwise protect your eyes when grinding.

If a grinding attachment is not available, hold the chisel or plane iron by hand on the tool rest at the proper angle. A general rule of thumb is that the width of the bevel is approximately twice the thickness of the blade (Fig. 12-16). Move the blade up on

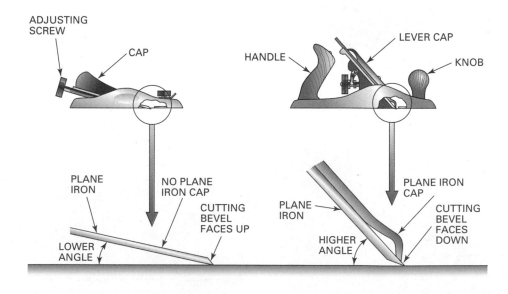

BLOCK PLANE   BENCH PLANE

Fig. 12-14 Difference in block and bench planes.

Fig. 12-15 Correct method of planing edges.

the grinding for a longer bevel and down for a shorter bevel.

Let the index finger of the hand holding the chisel ride against the outside edge of the tool rest as the chisel is moved back and forth across the revolving wheel. Dip the blade in water frequently to prevent overheating. Do not move the position of your index finger, making sure that the tool can be replaced on the wheel at exactly the same angle to

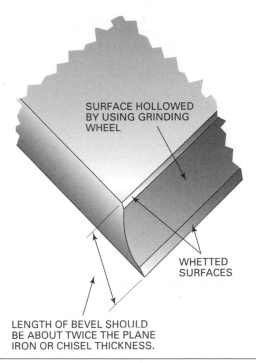

SURFACE HOLLOWED
BY USING GRINDING
WHEEL

WHETTED
SURFACES

LENGTH OF BEVEL SHOULD
BE ABOUT TWICE THE PLANE
IRON OR CHISEL THICKNESS.

Fig. 12-16 Details of the cutting edge of chisels and planes.

obtain a smooth hollow to the bevel. Grind the chisel or plane iron until an edge is formed (Fig. 12-17). A burr or *wire-edge* will be formed on the edge on the flat side. This can be felt by lightly rubbing your thumb along the flat side toward the edge.

Grinding does not sharpen the tool; it only shapes it. To produce a keen edge, it must be *whetted* (sharpened) using an oilstone. Hold the tool on a well-oiled stone so that the bevel rests flat on it. Move the tool back and forth across the stone for a few strokes. Then, make a few strokes with the flat side of the chisel or plane iron held absolutely flat on the stone. Continue whetting in this manner un-

Fig. 12-17 Grinding a plane iron.

til as keen an edge as possible is obtained. To obtain a keener edge, repeat the procedure on a finer stone or on a piece of leather. The edge is sharp when, after having whetted the bevel and before turning it over, no wire edge can be felt on the flat side (Fig. 12-18).

Chisels and plane irons do not have to be ground each time they need sharpening. Grinding is necessary only when the bevel has lost its concave shape by repeated whettings, the edge is badly nicked, or the bevel has become too short and blunt. The edge of a blade may be whetted many times before it needs grinding.

STEP BY STEP

P R O C E D U R E S

WHETTING THE BEVEL SIDE

WHETTING THE STRAIGHT SIDE

HONING ON A LEATHER STROP

Fig. 12-18 Whetting a plane iron.

Fig. 12-19 Straight snips.

Snips. *Straight tin snips* (Fig. 12-19) are generally used to cut straight lines on thin metal, such as roof flashing and metal roof edging. Three styles of *aviation snips* (Fig. 12-20) are available for straight metal cutting and for left and right curved cuts.

Hatchets. For wood shingling of side walls and roofs, among other purposes, the *shingling hatchet* is used. In addition to the shingling hatchet, many carpenters carry a slightly heavier hatchet for such uses as pointing stakes or otherwise tapering rough stock. A special drywall hatchet is also used for the installation of gypsum board. (Fig. 12-21)

Fig. 12-20 Right- and left-cutting aviation snips.

Fig. 12-21 Shingling hatchet.

> **CAUTION** When using hatchets for driving fasteners, make sure there are no workers in the path of the backswing.

Knives. Carpenters usually have a jackknife of good quality. The jackknife is used mostly for sharpening pencils and for laying out recessed cuts for some types of finish hardware, such as door hinges. The jackknife is used for laying out this type of work because a finer line can be obtained with it than with a pencil. In addition to marking, it also scores the layout line that is helpful when chiseling the recess (Fig. 12-22).

The *utility knife* (Fig. 12-23) is frequently used for such things as cutting gypsum board and softboards. Replacement blades are carried inside the handle.

Fig. 12-23 Utility knife.

Fig. 12-24 Hand scraper. (Courtesy of Stanley Tools)

Scrapers. The *hand scraper* (Fig. 12-24) is very useful for removing old paint, dried glue, pencil, crayon, and other marks from wood surfaces. The scraper blades are reversible, removable, and replaceable. They dull quickly, but can be easily sharpened by filing on the bevel and against the cutting edge (Fig. 12-25).

> **CAUTION** Be careful when filing so that the filing hand does not come in contact with the cutting edge. Also, care should be taken not to hollow out the center of the blade or the outside corners will dig into the work.

HELPFUL HINTS

Score layout lines with a knife for more accurate chiseling of recesses.

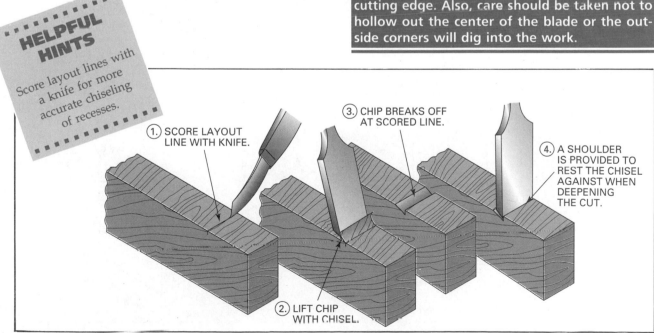

1. SCORE LAYOUT LINE WITH KNIFE.

2. LIFT CHIP WITH CHISEL.

3. CHIP BREAKS OFF AT SCORED LINE.

4. A SHOULDER IS PROVIDED TO REST THE CHISEL AGAINST WHEN DEEPENING THE CUT.

Fig. 12-22

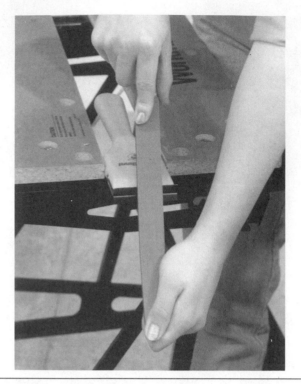

Fig. 12-25 Filing a hand scraper to a sharp edge.

Tooth-cutting Tools

The carpenter uses several kinds of saws to cut wood, metal, and other material. Each one is designed for a particular purpose.

Handsaws

Handsaws (Fig. 12-26) used to cut across the grain of lumber are called *crosscut saws.* To cut with the grain, *ripsaws* are used. The difference in the cutting action is in the shape of the teeth. The crosscut saw has teeth shaped like knives. These teeth cut through the wood fibers to give a smoother action and surface when cutting across the grain. The ripsaw has teeth shaped like rows of tiny chisels that cut the wood ahead of them. This results in a faster sawing action (Fig. 12-27). To keep the saw from binding, the teeth are *set,* that is, alternately bent, to make the saw cut or *kerf* wide enough to give clearance for the blade.

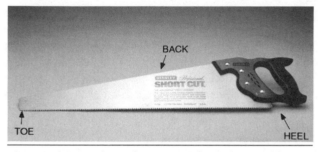

Fig. 12-26 Handsaw. (Courtesy of Stanley Tools)

CROSS SECTION OF CROSSCUT TEETH

HOW A CROSSCUT SAW CUTS

CROSS SECTION OF RIP TEETH

HOW A RIP SAW CUTS

Fig. 12-27 Cutting action of rip and crosscut saws. (Courtesy of Disston)

The carpenter ordinarily uses two crosscut saws. A 7-point or 8-point (number of tooth points to the inch) saw is used to cut across the grain of framing and other rough lumber. An 11-point or 12-point saw is used for fine cross cuts on finish work. One ripsaw, usually 5 1/2 points, is needed to cut with the grain of wood. The number of points is usually stamped on the saw blade at the heel.

Using Handsaws. Stock is handsawed with the face side up because the back side is splintered, along the cut, by the action of the saw going through the stock. This is not important on rough work. However, on finish work, it is essential to identify the face side of a piece and to make all layout lines and saw cuts with the face side up. It is possible, if no attention is given to cut one end from the face side and the other end from the back side. Then, no matter which way the piece is installed, one splintered end will be exposed to view.

The saw cut is made on the waste side of the layout line by cutting away part of the line and leaving the rest. This becomes difficult and takes practice, especially when it is important to make thin layout lines rather than broad, heavy ones. Press the blade of the saw against the thumb when starting a cut. Make sure the thumb is above the teeth; steady it with the index finger, with the rest of the hand on the work. Move the thumb until the saw is where desired and start the cut on the downstroke (Fig. 12-28). Move the hand away when the cut is deep enough. When handsawing, hold crosscut saws at about a 45 degree angle. Hold ripsaws, with the handle higher, at about 60 degrees.

CAUTION Do not use a ripsaw for cutting across the grain. It can jump at the start of the cut, possibly causing injury.

Most carpenters prefer to have their saws set and filed by sharpening shops. Sharpening handsaws requires special tools, much skill, and experience to do a professional job.

Special Purpose Saws

Compass and Keyhole Saws. The *compass saw* is used to make circular cuts in wood. The *keyhole saw* is similar to the compass saw except its blade is narrower for making curved cuts of smaller diameter (Fig. 12-29). To start the saw cut, a hole needs to be bored (except in soft material, when the point of the saw blade can be pushed through).

Coping Saws. The *coping saw* (Fig. 12-30) is used primarily to cut molding to make coped joints. A *coped joint* is made by cutting and fitting the end of a molding against the face of a similar piece. (Coping is explained in detail in a following unit.) The coping saw is also used to make any small, irregular curved cuts in wood or other soft material.

Use coping saw blades with fine teeth, and install with the teeth pointing away from the han-

Fig. 12-29 Compass and keyhole saws.

Fig. 12-28 Starting a cut with a handsaw.

Fig. 12-30 Coping saw.

dle. Cut on the downstroke with the handle above the work and the layout line.

Hacksaws. *Hacksaws* (Fig. 12-31) are used to saw metal. Hacksaw blades are available with 18, 24, and 32 points to the inch. Coarse-toothed blades are used for fast cutting in thick metal. Fine-toothed blades are used for smooth cutting of thin metal. At least three teeth of the blade should be in contact with thin metal or cutting will be difficult. Make sure that blades are installed with the teeth pointing away from the handle.

Wallboard Saws. The *wallboard saw* (Fig. 12-32) is similar to the compass saw, but is designed especially for gypsum board. The point is sharpened to make self-starting cuts for electric outlets, pipes, and other projections. Another type with a hand-saw handle is also used frequently. (Their use is discussed in detail in a following unit on drywall construction.)

Miter Boxes. The *miter box* (Fig. 12-33) is used to cut angles of various degrees on finish lumber by swinging the saw to the desired angle and locking it in place. These cuts are called *miters*. The joint between the pieces cut at these angles is called a *mitered joint*.

A mitered joint is made by cutting each piece at half the angle at which it is to be joined to another piece (Fig. 12–34). For instance, if two pieces are to be joined with a mitered joint at 90 degress to each other, each piece is cut at a 45 degree angle.

The miter box has built-in stops to locate the saw to cut 90, 67 1/2, 60, and 45 degree angles, which are commonly used miter angles. The miter box saw should only be used in the miter box and for no other purpose.

Fig. 12-32 Wallboard saws.

Fig. 12-33 Miter box and miter box saw.

Fig. 12-31 Hacksaw.

HELPFUL HINTS

How to find the miter cut for any angle.

LINE 1

STEP 1– PLACE A WIDE STRAIGHTEDGE AGAINST ONE SIDE OF THE ANGLE TO BE BISECTED AND DRAW A LINE ALONG THE OUTSIDE EDGE. (LINE 1)

LINE 1   LINE 2

STEP 2– PLACE THE SAME STRAIGHTEDGE AGAINST THE OTHER SIDE AND DRAW ANOTHER LINE ALONG ITS OUTSIDE EDGE. (LINE 2)

MITER ANGLE

LINE 1   LINE 2

STEP 3– DRAW A LINE FROM THE CORNER TO THE POINT WHERE LINE 1 AND LINE 2 INTERSECT. THIS IS THE MITER ANGLE.

Fig. 12-34

CHAPTER 13

FASTENING AND DISMANTLING TOOLS

iscussed in this chapter are those tools used to drive nails and turn screws and other fasteners. Tools used to clamp, hold, pry, and dismantle workpieces are also included.

Fastening Tools

The carpenter must decide which fastening tool to select and be able to use it competently and safely for the job at hand.

Hammers

The carpenter's *claw hammer* is available in a number of styles and weights. The claws may be straight or curved. Weights range from 7 to 32 ounces. Most popular for general work is the 16-ounce, curved claw hammer (Fig. 13-1). For rough work, a 20- or 22-ounce *framing hammer*, which has a longer handle (Fig. 13-2) is often used. In some areas, a 28- or 32-ounce, straight claw framing hammer is preferred for extra driving power.

Nail Sets. *Nail sets* (Fig. 13-3) are used to set nail heads below the surface. The most common sizes are 1/32, 2/32, and 3/32 inch. The size refers to the diameter of the tip. The surface of the tip is concave to prevent it from slipping off the nail head. If the tip becomes flattened, the nail set has lost its usefulness.

CAUTION Do not set hardened nails or hit the tip of the nail set with a hammer. This will cause the tip to flatten and may produce chips of flying metal. If it is necessary to drive hardened steel nails, wear eye protection (Fig. 13-4). Do not hit the side of the nail set for setting nails in tongue and groove flooring.

Fig. 13-3 Nail set.

Fig. 13-1 Curved claw hammer.

Fig. 13-2 Framing hammer.

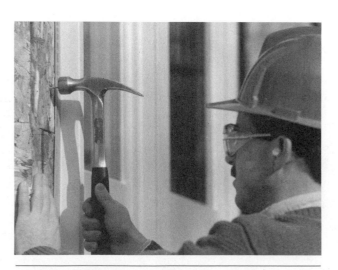

Fig. 13-4 Wear eye protection when driving hardened nails.

Fig. 13-5

Nailing Techniques. Hold the hammer firmly, close to the end of the handle, and hit the nail squarely. If the hammer frequently glances off the nail head, try cleaning the hammer face (Fig. 13-5). As a general rule, use nails that are three times longer than the thickness of the material being fastened.

Toenailing is the technique of driving nails at an angle to fasten the end of one piece to another (Fig. 13-6). It is used when nails cannot be driven into the end. Start the nail about 3/4 to one inch from the end and at an angle of about 30 degrees from the surface.

Drive finish nails almost home. Then set the nail below the surface with a nail set to avoid making hammer marks on the surface. Set finish nails at least 1/8 inch deep so the filler will not fall out.

In hardwood, or close to edges or ends, drill a hole slightly smaller than the nail shank to prevent the wood from splitting or the nail from bending. If a twist drill of the desired size is not available, cut the head off a finish nail of the same gauge and use it for making the hole.

Blunting or cutting off the point of the nail also helps prevent splitting the wood (Fig. 13-7). The point spreads the wood fibers as the nail is driven, while the blunt end pushes the fibers ahead of it and reduces the possibility of splitting. A little *paraffin* applied to the nail shank makes driving the nail easier. The end of a used candle is a handy thing to carry in a tool box for this and other lubricating purposes.

Holding the nail tightly with the thumb, and as many fingers as possible, while driving the nail in hardwood helps prevent bending the nail. Of course, hold the nail in this manner only as long as possible. Be careful not to glance the hammer off the nail and hit the fingers.

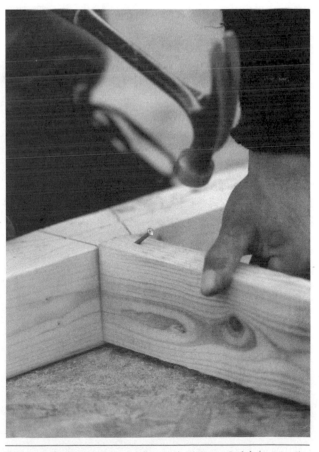

Fig. 13-6 Toenailing is the technique of driving nails at an angle.

HAMMER

BLUNTED
NAIL
POINT

SOMETIMES BLUNTING THE
POINT OF A NAIL WILL PREVENT
SPLITTING THE STOCK. TAP
THE POINT LIGHTLY SO AS
NOT TO BEND THE NAIL.

COMBINATION
PLIERS

IF A TWIST DRILL IS NOT HANDY,
CUT THE HEAD OFF A NAIL OF
THE SAME GAUGE AS THE NAILS
BEING USED, AND USE IT TO
DRILL HOLES.

Fig. 13-7 Methods to avoid splitting wood.

When nailing along the length of a piece, stagger the nails from edge to edge, rather than in a straight line. This avoids splitting and provides greater strength (Fig. 13-8). Drive nails at an angle into end grain for greater holding power. When fastening pieces side to side, nails are driven at an angle for greater strength. In addition, this may keep the nail points from protruding (Fig. 13-9) if using 16d spikes to fasten 2-inch nominal stock together.

Fig. 13-9 Driving nails at an angle increases holding power.

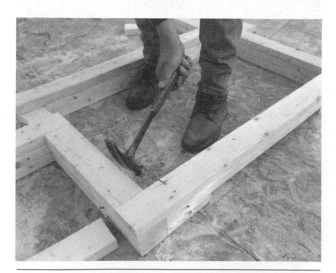

Fig. 13-8 Stagger nails for greater strength and to avoid splitting the stock.

When it is necessary to start a nail higher than you can hold it, press the nail tightly between the claws of the hammer with the head of the nail against the handle (Fig. 13-10). Turn the claws of the hammer toward the surface. Reach up and swing the hammer to start and hold the nail in the stock. Pull the hammer claws away from the nail, turn the hammer around, and drive the nail home.

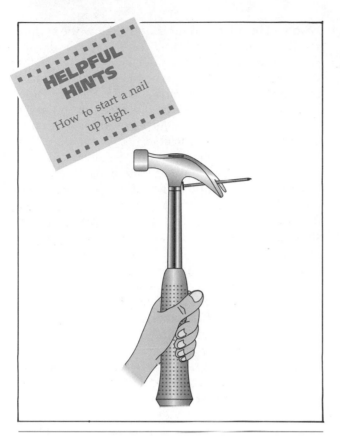

HELPFUL HINTS

How to start a nail up high.

Fig. 13-10

Fig. 13-12 Phillips screwdriver.

Fig. 13-13 Spiral screwdriver.

Screwdrivers

Screwdrivers are manufactured to fit all types of screw slots. The carpenter generally uses only the *common* screwdriver (Fig. 13-11), which has a straight tip to drive common screws, and the *Phillips* screwdriver, which has a cross-shaped tip (Fig. 13-12). In addition, *spiral* screwdrivers, with interchangeable tips (Fig. 13-13), are used to drive screws at a faster rate with greater ease.

Screwdriver Sizes. Common screwdrivers are sized by the length of the blade and by the type.

Fig. 13-11 Common screwdriver.

Lengths generally run from 3 to 12 inches. Phillips screwdrivers are sized by their length and point size. Commonly used sizes are lengths that run the same as common screwdriver and points that come in numbers 0, 1, 2, 3 and 4. The higher number indicates a point with a larger diameter.

Screwdrivers should fit snugly, without play, into the slot of the screw being driven. The screwdriver tip should not be wider than the screwhead, nor should it be too narrow (Fig. 13-14). The correct

TOO THIN   TOO THICK   TOO WIDE   TOO NARROW

CORRECT THICKNESS   CORRECT WIDTH

Fig. 13-14 Select the correct size screwdriver for the screw being driven.

size screwdriver helps assure that the screw will be driven without slipping out of the slot. When seated, the screwdriver slot should look the same as before the screw was driven, with no burred edges.

Screwdriver Bits. A *screwdriver bit* (Fig. 13-15) may be used with a bit brace to drive large screws with less effort. These bits are available in a number of sizes to accommodate a variety of screws.

Screwdriving Techniques. If possible, select screws so that 2/3 of their length penetrates the piece in which they are gripping. In preparation for driving a screw a *shank* hole and a *pilot hole* must be drilled. In addition, if the screw has a flat head, the shank hole must be countersunk so that the screw-

Fig. 13-15 Screwdriver bit. (Courtesy of Stanley Tools)

head will be flush with or slightly below the surface when driven (Fig. 13-16).

To select a drill for the shank hole, hold the drill bit against the shank of the screw and determine by eye if it is the same size. For the pilot hole, hold the drill bit against the threaded portion of the screw so that the drill covers the solid center section but leaves the threads visible.

Select drills with great care. Smaller drills may be used for a pilot hole in softwoods. Some may advocate that in softwood no pilot hole is necessary, but it is wise to drill them anyhow. It doesn't take that much more time, and the screw can be driven straight and easier. Without a pilot hole, the screw may follow the grain and go in at an undesirable angle.

In hardwoods if the pilot hole is too small or not deep enough, difficulty may be encountered in driving the screw. This causes slipping and damage to the screw slot. Also, if too much pressure is applied when driving the screw, the head may be twisted off. This is particularly true when driving screws of soft metal, such as aluminum or brass. It might be wise to first drive a steel screw, remove it, and then drive the screw of softer metal. Rub some wax (paraffin) on the threads of the screw to make driving easier. Remember that if the pilot hole is too large, the screw will not grip.

STEP BY STEP
P R O C E D U R E S

Fig. 13-16 Preparation needed to drive a screw.

CHUCK

BIT

WOOD STOP BLOCK

SCREWDRIVER

FLAT PLUG OR BUNG (GRAIN IS HORIZONTAL)

TAPERED PLUG

RABBETED PLUG

(1.) DRILL COUNTER BORED HOLE OF DESIRED DIAMETER. USE A STOP BLOCK TO ASSURE DESIRED DEPTH.

(2.) DRILL SHANK AND PILOT HOLES AS DESCRIBED PREVIOUSLY. DRIVE SCREW UNTIL WELL SEATED.

(3.) COUNTERBORED HOLES MAY BE PLUGGED WITH VARIOUS SHAPED PLUGS AS SHOWN ABOVE.

Fig. 13-17 It is sometimes desirable to counterbore when driving screws.

Use a stop when drilling the shank hole to make sure it will not be drilled too deep. This will prevent it from going through the material when drilling the pilot hole. A simple stop can be made by drilling a hole lengthwise through a piece of nominal 1 × 1 stock, cut to the desired length, and inserted on the twist drill against the chuck.

If the material to be fastened is thick, the screw may be set below the surface by *counterboring* to gain additional penetration without resorting to a longer screw. To set the screwhead below the surface, bore the counterbored hole first. Use a stop block to assure desired depth. The diameter of the hole should be equal to or slightly larger than the diameter of the screwhead. Next, drill the shank hole; then, the pilot hole (Fig. 13-17). Drilling shank and pilot holes first leaves no stock for the center point of the bit used to make the counterbored hole.

Dismantling Tools

Dismantling tools are used to take down staging and scaffolding, concrete forms, and other temporary structures. In addition, they are used for tearing out sections of a building when remodeling. Carpenters must be skilled in the use of the tools, and in the work, so that the dismantled members are not damaged any more than necessary.

Hammers

In addition to fastening, hammers are often used for pulling nails to dismantle parts. To increase leverage and make nail pulling easier, place a small block of wood under the hammer head (Fig. 13-18).

Fig. 13-18 Pull a nail more easily by placing a block of wood under the hammer.

Bars and Pullers

The *wrecking bar* (Fig. 13-19) is used to withdraw spikes and to pry when dismantling parts of a structure (Fig. 13-20). They are available in lengths from 12 to 36 inches, with the 30-inch size preferred for construction work.

Carpenters need a small *pry bar*, similar to that shown in Fig. 13–21, to pry small work and pull

Fig. 13-22 Nail claw.

small nails. To extract nails that have been driven home (all the way in) a *nail claw*, commonly called a *cat's paw,*)is used (Fig. 13-22).

Holding Tools

To turn nuts, lag screws, bolts, and other objects, an *adjustable wrench* is often used (Fig. 13-23). The wrench is sized by its overall length. The 10-inch adjustable wrench is the one most widely used.

For extracting, turning, and holding objects, a pair of pliers is often used. Many kinds are manufactured, but the *combination pliers* (Fig. 13-24) is designed for general use.

Fig. 13-19 Wrecking bar.

Fig. 13-20 Using a wrecking bar to pry stock loose.

Fig. 13-23 Adjustable wrench.

Fig. 13-21 Pry bar.

Fig. 13-24 Combination pliers.

C clamps (Fig. 13-25) are very useful for holding objects together while they are being fastened, holding temporary guides, applying pressure to glued joints, and many other purposes. The size is designated by the throat opening. C clamps are available in a number of styles. The carpenter will find a few 3- or 4-inch light C clamps to be very useful.

Fig. 13-25 C clamp.

Review Questions

Select the most appropriate answer.

1. When stretching a steel tape to lay out a measurement, place the ring on a nail so the
   a. one-inch mark is on the starting line
   b. end of the tape is on the starting line
   c. inside of the ring is on the starting line
   d. outside of the ring is on the starting line

2. In construction, the term *plumb* means
   a. horizontal.
   b. level.
   c. straight.
   d. vertical.

3. In addition to laying out circles and equal spaces, the wing dividers may be used for
   a. scaling.
   b. scoring.
   c. scribing.
   d. squaring.

4. A #8 auger bit will bore a hole
   a. 1/4 inch in diameter.
   b. 1/2 inch in diameter.
   c. 3/4 inch in diameter.
   d. one inch in diameter.

5. Countersinks are tools used to
   a. make sink cutouts in countertops.
   b. drive large screws.
   c. form recesses for flat head screws.
   d. set nails below the surface.

6. The longest of the bench planes is called a
   a. fore plane.
   b. jack plane.
   c. jointer plane.
   d. smooth plane.

7. For rough work, the carpenter uses a crosscut saw with
   a. 5 1/2 points.
   b. 7 to 8 points.
   c. 11 to 12 points.
   d. 16 to 18 points.

8. The kind of joint made by cutting one piece of molding to fit against the irregular face of a similar piece is made by using a
   a. chisel.
   b. coping saw.
   c. compass saw.
   d. countersink.

9. To prevent a nail set from slipping off the nail head, its tip is
   a. flat and smooth.
   b. convex.
   c. concave.
   d. checkered.

10. Grind the bevel on a chisel so it
    a. is hollow.
    b. has a wire-edge.
    c. is sharp.
    d. is smooth.

■ B U I L D I N G F O R S U C C E S S ■

Tool Care and Maintenance

It is possible to make some assumptions about a craftsperson's quality of work by inspecting his or her tools and equipment. In order to do quality work, skilled workers must properly maintain their entire tool inventory. It is essential for a good carpenter, electrician, mason, or plumber to develop a regular tool maintenance program.

All hand and power tools are to be properly used and cared for, regardless of the owner. The contractor's tools and equipment used by the worker on a daily basis must carry the same value as those of the worker. A carpenter must be ready to perform maintenance on the job if any equipment malfunctions or needs repair. Tasks such as honing a cutting edge, lubricating, adjusting, mending, and repairing can be completed successfully on the job. The majority of required maintenance will be conducted after work and on Saturdays.

An ongoing tool maintenance program may be a one-person operation or require the services of a professional repair person. Regardless of the situation, the scheduling of routine or special repair is required. An efficient tool maintenance program must begin with the proper storing and use of all hand and power tools. The tool inventory must be immediately accessible to be functional. A nail gun will not function properly if it is tossed into the back of a pickup.

An experienced craftsperson knows which tools call for daily, weekly, or monthly maintenance. The most frequently used tools, such as chisels, saws, planes, and router bits, may require daily inspection. The habit of storing tools in a protected manner such as tool boxes, bags, cases, or on shelves will prolong their service life.

The craftsperson who knows building materials, tools, and equipment will select the best tool for the job. When the hand or power tool is needed, it should be in proper working order. Potential accidents resulting from dull or malfunctioning equipment will be reduced. If a tool must be forced to perform, it has not been properly maintained.

When properly trained, a carpenter can sharpen a noncarbide saw blade at the job site in just a few minutes. A simple homemade sharpening jig, a sawhorse, and two C clamps will quickly bring a blade back into use. A plane iron can be honed to regain a sharp cutting edge in a very short time. At times, electric cords can be mended or replaced on the job. Blades and spare parts such as brushes can be carried in a tool box.

As repairs are made on tools and equipment, caution should be taken to assure the worker's safety. All power tools must be disconnected from their power source before any attempt is made to analyze the problem. When the problem appears to be beyond the repair capability of the worker, the equipment must be taken to an experienced tool repair facility.

One of the best assurances for having hand and power tools ready for use is to create and use an effective maintenance program. The following suggestions will help guarantee that hand and power tools are properly cared for.

1. Protect your tools and equipment during daily transport and use.

2. To prolong the life of your equipment, store it properly in tool boxes, carrying cases, racks, shelving, or special containers.

3. Carry an inventory of the most common replacement parts, such as cords, brushes, batteries, and gas cylinders (nail guns), for on-the-job repairs.

4. Avoid leaving tools and equipment in unprotected areas, on floors, on the ground (in water or mud), on ladders or scaffolding, on roofs, or near heavy traffic areas.

5. Sideline tools that have malfunctioned and schedule them for immediate repair.

6. Have a maintenance system in operation at the job site and schedule regular daily, weekly or monthly maintenance.

7. Establish and use as needed the services of a reputable maintenance service center.

8. Read and keep the service/use manual that accompanies each hand and power tool in your inventory. File each manual where it is easily accessible.

9. Clean, check for malfunctions, and properly store your equipment on a daily basis.

10. Schedule a brief period of time at the end of each week to perform maintenance in preparation for the next work week.

Remember that the tools and equipment in your inventory play a major role in generating your income. Don't assume that they will continue to operate without regular care and maintenance.

FOCUS QUESTIONS: For individual or group discussion.

1. In what ways can a personal tool maintenance process benefit you as a carpenter or other skilled worker?

2. What are some of the outcomes of not having a tool or machine maintenance program in place?

3. Assuming that you are working out of a vehicle, how would you transport your tools and machines that are used practically every day?

4. Do you know where you can take tools and machines for repairs within your community? How do you locate and make initial contact with a tool service center?

U N I T

6

Portable Power Tools

OBJECTIVES

After completing this unit, the student will be able to:

- state general safety rules for operating portable power tools

- identify, describe, and safely use the following portable power tools: circular saws, saber saws, reciprocating saws, drills, hammer-drills, screwdrivers, planes, routers, sanders, staplers, nailers, and powder-actuated drivers

CONTENTS

Portable power tools are widely used in the construction industry. They enable the carpenter to do more work in shorter time with less effort. However, they can cause serious injury if operated improperly. Most accidents are caused by a complete disregard for common sense safety precautions. Be aware of the dangers of operating portable power tools, and take steps to avoid these dangers and prevent accidents.

— GENERAL SAFETY RULES —

CAUTION Have a complete understanding of a tool before attempting to operate it.

Electrical shock has caused many fatal accidents. Make sure the tool is properly grounded. A *ground fault circuit interrupter (GFCI)* must be in the circuit being used. Even though the circuit is grounded, it is still possible for an operator of a portable power tool to be electrocuted should a short circuit occur without a GFCI in the circuit. The GFCI will trip at about 5 milliamperes, turning off the power before any chance of electrical shock or electrocution occurs.

Do not allow your attention to be distracted when operating power tools.

Never use a power tool that has a worn or frayed cord.

Use the proper size extension cord. Do not use one any longer than necessary to prevent excessive voltage drop.

Be careful to place extension cords where they will not be damaged, cut, caught in the tool, or tripped on.

Unplug tools when making adjustments or changing cutters.

Do not wear loose clothing that might become caught in the tool.

Wear safety goggles during all operations that cause particles to fly.

Make sure the material being worked on is securely held and supported. Use sharp tools and all safety guards supplied.

Stay alert and develop an attitude of care and concern for yourself and other workers.

CHAPTER **14**   SAWS, DRILLS, AND DRIVERS

Saws

The carpenter uses several kinds of portable saws for crosscutting and ripping, for making circular cuts, and for cutting openings in floors, walls, and ceilings.

Electric Circular Saws

Commonly called the *skilsaw,* the portable electric circular saw (Fig. 14-1) is used by the carpenter more than any other portable power tool. The circular saw blade is driven by an electric motor. The saw has a base that rests on the work to be cut. A handle with a trigger switch is provided for the operator to control the tool. The saw is adjustable for depth of cut. A retractable safety guard is provided over the blade, extending under the base.

The base may be tilted for making *bevel* and *miter* cuts. Bevel cuts are those at an angle through the thickness and with the grain or along the length. Edge or end miters are angle cuts through the thickness across the grain. Flat miters are cuts at an angle across the width. *Compound miters* are angle cuts across the width and also through the thickness (Fig. 14-2).

Fig. 14-1 Using a portable electric circular saw to cut compound angles on roof rafters.

Fig. 14-2 Edge, flat, and compound miters.

Saws are manufactured in many styles and sizes. They are available from very light- to heavy-duty models ranging from 1/6 hp to 1 1/2 hp. The size is determined by the diameter of the blade, which ranges from 4 1/2 to 12 inches. The handle and switch may be located on the top or in back. The blade may be driven directly by the motor or through a worm gear (Fig. 14-3).

The forward end of the base is notched in two places to serve as guides for following layout lines.

One notch is used to follow layout lines when the base is tilted to 45 degrees and the other when the base is not tilted.

> **CAUTION** Make sure the saw blade is installed with the teeth pointing in the correct direction. The teeth of the saw blade projecting below the base should be pointing away from the operator.

To loosen the bolt that holds the blade in place, insert a spike under the base and through the hole provided in the blade. Turn the bolt in the same direction as the rotation of the blade. To tighten, turn the bolt in a direction opposite to the rotation of the blade. Some saws have an arrangement that locks the spindle while loosening or tightening the bolt.

On most models an adjustable attachment that fits into the base is used for ripping narrow pieces parallel with the edge. The saw may also be guided by tacking or clamping a straightedge to the material and running the edge of the saw base against it. Allowance must be made for the distance from the saw blade to the edge of the saw base when positioning the straightedge.

Circular Saw Blades. Circular saw blades are available in a number of styles. The shape and number of teeth per inch determine their cutting action. Carbide-tipped blades are used more than high speed steel blades. They stay sharper longer when cutting material that dulls ordinary blades quickly. (More complete information on saw blades can be found in Unit 7.)

Fig. 14-3 Direct drive and worm-gear drive portable electric circular saws. (*Courtesy of Skil Power Tools*)

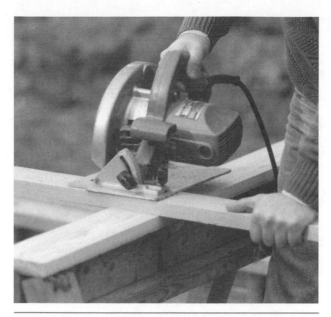

Fig. 14-4 Saw cuts are made over the end of supports so the waste will fall clear and not bind the blade.

Using the Portable Circular Saw. Safe and efficient cutting follows an established method:

■ Make sure the work is securely held and that the waste will fall clear and not bind the saw blade (Fig. 14-4).

■ Adjust the depth of cut so that the blade just cuts through the work. Never expose the blade any more than is necessary (Fig. 14-5).

CAUTION Make sure the guard operates properly. Be aware that the guard may possibly stick in the open position. *Never* wedge the guard back in an open position.

■ Mark the stock. Put on safety goggles. Rest the forward end of the base on the work. With the blade clear of the material, start the saw.

■ Advance the saw into the work when it has reached full speed. Make sure to observe the line to be followed. With the saw cut in the waste, cut as close to the line as possible for a short distance. Then observe the notch in the forward end of the base. Follow the line by using the notch on the saw base.

■ Follow the line closely.

CAUTION Any deviation from the line may cause the saw to bind and kick back. Do not force the saw forward. In case the saw does bind, stop the motor and bring the saw back to where it will run free. Continue following the line closely.

■ Near the end of the cut, the forward end of the base will go off the work. Guide the saw by observing the line at the saw blade and finish the cut. The saw may also be guided by watching the layout line at the saw cut for the whole length. Let the waste drop clear and release the switch.

CAUTION Keep the saw clear of your body until the saw blade has completely stopped.

■ When starting cuts across stock at an angle, it may be necessary to retract the guard by hand. A handle is provided for this purpose (Fig. 14-6). Release the handle after the cut has been started and continue as above.

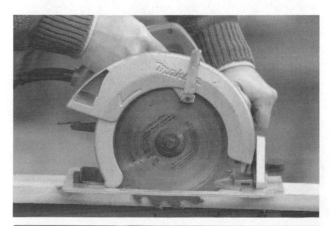

Fig. 14-5 The blade of the saw is adjusted for depth only enough to cut through the work.

Fig. 14-6 Retracting the guard of the portable circular saw by hand.

■ Compound miter cuts may be made by cutting across the stock at an angle with the base tilted.

Portable circular saws cut on the upstroke. The saw blade rotates upward through the material. As the teeth of the saw blade come through the top surface, splintering of the stock occurs at the layout line. The severity of the splintering depends of the kind of blade used, kind and thickness of the material being cut, and other factors. More splintering occurs when cutting across grain than with the grain.

On *finish work,* that is, work that will ultimately be exposed to view, any splintering along the cut is unacceptable. One way to prevent this is to mark layout lines on, and make cuts from, the back side. If it is not possible to cut from the back side, or if both sides may be exposed to view, mark the layout lines on the face side. Then score along the layout lines with a sharp knife. Make the cuts just outside the scored lines.

Making Plunge Cuts. Many times it is necessary to make internal cuts in the material such as for sinks in countertops or openings in floors and walls. To make these cuts with a portable electric circular saw, the saw must be plunged into the material:

■ Accurately lay out the cut to be made. Adjust the saw for depth of cut.

■ Wearing safety goggles, hold the guard open and tilt the saw up with the front edge of the base resting on the work. Have the saw blade over, and in line with, the cut to be made (Fig. 14-7.)

■ Make sure the teeth of the blade are clear of the work and start the saw. Lower the blade slowly into the work, following the line carefully, until the entire base rests squarely on the material.

■ Advance the saw into the corner. Release the switch and wait until the saw stops before removing it from the cut.

■ Reverse the direction and cut into the corner. Again, wait until the saw stops before removing it from the cut.

■ Proceed in like manner to cut the other sides. Cut into the corners with a handsaw or saber saw.

Saber Saws

The *saber saw* (Fig. 14-8) is sometimes called a *jigsaw* or *bayonet saw.* It is widely used to make curved cuts. The teeth of the blade point upward when installed, so the saw cuts on the upstroke. To produce a splinter-free cut on the face side, it is best to cut on the side opposite the face of the work, if possible, or to score with a knife along the layout line.

There are many styles and varieties of saber saws. The length of the stroke along with the amperage of the motor determines its size and quality. Strokes range from 1/2 to one inch. The longest stroke is the best for faster and easier cutting. Some saws can be switched from straight up-and-down strokes to several orbital (circular) motions to provide the most effective cutting action for various materials (Fig. 14-9). The base of the saw may be tilted to make bevel cuts.

Fig. 14-7 Making a plunge cut with a portable circular saw.

Fig. 14-8 The saber saw may be used either in straight or orbital cutting actions.

Straight line: cutting action for thick mild steel, very hard wood, ceramic tile, glass, scroll cuts and maximum precision.

Smallest orbit: for faster cutting on wood and hardboard and more aggressive cutting in mild steel and aluminum.

Medium orbit: additional aggressive cutting action for most plastics, fiberglass, composition board.

Largest orbit: for rough high speed cuts in wood, plasterboard, countertops; the most aggressive cuttting action.

Fig. 14-9 Guide to the selection of saber saw blade action.

Blade Selection. Many blades are available for fine or coarse cutting in wood or metal (Fig. 14-10). Wood cutting blades have teeth that are from 6 to 12

points to the inch. Blades with coarse teeth (less points to the inch) cut faster, but rougher. Blades with more teeth to the inch may cut slower, but produce a smoother cut surface. They do not splinter the work as much. Proper blade selection and use will produce the best cuts. The blade will last longer and the saber saw will not be subjected to harmful abuse by applying excessive pressure when cutting.

Using the Saber Saw. Follow a safe and established procedure:

- Outline the cut to be made. Secure the work either by hand, tacking, clamping, or by some other method.

- Using safety goggles, hold the base of the saw firmly on the work. With the blade clear, pull the trigger.

APPLICATION	PART NUMBER	TEETH PER INCH	OVERALL LENGTH (INCHES)	WIDTH (INCHES)
Metal cutting (High speed steel)				
Non-ferrous metal cutting ¼" to ¾" thick	94612	8	3⅝	5/16
Metal cutting over ⅛" thick	94613	12	2¾	5/16
Metal cutting up to ⅛" thick	94614	21	2¾	5/16
Metal cutting up to 1/16" thick	94615	36	2¾	5/16
Wood cutting (Taper ground, high speed steel)				
Fast, smooth curve cutting in wood up to 2" thick	94618	6	3⅝	¼
Fast, smooth cutting in wood up to 2" thick	94619	6	3⅝	5/16
Very smooth curve cutting in wood up to 1" thick	94622	10	3⅝	¼
Very smooth cutting in wood up to 1" thick	94623	10	3⅝	5/16
Wood cutting (Alternate set — ground, high carbon steel)				
Fast, medium curve cutting in wood up to 2" thick	94627	6	3⅝	¼
Fast, medium cutting in wood up to 2" thick	94628	6	3⅝	5/16
Wood cutting (Alternate set — high carbon steel)				
Fast, rough cutting in wood up to 2" thick	94631	6	3⅝	5/16
Medium cutting in wood up to 2" thick	94634	8	3⅝	5/16
General purpose (Alternate set — high speed steel)				
General purpose cutting in wood, plastic, metal, etc.	94637	12	3⅝	5/16
Wood scroll cutting (Taper ground, high carbon steel)				
Smooth, intricate scroll cuts in wood up to 1¾" thick	94640	20	2¾	3/16

Fig. 14-10 Saber saw blade selection guide.

■ Push the saw into the work, following the line closely. Make the saw cut into the waste, and cut as close to the line as possible without completely removing it.

■ Keep the saw moving forward, holding the base down firmly on the work. Turn the saw as necessary in order to follow the line to be cut. Feeding the saw into the work as fast as it will cut, but not forcing it, finish the cut. Keep the saw clear of your body until it has stopped.

Making Plunge Cuts. Plunge cuts are made with the saber saw in a manner similar to that used with the circular saw:

■ Tilt the saw up on the forward end of its base with the blade in line and clear of the work (Fig. 14-11).

■ Start the motor, holding the base steady. Very gradually and slowly, lower the saw until the blade penetrates the work and the base rests firmly on it.

CAUTION Hold the saw firmly to prevent it from jumping when the blade makes contact with the material and to make a successful plunge cut.

■ Cut along the line into the corner. Back up for about an inch, turn the corner, and cut along the other side and into the corner.

■ Continue in this manner until all the sides of the opening are cut.

Fig. 14-11 Making an internal cut by plunging the saber saw.

Fig. 14-12 Using the reciprocating saw to cut an opening in a wall.

CAUTION Make sure the tool comes to a complete stop before withdrawing it from the material being cut.

■ Turn the saw around and cut in the opposite direction to cut out the corners.

Reciprocating Saws

The *reciprocating saw* (Fig. 14-12), sometimes called a *sawzall*, is used primarily for *roughing in* work. This work consists of cutting holes and openings for such things as pipes, heating and cooling ducts, and roof vents. It can be likened to a powered compass saw.

Most models have a variable speed of from 0 to 2,400 strokes per minute. Like saber saws, some models may be switched to several *orbital* cutting strokes from a straight back and forth (reciprocal) cutting action. Ordinarily, the orbital cutting mode is used for fast cutting in wood. The reciprocating stroke is used for cutting metal.

Reciprocating Saw Blades. Common blade lengths run from 4 to 12 inches. They are available for cutting practically any type of material. Blades are available to cut wood, metal, plaster, fiberglass, ceramics, and other material.

Using the Reciprocating Saw. The reciprocating saw is used in a manner similar to the saber saw. The difference is that the reciprocating saw is heavier and more powerful. It can be used more efficiently to cut through rough, thick material, such as walls when remodeling. With a long blade, it can be used to cut flush with a floor or along the side of a stud.

Fig. 14-13 Portable power drills are available in a number of styles.

To use the saw, lines must be laid out and followed. The base or shoe of the saw is held firmly against the work whenever possible. If the blade can be inserted in a hole in the material, the cut may be started by starting the motor and following the layout lines. The blades can be reversed for cutting in confined areas.

Making Plunge Cuts. Plunge cuts are made in a manner opposite to that used with a saber saw. Check for location of electrical wiring and plumbing before making the cut.

■ Tilt the saw *down* on the bottom edge of its shoe. Hold the blade above the layout line and clear of the work.

■ Start the saw. Gradually lift the handle upward while keeping the shoe firmly on the work. Let the blade slowly penetrate the work and continue the cut.

CAUTION The saw must be held securely to keep it from jumping when the saw blade starts to enter the stock. Remove the blade from the work only after it has completely stopped.

Drills and Drivers

Portable power drills, manufactured in a great number of styles and sizes, are widely used to drill holes and drive fasteners in all kinds of construction materials. Battery-powered cordless models are used because of their convenience or when electrical power is not available.

Drills

The drills most generally used in the construction industry are classified as light-duty or heavy-duty. Light-duty drills usually have a *pistol-grip* handle. Heavy-duty drills may have a *spade-shaped* or *D-shaped* handle (Fig. 14-13).

Drill Sizes. The size of a drill is determined by the capacity of the *chuck,* that is, its maximum opening. The chuck is that part of the drill that holds the cutting tool. The most popular sizes for light-duty models are 1/4 and 3/8 inch. Heavy-duty drills with a 1/2-inch chuck are generally used.

Drill Speed and Rotation. Most drills have *variable speed* and *reversible* controls. Speed of rotation can be controlled from 0 to maximum RPM (revolutions per minute) by varying the pressure on the trigger switch. Slow speeds are desirable for drilling larger holes or for driving fasteners. Faster speeds are used for drilling smaller holes. A reversing switch changes direction of the rotation for removing screws or withdrawing bits and drills from holes.

Bits and Twist Drills. Twist drills are used in electric drills to make small holes in wood or metal (Fig. 14-14). For larger holes in wood, plastics, and composition materials, a variety of wood-cutting bits are used.

For boring holes in rough work, the *spade bit* is commonly used. For a hole with a cleaner edge in finish work, the *Power Bore* bit may be used (Fig. 14-15). Notice that none of these types has a center point that is threaded.

> **CAUTION** Never use bits with threaded center points in electric drills when boring deep holes. A threaded center will draw the bit into the work. This makes it difficult to withdraw the bit, and may cause the operator to lose control of the drill.

Other Drill Accessories. Occasionally, carpenters may use *hole saws*. These saws cut holes through thin material from 5/8 inch to 6 inches in diameter. One disadvantage of the hole saw is that a hole cannot be made partially through the material. Only the circumference is cut and the waste is not expelled.

(1) Power Bore Bit   (2) Spade Bit

Fig. 14-14 Twist drills are used to drill small holes. (*Courtesy of DeWalt Industrial Tool Company*)

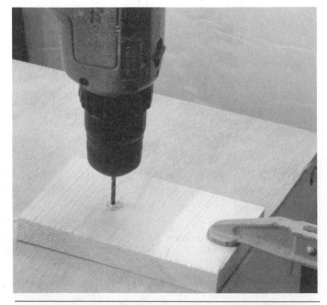

Fig. 14-15 Spade and Power Bore bits are used to drill larger holes in wood and similar material.

The *combination drill and countersink,* discussed in Chapter 12, is frequently used with power drills for fast drilling, countersinking, and counterboring in one operation when needed.

Masonry drill bits have carbide tips for drilling holes in concrete, brick, tile, and other masonry. They are frequently used in portable power drills. They are more efficiently used in *hammer-drills.*

Using Portable Electric Drills. Select the proper size bit or twist drill and insert it. Tighten the chuck with the chuck key. For accuracy, holes in wood are center-punched. Holes in metal must be center-punched because the drill will wander off center.

> **CAUTION** Hold small pieces securely by clamping or other means. When drilling through metal, especially, the drill has a tendency to hang up when it penetrates the underside. If the piece is not held securely, the hangup will cause the piece to rotate with the drill. It could then hit anything in its path and possibly cause serious injury to a person before power to the drill can be shut off.

Place the bit on the center of the hole to be drilled. Hold the drill at the desired angle. (Fig. 14-16) and start the motor. Apply pressure as required. Drill into the stock, being careful not to wobble the drill. Failure to hold the drill steady may result in breakage of small twist drills.

> **CAUTION** Remove the bit from the hole frequently to clear the chips. Failure to do this may result in the drill binding and twisting from the operator's hands. Be ready to instantly release the trigger switch if the drill does bind.

Hammer-drills

Hammer-drills (Fig. 14-17) are similar to other drills. However, they can be changed to hammers as they

Fig. 14-17 The hammer-drill is used to make holes in concrete.

PERPENDICULAR TO SURFACE

AT AN ANGLE TO SURFACE

GUIDE BLOCK

SIGHT ALONG THE EDGE OF THE GUIDE BLOCKS AS THE HOLE IS DEEPENED.

DRILL BIT

GUIDE BLOCK

HELPFUL HINTS
How to guide a bit to drill a hole at an angle.

Fig. 14-16

drill, quickly making holes in concrete or other masonry. Some models deliver as much as 50,000 hammer blows per minute. Most popular are the 3/8- and 1/2-inch sizes.

A depth stop is usually attached to the side of the hammer-drill. It can be converted to a conventional drill by a quick-change mechanism. Most models have a variable speed of from 0 up to 2,600 rpm. The hammer-drill has the same type chuck and is used in the same manner as conventional portable power drills.

Portable Electric Screwdrivers

Portable electric screwdrivers are commonly called *screwguns* (Fig. 14-18). They are similar in appearance to light-duty drills, except for the chuck. Most screwdrivers have a pistol-type grip for one-hand operation and controls for variable speed and reversing rotation.

The chuck is made to receive special screwdriver bits of various shapes and sizes. Most models have a positive clutch for consistent driving depth. Other models have an adjustable clutch that can be set to various torques.

Using the Portable Electric Screwdriver. In
the positive clutch model, the screw is turned when the trigger is pulled and pressure is applied to the screw. The amount of tightening is dependent on

the operator. As soon as pressure is released, the screwdriver stops turning. This model is preferred when fastening different kinds of materials or driving different size screws. It is the type most widely used.

For uniform fastening using identical screws in the same type material, the adjustable clutch model may be used. The clutch can be adjusted for different amounts of torque. The screwdriver automatically stops turning when the set torque is reached.

Drywall Drivers. *Drywall drivers* are used extensively for fastening gypsum board to walls and ceilings with screws (Fig. 14-19). They are similar to other screwdrivers except that an adjustable nosepiece, which surrounds the screwdriver bit, can be set to drive screws to the desired depth. When the forward end of the nosepiece touches the surface,

SCREW DEPTH ADJUSTMENT

NOSE PIECE

Fig. 14-18 The portable electric screwdriver.

Fig. 14-19 The drywall driver is used to fasten wall board with screws. (A) (*Courtesy of Skil Power Tools*), (B) (*Courtesy of U.S. Gypsum Corporation*)

DEPTH GAUGE

MASONRY DRILL BIT

HAMMER DRILL

CHARGER

CHARGER EXTRA BATTERY

CORDLESS SCREWDRIVER

Fig. 14-20 Cordless drill, screwdriver, and hammer-drill. (A) (*Courtesy of Skil Power Tools*), (B) (*Courtesy of Robert Bosch Power Tool Company*)

the bit can go no farther than the depth for which it is set and slips out of the screw slot.

Accessories are available for driving hex head as well as screws with slotted heads. Hex head nutsetters are available in magnetic or nonmagnetic styles, in sizes ranging from 3/16 to 3/8 inch. Screwdriver bits include slotted, Phillips, Pozidrive, Torx, clutch head, and Robertson.

Cordless Drills and Drivers

Cordless power drills and drivers are widely used when power is not available or inconvenient to use (Fig. 14-20). Power comes from a removable battery, usually attached to the handle of the tool. The battery has to be recharged when it has run down.

Downtime is practically eliminated with spare batteries and improvements in the chargers, which can quick charge in as little as thirteen minutes (Fig. 14-21). Improved chargers can charge batteries of various voltages. They stop charging when the volt-

Fig. 14-21 Cordless tool battery charger. (*Courtesy of DeWalt Industrial Tool Co.*)

age level is reached, before the battery begins to heat. This prolongs the life of the battery.

Despite some of the disadvantages, cordless tools are becoming increasingly popular and are used in many applications (Fig. 14-22).

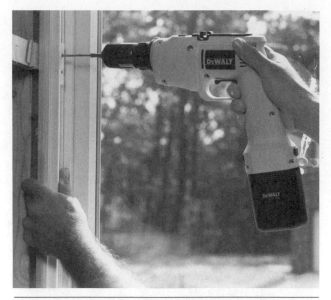

Fig. 14-22 Cordless screwdriver being used to install a window. (*Courtesy of DeWalt Industrial Tool Co.*)

CHAPTER 15

PLANES, ROUTERS, AND SANDERS

Portable Power Planes

Portable power planes make some planing jobs much easier for the carpenter. Planing the ends and edges of hardwood doors and stair treads, for example, takes considerable effort with hand planes, even with razor-sharp cutting edges.

Jointer Planes

The *jointer plane* is used primarily to smooth and straighten long edges, such as fitting doors in openings (Fig. 15-1). It is manufactured in lengths up to 18 inches. The electric motor powers a cutter head that may measure up to 3 3/4 inches wide. The

Fig. 15-1 Using the portable electric jointer plane to fit a door. (*Figure A Courtesy of Skil Corp.*)

planing depth, or the amount that can be taken off with one pass, can be set for 0 up to 1/8 inch.

An adjustable fence allows planing square, beveled edges to 45 degrees, or *chamfers* (Fig. 15-2). A rabbeting guide is used to cut rabbets up to 7/8 inch deep.

Block Planes

The *block plane* is about 7 inches long. It is ordinarily operated with one hand. It is used to smooth the ends and edges of small pieces of stock (Fig. 15-3). Like the jointer plane, a motor powers the cutter head. The toe of the shoe can be adjusted for depth of cut. The fence may be used for guiding the block plane in the same manner as for jointer planes.

Abrasive Planes

At least one manufacturer has replaced the cutter head of the block plane with a heavy-duty sleeve. This tool is called an *abrasive plane*. Its maximum depth of cut is 1/64 inch. The plane leaves the surface sanded and prepared for finishing. With carbide abrasive sleeves, the tool can be used to sand masonry and metal, in addition to wood and plastics.

BEVEL   CHAMFER

A BEVEL IS A CUT AT AN ANGLE THROUGH THE TOTAL THICKNESS

A CHAMFER IS AN ANGLED CUT PART-WAY THROUGH THE THICKNESS

Fig. 15-2 A bevel and a chamfer.

Fig. 15-3 Using a power block plane requires only one hand.

HELPFUL HINTS
How to plane a taper.

1ST PASS

2ND PASS

3RD PASS

4TH PASS

FINAL PASS

DEPTH OF CUT EXAGGERATED FOR EFFECT

FINAL PASS FROM ONE END TO THE OTHER

Fig. 15-4

Operating Power Planes

> **CAUTION** Extreme care must be taken when operating power planes. There is no retractable guard, and the high-speed cutterhead is exposed on the bottom of the plane. Keep the tool clear of your body until it has completely stopped. Keep extension cords clear of the tool.

- Set the side guide to the desired angle, and adjust the depth of cut.
- Hold the toe (front) firmly on the work, with the plane cutterhead clear of the work.

Fig. 15-5 Using a portable electric router.

Fig. 15-6 Router bit selection guide. (*Courtesy of Stanley Tools*)

■ Start the motor. With steady, even pressure make the cut through the work for the entire length. Guide the angle of the cut by holding the guide against the side of the stock. Apply pressure to the toe of the plane at the beginning of the cut. Apply pressure to the heel (back) at the end of the cut to prevent tipping the plane over the ends of the work.

■ To plane a *taper,* that is, to take more stock off one end than the other, make a number of passes. Each pass should be shorter than the preceding one. Lift the plane clear of the stock at the end of the pass. Make the last pass completely from one end to the other (Fig. 15-4 page 121).

Portable Electric Routers

One of the most versatile portable tools used in the construction industry is the *router* (Fig. 15-5). It is available in many models, ranging from 1/4 hp to over 3 hp with speeds of from 18,000 to 30,000 rpm. These tools have high-speed motors that enable the operator to make clean, smooth-cut edges.

The motor powers a chuck in which cutting bits of various sizes and shapes are held (Fig. 15-6). An adjustable base is provided to control the depth of cut. A trigger or toggle switch controls the motor.

The router is used to make many different cuts, including grooves, dadoes, rabbets, and *dovetails* (Fig. 15-7). It is also used to shape edges and make cutouts, such as for sinks in countertops. It is extensively used with accessories, called *templates,* to cut recesses for hinges in door edges and door frames.

Laminate Trimmers. A light-duty specialized type of router is called a *laminate trimmer.* It is used almost exclusively for trimming the edges of *plastic laminates* (Fig. 15-8). Plastic laminate is a thin, hard material used primarily as a decorative covering for kitchen and bathroom cabinets and countertops. (The installation of this material is described in a following chapter.)

LAMINATE TRIMMER

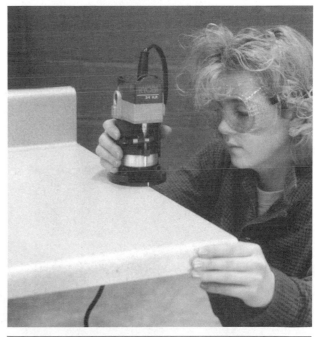

Fig. 15-8 The laminate trimmer is used to trim plastic laminates. (*Courtesy of Rockwell, International*)

Guiding the Router

Controlling the sideways motion of the router is accomplished by the following methods:

■ By using a router bit with a pilot (guide) (Fig. 15-9). The pilot may be solid and rotate with

Fig. 15-7 A dovetail cut is easily made with a router.

ROUTER BIT

PILOT

Fig. 15-9 Guiding the router with a pilot on the bit for a guide.

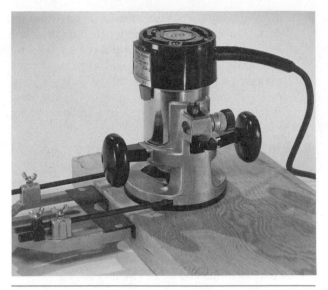

Fig. 15-11 A guide attached to the base of the router rides along the edge of the stock and controls the sideways motion of the router. (*Courtesy of Black & Decker*)

the bit. Other bits may have ball-bearing pilots.

■ By guiding the edge of the router base against a straightedge (Fig. 15-10).

■ By using an adjustable guide attached to the base of the router (Fig. 15-11). The guide rides along the edge of the stock. Make sure the edge is in good condition.

■ By using a template (pattern) with template guides attached to the base of the router (Fig. 15-12). This is the method widely used for cutting recesses for door hinges. (It is explained in detail in a following unit on installing doors.)

■ By freehand routing, in which the sideways motion of the router is controlled by the operator only (Fig. 15-13).

STRAIGHTEDGE IS ON OPERATOR'S RIGHT AS ROUTER IS PULLED. ROTATION OF ROUTER TENDS TO KEEP IT AGAINST STRAIGHTEDGE. IF STRAIGHTEDGE WERE ON LEFT SIDE, ROUTER WOULD HAVE TENDENCY TO PULL AWAY FROM THE STRAIGHTEDGE.

ROUTER

DIRECTION ROUTER IS PULLED

Fig. 15-10 Using a straightedge to guide the router.

Fig. 15-12 Guiding the router by means of a template and template guide.

■ To make *circular* cuts, remove the subbase. Replace it, using the same screwholes, with a custom-made one in which one side extends to any desired length. Along a centerline make a series of holes to fasten the newly made subbase to the center of the desired arc (Fig. 15-14).

Using the Router

Before using the router, make sure power is disconnected. Follow the method outlined:

■ Select the correct bit for the type of cut to be made.

Fig. 15-13 Freehand routing. (*Courtesy of Black & Decker*)

HELPFUL HINTS

How to guide the router to make circular cuts.

REPLACE THE ROUTER SUB-BASE WITH A CUSTOM-MADE ELONGATED BASE. DRILL SEVERAL HOLES TO BE USED AS CENTERS.

PIVOT THE ROUTER ON A CENTER TO MAKE CIRCULAR CUT.

Fig. 15-14

■ Insert the bit into the chuck. Make sure the chuck grabs at least 1/2 inch of the bit. Adjust the depth of cut.

■ Control the sideways motion of the router by one of the methods previously described.

■ Clamp the work securely in position. Plug in the cord.

■ Lay the base of the router on the work with the router bit clear of the work. Start the motor.

■ Advance the bit into the cut, pulling the router in a direction that is against the rotation of the bit. On outside edges and ends, the router is moved counterclockwise around the piece. When making internal cuts, the router is moved in a clockwise direction.

CAUTION Finish the cut, keeping the router clear of your body until it has stopped. Be aware that the router bit is unguarded.

Portable Electric Sanders

Interior trim, cabinets, and other finish should be sanded before any finishing coats of paint, stain, polyurethane, or other material is applied. It is shoddy workmanship to coat finish work without sanding. In too many cases, expediency seems to take precedence over quality. Trim needs to be sanded because the grain probably has been *raised.* This happens because the stock has been exposed to moisture in the air between the time it was planed and the time of installation. Also, rotary planing of lumber leaves small ripples in the surface. Although hardly visible before a finish is applied, they become very noticeable later.

Some finishing coats require more sanding than others. If a penetrating stain is to be applied, extreme care must be taken to provide a surface that is evenly sanded with no cross-grained scratches. If paints or transparent coatings are to be applied, surfaces need not be sanded as thoroughly. Portable sanders make sanding jobs less tedious.

Portable Electric Belt Sanders

The *belt sander* is used frequently for sanding cabinetwork and interior finish (Fig. 15-15). The size of the belt determines the size of the sander. Belt widths range from 2 to 4 inches. Belt lengths vary from 21 to 24 inches. The 3 inch by 21 inch belt sander is a popular, lightweight model. Some sanders have a bag to collect the sanding dust.

3" X 21" BELT SANDER

Fig. 15-15 Using a portable electric belt sander.

Installing Sanding Belts. Sanding belts are usually installed by retracting the forward roller of the belt sander. An arrow is stamped on the inside of some sanding belts to indicate the direction in which the belt should run. The sanding belt should run *with,* not against, the lap of the joint (Fig. 15-16). Sanding belts joined with butt joints may be installed in either direction. Install the belt over the rollers. Then release the forward roller to its operating position.

The forward roller can be tilted to keep the sanding belt centered as it is rotating. Stand the sander on its back end. Hold it securely, and start it. Turn the adjusting screw one way or the other to track the belt and center it on the roller (Fig. 15-17).

Using the Belt Sander. There has been probably more work ruined by improper use of the portable belt sander than any other tool. It is wise to practice on scrap stock until enough experience in its use will assure an acceptable sanded surface. Care must be taken to sand squarely on the sander's pad. Allowing the sander to tilt sideways or to ride on either roller results in a gouged surface.

Fig. 15-16 Sanding belts must be installed in the proper direction.

Fig. 15-17 The belt sander should be adjusted to keep the sanding belt centered on its rollers.

CAUTION Make sure the switch of the belt sander is off before plugging the cord into a power outlet. Some trigger switches can be locked in the "ON" position. If the tool is plugged in when the switched is locked in this position, the sander will travel at high speed across the surface. This could cause damage to the work and/or injury to anyone in its path.

- Secure the work to be sanded. Make sure the belt is centered on the rollers and tracking properly.
- Hold the tool with both hands so that the edge of the sanding belt can be clearly seen (Fig. 15-18).
- Start the machine. Place the pad of the sander flat on the work. Pull the sander back and lift it just clear of the work at the end of the stroke.

CAUTION Be careful to keep the electrical cord clear of the tool. Because of the constant movement of the sander, the cord may easily get tangled in the sander if the operator is not alert.

- Bring the sander forward. Continue sanding, using a skimming motion that lifts the sander just clear of the work at the end of every stroke. Sanding in this manner prevents overheating the sander, the belt, and the material being sanded. It allows debris to be cleared from the work. The operator can also see what has been done.
- Do not sand in one spot too long. Be careful not to tilt the sander in any direction. Always sand with the pad flat on the work. Do not exert excessive pressure. The weight of the sander is enough. Always sand with the grain to produce a smooth finish.
- Make sure the sander has stopped before setting it down. It is a good idea to lay it on its side to prevent accidental traveling.

Fig. 15-18 Hold the belt sander with both hands so the edge of the sanding belt can be clearly seen.

Finishing Sanders

The finishing sander (Fig. 15-19) is used for the final sanding of interior work. These tools are manufactured in many styles and sizes. They are available in cordless models.

Finishing sanders either have an orbital motion, an oscillating (straight back and forth) motion, or a combination of motions controlled by a switch. The orbital motion has faster action, but leaves scratches across the grain. The straight line motion is slower, but produces no cross-grain scratches on the surface.

Most sanders take 1/3 or 1/2 sheet of sandpaper. It is usually attached to the pad by some type of friction or spring device. Some sanding sheets come precut with an adhesive backing for easy attachment to the sander pad.

To use the finishing sander, proceed as follows:

■ Select the desired grit sandpaper. Attach it to the pad, making sure it is tight. A loose sheet will tear easily.

■ Start the motor and sand the surface evenly, *slowly* pushing and pulling the sander with the grain. Let the action of the sander do the work. Do not use excessive pressure. This may overload the machine and burn out the motor. Always hold the sander flat on its pad.

1/4 SHEET PALM GRIP SANDER

CORDLESS BLOCK SANDER

Fig. 15-19 A portable electric finishing sander.

Coated Abrasives

Sandpaper is the common name for *coated abrasives.* Sandpaper does not use sand as an abrasive. Nor is its backing necessarily paper. It comes in the form of sheets, belts, discs, and sleeves, among others. The quality of the coated abrasive depends on the kind of abrasive, the backing on which the abrasive is coated, and the type of adhesive used to secure the abrasive.

DESCRIPTION	GRIT NO.	0 SERIES
	400	10/0
VERY FINE	360	—
	320	9/0
	280	8/0
	240	7/0
FINE	220	6/0
	180	5/0
	150	4/0
MEDIUM	120	3/0
	100	2/0
	80	1/0
COARSE	60	1/2
	50	1
	40	1 -1/2
	36	2
	30	2 - 1/2
VERY COARSE	24	3
	20	3 - 1/2
	16	4
	12	4 - 1/2

Fig. 15-20 Grits of coated abrasives.

Abrasives. The quality of the abrasive is determined by the length of time it is able to retain its sharp cutting edges. *Flint* and *garnet* are natural minerals used as abrasives. Although sandpaper made with flint is less expensive, it does not last as long as garnet. Synthetic (man-made) abrasives include *aluminum oxide* and *silicon carbide*. Sandpaper coated with aluminum oxide is probably the most widely used for wood.

Grits. Sandpaper *grit* refers to the size of the abrasive particles. Sandpaper with large abrasive particles is considered coarse. Small abrasive particles are used to make fine sandpaper.

Sandpaper grits are designated either by a grit number or by an equivalent "0" series number. The grit numbers range from No. 12 (coarsest) to No. 600 (finest). The "0" series number 10/0 is equivalent to a 400 grit number. A 1/0 is the same as 80 grit (Fig. 15-20). Commonly used grits are 60 or 80 for rough sanding, and 100 or 120 for finish sanding.

Sand with a coarser grit until a surface is uniformly sanded. Do not switch to a finer grit too soon. Do not use worn or clogged abrasives. Their use causes the surface to become glazed or burned.

CHAPTER 16 FASTENING TOOLS

Portable fastening tools included in this chapter are those called *pneumatic* (powered by compressed air) and *powder-actuated* (drive fasteners with explosive powder cartridges). They are widely used throughout the construction industry for practically every fastening job from foundation to roof.

Pneumatic Staplers and Nailers

Pneumatic staplers and *nailers* are commonly called *guns* (Fig. 16-1). They are used widely for quick fastening of framing, subfloors, wall and roof sheathing, roof shingles, exterior finish, and interior trim. A number of manufacturers make a variety of models in several sizes for special fastening jobs. For instance, a *framing gun,* although used for many fastening jobs, is not used to apply interior trim.

Older models required frequent oiling, either by hand or by an oiler installed in the air line. Due to improvements in design, some newer models require no oiling at all.

Nailing Guns

One particular heavy-duty *framing gun* (Fig. 16-2) drives smooth-shank, headed nails up to 3 1/4

Fig. 16-1 Pneumatic nailers and staplers are widely used to fasten building parts. (*Courtesy of ITW Paslode*)

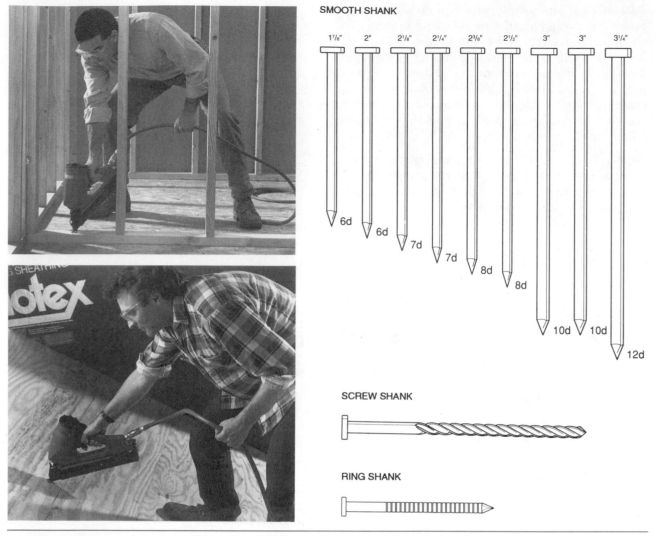

Fig. 16-2 Heavy-duty framing nailers are used for floor, wall, and roof framing. (*Courtesy of Senco Products, Inc.*)

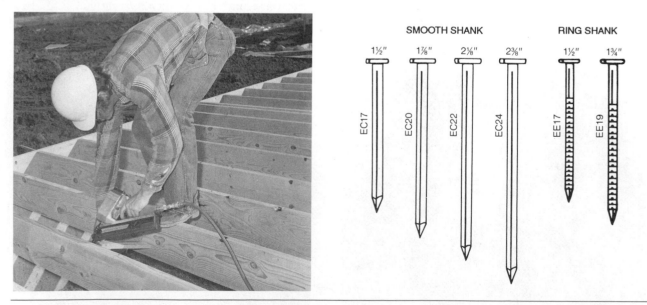

Fig. 16-3 A light-duty nailer is used to fasten light framing, subfloors, and sheathing. (*Courtesy of Senco Products, Inc.*)

inches, ring-shank nails 2 3/8 inches and screw shank nails 3 inches to fasten all types of framing. A light-duty version (Fig. 16-3) drives smooth-shank nails up to 2 3/8 inches and ring-shank nails up to 1 3/4 inches to fasten light framing, subfloor, sheathing, and similar components of a building. Nails come glued in strips for easy insertion into the magazine of the gun (Fig. 16-4).

The *finish nailer* (Fig. 16-5) drives finish nails from one to 2 inches long. It can be used for the application of practically all kinds of exterior and interior finish work. It sets or flush drives nails as desired. A nail set is not required, and the possibility of marring the wood is avoided.

The *brad nailer* (Fig. 16-6) drives both slight-headed and medium-headed brads ranging in

Fig. 16-4 Both headed and finish nails used in nailing guns come glued together in strips. (*Courtesy of Senco Products, Inc.*)

Fig. 16-5 The finish nailer is used to fasten all kinds of interior trim. (*Courtesy of Senco Products, Inc.*)

Fig. 16-6 A light-duty brad nailer is used to fasten thin molding and trim. (*Courtesy of Senco Products, Inc.*)

Fig. 16-8 Roofing nails come in coils for use in the roofing nailer. (*Courtesy of Senco Products, Inc.*)

COIL ROOFING NAILS

Fig. 16-7 A coil roofing nailer is used to fasten asphalt roof shingles. (*Courtesy of Senco Products, Inc.*)

length from 5/8 inch to 1 5/8 inches. It is used to fasten small moldings and trim, cabinet door frames and panels, and other miscellaneous finish carpentry.

The *roofing nailer* (Fig. 16-7) is designed for fastening asphalt and fiberglass roof shingles. It drives five different size, wide, round-headed roofing nails from 7/8 to 1 3/4 inches. The nails come in coils of 120 (Fig. 16-8), which are easily loaded in a nail canister.

Staplers

Like nailing guns, *staplers* are manufactured in a number of models and sizes. Although they differ somewhat, they are classified as light-, medium-, and heavy-duty.

A popular tool is the *roofing stapler* (Fig. 16-9), which is used to fasten roofing shingles. It comes in

Fig. 16-9 The wide-crown stapler, being used to fasten roof shingles, may be used to fasten a variety of materials. (*Courtesy of Senco Products, Inc.*)

several models and drives one-inch wide-crown staples in lengths from 3/4 inch to 1 1/2 inches. It can also be used for fastening other material, such as lath wire, insulation, furniture, and cabinets. The staples, like nails, come glued together in strips (Fig. 16-10) for quick and easy reloading. Most stapling guns can hold up to 150 staples.

No single model stapler drives all widths and lengths of staples. Models are made to drive narrow-crown and intermediate-crown, in addition to wide-crown staples. A popular model drives 3/8-inch intermediate-crown staples from 3/4 inch to 2 inches in length.

Cordless Guns

Conventional pneumatic staplers and nailers are powered by compressed air. The air is supplied by an air compressor (Fig. 16-11) through long lengths of air hose stretched over the construction site. The development of *cordless* nailing and stapling guns (Fig. 16-12) eliminates the need for air compressors

Fig. 16-12 Each cordless gun comes in its own case with battery, battery charger, safety glasses, instructions, and storage for fuel cells. (*Courtesy of ITW Paslode*)

Fig. 16-10 Staples, like nails, come glued together in strips for use in stapling guns. (*Courtesy of Senco Products, Inc.*)

Fig. 16-11 The development of cordless nailers and staplers eliminates the need for air compressors power conventional models.

and hoses. The cordless gun utilizes a disposable fuel cell. A battery and spark plug power an internal combustion engine that forces a piston down to drive the fastener. Another advantage is the time saved to set up the compressor, draining it at the end of the day, coiling the hoses, and storing the equipment.

The *cordless framing nailer* (Fig. 16-13) drives nails from 2 to 3 1/4 inches in length. Each fuel cell will deliver energy to drive about 1,200 nails. The battery will last long enough to drive about 4,000 nails before recharging is required.

The *cordless finish nailer* (Fig. 16-14) drives finish nails from 3/4 in to 2 1/2 inches in length. It will drive about 2,500 nails before a new fuel cell is needed and about 8,000 nails before the battery has to be recharged.

The *cordless stapler* (Fig. 16-15) drives intermediate-crown staples from 3/4 inch to 2 inches in length. It will drive about 2,500 staples with each fuel cell and about 8,000 staples with each charge of the battery.

Fig. 16-13 The cordless framing nailer eliminates the nuisance of bothersome air hoses. (*Courtesy of ITW Paslode*)

Using Staplers and Nailers

Because of the many designs and sizes of staplers and nailers, you should study the manufacturer's directions and follow them carefully. Use the right nailer or stapler for the job at hand. Make sure all safety devices are working properly. A work contact element allows the tool to operate only when this device is firmly depressed on a work surface and the trigger is pulled, promoting safe tool operation.

■   Load the magazine with the desired size staples or nails.

■   Connect the air supply to the tool. For those guns that require it, make sure there is an oiler in the air supply line, adequate oil to keep the gun lubricated during operation, and an air filter to keep moisture from damaging the gun. Use the recommended air pressure.

> **CAUTION** Exceeding the recommended air pressure may cause damage to the gun or burst air hoses, possibly causing injury to workers.

■   Press the trigger and tap the nose of the gun to the work. When the trigger is depressed, a fastener is driven each time the nose of the gun is tapped to the work. The fastener may also be safely driven by first pressing the nose of the gun to the surface and then pulling the trigger.

■   Upon completion of fastening, disconnect the air supply.

> **CAUTION** Never leave an unattended gun with the air supply connected. Always keep the gun pointed toward the work. Never point it at other workers or fire a staple except into the work. A serious injury can result from horseplay with the tool.

Powder-actuated Drivers

Powder-actuated drivers (Fig. 16-16) are used to drive specially designed pins into masonry or steel. They are used in a manner similar to firing a gun. Powder charges of various strengths drive the pin when detonated.

Drive Pins

Drivepins are available in a variety of sizes. Three styles are commonly used. The *headed* type is used for fastening material. The *threaded* type is used to

APPLICATION	DIMENSION
PANELING, BED MOLDING	3/4" X 16 GA.
PANELING, CAP/SHOE MOLDING	1" X 16 GA.
CAP/SHOE MOLDING	1 1/4" X 16 GA.
BASEBOARD, CROWN, CHAIR	1 1/2" X 16 GA.
BASEBOARD, CROWN, CHAIR	1 3/4" X 16 GA.
BASEBOARD, CROWN, CASINGS	2" X 16 GA.
CASINGS	2 1/4" X 16 GA.
CASINGS	2 1/2" X 16 GA.

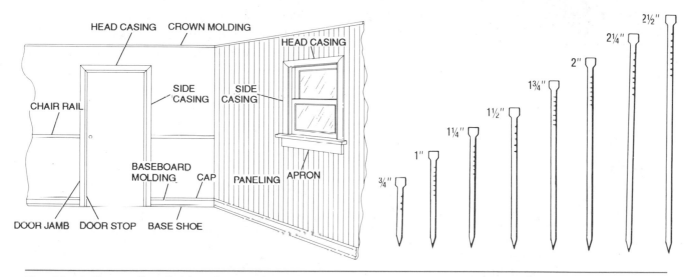

Fig. 16-14 The cordless finish nailer can drive about 2,500 nails with one fuel cell (*Courtesy of ITW Paslode*)

Fig. 16-15 The cordless stapler can drive 2500 intermediate- crown staples from 3/4 inch up to 2 inches without changing fuel cells. (*Courtesy of ITW Paslode*)

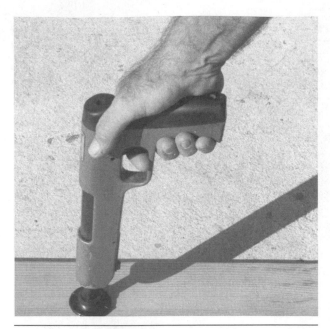

Fig. 16-16 Powder-actuated drivers are used for fastening into masonry or steel.

bolt an object after the pin is driven. The *eyelet* type is used when attachments are to be made with wire.

Powder Charges

Powder charges are color-coded according to strength. Learn the color codes for immediate recognition of the strength of the charge. A stronger charge is needed for deeper penetration or when driving into hard material. The strength of the charge must be selected with great care.

Because of the danger in operating these guns, many states require certification of the operator.

Certificates may be obtained from the manufacturer's representative after a brief training course.

Using Powder-actuated Drivers

■ Study the manufacturer's directions for safe and proper use of the gun.

■ Make sure the drivepin will not penetrate completely through the material into which it is driven. This has been the cause of fatal accidents.

■ To prevent ricochet hazard, make sure the recommended shield is in place on the nose of the gun. A number of different shields are available for special fastening jobs.

■ Select the proper fastener for the job. Consult the manufacturer's drivepin selection chart to determine the correct fastener size and style.

■ Select a powder charge of necessary strength. Always use the weakest charge that will do the job. Load the driver with the pin first and the cartridge second.

■ Keep the tool pointed at the work. Wear safety goggles. Press hard against the work surface, and pull the trigger. The resulting explosion drives the pin. Eject the spent cartridge.

> **CAUTION** If the gun does not fire, hold it against the work surface for at least thirty seconds. Then remove the cartridge according to the manufacturer's directions. Do not attempt to pry out the cartridge with a knife or screwdriver; most cartridges are rim-fired and could explode.

Review Questions

Select the most appropriate answer.

1. To prevent electrical shock, make sure the tool is properly
   a. connected.　　　c. insulated.
   b. grounded.　　　d. wired.

2. The guard of the portable electric saw should never be
   a. lubricated.　　　c. retracted by hand.
   b. adjusted.　　　d. wedged open.

3. The saber saw is used primarily for making
   a. curved cuts.　　　c. internal cuts.
   b. holes.　　　d. straight cuts.

4. The reciprocating saw is most similar to a
   a. hammer-drill.
   b. powered compass saw.
   c. saber saw.
   d. vibrating sander.

5. The size of an electric drill is determined by its
   a. ampere rating.　　　c. weight.
   b. horsepower.　　　d. chuck capacity.

6. In power drills, never use bits with
a. a double twist.
b. no center.
c. a single spur.
d. a threaded center.

7. To prevent a drill bit from binding in a hole
a. clear chips frequently.
b. drill at a slow speed.
c. drill at a high speed.
d. use bits with a twist.

8. The amount of screw tightening in the positive clutch model electric screwdriver is dependent on the
a. adjustment of clutch torque.
b. release of pressure.

c. size of the screw being driven.
d. ampere rating of the screwdriver.

9. The portable electric jointer plane is well suited to
a. nose the edge of closet shelves.
b. plane miter joints to fit.
c. trim end grain of stair treads.
d. plane doors to fit in openings.

10. When using the router to shape external edges and ends, the router is pulled in a
a. direction with the grain.
b. clockwise direction.
c. counterclockwise direction.
d. direction with the rotation of the bit.

■ BUILDING FOR SUCCESS ■

The Importance of Following Instructions

How many times have we all rushed through something only to find that the results did not meet our expectations? Perhaps the time required to read and follow instructions would have produced better results.

Our nation's construction workforce continues to be very active and geared to be time conscious. We move through a work day at rates that minimize time for planning, thinking, and organizing properly. Some workers believe that taking time to create a plan of action only slows them down. A short pause to assess the situation will usually help us obtain better results and meet our expected goals.

Almost everyone has purchased or received as a gift something that had instructions for assembling or processing. The instructions were supplied by the manufacturer to assist each person in utilizing the product. Some instructions are lengthy; others are brief but adequate to meet the objectives. Instructions should focus on the goals and lead users through the assembly process successfully.

The key to following any instructions is to believe they are necessary for success. With that understanding, each person will adequately reach his or her goals.

As a society, we tend to be more audio- or visual-oriented than reading-oriented. For the carpenter, reading skills are crucial to following blueprints, codes, and specifications. Each carpenter is reading, listening to, or observing instructions daily.

A person must believe that he or she can benefit from following the instructions. As instructions are followed, a learning process takes place. By repeating the procedures at regular intervals, strict reliance on the written instructions is lessened. Once the procedures are memorized or become automatic, reliance for adherence is transferred more to memory.

How should we follow instructions?

1. Observe, listen to, and read instructions before attempting to follow through on the process. Work with an experienced person whenever possible.

2. Prepare the necessary tools and materials called for in the instructions. This may call for preparation a few days in advance.

3. Begin with the proverbial "step one," then proceed sequentially through to the next step. Do not skip steps.

4. Ask questions along the way that will help complete a step or procedure.

5. Repeat any step that did not bring the expected results. The next step will be

based on completion of the previous step.

6. As the project or process is completed, maintain the instructions for future reference. If the instructions include maintenance procedures, they must be forwarded to whomever has that responsibility.

A carpenter working with a quality product will want to assure the new homeowner that the installation instructions have been followed. Deviation from the manufacturer's instructions may very well affect the warranty of the product. For example, the application of "Colorlok" siding (Masonite Corporation) will call for joint moldings, 1/8-inch expansion gaps at the ends where siding pieces butt against trim and corners, and nailing to be on each stud 5/8 inch down from the top edge of each course. If the expansion gap is not allowed, buckling will occur. If joint moldings are not installed, weathering can take place at the ends of each exposed piece of siding. If the nails are exposed below the buttline of the succeeding course, undue weather of the painted surface will shorten the life of the product.

As the result of improper installation, the carpenter may be responsible for product failure, not the manufacturer. Any disregard for instructions could affect the carpenter's profit margin on a job.

In conclusion, a carpenter must be prepared to observe all instructions to avoid possible setbacks and financial losses. The attitude for success understands the need for following instructions.

FOCUS QUESTIONS: For individual or group discussion.

1. Who will be affected by the decision to follow or disregard instructions? How can they be affected?

2. What might be some good reasons to develop the habit of following directions?

3. What might be the attitudes possessed by workers who disregard installation instructions?

4. In what way can a skilled worker such as a carpenter maintain a volume of the manufacturer's installation instructions?

Stationary Power Tools

Many kinds of stationary power woodworking tools are used in wood mills and cabinet shops for specialized work. On the building site, usually only the radial arm, table, and miter saws are available for use by carpenters. These are not heavy-duty machines. They must be light enough to be transported from one job to another. These saws are ordinarily furnished by the contractor.

OBJECTIVES

After completing this unit, the student will be able to:

- describe different types of circular saw blades and select the proper blade for the job at hand

- describe, adjust, and operate the radial arm saw and table saw safely to crosscut lumber to length, rip to width, and make miters, compound miters, dadoes, grooves, and rabbets

- operate the table saw in a safe manner to taper rip and to make cove cuts

- describe, adjust, and operate the power miter saw to crosscut to length, making square and miter cuts safely and accurately

CONTENTS

GENERAL SAFETY RULES

- Be trained and competent in the use of stationary power tools before attempting to operate them without supervision.
- Make sure power is disconnected when making adjustments to machines.
- Make sure saw blades are sharp and suitable for the operation. Ensure that safety guards are in place, and that all guides are in proper alignment and secured.
- Wear eye protection and appropriate, properly fitted clothing.

- Keep the work area clear of scrap that might present a tripping hazard.
- Keep stock clear of saw blades before starting a machine.
- Do not allow your attention to be distracted while operating power tools.
- Turn off the power and make sure a machine has stopped before leaving the area.

Safety precautions that apply to specific operations are given when those operations are described in this unit.

CHAPTER 17   CIRCULAR SAW BLADES

All the stationary power tools described in this unit use circular saw blades. To ensure safe and efficient saw operation, it is important to know which type to select for a particular purpose.

Circular Saw Blades

The more teeth a saw blade has, the smoother the cut. However, a fine-toothed blade does not cut as fast. This means that the stock must be fed more slowly. A coarse-toothed saw blade leaves a rough surface, but cuts more rapidly. Thus, the stock can be fed faster by the operator.

If the feed is too slow, the blade may overheat. This will cause it to lose its shape and wobble at high speed. This is a dangerous condition that must be avoided. An overheated saw will probably start to bind in the cut, possibly causing kickback and serious operator injury.

The same results occur when trying to cut material with a dull blade. Always use a sharp blade. Use fine-toothed blades for cutting thin, dry material, and coarse-toothed blades for cutting heavy, rough lumber.

Types of Circular Saw Blades

Saw blades are classified into two general groups. In one group, the saw blade is made with high-speed steel. In the other group, a steel blade has *tungsten carbide cutting tips.* Tungsten carbide is such a hard metal that it is sharpened with diamond-impregnated grinding wheels. The carbide-tipped blades stay sharp longer. They are used for cutting material that contains adhesives, or other foreign material, such as particleboard, plywood, hard-board, plastics, and engineered lumber that rapidly dull ordinary steel blades.

In both groups of saw blades, the number and shape of the teeth vary to give different cutting actions according to the kind, size, and condition of the material to be cut. Most of the saw blades being used today are carbide-tipped. However, they have not entirely replaced high-speed steel blades.

High-speed Circular Saw Blades

High-speed steel blades are classified as *rip, crosscut,* or *combination* blades (Fig. 17-1). They may be given other names, such as plywood, panel, or flooring blades, but they are still in one of the three classifications just named.

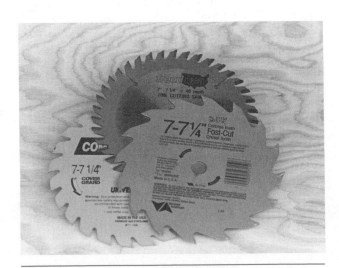

Fig. 17-1 High-speed circular saw blades.

Fig. 17-2 The teeth of a ripsaw have square edge cutting tips.

Fig. 17-4 The teeth of a combination blade.

Fig. 17-5 The teeth of a chisel-point combination blade.

Ripsaws. The ripsaw blade (Fig. 17-2) usually has less teeth than crosscut or combination blades. As in the hand ripsaw, every tooth of the circular ripsaw blade is filed or ground at right angles to the face of the blade. This produces teeth with a cutting edge all the way across the tip of the tooth. These teeth act like a series of small chisels that cut and clear the stock ahead of them. Look for these cutting edges to identify a ripsaw.

Use a ripsaw for cutting solid lumber with the grain when a smooth edge is not necessary. Also, use a ripsaw when cutting unseasoned or green lumber and lumber of heavy dimension with the grain.

Crosscut Saws. The teeth of a crosscut circular saw blade are shaped like those of the crosscut handsaw (Fig. 17-3). The sides of the teeth are alternately filed or ground on a bevel. This produces teeth that come up to points instead of edges. Look for these beveled sides and points on the teeth to determine a crosscut blade.

Crosscut teeth slice through the wood fibers smoothly. A crosscut blade is an ideal blade for cutting across the grain of solid lumber. It also cuts plywood satisfactorily with little splintering of the cut edge.

Combination Saws. The combination blade is used when a variety of ripping and crosscutting is to be done. It eliminates the need to change blades for different operations.

There are several types of combination blades. One type has groups of teeth around its circumference (Fig. 17-4). The leading tooth in each group is a rip tooth. The ones following are crosscut teeth.

The *chisel-point* combination blade (Fig. 17-5) is generally available with from twenty to forty teeth per blade. The number of teeth depends on the size of the blade. Each tooth is the same shape. Ripsaw teeth are given a slight alternate bevel to produce combination crosscut and rip cutting action.

Providing Clearance in the Saw Cut

All saw blades must have some provision for clearance in the saw cut. Without this clearance, the saw blade, no matter how sharp will bind in the cut and overheat. The teeth of the blades just described are set in a manner similar to that used for handsaws to provide this clearance. Special circular saw blade setting tools are needed for this purpose (Fig. 17-6).

Another type of blade provides this clearance by being taper ground. A *taper ground blade* is thicker at the tips of the teeth and is thinner toward its center (Fig. 17-7). Its teeth have no set. The taper of the blade provides the clearance in the saw cut. A taper ground blade can be identified by the ridge of thicker metal around the arbor hole.

Fig. 17-3 The teeth of a crosscut blade have beveled sides and pointed tips.

Fig. 17-6 Special tools are needed to set the teeth of high-speed circular saw blades.

CARBIDE-TIP

SAW BLADE

SQUARE GRIND   ALTERNATE BEVEL   TRIPLE CHIP

(A)

(B)

Fig. 17-8 Three main styles of carbide-tipped saw blades are commonly used. (*Figure B Courtesy of Dewalt*)

SAW TEETH HAVE NO SET

BLADE TAPERS THINNER FROM TEETH TOWARDS CENTER

HUB – SAME THICKNESS AS OUTER CIRCUMFERENCE

ARBOR HOLE

Fig. 17-7 Cross-section of a taper ground circular saw planer blade.

The reason for using a taper ground blade is to obtain an extremely smooth cut surface. Taper ground blades usually are a combination type. They are called *planer* blades. Use planer blades only on straight, dry lumber of relatively small dimension. Taper ground blades do not provide much clearance. The slightest twist in the stock will cause it to bind against the blade. Binding causes overheating and dangerous conditions. Binding also produces black, burned areas on the cut surface.

Carbide-tipped Blades

The carbide tips welded on a circular saw blade are very hard. They are also very *brittle*. For this reason, extreme care must be taken when handling carbide-tipped blades. Be careful not to drop the blades. Some of the tips could shatter like glass. Although the tips are shaped in many ways for various purposes, three main styles of teeth are commonly used (Fig. 17-8).

Square Grind. The *square grind* is similar to the rip teeth in a steel blade. It is used primarily to cut solid wood with the grain. It can also be used on composition boards when the quality of the cut surface is not important.

Alternate Top Bevel. The *alternate top bevel* grind is used with excellent results for crosscutting solid lumber, plywood, hardboard, particleboard, fiberboard, and other wood composite products. It can also be used for ripping operations. However, the feed is slower than square or combination grinds.

Triple Chip. The *triple chip grind* is designed for cutting brittle material without splintering or chipping the surface. It is particularly useful for cutting an extremely smooth edge on plastic laminated material, such as countertops. It can also be used like a planer blade to produce a smooth cut surface on straight, dry lumber of small dimension.

Combination. Carbide-tipped saws also come with a combination of teeth. The leading tooth in each set is square ground. The following teeth are ground at alternate bevels. These saws are ideally suited for ripping and crosscutting solid lumber and all kinds of engineered wood products. Because of their versatility, these saw blades are probably the most widely used by carpenters.

Carbide-tipped teeth are not set. The carbide tips are slightly thicker than the saw blade itself. Therefore, they provide clearance for the blade in the saw cut. In addition, the sides of the carbide tips are slightly beveled back to provide clearance for the tip.

Removing and Replacing Circular Saw Blades

Saw arbor shafts may have a right- or left-hand thread for the nut, depending on which side of the blade it is located. No matter what direction the arbor shaft is threaded, the arbor nut is loosened in the same direction in which the saw blade rotates (Fig. 17-9). The arbor nut is always tightened against the rotation of the saw blade. This design prevents the arbor nut from loosening during operation.

Sharpening Circular Saw Blades

Carbide-tipped blades are sent out to shops that have specialized sharpening equipment. Usually high-speed steel blades are also sent to sharpening shops. However, in an emergency, it is sometimes necessary to bring these blades back into condition on the job.

Jointing. To sharpen a steel circular saw blade, it first must be *jointed*. Jointing is the process of making all the teeth the same height. With the blade on the table saw, lower it until the teeth are below the surface of the table. Hold an oilstone over the teeth of the blade. Start the machine. *Slowly* raise the blade until the teeth come in contact with the stone

Fig. 17-9 The saw arbor nut is loosened in the same direction in which the blade rotates.

Fig. 17-10 Jointing a circular saw blade.

(Fig. 17-10). Stop the machine and inspect the teeth to see if all or most of them have been touched by the stone. Do not joint any more than is necessary.

Gumming. *Gumming* is the process of deepening the *gullets* (the valleys between teeth) of a saw blade. With the blade still mounted on the table saw and the power disconnected, raise the blade until the point of a pencil held flat on the table and against the side of the blade is just below the deepest gullet. Rotate the blade by hand so that a pencil mark is made around it (Fig. 17-11). This will be a guide for grinding the gullets to their proper depth.

Remove the blade from the machine. Grind the gullets using a bench grinder with a narrow, round-edge grinding wheel (Fig. 17-12). Grind down to the pencil mark, maintaining the original shape of the gullet. Be careful not to overheat the teeth. If necessary, a *round file* may be used to gum the blade instead of a grinder.

CAUTION Use eye protection when operating the grinder.

Fig. 17-11 Marking the saw blade for gumming.

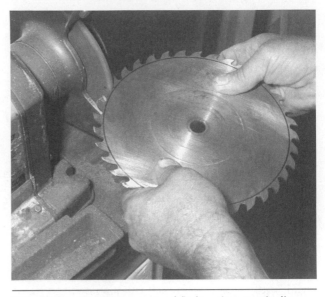

Fig. 17-12 Gumming a saw blade using a grinding wheel.

Setting. The teeth of the blade are set by using a circular saw set. The saw teeth are laid on a beveled anvil so that the top half of the tooth is bent alternately outward.

Filing. For filing, hold the saw blade in a *circular* saw vise. The teeth are filed in the same manner as for handsaws. Try to maintain a uniform shape. Use *flat* or *triangular* files as necessary. File only until the shiny tip of the tooth disappears (Fig. 17-13).

Fig. 17-13 File saw teeth only until the shiny tip disappears.

CHAPTER 18   RADIAL ARM AND MITER SAWS

The radial arm saw and the miter saw are similar in purpose. While many other operations can be performed with the radial arm saw, it is used primarily for crosscutting framing members. The miter saw is specifically designed to crosscut interior and exterior trim.

Radial Arm Saws

The *radial arm saw* (Fig. 18-1) is ideally suited for crosscutting operations. The stock remains stationary while the saw moves across it. The operator does not have to push or pull the stock through the blade. The end of long lengths are easily cut squared or mitered. The cut is made above the work and all layout lines are clearly visible. The radial arm saw may also be used for ripping operations. In many cases, this saw is brought to the job site when construction begins.

The size of the radial arm saw is determined by the diameter of the largest blade that can be

Fig. 18-2 The arm of the radial saw moves horizontally in a complete circle. (*Courtesy of Delta International*)

used. The arm of the saw moves horizontally in a complete circle (Fig. 18-2). The motor unit tilts to any desired angle and also rotates in a complete circle. The depth of cut is controlled by raising or lowering the arm. This flexibility allows practically any kind of cut to be made.

Crosscutting

For straight crosscutting, make sure the arm is at right angles to the fence. Adjust the depth of cut so that the teeth of the saw blade are about 1/16 inch below the surface of the table. With the saw all the way back and all guards in place, hold the stock against the fence. Make the cut by bringing the saw forward and cutting to the layout line (Fig. 18-3).

CAUTION When holding the stock on the left side of the blade, hold with the left hand and pull the saw with the right hand. When holding the stock on the right side of the blade, hold with the right hand and pull the saw with the left hand. Do not cross your hands over each other.

Fig. 18-1 The radial arm saw is well suited for crosscutting operations. (*Courtesy of DeWalt*)

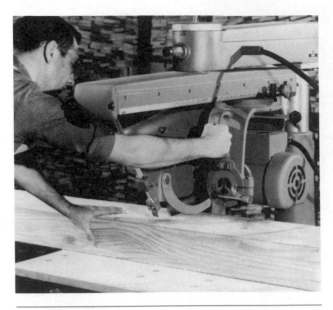

Fig. 18-3 Straight crosscutting using the radial arm saw. (*Courtesy of Rockwell, International*)

Fig. 18-4 Cutting identical lengths.

When crosscutting stock thicker than the capacity of the saw, cut through half the thickness. Then turn the stock over and make another cut.

> **CAUTION** Do not pull the saw quickly through the work. Forcing the saw may cause it to jam in the work or ride over it. This could cause injury to the operator. Pull the saw slowly through the work or even hold it back somewhat, especially when cutting thick material. If the stock is bowed, crosscut with the crown of the bow down on the table. If the cut is made with the crown up, the saw will probably bind as the stock flattens out on the table surface.

When the cut is complete, return the saw to the starting position, behind the fence. Turn off the power. Some saws come with devices that pull the motor unit back after the cut is made. In some areas, such a device is a required safety attachment.

Cutting Identical Lengths

When many pieces of identical length need to be cut, clamp, or otherwise fasten, a *stop block* to the table in the desired location. Cut one end of the stock. Slide the cut end against the stop block. Then cut the other end (Fig. 18-4). Continue in this manner until the desired number of pieces is cut.

Be careful not to bring the stock against the stop block with too much force. Doing so will cause the stop block to move. Any movement results in pieces of unequal length. It is helpful, when cutting

many identical lengths, to mark the location of the stop block. Occasionally, look to see if it has moved.

Be careful that no sawdust or wood chips are trapped between the stock and the stop block. This results in pieces shorter than desired. A stop block with a rabbeted end helps prevent sawdust or chips from being trapped between the stop block and the piece being cut (Fig. 18-5).

> **CAUTION** Replace the fence of the radial arm saw when saw cuts in it become too wide or too close together. Wide spaces in the fence allow small cut ends to shoot out toward the rear. The pieces can bounce back and cause injury to the operator.

Mitering

Flat miters. To cut a *flat miter*, the arm of the saw is rotated to the right or left to the desired angle. Usually the arm is swung to the right for more advantageous miter cutting, where the motor does not obscure the view or get in the way. The *miter latch* locates the arm in the 90 or 45 degree angle position.

For any other angle, the arm may be located by viewing the degree scale. This method may not be accurate, however. The pointer of the scale may have been bent or moved. In addition, the gradations on the scale are so fine that it is difficult to rely on the accuracy of the setting. Test pieces of known accuracy should be used to adjust the saw to cut miters other than 45 degrees.

In both cases, the arm is locked in position by the *miter clamp* after being adjusted. Miter cuts are made in the same manner as straight crosscutting (Fig. 18-6).

RADIAL ARM SAW FENCE

C CLAMP

STOP BLOCK

DUPLICATE LENGTH TO BE CUT

RABBET ON END OF STOP BLOCK PREVENTS SAW DUST AND CHIPS INTERFERING WITH LENGTH TO BE CUT.

Fig. 18-5

Fig. 18-6 Making miter cuts.

Mitering Jig. Flat miters may also be cut with the use of a *mitering jig*. One type of mitering jig consists of a piece of plywood, or similar material, to which strips are fastened at 45 degree angles to its edge. The jig is clamped to the radial arm table surface.

The pieces are cut by holding them at a 45 angle against the strips with the radial arm saw in its straight crosscutting position (Fig. 18-7). Using this jig allows both right and left miters, such as for door and window casings, to be cut without swinging the arm of the saw. Mitering jigs may also be constructed to cut miters other than 45 degrees.

End Miters. An *end miter* is cut in the same manner, except that the saw blade is tilted to the desired angle (Fig. 18-8). Stops locate the blade angle at 90, 45, and 0 degrees. Any other angle is located by viewing the angle degree scale or by using test blocks. When the blade is tilted to the desired angle, it is locked in position with the clamp.

Making Compound Miters. *Compound miters* are frequently made on some types of roof rafters and moldings. The compound miter is made by swinging the arm and also tilting the motor to the desired angles (Fig. 18-9).

Ripping

For ripping operations, the arm is locked in the straight crosscutting position at right angles to the fence. The motor unit is rotated horizontally to either the in-rip or the out-rip position (Fig. 18-10). The out-rip position is used to rip wider material than the in-rip position. The ripping width may be measured with a rule or tape from the fence to the

STEP BY STEP
P R O C E D U R E S

SHALLOW SAW CUT

1. USE 1/2" PLYWOOD AS WIDE & LONG AS THE RADIAL ARM SAW TABLE. PLACE ONE EDGE AGAINST THE FENCE & MAKE A SHALLOW CUT ABOUT 1/16" DEEP PART-WAY ACROSS THE PLYWOOD. REMOVE IT FROM THE SAW.

SAME DISTANCE ON BOTH SIDES
FENCE EDGE
FRAMING SQUARE
LAYOUT LINE
12"   12"

2. LAYOUT A STRAIGHT LINE 12" FROM THE FENCE EDGE. HOLD A FRAMING SQUARE SO ITS HEEL IS CENTERED ON THE SAW CUT, FLUSH WITH THE FENCE EDGE, & WITH EQUAL DISTANCES TO THE LAYOUT LINE ALONG BOTH TONGUE & BLADE. MARK LAYOUT LINES ALONG THE OUTSIDE EDGES OF THE SQUARE.

1" x 2" GUIDE STRIPS

3. ATTACH 1" x 2" GUIDE STRIPS SO THEIR INSIDE EDGES ARE TO THE LAYOUT LINES.

4. THE COMPLETED MITER JIG. TO USE, PLACE ON TABLE AGAINST FENCE SO SAW CUT LINES UP WITH SAW BLADE. CLAMP OR, OTHERWISE FASTEN, TO TABLE SURFACE. PIECES TO BE MITERED ARE HELD AGAINST INSIDE EDGES OF GUIDE STRIPS.

Fig. 18-7A

Fig. 18-7B Using a mitering jig on the radial arm saw.

Fig. 18-8 Cutting an end miter.

Fig. 18-9 Making a compound miter cut.

Fig. 18-10 In-rip and out-rip positions of the radial arm saw.

blade. It may also be set by reading the rip scale built in on the arm of the saw (Fig. 18-11).

Lock the saw carriage in the desired position by tightening the *rip lock* against the side of the arm. Lower the motor until the saw teeth are just below the table surface.

Adjust the safety guard so that the in-feed end almost touches the material to be cut. This is done to prevent sawdust from flying at the operator. It

Fig. 18-11 The rip scale indicates the width of cut for both the in-rip and out-rip positions.

also holds the stock to be ripped down on the table surface.

Lower the kickback assembly so that the kickback fingers are about 1/8 inch lower than the material to be ripped. Using the fence as a guide, feed the stock evenly into the saw blade (Fig. 18-12).

> **CAUTION** Do not feed stock from the kickback end of the guard. *Feed the stock against the rotation of the blade.* Feeding stock in the wrong direction may cause it to be pulled from the operator's hands and through the saw with great force. This could cause serious or fatal injury to anyone in its path.

Do not force the stock into the saw. Feed it with a continuous motion. When ripping narrow stock, use a *push stick* on the end of the stock to push it between the saw blade and the fence.

Bevel Ripping. *Bevel ripping* is done in the same manner as straight ripping, except that the blade is tilted to the desired angle (Fig. 18-13). A V-groove

Fig. 18-12 Using the radial arm saw to rip lumber (note direction of feed in relation to blade rotation).

Fig. 18-13 Ripping stock at a bevel.

may be made in this position by cutting partway through the stock. Then turn it around, end for end, and run it through again. The depth, width, and angle of the groove may be varied with different adjustments.

Dadoing, Grooving, and Rabbeting

Dadoes, grooves, and *rabbets* can be cut using a single saw blade. A *dado head* is commonly used to make them faster, with one pass of the stock through the cutting tool. One type of dado head consists of two outside circular saw blades with several *chippers* of different thicknesses placed in between (Fig. 18-14). Most dado heads make cuts of from 1/4 to 13/16 inch wide. Wider cuts are obtained by making two or more passes through the saw.

When installing this type of dado head, make sure the tips of the chippers are opposite the gullets and not against the side of the blade. Chipper tips are *swaged* (made wider than the body of the chipper) to assure a clean cut across the width of the dado or groove.

Another type is the one-unit *adjustable dado head* (Fig. 18-15), commonly called a *wobbler head.*

This type can be adjusted, for width of dado, without removing it from the saw arbor. Both types are available in high-speed steel or carbide-tipped styles.

Make dadoes and end rabbets in the same manner as for crosscutting. Edge rabbets and grooves are made with the saw in the ripping position. V-grooves can be made by tilting the saw blade (Fig. 18-16).

Fig. 18-14 The dado head consists of two blades and several chippers of different thicknesses.

Fig. 18-15 The one-unit adjustable dado head.

Fig. 18-16 Cutting dadoes and grooves.

Miter Saw

The *power miter saw* is also called a *power miter box* or a *chop saw*. Usual sizes, designated by the blade diameter, are 10 and 12 inches (Fig. 18-17). It is not considered a portable power tool. However, it is light enough to be moved around wherever needed on the job.

A circular saw blade, mounted above the base, is pushed down using a chopping action to make the cut. The saw blade may be turned 45 degrees to the right or left. Positive stops are located at 90, 67 1/2, 60, and 45 degrees. Other angles are cut by reading the degree scale and locking in the desired position.

In some models, the blade can be tilted to make compound miter cuts. The 12-inch *compound miter saw* can make compound miters in crown molding up to 6 5/8 inches wide and other compound cuts up to 45 degrees by 45 degrees (Fig. 18-18). This saw, although used mostly for finish work, can also easily cut a 2×8 at 90 degrees and a 2×6 or 4×4 at 45 degrees.

To make cuts, the stock is held firmly on the table and against the fence. The saw is adjusted to the desired angle and brought down through the stock. A retractable blade guard provides operator protection and should be in place when the saw is being used (Fig. 18-19).

Fig. 18-18 The blade of the compound miter saw may be tilted.

10" MITER SAW

Fig. 18-17 The power miter saw is widely used to cut miters on interior trim.

Fig. 18-19 Using the compound miter saw to miter a wide molding.

CHAPTER 19   TABLE SAWS

The table saw is one of the most frequently used woodworking power tools. In many cases, it is brought to the job site when the interior finish work begins. It is a very useful tool because so many kinds of work can be performed with it. Common table saw operations, with different jigs to aid the process, are discussed in this chapter.

Table Saws

The size of the *table saw* (Fig. 19-1) is determined by the diameter of the saw blade. It may measure up to 16 inches or more. A commonly used table saw on the construction site is the 10-inch model.

The blade is adjusted for depth of cut and tilted up to 45 degrees by means of handwheels. A *rip fence* guides the work during ripping operations. A *miter gauge* is used to guide the work when cutting square ends and miters. The miter gauge slides in grooves in the table surface. It may be turned and locked in any position up to 45 degrees.

Ripping Operations

For *ripping* operations, the table saw is easier to use and safer than the radial arm saw. To rip stock to width, measure from the *rip fence* to the point of a saw tooth set closest to the fence. Lock the fence in place. Adjust the height of the blade to about 1/4 inch above the stock to be cut. With the stock clear of the blade, turn on the power.

CAUTION Stand to the left of the blade. Never stand directly in back of the saw blade. Make sure no one else is in line with the saw blade in case of kickback.

Hold the stock against the fence with the left hand. Push the stock forward with the right hand, holding the end of the stock (Fig. 19-2). As the end approaches the saw blade, let it slip through the left hand. Remove the left hand from the work. Push the end all the way through the saw blade with the right hand, if the stock is of sufficient width (at least 5 inches wide). If the stock is not wide enough, use a *push stick* (Fig. 19-3).

Fig. 19-2 Using the table saw to rip lumber.

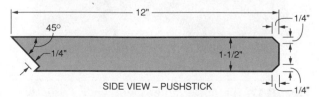

NOTE: PUSH STICK MUST BE THIN ENOUGH TO PASS BETWEEN THE RIP FENCE AND SAW SAW BLADE WITH CLEARANCE.

Fig. 19-3 A. Use a push stick to rip narrow pieces (guard has been removed for clarity). B. Push stick design.

Fig. 19-1 The 10-inch table saw is frequently used on the job site.

CAUTION Make sure the stock is pushed all the way through the saw blade. Leaving the cut stock between the fence and a running saw blade may cause a kickback, injuring anyone in its path. Use a push stick when ripping narrow pieces. Do not pick small waste pieces or strips from the saw table when the saw is running. Remove them with a stick or wait until the saw has stopped. Always use the rip fence for ripping operations. Never make freehand cuts. Never reach over a running saw blade.

Fig. 19-5 Using a taper ripping jig to cut identical wedges (guard has been removed for clarity).

Fig. 19-4 When bevel ripping, the blade is tilted away from the fence (guard has been removed for clarity).

Bevel Ripping

Ripping on a *bevel* is done in the same manner as straight ripping except that the blade is tilted. For safer and more efficient bevel ripping, tilt the blade away from the fence and not toward it (Fig. 19-4).

Taper Ripping

Tapered pieces (one end narrower than the other) can be made with the table saw by using a *taper ripping jig*. The jig consists of a wide board with the length and amount of taper cut out of one edge. The other edge is held against the rip fence. The stock to be tapered is held in the cutout of the jig. The taper is cut by holding the stock in the jig as both are passed through the blade (Fig. 19-5).

By using taper ripping jigs, small tapered pieces, such as wedges for concrete forms, or large ones, such as rafter tails, may be cut according to the design of the jig. A handle on the jig makes it safer to use. Also, if the jig is the same thickness as

the stock to be cut, then the cutout section of the jig can be covered with a thin strip of wood to prevent the stock from flying out of the jig and back toward the operator.

Making Cove Cuts

At times it may be necessary to make a *cove* cut, such as the concave shape of a large cove molding, to match an existing piece. The table saw can be used to make these cuts. Feed the stock across the saw blade at an angle.

A straightedge clamped to the table surface guides the stock. The angle at which the straightedge is clamped determines the radius of the cove (Fig. 19-6). A number of light cuts must be made because of the side stress placed on the saw blade. Make a number of passes. Take off about 1/16 inch with each pass until the desired depth is reached.

Rabbeting and Grooving

Making *rabbets* and narrow *grooves* may be done using a single saw blade in a ripping fashion. To make a rabbet, usually two settings of the rip fence and two passes through the saw blade are required (Fig. 19-7). For narrow grooves, one or more passes are required. Move the rip fence slightly with each pass until the desired width of groove is obtained.

To make these cuts with fewer passes, a dado head may be used (Fig. 19-8).

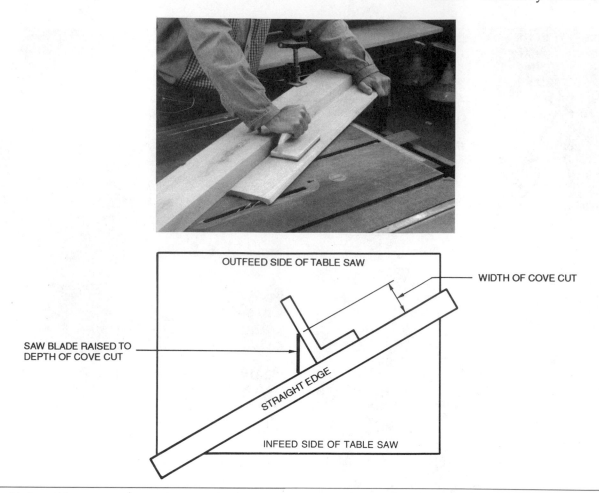

Fig. 19-6 Making a cove cut with the table saw (guard has been removed for clarity).

Fig. 19-7 Rabbeting an edge using a single saw blade (guard has been removed for clarity).

Crosscutting Operations

For most crosscutting operations, the *miter gauge* is used. To cut stock to length with both ends squared, first check the miter gauge for accuracy. Hold a framing square against it and the side of the saw

blade. Usually the miter gauge is operated in the left-hand groove. The right-hand groove is used only when it is more convenient.

Square one end of the stock by holding the work firmly against the miter gauge with one hand

Fig. 19-8 Cutting a groove with a dado head installed on the table saw (guard has been removed for clarity).

Fig. 19-10 Cutting identical lengths using a stop block on an auxiliary fence of the miter gauge (guard has been removed for clarity).

while pushing the miter gauge forward with the other hand. Measure the desired distance from the squared end. Mark on the front edge of the stock. Repeat the procedure, cutting to the layout line (Fig. 19-9).

Cutting Identical Lengths

When a number of identical lengths need to be cut, first square one end of each piece. Clamp a stop to an *auxiliary wood fence* installed on the miter gauge. Place the square end of the stock against the stop block. Then make the cut. Slide the remaining stock across the table until its end comes in contact with the stop block. Then make another cut. Continue in this manner until the desired number of pieces is cut (Fig. 19-10).

When short pieces of identical length are to be cut, a block is clamped to the rip fence in a location so that once the cut is made, there is clearance between the cut piece and the rip fence. The fence is adjusted so that the desired distance is between the face of the stop block and saw blade.

Square one end of the stock. Slide the squared end against the stop block. Then make a cut. Continue making cuts in this manner until the desired number of pieces is obtained (Fig. 19-11).

Mitering

Flat miters are cut in the same manner as square ends, except the miter gauge is turned to the desired angle. *End miters* are made by adjusting the

Fig. 19-9 Using the miter gauge as a guide to crosscut.

Fig. 19-11 Using the rip fence as a stop to cut short, identical lengths (guard has been removed for clarity).

miter gauge to a square position and making the cut with the blade tilted to the desired angle.

Compound miters are cut with the miter gauge turned and the blade tilted to the desired angles (Fig. 19-12).

A *mitering jig* similar to that used on the radial arm saw can be used with the table saw. Figure 19-13 shows the construction and use of the jig. Use of such a jig eliminates turning the miter gauge each time for left- and right-hand miters.

Fig. 19-12 Making a compound miter cut (guard has been removed for clarity).

1½" PLYWOOD OR SIMILAR

WOOD STRIPS SIZED TO FIT IN MITER GAGE GROOVES

1. FASTEN WOOD STRIPS, SIZED, SPACED, AND LOCATED TO FIT IN MITER GAGE GROOVES, TO UNDER SIDE OF PLYWOOD. RUB WAX ON PLYWOOD AND WOOD STRIPS.

MITER GAGE GROOVES

TABLE SAW TOP– RUB WITH WAX

MITERING JIG

SAW CUT

2. WITH SAW BLADE BELOW TABLE SURFACE PLACE JIG SO WOOD STRIPS FIT IN GROOVES AND SLIDE BACK AND FORTH. JIG SHOULD MOVE EASILY, BUT WITH NO PLAY. MAKE A SAW CUT ABOUT HALF-WAY ACROSS THE WIDTH.

SAW CUT

FRAMING SQUARE

INFEED EDGE OF JIG

3. HOLD FRAMING SQUARE AT 12 AND 12 ON INFEED EDGE WITH HEEL CENTERED ON THE SAW CUT. MARK LAYOUT LINES ON OUTSIDE EDGES OF SQUARE.

NOTE: USE SAW GUARD WHEN MAKING CUTS– GUARD IS NOT SHOWN FOR CLARITY.

HOLD PIECES TO BE MITERED AGAINST OUTSIDE EDGES OF GUIDE STRIP.

4. ATTACH GUIDE STRIPS SO EDGES ARE TO LAYOUT LINES.

JIG

STOCK

Fig. 19-13

Dadoing

Dadoing is done in a similar manner as crosscutting except, with the use of a dado head (Fig. 19-14). The dado head is only used to cut *partway* through the stock thickness.

Table Saw Aids

A very useful aid is an *auxiliary fence*. A straight piece of 3/4-inch plywood about 12 inches wide and as long as the ripping fence is screwed or bolted to the metal fence. When cuts must be made close to the fence, the use of an auxiliary wood fence prevents the saw blade from cutting into the metal fence. Also, the additional height provided by the fence gives a broader surface to steady wide work when its edge is being cut. The auxiliary wood fence also provides a surface on which to clamp feather boards. An auxiliary fence is also useful when attached to the miter gauge.

Feather boards are useful aids to hold work against the fence as well as down on the table surface during ripping operations (Fig. 19-15). Feather boards may be made of 1×6 nominal lumber, with one end cut at a 45 degree angle. Saw cuts are made in this end about 1/4 inch apart. This gives the end some spring and allows pressure to be applied to the piece being ripped.

An *auxiliary table top* serves many useful purposes. With the saw blade below the table surface, a wide piece of thin plywood, or similar material, as long as the saw table top is clamped in place after the rip fence is adjusted (Fig. 19-16). The saw motor is started. The blade is slowly raised to the desired height through the auxiliary top. The top prevents narrow rippings from slipping through the space

between the saw blade and the table insert. It also prevents thin material from slipping under the rip fence. Use of the auxiliary table top can prevent accidents and serious injury.

Fig. 19-15 A. Feather boards are useful aids to hold work during table saw operations (guard has been removed for clarity). B. Feather board design.

Fig. 19-14 Cutting dadoes using a dado head (guard has been removed for clarity).

Fig. 19-16 An auxiliary table serves many useful purposes (guard has been removed for clarity).

Review Questions

Select the most appropriate answer.

1. To cut lumber with the grain use a
   a. carbide-tipped blade.
   b. ripsaw blade.
   c. high-speed steel blade.
   d. blade with alternate beveled teeth.

2. The most frequently used blade in general carpentry is the
   a. combination planer blade.
   b. combination carbide-tipped blade.
   c. square-grind carbide-tipped blade.
   d. ripsaw blade.

3. A high-speed steel circular saw blade that has no set is the
   a. combination blade.   c. ripsaw blade.
   b. crosscut blade.   d. taper ground blade.

4. The alternate top bevel grind, carbide-tipped circular saw
   blade is designed to
   a. rip solid lumber.
   b. crosscut solid lumber.
   c. cut in either direction of the grain.
   d. cut green lumber of heavy dimension.

5. Saw arbor nuts are loosened
   a. clockwise.
   b. with the rotation of the blade.
   c. counterclockwise.
   d. against the rotation of the blade.

6. When using the radial arm saw for ripping operations, the in-feed end of the blade guard is lowered close to the work to
   a. protect the operator from the saw blade.
   b. prevent sawdust from flying out.
   c. hold the work down on the table surface.
   d. all of the above.

7. The table saw guide used for cutting with the grain is called
   a. rip fence.   c. tilting arbor.
   b. miter gauge.   d. ripping jig.

8. When using the table saw for bevel ripping, the fence is
   placed on the
   a. right side.
   b. side away from the blade tilt.
   c. left side.
   d. side toward the blade tilt.

9. A dado is a wide cut partway through the thickness of the material and
   a. across the grain.
   b. with the grain.
   c. in either direction of the grain.
   d. close to the edge.

10. When using the table saw for ripping operations, use a push stick if the ripped width is narrower than
   a. 3 inches
   b. 4 inches
   c. 5 inches
   d. 6 inches

■ BUILDING FOR SUCCESS ■

Implementing a Safety Program

At the job site or in the school construction lab, no objective should neglect to take into consideration the safety and welfare of workers and students. Completion dates, schedules, quality, and profits are important goals in the construction industry. However, none should be attained at the expense of people's lives. The highest priority on the job has to be an effective plan to ensure worker safety. The safety program has to be ongoing from day one to be effective. The responsibility of this program is everyone's. Both skilled and unskilled men and women who subject themselves to numerous hazards and potential accidents while working deserve the best protection available.

In most cases the contractor has informed his or her employees of appropriate job safety. Acceptable safety practices are reinforced by insurance companies and local, state, and national agencies such as the Occupational Safety and Health Administration (OSHA).

A company safety program must be well researched, prepared, and presented to the employees so they see it is for their benefit. Implementation must become part of everyone's daily tasks. The basic premise should be to greatly reduce the probability of construction related accidents. Safety research should be conducted using all available resources, including statistics depicting the types and causes of construction accidents. The collected information should then be used to structure and implement the safety program.

Much research information may be obtained from insurance companies, labor statistics, OSHA, and the National Safety Council. A desired result of the research is to prepare a safety program that the employees will accept. The focus will center on the workers' welfare. Realistically this concentration should be a high priority with all workers.

Employees have to be involved from the beginning. The success of the program depends on how much the workers themselves believe in staying healthy. It has to be a program that automatically kicks in on a daily basis. A safe working environment has to be important to every worker or student on the job. This calls for influence on the part of the instructor or contractor. He or she will have to "sell" the safety plan to employees. Almost everyone has listened to radio station "WIIFM" (What's in It for Me?). That will be the key for acceptance. This connection has to be made if the workers are to accept the plan as a part of their job and security.

The annual number of injuries and fatalities in the construction industry share the lead with those in mining and farming. The very nature of these professions exposes people to a high risk for accidents. Each trade represented in construction has specific high-risk tasks that endanger workers. The overall safety program should become an element in all construction jobs.

Discussions, videos, speakers, demonstrations, and literature can be a vital part of an effective safety program. Each student or worker can be given a part in the presentation, application, or feedback report of a company safety program. As potential accidents are discovered and studied by workers, attention can be drawn to accident prevention.

Hazards and dangerous working conditions can be eliminated from the work site through employee involvement. Potential subjects for good safety programs are numerous due to the multitude of construction related jobs. Concentrations can be on excavating, shoring, hauling, lifting, cutting, climbing, driving, assembling, electrical grounding, or a myriad of other activities. Ideally the studies should center on tasks that are carried out by the people on a daily basis, then go on to other more infrequent tasks.

The largest obstacles may be researching the safety program, informing workers, or keeping the program ongoing. Regardless, the program should have the safety of the workers in mind. Many construction companies take pride in the length of time between accidents on the job. An incentive program helps everyone become strongly committed to worker safety.

A good safety program will answer the important questions of Who, Why, What, When, Where, and How (in no particular order). To answer these questions: (1) identify the need and target population (Who); (2) justify the safety program with reason (Why); (3) relate the safety program concept (What); (4) establish a time frame for the program (When); (5) list the various work sites for implementation (Where); and (6) clarify the implementation process for everyone (How). Then be prepared to document, report, and reinforce the use of the safety program. A successful safety program will always involve the target population in the planning process. Ownership at this point is extremely important if goals are to be reached. The business owners and/or managers may see how a good safety record can ensure staying on schedule or assure profits for the company, but the employees may view it differently. The workers will be interested in assisting the company with their goals of quality and profits. They will become more highly motivated, however, by a safety program that centers on their health and welfare. That's when the buy-in takes place.

As employers reflect on the time and cost of an effective safety program, they will see and feel the benefits in satisfied workers, lower insurance rates, fewer accidents, less downtime, and healthier work climates. The employees in turn will feel they are a valued part of "their" company.

FOCUS QUESTIONS: For individual or group discussion

1. What might be the reasons for implementing a safety program at school or on the job?

2. In order for a safety program to be effective, people must see how it can benefit them. How could you as a contractor persuade them to "buy in" to this type of program?

3. As you introduce a safety program, how might you answer the questions Who, Why, What, When, Where, and How.

U N I T

8

Blueprints and Building Codes

OBJECTIVES

After completing this unit, the student will be able to:

- describe and explain the function of the various kinds of drawings contained in a set of blueprints

- demonstrate how specifications are used

- read and use an architect's scale

- identify various types of lines and read dimensions

- identify and explain the meaning of symbols and abbreviations used on blueprints

- read and interpret plot, foundation, floor, and framing plans

- locate and explain information found in exterior and interior elevations

- identify and utilize information from sections and details

- use schedules to identify and determine location of windows, doors, and interior finish

- define and explain the purpose of building codes and zoning laws

- explain the requirements for obtaining a building permit and the duties of a building inspector

The ability to interpret blueprints and to understand building codes, zoning ordinances, building permits, and inspection procedures is necessary for promotion on the job. With this ability, the carpenter has the competence needed to handle jobs of a supervisory nature, including foreman, superintendent, and contractor.

CONTENTS

Blueprints

An *architect* designs buildings, supervises construction, and manages projects. The *drafter* makes working drawings from the architect's designs. These drawings use various kinds of lines, dimensions, notes, symbols, and abbreviations to describe a structure to be built.

Standards for architectural drawings are more flexible than, for instance, machine drafting. Therefore, the method of drawing, lettering, and dimensioning may vary slightly with the drafter.

Many copies of the drawings must be made for use by clients, contractors, and subcontractors, for bidding purposes, for obtaining building permits, and for use on the job.

Making Blueprints

Original black line drawings are made with graphite or ink on translucent paper, cloth, or polyester film. Copies are usually made with a *diazo* printer. They are called diazo or *ozalid* prints. The copying process utilizes an ultraviolet light that passes through the paper, but not the lines. The light exposes a chemical on the copy paper, except where the lines are drawn. The copy paper is then exposed to ammonia vapor. This vapor develops the remaining chemical into blue, black, or brown lines against a white background. The diazo printing process is faster and less expensive than older methods.

Blueprinting is an older method of making copies. It uses a water wash to produce prints with white lines against a dark blue background. Although these are true blueprints, the process is seldom used today. The word *blueprint* remains in use, however, to mean any copy of the original drawing.

Kinds of Drawings

Different kinds of drawings are required for the construction of a building. These are called *plans, elevations, sections,* and *details.* When put together, they constitute a set of blueprints.

Types of Views

It may be necessary to use several views to describe an object with clarity. *Pictorial drawings* are usually three-dimensional *perspective* or *isometric* views. Pictorials do not give much information for building.

Multiview, also called *orthographic,* drawings are two-dimensional. They are used to convey most of the information needed for construction. They constitute the bulk of a set of blueprints (Fig. 20-1).

The lines in a perspective diminish in size as they converge toward vanishing points on a line

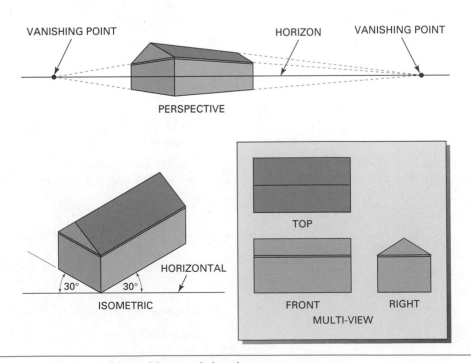

Fig. 20-1 Different views are used in architectural drawings.

Fig. 20-2 A perspective view is used for a presentation drawing.

Presentation Drawings

A *presentation drawing* is usually a perspective type (Fig. 20-2). It shows the building from a desirable vantage point to display its most interesting features. Walks, streets, shrubs, trees, vehicles, and even people may be drawn. Many times, the drawing is colored for greater eye appeal. Presentation drawings provide no information for construction purposes. They are usually shown to clients to let them see how the completed building will look.

Plans

- The *floor plan* is a drawing of the structure viewed from above. Imagine a horizontal cut made about 4 to 5 feet above the floor. After the material above the cut is removed, the floor plan remains (Fig. 20-3).

- The *foundation plan* (Fig. 20-4) shows a horizontal cut through the walls. It is similar to a floor plan. It shows the shape and dimensions of the foundation, among other things.

called a *horizon*. In an isometric drawing, the horizontal lines are drawn at 30 degree angles. All lines are drawn to actual scale. They do not diminish or converge as in perspective drawings.

In multiview projections, views of the front, right side, left side, and rear are called *elevations*. Views from the top are called *plans*, of which there are several kinds. The bottom view is never used in architectural drawings.

Fig. 20-3 A floor plan is a horizontal cut through the building.

Fig. 20-4 A foundation plan. (*Courtesy of Pinellas Voc-Tech Institute*)

The *plot plan* (Fig. 20-5) shows information about the lot, such as the location of the building, walks, and driveways. The drawing simulates a view looking down from a considerable height. It is made at a small scale because of the relatively large area it represents.

Framing plans are not always found in a set of prints. They include floor framing and roof framing. They show the direction and spacing of the framing members (Fig. 20-6.)

PLOT PLAN

Fig. 20-5 A plot plan. (*Courtesy of Pinellas Voc-Tech Institute*)

ROOF FRAMING PLAN

Fig. 20-6 A roof framing plan. (*Courtesy of Pinellas Voc-Tech Institute*)

Elevations

Elevations are a group of drawings (Fig. 20-7) that show the shape and finishes of all sides of the exterior of a building.

Interior elevations are drawings of certain interior walls. The most common are kitchen and bath-room wall elevations. They show the design and size of cabinets built on the wall (Fig. 20-8). Other walls that have special features, such as a fireplace, may require an elevation drawing.

Occasionally found in some sets of prints are *framing elevations*. Similar to framing plans, they

Fig. 20-7 Elevations. (*Courtesy of Pinellas Voc-Tech Institute*)

show the spacing, location, and sizes of wall framing members. No further description of framing drawings is required to be able to interpret them.

Sections

A *section* is a drawing that shows a vertical cut through all or part of a construction (Fig. 20-9). A section reference line is found in the plans or elevations to identify the section being viewed.

Details

To make parts of the construction more clear, it is usually necessary to draw *details*. Details are small parts drawn at a very large scale, even full-size. The detail shown in Figure 20-10 shows a portion of the construction more clearly.

Other Drawings

Drawings relating to electrical work, plumbing, heating, and ventilating may be on separate sheets

ELEVATION Ⓐ   ELEVATION Ⓑ

Fig. 20-8 Interior wall elevations. (*Courtesy of Pinellas Voc-Tech Institute*)

Fig. 20-9 A section is a view of a vertical cut through part of the construction. (*Courtesy of Pinellas Voc-Tech Institute*)

in a set of prints. For smaller projects, separate plans are not always needed. All necessary information can be usually found on the floor plan.

The carpenter is responsible for building to accommodate wiring, pipes, and ducts. He or she should be able to read these plans with some degree of proficiency to understand the work involved.

Fig. 20-10 Detail of a foundation wall.

Schedules

Besides drawings, printed instructions are included in a set of drawings. *Window schedules* and *door schedules* (Fig. 20-11) give information about the location, size, and kind of windows and doors to be installed in the building. Each of the different units is given a number or letter. A corresponding number or letter is found on the floor plan to show the location of the unit. Windows may be identified by letters and doors by numbers. The letters and numbers may be framed with various geometric figures, such as circles and triangles.

A *finish schedule* (Fig. 20-12) may also be included in a set of prints. This schedule gives information on the kind of finish material to be used on the floors, walls, and ceilings of the individual rooms.

Window, door, and finish schedules are used for easy understanding and to conserve space on floor plans.

Specifications

Specifications, commonly called *specs*, are written to give information that cannot be completely provided in the drawings or schedules. They supplement the working drawings with more complete descriptions of the methods, materials, and quality

WINDOW SCHEDULE (SIZE OF OPENING FOR FRAME)				
TYPE	HEIGHT	WIDTH	STYLE	MATERIAL
1	4^0	2^0	CASEMENT	WOOD
2	6^0	4^0	CASEMENT	WOOD
3	6^0	3^0	CASEMENT	WOOD
4	3^0	4^0	CASEMENT	WOOD
5	5^0	3^0	CASEMENT	WOOD
6	6^0	8^0	CASEMENT	WOOD
7	4^0	3^0	CASEMENT	WOOD

DOOR SCHEDULE			
TYPE	SIZE	STYLE	MATERIAL
A	$2^8 \times 6^8$	H.C.	WOOD
B	$(2)2^0 \times 6^8$	H.C. DOUBLE (LOUVER)	WOOD
C	$(2)2^0 \times 6^8$	S.C. DOUBLE (SWINGING)	WOOD
D	$(2)2^0 \times 6^8$	S.C. FRENCH DOORS	WOOD & GLASS
E	$3^0 \times 6^8$	S.C.	WOOD
F	$3^0 \times 6^8$	H.C.	WOOD
G	$5^0 \times 6^8$	DOUBLE SLIDING GLASS	GLASS & ALUM.
H	$3^0 \times 6^8$	S.C. (ONE — LIGHT)	WOOD

Fig. 20-11 A typical window and door schedule. (*Courtesy of Pinellas Voc-Tech Institute*)

FINISH SCHEDULE							
ROOM	WALLS	DADO	BASE	FLOOR	CEILING	CORNICE	REMARKS
LIV. RM.	PLASTER.	CANVAS	WOOD	OAK	PLASTER	WOOD	BOOKCASE
DIN. RM	"	"	"	"	"	PICT MLDG	CLIPBD.
KITCHEN	"	4'-0' TILE	TILE	LINOLEUM	"	—	—
HALL	"	—	WOOD	OAK	"	WOOD	SEE DTL.
ENTRY	"	—	"	"	"	—	—

Fig. 20-12 A typical finish schedule.

of construction. If there is a conflict, the specifications take precedence over the drawings. Any conflict should be pointed out to the architect so corrections can be made.

The amount of detail contained in the specs will vary, depending on the size of the project. On small jobs, they may be written by the architect. On larger jobs, a *specifications writer*, trained in the construction process, may be required. For complex commercial projects, a specifications guide, used by spec writers, has been developed by the Construction Specifications Institute (CSI). The guide has sixteen major divisions, each containing a number of subdivisions (Fig. 20-13.)

Using the specification guide, under Division 6—WOOD AND PLASTICS, Section 06200, an example of the content and the manner in which specifications are written is shown in Figure 20-14.

For some light commercial and residential construction, many sections of the spec guide would not apply. A shortened version is then used, eliminating divisions 12,000, 13,000, and 14,000. On

DIVISION 1 – GENERAL REQUIREMENTS
01010 SUMMARY OF WORK
01100 ALTERNATIVES
01150 MEASUREMENT & PAYMENT
01200 PROJECT MEETINGS
01300 SUBMITTALS
01400 QUALITY CONTROL
01500 TEMPORARY FACILITIES &
       CONTROLS
01600 MATERIAL & EQUIPMENT
01700 PROJECT CLOSEOUT

DIVISION 2 – SITE WORK
02010 SUBSURFACE EXPLORATION
02100 CLEARING
02110 DEMOLITION
02200 EARTHWORK
02250 SOIL TREATMENT
02300 PILE FOUNDATIONS
02350 CAISSONS
02400 SHORING
02500 SITE DRAINAGE
02550 SITE UTILITIES
02600 PAVING & SURFACING
02700 SITE IMPROVEMENTS
02800 LANDSCAPING
02850 RAILROAD WORK
02900 MARINE WORK
02950 TUNNELING

DIVISION 3 – CONCRETE
03100 CONCRETE FORMWORK
03150 FORMS
03200 CONCRETE REINFORCEMENT
03250 CONCRETE ACCESSORIES
03300 CAST-IN-PLACE CONCRETE
03350 SPECIALLY FINISHED
       (ARCHITECTURAL) CONCRETE
03360 SPECIALLY PLACED CONCRETE
03400 PRECAST CONCRETE
03500 CEMENTITIOUS DECKS
03600 GROUT

DIVISION 4 – MASONRY
04100 MORTAR
04150 MASONRY ACCESSORIES
04200 UNIT MASONRY
04400 STONE
04500 MASONRY RESTORATION &
       CLEANING
04550 REFRACTORIES

DIVISION 5 – METALS
05100 STRUCTURAL METAL FRAMING
05200 METAL JOISTS
05300 METAL DECKING
05400 LIGHTGAGE METAL FRAMING
05500 METAL FABRICATIONS
05700 ORNAMENTAL METAL
05800 EXPANSION CONTROL

DIVISION 6 – WOOD & PLASTICS
06100 ROUGH CARPENTRY
06130 HEAVY TIMBER CONSTRUCTION
06150 TRESTLES
06170 PREFABRICATED STRUCTURAL
       WOOD
06200 FINISH CARPENTRY
06300 WOOD TREATMENT
06400 ARCHITECTURAL WOODWORK
06500 PREFABRICATED STRUCTURAL
       PLASTICS
06600 PLASTIC FABRICATIONS

DIVISION 7 – THERMAL & MOISTURE
PROTECTION
07100 WATERPROOFING

07150 DAMPPROOFING
07200 INSULATION
07300 SHINGLES & ROOFING TILES
07400 PREFORMED ROOFING & SIDING
07500 MEMBRANE ROOFING
07570 TRAFFIC TOPPING
07600 FLASHING & SHEET METAL
07800 ROOF ACCESSORIES
07900 SEALANTS

DIVISION 8 – DOOR & WINDOWS
08100 METAL DOORS & FRAMES
08200 WOOD & PLASTIC DOORS
08300 SPECIAL DOORS
08400 ENTRANCES & STOREFRONTS
08500 METAL WINDOWS
08600 WOOD & PLASTIC WINDOWS
08650 SPECIAL WINDOWS
08700 HARDWARE & SPECIALTIES
08800 GLAZING
08900 WINDOW WALLS/CURTAIN WALLS

DIVISION 9 – FINISHES
09100 LATH & PLASTER
09250 GYPSUM WALLBOARD
09300 TILE
09400 TERRAZZO
09500 ACOUSTICAL TREATMENT
09540 CEILING SUSPENSION SYSTEMS
09550 WOOD FLOORING
09650 RESILIENT FLOORING
09680 CARPETING
09700 SPECIAL FLOORING
09760 FLOOR TREATMENT
09800 SPECIAL COATINGS
09900 PAINTING
09950 WALL COVERING

DIVISION 10 – SPECIALTIES
10100 CHALKBOARDS & TACKBOARDS
10150 COMPARTMENTS & CUBICLES
10200 LOUVERS & VENTS
10240 GRILLES & SCREENS
10260 WALL & CORNER GUARDS
10270 ACCESS FLOORING
10280 SPECIALTY MODULES
10290 PEST CONTROL
10300 FIREPLACES
10350 FLAGPOLES
10400 IDENTIFYING DEVICES
10450 PEDESTRIAN CONTROL DEVICES
10500 LOCKERS
10530 PROTECTIVE COVERS
10550 POSTAL SPECIALTIES
10600 PARTITIONS
10650 SCALES
10670 STORAGE SHELVING
10700 SUNCONTROLDEVICES(EXTERIOR)
10750 TELEPHONE ENCLOSURES
10800 TOILET & BATH ACCESSORIES
10900 WARDROBE SPECIALTIES

DIVISION 11 – EQUIPMENT
11050 BUILT-IN MAINTENANCE
       EQUIPMENT
11100 BANK & VAULT EQUIPMENT
11150 COMMERCIAL EQUIPMENT
11170 CHECKROOM EQUIPMENT
11180 DARKROOM EQUIPMENT
11200 ECCLESIASTICAL EQUIPMENT
11300 EDUCATIONAL EQUIPMENT
11400 FOOD SERVICE EQUIPMENT
11480 VENDING EQUIPMENT
11500 ATHLETIC EQUIPMENT
11550 INDUSTRIAL EQUIPMENT
11600 LABORATORY EQUIPMENT
11630 LAUNDRY EQUIPMENT
11650 LIBRARY EQUIPMENT

11700 MEDICAL EQUIPMENT
11800 MORTUARY EQUIPMENT
11830 MUSICAL EQUIPMENT
11850 PARKING EQUIPMENT
11860 WASTE HANDLING EQUIPMENT
11870 LOADING DOCK EQUIPMENT
11880 DETENTION EQUIPMENT
11900 RESIDENTIAL EQUIPMENT
11970 THEATER & STAGE EQUIPMENT
11990 REGISTRATION EQUIPMENT

DIVISION 12 – FURNISHINGS
12100 ARTWORK
12300 CABINETS & STORAGE
12500 WINDOW TREATMENT
12550 FABRICS
12600 FURNITURE
12670 RUGS & MATS
12700 SEATING
12800 FURNISHING ACCESSORIES

DIVISION 13 – SPECIAL CONSTRUCTION
13010 AIR SUPPORTED STRUCTURES
13050 INTEGRATED ASSEMBLIES
13100 AUDIOMETRIC ROOM
13250 CLEAN ROOM
13350 HYPERBARIC ROOM
13400 INCINERATORS
13440 INSTRUMENTATION
13450 INSULATED ROOM
13500 INTEGRATED CEILING
13540 NUCLEAR REACTORS
13550 OBSERVATORY
13600 PREFABRICATED STRUCTURES
13700 SPECIAL PURPOSE ROOMS &
       BUILDINGS
13750 RADIATION PROTECTION
13770 SOUND & VIBRATION CONTROL
13800 VAULTS
13850 SWIMMING POOLS

DIVISION 14 – CONVEYING SYSTEMS
14100 DUMBWAITERS
14200 ELEVATORS
14300 HOISTS & CRANES
14400 LIFTS
14500 MATERIAL HANDLING SYSTEMS
14570 TURNTABLES
14600 MOVING STAIRS & WALKS
14700 TUBE SYSTEMS
14800 POWERED SCAFFOLDING

DIVISION 15 - MECHANICAL
15010 GENERAL PROVISIONS
15050 BASIC MATERIALS & METHODS
15180 INSULATION
15200 WATER SUPPLY & TREATMENT
15300 WASTE WATER DISPOSAL &
       TREATMENT
15400 PLUMBING
15500 FIRE PROTECTION
15600 POWER OR HEAT GENERATION
15650 REFRIGERATION
15700 LIQUID HEAT TRANSFER
15800 AIR DISTRIBUTION
15900 CONTROLS & INSTRUMENTATION

DIVISION 16 - ELECTRICAL
16010 GENERAL PROVISIONS
16100 BASIC MATERIALS & METHODS
16200 POWER GENERATION
16300 POWER TRANSMISSION
16400 SERVICE & DISTRIBUTION
16500 LIGHTING
16600 SPECIAL SYSTEMS
16700 COMMUNICATIONS
16850 HEATING & COOLING
16900 CONTROLS & INSTRUMENTATION

Fig. 20-13 CSI format for specifications.

simpler plans, notations made on the same sheets as the drawings may take the place of specifications. The ability to read notations and specifications accurately is essential in order to conform to the architect's design.

Blueprint Language

Carpenters must be able to read and understand the combination of lines, dimensions, symbols, and notations on the drawings. Only then can they build exactly as the architect has designed the construc-

DIVISION 6—WOOD AND PLASTICS

Section 06200—FINISH CARPENTRY

General: This section covers all finish woodwork and related items not covered elsewhere in these specifications. The contractor shall furnish all materials, labor, and equipment necessary to complete the work, including rough hardware, finish hardware, and specialty items.

Protection of Materials: All millwork (finish woodwork*) and trim is to be delivered in a clean and dry condition and shall be stored to insure proper ventilation and protection from dampness. Do not install finish woodwork until concrete, masonry, plaster, and related work is dry.

Materials: All materials are to be the best of their respective kind. Lumber shall bear the mark and grade of the association under whose rules it is produced. All millwork shall be kiln dried to a maximum moisture content of 12%.

1. Exterior trim shall be select grade white pine, S4S.
2. Interior trim and millwork shall be select grade white pine, thoroughly sanded at the time of installation.

Installation: All millwork and trim shall be installed with tight fitting joints and formed to conceal future shrinkage due to drying. Interior woodwork shall be mitered or coped at corners (cut in a special way to form neat joints*). All nails are to be set below the surface of the wood and concealed with an approved putty or filler.

*(explanations in parentheses have been added to aid the student.)

Fig. 20-14 Sample specifications following the CSI format. (*From Huth,* Construction Technology, *1989, by Delmar Publications, Inc.)*

tion (Fig. 20-15). No deviation from the blueprints may be made without the approval of the architect.

Scales

It would be inconvenient and impractical to make full-sized drawings of a building. Therefore, they are made *to scale.* This means that each line in the drawing is reduced proportionally to a size that clearly shows the information and can be handled conveniently. Not all drawings in a set of prints, or even on the same page, are drawn at the same scale. The scale of a drawing is stated in the title block of the page. It can also be directly below a drawing.

Architect's Scale. The triangular *architect's scale* is commonly used to scale lines when making drawings (Fig. 20-16). Both sides of the three edges

Fig. 20-15 The carpenter must be able to read blueprints.

Fig. 20-16 The architect's scale.

are divided into different size gradations. The side of one edge is divided into inch and 1/16-inch increments. Along each edge of the remaining five sides are two scales. One scale is twice as large as the other. One is read from left to right. The other scale is read from right to left.

The scales are paired as follows:

$$3'' = 1'\text{-}0'' \text{ and } 1\ 1/2'' = 1'\text{-}0''$$
$$1'' = 1'\text{-}0'' \text{ and } 1/2'' = 1'\text{-}0''$$
$$3/4'' = 1'\text{-}0'' \text{ and } 3/8'' = 1'\text{-}0''$$
$$1/4'' = 1'\text{-}0'' \text{ and } 1/8'' = 1'\text{-}0''$$
$$3/16'' = 1'\text{-}0'' \text{ and } 3/32'' = 1'\text{-}0''$$

When reading a scale in one direction be careful not to confuse it with the one running in the opposite direction. On the end of each scale, a space representing one foot is divided into inches. In some cases, it is divided into fractions of an inch. This space is used when scaling off fractions of a foot (Fig. 20-17).

Commonly Used Scales. Probably the most commonly used scale found on blueprints is 1/4 inch equals 1 foot. This is indicated as 1/4″ = 1'-0″. It is often referred to as a "quarter-inch scale." This means that every 1/4 inch on the drawing will

Fig. 20-17 Various architect's scales. (*From Jefferis and Madsen,* Architectural Drafting & Design, *1991, by Delmar Publishers, Inc.*)

equal one foot in the building. Floor plans and exterior elevations for most residential buildings are drawn at this scale.

To show the location of a building on a lot and other details of the site, the architect may use a scale of 1/16" = 1'-0". this reduces the size of the drawing to fit it on the paper. To show certain details more clearly, larger scales of 1 1/2" = 1'-0" or 3" = 1'-0" are used. Complicated details may be drawn full size or half size. Other scales are used when appropriate. Views showing the elevation of interior walls are often drawn at 1/2" = 1'-0" or 3/4" = 1'-0".

Drawing blueprints to scale is important. The building and its parts are shown in true proportion, making it easier for the builder to visualize the construction. However, *the use of a scale rule to determine a dimension should be a last resort.* Dimensions on blueprints should be determined either by reading the dimension or adding and subtracting other dimensions to determine it. The use of a scale rule to determine a dimension results in inaccuracies.

Types of Lines

Some lines in an architectural drawings look darker than others. They are broader so they stand out clearly from other lines. This variation in width is called *line contrast.* This technique, like all architectural drafting standards, is used to make the drawing easier to read and understand (Fig. 20-18).

- Lines that outline the object being viewed are broad, solid lines called *object lines.*

- To indicate an object not visible in the view, a *hidden line* consisting of short, fine, uniform dashes is used. Hidden lines are used only when necessary. Otherwise the drawing becomes confusing to read.

- *Centerlines* are indicated by a fine, long dash, then a short dash, then a long dash, and so on.

Fig. 20-18 Types of lines on architectural drawings.

They show the centers of doors, windows, partitions, and similar parts of the construction.

- A *section reference* or *cutting-plane line* is, sometimes, a broad line consisting of a long dash followed by two short dashes. At its ends are arrows. The arrows show the direction in which the cross-section is viewed. Letters identify the cross-sectional view of that specific part of the building. More elaborate methods of labeling section reference lines are used in larger, more complicated sets of plans (Fig. 20-19). The sectional drawings may be on the same page as the reference line or on other pages.

- A *break line* is used in a drawing to terminate part of an object that, in actuality, continues. It can only be used when there is no change in the drawing at the break. Its purpose is to shorten the drawing to utilize space.

- A *dimension line* is a fine, solid line used to indicate the location, length, width, or thickness of an object. It is terminated with arrowheads, dots, or slashes (Fig. 20-20.)

- *Extension lines* are fine, solid lines projecting from an object to show the extent of a dimension.

- A *leader line* is a fine solid line. It terminates with an arrowhead, and points to an object from a notation.

Dimensions

Dimension lines on a blueprint are generally drawn as continuous lines. The dimension appears above and near the center of the line. All dimensions on vertical lines should appear above the line when the print is rotated 1/4 turn clockwise. Extension lines are drawn from the object so that the end point of the dimension is clearly defined. When the space is

Fig. 20-19 Several ways of labeling section reference lines.

Fig. 20-20 Several methods of terminating dimension lines.

too small to permit dimensions to be shown clearly, they may be drawn as shown in Figure 20-21.

Kinds of Dimensions. Dimensions on architectural blueprints are given in feet and inches, such as 3'-6", 4'-8", and 13'-7". A dash is always used to separate the foot measurement from the inch measurement. When the dimension is a whole number of feet with no inches, the dimension is written with zero inches, as 14'-0". The use of the dash prevents mistakes in reading dimensions.

Dimensions of one foot and under are given in inches, as 10", 8", and so on. Dimensions involving fractions of an inch are shown, for example, as 1'-0 1/2", 2'-3 3/4", or 6 1/2". If there is a difference between a written dimension and a scaled dimension of the same distance, the written dimension should be followed.

Modular Measure

In recent years, *modular measurement* has been used extensively. A grid with a unit of 4 inches is used in designing buildings (Fig. 20-22). The idea is to draw the plans to use material manufactured to fit the grid spaces. Drawing plans to a modular measure enables the builder to use manufactured component parts with less waste, such as 4x8 sheet materials and manufactured wall, floor, and roof sections that fit together with greater precision.

Fig. 20-21 Dimensioning small spaces.

PRINCIPLES OF MODULAR PLANNING,
SHOWING THE PLANNING GRID.

PRE-PLANNING SAVES
BOTH MATERIALS AND LABOR

24″ MINOR MODULE

48″ MAJOR MODULE

48″ MODULE

EXTERIOR WALL, DOOR AND
WINDOW SECTIONS SHOW THE
NEED FOR MODULAR PLANNING
OF THESE COMPONENTS. A $\frac{1}{8}$—INCH
TOLERANCE IS PROVIDED AT EACH
SEPARATION.

Fig. 20-22 Modular measurement uses a grid of 4 inches.

The spacing of framing members and the location and size of windows and doors adhering to the concept of modular measurement cut down cost and conserve materials.

Symbols

Symbols are used on drawings to represent objects in the building, such as doors, windows, cabinets, plumbing, and electrical fixtures. Others are used in regard to the construction, such as for walls, stairs, fireplaces, and electrical circuits. They may be used for identification purposes, such as those used for section reference lines. The symbols for various construction materials, such as lumber, concrete, sand, and earth (Fig. 20-23), are used when they make the drawing easier to read. (More detailed

illustration, description, and use of architectural symbols are presented in following units where appropriate.)

Abbreviations

Architects find it necessary to use abbreviations on drawings to conserve space. Only capital letters, such as DR for door, are used. Abbreviations that make an actual word, such as FIN. for finish, are followed by a period. Several words may use the same abbreviation, such as BR for bearing, bedroom, or brass and, also, W for west, width, or with. The *location* of these abbreviations is the key to their meaning. A list of commonly used abbreviations is shown in Figure 20-24.

Fig. 20-23 Symbols for commonly used construction materials.

Access Panel . . . . AP	Dressed and Matched . . . . D & M	Plate . . . . PL
Acoustic . . . . ACST	Dryer .D	Plate Glass . . . . .PL GL
Acoustical Tile . . . . .AT	Electric Panel . . . . EP	Platform . . . . .PLAT
Aggregate . . . . .AGGR	End to End . . . . .E to E	Plumbing . . . . .PLBG
Air Conditioning . . . . . .AIR COND	Excavate . . . . EXC	Plywood . . . . .PLY
Aluminum . . . . .AL	Expansion Joint . . . . EXP JT	Porch .P
Anchor Bolt . . . . .AB	Exterior . . . . .EXT	Precast . . . . .PRCST
Angle	Finish . . . . FIN	Prefabricated . . . . PREFAB
Apartment . . . . .APT	Finished Floor . . . . FIN FL	Pull Switch . . . . PS
Approximate . . . . .APPROX	Firebrick . . . . .FBRK	Quarry Tile Floor . . . . .QTF
Architectural . . . . .ARCH	Fireplace . . . . FP	Radiator . . . . .RAD
Area .A	Fireproof . . . . .FPRF	Random . . . . RDM
Area Drain . . . . .AD	Fixture . . . . FIX	Range . . . . .R
Asbestos . . . . ASB	Flashing . . . . FL	Recessed . . . . .REC
Asbestos Board . . . . .AB	Floor . . . . FL	Refrigerator . . . . .REF
Asphalt . . . . ASPH	Floor Drain . . . . FD	Register . . . . .REG
Asphalt Tile . . . . .AT	Flooring . . . . FLG	Reinforce or Reinforcing . . . . REINF
Basement . . . . BSMT	Fluorescent . . . . .FLUOR	Revision . . . . .REV
Bathroom . . . . .B	Flush . . . . FL	Riser .R
Bathtub . . . . .BT	Footing . . . . .FTG	Roof . . . . .RF
Beam . . . . .BM	Foundation . . . . .FND	Roof Drain . . . . .RD
Bearing Plate . . . . BRG PL	Frame . . . . FR	Room . . . . RM or R
Bedroom . . . . .BR	Full Size . . . . FS	Rough . . . . .RGH
Blocking . . . . BLKG	Furring . . . . FUR	Rough Opening . . . . .RO
Blueprint . . . . .BP	Galvanized Iron . . . . GI	Rubber Tile . . . . R TILE
Boiler . . . . .BLR	Garage . . . . .GAR	Scale . . . . .SC
Book Shelves . . . . .BK SH	Gas . . . . .G	Schedule . . . . .SCH
Brass . . . . BRS	Glass . . . . .GL	Screen . . . . .SCR
Brick . . . . .BRK	Glass Block . . . . .GL BL	Scuttle . . . . .S
Bronze . . . . BRZ	Grille . . . . .G	Section . . . . SECT
Broom Closet . . . . .BC	Gypsum . . . . .GYP	Select . . . . SEL
Building . . . . BLDG	Hardware . . . . .HDW	Service . . . . .SERV
Building Line . . . . .BL	Hollow Metal Door . . . . .HMD	Sewer . . . . .SEW
Cabinet . . . . .CAB	Hose Bib . . . . .HB	Sheathing . . . . .SHTHG
Calking . . . . .CLKG	Hot Air . . . . .HA	Sheet . . . . SH
Casing . . . . .CSG	Hot Water . . . . HW	Shelf and Rod . . . . SH & RD
Cast Iron CI	Hot Water Heater . . . . .HWH	Shelving . . . . SHELV
Cast Stone . . . . .CS	I Beam I	Shower . . . . .SH
Catch Basin . . . . .CB	Inside Diameter . . . . ID	Sill Cock . . . . SC
Cellar . . . . CEL	Insulation . . . . INS	Single Strength Glass . . . . .SSG
Cement . . . . CEM	Interior . . . . INT	Sink . . . . SK or S
Cement Asbestos Board . . . . CEM AB	Iron I	Soil Pipe . . . . SP
Cement Floor . . . . .CEM FL	Jamb . . . . JB	Specification . . . . .SPEC
Cement Mortar . . . . CEM MORT	Kitchen . . . . .K	Square Feet . . . . .SQ FT
Center . . . . .CTR	Landing . . . . LDG	Stained . . . . .STN
Center to Center . . . . .C to C	Lath . . . . LTH	Stairs . . . . ST
Center Line . . . . or CL	Laundry . . . . LAU	Stairway . . . . STWY
Center Matched . . . . .CM	Laundry Tray . . . . LT	Standard . . . . .STD
Ceramic . . . . CER	Lavatory . . . . .LAV	Steel . . . . ST or STL
Channel . . . . CHAN	Leader L	Steel Sash . . . . SS
Cinder Block . . . . CIN BL	Length . . . . L, LG, or LNG	Storage . . . . .STG
Circuit Breaker . . . . CIR BKR	Library . . . . LIB	Switch . . . . SW or S
Cleanout . . . . .CO	Light . . . . LT	Telephone . . . . TEL
Cleanout Door . . . . .COD	Limestone . . . . LS	Terra Cotta . . . . TC
Clear Glass . . . . .CL GL	Linen Closet . . . . .L CL	Terrazzo . . . . TER
Closet . . . . C, CL, or CLO	Lining . . . . LN	Thermostat . . . . .THERMO
Cold Air . . . . CA	Living Room . . . . LR	Threshold . . . . .TH
Cold Water . . . . .CW	Louver LV	Toilet . . . . .T
Collar Beam . . . . COL B	Main . . . . MN	Tongue and Groove . . . . T & G
Concrete . . . . CONC	Marble . . . . MR	Tread . . . . .TR or T
Concrete Block . . . . .CONC B	Masonry Opening . . . . MO	Typical . . . . TYP
Concrete Floor . . . . .CONC FL	Material . . . . .MATL	Unfinished . . . . UNF
Conduit . . . . CND	Maximum . . . . MAX	Unexcavated . . . . .UNEXC
Construction . . . . CONST	Medicine Cabinet . . . . .MC	Utility Room . . . . URM
Contract . . . . CONT	Minimum . . . . MIN	Vent V
Copper . . . . COP	Miscellaneous . . . . MISC	Vent Stack . . . . .VS
Counter . . . . CTR	Mixture . . . . MIX	Vinyl Tile . . . . V TILE
Cubic Feet . . . . .CU FT	Modular . . . . .MOD	Warm Air . . . . WA
Cut Out . . . . CO	Mortar . . . . MOR	Washing Machine . . . . .WM
Detail . . . . DET	Moulding . . . . .MLDG	Water . . . . W
Diagram . . . . DIAG	Nosing . . . . .NOS	Water Closet . . . . .WC
Dimension . . . . DIM	Obscure Glass . . . . OBSC sL	Water Heater . . . . .WH
Dining Room . . . . .DR	On Center . . . . .OC	Waterproof . . . . .WP
Dishwasher . . . . DW	Opening . . . . .OPNG	Weather Stripping . . . . .WS
Ditto . . . . DO	Outlet . . . . .OUT	Weephole . . . . .WH
Double-Acting . . . . .DA	Overall . . . . OA	White Pine . . . . .WP
Double Strength Glass . . . . . . DSG	Overhead . . . . .OVHD	Wide Flange . . . . .WF
Down . . . . DN	Pantry . . . . .PAN	Wood . . . . .WD
Downspout . . . . .DS	Partition . . . . PTN	Wood Frame . . . . .WF
Drain . . . . .D or DR	Plaster . . . . .PL or PLAS	Yellow Pine . . . . .YP
Drawing . . . . DWG	Plastered Opening . . . . .PO	

Fig. 20-24 Commonly used abbreviations.

CHAPTER 21   *FLOOR PLANS*

The chapters in this unit are based on a set of drawings for a two-story, three-bedroom residence (Fig. 21-1). Part of the basement is an activity room.

Floor Plans

Floor plans (Fig. 21-2) contain a substantial amount of information. They are used more than any other kind of drawing. After consideration of many factors that determine the size and shape of the building, floor plans are drawn first. Others, such as the foundation plan and elevations, are derived from it. They are generally drawn at a scale of $1/4'' = 1'-0''$. A separate plan is made for each floor of buildings with more than one story.

Floor Plan Symbols

In order to make the plan as uncluttered as possible, numerous *symbols* are used. Recognition of commonly used symbols makes it easier to read the floor plan as well as other plans that use the same symbols. (Symbols used in elevation and section drawings are different from plan symbols. They are described in following chapters.)

Door Symbols. Symbols for exterior doors are drawn with a line representing the outside edge of the sill. Interior door symbols show no sill line. The symbols in Figure 21-3 identify the *swing* and show on which side of the opening to hang the door.

Similarly, exterior *sliding* door symbols show the sill line. The symbols for interior sliding doors, called *bypass* doors, show none. *Pocket doors* slide inside the wall (Fig. 21-4).

Bifold and *accordian* doors open to almost the full width of a closet opening. They are used when complete access to the closet is desired. The sections or panels of the doors are clearly seen in the symbols (Fig. 21-5).

Window Symbols. The inside and outside lines of window symbols represent the edges of the window sill. In between, other lines are drawn for the panes of glass. A window with a fixed, single sash

Fig. 21-1 The plans for this residence are used throughout this unit. (*From Jefferis and Madsen,* Architectural Drafting & Design, *1991, by Delmar Publishers, Inc.*)

MAIN FLOOR PLAN
SCALE : ¼" = 1'-0"

NOTES:
1. ALL FRAMING LUMBER TO BE DFL 2.
2. ALL EXT. STUDS TO BE 2x6 @16" O.C.
   *HEATED AREAS.
3. ALL METAL CONNECTORS TO BE SIMPSON
   CO. OR EQUAL.
4. VENT ALL FANS TO OUTSIDE AIR.

Fig. 21-2 First and second floor plans. (*From Jefferis and Madsen,* Architectural Drafting & Design, *1991, by Delmar Publishers, Inc.*)

UPPER FLOOR PLAN

SCALE : ¼" = 1'-0"

Fig. 21-2 (*continued*)

LOWER FLOOR PLAN

SCALE : ¼" = 1'-0"

WINDOW SCHEDULE

SYM	SIZE	MODEL	ROUGH OPEN	QUAN
A	$1^0 \times 5^0$	JOB-BUILT		2
B	$8^0 \times 5^0$	W4N5 CSM	8'-2¾"×5'-5⅛"	1
C	$4^0 \times 5^0$	W2N5 CSM	4'-2¼"×5'-5⅛"	2
D	$4^0 \times 3^6$	W2N3 CSM	4'-2¼"×3'-5½"	2
E	$3^6 \times 3^6$	2N3 CSM	3'-5¼"×3'-5½"	2
F	$6^0 \times 4^0$	G64 SLDG	6'-0½"×4'-0½"	1
G	$5^0 \times 3^6$	G536 SLDG	5'-0½"×3'-6½"	4
H	$4^0 \times 3^6$	G436 SLDG	4'-0½"×3'-6½"	1
J	$4^0 \times 2^0$	A41 AWN	4-0½"×2'-0⅝"	3
K	$4^0 \times 2^0$	G42 SLDG	4'-0½"×2'-0½"	3

DOOR SCHEDULE

SYM	SIZE	TYPE	QUAN
1	$3^0 \times 6^8$	S.C. R.P. METAL INSULATED	1
2	$3^0 \times 6^8$	S.C.-FLUSH-METAL INSUL	2
3	$2^8 \times 6^8$	S.C.-SELF CLOSING	2
4	$2^8 \times 6^8$	H.C.	5
5	$2^6 \times 6^8$	H.C.	5
6	$2^6 \times 6^8$	POCKET	2
7	$2^4 \times 6^8$	POCKET	1
8	PR $2^6 \times 6^8$	H.C.	1
9	$5^0 \times 6^8$	BI-PASS	2
10	$3^0 \times 6^8$	BI-FOLD	1
11	$4^0 \times 6^8$	BI-FOLD	1
12	$2^0 \times 6^0$	SHATTER PROOF	1
13	$6^0 \times 6^8$	WOOD FRAME-TEMP. SLDG. GL.	1
14	$9^0 \times 7^0$	OVERHEAD GARAGE	2

Fig. 21-2 *(continued)*

DOUBLE

SINGLE

FLOOR PLAN SYMBOLS FOR DOUBLE-ACTING DOORS

FLOOR PLAN SYMBOL FOR AN EXTERIOR DOOR

DOUBLE-ACTING DOORS PICTORIAL

EXTERIOR DOOR PICTORIAL

EXTERIOR WALL

FLOOR PLAN SYMBOL FOR AN EXTERIOR DUTCH DOOR

FLOOR PLAN SYMBOL FOR AN INTERIOR SWINGING DOOR

DUTCH DOOR PICTORIAL

INTERIOR SWINGING DOOR PICTORIAL

Fig. 21-3 Floor plan symbols for exterior and interior swinging doors.

EXTERIOR WALL

FLOOR PLAN SYMBOL FOR EXTERIOR SLIDING DOOR

EXTERIOR SLIDING DOOR PICTORIAL

CLOSET

FLOOR PLAN SYMBOL FOR BY-PASSING DOORS

BATH

FLOOR PLAN SYMBOL FOR A POCKET DOOR

BY-PASSING DOORS PICTORIAL

POCKET DOOR PICTORIAL

Fig. 21-4 Symbols for exterior and interior sliding doors.

FLOOR PLAN SYMBOL FOR BIFOLD DOOR

FLOOR PLAN SYMBOL FOR ACCORDIAN DOOR

BIFOLD DOOR PICTORIAL

ACCORDIAN DOOR PICTORIAL

Fig. 21-5 Bifold and accordian doors are used on closets and wardrobes.

is indicated by one line. Because the *double-hung* window has two sash that slide vertically, its symbol shows two lines (Fig. 21-6).

The *casement* window, which swings outward, is depicted by a symbol similar to that for a swinging door. They may be shown having two or more

WINDOW WITH FIXED SASH

DOUBLE-HUNG WINDOW

FLOOR PLAN SYMBOLS

WINDOW WITH FIXED SASH

DOUBLE-HUNG WINDOW

PICTORIAL

Fig. 21-6 Fixed sash and double-hung window symbols.

Fig. 21-7 Symbols for casement windows.

units in each window (Fig. 21-7). An *awning* window is similar to a casement except it swings outward from the top. Its open position is indicated by dashed lines (Fig. 21-8).

The symbol for a *sliding window* is similar to that for a sliding door except for a line indicating the inside edge of the sill (Fig. 21-9).

Fig. 21-8 The open position of an awning window is shown with dashed lines.

FLOOR PLAN SYMBOL

PICTORIAL

Fig. 21-9 Sliding window floor plan symbol.

FLOOR PLAN SYMBOL

PICTORIAL

Fig. 21-10 A window consisting of a fixed sash and casement units.

Different kinds of windows may be used in combination. Figure 21-10 shows a window with a fixed sash, with casements on both sides. Main entrances may consist of a door with a *sidelight* on one or both sides (Fig. 21-11).

Close to the window and door symbols are letters and numbers that identify the units in the window and door schedules.

Structural Members. Openings, without doors, in interior walls for passage from one area to another are indicated by dashed lines. Notations are given if the opening is to be *cased* (trimmed with molding), and if the top is to be *arched* or treated in any other manner (Fig. 21-12).

The location of garage door *headers* may be shown by a series of dashes. Their size is usually indicated with a notation (Fig. 21-13). Window and other headers are shown in the same manner.

CASED OPENING

FLOOR PLAN SYMBOL

PICTORIAL

Fig. 21-12 Dashed lines indicate an interior wall opening without a door.

FLOOR PLAN SYMBOL

PICTORIAL
AN ENTRANCE DOOR FLANKED BY SIDELIGHTS

Fig. 21-11 An entrance door flanked by sidelights.

Fig. 21-13 The symbol for a garage door header is a dashed line.

Fig. 21-14 Symbols for ceiling beams and ceiling joists.

Ceiling beams, above the cutting plane of the floor, which support the ceiling joists are also represented by a series of dashes. *Ceiling joists* or *trusses* are identified in the floor plan with a double-ended arrow showing the direction in which they run. Their size and spacing are noted alongside the arrow (Fig. 21-14).

Kitchen, Bath, and Utility Room. The location of *bathroom* and *kitchen fixtures,* such as sinks, tubs, refrigerators, stoves, washers, and dryers, is shown by obvious symbols, abbreviations, and notations. The extent of the *base cabinets* is indicated by a line indicating the edge of the countertop. Objects such as dishwashers, trash compactors, and lazy susans are shown by a dashed line and notations or abbreviations. The *upper cabinets* are symbolized by dashed lines (Fig. 21-15).

Fig. 21-15 Floor plan symbols for kitchen and bath cabinets.

Other Floor Plan Symbols. The floor plan also shows the location of *stairways*. Lines indicate the outside edges of the *treads*. Also shown is the direction of travel and the number of *risers* (vertical distance from tread to tread) in the staircase (Fig. 21-16).

The location and style of *chimneys, fireplaces,* and *hearths* are shown by the use of appropriate symbols. Fireplace dimensions are generally not given. The size of the chimney flue and the hearth is usually indicated. The fireplace material may be shown by symbols according to the material specified (Fig. 21-17). The kind of material may also be identified with a notation.

An *attic access*, also called a *scuttle,* is usually located in a closet or hall. It is outlined with dashed

Fig. 21-16 Stair symbols vary according to the style of the staircase.

Fig. 21-17 Symbols for fireplace and hearth.

AN ATTIC ACCESS IS OUTLINED WITH DASHED LINES
AND NOTED WITHIN

Fig. 21-18 Attic access is outlined with dashed lines.

Fig. 21-20 Hose bibb.

lines. It may also be identified by a notation (Fig. 21-18).

The floor plan symbol for a *floor drain* is a small circle, square, or circle within a square. The slope of the floor is shown by straight lines from the corners of the floor to the center of the drain. Floor drains are appropriately installed in utility rooms where washers, dryers, laundry tubs, and water heaters are located (Fig. 21-19).

The location of outdoor water faucets, called *hose bibbs*, is shown by a symbol (Fig. 21-20) projecting from exterior walls where desired. For clarity, the symbol is labeled.

If not distracting, *electrical wiring* may be shown on the floor plan of simpler structures by the use of curved, dashed lines running to *switches*, *outlets*, and *fixtures* (Fig. 21-21). The symbols for these electrical components are shown in Figure 21-22.

Complex buildings require separate electrical plans, as well as prints for plumbing and for heating and ventilation.

Dimensions

Dimensions are placed and printed in a manner to make them as easy to read as possible. Read dimensions carefully. A mistake in reading a dimension early in the construction process could have serious consequences later.

Exterior Dimensions. On floor plans, the *overall* dimensions of the building are found on the outer dimension lines. In a *wood frame*, the dimensions are to the outside face of the frame. *Concrete block* walls are dimensioned to their outside face. *Brick veneer* walls are dimensioned to the outside face of

SOME SYMBOLS FOUND IN A UTILITY ROOM

Fig. 21-19 Some symbols found in a utility room.

Fig. 21-21 Part of a typical electrical floor plan. (*From Jefferis and Madsen,* Architectural Drafting & Design, *1991, by Delmar Publishers, Inc.*)

the wood frame, with an added dimension and notation for the veneer (Fig. 21-23).

Dimension lines closest to the exterior walls are used to show the location of *windows* and *doors*. In a wood frame, they are dimensioned to their centerline. In concrete block walls, the dimensions are to the edges of the openings and also show the opening width.

The second dimension line from the exterior wall is used to locate the centerline of *interior partitions*, which intersect the exterior wall.

Interior Dimensions. Dimensions are given from the outside of a wood frame or from the inside of concrete block walls, to the centerlines of *partitions*. Interior *doors* and other openings are dimen-

sioned to their centerline similar to exterior walls (Fig. 21-24).

Not all interior dimensions are required. Some may be assumed. For instance, it can be clearly seen if a door is centered between two walls of a closet or hallway (Fig. 21-25).

If a door is to be located a minimum distance from a wall, place the door opening near enough to the wall so that any door casing later applied can be scribed (fitted) to the wall. Or place the opening far enough away to allow convenient finishing (painting, wallpapering) between the casing and the wall (Fig. 21-26). A small space (one inch or less) between the outside edge of side casing and a wall is very difficult to finish.

Fig. 21-22 Electrical symbols. (*From Jefferis and Madsen,* Architectural Drafting & Design, *1991, by Delmar Publishers, Inc.*)

Fig. 21-23 Overall dimensions.

Fig. 21-24 Dimensioning windows, doors, and partitions.

Fig. 21-25 Some dimensions are obvious and are not given.

Locating a door opening a minimum distance from a wall.

SCRIBE
DOOR TRIM
AGAINST WALL

DOOR TRIM
FAR ENOUGH
AWAY FROM
WALL TO ALLOW
ADEQUATE
ROOM FOR
FINISHING WALL

Fig. 21-26

CHAPTER 22　　Sections and Elevations

Sections

Floor plans are views of a horizontal cut. *Sections* show *vertical* cuts called for on the floor, framing, and foundation plans (Fig. 22-1). Sections provide information not shown on other drawings. The number and type of section drawings in a set of prints depend on what is required for a complete understanding of the construction. They are usually drawn at a scale of $3/8'' = 1'-0''$.

Kinds of Sections

Full sections cut across the width or through the length of the entire building (Fig. 22-2). For a small residence, only one full section may be required to fully understand the construction. Commercial

Fig. 22-1 Sections are views of vertical cutting planes across the width or through the length of a building. Section reference lines identify the location of the section and the direction from which it is being viewed. (*From Jefferis and Madsen,* Architectural Drafting & Design, *1991, by Delmar Publishers, Inc.*)

structures may require several full and many partial sections for complete understanding.

A *partial section* shows the vertical relationship of the parts of a small portion of the building. Partial sections through exterior walls are often used to give information about materials from foundation to roof (Fig. 22-3). They are drawn at a larger scale. The information given in one wall section does not necessarily apply for all walls. It may not apply, in fact, for all parts of the same wall. The wall section, or any other section being viewed, applies only to that part of the construction located by the section reference lines. Because the construction changes throughout the building, many section views are needed to provide clear and accurate information.

An enlargement of part of a section is required when enough information cannot be given in the space of a smaller scale drawing. These large-scale drawings are called *details*. An example is the foundation detail shown in Figure 22-4.

Reading Sections

From a typical exterior wall section of a residence, shown in Figure 22-3, much necessary information is provided. If a desired dimension is not given, enough others are provided to calculate it. Starting from the bottom and working upward to the roof, the following is an example of data that may be read from a full cross-section.

Foundation

■ The height, width, and shape of the foundation footings, including the footing keyway. The material used for the footing, and the lo-

Fig. 22-2 A typical full cross-section of a residence. (*From Jefferis and Madsen,* Architectural Drafting & Design, *1991, by Delmar Publishers, Inc.*)

Fig. 22-3 A partial section through an exterior wall. (*From Jefferis and Madsen,* Architectural Drafting & Design, *1991, by Delmar Publishers, Inc.*)

3 1/2" FIBER GLASS INSULATION

POLYETHYLENE VAPOR BARRIER

STUCCO OVER
15# FELT
& 26 GA LINEWIRE

1/2" GYP BD.

2 × 4 STUDS 16" O.C.

5/8" × 10" ANCHOR BOLT 6'-0" O.C.

4" CONC. SLAB

6" × 6" × #10 W.W.M.

2 × 4 PT

VAPOR BARRIER

6"

GRADE

12"

6"

4" GRAVEL

#4 φ CONT

4"

12"

A DETAIL IS A SMALL PART OF A SECTION DRAWN
AT A LARGER SCALE.

Fig. 22-4 A detail is a small part of a section drawn at a large scale.

cation and size of any steel reinforcement. The location of the finished grade and the depth of the footings below the grade line.

■ The height and thickness of the foundation wall and the material used to construct it. The shape, size, and location of a wall ledge for brick veneer. The location and size of reinforcing rods and anchor bolts in the foundation wall. The type, size, and location of below grade foundation wall insulation. The kind and surface coverage of foundation wall waterproofing. The location of foundation drain pipe and the kind and amount of drain bed material.

■ The thickness of the basement floor slab. The floor slab material and the size, gauge, and kind of slab reinforcement. The thickness and kind of material directly under the slab. The kind, size, and location of a vapor barrier.

Column footings material, size, and depth below grade. The location, size, and spacing of steel reinforcing rods in the column footings. The type and size of columns are given, but the number and spacing are found in a longitudinal section or a foundation plan.

Floor

■ The thickness, width, location, and kind of sill plate, The width, depth, location, and kind of girder. The kind, thickness, width, length, and spacing of floor joists. The header joist or rim joist material, thickness, and width. The size, type, location, and kind of bridging. The kind and thickness of the subfloor and finish floor. The kind, size, and location of floor insulation. The type, location, and spacing of framing anchors, connectors, and hold-downs. The location, shape, and material used for flashings and termite shields.

Walls

■ The kind, location, thickness, and width of sole and top plates. The kind, width, and spacing of the studs. The type and thickness of wall insulation. The thickness and width of certain window and door opening headers, and their height from the finish floor. The kind and thickness of exterior wall sheathing, and finish, including type of building paper or housewrap. The location of certain partitions. The kind and thickness of interior wall finish.

The type, shape, and size of the base and base shoe. The finish floor to ceiling height.

Ceiling and Roof

■ The kind, width, thickness, length, and spacing of ceiling joists. The kind, size, and spacing of ceiling furring strips. The type and thickness of ceiling insulation and ceiling finish. The kind, size, spacing, and pitch of roof rafters or trusses. The thickness and width of the ridgeboard. The size and location of collar ties. The type and thickness of roof sheathing and the type of roofing felt and roof finish. The amount of roof overhang. The kind, thickness, and width of cornice finish; soffit, fascia, and frieze. The type and location of soffit and roof vents, including insulation baffles for air circulation. Kind and location of framing ties and connectors.

Elevations

Elevations are orthographic drawings. They are usually drawn at the same scale as the floor plan. They show each side of the building as viewed from the outside. Generally four elevations, one for each side, are included in a set of drawings. They are titled Front, Rear, Left Side, and Right Side. They may also be titled according to the compass direction that they face, for instance, North, South, East, and West. From the exterior elevations, the general shape and design of the building can be determined (Fig. 22-5).

Fig. 22-5 Elevations show the exterior of the building. (*From Jefferis and Madsen,* Architectural Drafting & Design, *1991, by Delmar Publishers, Inc.*)

Fig. 22-5 (*continued*)

Symbols.
Elevation symbols are different than floor plan symbols for the same object. In elevation drawings, the symbols represent, as closely as possible, what they would appear to the eye. To make the drawing more clear, the symbols are usually identified with a notation.

Interpreting Elevations

- The location of any steps, porches, dormers, skylights, and chimneys, although not dimensioned, can be seen.

- Foundation footings and walls below the grade level may be shown with hidden lines.

- The kind and size of exterior siding, railings, entrances, and special treatment around doors and windows are shown (Fig. 22-6).

The elevations show the windows and doors in their exact location. Other openings, such as vents and louvers, are shown in place. Their style and size are identified by appropriate symbols and notations (Fig. 22-7).

The type of roofing material, the roof pitch, and the cornice style may also be determined from the exterior elevations (Fig. 22-8).

Dimensions. In relation to other drawings, elevations have few dimensions. Some dimensions usu-

Fig. 22-6 Symbols for siding, railings, and shutters.

Fig. 22-7 Symbols for windows, doors, vents, and louvers.

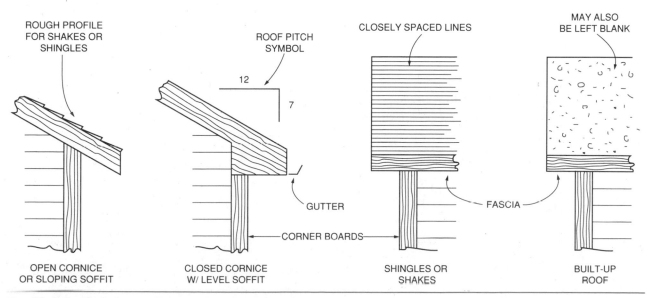

Fig. 22-8 Symbols for roofing, roof pitch, and cornice style.

ally given are floor to floor heights, distance from grade level to finished floor, height of window openings from the finished floor, and distance from the ridge to the top of the chimney.

A number of other things may be shown on exterior elevations, depending on the complexity of the structure. Little information is given in elevations that cannot be seen in more detail in plans and sections. However, elevations serve an important purpose in making the total construction easier to visualize.

CHAPTER 23 PLOT AND FOUNDATION PLANS

Plot Plans

A *plot plan* is a map of a section of land used to show the proposed construction (Fig. 23-1). Depending on the size, the scale of plot plans may vary from 1″ = 10′ to 1″=200′. It is a required drawing when applying for a permit to build in prac-

tically every community. It is a necessary drawing to plan construction that may be affected by various features of the land. The plan must show compliance with zoning and health regulations. Although plot plan requirements may vary with localities, certain items in the plan are standard.

Fig. 23-1 A typical plot plan.

Property Lines

The property line *measurements* and *bearings*, known as *metes* and *bounds*, show the shape and size of the parcel. They are standard in every plot plan.

Measurements. The boundary lines are measured in *feet, yards, rods,* or *chains.*

3 feet equals one yard.

16 1/2 feet or 5 1/2 yards equal one rod.

66 feet or 22 yards or 4 rods equal one chain.

Parts of measurement units are expressed as decimals. For instance, a boundary line dimension is expressed as 100.50 feet, not 100 feet, 6 inches. It is shown centered on, close to, and inside the line.

North. The North compass direction is clearly marked on every plot plan. In a clear space, an arrow of any style labeled with the letter "N" is pointed in the north direction (Fig. 23-2).

Bearings. In addition to the length of the boundary line, its *bearing* is shown. The bearing is a compass direction given in relation to a *quadrant* of a circle. There are 360 degrees in a circle and 90 degrees in each quadrant. Degrees are divided into *minutes* and *seconds.*

One degree equals 60 minutes (60')

One minute equal 60 seconds (60")

The boundary line bearing is expressed as a certain number of degrees clockwise or counterclockwise from either North or South. For instance, a bearing may be shown as N 30° E, N 60° W, S 45° E, or S 75° W (Fig. 23-3). No bearings begin with East or West as a direction. The bearing is shown centered close to, and outside the boundary line opposite its length.

Point of Beginning. An object that is unlikely to be moved easily, such as a large rock, tree, or iron rod driven into the ground, is used for a *point of beginning.* To denote this point on the plot plan, one corner of the lot may be marked with the abbreviation POB. It is from this point that the lot is laid out and drawn (Fig. 23-4).

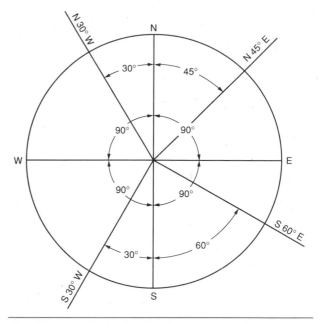

Fig. 23-3 Method of indicating property line bearings.

Fig. 23-4 Measurements, bearings, and legal description of a parcel of land.

Topography

Topography is the detailed description of the land surface. It includes any outstanding physical features and differences in *elevation* of the building site. Elevation is the height of a surface above sea level. It is expressed in feet and 1/100 of a foot.

Contour Lines. *Contour lines* are irregular, curved lines connecting points of the same elevation of the land. The vertical distance between contour lines is called the *contour interval.* It may vary depending on how specifically the contour of the land needs to be shown on the plot plan.

When contour lines are close together, the slope is steep. Widely spaced contour lines indicate a gradual slope. At intervals, the contour lines are broken and the elevation of the line inserted in the

Fig. 23-2 Typical North direction symbols.

Fig. 23-5 Contour lines show the elevation and slope of the land.

space (Fig. 23-5). On some plans, dashed contour lines indicate the existing grade and solid lines depict the new grade. Topography is not always a requirement on plot plans. This is especially true for sites where there is little or no difference in elevation of the land surface. The slope of the finished grade may be shown by arrows instead of contour lines (Fig. 23-6).

Fig. 23-6 Arrows are sometimes used in place of contour lines to show the slope of the land.

Elevations

The height of several parts of the site and the construction are indicated on the plot plan. It is necessary to know these elevations for grading the lot and construction of the building and accessories.

Benchmark. Before construction begins, a reference point, called a *benchmark*, is established on or close to the site. It is used for conveniently determining differences in elevation of various land and building surfaces.

The benchmark is established on some permanent object, which will not be moved or destroyed, at least until the construction is complete. It may be the actual elevation in relation to sea level. It may also be given an arbitrary elevation of 100.00 feet. All points on the lot, therefore, would either be above, level with, or below the benchmark (Fig. 23-7). The location of the benchmark is clearly shown on the plot plan with the abbreviation BM.

Finish Floor. The elevation of finished floor levels is clearly shown and noted on the plot plan. Building lines are laid out to the same height as the

Fig. 23-7 A benchmark is a reference point used for determining differences in elevation.

top of the foundation wall before any floors are built. Therefore, the distance from the finish floor to the top of the foundation wall must be found. It is then subtracted from the finish floor elevation to determine the elevation of the top of the foundation wall. This distance can be calculated with information from an exterior wall section (Fig. 23-8).

Converting Decimals to Fractions. Calculations, especially those dealing with elevations and roof framing, require the carpenter to convert decimals of a foot, to feet, inches, and 16ths of an inch as found on the rule or tape. To convert, use the following method:

■ Multiply a decimal of a foot by 12 (the number of inches in a foot) to get inches.

Fig. 23-8 The distance from the finish floor elevation to the top of the foundation wall must be calculated.

- Multiply any remainder decimal of an inch by 16 (the number of 16ths in an inch) to get 16ths of an inch. Round off any remainder to the nearest 16th of an inch.
- Combine whole feet, whole inches, and 16ths of an inch to make the conversion.

Example: Convert the finish floor elevation of 104.65 feet to feet, inches, and 16ths of an inch.

1. Multiply .65 ft. × 12 = 7.80 inches
2. Multiply .80 inches × 16 = 12.8 16ths of an inch.
3. Round off 12.8 16ths of an inch to 13 16ths of an inch.
4. Combine feet, inches, and sixteenths = 104'-7 13/16".

It is more desirable to remember the method of conversion rather than use conversion tables. Reliance on conversion tables requires access to the tables and knowledge of their use, encourages dependence on them, and results in helplessness without them.

Other Elevations

- In addition to contour lines, the elevation of each corner of the property is noted on the plot plan (Fig. 23-9). The top of one of the boundary corner markers makes an excellent benchmark.
- Existing and proposed roads adjacent to the property are shown as well as any *easements*. Easements are right-of-way strips running through the property. They are granted for various purposes, such as access to other property, storm drains, or utilities. Elevations of a street at a driveway and at its centerline are usually required. One of these is sometimes used as the benchmark.

The Structure

The shape and location of the building are shown. Distances, called *setbacks*, are dimensioned from the boundary lines to the building.

The shape, width, and location of patios, walks, driveways, and parking areas are also found on plot plans. Details of their construction are found in another drawing.

A plot plan may also show any *retaining walls.* These walls are used to hold back earth to make more level surfaces instead of steep slopes.

Utilities. The water supply and public sewer connections are shown by noted lines from the structure to the appropriate boundary line. If a private sewer disposal system is planned, it is shown on the plot plan (Fig. 23-10). Although there are many kinds of sewer disposal systems, those most commonly used consist of a septic tank and leaching field. There are usually strict regulations in regard to location and construction.

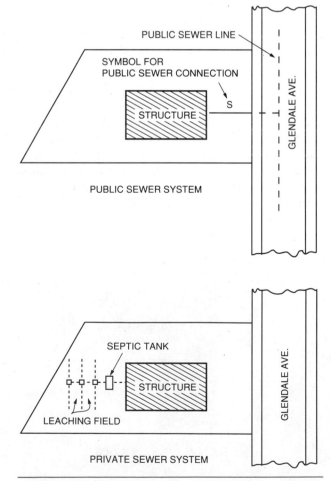

Fig. 23-10 Sewage disposal systems are shown on the plot plan.

Fig. 23-9 The elevations of property line corners are indicated on the plot plan.

DECIDUOUS
TREES

CONIFEROUS
TREES

PALM
TREES

Fig. 23-11 Various symbols are used to indicate different kinds of trees.

The location of gas lines is shown, if applicable. Sometimes the location of the nearest utility pole is given. It is shown using a small solid circle as a symbol and is noted. Foundation drain lines leading to a storm drain, drywell, or other drainage may be shown and labeled.

Landscaping. The location and kind of existing and proposed trees are shown on the plot plan. Existing trees are noted, whether they are to be saved or removed. Those that are to be saved are protected by barriers during the construction process. Various symbols are used to show different kinds of trees (Fig. 23-11).

Identification. Included on the plot plan are the name and address of the property owner, the title and scale of the drawing, and a legal description of the property (Fig. 23-12).

Foundation Plans

The foundation plan is drawn at the same scale as the floor plan. It is a view from above of a horizontal cut through the foundation. Great care must be taken when reading the foundation plan so no mistakes are made. A mistake in the foundation affects the whole structure, and generally requires adjustments throughout the construction process.

Two commonly used types of foundations are those having a crawl space, or basement below grade, and those with a concrete slab floor at grade level (Fig. 23-13).

Crawl Space and Basement Foundations

The *crawl space* is the area enclosed by the foundation between the ground and the floor above. A minimum distance of 18 inches from the ground to the floor and 12 inches from the ground to the bottom of any beam is required. The ground is covered with a plastic sheet, called a vapor barrier, to prevent moisture rising from the ground from penetrating into the floor frame above.

A foundation enclosing a *basement* is similar to that of a crawl space except the walls are higher, windows and doors may be installed, and a concrete floor is provided below grade. The basement may be used for additional living area, garage, utility room, or workshop.

Reading Plans. Whether the foundation supports a floor using closely spaced floor joists or more widely spaced post-and-beam construction, the information given in the foundation plan is similar. A typical foundation plan is shown in Figure 23-14. Its usual components are as follows:

- The inside and outside of the foundation wall are clearly outlined. Dashed lines on both sides of the wall show the location of the foundation footing. The type, size, and spacing of anchor bolts are shown by a notation.

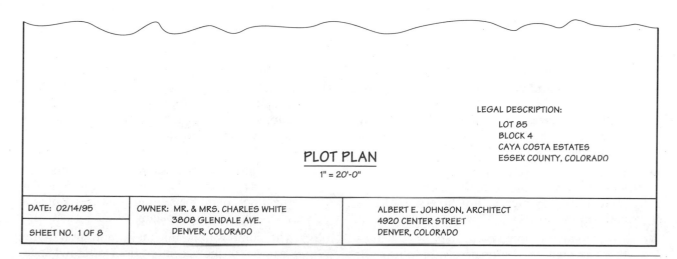

LEGAL DESCRIPTION:

LOT 85
BLOCK 4
CAYA COSTA ESTATES
ESSEX COUNTY, COLORADO

PLOT PLAN
1" = 20'-0"

DATE: 02/14/95

SHEET NO. 1 OF 8

OWNER: MR. & MRS. CHARLES WHITE
3808 GLENDALE AVE.
DENVER, COLORADO

ALBERT E. JOHNSON, ARCHITECT
4920 CENTER STREET
DENVER, COLORADO

Fig. 23-12 Certain identification items are needed on the plot plan.

Fig. 23-13 Commonly used foundations.

■ Wall openings for windows, doors, or crawl space access and vents are shown with appropriate symbols and noted. Small retaining walls of concrete or metal, called *areaways*, to hold earth away from windows that are below grade may be shown (Fig. 23-15).

■ Walls for *stoops* (platforms for entrances) are shown. A notation is made in regard to the material with which to fill the enclosed area and cap the surface.

■ Other footings shown by dashed lines include those for chimneys, fireplaces, and columns or posts. Columns or posts support girders (shown by a series of long and short dashes directly over the center of the columns). A recess in the foundation wall, called a *pocket*, to support the ends of the girder is shown. Notations are made to identify all of these items (Fig. 23-16).

■ Floor joist direction is shown by a line with arrows on both ends similar to those shown in the floor plan for ceiling joists. A notation gives the size and spacing of the joists.

■ The composition, thickness, and underlying material of the basement floor or crawl space surface are noted.

Fig. 23-14 A foundation plan for a partial basement. (*From Jefferis and Madsen,* Architectural Drafting & Design, *1991, by Delmar Publishers, Inc.*)

Fig. 23-15 Partial plan of a foundation wall with various items.

■ The location of a stairway to the basement is shown with the same symbol as used in the floor plan.

■ Although it may be stated in the specifications, the strength of concrete used for various parts of the foundation, and the wood type and grade, in addition to size, may be specified by a notation.

■ Plans for foundations with basements may show the location, with appropriate notations, of furnaces and other items generally found in the floor plan if the basement is used as part of the living area.

Dimensions. It is important to understand how parts of the foundation are dimensioned. Foundation walls are dimensioned face to face. Footings, columns, posts, girders, and beams are dimensioned to their centerline (Fig. 23-17).

Slab-on-Grade Foundation

The *slab-on-grade foundation* is used in many residential and commercial buildings. It takes less labor and material than foundations to support beam and joist floor framing. Often the concrete for the footing, foundation, and slab can be placed at the same time. There are several kinds, but slab-on-grade foundation plans (Fig. 23-18) show common components:

■ The shape and size of the slab are shown with solid lines, as are patios and similar areas. Changes in floor level, such as for a fireplace hearth or a sunken living area, are indicated by solid lines. A notation gives the depth of the recess.

■ Exterior and interior footing locations ordinarily below grade are indicated with dashed

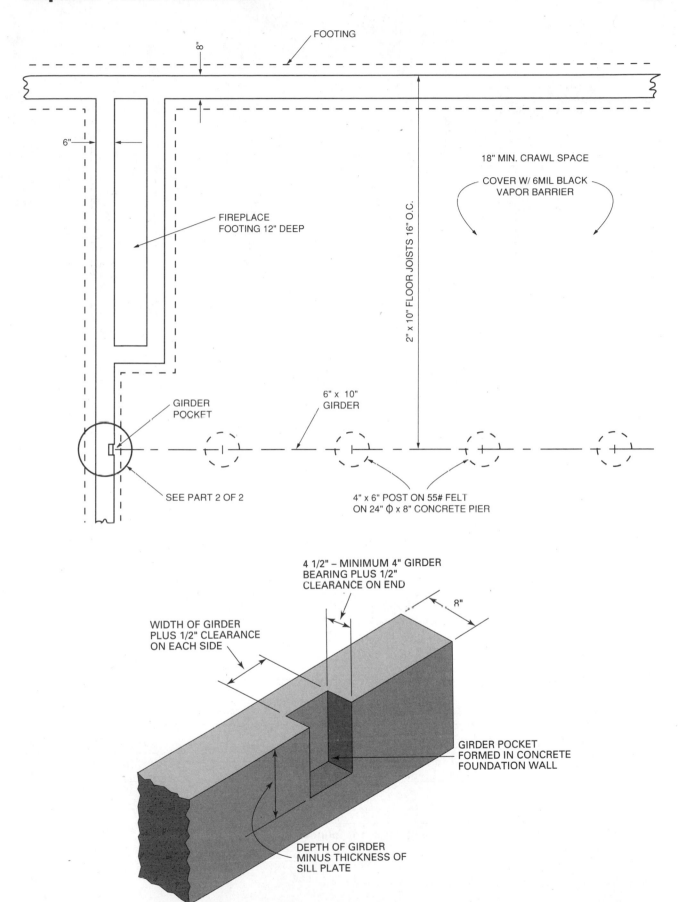

FOOTING

8"

6"

FIREPLACE
FOOTING 12" DEEP

18" MIN. CRAWL SPACE

COVER W/ 6MIL BLACK
VAPOR BARRIER

2" x 10" FLOOR JOISTS 16" O.C.

GIRDER
POCKET

6" x 10"
GIRDER

SEE PART 2 OF 2

4" x 6" POST ON 55# FELT
ON 24" ⌀ x 8" CONCRETE PIER

4 1/2" – MINIMUM 4" GIRDER
BEARING PLUS 1/2"
CLEARANCE ON END

8"

WIDTH OF GIRDER
PLUS 1/2" CLEARANCE
ON EACH SIDE

GIRDER POCKET
FORMED IN CONCRETE
FOUNDATION WALL

DEPTH OF GIRDER
MINUS THICKNESS OF
SILL PLATE

Fig. 23-16 Enlarged view of foundation plan showing girder and column details.

Fig. 23-17 Dimensioning of crawl space and basement foundations.

Fig. 23-18 Slab-on-grade foundation plan. (*From Jefferis and Madsen,* Architectural Drafting & Design, *1991, by Delmar Publishers, Inc.*)

lines. Footings outside the slab and ordinarily above grade are shown with solid lines.

■ Appropriate symbols and notations are used for fireplaces, floor drains, and ductwork for heating and ventilation (Fig. 23-19). Note: Blueprints for commercial work usually have separate drawings for electrical, plumbing, and mechanical work.

OUTSIDE
EDGE OF
FOOTING

EDGE OF RECESS

INTERIOR FOOTING

OUTSIDE
EDGE OF
SLAB

RECESS SLAB 6" FOR
LIVING ROOM FLOOR

4" CONCRETE SLAB OVER
4" SAND FILL WITH .006
VAPOR BARRIER & 2"x24"
RIGID INSULATION AROUND
ENTIRE PERIMETER.

Fig. 23-19 Some symbols used in slab-on-grade foundation plans.

DIMENSION TO
OUTSIDE EDGE
OF SLAB

DIMENSIONS TO CENTER
LINE OF INTERIOR FOOTINGS

DIMENSIONS TO CENTER
LINE OF COLUMNS
& COLUMN FOOTINGS

Fig. 23-20 Dimensioning slabs on-grade foundation plans.

- Notations are made for the slab thickness, wire mesh reinforcing, fill material, and vapor barrier under the slab.
- Interior footing, mudsill, reinforcing steel, and anchor bolt size, location, and spacing are written.

Dimensions. Overall dimensions are to the outside of the slab. Interior piers are located to their centerline. Door openings are dimensioned to their sides (Fig. 23-20).

CHAPTER 24 BUILDING CODES AND ZONING REGULATIONS

Cities and towns have laws governing many aspects of new construction and remodeling. These laws protect the consumer and the community. Codes and regulations provide for safe, properly designed buildings in a planned environment. Contractors and carpenters should have knowledge of zoning regulations and building codes.

Zoning Regulations

Zoning regulations deal, generally speaking, with keeping buildings of similar size and purpose in areas for which they have been planned. They also regulate the space in each of the areas. The community is divided into areas called *zones*, shown on *zoning maps.*

Zones. The names given to different zones vary from community to community. The zones are usually abbreviated with letters or a combination of letters and numbers. A large city may have thirty or more zoning districts.

There may be several *single family residential zones.* Some zones have less strict requirements than others. Other areas may be zoned as *multifamily residential.* They may be further subdivided into areas according to the number of apartments. Other residential zones may be set aside for *mobile home parks,* and those that allow a combination of *residences, retail stores, and offices.*

Other zones may be designated for the *central business district,* various kinds of *commercial districts* and different *industrial* zones.

Lots. Zoning laws regulate buildings and building sites. Most cities specify a *minimum lot size* for each zone and a *maximum ground coverage* by the structure. The *maximum height* of the building for each zoning district is stipulated. A *minimum lot width* is usually specified as well as *minimum yards.*

Minimum yard refers to the distance buildings must be kept from property lines. These distances are called *setbacks.* They are usually different for front, rear, and side.

Some communities require a certain amount of landscaped area, called *green space,* to enhance the site. In some residential zones, as much as half the lot must be reserved for green space. In a central business area, only 5 to 10 percent may be required.

In most zones, off-street parking is required. For instance, in single-family residential zones, room for two parking spaces on the lot is required.

Nonconforming Buildings. Because some cities were in existence before the advent of zoning laws, many buildings and businesses may not be in their proper zone. They are called *nonconforming.* It would be unfair to require that buildings be torn down, or to stop businesses, in order to meet the requirements of zoning regulations.

Nonconforming businesses or buildings are allowed to remain. However, restrictions are placed on rebuilding. If partially destroyed, they may be allowed to rebuild, depending on the amount of destruction. If 75 percent or more is destroyed, they are not usually allowed to rebuild in the same manner or for the same purpose in the same zone.

Any hardships imposed by zoning regulations may be relieved by a *variance.* Variances are granted by a Zoning Board of Appeals within each community. A public hearing is held after a certain period of time. The general public, and, in particular, those abutting the property are notified. The petitioner must prove certain types of hardship specified in the zoning laws before the zoning variance can be granted.

Building Codes

Building codes regulate the design and construction of buildings by establishing minimum safety standards. They prevent such things as roofs being ripped off by high winds, floors collapsing from inadequate support, buildings settling because of a poor foundation, and tragic deaths from fire due to lack of sufficient exits from buildings. In addition to building codes, other codes govern the mechanical, electrical, and plumbing trades.

Some communities have no building codes. Some write their own. Some have codes, but exempt residential construction. Some have adopted one of three national building codes. Some use one of the national codes supplemented with their own.

It is important to have a general knowledge of the building code used by a particular community. Construction superintendents and contractors must have extensive knowledge of the codes.

National Building Codes. The *Basic National Building Code* (BNBC) is used primarily in the Northeast and Midwest. The *Uniform Building Code (UBC)* has been adopted in the West and Southwest. The *Standard Building Code* is used by states in the Southeast.

Many communities use the Council of American Building Officials (CABO) One- and Two-Family Dwelling Code for residential work. They then keep one of the national codes for commercial work. For residential work, other communities may use *Dwelling Construction under the Uniform Building Code.*

Use of Residential Codes. In addition to structural requirements, major areas of residential codes include:

- Exit facilities, such as doors, halls, stairs, and windows as emergency exits, and smoke detectors.

- Room dimensions, such as ceiling height and minimum area.

- Light, ventilation, and sanitation, such as window size and placement, maximum limits of glass area, fans vented to the outside, requirements for baths, kitchens, and hot and cold water.

Use of Commercial Codes. Codes for commercial work are much more complicated than those for residential work. The structure must first be defined for code purposes. In order to define the structure, six classifications must be used.

1. The *occupany group* classifies the structure by how and whom it will be used. The classification is designated by a letter, such as R, which includes not only single-family homes but apartments and hotels.

2. The size and location of the building.

3. The type of construction. Five general types are given numbers 1 through 5. Types 1 and 2 require that all structural parts be noncombustible. Construction in types 3, 4, or 5 can be made of either masonry, steel, or wood.

4. The floor area of the building.

5. The height of the building. Zoning regulations may also affect the height.

6. The number of people who will use the building, called the *occupant load,* determines such things as the number and location of exits.

Once the structure is defined, the code requirements may be studied.

Building Permits

A *building permit* is needed before construction can begin. Application is made by the contractor to the office of the local building official. The building permit application form (Fig. 24-1) requires a general description of the construction, legal description and location of the property, estimated cost of construction, and information about the applicant.

Drawings of the proposed construction are submitted with the application. The type and kind of drawings required depend on the complexity of the building. For commercial work, usually five sets of plot plans and two sets of other drawings are required. The drawings are reviewed by the building inspection department. If all is in order, a permit (Fig. 24-2) is granted upon payment of a fee. The fee is usually based on the estimated cost of the construction. Electrical, mechanical, plumbing, water, and sewer permits are usually obtained by subcontractors. The permit card must be displayed on the site in a conspicuous place until the construction is completed.

Inspections

Building inspectors visit the job site to perform code inspections at various intervals:

1. A *foundation inspection* takes place after the trenches have been excavated and forms erected and ready for the placement of concrete. No reinforcing steel or structural framework of any part of any building may be covered without an inspection and a release.

2. A *frame inspection* takes place after the roof, framing, fire blocking, and bracing are in place, and all concealed wiring, pipes, chimneys, ducts, and vents are complete.

3. The *final inspection* occurs when the building is finished. A Certificate of Occupancy or Completion is then granted.

It is the responsibility of the contractor to notify the building official when the construction is ready for a scheduled inspection. If all is in order, the inspector signs the permit card in the appropriate space and construction continues. If the inspector finds a code violation, it is brought to the attention of the contractor or architect for compliance.

These inspections ensure that construction is proceeding according to approved plans. They also

CITY OF ST. PETERSBURG
APPLICATION FOR BUILDING PERMIT

RADON GAS FEE _____

```
|                    FOR OFFICE USE ONLY                    |
```

Permit Type _____   Permit # _____

Permit Class of Work _____   Log # _____

Permit Use Code _____   Issue Date _____

Lot _____ Block _____ Sub _____   Permit Cost _____

Fire Zone: IN _____   OUT _____   Zone: _____   T.I.F. Due (Y/N or NA) _____

Utility Notification   1. FL Power _____   B of A (Y/N) _____ Case No.
                              2. Peoples Gas _____   E.D.C. (Y/N) _____ Case No.
NOTE: Items with*   3. Water Dept. _____   C.R.A. (Y/N) _____ Case No.
must be entered in computer.   H.P.C. (Y/N) _____ Case No.

*Plat Page _____   *Sec _____   *Township _____   *Range _____   Zone _____

*Dept of Commerce Code _____   *Const. Type _____ Protected _____ Unprotected _____

*Additional Permits Required:
   Building _____   Plumbing _____   No. of W.C. _____   No. of Meters _____

   Electrical _____   Mechanical _____   Gas _____   Fire Sprk. _____   Landscape _____

   Park/Paving _____   Total Spaces _____   Handicap _____

*Flood Zone _____   *Setbacks: Front _____   Left Side _____   Right Side _____
                                              Rear _____   Other Requirements _____
Threshold Building YES _____   NO _____   _____

Special Notes/Comments to Inspector: _____
_____
_____

```
|              APPLICANT PLEASE FILL OUT THIS SECTION              |
```

JOB ADDRESS _____ Suite or Apt. No. _____

CONTRACTOR _____ Cert./Reg. No. _____ Telephone _____

PROPERTY OWNER'S Name _____ Address _____

              City _____ State _____ Zip _____ Telephone _____

Building Description: Total Sq. Ft. _____   Estimated Job Value _____
   (LF-SF or Dimensions _____   Building Use _____
   Valuation of Work _____   Former Use _____
   No. of Units _____ No. of Suites _____   No. of Stories _____

Special Notes or Comments: _____
_____

PHONE 893-7386 FOR ALL INSPECTIONS   HCS-12 Rev. 6-1-88

(OVER)

Fig. 24-1 A form used to apply for a building permit.

CITY OF ST. PETERSBURG
DEPARTMENT OF HOUSING & CONSTRUCTION SERVICES
BUILDING PERMIT

THIS PERMIT BECOMES INVALID IF NO INSPECTIONS HAVE BEEN MADE DURING ANY 3 MONTH PERIOD.

Flood Elevation - _____ Lowest Floor Minimum Required

☐ New Construction   ☐ Moving   ☐ Siding
☐ Grounds Improvements   ☐ Fences   ☐ Walls
☐ Utility Building   ☐ Pool
☐ Reroofing   ☐ Other _____

Permit No. _____ (ZONE) _____

Job Address _____

Lot _____   **Blk.** _____   **Sub.** _____

Date _____

This permit covers building construction only. Additional permits are required for electric, plumbing, gas and/or mechanical installations.

BUILDING			ELECTRICAL			PLUMBING			MECHANICAL-GAS		
Type of Inspection	Date	Inspector	Type of Inspection	Date	Inspector	Type of Inspection	Date	Inspector	Type of Inspection	Date	Inspector

NOTE: Building, Electrical, Plumbing and Mech/Gas Inspections shall be dated and initialed by inspectors before walls and ceilings are covered.

BUILDING OK TO COVER		ELECTRICAL OK TO COVER		PLUMBING OK TO COVER		MECH/GAS OK TO COVER	
Date	Inspector	Date	Inspector	Date	Inspector	Date	Inspector

NOTE: This card shall remain posted at the job site until all final inspections have been dated and initialed by inspectors.

BUILDING FINAL OK		ELECTRICAL FINAL OK		PLUMBING FINAL OK		MECH/GAS FINAL OK	
Date	Inspector	Date	Inspector	Date	Inspector	Date	Inspector

THIS CARD MUST BE POSTED IN AN EASILY SEEN LOCATION.

For INSPECTIONS, call 893-7386. ■ For other information, call 893-7388.

Fig. 24-2 A building permit. Communities use different kinds of forms.

make sure construction is meeting code requirements. This protects the future occupants of the building and the general public. In most cases, a good rapport exists between inspectors and builders, enabling construction to proceed smoothly and on schedule.

Review Questions

Select the most appropriate answer.

1. A view looking from the top downward is called
   a. an elevation.　　c. a plan.
   b. a perspective.　　d. a section.

2. A view showing a vertical cut through the construction is called
   a. an elevation.　　c. a plan.
   b. a detail.　　　　d. a section.

3. The most commonly used scale for floor plans is
   a. 1/4″ = 1′-0″.　　c. 1 1/2″ = 1′-0″.
   b. 3/4″ = 1′-0″.　　d. 3″ = 1′-0″.

4. To determine a missing dimension,
   a. use the scale rule.
   b. calculate it.
   c. read the specifications.
   d. find a typical wall section.

5. Centerlines are indicated by a
   a. series of short, uniform dashes.
   b. a series of long and short dashes.
   c. a long dash followed by two short dashes.
   d. a solid, broad, dark line.

6. Which of the dimensions below would be found on a blueprint?
   a. 3′.　　　　c. 3′-0″.
   b. 3 ft.　　　d. 36″.

7. On what drawing would the setback of the building from the property lines be found?
   a. Floor plan.　　c. Elevation.
   b. Plot plan.　　d. Foundation plan.

8. To find out which edge of doors are to be hinged, look at the
   a. elevations.　　c. floor plan.
   b. framing plan.　　d. wall section.

9. The elevation of the finished first floor is usually found on the
   a. plot plan.　　　c. foundation plan.
   b. floor plan.　　d. framing plan.

10. An exterior wall stud height can best be determined from the
   a. floor plan.　　　c. wall section.
   b. framing elevation.　　d. specifications.

■ BUILDING FOR SUCCESS ■

The Role of Construction Documents

The construction student who desires to excel as a carpenter and eventually as a builder will need to be proficient in interpreting plans and reading specifications. Units 8 and 9 have provided essential information about reading plans, codes, and specifications. With this information, you will want to combine your basic understanding of how all of this documentation fits together. All of the information contained in the construction documents will be implemented, at various intervals, either before, during, or after the construction process is completed.

Construction documents vary. However, they normally consist of at least the working drawings and specifications. Part of the package may also involve local codes, ordinances, bidding information and forms, conditions of the contract, and any other pertinent agreements. The complexity of the project will often dictate the exact makeup of the document package. Generally residential projects do not contain the extensive documentation that commercial construction projects do. The full purpose of this information is to assure that the construction process is completed as proposed. Additionally, having a structure that is safe for occupancy and completed on schedule are important goals included in the document group. By law the contracts become binding, and may require legal counsel to clarify discrepancies or disputes.

During the construction planning process, the builder will use the blueprints, specifications, codes, and local regulations to prepare an accurate cost estimate of the job and materials needed to complete the project. The cost estimate in turn will be submitted for consideration by the client. Until the contractor submits the bid, the opportunity exists for questions to be answered that will affect his or her estimated bid. These questions may center on a variety of things, such as materials, assemblies, time schedules, responsibilities of all parties, or other factors affecting the project.

Throughout the construction process, everyone involved will be required to adhere to the construction documents and all agreements. Once construction is underway, the client, contractor, subcontractors, inspectors, and architects will refer to the documents for accuracy and accountability. The outcomes must be as proposed in the beginning. Payment for services will be made by the owner only if these services comply with the stipulations contained in the construction documents.

The successful conclusion of the project will be based on how well the construction requirements were met. Without a clear understanding of the stipulations surrounding the construction project, many setbacks will occur. Any delays or unexpected deviations from the proposed plan will greatly affect the schedules and profit margin for the builder. Additionally, deviations not planned will affect the relationship of the builder, subcontractors, and client.

Construction documents serve as the focal point for the entire construction process. Therefore, they must be understood by everyone. In many cases involving disputes, the specifications and contract will have precedence over the blueprints. Historically, successful construction projects have been associated with clear understandings centered on the construction document—*before* the construction began.

FOCUS QUESTIONS: For individual or group discussion.

1. What do you think are included in the term *construction documents*?

2. What bearing should these various written stipulations and drawings have on the outcome of a new custom-built home?

3. How can these documents play a role in the settling of disputes that may come up during or following the construction of a new commercial structure such as a hotel?

4. Is there a need to maintain the construction documents for future reference? Think of a case in which there was structural failure that affected people's lives.

STUDENT ACTIVITIES for

Unit 1 Wood and Lumber

1. Using a tree cross-section, identify the various parts. Explain the function of each in the growth process.
2. Identify and present eight wood defects. Explain how these defects affect the wood's use and grading.
3. Present and explain a lumber grade stamp in detail.
4. Identify and show the difference between softwood lumber that is plain-sawed and quarter-sawed.
5. Using boards, define and show the types of warp. Using boards, define and show the types of warp.
6. Calculate the quantity of board feet and cost of the following dimensional lumber:
   a. $14 - 2'' \times 10'' \times 16'$ @ \$525/MBF
   b. $84 - 1'' \times 8'' \times 14'$ @ \$754/MBF
   c. $450 - 2'' \times 4'' \times 8'$ @ \$485/MBF

Unit 2 Engineered Panels

1. Select a plywood or composite panel grade stamp. Interpret the stamp markings. Then state the possible uses for that panel in construction.
2. Select three nonstructural panels. State their manufacturing processes and uses in construction.
3. Using a residential blueprint, calculate the number of sheets of 4×8 floor, wall, and roof sheathing needed to cover the exterior of the house. (Make no deductions for openings.)

Unit 3 Engineered Lumber

1. Select a laminated veneer lumber (LVL) product. Explain the manufacturing process. Using an example from the manufacturer's product guide, explain how the sizing is determined for a floor beam product.
2. Using a manufacturer's guide and design manual (such as Boise Cascade, BCI Joist;

Louisiana Pacific, Inner Seal I, Joists 23 series, or Truss Joist MacMillan), explain how an "I" joist system can be sized for a residential floor and roof. Explain how to use the span charts with an example house plan.
3. Using the same product guide and manual, explain the process of web hole location and sizing in the "I" joist products.
4. Explain in detail the fastening systems for "I" floor and roof assemblies by using framing connectors. Give examples of proper sizing, placing, and nailing with the connectors.

Unit 4 Fasteners

1. Prepare and present a display board showing a large variety of nails used in residential and light commercial construction. Explain the names, available sizes, various surface treatments, uses, approximate cost, and in what quantities they may be purchased.
2. Prepare and present a display board showing a large variety of screws, bolts, and other construction fasteners. Include the same information as in the nail display.
3. Create a display board assortment of the various construction adhesives. Exhibit products, empty containers, product labels, literature, or samples that illustrate a large array available for use. Classify the adhesives as mastics, glues, and so forth. State their safe uses.

Unit 5 Hand Tools

1. Orally present the fractions of an inch, by 16ths (begin with 1/16, 1/8, 3/16, etc.)
2. Using a grinder and a hone, demonstrate the proper procedure for sharpening a chisel and plane iron.
3. Using a block plane, a combination square, and a piece of softwood approximately $1'' \times 4'' \times 12''$, with a 4- to 5-degree bevel on one edge, reestablish a square edge to replace the bevel.

4. Using a hammer, fasten together two pieces of 2″ × 4″ × 2′, with two 16d box nails. Set the nails flush with the surface. Do not leave any hammer marks. Then fasten a small one-inch board (fascia) to the surface of the 2″ × 4″ assembly, using 8d box nails. Leave no manner marks.

5. On a piece of 2″ × 4″ lay out a 90 degree angle across the surface and on the edge. using an 8- to 10-pt. hand saw, make a perpendicular cut through the lumber. Repeat the procedure using a 30 degree angle across the surface of the lumber. Cut to the line.

Unit 6 Portable Power Tools

The following exercises should only be attempted after proper instruction has been given by the instructor or qualified tool representative. Use safety glasses or goggles at all times.

1. Using saw horses and a portable electric saw, safely cut the following on the marked lines.

   a. Using a 6-foot 2″ × 4″, cut off four or five one-inch marked increments at 90 degrees, freehand. Then cut four or five using a saw protractor. Repeat the process, making a 45 degree compound miter across the end.

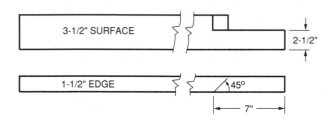

   b. Using the rip guide and protractor, make a notch in the 2″ × 4″ as shown. do not overcut.

   c. cut a notch in the lumber as shown. A hammer and chisel may be used to complete the notch.

   d. Using a piece of 3/4-inch plywood or waferboard, make a plunge cut as shown.

2. Using a portable electric jointer plane, and a 6-foot 2″ × 4″, plane a 3 degree bevel on the edge. The completed bevel should show no variations from a straight line or 3 degree angle. Check with a straightedge and a sliding-T-bevel.

3. Select any three router bits. Safely demonstrate the setup and cutting process on a piece of plywood or lumber. Relate the names, type of cut made, and where they would be used in carpentry.

4. Demonstrate the safe procedure for setting up a portable compressor and pneumatic nailer for nailing dimensional lumber and sheet materials. Include: compressor / nailer preparation, airline connections, safe nailing procedures, disconnections, and proper storage.

5. Following a demonstration and training course by a certified field representative, and personal experience, demonstrate the safe and proper use of a powder actuated driver fastening system. Explain and demonstrate how to safely fasten wood to concrete and wood to steel. Explain the powder charge color-coded system.

Unit 7 Stationary Power Tools

Upon receiving safe tool operation instruction by an instructor, complete the following.

1. Properly change the blade on a table saw and radial arm saw.
2. Using the appropriate saw (table saw, radial arm saw, or power miter box) make the following cuts: crosscut, rip, miter, compound miter, dado, plough, rabbet, bevel rip, taper rip, cove cut, duplicate length cuts.
3. Using the jointer, cut a 90 degree edge joint, end grain joint, taper cut, and 45 degree edge bevel joint.

Unit 8 Blueprints and Building Codes

Using a full set of residential blueprints and specifications (if included), a building code, architect's scale, and calculator, assemble a take-off list of construction materials for the concrete (foundation, floors, driveway, and walks), frame, exterior, and interior finish. The completed list should contain quantities and volumes of the construction materials. (This estimating exercise can be divided into various individual assignments.)

Rough Carpentry

UNIT
9

Building Layout

OBJECTIVES

After completing this unit, the student will be able to:

- establish level points across a building area using a water level and by using a carpenter's hand spirit level in combination with a straightedge.

- accurately set up and use the builder's level, transit-level, and laser level for leveling, determining and establishing elevations, and laying out angles.

- lay out building lines by using the 6-8-10 method for squaring corners and check the layout for accuracy.

- build batter boards, and accurately establish layout lines for a building using building layout instruments.

CONTENTS

Before construction begins, lines must be laid out to show the location and elevation of the building foundation. Accuracy in laying out these lines is essential in order to comply with local zoning ordinances. In addition, accurate layout lines provide for a foundation that is level and to specified dimensions. Accuracy in the beginning makes the work of the carpenter and other construction workers easier later.

CHAPTER 25

LEVELING AND LAYOUT TOOLS

Building layout requires leveling lines as well as laying out various angles over the length and width of the structure. The carpenter must be able to set up, adjust, and use a variety of leveling and layout tools.

Leveling Tools

Several tools, ranging from simple to state-of-the-art, are used to level the layout. More sophisticated leveling and layout tools, although preferred, are not always available.

HELPFUL HINTS

Make a pilot hole in the earth for straight and easy driving of stakes.

CROW BAR

DRIVE BAR REPEATEDLY INTO GROUND. WIGGLE HANDLE EACH TIME BAR IS DRIVEN.
CONTINUE UNTIL DESIRED DEPTH IS REACHED.

WRECKING BAR

GRADE

Fig. 25-1

Levels and Straightedges

If no other tools are available, a *carpenter's hand level* and a long *straightedge* may be used to level across the building area. Select an 8- to 10-foot length of lumber. Make sure it has a straight edge and is wide enough that it will not sag when placed on edge and supported only on its ends. (Use the method previously described in this book to determine if the edge is straight.)

Hold the straightedge on edge. Place one end on the surface to be leveled and the other end on the top of a stake driven in the ground so that the straightedge is level. Determine levelness by holding the spirit level against the bottom edge.

Move the straightedge and place the beginning end on the driven stake. Place the other end on another driven stake until the straightedge is again level. Stakes can be driven straight with relative ease if a pilot hole is first made in the earth (Fig. 25-1). Continue moving the straightedge from stake to stake until the desired distance is leveled (Fig. 25-2). This is an accurate, although time-consuming, method of leveling over a long distance. It can be done by one person.

If you want to level to the corners of building layouts, start by driving a stake near the center so its top is to the desired height. Level from the center stake to each corner in the manner described (Fig. 25-3).

Water Levels

A *water level* is a very accurate tool, dating back centuries. It is used for leveling from one point to another. Its accuracy, within a pencil point, is based on the principle that water seeks its own level (Fig. 25-4).

One commercial model consists of 50 feet of small diameter, clear vinyl tubing and a small tube storage container. A built-in reservoir holds the colored water that fills the tube. One end is held to the starting point. The other end is moved down until the water level is seen and marked on the surface to be leveled (Fig. 25-5).

Although highly accurate, the water level is somewhat limited by the length of the plastic tube. However, extension tubings are available. Also, though slightly inconvenient, the water level may be moved from point to point.

A water level is accurate only if both ends of the tube are open to the air and there are no air

1st POSITION

2nd POSITION

2" x 6" STRAIGHTEDGE
8'–10' LONG

HAND LEVEL

2" x 2" STAKE

Fig. 25-2 Leveling with a straightedge from stake to stake.

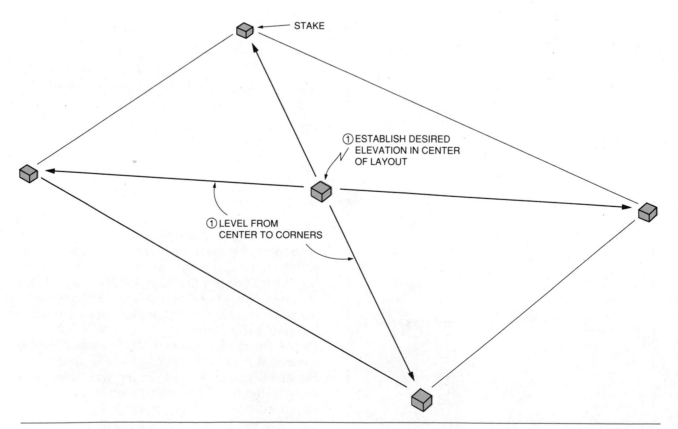

STAKE

① ESTABLISH DESIRED
ELEVATION IN CENTER
OF LAYOUT

① LEVEL FROM
CENTER TO CORNERS

Fig. 25-3 Leveling from the center of a building layout to the corners.

Fig. 25-4 Water seeks its own level. Both ends of the water level must be open to the atmosphere.

Builder's Levels

The *builder's level* (Fig. 25-6) consists of a *telescope* to which a *spirit level* is mounted. The telescope is fixed in a horizontal position. It can rotate 360 degrees for measuring horizontal angles, but cannot be tilted up or down.

Transit-levels

The *transit-level* (Fig. 25-7) is similar to the builder's level. However, its telescope can be moved up and down 45 degrees in each direction. This feature enables it to be used more effectively than the builder's level.

bubbles in the length of the tube. Because both ends must be open, there may occasionally be some loss of liquid. However, this is replenished by the reservoir. Any air bubbles can be easily seen with the use of colored water.

In spite of these drawbacks, the water level is an extremely useful, inexpensive, simple tool for leveling from room to room, where walls obstruct views, down in a hole, or around obstructions. Another advantage is that leveling can be done by one person.

Optical Levels

The most commonly used optical instruments for leveling, plumbing, and angle layout are the *builder's level* and the *transit-level*.

Fig. 25-6 The builder's level. (*Courtesy of David White*)

Fig. 25-5 The water level is a simple tool to use.

Fig. 25-7 The telescope of the transit-level may be moved up and down 45 degrees each way. (*Courtesy of David White*)

Automatic Levels

Automatic levels and *automatic transit-levels* (Fig. 25-8) are similar to those previously described except that they have an internal *compensator*. This compensator uses gravity to maintain a true level line of sight. Even if the instrument is jarred, the line of sight stays true because gravity does not change.

AUTOMATIC TRANSIT–LEVEL

AUTOMATIC BUILDER'S LEVEL

THE CORRECTION RANGE ± 15 MINUTES OF ANGLE. THIS AMOUNTS TO AN ERROR CORRECTION CAPABILITY OF UP TO 5 INCHES ON A LEVELING ROD READING AT 100 FEET.

1 Conventional instrument correctly leveled. Rod reading is 3'-3".

2 Conventional instrument slightly out of level. Vial bubble is off center and incorrect rod reading is 3'-1 1/2".

3 Automatic level-transit correctly leveled. Rod reading is 3'-3".

4 Automatic level-transit slightly out of level. Circular bubble is off center, but the compensator corrects for the variation from level and maintains a correct rod reading of 3'-3".

Fig. 25-8 Automatic levels and transit-levels. (*Courtesy of David White*)

There are many models of leveling instruments available. To become familiar with more sophisticated levels, study the manufacturers' literature. No matter what type of level is used, the basic procedures are the same.

Using Optical Levels

Before the level can be used, it must be placed on a *tripod* or some other solid support and leveled.

Setting Up and Adjusting the Level

The telescope is adjusted to a level position by means of four *leveling screws* that rest on a *base leveling plate*. In higher-quality levels, the base plate is part of the instrument. In less expensive models, the base plate is part of the tripod.

Open and adjust the legs of the tripod to a convenient height. Spread the legs of the tripod well apart, and firmly place its feet into the ground.

CAUTION On a smooth surface it is essential that the points on the feet hold without slipping. Make small holes or depressions for the tripod points to fit into. Or, insert screw eyes at the lower inside of the tripod legs and attach wire or light chain to the three screw eyes (Fig. 25-9).

When set up, the top of the tripod should be as level as possible. Sight by eye and tighten the tripod wing nuts. With the top of the tripod close to level, adjustment of the instrument is made easier.

Lift the instrument from its case by the frame (do not hold on to the telescope to do so). Note how it is stored so it can be replaced in the case in the same position. Make sure the horizontal clamp screw is loose so the telescope revolves freely. While holding onto the frame, secure the instrument to the tripod. (The attaching mechanism varies according to the manufacturer and the quality of the instrument.)

CAUTION Care must be taken not to damage the instrument. Never use force on any parts of the instrument. All moving parts turn freely and easily by hand. Excessive pressure on the leveling screws may damage the threads or the base plate. Unequal tension on the screws will cause the instrument to wobble on the base plate resulting in leveling errors. Periodically use a toothbrush dipped in light instrument oil to clean and lubricate threads of the adjusting screws.

TRIPOD

LIGHT CHAIN

SCREWEYE

Fig. 25-9 Make sure the feet of the tripod do not slip on smooth or hard surfaces.

Accurate leveling of the builder's level is important. Line up the telescope directly over two opposite leveling screws. Turn the screws in opposite directions with forefingers and thumbs. Move the thumbs toward or away from each other, as the case may be, to center the bubble in the spirit level (Fig. 25-10).

Rotate the telescope 90 degrees over the other two opposite leveling screws and repeat the procedure. Make sure each of the screws have the same, but not too much, tension. Return to the original position, check, and make minor adjustments. Continue adjustments until the bubble remains exactly centered when the instrument is revolved in a complete circle.

CAUTION Do not leave a set-up instrument unattended near moving equipment.

Fig. 25-10 A. Level the instrument by moving thumbs toward or away from each other. B. The instrument is level when the bubble remains centered as the telescope is revolved in a complete circle.

Sighting the Level

To sight an object, rotate the telescope and sight over its top, aiming it at the object. Look through the telescope. Focus it by turning the focusing knob

one way or the other, until the object becomes clear. Keep both eyes open. This eliminates squinting, does not tire the eyes, and gives the best view through the telescope (Fig. 25-11).

CAUTION If the lenses need cleaning, dust them with a soft brush or rag. Do *not* rub the dirt off. Rubbing may scratch the lens coating.

When looking into the telescope, vertical and horizontal *cross-hairs* are seen. They enable the target to be centered properly (Fig. 25-12). The cross-hairs can be brought into focus by turning the

Fig. 25-11 When sighting through the instrument, keep both eyes open. The worker is focusing the instrument.

VERTICAL CROSS-HAIR

TELESCOPE

NAIL

STAKE

HORIZONTAL CROSS-HAIR

THE CROSS-HAIRS MAY BE FOCUSED BY TURNING THE EYE PIECE

Fig. 25-12 When looking in the telescope, vertical and horizontal cross-hairs are seen.

eyepiece one way or the other. Center the cross-hairs on the object by moving the telescope left or right. A fine adjustment can be made by tightening the horizontal clamp screw and turning the horizontal tangent screw one way or the other. The horizontal cross-hair is used for reading elevations. The vertical cross-hair is used when laying out angles and aligning vertical objects.

Leveling

When the instrument is leveled, a given point on the line of sight is exactly level with any other point. Any line whose points are the same distance below or above the line of sight is also level (Fig. 25-13). To level one point with another, a helper must hold a *target* on the point to be leveled. A reading is taken. The target is then moved to selected points that are brought to the same elevation by moving those points up or down to get the same reading.

Targets. A *folding rule* is often used as a target (Fig. 25-14). The end of the rule is placed on the point to be leveled. The rule is then moved up or down until the same mark is read on the rule as was read at the starting point.

A tape may also be used as a target. Because of its flexibility, it may need to be backed up by a strip of wood to hold it rigid (Fig. 25-15). It is then used in the same manner as the folding rule.

Fig. 25-14 Leveling the top of a concrete form by sighting at a folding rule.

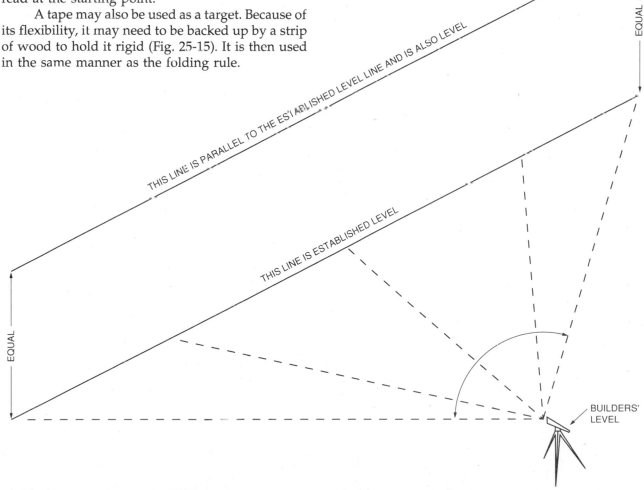

Fig. 25-13 Any line parallel to the established level line is also level.

Fig. 25-15 A tape may need to be backed up by a strip of wood if used as a target.

The simplest target is a plain 1x2 strip of wood. The end of the stick is held on the starting point of desired elevation. The line of sight is marked on the stick. The end of the stick is then placed on top of various points. They are moved up or down to bring the mark to the same height as the line of sight (Fig. 25-16). A stick of practically any length can be used. If a greater length is required, two or more strips may be fastened together.

Cut the stick to a length so that the mark to be sighted is a noticeable distance off center of its length. It is immediately noticeable if the stick is inadvertently turned upside down (Fig. 25-17). If, for some reason, it is not desirable to cut the stick, clearly mark the top and bottom ends.

Leveling Rods. For longer sightings, the *leveling* rod is used because of its clearer graduations. A variety of rods are manufactured of wood or fiber glass for several leveling purposes. They are made with two or more sections that extend easily and lock into place. Rods vary in length—from two-section rods extending 9'-0" up to seven-section rods extending 25'-0".

Fig. 25-16 Using a stick for a target.

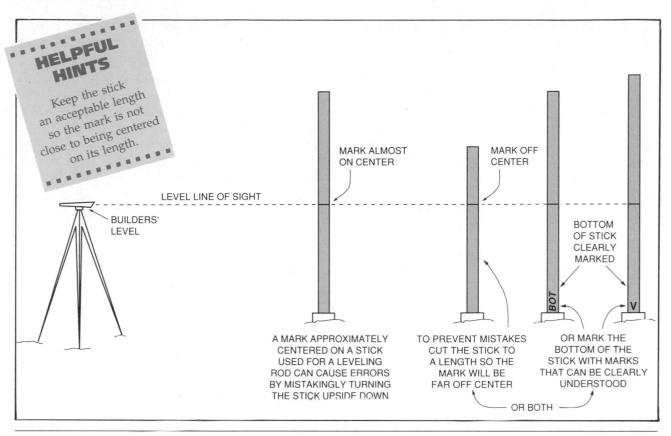

Fig. 25-17

The builder's rod has feet, inches, and 8ths of an inch. The graduations are 1/8 inch wide and 1/8 inch apart. They are designed for easy reading. An oval-shaped, red and white, movable target is available to fit on any rod for easy readings (Fig. 25-18). Other types of rods are graduated in feet, 10ths, and

Fig. 25-18 The builder's leveling rod is marked in feet, inches, and 8ths of an inch. (*Courtesy of David White*)

100ths of a foot for surveying work. Some rods have metric gradations.

Communication. A responsible rod operator holds the rod vertical and faces the instrument so it can be read with ease and accuracy. Sighting distances are not usually over 100 to 150 feet. Sometimes voice commands cannot be used. Hand signals are then given to the rod operator to move the target as desired by the instrument operator. Usually appropriate hand signals are given even when distances are not great. Shouting on the job site is unnecessary, unprofessional, and creates more confusion than it eliminates.

Establishing Elevations

Many points on the job site, such as the depth of excavations, the height of foundation footing and walls, and the elevation finish floors, are required to be set at specified elevations or grades. These elevations are established by starting from the *benchmark.* The benchmark is a point of designated elevation,

usually 100.00 feet. The instrument operator records elevations and rod readings in a notebook to make calculations.

Elevation of the Instrument (EI). When it is necessary to set a point at some definite elevation, determine the *elevation of the instrument (EI).* To find EI, place the rod on the benchmark and add the reading to the elevation of the benchmark (Fig. 25-19). For instance, if the benchmark has an elevation of 100.00 feet and the rod reads 5'-8", then the EI is 105'-8".

Grade Rod. What must be read on the rod when its base is at the desired elevation is called the *grade rod.* This is found by subtracting the desired elevation from the elevation of the instrument (EI). For instance, if the elevation to be established is 102'-0", subtract it from 105'-8" (EI) to get 3'-8" (the grade rod). The rod operator places the rod at the desired point. He or she then moves it up or down, at the direction of the instrument operator, until the grade rod of 3'-8" is read on the rod. The base of the build-

Fig. 25-19 Determining the elevation of the instrument (EI).

er's rod is then at the desired elevation (Fig. 25-20). A mark, drawn at the base of the rod on a stake or other object, establishes the elevation.

Base plates for bridges and large buildings must be accurately set to specified elevations. The concrete pier or foundation is formed slightly below grade. Four wedges, one on each corner, are placed upon the concrete foundation. The base plate is then lowered into place, resting on the wedges. The grade rod is calculated, and the rod is placed on one corner. The wedge at that corner is slowly driven until the correct reading and elevation are obtained. This procedure is repeated at the four corners and then rechecked. The space between the top of the foundation and the bottom of the base plate is then filled with concrete grout.

Determining Differences in Elevation

Differences in elevation need to be determined for such things as grading driveways, sidewalks, and parking areas, laying out drainage ditches, plotting contour lines, and estimating cut and fill requirements. The difference in elevation of two or more points is easily determined with the use of the builder's level or transit-level.

Determining Difference in Elevation Requiring Only One Setup. To find the difference in elevation between two points, set up the instrument about midway between them. Place the rod on the first point. Take a reading, and record it. This is called a *backsight*. All backsights are *plus (+) sights*.

Place the rod on the second point. Take a reading, and record it. This is called a *forward sight*. All forward sights are *minus (−) sights*.

Add the plus and minus sights to get the difference in elevation (Fig. 25-21). For any sight, the plus or minus is reversed if readings are taken with the rod upside down (Fig. 25-22).

Determining Difference in Elevation Requiring More Than One Setup. Sometimes the difference in elevation of two points is too great or the distance is too far apart. Then it is necessary to make more than one instrument setup to determine the difference. The procedure is similar to a series of one setup operation until the final point is reached.

FROM THE ELEVATION OF THE INSTRUMENT (E.I.) ⟶ 105' - 8"
SUBTRACT THE GRADE TO BE ESTABLISHED
(FINISH FLOOR GRADE, FOR INSTANCE) ⟶ 102' - 0"

TO GET THE GRADE ROD ⟶ 3' - 8"

THE "GRADE ROD" IS WHAT THE ROD MUST READ WHEN ITS BASE IS
AT THE DESIRED GRADE OR ELEVATION

Fig. 25-20 Calculating the grade rod and establishing a desired elevation.

SETUP #	PLUS SIGHT +	MINUS SIGHT –
1	5' - 6-1/2"	

NOTE BOOK

BACK SIGHT IS A PLUS SIGHT

SETUP #	PLUS SIGHT +	MINUS SIGHT –
1	5' - 6-1/2"	1' - 2-1/2"

NOTE BOOK

FORWARD SIGHT IS A MINUS SIGHT

THE DIFFERENCE BETWEEN THE PLUS SIGHT & MINUS SIGHT IS 4' - 4"
THE PLUS SIGHT IS LARGER, SO POINT "B" IS 4' - 4" HIGHER THAN POINT "A".

Fig. 25-21 Determining a difference of elevation between two points requiring only one setup.

SETUP #	PLUS SIGHT +	MINUS SIGHT –
1	4' - 6-1/2"	
	7' - 9-3/8"	

NOTE BOOK

A MINUS SIGHT BECOMES A PLUS SIGHT IF THE ROD IS READ UPSIDE DOWN.
POINT "B" IS 12' - 3-7/8" HIGHER THAN POINT "A".

Fig. 25-22 Plus and minus sightings are reversed for any readings taken with the rod upside down.

CAUTION When carrying a tripod-mounted instrument, handle with care. Carry it in an upright position. Do not carry it over the shoulder or in a horizontal position. Be careful when going through buildings or close quarters not to bump the instrument.

Record all backsights as plus sights and all forward sights as minus sights, unless the rod is upside down when read (Fig. 25-23). Find the sum of all minus sights and all plus sights. The difference between them is the difference in elevation of the beginning and ending points. If the sum of the plus sights is larger, then the end point is higher than the starting point. If the sum of the minus sights is larger, then the end point is lower than the starting point.

Measuring and Laying Out Angles

To measure or lay out angles, the instrument must be set over a point on the ground. A hook, centered below the instrument, is provided for suspending a plumb bob. The plumb bob is used to place the level directly over a particular point. In more sophisticated instruments, a built-in *optical plumb* allows the operator to sight to a point below, exactly plumb with the center of the instrument. This enables quick and accurate setups over a point (Fig. 25-24).

Setting Up over a Point. Suspend the plumb bob from the instrument. Secure it with a slip knot. Move the tripod and instrument so that the plumb bob appears to be over the point.

Press the legs of the tripod into the ground. Lower the plumb bob, by moving the slip knot, until it is about 1/4 inch above the point on the ground. The final centering of the instrument can be made by loosening any two adjacent leveling screws and slowly shifting the instrument until the plumb bob is directly over the point (Fig. 25-25). Retighten the same two leveling screws that were previously loosened, and level the instrument. Shift the instrument on the base plate until the plumb bob is directly over the point. Check the levelness of the instrument. Adjust, if necessary.

Circle Scale and Index. A *horizontal circle scale* (outside ring) is divided into 90 degree quadrants. A pointer, or *index*, turns with the telescope. The circle scale remains stationary, and indicates the number of degrees the telescope is turned. When desired, the horizontal circle may be turned by hand for setting to zero degrees, no matter which way the telescope is pointing. Starting at zero and rotating the telescope on it, any horizontal angle can be easily measured (Fig. 25-26).

Reading the Horizontal Vernier. For more precise readings, the *horizontal vernier* is used to read minutes of a degree (Fig. 25-27). The vernier is

Fig. 25-23 Determining a difference of elevation between two points requiring more than one setup.

Fig. 25-24 Some instruments have a device called an optical plumb for setting the instrument directly over a point. (*Courtesy of David White*)

HANGING THE
PLUMB BOB

TO HANG THE
PLUMB BOB,
ATTACH CORD TO
THE PLUMB BOB
HOOK ON THE
TRIPOD AND KNOT
THE CORD AS
ILLUSTRATED.

Fig. 25-25 To locate the instrument directly over a point, a plumb bob is suspended from the level.

actually two verniers, one on each side of the vernier zero index. This makes it possible to read any angle, whether turned to the right or to the left.

The vernier scale turns with the telescope. If the eye end of the telescope is turned to the left (clockwise), the vernier scale on the left side of the zero index is used. If the eye end of the telescope is turned to the right (counterclockwise), the vernier scale on the right side of the zero index is used. Use either the left or right vernier scale according to the direction in which the eye end of the telescope is turned when measuring or laying out angles (Fig. 25-28).

If the circle index zero lines up between degree marks on the circle scale, notice the last degree mark passed. On the vernier scale, locate a vernier mark that coincides with a circle mark. Add the reading in minutes to the degrees noted on the circle scale (Fig. 25-29). The carpenter is not usually concerned with any finer readings.

THE HORIZONTAL CIRCLE SCALE IS DIVIDED INTO QUADRANTS OF 90° EACH, AND REMAINS STATIONARY AS THE TELESCOPE IS TURNED.

IT IS GRADUATED IN DEGREES AND NUMBERED EVERY 10 DEGREES.

IT MAY BE ROTATED BY HAND.

THE HORIZONTAL CIRCLE SCALE

THE ZERO INDEX

THE VERNIER SCALE

QUADRANT

75 DEGREES

THE HORIZONTAL CIRCLE SCALE READS 75 DEGREES

Fig. 25-26 Reading the horizontal circle scale.

THE VERNIER ROTATES WITH THE TELESCOPE

VERNIER

ZERO INDEX

HORIZONTAL CIRCLE SCALE

THE VERNIER IS GRADUATED AND NUMBERED INTO 15 MINUTE INTERVALS AND IS ACTUALLY A DOUBLE VERNIER - ONE ON EACH SIDE OF THE ZERO INDEX

Fig. 25-27 The horizontal vernier scale is used for reading minutes of a degree.

Measuring a Horizontal Angle. After leveling the instrument over the point of an angle, called its *vertex*, loosen the horizontal clamp screw. Rotate the instrument until the vertical cross-hair is nearly in line with a distant point on one side of the angle. Tighten the clamp screw. Then turn the tangent screw to line up the vertical cross-hair exactly with the point.

If the point is above or below the line of sight, and a transit-level is not available, sight a straight-edge held plumb from the point. Or, sight the line of a plumb bob line suspended over the point (Fig. 25-30). If using a transit-level, release the locking lever and tilt the telescope to sight the point.

By hand, turn the horizontal circle scale to zero. Loosen the clamp screw. Swing the telescope until the vertical cross-hair lines up with a point on

Fig. 25-28 Method of determining which vernier scale to read.

75 DEGREES, 45 MINUTES

Fig. 25-29 Add the reading in minutes from the vernier scale to the reading in degrees from the circle scale.

the other side of the angle. Tighten the horizontal clamp screw. Then turn the tangent screw for a fine adjustment, if necessary. Read the degrees on the circle scale and minutes on the vernier scale (Fig. 25-31).

Laying Out a Horizontal Angle. Center and level the instrument over the vertex of the angle to be laid out. Sight the telescope on a distant point on

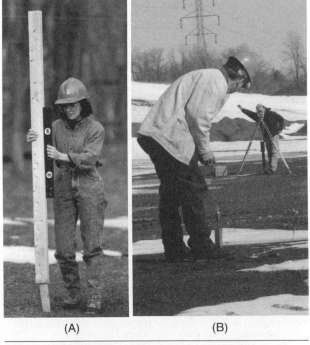

(A)   (B)

Fig. 25-30 Locating a point below the line of sight.
a. Using a hand level and straightedge
b. Using a plumb bob

① TO MEASURE ANGLE ABC, SET UP THE TRANSIT-LEVEL OVER POINT B AND SIGHT TO POINT A. TIGHTEN HORIZONTAL CLAMP SCREW. BY HAND, SET THE HORIZONTAL CIRCLE SCALE TO ZERO.

② LOOSEN THE HORIZONTAL CLAMP SCREW AND ROTATE THE TELESCOPE TO SIGHT POINT C. TIGHTEN THE HORIZONTAL CLAMP SCREW. USE THE TANGENT SCREW, IF NECESSARY. READ THE NUMBER OF DEGREES ON THE CIRCLE SCALE. USE LEFT VERNIER, BECAUSE EYEPIECE WAS TURNED TO THE LEFT, TO GET MINUTES OF A DEGREE.

Fig. 25-31 Measuring an angle.

one side of the angle. Tighten the horizontal clamp, and set the circle scale to zero. Loosen the clamp. Then turn the telescope until close to the desired number of degrees. Tighten the clamp. Use the tan-gent screw and make a fine adjustment until the index reads exactly the desired number of degrees. Sight the vertical cross-hair to lay out the other side of the angle (Fig. 25-32).

① TO LAY OUT A 90° ANGLE FROM LINE EF, SET UP THE TRANSIT-LEVEL OVER POINT E AND SIGHT TO POINT F. TIGHTEN THE HORIZONTALCLAMP SCREW. TURN THE HORIZONTAL CIRCLE SCALE TO ZERO.

② LOOSEN THE HORIZONTAL CLAMP SCREW AND ROTATE THE TELESCOPE UNTIL 90° IS READ ON THE HORIZONTAL CIRCLE SCALE. TIGHTEN THE HORIZONTAL CLAMP SCREW. DRIVE A STAKE WITH A NAIL CENTERED IN ITS END TO SIGHT POINT D. AN ANGLE OF 90° IS LAID OUT.

Fig. 25-32 Laying out an angle.

Special Uses of the Transit-level

The transit-level can be used efficiently for all the operations described previously in this chapter. In addition, it is a valuable tool for such things as measuring vertical angles, setting points in a line, and plumbing posts, columns, walls, and similar objects.

Measuring Vertical Angles. The *vertical arc* scale is attached to the telescope. It measures vertical angles, to 45 degrees, above and below the horizontal. By tilting and rotating the telescope, set the horizontal cross-hair on the points of the vertical angle being measured. Tighten the vertical clamp. Then turn the tangent screw for a fine adjustment to

place the cross-hair exactly on the point. Vertical angles are read by means of the vertical arc scale and the obvious vernier similar to the reading of horizontal angles (Fig. 25-33).

Setting Points in a Straight Line. It may be necessary to set points in a straight line, such as on a property boundary line. Set up the instrument over the one point. Rotate and tilt the telescope to sight the other point fairly close. Tighten the horizontal clamp. Then turn the tangent screw until the vertical cross-hair is exactly on the far point. With the horizontal clamp tight, release the lock levers and depress the telescope, as required, to sight points between the corners and along the boundary line (Fig. 25-34).

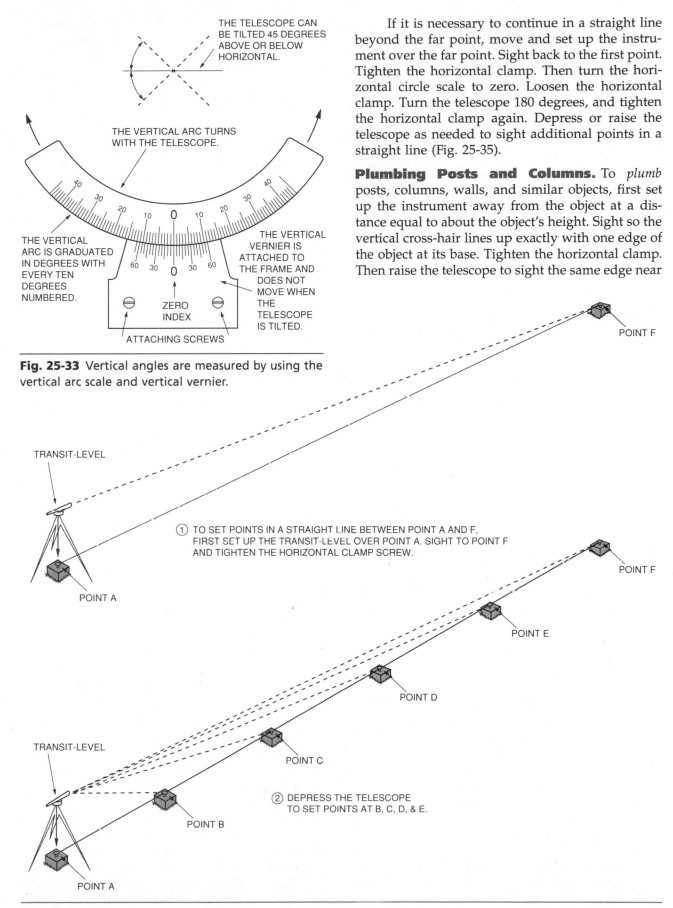

If it is necessary to continue in a straight line beyond the far point, move and set up the instrument over the far point. Sight back to the first point. Tighten the horizontal clamp. Then turn the horizontal circle scale to zero. Loosen the horizontal clamp. Turn the telescope 180 degrees, and tighten the horizontal clamp again. Depress or raise the telescope as needed to sight additional points in a straight line (Fig. 25-35).

Plumbing Posts and Columns. To *plumb* posts, columns, walls, and similar objects, first set up the instrument away from the object at a distance equal to about the object's height. Sight so the vertical cross-hair lines up exactly with one edge of the object at its base. Tighten the horizontal clamp. Then raise the telescope to sight the same edge near

THE TELESCOPE CAN BE TILTED 45 DEGREES ABOVE OR BELOW HORIZONTAL.

THE VERTICAL ARC TURNS WITH THE TELESCOPE.

THE VERTICAL ARC IS GRADUATED IN DEGREES WITH EVERY TEN DEGREES NUMBERED.

THE VERTICAL VERNIER IS ATTACHED TO THE FRAME AND DOES NOT MOVE WHEN THE TELESCOPE IS TILTED.

ZERO INDEX

ATTACHING SCREWS

Fig. 25-33 Vertical angles are measured by using the vertical arc scale and vertical vernier.

TRANSIT-LEVEL

POINT F

POINT A

① TO SET POINTS IN A STRAIGHT LINE BETWEEN POINT A AND F, FIRST SET UP THE TRANSIT-LEVEL OVER POINT A. SIGHT TO POINT F AND TIGHTEN THE HORIZONTAL CLAMP SCREW.

POINT F

POINT E

POINT D

POINT C

TRANSIT-LEVEL

POINT B

② DEPRESS THE TELESCOPE TO SET POINTS AT B, C, D, & E.

POINT A

Fig. 25-34 Setting points in a straight line.

③ TURN 180° & SIGHT POINT L

① SET UP OVER POINT F

POINT F

② SIGHT BACK TO POINT A

POINT L

K

J

I

H

G

E

D

C

B

POINT A

TO CONTINUE POINTS IN A STRAIGHT LINE:

① SET TRANSET LEVEL OVER POINT F. THE FARTHEST POINT SIGHTED FROM POINT A.

② SIGHT BACK TO POINT A. - TIGHTEN HORIZONTAL CLAMP. TURN CIRCLE SCALE TO ZERO.

③ LOOSEN HORIZONTAL CLAMP AND TURN TELESCOPE 180 DEGREES TO SIGHT POINT L. DEPRESS TELESCOPE TO SIGHT INTERMEDIATE POINTS G, H, I, J, & K.

Fig. 25-35 Continuing a straight line.

the top of the object. If plumb, the same edge will line up exactly as it did at the base. If it does not line up, move the top of the object until it does and brace in position.

To complete the plumbing process, move and set the instrument up at a position about 90 degrees from the first position. Then repeat the procedure (Fig. 25-36).

Storing the Instrument. If the instrument gets wet, dry it before returning it to its case. Keep it in its carrying case when it is not being used or when being transported in a vehicle over long distances.

Laser Levels

A *laser* is a device in which energy is released in a narrow beam of light. The light beam is absolutely straight. Unless interrupted by an obstruction or otherwise disturbed, the light beam can be seen for a long distance.

Lasers have many applications in space, medicine, agriculture, and engineering. The *laser level* has been developed for the construction industry to provide more efficient layout work (Fig. 25-37).

Kinds and Uses of Laser Levels

Several manufacturers make laser levels in a number of different models. The least expensive models are the least sophisticated. A low-price unit is leveled and adjusted manually. More expensive ones are automatically adjusted to and maintained in level. Power sources range from four C-cell batteries, a detachable, rechargeable battery pack, a 12-volt battery, or a AC/DC converter for 110 or 220 volts.

Establishing and Determining Elevations. A simple, easy-to-use model is mounted on a tripod or solid flat surface and leveled like manual or automatic optical instruments. The laser is turned on. It will emit a red beam, usually 3/8 inch in diameter. The beam rotates through a full 360 degrees, creating a level *plane* of light. As it rotates, it establishes equal points of elevation over the entire job site, similar to a line of sight being rotated by the telescope of an optical instrument (Fig. 25-38).

Depending on the quality of the instrument, the laser head may rotate at various revolutions per second (RPS), up to 40 RPS. Its quality also deter-

① SET UP TRANSIT-LEVEL AT "A". SIGHT OUTSIDE EDGE OF POST AT BOTTOM. TIGHTEN HORIZONTAL CLAMP. USE TANGENT SCREW FOR FINE ADJUSTMENT, IF NECESSARY.

② RAISE LINE OF SIGHT TO TOP OF POST. MOVE POST TO BRING OUTSIDE EDGE TO LINE OF SIGHT. BRACE POST.

③ MOVE INSTRUMENT 90° TO POSITION "B". REPEAT PROCEDURE.

Fig. 25-36 Plumbing a post.

mines its working range. This may vary from a 75- to 1,000-foot radius.

Some laser beams cannot be seen by the human eye. Those that can are difficult to see outdoors

Fig. 25-37 Laser levels have been developed for use in the construction industry. (*Courtesy of Laser Alignment, Inc.*) (*Courtesy of Spectra-Physics*)

Fig. 25-38 The laser beam rotates 360 degrees, creating a level plane of light. (*Courtesy of Laser Alignment, Inc.*)

in bright sunlight. To detect the beam, a battery-powered electronic *sensor* target, also called a *receiver* or *detector,* is attached to the leveling rod or stick. Most sensors have a visual display with a selectable audio to indicate when it is close to or on the beam (Fig. 25-39). In addition to electronic sensor targets, specially designed targets are used for interior work, such as installing ceiling grids and leveling computer floors.

The procedures for establishing and determining elevations with laser levels are similar to those with optical instruments. To establish elevations, the sensor is attached to the grade rod. The grade rod is moved up or down until the beam indicates that the base of the rod is at grade.

To determine elevations, the base of the rod is held to the elevation to be determined. The sensor is moved up or down on the rod until the display indicates it is centered on the beam. The reading from the rod is then recorded.

One of the great advantages of using laser levels is that, in most cases, only one person is needed to do the operations (Fig. 25-40). Another advantage is that certain operations are accomplished more easily in less time.

Fig. 25-40 When using laser levels, only one person is required for leveling operations. (*Courtesy of Laser Alignment, Inc.*)

Selectable Audio provides tones to match visual display.

▼ **HIGH** .... FAST BEEPING

ON-GRADE ...... SOLID TONE

▲ **LOW** .... SLOW BEEPING

Fig. 25-39 An electronic target senses the laser beam. An audio provides tones to match the visual display. (*Courtesty of Laser Alignment, Inc.*)

Special Horizontal Operations. For leveling *suspended ceiling grids,* the laser level is mounted to an adjustable grid mount bracket. Once the first strip of angle trim has been installed, the laser and bracket can easily be attached and clamped into place.

The unit is leveled. The height of the laser beam is then adjusted for use with a special type magnetic or clip-on target. Magnetic targets are used on steel grids. Clip-on targets are used on aluminum or steel. Once the laser is set up, the ceiling grid is quickly and easily leveled using the rotating beam as a reference and viewing the beam through the target (Fig. 25-41). Sprinkler heads, ceiling outlets, and similar objects can be set in a similar manner.

For leveling *raised access computer floors,* the laser unit is mounted on a *mini tripod* and leveled.

Fig. 25-41 Leveling suspended ceiling grids. The beam is viewed through the target. (*Courtesy of Spectra-Physics*)

Adjust the special computer floor target to the elevation of the beam by placing it on a floor pedestal set at the desired elevation. Move the target to each floor pedestal and set to the correct height indicated when the beam hits the target (Fig. 25-42). In addition to precise leveling jobs, like computer floors, European-style home and office cabinets and similar objects can also be leveled.

Special Vertical Operations. Most laser levels are designed to work laying on their side using special brackets or feet. The unit is placed on the floor and leveled. To *lay out a partition*, rotate the head of the laser downward. Position the laser beam over one end of the partition. Align the beam toward the far point by turning the head toward

Fig. 25-42 Leveling pedestals for a computer floor. (*Courtesy of Spectra-Physics*)

and adjusting to the second point. Recheck the alignment and turn on to rotate the beam. The beam will be displayed as continuous straight and plumb lines on the floor, ceiling, and walls to both align and plumb the partition at the same time (Fig. 25-43). Mounted on its side on a tripod, the laser can be used for aligning and plumbing tilt-ups, curtain walls, and a variety of similar work (Fig. 25-44).

Fig. 25-43 Laid on its side, the laser level is used to lay out and align partitions and walls. (*Courtesy of Laser Alignment, Inc.*)

Fig. 25-44 On exterior applications, the laser unit is mounted on its side on a tripod to plumb and align formwork and similar objects. (*Courtesy of Laser Alignment, Inc.*)

Layout Operations. Some laser levels emit a plumb line of light projecting upward from the top at a right angle (90 degrees) to the plane of the rotating beam. The plumb reference beam allows one person to lay out 90 degree cross-walls or building lines. Set the unit on its side as for laying out a partition. The reference beam establishes a 90 degree corner with the rotating laser beam (Fig. 25-45).

Plumbing Operations. The vertical beam provides a ready reference for plumbing posts, columns, elevator shafts, slip forming, and wherever a plumb reference beam is required.

Mount the laser unit on a tripod over a point of known offset from the work to be plumbed. Suspend a plumb bob from the center of the tripod directly over the point. Apply power to the unit. The vertical beam that is projected is ready for use as a reference. Move the top of the object until it is offset from the beam the same distance as the bottom point.

The heads on some laser units can be tilted so the rotating beam produces a plane of light at an angle to the horizontal. Such units are used for laying out slopes. Other units are manufactured for special purposes such as pipelaying and tunnel guidance. Marine laser units, with ranges up to 10 miles, are used in port and pier construction and offshore work.

Laser Safety

With a little common sense, the laser can be used safely. All laser instruments are required to have warning labels attached (Fig. 25-46). The following are safety precautions for laser use:

- Only trained persons should set up and operate laser instruments.
- Never stare directly into the laser beam or view it with optical instruments.
- When possible, set the laser up so it is above or below eye level.
- Turn the laser off when not in use.
- Do not point the laser at others.

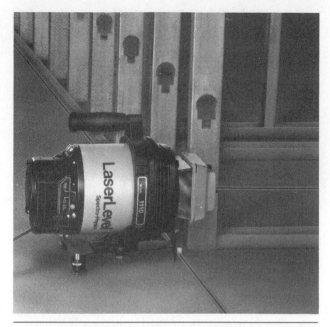

Fig. 25-45 A reference beam at right angles to the plane of the rotating beam is used for 90 degree layout work. (*Courtesy of Spectra-Physics*)

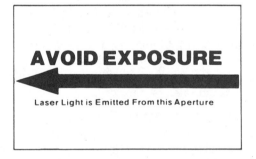

AVOID EXPOSURE

Laser Light is Emitted From this Aperture

APERTURE LABEL

DANGER

LASER LIGHT - AVOID DIRECT EYE EXPOSURE.

HELIUM NEON LASER
5 MW MAXIMUM OUTPUT
CLASS III a. LASER PRODUCT

WARNING LABEL

Fig. 25-46 Warning labels must be attached to every laser instrument. (*Courtesy of Laser Alignment, Inc.*)

CHAPTER 26

LAYING OUT FOUNDATION LINES

Laying Out Foundation Lines

Before any layout can be made, the carpenter must determine the dimensions of the building and its location on the site from the plot plan. It is usually the carpenter's responsibility to lay out building lines.

Staking the Building

Find the survey rods that mark the corners of the property. Do not guess where the property lines are. Stretch and secure lines between each corner, laying out all the property boundary lines.

Measure in on each side from the front property line the specified front setback. Drive stakes on each end. Stretch a line between the stakes to mark the front line of the building (Fig. 26-1).

Measure back from the front building line the width of the building. Drive stakes. Then lay out the rear building line in a manner similar to that used when laying out the front building line (Fig. 26-2).

Along the front building line, measure in from the side property line the specified side setback of the building. Drive a stake. Drive a nail in the top of the stake to the exact side setback. From this nail, measure the dimension of the building along the front building line. Drive another stake. Drive a nail in the top of the stake marking the exact length of the building (Fig. 26-3).

Set up a level, for 90 degree layout, over the nail in one of the stakes marking the front corner of the building. A transit-level or a laser level with a reference beam is the preferred choice. Sight along the front property line to the nailhead in the stake of the opposite front corner.

Swing the telescope 90 degrees. Sight, locate, and drive a stake for the rear corner of the building. Drive a nail in the top of the stake marking the rear corner exactly (Fig. 26-4).

After squaring the corner, locate the stake for the other corner of the building by measuring along the established rear building line the same distance

STEP BY STEP
PROCEDURES

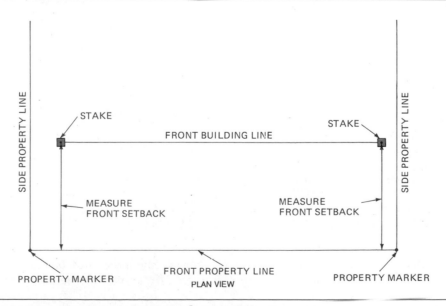

Fig. 26-1 Step 1—Locate, stretch, and secure the front building line.

. . . .

STEP BY STEP
P R O C E D U R E S

(continued)

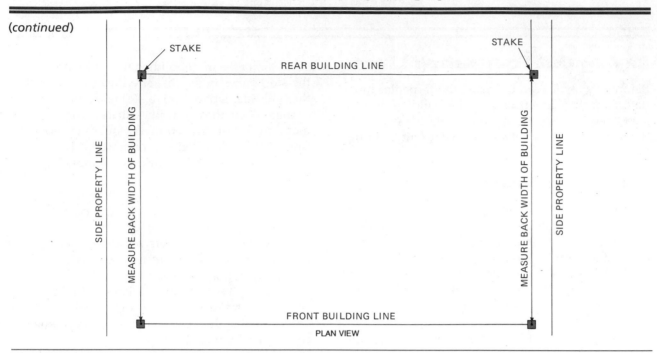

Fig. 26-2 Step 2—Locate, stretch, and secure the rear building line.

Fig. 26-3 Step 3—Drive stakes to lay out the front corners of the building.

as in the front. Locate a stake at that point. Drive a nail in its top to mark exactly the other rear corner (Fig. 26-5).

Check the accuracy of the work by measuring diagonally from corner to corner. The diagonal measurements should be the same. Make adjustments if the measurements differ (Fig. 26-6).

Laying Out a Square Corner by the 6-8-10 Method. A right angle corner may be laid out

STEP BY STEP
P R O C E D U R E S

Fig. 26-4 Step 4—Lay out a 90 degree angle to locate a rear corner stake.

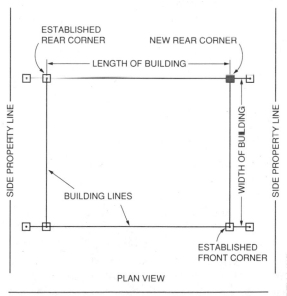

Fig. 26-5 Step 5—Measure from established front and rear corners to locate the other rear corner of the building.

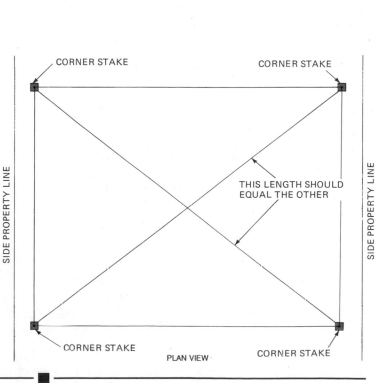

Fig. 26-6 Step 6—If the diagonal measurements of a rectangular layout are equal, the corners are square.

using the 6-8-10 method in the absence of a builder's level or transit-level. Measure 6 feet from the corner stake along the front building line. Mark the line by pushing a small brad through it. Measure and mark 8 feet on the side building line. Move the rear end of the side line until the distance between the marks is 10 feet. The corner is then square (Fig. 26-7). Secure the side layout line.

This method is based on the fact that any triangle with sides of 3, 4, and 5, or multiples thereof, is a right triangle. For example, a triangle with sides of 12, 16, and 20 is also a right triangle. Carpenters use measurements of 6, 8, and 10 feet because they are more convenient. Carpenters use this method of laying out a square corner for various installations in the construction.

When taking measurements over layout lines, steady the tape by holding it against a solid object such as a rock, another stake, or some part of the body. Take readings from the one- or 2-foot mark. Remember to add that amount to the reading on the other end (Fig. 26-8).

All measurements must be made on the level. If the land slopes, the tape is held level with a plumb bob suspended from it (Fig. 26-9).

Irregular-shaped buildings are laid out using the fundamental principles outlined above. All cor-

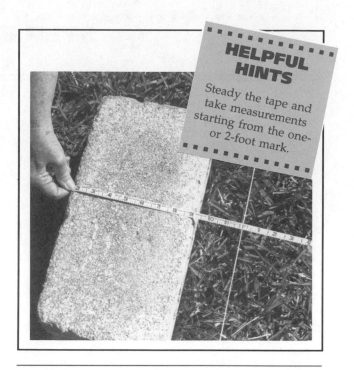

HELPFUL HINTS

Steady the tape and take measurements starting from the one- or 2-foot mark.

Fig. 26-8

ner stakes are located by measuring from the established front and side building lines (Fig. 26-10).

Stakes, other than those used for the front corners, may have to be moved slightly for accuracy. Also, staking the building is a temporary aid for erecting *batter boards*. Consequently, mark the corners by driving a long spike into the ground through a small square of thin cardboard (Fig. 26-11).

Erecting Batter Boards

Batter boards are wood frames to which building layout lines are secured. Batter boards consist of horizontal members, usually 1×6, called *ledgers*. These are attached to 2×4 stakes driven into the ground. The ledgers are fastened in a level position to the stakes, usually at the same height as the foundation wall (Fig. 26-12). Batter boards are built in the same way for both residential or commercial construction.

Fig. 26-7 Laying out a square corner by the 6-8-10 method.

Fig. 26-9 For building layouts on sloping land, measurements must be taken on the level.

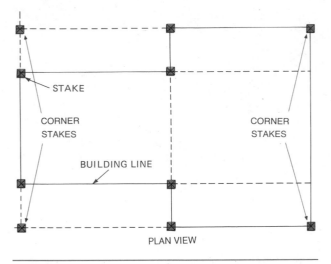

Fig. 26-10 Locate and drive corner stakes first, then intermediate ones, when laying out an irregular building.

Batter boards are erected in such a manner that they will not be disturbed during excavation. Drive 2×4 stakes into the ground a minimum of 4 feet outside the building lines at each corner to construct right angle batter boards. When setting batter boards for large construction, make sure to leave room for heavy excavating equipment to operate without disturbing them. Straight batter boards are used for long setbacks. In loose soil or when stakes are higher than 3 feet, they must be braced (Fig. 26-13).

Set up the builder's level about center on the building location. Sight to the benchmark, and record the sighting. Determine the difference between the benchmark sighting and the height of the ledgers. Sight and mark each corner stake at the specified elevation. Attach ledgers to the stakes so that the top edge of each ledger is on the mark. Brace the batter boards for strength, if necessary.

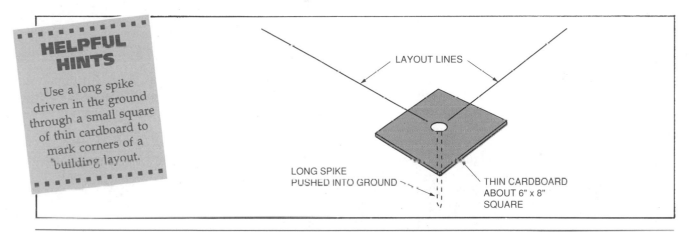

HELPFUL HINTS

Use a long spike driven in the ground through a small square of thin cardboard to mark corners of a building layout.

Fig. 26-11

Fig. 26-12 Batter boards are erected for a large commercial building.

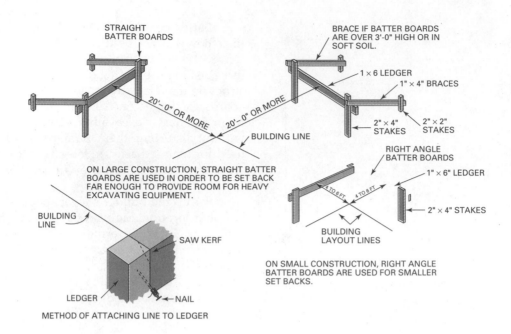

Fig. 26-13 Batter boards are placed back far enough so they will not be disturbed during excavation operations.

Stretch lines between batter boards directly over the nailheads in the original corner stakes. Locate the position of the lines by suspending a plumb bob directly over the nailheads. When the lines are accurately located, make a saw cut on the outside corner of the top edge of the ledger. This prevents the layout lines from moving when stretched and secured. Be careful not to make the saw cut below the top edge (Fig. 26-14).

Check the accuracy of the layout by again measuring the diagonals to see if they are equal. If not, make the necessary adjustment until they are equal (Fig. 26-15).

Saw cuts are also often made on batter boards to mark the location of the foundation footing. The footing width usually extends outside and inside of the foundation wall.

Fig. 26-14 Typical batter board construction and method of locating layout lines on batter boards by suspending a plumb bob directly over corner stakes.

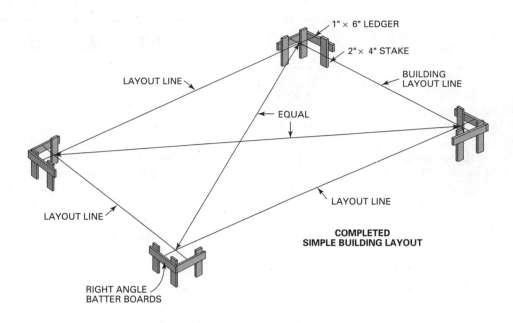

Fig. 26-15 Check the diagonals for accuracy of the building layout.

Review Questions

Select the most appropriate answer.

1. The location of the building on the lot is determined from the
   a. foundation plan.   c. architect.
   b. floor plan.   d. plot plan.

2. The builder's level is ordinarily used for
   a. laying out straight lines.
   b. reading elevations.
   c. plumbing walls.
   d. all of these.

3. When sighting objects through the telescope of the builder's level
   a. use a sun shade.
   b. keep both eyes open.
   c. close one eye.
   d. have the cross-hairs centered on the object.

4. The horizontal cross-hair is used for
   a. laying out angles.
   b. laying out straight lines.
   c. plumbing walls and posts.
   d. reading elevations.

5. The graduations of a leveling rod are
   a. 1/8 inch wide and 1/8 inch apart.
   b. 1/4 inch wide and 1/4 inch apart.
   c. 1/16 inch wide and 1/16 inch apart.
   d. 1/8 inch wide and 1/4 inch apart.

6. If the rod reading at point A is 6'-0" and the reading at point B is 3'-6" then
   a. the elevation at point B is 9'-6".
   b. point A is higher than point B.
   c. point A has an elevation of 2'-4".
   d. point B is higher than point A.

7. The reference for establishing elevations on a construction site is called
   a. starting point.   c. benchmark.
   b. reference point.   d. sight mark.

8. Starting from point A, backsight rod readings are 36", 48", and 59". Forward sight rod readings to point B are 28", 42", and 60".
   a. point A is 13" higher than point B.
   b. point A is 13" lower than point B.
   c. point A is 23" higher than point B.
   d. point A is 23" lower than Point B.

9. To read minutes of a degree, when laying out horizontal angles with the builder's level, use the
   a. base plate.   c. circle scale.
   b. index.   d. vernier scale.

10. A right-angle triangle has sides of
   a. 2, 4, and 6 feet.
   b. 3, 5, and 7 feet.
   c. 6, 8, and 12 feet.
   d. 12, 16, and 20 feet.

■ BUILDING FOR SUCCESS ■

Building Site Layout

Accurate building site layout requires placing the structure in the correct location and being in compliance with all local zoning regulations. The establishment of any required elevations will be determined in the layout process. An adequate working knowledge of site layout, instrument use, calculations, site plan reading, and local zoning ordinances should be possessed by the carpenter.

This phase of the construction process calls for workers to use their acquired skills in math, vernier and elevation reading, staking, and batter board construction. The layout must be squared. Then batter boards must be marked for building lines, footing lines, and excavation lines. If the process is followed as explained in Unit 9, the results will be accurate.

Site layout can pose many challenges for the carpenter. As site plans are read and interpreted, the layout person will be solving math problems in the process. Once the site layout is underway, it is wise to recheck the measurements. This will minimize errors. This is time to check for any easements that have been established by the local authorities. With each additional site layout, the need for extensive double checking will diminish.

Site layout is most convenient with two or three people. One person working alone must improvise throughout the process for accurate results. The site layout process takes on the characteristics of surveying and calls for a team approach to obtain the best results. Accuracy in the layout process also depends on how well the equipment is maintained and transported. Abused instruments will not provide accurate readings. Care should be taken to wipe or clean the instruments, leveling rods, tapes, tripods, and other tools after each layout is complete. The storing and transporting of the equipment should be done with care.

Weather variations will call for a variety of methods to be used during the site layout process. For example, steel stakes may be more appropriate as batter boards when the ground is frozen. Excessive winds may affect the string lines and plumb bobs. Differing temperatures and climatic conditions may af-

fect the layout instruments. A warm builder's level or transit-level may have to adjust to the cold temperatures in the winter. Expansions and contractions in the metal housing can cause instruments to give incorrect readings. Rain, snow, and wind will interfere with the layout process. Rain may cause delays that last for days. Instruments should be covered if exposed to rain during use. Snow may have to be removed from the site in order to expose the soil. Wind may cause string lines to vibrate extensively.

Some of the more important aspects of accurate building site layout are:

1. Follow the site plan precisely. Confirm all measurements, setbacks, utility placements, and easements.
2. Properly interpret all local zoning ordinances. Ask local authorities for clarification.
3. Rely on the most recent survey to locate the property lines. If any property lines cannot be located, consult the surveyor.
4. Be sure the builder's level or transit-level is not damaged. This could give inaccurate readings.
5. Recheck measurements, elevation readings, and calculations as needed to confirm accuracy.
6. Construct durable batter boards and notch all necessary lines (excavation, footing, foundation), for future reference.

As experience is gained in site layout, confidence is increased. That confidence will in turn, allow each person to reach a high level of accuracy and efficiency.

FOCUS QUESTIONS: For individual or group discussion.

1. What might be some reasons for accurate site layout?
2. Once the site layout process is understood and practiced, what would be a good procedure to follow?
3. What could be some of the results of an inaccurate building site layout? How could they affect the preestablished time schedule and goals of the project?

Concrete Form Construction

The construction of concrete forms is the responsibility of the carpenter. The forms must meet specified dimensions and be strong enough to withstand tremendous pressure. If a form fails to hold, costly labor and materials are lost. Time must be spent in dismantling and cleanup. The formwork must be erected all over again. Cleanup is difficult because the concrete has probably set up by that time. If reinforcing rods are used in the concrete, the cleanup and rebuilding are even more costly. The carpenter must know that the forms will hold before the concrete is placed.

OBJECTIVES

After completing this unit, the student will be able to:

■ describe the composition of concrete and factors affecting its strength, durability, and workability.

■ explain the reasons for reinforcing concrete and describe the materials used.

■ job-mix a batch of concrete and explain the method of and reasons for making a slump test.

■ explain techniques used for the proper placement and curing of concrete.

■ construct forms for footings, slabs, walks, and driveways.

■ construct concrete forms for foundation walls.

■ construct concrete column forms.

■ lay out and build concrete forms for stairs.

■ estimate quantities of concrete.

CONTENTS

CHAPTER 27 CHARACTERISTICS OF CONCRETE

Understanding the characteristics of concrete is essential for the construction of reliable concrete forms, the correct handling of freshly mixed material, and the final quality of hardened concrete.

Concrete

Concrete is the most widely used of all building materials. It can be formed into practically any shape for construction of buildings, bridges, dams, and roads. Improvements over the years have created a product that is strong, durable, and versatile (Fig. 27-1).

Composition of Concrete

Concrete is a mixture of *portland cement*, fine and coarse *aggregate*, water, and various *admixtures*. Aggregates are fillers, usually sand, gravel, or stone. Admixtures are materials added to the mix to achieve certain desired qualities.

Portland Cement. In 1824 Joseph Aspdin, an Englishman, developed an improved type of cement. He named it *portland cement* because it produced a concrete that resembled stone found on the Isle of Portland, England. Portland cement is a fine gray powder. When mixed with water it forms a paste. A chemical reaction, called *hydration,* takes places that causes the cement to first set. It then hardens to bind the aggregates together into a strong, durable, and watertight mass called concrete.

When moisture is present, hydration and, consequently, hardening of the concrete, continues for many years. Rapid evaporation or freezing of the water in concrete can cause the cement hydration process and the hardening to stop.

Portland cement is usually sold in paper bags that hold one cubic foot and weigh 94 pounds each. It can still be used even after a long period of time as long as it remains dry. Eventually, however, it will absorb moisture from the air. It should not be used if it contains lumps that cannot be broken up easily.

Types of Portland Cement. New developments have produced several types of portland cement. Type I is the familiar gray kind most generally purchased and used. Type IA is an *air-entraining* cement that is more resistant to freezing and thawing. Types II through V have special uses, such as for low or high heat generation, high early strength, and resistance to severe frost. White portland cement, used for decorative purposes, differs only in color from gray cement. There are many other types, each with its specific purpose—ranging from underwater use to sealing oil wells.

Water. Any water used for drinking is suitable for use to make concrete. Other water may be used, but should be tested first to make sure it is acceptable.

The amount of water used to make concrete in relation to the amount of portland cement is called the *water-cement ratio.* The amount of water used

Fig. 27-1 Concrete is the most widely used material in the construction industry.

largely determines the quality of the concrete. A common practice of adding more water to a mix with the proper ratio to make it more fluid and to allow it to flow easier alters the water-cement ratio and weakens the concrete.

Aggregates. *Aggregates* have no cementing value of their own. They serve only as a filler, but are an important ingredient. They constitute from 60 to 80 percent of the concrete volume.

Fine aggregate consists of particles 1/4 inch or less in diameter. Sand is the most commonly used fine aggregate. *Coarse* aggregate, usually gravel or crushed stone, comes usually in sizes of 3/8, 1/2, 3/4, 1, 1 1/2, and 2 inches. When strength is the only consideration, 3/4 inch is the optimum size for most aggregates. Large-size aggregate uses less water and cement. Therefore, it is more economical. However, the maximum aggregate size must not exceed the following:

- 1/5 the smallest dimension of the unreinforced concrete.

- 3/4 the clear spacing between reinforcing bars or between the bars and the form.

- 1/3 the depth of unreinforced slabs on the ground (Fig. 27-2).

Fine and coarse particles must be in a proportion that allows the finer particles to fill the spaces between the larger particles. The aggregate should be clean and free of dust, loam, clay, or vegetable matter.

Air. An important advance was made with the development of *air-entrained* concrete. It is produced by using air-entraining portland cement or an admixture. The intentionally made air bubbles are very small. Billions of them are contained in a cubic yard of concrete.

The introduction of air into concrete was designed to increase its resistance to freezing and thawing. It also has other benefits: less water and sand are required, the workability is improved, separation of the water from the paste is reduced, and the concrete can be finished sooner and is more watertight than ordinary concrete. Air-entrained concrete is now recommended for almost all concrete projects.

Admixtures. Admixtures are available to quicken or retard setting time, develop early strength, inhibit corrosion, retard moisture, control bacteria and fungus, improve pumping, and color concrete, among many other purposes.

Mixing Concrete

Concrete mixtures are designed to achieve the desired qualities of strength, durability, and workability in the most economical manner. The mix will vary according to the strength and other desired qualities, plus other factors such as the method and time of curing.

For very large jobs, such as bridges and dams, concrete mixing plants are built on the job site. This

Fig. 27-2 The maximum aggregate size allowed in the concrete mixture.

is because freshly mixed concrete is needed on a round-the-clock basis. The construction engineer calculates the proportions of the mix and supervises tests of the concrete.

Ready-Mixed Concrete. Most concrete used in construction is *ready-mixed* concrete. It is sold by the cubic yard. There are 27 cubic feet to the cubic yard. The purchaser usually specifies the amount and the desired strength of the concrete. The concrete supplier is then responsible for mixing the ingredients in the correct proportion to yield the desired strength.

Sometimes the purchaser specifies the proportion of the ingredients. He or she then assumes responsibility for the design of the mixture.

The ingredients are accurately measured at the plant, often with computerized equipment. The mixture is delivered to the job site, usually in *transit-mix trucks* (Fig. 27-3). The mixer truck contains a large revolving drum, capable of holding from one to 12 cubic yards. There is a separate water tank with a water measuring device. The drum rotates to mix the concrete as the truck is driven to the construction site.

Specifications require that each batch of concrete be delivered within 1 1/2 hours after water has been added to the mix. If the job site is a short distance away, water is added to the cement and aggregates at the plant. If the job site is farther away, the proper amount of water is added to the dry mix from the water tank on the truck as it approaches or arrives at the job.

Job-Mixed Concrete. A small job may require that the concrete be mixed on the site either by hand or with a powered cement mixer. All materials must be measured accurately. For measuring purposes, remember that a cubic foot of water weighs about 62 1/2 pounds and contains approximately 7 1/2 gallons.

Fig. 27-3 A newer, over-the-cab, front end discharge design of transit-mix trucks makes it easier to deliver ready-mix concrete.

WATER GALS	PORTLAND CEMENT BAGS	AGGREGATE (FINE) CU. FT.	AGGREGATE (COARSE) CU. FT.
5 1/2	1	3	4
5	1	2 1/4	3
4	1	2	2 1/4

TRIAL BATCHES FOR SLUMP & WORKABILITY

Fig. 27-4 Formulas for several concrete mixtures.

All ingredients should be thoroughly mixed according to the proportions shown in Figure 27-4. A little water should be put in the mixer before the dry materials are added. The water is then added uniformly while the ingredients are mixed from one to three minutes.

Concrete Reinforcement

Concrete has high *compressive strength*. This means that it resists being crushed. It has, however, low *tensile* strength. It is not as resistant to bending or pulling apart. Steel bars, called *rebars*, are used in concrete to increase its tensile strength. Concrete is then called *reinforced concrete*.

Rebars. Rebars used in construction are usually *deformed*. Their surface has ridges that increase the bond between the concrete and the steel. They come in standard sizes, identified by numbers that indicate the diameter in 8ths. For instance, a #6 rebar has a diameter of 6/8 or 3/4 inch (Fig. 27-5).

The size, location, and spacing of rebars are determined by concrete engineers and shown on the plans. The concrete worker places the rebars inside the form before the concrete is placed. Cutting, bending, placing, and tying rebars require workers who have been trained in that trade.

Wire Mesh. Welded wire mesh is used to reinforce concrete floor slabs resting on the ground, driveways, and walks. It is identified by the gauge

BAR #	BAR Ø INCHES	BAR WEIGHT LBS PER 100 LIN FT.
2	1/4	17
3	3/8	38
4	1/2	67
5	5/8	104
6	3/4	150
7	7/8	204
8	1	267

Fig. 27-5 Numbers and sizes of commonly used reinforcing steel bars.

MESH SIZE INCHES	MESH GAGE	MESH WEIGHT LBS PER 100 SQ. FT.
6X6	#6	42
6X6	#8	30
6X6	# 10	21

Fig. 27-6 Size, gauge, and weight of commonly used welded wire mesh.

and spacing of the wire. Common gauges are #6, #8, and #10. The wire is usually spaced to make 6-inch squares (Fig. 27-6).

Welded wire mesh is laid in the form above the vapor barrier, if used, before the concrete is placed. It is spliced by lapping one full square plus 2 inches.

Placing Concrete

Sawdust, nails, and other debris should be removed from the forms. The inside surfaces are brushed with oil to make form removal easier. Also, before concrete is placed, the forms and subgrade are moistened with water. This is done to prevent rapid absorbing of water from the concrete.

Concrete is *placed,* not poured. Water should never be added so that concrete flows into forms without working it. Adding water alters the water-cement ratio on which the quality of the concrete depends.

Slump Test. *Slump tests* are made by supervisors on the job to determine the consistency of the concrete. The concrete sample for a test should be taken just before the concrete is placed. The *slump cone* is first dampened. It is placed on a flat surface and filled to about one-third of its capacity with the fresh concrete. The concrete is then *rodded* by moving a metal rod up and down twenty-five times over the entire surface. Two more approximately

equal layers are added to the cone. Each layer is rodded in a similar manner with the rod penetrating the layer below. The excess is screeded from the top. The cone is removed by carefully lifting it vertically in three to seven seconds. The cone is gently placed beside the concrete, and the amount of slump measured (Fig. 27-7). The test should be completed within two to three minutes.

Changes in slump should be corrected immediately. The table shown in Figure 27-8 shows recommended slumps for various types of construction. Concrete with a slump greater than 6 inches should not be used.

Placing in Forms. The concrete truck should get as close as possible. Concrete is placed by chutes where needed. It should not be pushed or dragged any more than necessary. It should not be dropped more than 4 to 6 feet, and should be dropped vertically and not angled. Drop chutes should be used in high forms to prevent the buildup of dry concrete

Fig. 27-7 A slump test shows the wetness or dryness of a concrete mix.

TYPES OF CONSTRUCTION	SLUMP IN INCHES	
	MAXIMUM	MINIMUM
REINFORCED FOUNDATION WALLS & FOOTINGS	3	1
PLAIN FOOTINGS, CAISSONS & SUBSTRUCTURE WALLS	3	1
BEAMS & REINFORCED WALLS	4	1
BUILDING COLUMNS	4	1
PAVEMENTS & SLABS	3	1
HEAVY MASS CONCRETE	2	1

Fig. 27-8 Recommended slumps for different kinds of construction.

on the side of the form or reinforcing bars above the level of the placement. Drop chutes also prevent separation caused by concrete striking and bouncing off the side of the form. Pumps are used on large jobs that need concrete continuously over long distances or to heights up to 500 feet.

Rate of Placement. Concrete must not be placed at a rapid rate, especially in high forms. The amount of pressure at any point on the form is determined by the height and weight of the concrete

above it. Pressure is not affected by the thickness of the wall (Fig. 27-9).

A slow rate of placement allows the concrete nearer the bottom to begin to harden. Once concrete hardens it cannot exert more pressure on the forms even though liquid concrete continues to be placed above it (Fig. 27-10). The use of stiff concrete with a low slump, which acts less like a liquid, will transmit less pressure. Rapid placing leaves the concrete in the bottom still in a fluid state. It will exert great lateral pressure on the forms at the bottom. This may cause the form ties to fail or the form to deflect excessively.

Concrete should be placed in forms in layers of not more than 12 to 18 inches thick. Each layer should be placed before setting occurs in the previous layer. The layers should be thoroughly consolidated (Fig. 27-11). The rate of concrete placement in high forms should be carefully controlled.

Consolidation of Concrete. To eliminate voids or honeycombs in the concrete, it should be thoroughly worked by hand spading or vibrated after it goes into the form. Vibrators make it possible to use a stiff mixture that would be difficult to consolidate by hand.

An immersion type vibrator, called a *spud vibrator*, has a metal tube on its end. This tube vi-

RATE (FT/HR)	PRESSURES OF VIBRATED CONCRETE (PSF)			
	50° F		70° F	
	COLUMNS	WALLS	COLUMNS	WALLS
1	330	330	280	280
2	510	510	410	410
3	690	690	540	540
4	870	870	660	660
5	1050	1050	790	790
6	1230	1230	920	920
7	1410	1410	1050	1050
8	1590	1470	1180	1090
9	1770	1520	1310	1130
10	1950	1580	1440	1170

PRESSURE INCREASES AS PLACEMENT RATE INCREASES.
PRESSURE INCREASES AT LOWER TEMPERATURES.

PSF = POUNDS PER SQUARE FOOT

Fig. 27-9 The amount of pressure against the form is determined by the height and weight of the concrete before it hardens. Pressure is not affected by the thickness of the wall.

Fig. 27-10 Once concrete sets up and starts to harden it cannot exert more pressure on the form even though liquid concrete continues to be placed above it.

Fig. 27-11 A safe, consistent pour rate is accomplished using drop chutes to place concrete in approximately 12- to 18-inch internally vibrated layers.

brates at a rapid rate. It is commonly used in construction to vibrate and consolidate concrete. The operator should be skilled in the use of the vibrator. It is important to keep it moving up and down to consolidate layers of concrete and to prevent overvibrating. Excessive vibration makes concrete more liquid. This causes more pressure on forms and possible failure (Fig. 27-12). Concrete with a high slump that is more fluid requires little or no vibration.

Curing Concrete

Concrete hardens and gains strength because of a chemical reaction called hydration. All the desirable properties of concrete are improved the longer this process takes place. A rapid loss of water from fresh concrete can stop the hydration process too soon, weakening the concrete. Curing prevents loss of

moisture, allowing the process to continue so that the concrete can gain strength.

Concrete is *cured* either by keeping it moist or by preventing loss of its moisture for a period of time. For instance, if cured for seven days, it is about 50 percent stronger. If kept moist for a month, its strength is doubled. In addition, a rapid loss of moisture causes the concrete to shrink, resulting in cracks. Curing should be started at soon as the surface is hard enough to resist marring.

Methods of Curing. Flooding or constant sprinkling of the surface with water is the most effective method of curing concrete. Curing can also be accomplished by covering the concrete with burlap, straw, sand, or other material that retains water, and wetting it continuously.

In hot weather, the main concern is to prevent rapid evaporation of moisture. Sunshades or wind-

RIGHT

Liquid Concrete

Concrete has
Stiffened Sufficiently
so Lateral Pressure
is Reduced

WRONG

Liquid Concrete
Full Height of
the Form

Do not over-vibrate the concrete as excessive vibration will cause concrete at the bottom of the form to remain in a liquid state longer than expected. This will result in higher than expected lateral form pressures and may cause a premature failure. Depth of vibration should be limited to within 4 feet of the top of the fresh concrete.

Fig. 27-12 Avoid excessive vibration of concrete. (*Courtesy of Dayton Superior*)

breaks may need to be erected. The formwork may be allowed to stay in place or the concrete surface may be covered with plastic film or other water-proof sheets. The edges of the sheets are overlapped and sealed with tape or covered with planks. Liquid curing chemicals may be used to seal in moisture and prevent evaporation. However, manufacturers' direction for their use should be carefully followed.

Curing Time. Concrete should be cured for as long as practical. The curing time depends on the temperature. At or above 70 degrees, curing should take place at least three days. At or above 50 degrees, concrete is cured at least five days. Near freezing, there is practically no hydration and concrete takes considerably longer to gain strength. There is no strength gain while concrete is frozen. When thawed, hydration resumes with appropriate curing. If concrete is frozen within the first twenty-four hours after being placed, permanent damage to the concrete is almost certain. In cold weather, *accelerators* that shorten the setting time are sometimes used. Protect concrete from freezing for at least four days after being placed by providing insulation or artificial heat, if necessary.

Forms may be removed after the concrete has set and hardened enough to maintain its shape. This time will vary depending on the mix, temperature, humidity, and other factors.

Precautions Using Concrete

Avoid prolonged contact with fresh concrete or wet cement because of possible skin irritation. Wear protective clothing when working with newly mixed concrete. Wash skin areas that have been exposed to wet concrete as soon as possible. If any material containing cement gets into the eyes, flush immediately with water and get medical help.

CHAPTER 28   Forms for Footings, Slabs, Walks, and Driveways

Concrete for footings, slabs, walks, and driveways is placed directly on the soil. The supporting soil must be suitable for the type of concrete work placed upon it.

Footing Forms

The *footing* for a foundation provides a base on which to spread the load of a structure over a wider area of the soil. For foundation walls, the most typical type is a *continuous* or *spread* footing. To provide support for columns and posts, *pier* footings of square, rectangular, circular, or tapered shape are used. In heavy construction, sometimes it is not practical to excavate deep enough to reach load-bearing soil. Then *piles* are driven and capped with a *grade beam* (Fig. 28-1).

MONOLITHIC SLAB

FOUNDATION WALL

CONTINUOUS OR SPREAD FOOTING

WOOD POST

DOWEL

CONCRETE PEDESTAL

METAL COLUMN

METAL PLATE

COLUMN FOOTING

POST

TAPERED PIER FOOTING

GRADE BEAM

FLARED FOOTING

PIER

Fig. 28-1 Several types of footings are constructed to support foundations.

Fig. 28-2 Typical footing for residential construction.

Continuous Wall Footings

In most cases, the footing is formed separately from the foundation wall. In residential construction, often the footing width is twice the wall thickness. The footing depth is equal to the wall thickness (Fig. 28-2). However, to be certain, consult local building codes.

For larger buildings, architects or engineers design the footings to carry the load imposed upon them. Usually these footings are strengthened by reinforcing rods of specified size and spacing.

Frost Line. Footings must be located below the *frost line*. The frost line is the point below the surface to which the ground freezes in winter. Because water expands when frozen, foundations whose footings are above the frost line will heave and buckle when the ground freezes. In extreme northern climates, footings must be placed as much as 6 feet below the surface (Fig. 28-3). In tropical climates, footings only need to reach solid soil, with no consideration given to frost.

In areas where the soil is stable, no formwork is necessary. A trench is dug to the width and depth of the footing. The concrete is carefully placed in the trench (Fig. 28-4). In other cases, forms need to be built for the footing.

Locating Footings

To locate the footings, stretch lines on the batter boards in line with the outside of the footing. Suspend a plumb bob from the batter board lines at each corner. Drive stakes. Stretch lines between the stakes to the correct elevation of the top of the footing (Fig. 28-5).

Building Wall Footing Forms

When building wall footing forms, erect the outside form first. Usually 2×8 lumber is used for the side of the form, and 1×4 stakes are used to hold the

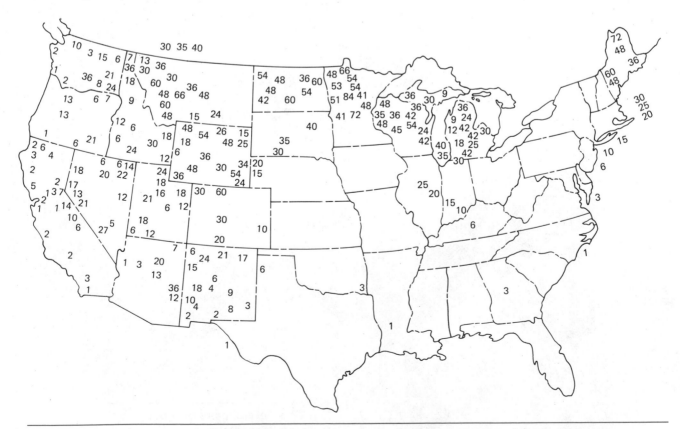

Fig. 28-3 Frost line penetration in the United States. (*Courtesy of U.S. Department of Commerce*)

Fig. 28-4 In stable soils, no footing formwork is necessary.

sides in position. Fasten the sides by driving nails through the stakes. Use duplex nails for easy removal. Steel stakes, spreaders, and braces are manufactured for use in building footing forms and other edge formwork. They come with nail holes for easy fastening (Fig. 28-6).

STEEL STAKES

LUMBER FORM SPREADER
AND LUMBER FORM BRACE

Fig. 28-6 Steel stakes, spreaders, and braces are manufactured for use in building footing forms and other edge formwork. (*Courtesy of Symons Corporation*)

BATTER BOARD

SAW KERF

CHALKLINE

TOPSOIL

BACK SLOPE

PLUMB LINE

WALL LINE

SPACE FOR TILING OR WATERPROOFING WALL

FOOTING TRENCH, SHARP CUT OR USE FORM BOARDS

NO FILL

Fig. 28-5 The footing is located by suspending a plumb bob from the batter board lines.

Keep the top inside corner of the form as close as possible to the line *without touching it.* Be sure the form does not touch the line. If the form touches the line at any point, the line is moved and is no longer straight. Form the outside of the footing in this manner all around. Space stakes 4 to 6 feet apart or as necessary to hold the form straight.

Before erecting the inside forms, cut a number of 1×3 *spreaders*. These are nailed to the top edges of the form. They tie the two sides together and keep them the correct distance apart. Erect the inside forms in a manner similar to that used in erecting the outside forms. Place stakes for the inside forms opposite those holding the outside form. Level across from the outside form to determine the height of the inside form. Fasten the spreaders across the form at intervals necessary to hold the form the correct distance apart.

Brace the stakes where necessary to hold the forms straight. In many cases, no bracing is necessary. Footing forms are sometimes braced by shoveling earth or placing large stones against the outside of the forms. Figure 28-7 shows the steps in the construction of a typical wall footing form.

Keyways. A keyway is formed in the footing by pressing 2×4 lumber into the fresh concrete (Fig. 28-8). The keyway form is beveled on both edges for easy removal after the concrete has set. The purpose of a keyway is to provide a lock between the footing and the foundation wall. This joint helps the foundation wall resist the pressure of the back-filled earth against it. It also helps to prevent seepage of water into the basement. In some cases, where the design of the keyway is not so important, 2×4

STEP BY STEP
P R O C E D U R E S

STEP 1 – SUSPEND PLUMB BOB TO LOCATE OUTSIDE OF FOOTING. DRIVE STAKES AT EACH CORNER & STRETCH LINES FROM CORNER TO CORNER AT ELEVATION OF TOP OF FOOTING.

STEP 2 – DRIVE INTERMEDIATE STAKES OUTSIDE OF STRETCHED LINE BY THE THICKNESS OF THE EDGE FORM.

STEP 3 – SNAP A CHALK LINE ON THE INSIDE FACE OF ALL STAKES. THIS LINE REPRESENTS THE TOP OF THE FORM.

Fig. 28-7 Steps in the construction of a form for a typical continuous footing.

STEP BY STEP
P R O C E D U R E S

(*continued*)

STEP 4 – FASTEN OUTSIDE EDGE FORM TO STAKES, KEEPING TOP EDGE TO CHALK LINE.

STEP 5 – LOCATE STAKES FOR INNER EDGE FORM. FASTEN INNER EDGE FORM TO STAKES. LEVEL ACROSS FROM OUTSIDE FORM FOR PROPER HEIGHT.

STEP 6 – INSTALL SPREADERS AND BRACE AS REQUIRED

pieces are not beveled on the edges, but are pressed into the fresh concrete at an angle.

Stepped Wall Footings

When the foundation is to be built on sloped land, it is sometimes necessary to *step* the footing. The footing is formed at different levels, to save material. In building stepped footing forms, the thickness of the footing must be maintained. The vertical part of each step must not exceed the footing thickness. The horizontal part of the step must be at least twice the vertical part (Fig. 28-9). Vertical boards are placed between the forms to retain the concrete at each step.

SECTIONS THROUGH FOOTING

2" X 4" WITH BEVELED EDGES PRESSED INTO FRESH CONCRETE

2" X 4" PRESSED INTO FRESH CONCRETE AT AN ANGLE

Fig. 28-8 Methods of forming keyways in the footing.

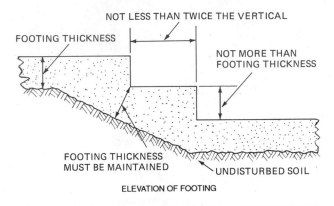

NOT LESS THAN TWICE THE VERTICAL

FOOTING THICKNESS

NOT MORE THAN FOOTING THICKNESS

FOOTING THICKNESS MUST BE MAINTAINED

UNDISTURBED SOIL

ELEVATION OF FOOTING

Fig. 28-9 Dimensions of a stepped footing.

Column Footings

Concrete for footings, supporting columns, posts, fireplaces, chimneys, and similar objects is usually placed at the same time as the wall footings. The size and shape of the column footing vary according to what it has to support. The dimensions are determined from the foundation plan.

In residential construction, the forms for this type of footing are usually built by spiking 2×8 pieces together in square, rectangular or tapered shapes to the specified size (Fig. 28-10).

Measurements are laid out on the wall footing forms to locate the column footings. Lines are stretched from opposite sides of the wall footing forms to locate the position of the forms. They are laid in position corresponding to the stretched lines

(Fig. 28-11). Stakes are driven. Forms are usually fastened in a position so that the top edges are level with the wall footing forms.

Forms for Slabs

Building forms for slabs, walks, and driveways is similar to building continuous footing forms. The sides of the form are held in place by stakes driven into the ground. Forms for floor slabs are built level. Walks and driveways are formed to shed water. Usually 2×4 or 2×6 lumber is used for the sides of the form.

Slab-on-Grade

In warm climates, where frost penetration into the ground is not very deep, little excavation is necessary. The first floor may be a concrete slab placed directly on the ground. This is commonly called *slab-on-grade* construction (Fig. 28-12). With improvements in the methods of construction, the

INTERIOR FOOTING FORM

LINES

FOOTING FORM

Fig. 28-11 Locating interior footings.

LET ENDS RUN BY. NO NEED TO SPEND TIME CUTTING AND FITTING

A RECTANGULAR FORM FOR A COLUMN FOOTING

Fig. 28-10 Construction of pier forms.

Fig. 28-12 Slab-on-grade foundation. (*From Mark W. Huth*, Understanding Construction Drawings, © *1983 by Delmar Publishers Inc.*)

need for lower construction costs, and the desire to give the structure a lower profile, slabs-on-grade are being used more often in all climates.

Basic Requirements

The construction of concrete floor slabs should meet certain basic requirements:

1. The finished floor level must be high enough so that the finish grade around the slab can be sloped away for good drainage. The top of the slab should be no less than 8 inches above the finish grade.

2. All topsoil in the area in which the slab is to be placed must be removed. A base for the slab consisting of 4 to 6 inches of gravel, crushed stone, or other approved material must be well compacted in place.

3. The soil under the slab may be treated with chemicals for control of termites, but caution is advised. Such treatment should be done only by those thoroughly trained in the use of these chemicals.

4. All water and sewer lines, heating ducts, and other utilities that are to run under the slab must be installed.

5. A vapor barrier must be placed under the concrete slab to prevent soil moisture from rising through the slab. The vapor barrier should be a heavy plastic film, such as 6-mil polyethylene or other material having equal or superior resistance to the passage of vapor. It should be strong enough to resist puncturing during the placing of the concrete. Joints in the vapor barrier must be lapped at least 4 inches and sealed.

6. Where necessary, to prevent heat loss through the floor and foundation walls, a waterproof, rigid insulation is installed around the perimeter of the slab.

7. The slab should be reinforced with 6 × 6 inch, #10 welded wire mesh, or by other means to provide equal or superior reinforcing. The concrete slab must be at least 4 inches thick and *haunched* (made thicker) under load-bearing walls (Fig. 28-13).

Monolithic Slabs

In warm climates, where there is little or no frost penetration into the ground, a combined slab and foundation, called a *monolithic slab,* is commonly used (Fig. 28-14). This type of slab is also referred to as a *thickened edge slab.* It consists of a shallow footing around the perimeter. The perimeter is placed at the same time as the slab. The slab and footing make up a one-piece integral unit. The bottom of the footing must be at least one foot below the finish grade, unless local building codes dictate otherwise.

Fig. 28-13 The slab is haunched under load-bearing walls. (*From Mark W. Huth*, Understanding Construction Drawings, © *1983 by Delmar Publishers Inc.*)

Fig. 28-14 Monolithic slab-on-grade.

Forms for monolithic slabs are constructed using stakes and edge form boards, plank, or steel manufactured especially for forming slabs and similar objects.

■ Locate corner stakes by plumbing down from batter board lines. Drive the stakes outside of the building line by the thickness of the edge form material. Stakes should be long enough to be driven below the edge thickness of the slab.

■ Stretch lines at desired elevation between the corner stakes. Drive intermediate stakes. Snap lines from corner stakes on to intermediate stakes. Fasten edge form boards to the tops and to the snapped lines.

■ Tie all corners and joints together. Sight the forms for straightness. Brace by methods described earlier for continuous footing forms (Fig. 28-15).

STEP BY STEP
P R O C E D U R E S

STEP 1– SUSPEND A PLUMB BOB FROM THE INTERSECTION OF BUILDING LINES AND MARK LOCATION OF OUTSIDE CORNER OF SLAB ON A STAKE.

STEP 2– DRIVE STAKES OFFSET FROM EACH CORNER STAKE BY THE THICKNESS OF THE EDGE FORM.

Fig. 28-15 Steps in forming a monolithic slab.

STEP BY STEP
P R O C E D U R E S

(continued)

STEP 3 – STRETCH LINES FROM CORNERS AND DRIVE INTERMEDIATE STAKES.
SNAP A LINE ON THE STAKES AT THE SPECIFIED ELEVATION.

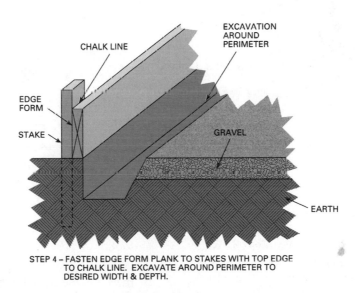

STEP 4 – FASTEN EDGE FORM PLANK TO STAKES WITH TOP EDGE
TO CHALK LINE. EXCAVATE AROUND PERIMETER TO
DESIRED WIDTH & DEPTH.

■ Excavate around the perimeter for the thickened edge and in the interior for any load bearing walls. Temporary screeds will need to be placed when the area is ready for concrete.

Independent Slabs

In areas where the ground freezes to any appreciable depth during winter, the footing for the walls of the structure must extend below the frost line. If slab-on-grade construction is desired in these areas, the concrete slab and foundation wall are separate. This type of slab-on-grade is called an *independent slab*. It may be constructed in a number of ways according to conditions (Fig. 28-16).

When concrete blocks are used for the foundation wall, special blocks are used for the top course. They act as a form for the concrete slab. The top of a solid concrete wall may be formed with a shoulder for the same purpose.

If the frost line is not too deep, the footing and wall may be an integral unit and placed at the same

14' FOUNDATION FOR AN INDEPENDENT SLAB

Fig. 28-16 Independent slabs are constructed in a number of ways.

time. In colder climates, the foundation wall and footing are formed and placed separately. The wall may be set on piles or on a continuous footing, the forming of which has been described previously.

Slab Insulation Requirements

Rigid insulation is placed between the foundation wall and the slab edge to provide for contraction and expansion of the slab. Insulation also prevents heat loss.

Perimeter insulation is required to prevent heat loss, except in warm climates. Two general rules to follow when determining insulation requirements are:

1. When average winter low temperatures are 0°F and higher, the R factor should be about 2.0. The depth of the insulation or the width

under the slab should not be less than one foot.

2. When average winter temperatures are lower than 0°F, the R factor should be about 3.0 without floor heating. The depth or width of the insulation should be no less than 2 feet.

Forms for Walks and Driveways

Forms for walks and driveways are usually built so water will drain from the surface of the concrete. In these cases, grade stakes must be established and grade lines carefully followed (Fig. 28-17).

Usually a combination of 1×4 boards and 2×4 stakes are used or a combination of 2×4 edging and 1×4 stakes. Establish the grade of the walk or driveway on stakes at both ends. Stretch lines tightly

Fig. 28-17 Forms for walks are built so water drains from the surface.

between the end stakes. Drive intermediate stakes. Fasten the edge pieces to the stakes following the line in a manner similar to making continuous footing forms.

Forms for Curved Walks and Driveways

In many instances, walks and driveways are curved. To form the curve, 1/4-inch plywood or hardboard is used for small radius curves. If using plywood, install it with the grain vertical for easier bending without breaking. Wetting the stock sometimes helps the bending process.

For curves of a long radius, 1×4 lumber may be used and satisfactorily bent if the curve is not too tight. Lumber may also be curved by making saw *kerfs* (cuts) spaced close together. The lumber is then bent until the saw kerfs close (Fig. 28-18). A method of determining the spacing of the saw kerfs to obtain a desired curve is shown in Figure 28-19.

SMOOTH INNER FORM FACE

$\frac{1}{4}''$ PLYWOOD

2 X 4 FORM

2 X 2 STAKE

A) SUGGESTED DETAIL AT JOINT BETWEEN STRAIGHT AND CURVED CURVED FORMS

1 X 4 FORM

C) USE 1'' LUMBER FOR LONG RADIUS CURVES

STAKES AT 2 TO 3 FT INTERVALS

PLYWOOD GRAIN VERTICAL

B) USE $\frac{1}{4}''$ PLYWOOD OR HARDBOARD FOR SHORT RADIUS CURVES.

STAKES AT 1 TO 2 FT INTERVALS

2 X 4 FORM

D) USE SAW KERFING TO BEND 2'' LUMBER. BEND SO THAT CUTS CLOSE.

T

CUT $\frac{1}{2}$ TO $\frac{2}{3}$ T

Fig. 28-18 Forming for curved walks and driveways.

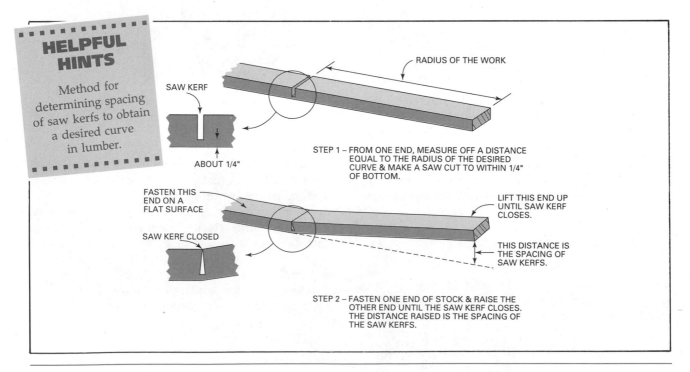

HELPFUL HINTS

Method for determining spacing of saw kerfs to obtain a desired curve in lumber.

SAW KERF

ABOUT 1/4"

RADIUS OF THE WORK

STEP 1 – FROM ONE END, MEASURE OFF A DISTANCE EQUAL TO THE RADIUS OF THE DESIRED CURVE & MAKE A SAW CUT TO WITHIN 1/4" OF BOTTOM.

FASTEN THIS END ON A FLAT SURFACE

SAW KERF CLOSED

LIFT THIS END UP UNTIL SAW KERF CLOSES.

THIS DISTANCE IS THE SPACING OF SAW KERFS.

STEP 2 – FASTEN ONE END OF STOCK & RAISE THE OTHER END UNTIL THE SAW KERF CLOSES. THE DISTANCE RAISED IS THE SPACING OF THE SAW KERFS.

Fig. 28-19

Foundation walls and columns are usually formed by using *panels* rather than building forms in place, piece by piece. Panel construction simplifies the erection and stripping of formwork.

Wall Forms

Wall Form Components

Various kinds of panels and panel systems are used. Some concrete panel systems are manufactured of wood, aluminum, or steel. Specially designed hardware is used for joining, spacing, aligning, and bracing the panels. Figure 29-1 shows the components of one sytem used for light formwork. (Other forming systems are described later in this chapter.)

THREE STANDARD BAR CONFIGURATIONS ARE AVAILABLE IN HEIGHTS OF 2' TO 10', AND WIDTHS OF 4" TO 24" IN 1" INCREMENTS

4 BAR   5 BAR   6 BAR

8' PANELS

TIES
CONSISTENT STRENGTH (5400 LB. ULT. CAP.)
STRAIGHT OR BENT TIES AVAILABLE
TIE LENGTHS UP TO 48"

BREAK BACK TOOL

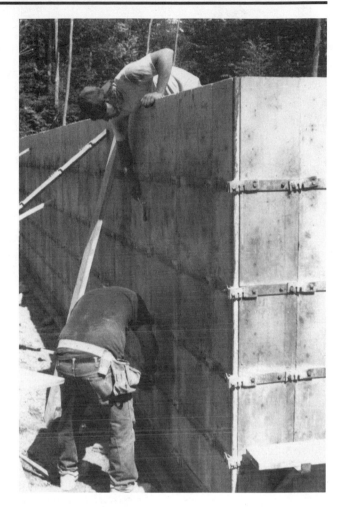

WALER HARDWARE

WALER BRACKETS
    2" x 4" TOP WALER BRACKET
    2" x 6" WALER BRACKET
    2" x 8" WALER BRACKET
    2" x 10" WALER BRACKET
    2" x 12" WALER BRACKET
WALERS ARE DESIGNED FOR ALIGNMENT ONLY, AND ADD NO STRUCTURAL INTEGRITY TO THE FORM. THEREFORE, ONLY ONE SIDE OF THE FORMWORK NEED BE WALED.

Fig. 29-1 Components of a panel system used for light formwork. (*Courtesy of Symonds Corporation*)

OUTSIDE CORNERS

INSIDE CORNER

AVAILABLE IN
4" x 4", 4" x 6" AND
6" x 6" WIDTHS FOR
QUICK AND
EASY ASSEMBLY

THE 4" x 6" CORNER
CONVERTS TO
6" x 4" SIMPLY
BY TURNING END
FOR END AND
MOVING THE
HARDWARE

HINGED CORNER

4" x 4" (MINIMUM
ANGLE = 90°)

8" x 8" (MINIMUM
ANGLE = 45°)

HINGED CORNER MAY
BE USED AS INSIDE,
OUTSIDE OR BAY
CORNER.

BEAM POCKET

TAPERED FOR QUICK
REMOVAL, PROVIDES
FAST BEAM BOX-OUTS.

1" x 1"
STANDARD

2" x 2"

Fig. 29-1 (*continued*)

Form Panels. Panels built of special form plywood backed by 2×4 *studs* (vertical framing members) are often used. They can be purchased or built on the job. They are placed side by side to form the inside and outside of the foundation walls. A usual panel size is 36 inches wide by 8 feet high. Any larger size panel cannot be handled easily by one person. Narrower panels of several widths are used for fillers when the end space is too narrow for a standard size panel. The edges of the panels are *dadoed* (grooved across the grain) and holes are bored to receive *snap ties* (Fig. 29-2).

Snap Ties. Snap ties hold the wall forms together at the desired distance apart. They support both sides against the lateral pressure of the concrete (Fig. 29-3). These ties reduce the need for external bracing and greatly simplify the erection of wall forms. The design of a form for a particular job, including the spacing of the ties and studs, is decided by a construction engineer.

These ties are called snap ties because projecting ends are snapped off slightly inside the concrete after removal of the forms. A special snap tie wrench is used to break back the ties (Fig. 29-4). The small remaining holes are easily filled.

Because of the great variation in the size and shape of concrete forms, there are a large number of styles used. For instance, *flat ties* of various styles are used with some manufactured panels. For heavier formwork, *coil ties* and reuseable *coil bolts* are used (Fig. 29-5). For each kind of tie, there are also several sizes and styles. There are hundreds of kinds of

PLYWOOD FACE

HOLES FOR
SNAP TIES

SNAP TIE

DADOES FOR
SNAP TIES

2 x 4 STUDS
16" O.C. OR AS
SPECIFIED

Fig. 29-2 Typical plywood panel used for formwork. Edges are dadoed and holes are bored for snap ties.

CRIMP TO PREVENT TIE
FROM TURNING IN THE
CONCRETE

BREAKBACK POINT

HUMP HOLDS SPREADER
IN PROPER POSITION

Fig. 29-3 A snap tie for light formwork. (*Courtesy of Dayton Superior*)

1. SLIDE THE SNAP TIE WRENCH UP AGAINST THE TIE SO
THAT THE FRONT OF THE WRENCH IS TOUCHING THE CONCRETE.

2. KEEPING THE FRONT OF THE WRENCH TIGHT AGAINST THE CONCRETE,
PUSH THE HANDLE END TOWARDS THE CONCRETE WALL SO THAT THE TIE
IS BENT OVER AT APPROXIMATELY A 90° ANGLE.

3. ROTATE THE WRENCH AND TIE END 1/4 TO 1/2 TURN
BREAKING OFF THE TIE END.

Fig. 29-4 Method of breaking back snap ties.
(*Courtesy of Dayton Superior*)

form hardware. To become better acquainted with form hardware, study manufacturers' catalogs.

Walers. The snap ties run through and are wedged against form members called *walers*. Walers may be 4×4 lumber with holes bored for the snap ties or doubled 2×4 pieces with spaces between them. They may be horizontal or vertical. Walers are spaced at right angles to the panel frame members. The number and spacing depend on the pressure exerted on the form. Figure 29-6 shows the construction of a typical wall form.

BEAD LOCATES SPREADER WASHER

HOT FORGED HEAD ON STOCK SIZES OF TIES

ANTI-TURN FEATURE (FLAT OR CRIMP)

BREAK BACK

15/16″ DIAMETER METAL SPREADER WASHER

BEAD LOCATES SPREADER WASHER

COLD FORMED HEAD ON NON-STOCK SIZES OF TIES

ANTI-TURN FEATURE (FLAT OR CRIMP)

BREAK BACK

15/16″ DIAMETER METAL SPREADER WASHER

ANTI-TURN FEATURE (CRIMP OR FLAT)

HOT FORGED ½″HEX HEAD ON STOCK SIZES OF TIES

1″ x 1″ PLASTIC CONE WITH BREAK BACK INSIDE CONE

WIRE SNAP TIES

1 5/8″ TYP.

WALL THICKNESS

1/4″ BREAK BACK

6″

6″

MODULAR TYPE FORM

WALL THICKNESS

BREAK BACK

1/4″ X 1″ SLOTS

WELL CRIMPED NAIL

$5/16″$ x 1″ x 4′-0″, 6′-0″ or 8′-0″ LONG BAR AS REQUIRED

Fig. 29-5 A large variety of snap ties are manufactured. (*Courtesy of Dayton Superior*)

STEP BY STEP
P R O C E D U R E S

STEP 1 – SUSPEND PLUMB BOB FROM BUILDING LINES & MARK CORNERS OF WALL.
SNAP CHALK LINE ON TOP OF FOOTING BETWEEN CORNERS.

STEP 2 – APPLY PLATES AROUND PERIMETER OF FOOTING.

STEP 3 – START IN CORNER AND PLACE PANELS ON PLATE. FASTEN PANELS TO
PLATE WITH DUPLEX NAILS. KEEP FACE OF PANELS FLUSH WITH INSIDE
EDGE OF PLATE. INSERT SNAP TIES BETWEEN PANELS. BRACE PANELS
WITH TEMPORARY BRACE.

Fig. 29-6 Construction of a typical wall form using plywood panels.

STEP BY STEP

P R O C E D U R E S

(continued)

OUTSIDE PANELS

INSIDE PANELS

SNAP TIE

FOOTING

PLATE

STEP 4 – INSERT REMAINDER OF TIES THROUGH OUTSIDE PANELS. INSTALL INSIDE
PANELS OVER SNAP TIE ENDS. TILT TOP OF PANEL AND INSERT SNAP TIE ENDS
STARTING FROM BOTTOM.

PANEL

WALER

SNAP TIE
& WEDGE

FOOTING

PLATE

YOKE

STEP 5 – INSTALL WALERS. INSTALL SNAP TIE WEDGES & FASTEN.
YOKE CORNERS.

The vertical spacing of the snap ties and walers depends on the thickness and height of the concrete wall. The vertical spacing is closer together near the bottom. This is because there is more lateral pressure from the concrete there than at the top (Fig. 29-7).

For low wall forms less than 4 feet in height, the panels may be laid horizontal with vertical walers spaced as required (Fig. 29-8).

Erecting Wall Forms

After the concrete placed in the footing has hardened sufficiently (usually at least three days), the forms are removed and cleaned. The salvaged forms are reused.

Locating the Forms. Lines are stretched on the batter boards in line with the outside of the foundation wall. A plumb bob is suspended from the layout lines to the footing. Marks that are plumb with the layout lines are placed on the footing at each corner. A chalk line is snapped on the top of the footing between the corner marks outlining the outside of the foundation wall.

Installing Plates. Panels can be set directly on the concrete footing to the chalk line or on 2×4 or 2×6 lumber plates. Plates are recommended because they provide a positive on-line wall pattern.

Fig. 29-8 When constructing low wall forms, the panels are laid horizontal.

They also level out rough areas on the footing. Plates function to locate the position and size of pilasters, changes in wall thickness, and corners (Fig. 29-9). Secure plates to the footing using masonry nails or pins driven by a powder-actuated tool.

Erecting Panels. Stack the number of panels necessary to form the inside of the wall in the center of the excavation. Lay the panels needed for the outside of the wall around the walls of the excavation. The face of all panels should be oiled or treated with a chemical releasing agent. This provides a smooth face to the hardened concrete and makes stripping of the forms easy.

Erect the outside wall forms first. Set all corner panels in place by nailing into the plate with duplex nails. Make sure the corners are plumb by testing with a hand level.

Fill in between the corners with panels, keeping the same width panels opposite each other. Place snap ties in the dadoes between panels as work progresses. Tie panels together by driving U-shaped clamps over the edge 2×4s or by spiking them together with duplex nails. Use filler panels as necessary to complete each wall section. Brace the wall temporarily as needed.

Placing Snap Ties. After the panels for the outside of the wall have been erected, place snap ties in the intermediate holes. Be careful not to leave out any snap ties. Erect the panels for the inside of the wall. Keep joints between panels opposite to those for the outside of the wall. Insert the other end of the snap ties between panels and in intermediate holes as panels are erected (Fig. 29-10). If the concrete is to be reinforced, the rebars are tied in place before the inside panels are erected.

Installing Walers. When all panels are in place, install the walers. Let the snap ties come through them and wedge into place. Care must be taken

Fig. 29-7 Horizontal stiffeners are placed closer together near the bottom than they are at the top. Dimensions are for purposes of illustration only.

2 × 4 PLATES

2 × 4 PLATES

FOOTING

KEYWAY

PLATES

PLASTER

PLAN OF FOOTING
SHOWING PLATES

KEYWAY

LET PLATES RUN BY CORNERS.
NOT NECESSARY TO FIT.

Fig. 29-9 Plates should be used when erecting wall forms.

Fig. 29-10 Erecting the walls for the inside of the form.

when installing and driving snap tie wedges (Fig. 29-11). Let the ends of the walers extend by the corners of the formwork. Reinforce the corners with vertical 2×4s as shown in Figure 29-12. This is called *yoking* the corners.

Forming Pilasters. The wall may be formed at intervals for the construction of *pilasters* (thickened portions of the wall). They may strengthen the wall or provide support for beams. They may be constructed on the inside or outside of the wall. They

RIGHT   WRONG

Do not attempt to draw up warped wales with a wedge or overtighten the wedge in any manner. Overtightening will cause the metal spreader washers to bend out of shape or will shatter the plastic cones. resulting in a decreased and incorrect wall thickness.

RIGHT   WRONG

The optimimum wedge position is when the snap tie head is at the midpoint of the wedge. You may place the snap tie head higher on the wedge, as long as it is not overtightened. However, the snap tie head must not be positioned lower than the midpoint, as this will place it on a section of the wedge that has not been designed to carry the rated load.

RIGHT   WRONG

As the concrete is consolidated, internal vibrations may cause the steel wedges that have not been nailed into position to loosen, bounce around and eventually fall off resulting in premature form failure.

Fig. 29-11 Care must be taken when installing snap tie wedges. (*Courtesy of Dayton Superior*)

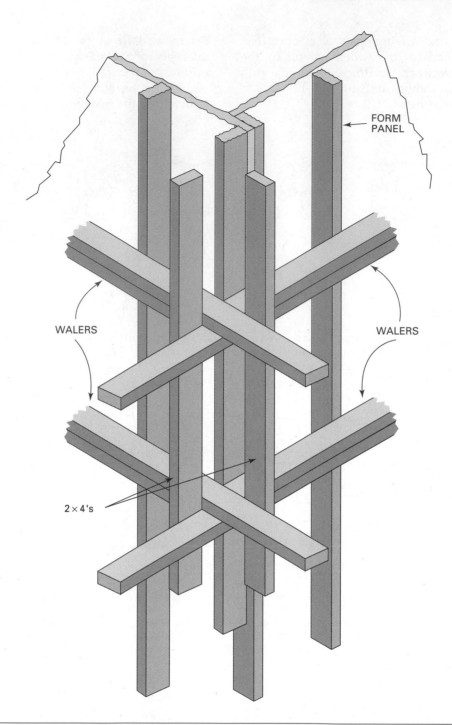

FORM
PANEL

WALERS

WALERS

2 × 4's

Fig. 29-12 The outside corners of wall forms are yoked.

can be formed with job-built panels, making inside and outside corners in the usual manner. In the pilaster area longer snap ties are necessary (Fig. 29-13).

Straightening and Bracing. Brace the walls inside and outside as necessary to straighten them. Wall forms are easily straightened by sighting by eye along the top edge from corner to corner. Another method of straightening is by stretching a line from corner to corner at the top of the form over

two blocks of the same thickness. Move the forms until a test block of equal thickness passes just under the line (Fig. 29-14).

Braces are cut with square ends and spiked into place against *strongbacks.* Strongbacks are placed across walers at right angles wherever braces are needed. The sharp corner of the square ends helps hold the braces in place. It also allows easy prying with a bar to tighten the braces and move the forms (Fig. 29-15). There is no need to make a bevel cut on the ends of braces. A special

Fig. 29-13 Formwork for a pilaster.

ALL THREE BLOCKS ARE OF EQUAL THICKNESS

TEST BLOCK GOES UNDER LINE WITHOUT TOUCHING IT

LINE

Fig. 29-14 Straightening the wall form with a line and test block.

adjustable form brace and aligner is sometimes used for positioning and holding wall forms (Fig. 29-16).

Screeding. After the wall forms have been straightened and braced, chalk lines are snapped on the inside of the form for the height of the foundation wall. Nails may be driven partway in at intervals along the chalk line as a guide for *screeding* (leveling) the top of the wall. If the tops of the pan-els are level with each other, a short piece of stock notched at both ends can be run along the panel tops to screed the concrete. Another method is to fasten strips on the inside walls along the chalk line and use a similar screeding board, notched to go over the strips (Fig. 29-17).

Setting Anchor Bolts. As soon as the wall is screeded, anchor bolts are set in the fresh concrete. A number of various styles and sizes are manufac-

Fig. 29-15 Cut braces with square ends. This allows for easy prying when straightening the formwork.

Fig. 29-16 Manufactured wall brace and aligner. (*Courtesy of Dayton Superior*)

SCREEDING BLOCK

THE TOP OF THE FORM MAY BE USED TO SCREED THE CONCRETE WALL.

SCREEDING BLOCK

SCREED

STRIPS MAY BE FASTENED TO BOTH SIDES OF THE FORM TO SCREED THE TOP OF THE CONCRETE WALL.

CHALK LINE

Fig. 29-17 Nails are driven partway in along a chalk line as a guide to screed the top of the wall. Walls are screeded with other methods.

tured (Fig. 29-18). The type is usually specified on the foundation plan. Care must be taken to set the anchor bolts at the correct height and at specified locations. An anchor bolt *template* is sometimes used to accurately place the bolts (Fig. 29-19).

Openings in Concrete Walls

In many cases, openings must be formed in concrete foundation walls for such things as windows, doors, ducts, pipes, and beams. The forms used for providing the larger openings are called *bucks*.

Constructing Bucks. The buck is usually made of 2-inch stock. Its width is the same as the thickness of the foundation wall. Nailing blocks or strips are often fastened to the outside of the bucks. These are beveled on both edges to lock them into the concrete when the form is stripped. They provide for the fastening of window and door frames in the openings. Intermediate pieces may be necessary in bucks for large openings to withstand the pressure of the concrete against them (Fig. 29-20).

Bucks are made to the specified dimension. They are installed against the inside face of the out-

side panels. Duplex nails through the outside panels hold the bucks in place. The inside wall panels are then installed against the other side of the bucks. Nails are driven through the inside wall panels to secure the bucks on the inside.

ANCHOR TYPE B
HEX-HEAD BOLT

ANCHOR TYPE A
L-BOLT

Fig. 29-18 Various styles of anchor bolts. (*Courtesy of Simpson Strong-Tie Co.*)

Fig. 29-19 Templates are sometimes used to accurately place anchor bolts in fresh concrete.

Fig. 29-20 Construction of a typical window buck.

Fig. 29-21 A small box is used to form a girder pocket in a foundation wall.

Girder Pockets. Recesses in the top of the foundation wall are called *girder pockets*. They are sometimes required to receive the ends of girders (beams). A box of the size needed is made and fastened to the inside at the specified location (Fig. 29-21). Since it is near the top of the form, not much pressure from the liquid concrete is exerted against it.

Column Forms

To form columns, like all other kinds of formwork, as much use as possible is made of panels, manufactured or job-built, to simplify erection and stripping. Columns may be formed in square, rectangular, circular, or a number of other shapes.

Erecting Column Forms

The following is a procedure for forming a square or rectangular column. Stretch lines and mark the location of the column on the footing. Fasten two 2×4 pieces to the footing on opposite sides of the column and outside of the line by the panel thickness. Fasten the other two to the overlapping ends of the first pair in a similar location (Fig. 29-22).

Build and erect two panels to form the thickness and height of the column. A *cove* molding of desired size may be fastened on the edges of this panel to form a radius to the corners of the concrete column. A *quarter-round* molding may be used for a cove shape. Triangular-shaped strips of wood can

Fig. 29-22 A yoke is constructed on the column footing.

be used to form a chamfer on the corners of the column (Fig. 29-23). (A helpful hint is shown in Fig. 29–24 to rip these strips in a safe and efficient manner.)

The face of the column may decorated with

flutes by fastening vertical strips of *half-round* molding spaced on the panel faces. In addition, *form liners* are often used. They provide various wood, brick, stone, and many other textures in the face of the concrete wall or column.

Fig. 29-23 The corners of a concrete column may be formed in several ways.

SQUARE STOCK HELD AT ANGLE AND RUN THROUGH JIG. MAKES TWO CHAMFERING STRIPS WITH ONE PASS THROUGH SAW.

LINE OF CUT

RIP FENCE

TABLE SAW

CLAMP JIG TO TABLE TOP

TRIANGULAR (45° ANGLE) STRIPS FASTENED TO PLYWOOD BASE

1/4" PLYWOOD BASE

NOTE: **USE SAW GUARD** GUARD NOT SHOWN FOR CLARITY

C CLAMP

Fig. 29-24

Build panels for the opposite sides of the column to overlap the previously built panels. Plumb and nail the corners together with duplex nails. Install 2×4 *yokes* around the column forms, letting their ends extend beyond the corners. Spike them together where they overlap. Yoke the column closer together at the bottom. The number and spacing of yokes depend on the height of the column. These details are specified by the form designer. Install vertical 2×4s between the overlapping ends of the yokes (Fig. 29-25). Brace the formwork securely to hold it plumb. Triangular strips placed in the corners of the form are sometimes used to bevel the column corners.

Concrete Forming Systems

A concrete forming system consists of manufactured items for concrete form construction. Among these are specially designed panels of steel or a combination of steel and wood. The panels may be tied together with metal wedges (Fig. 29-26). Much forming system hardware is specially designed for use with these systems.

Small column forms can be erected quickly with less material. The corners are wedged and act as yokes (Fig. 29-27). Large columns can be formed in much less time because less material is needed

(Fig. 29-28). Circular columns and *column capitals* can be formed with reusable fiber glass and steel forms or with tubular fiber forms that are stripped by peeling them off (Fig. 29-29).

Walers are easily installed on wall forms when the forming system hardware is used (Fig. 29-30). Adjustable braces are available. Few are required to hold the wall straight (Fig. 29-31). Steel inside corners are used with wall panels for fast and easier forming of the inside corner (Fig. 29-32).

Other accessories are available, such as molded beam forms of reinforced fiberglass (Fig. 29-33) and Domeforms® for forming beams and slabs (Fig. 29-34).

Forming systems require a considerable initial outlay of funds. However, the investment is saved many times over because of low labor costs and reusable forming system components. A number of forming systems are available for various types of work. Systems are made for light residential work up to heavy-duty concrete formwork. With forming systems, fewer ties and materials are used. For special jobs, manufacturers also make custom-made formwork. The time saved by using custom-made formwork on a job is well worth the cost (Fig. 29-35).

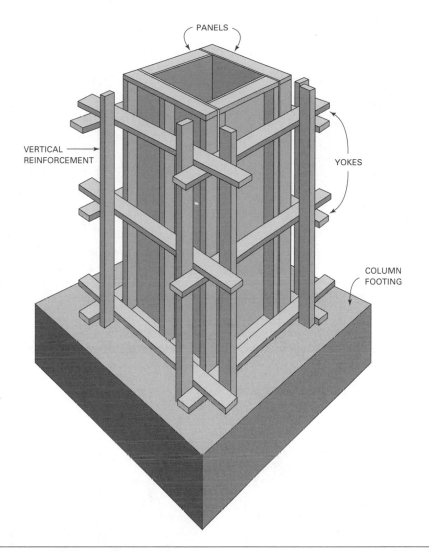

Fig. 29-25 A completed concrete column form.

Fig. 29-26 Only a hammer is needed to tie panels or corners together when metal wedges are used. (*Courtesy of Symons Corporation*)

Fig. 29-27 Yokes are not necessary to form columns when forming systems are used. (*Courtesy of Symons Corporation*)

COLUMN FORMS

COLUMN CAPITALS

Fig. 29-29 Manufactured circular column and column capital forms need only to be set in place and braced. (*Courtesy of Symons Corporation*)

Fig. 29-28 Large rectangular column form with walers and strongbacks attached. The forming system eliminates the need for yoking the corners. (*Courtesy of Symons Corporation*)

Fig. 29-30 Walers are easily installed on forming systems when special hardware is used. (*Courtesy of Symons Corporation*)

Fig. 29-31 Adjustable braces are available with forming systems. (*Courtesy of Symons Corporation*)

Fig. 29-32 Steel inside corners of a forming system make fast erection for formwork possible. (*Courtesy of Symons Corporation*)

Fig. 29-33 Slab and beam formed with a molded fiberglass beam form. (*Courtesy of Symons Corporation*)

Fig. 29-34 Domeforms® of fiber glass or steel form beams and floors easily. (*Courtesy of Symons Corporation*)

Fig. 29-35 Custom-made framework was reused for all seventy floors of this high-rise building. (*Courtesy of Symons Corporation*)

CHAPTER 30

CONCRETE STAIR FORMS

Stair Forms

It may be necessary to refer to later chapters on stair framing for definition of stair terms, types of stairs, stair layout, and methods of stair construction.

Concrete stairs may be suspended or supported by earth (Fig. 30-1). Each type may be constructed between walls or have open ends.

Forms for Earth Supported Stairs

Before placing concrete, stone, gravel, or other suitable fill is graded to provide proper thickness to the stairs. It should not be overly thick, or concrete will be wasted. It may be necessary to lay out the stairs before the supporting material is placed.

Forming between Existing Walls. When earth-supported stairs are formed between two existing walls, the *rise* and *run* of each step are laid out on the inside of the existing walls. *Planks* are ripped to width to correspond to the height of each riser. The planks are then beveled on their bottom edge and wedged securely in position to the layout lines.

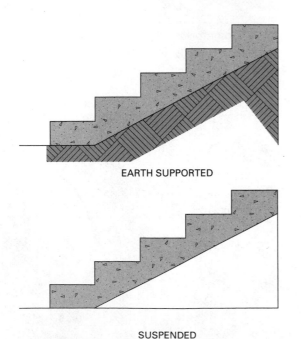

EARTH SUPPORTED

SUSPENDED

Fig. 30-1 Types of concrete stairs. Stairs may be formed with both ends closed or open, or with one end closed and the other end open.

The plank is wedged and secured in place with its inside face to the riser layout line. The top and bottom edges are to tread layout lines.

Beveling the bottom edge of the plank permits the mason to trowel the entire surface of the tread. Otherwise, the bottom edge of the riser form will leave its impression in the concrete tread. After the riser planks are secured in position, they are braced from top to bottom between the ends. This keeps them from bowing outward due to the pressure of the concrete (Fig. 30-2).

Forming Stairs with Open Ends. In cases where the ends of the stairs are to be open, panels are erected on each end. It does no harm if the panels are larger than needed. The panels must be plumb. The distance between them must be the desired width of the stairs. They are then braced firmly in position.

The risers and treads are laid out on the inside surfaces of the panels. *Cleats* (short strips of wood) are fastened at each riser location. Allowances must be made for the thickness of the riser form board. Screws or duplex nails should be used to make stripping the form easier. The space should then be filled with the supporting material to the proper level. Any necessary reinforcing should be installed (Fig. 30-3).

The boards or planks to form the risers are ripped to width and cut to length. The bottom edge, opposite the face side, is beveled to permit finishing

Fig. 30-2 Stairs formed between two existing walls. (*Courtesy of Portland Cement Association*)

SIDE FORM

FILL

CLEATS

Fig. 30-3 End panels are braced firmly in position. The stairs are laid out, cleats fastened, and the space filled with supporting material.

of the total tread width. They are then fastened with duplex nails through the side panels in a position against the cleats, with the top and bottom edges to the tread layout lines. The riser forms are then braced from top to bottom at intervals between its two ends (Fig. 30-4).

Forms for Suspended Stairs

Forms for suspended stairs are more difficult to build. Instead of earth support, a form needs to be built on the bottom to support the stair slab. With proper design and reinforcement, the stairs are

Fig. 30-4 Typical form construction for concrete stairs having open ends. (*Courtesy of Portland Cement Association*)

strong enough to support themselves in addition to the weight of the traffic.

Forming Suspended Stairs between Existing Concrete Walls.

A cross-section of the formwork for suspended stairs between existing walls, with the form members identified, is shown in Figure 30-5. First, lay out the treads, risers, and stair slab bottom on both of the walls between which the stairs run. This can be done by using a hand level, rule, and chalk line (Fig. 30-6).

The form for the bottom of the stair slab is then laid out. Allow for the thickness of the plywood deck, the width of the supporting joists, and the depth of the *horses*. Snap a line on both walls to

indicate their bottom and also the top of the supporting *shores*.

Allowing for a plank sill and wedges at the bottom end of the shores, cut them to length with the correct angle at the top. They should be cut a little short to allow for wedging because of the unevenness of the concrete floor. Install them at specified intervals on top of the plank sill. Brace in position, and place and fasten the horses on top with *scabs*. Scabs are short lengths of narrow boards fastened across a joint to strengthen it. Install joists at right angles across the horses at the specified spacing (Fig. 30-7). Fasten the form plywood in position on top of the joists. Use only as many fasteners as needed for easier stripping later. Wedge

Fig. 30-5 Cross-section of formwork for suspended stairs between walls.

Fig. 30-6 Make the layout for suspended stairs on the inside of the walls.

Fig. 30-7 The framework for the bottom of suspended stairs.

the shoring as necessary to bring the surface of the plywood to the layout line. Fasten the wedges in place so they will not move. The plywood should now be oiled to facilitate stripping. Install the reinforcing.

Rip the riser form planks to width. Cut to length, allowing for wedges. Bevel the outside bottom edge if the treads are to be finished. There is no need for beveling if some type of finish material is to be applied to the tread surface later. Install the riser form planks to the layout lines. Wedge in position.

Another method of securing riser form boards

is by fastening planks on the walls about an inch above the stair layout. Project the riser layout lines upward on the planks. Fasten cleats to the plank. Then brace the cleats and fasten the riser forms to them.

Forming Suspended Stairs with Open Ends.

A completed form for suspended stairs with open ends is shown in Figure 30-8. Because there are no walls upon which to lay out the stairs, the layout must be made on the side panels used for forming the ends of the stairs.

Laying Out the Side Forms. Lay out the risers,

Fig. 30-8 A completed form for suspended stairs with open ends.

treads, and stair slab bottom on the left and right side form panels. First, measure and snap a line up from the bottom a distance equal to the thickness of the slab. The intersection of the treads and risers lies at intervals along this line.

Make a *pitch board* by laying out the width of the tread and the height of the riser on a thin piece of plywood. Mark tread and riser locations by using the pitch board held to the slab line just snapped (Fig. 30-9).

Laying Out the Stair Horses. Measure down from the bottom of the side to form a distance equal to the combined joist and plywood deck thickness. Snap a line parallel to the bottom of the side form to locate the top of the stair horse. Extend the bottom floor line and the top plumb line on the side forms to the snapped line. Mark the length and cuts of the stair horse (Fig. 30-10). Cut one stair horse for each side to length at the angles indicated.

Installing Horses and Joists. Temporarily support and brace the horses in position. Measure, cut, and install shores in a manner similar to that of closed stairs. Brace shores and horses firmly in position. Remove temporary supports and braces. Fasten joists at designated intervals to the horses, leaving an adequate and uniform overhang on both sides. Fasten decking to joists using a minimum number of nails.

Installing Side and Riser Forms. Snap lines on the deck for both sides of the stairs. Stand the side forms up with their inside face to the chalk line. Fasten them through the deck into the joists with duplex nails. Brace the side forms at intervals so they are plumb for their entire length (Fig. 30-11). Apply form oil to the deck before reinforcing bars are installed.

Rip the riser form boards to width and cross-cut them to length. Install them on the riser layout line. Fasten them with duplex nails through the side forms into their ends. Install cleats against the riser boards on both ends for additional support. Install intermediate braces to the riser form boards from top to bottom, if needed.

Economy and Conservation in Form Building

Economical concrete construction depends on the reuse of forms. Forms should be designed and built to facilitate stripping and reuse. Use panels to build forms whenever possible. Use only as many nails as necessary to make stripping forms easier.

HELPFUL HINTS

How to make a tread and riser layout on the side form panel using a pitch board.

1. MAKE A PITCH BOARD AS SHOWN ABOVE.

2. HOLD PITCH BOARD TO A LINE ON THE SIDE FORM REPRESENTING THE TOP OF THE STAIR SLAB AND MARK BOTTOM RISER & TREAD.

3. MOVE PITCH BOARD ALONG LINE & MARK LOCATION OF RISERS & TREADS FOR ENTIRE LENGTH OF STAIRS.

Fig. 30-9

Fig. 30-10 Method of finding the length and cuts of the stair horses.

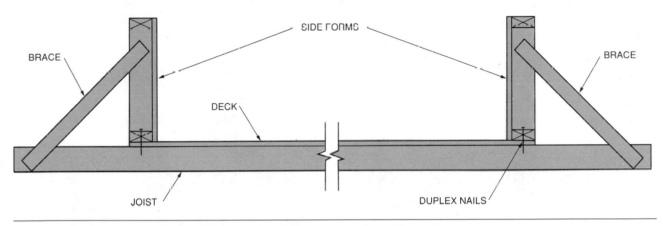

Fig. 30-11 Cross-section of the side form installed on the deck and braced.

Care must be taken when stripping forms to prevent damage to the panels so they can be reused. Stripped forms should be cleaned of all adhering concrete and stacked neatly.

CAUTION Remove all protruding nails to eliminate the danger of stepping on or brushing against them.

Long lengths of lumber can often be used without trimming. Random length walers can extend beyond the forms. There is no need to spend a lot of time cutting lumber to exact lengths. The important thing is to form the concrete to specified dimensions without spending too much time in unnecessary fitting.

Estimating Concrete

Sometimes it is the carpenter's responsibility to order concrete for the job. Ready-mix concrete is sold by the cubic yard. To determine the number of cubic yards of concrete needed for a job, find the number of cubic feet and divide by 27. For example, a wall 8 inches thick, 8 feet high, and 36 feet long requires 7.1 cubic yards of concrete:

$$2/3 \times 8 \times 36 \div 27 = 7.1$$

In this formula, always change all dimensions to feet or fractions of a foot. In the above example, the 8-inch thickness is changed to 2/3 of a foot.

Unit Summary

Concrete can be formed in practically as many shapes as the human mind can conceive (Fig. 30-12). Specialized formwork such as for lift slabs, and tilt-up construction, which were illustrated in Figure 27-1, is required for more efficient concrete construction. In addition, large, crane handled *gang, flying, slip, room tunnel, bridge,* and other special forms allow fast setting and quick stripping (Fig. 30-13).

Fig. 30-12 Concrete can be formed in many complex and pleasing shapes. (*Courtesy of American Plywood Association*)

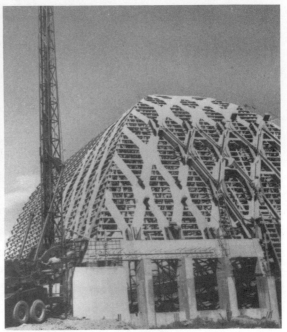

Information is this unit is provided to enable the student to build several basic types of concrete forms. The knowledge of concrete form construction acquired from this unit can be used as a base for building more complicated concrete formwork.

THE STRUCTURE PICTURED HERE WAS BUILT WITH A CLASSIC SLIPFORM SYSTEM.

THIS BUILDING IS A 15-STORY APARTMENT IN SAN FRANCISCO. THE 4-FOOT DEEP SLIPFORMS WERE ADVANCED 15 INCHES PER HOUR DURING THE SLIPPING PROCESS TO COMPLETE A STORY-HEIGHT IN 8 HOURS.

JACKROD

HYDRAULIC JACK

YOKE LEG

WORKING PLATFORM

2x8 JOISTS

3 PLY WALER
2x6 OR 2x8

2x4 STUD

2x6 VERTICAL AT
LIFTING POINTS

3/4" HDO PLYWOOD

STANDARD SLIPFORM
FOR STRAIGHT WALL

Fig. 30-13 Many methods are used to place and shape concrete. (*Courtesy of American Plywood Association*)

TWO FORMS, EACH CAPABLE OF FORMING A FOLDED PLATE 150 x 30 FEET, DEVELOPED THE
ROOF SYSTEM FOR THIS 300 x 480-FOOT AIR TERMINAL.
THE WORK SCHEDULE REQUIRED 17 REUSES OF THE SAME FORM IN A WELL-PLANNED
SERIES OF POURS. THE SAME 3/4-INCH PLYFORM SERVED THROUGH THE SERIES OF POURS,
THEN WAS SALVAGED FOR USE IN OTHER PHASES OF THE JOB.

Fig. 30-13 (*continued*)

Review Questions

Select the most appropriate answer.

1. The inside surfaces of forms are oiled to
   a. protect the forms from moisture.
   b. prevent the loss of moisture from concrete.
   c. strip the forms easier.
   d. prevent honeycombs in the concrete.

2. Rapid placing of concrete
   a. prevents adequate vibrating.
   b. may burst the forms.
   c. separates the aggregate.
   d. causes voids and honeycombs.

3. Unless footings are placed below the frost line,
   a. the foundation will settle.
   b. the foundation will heave and crack.
   c. excavation is difficult in winter.
   d. problems with water and moisture will
      result.

4. Spreaders for footing forms are used
   a. to allow easy placement of the concrete.
   b. to prevent forcing the form out of
      alignment.
   c. because they are easier to fasten.
   d. because they maintain the footing width.

5. The sides of a residential footing form are usually
   a. 2×4s.   c. 2×8s.
   b. 2×6s.   d. 2×10s.

6. The horizontal surface of a stepped footing
must be at least
   a. 4 feet.
   b. twice the vertical distance.
   c. the vertical distance.
   d. the thickness of the footing.

7. The usual size of a standard wall form panel is
a. 4 feet by 8 feet.
b. 3 feet by 8 feet.
c. 32 inches by 8 feet.
d. 2 feet by 8 feet.

8. Reinforcing the walers at the corners with vertical 2×4s is called
a. double locking the corners.
b. interlocking the corners.
c. yoking the corners.
d. tying the corners.

9. To screed the top of a foundation wall
a. space nails along a chalk line.
b. vibrate the top surface so it will flow level.
c. use a chalk line only.
d. add enough water to the concrete so it will flow level.

10. When erecting concrete formwork
a. drive extra nails for added strength.
b. drive all nails home.
c. use as few nails as possible.
d. use duplex nails throughout.

■ BUILDING FOR SUCCESS ■

Quality Concrete Forms

Undoubtedly, the carpenter will continue to play a critical role in the forming and placing of concrete. Carpenters may be assisted by concrete workers, cement finishers, or masons. However, the carpenter usually initiates, constructs, and disassembles concrete forms. Many small residential builders, remodelers, and some light commercial contractors do most of their own concrete work. Their lead carpenter will generally have extensive experience in building concrete forms of various types. Unit 10 has provided many of the important aspects and form building methods for footings, walls, columns, and floor systems. These methods and construction systems continue to be the most preferred in the industry.

"Hands-on" exercises help construction students develop the most meaningful skills. By digging a short footing by hand, the skill or "art" of using a shovel will be learned. Small form construction produces the skills and knowledge needed to build large ones. Understanding the basics of form building is a prerequisite for quality form construction. When the forms are plumb, level, square, inclined, or contoured properly, the correct size as called for on the prints, and in the right location, the criteria for quality forms have been met.

The site plan, footing and foundation plans, and perhaps an elevation plan will provide the necessary measurements and information for the forms. This is where excellent blueprint reading and measuring skills are a must. Accuracy will be the final goal in form construction. Much of the flatwork, such as sidewalks, driveways, and patios, will require certain inclines for proper drainage. This in turn calls for varying elevations to be maintained. Predetermined drops per lineal foot are required to divert water to designated areas, This will call for the carpenter to accurately read a builder's level, spirit level, layout lines, and grade stakes.

At times there is a tendency for the form carpenter to disregard accuracy and detail. This has proven to be a disaster for many concrete jobs. If water does not drain away from the structure, a multitude of complications surface soon after. The key to quality is attention to detail, code requirements, specifications, and plans. Shortcuts or unapproved alterations affect the appearance and safety of a job. For example, if the amount of shoring, bracing, and reinforcing of forms is minimized, there is no guarantee the forms will properly contain the concrete placed in them. The exterior stair form that does not allow water to drain from the treads will allow for ice to form and become potentially dangerous.

Today's regulations, uniform codes, ADA requirements, and other special considerations require a carpenter to be knowledgeable about all applicable requirements during the planning and construction process. Among other things, an acceptable concrete job must be well planned, meet its intended use, and be safe for people to use.

The versatility in skills required of a carpenter will continue indefinitely. Basic carpentry skills must be combined with problem-solving abilities and communication skills. A good form carpenter must place as much emphasis on the quality of his or her work as the trim carpenter. There can be little allowance for error or disregard for exactness anywhere in quality construction. After the project has been completed, people may not even notice how much care, talent, and skill went into forming the walls, columns, stairs, or floors. But when everyone can ascend or descend stairs with no difficulty, water drains away properly, and there are no slips or falls due to poor construction, then people notice and appreciate quality.

Quality work may go unnoticed, but a product lacking quality will become the center of attention. If the form carpenter sets quality work as a goal, he or she will not need to look for employment. Quality work is always recognized by builders throughout the country, and they are willing to pay for it.

FOCUS QUESTIONS: For individual or group discussion.

1. How can you identify a properly formed concrete job?

2. What are some problems that develop when water does not flow in the designated direction on concrete flatwork?

3. How can you draw a correlation between concrete forming and trim carpentry?

4. If you were supervising the forming of a concrete patio slab that is attached to a house, what are some of the forming considerations that assure accuracy?

Floor Framing

OBJECTIVES

After completing this unit, the student will be able to:

- describe platform, balloon, and post-and-beam framing, and identify framing members of each.

- describe several energy and material conservation framing methods.

- build and install girders, erect columns, and lay sills.

- lay out and install floor joists.

- frame openings in floors.

- lay out, cut, and install bridging.

- apply subflooring.

- describe how termites and fungi destroy wood and state some construction techniques used to prevent destruction by wood pests.

CONTENTS

Wood frame construction is used for residential and light commercial construction for important reasons of economy, durability, and variety. The cost for wood frame construction is generally less than for other types of construction. Fuel and air-conditioning expenses are reduced because wood frame construction provides better insulation.

Wood frame homes are very durable. If properly maintained, a wood frame building will last indefinitely. Many existing wood frame structures are hundreds of years old.

Because of the ease with which wood can be cut, shaped, fitted, and fastened, many different architectural styles are possible. In addition to single-family homes, wood frame construction is used for all kinds of low-rise buildings, such as apartments, condominiums, offices, motels, warehouses, and maunfacturing plants.

CHAPTER 31 TYPES OF FRAME CONSTRUCTION

There are several methods of framing a building. Some types are rarely used today but exist so knowledge of them is necessary when remodeling. Other types are relatively new and knowledge is not so widespread. Some wood frames are built using a combination of types. New designs utilizing engineered lumber are increasing the height to which wood frame structures can be built.

Platform Frame Construction

The *platform* frame, sometimes called the *western* frame, is most commonly used in residential construction (Fig. 31-1). In this type of construction, the floor is built and the walls are erected on top of it. When more than one story is built, the second-floor platform is erected on top of the walls of the first story.

A platform frame is easy to erect. At each floor level a flat surface is provided on which to work. A common practice is to assemble wall framing units on the floor and then tilt the units up in place.

Effects of Shrinkage

Lumber shrinks mostly across width and thickness. A disadvantage of the platform frame is the relatively large amount of settling caused by the shrinkage of the large number of horizontal load-bearing frame members. However, because of the equal amount of horizontal lumber, the shrinkage is more or less equal throughout the building. To reduce shrinkage, only framing lumber with the proper moisture content should be used.

Balloon Frame Construction

In the *balloon* frame, the wall *studs* and first-floor *joists* rest on the sill. The second-floor joists rest on a 1×4 *ribbon* that is cut in flush with the inside edges of the studs (Fig. 31-2). This type of construction is now rarely used, but a substantial number of structures built with this type of frame are still in use.

Effects of Shrinkage

There is so little shrinkage in lumber from one end to the other that it is not taken into consideration. In the balloon frame, settling caused by shrinkage of lumber is held to a minimum in the exterior walls. This is because the studs are continuous from sill to top plate. To prevent unequal settling of the frame due to shrinkage, the studs of *bearing partitions* rest directly on the *girder.*

This is a preferred frame when the exterior wall finish is to be brick veneer or stucco. There is little movement between the wood frame and the masonry veneer. Consequently, there are less cracks in the wall finish (Fig. 31-3).

In any type of frame, the closer the moisture content of lumber is to the *equilibrium moisture content,* the less effect shrinkage will have on the structure. The use of engineered lumber, manufactured with low moisture content, reduces shrinkage in the frame. Solid sawn lumber used for framing members should be, at least, S-Dry or, preferably at equilibrium moisture content of 8 to 12 percent, depending on the area.

Firestops

In a balloon frame, the studs run from sill to plate. *Firestops* must be installed in the walls in several locations. A firestop is an approved material used in the space between frame members to prevent the spread of fire for a certain period of time. In a wood frame, a firestop in a wall might consist of 2×4 blocking between studs. In the platform frame, the wall plates act as firestops.

Firestops must be installed in the following locations:

- In all stud walls, partitions, and furred spaces at ceiling and floor levels.

- Between stair *stringers* at the top and bottom. (Stringers are stair framing members. They are sometimes called stair *horses.*)

- Around chimneys, fireplaces, vents, pipes, at ceiling and floor levels with noncombustible material.

- The space between floor joists at the sill and girder.

- All other locations as required by building codes (Fig. 31-4, p. 313).

Post-and-Beam Frame Construction

The *post-and-beam* frame uses fewer but larger pieces. Large timbers, widely spaced, are used for joists, posts, and rafters. *Matched* planks (tongue

Fig. 31-1 Platform frame construction showing the location and names of its members. (*Part 2 is Courtesy of Western Wood Products Association*)

Fig. 31-2 The location of the members of a balloon frame.

and grooved) are often used for floors and roof sheathing (Fig. 31-5, p. 314).

Floors

Floor beams are sometimes spaced 4 feet on center to allow for the use of tongue and grooved plywood panels instead of matched planks (Fig. 31-6, p. 315). In addition to being nailed, the plywood panels are glued to the floor beams with construction adhesive applied with caulking guns. The use of matched planks allows the floor beams to be more widely spaced.

Walls

Exterior walls of a post-and-beam frame may be constructed with widely spaced posts. This allows

Fig. 31-3 Comparison of shrinkage in the platform and balloon frames.

wide expanses of glass to be used from floor to ceiling. Usually some sections between posts in the wall are studded at close intervals, as in platform framing. This provides for door openings, fastening for finish, and wall sheathing. In addition, close spacing of the studs permits the wall to be adequately braced (Fig. 31-7, p. 315).

Roofs

The post-and-beam frame roof is widely used. The exposed roof beams and sheathing on the underside are attractive. Usually the bottom surface of the roof planks is left exposed to serve as the finished ceiling. Roof planks come in 2-, 3-, and 4-inch nomi-

nal thicknesses. Some are end matched as well as edge matched. Some buildings may have a post-and-beam roof, while the walls and floors may be conventionally framed.

The post-and-beam roof may be constructed with a *longitudinal* frame. The beams run parallel to the ridge beam. Or they may have a *transverse* frame. The beams run at right angles to the ridge beam similar to roof rafters.

The ridge beam and longitudinal beams, if used, are supported at each end by posts in the end walls. They must also be supported at intervals along their length. This prevents the side walls from spreading and the roof from sagging (Fig. 31-8, p. 316). One of the disadvantages of a post-and-

SOFFIT

FIRESTOP

SOFFIT

DROP CEILING

FIRESTOP

DROP CEILING

COVE CEILING

FIRESTOP

COVE CEILING

FIRESTOPS

STAIRS

NONCOMBUSTIBLE
FIRESTOP

VENTS, PIPES, DUCTS

NONCOMBUSTIBLE
FIRESTOP

NONCOMBUSTIBLE
FIRESTOP

CHIMNEYS, FIREPLACES

Fig. 31-4 Firestops prevent the spread of fire for a period of time. (*Courtesy of Western Wood Products Association*)

Fig. 31-5 The post-and-beam frame. (*Part 2 is Courtesy of Bohemia Glulam manufactured by Williamette Industries*)

1/8" SPACING IS RECOMMENDED AT ALL EDGE AND END JOINTS UNLESS OTHERWISE INDICATED BY PANEL MANUFACTURER.

APA RATED STURD-I-FLOOR 48 oc

STAGGER END JOINTS

4x GIRDER

STRENGTH AXIS

12" MIN.

4 x 4 POST

2x BAND JOIST

2x SILL

NOTE: PROVIDE ADEQUATE VENTILATION AND USE GROUND COVER VAPOR BARRIER IN CRAWL SPACE. PANELS MUST BE DRY BEFORE APPLYING FINISH FLOOR.

T&G JOINT (OR BLOCK EDGES)

Fig. 31-6 Floor beams are sometimes spaced 4 feet OC when panels are used for a floor. (*Courtesy of American Plywood Association*)

CONVENTIONALLY FRAMED SECTION

POSTS

DOOR OPENING

WALL BRACE

Fig. 31-7 Sections of the exterior walls of a post-and-beam wall may need to be conventionally framed.

Fig. 31-8 Longitudinal and transverse post-and-beam roofs.

Fig. 31-9 Metal connectors are specially made to join glulam beams. (*Courtesy of Simpson Strong-Tie Company*)

beam roof is that interior partitions and other interior features must be planned around the supporting roof beam posts.

The roof is insulated on top of the deck in order not to spoil the appearance of the exposed beams and deck on the underside. A vapor barrier is first applied. The insulation is installed before the roof covering is put on. (Insulation is covered in detail in a following unit.)

Because of the fewer number of pieces used, a well-planned post-and-beam frame saves material and labor costs. Care must be taken when erecting the frame to protect the surfaces and make well-fitting joints on exposed posts and beams. Glulam beams are well suited for and frequently used in post-and-beam construction. A number of metal connectors are used to join members of the frame (Fig. 31-9).

Energy and Material Conservation Framing Methods

There has been much concern and thought about conserving energy and materials in building construction. Several systems have been devised that differ from conventional framing methods. They conserve energy and use less material and labor.

The 24-Inch Module Method

One of the new methods uses plywood over lumber framing spaced on a 24-inch module. All framing, floors, walls, and roofs are spaced 24 inches on center (OC). Joists, studs, and rafters line up with each other (Fig. 31-10).

Floors. For maximum savings, a single layer of 3/4-inch tongue and grooved plywood is used over joists spaced 24 inches OC. In-line floor joists are used to make installation of the plywood floor easier (Fig. 31-11). The use of adhesive when fastening the plywood floor is recommended. Gluing increases stiffness and prevents squeaky floors (Fig. 31-12).

Walls. Studs up to 10 feet long with 24-inch spacing can be used in single-story buildings. Stud height for two-story buildings should be limited to 8 feet. A single layer of plywood acts as both sheathing and exterior siding. In this case, the plywood must be at least 1/2-inch thick (Fig. 31-13). If two-layer construction is used, 3/8-inch plywood is acceptable.

Wall openings are planned to be located so that at least one side of the opening falls on the module. Whenever possible, window and door sizes are selected so that the rough opening width is a multiple of the module (Fig. 31-14). Also, locate partitions at normal wall stud positions if possible.

Roofs. Rafters or trusses are spaced 24 inches on center in line with the studs under 3/8-inch plywood sheathing. Increasing the spacing of the roof framing members and using a thinner roof sheathing conserves material.

House Depths

House depths that are not evenly divisible by four waste floor framing and sheathing. Lumber for floor joists is produced in increments of 2 feet. A 25-foot house depth uses the same total linear footage of standard floor joist lengths as a house depth of 28 feet. Full 48-inch-width plywood panels can be used without ripping for 24-, 28-, and 32-foot house depths.

Fig. 31-10 The 24-inch module system of framing uses less material and labor.

IN-LINE FLOOR JOISTS

SCAB JOINTS

FLOOR PANELS LAP BUTT JOINT TO PROVIDE TIE

Fig. 31-11 In-line floor joists make installation of plywood subflooring simpler.

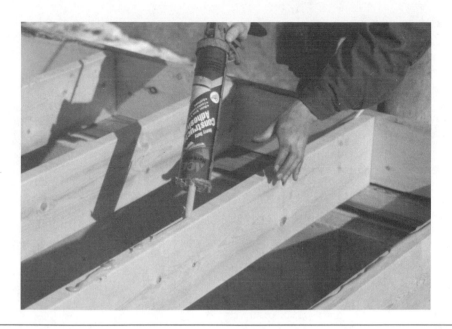

Fig. 31-12 Using adhesive when fastening subflooring makes the floor frame stiffer and stronger. (*Courtesy of American Plywood Association*)

SINGLE WALL
(PLYWOOD SIDING DIRECT TO STUDS)

GRADES AND
SPECIES AS
ACCEPTED BY
LOCAL CODE

EXTERIOR
PLYWOOD
SIDING

FOR VERTICAL BUTT
JOINTS, USE BATTENS
OR BUILDING PAPER.
(NO BATTEN OR BUILDING
PAPER NEEDED WITH
SHIPLAP JOINTS.)

BATTEN (OPTIONAL)

DOUBLE WALL
(SHEATHING AND SIDING)

GRADES AND
SPECIES AS
ACCEPTED BY
LOCAL CODE

HORIZONTAL
WOOD SIDING

3/8" PLYWOOD SHEATHING
(HORIZONTAL OR VERTICAL)

Fig. 31-13 Single-layer and double-layer exterior wall covering.

WINDOW ON MODULE

WINDOW OFF MODULE

6' WALL SECTION

6' WALL SECTION

STUDS   24 L.F.
JACKS   14 L.F.
CRIPPLES 9 L.F.
          47 L.F.

STUDS   32 L. F.
JACKS   14 L.F.
CRIPPLES 12 L.F.
          58 L.F.

(23% MORE
FRAMING
REQUIRED)

24''   24''   24''

24''   24''   24''

Fig. 31-14 To conserve materials, locate wall openings so they fall on the spacing module.

Reducing the Clear Span of Floor Joists

The need to select large-size floor joists can be avoided by reducing the clear span. This can be done by using wider sill plates and center-bearing plates (Fig. 31-15). When the use of wider plates makes it possible to use smaller joists, the added cost of the plates is little when compared to the savings achieved.

ONE-HALF HOUSE DEPTH (D/2)

CLEAR SPAN

BAND
JOIST

JOIST

SILL
PLATE

CENTER
BEARING
PLATE

SILL PLATE	CLEAR SPAN	CENTER BEARING PLATE
2 X 4	D/2 MINUS $5\frac{1}{4}''$	2 X 4
2 X 6	D/2 MINUS $7\frac{1}{4}''$	2 X 4
2 X 6	D/2 MINUS $8\frac{1}{4}''$	2 X 6
2 X 8	D/2 MINUS $10\frac{7}{8}''$	2 X 8

USE OF WIDER BEARING PLATES TO REDUCE CLEAR SPANS

Fig. 31-15 Smaller floor joists can sometimes be used if the clear span is reduced. (*Courtesy of Southern Pine Marketing Council*)

The Arkansas System

An energy saving construction system developed by the Arkansas Power and Light Company uses 2×6 wall studs spaced 24 inches OC. This permits using full 6-inch insulation in the exterior walls. A modified truss accommodates 12 inches of insulation in the ceiling without compressing it at the eaves (Fig. 31-16).

Fig. 31-16 Modified truss design accommodates 12-inch ceiling insulation without compressing at eaves.

CHAPTER 32

LAYOUT AND CONSTRUCTION OF THE FLOOR FRAME

A floor frame consists of members fastened together to support the loads a floor is expected to bear. The floor frame is started after the foundation has been placed and has hardened. A straight and level floor frame makes it easier to frame and finish the rest of the building.

Because platform framing is used more than any other type, this chapter describes how to lay out and construct its floor frame. The knowledge gained in this chapter can be used to lay out and construct any type of floor frame.

Description and Installation of Floor Frame Members

In the usual order of installation, the floor frame consists of *girders*, *posts* or *columns*, *sill plates*, *joists*, *bridging*, and *subflooring* (Fig. 32-1).

Description of Girders

Girders are heavy beams that support the inner ends of the floor joists. Several types are commonly used.

Kinds of Girders. Girders may be made of solid wood or built up of two or more 2-inch planks. Laminated veneer lumber or glulam beams may also be used as girders. Sometimes, wide flange, I-shaped steel beams are used (Fig. 32-2).

Determining the Size of Girders and Other Structural Members. The size of solid, built-up, engineered lumber, or steel girders, or any structural component, is best determined by professional architects or engineers. There are many factors to consider when determining structural lumber or steel sizes. Only people specifically trained should select the size of a structural member. Estimating or guessing the size can have serious consequences. If undersize, the member could sag. Or, worse, it could fail and possibly cause the section of building it supports to collapse. If oversize, it would not be economical and might take up more room than necessary. If the blueprints do not specify the kind or size, have the design checked by a professional engineer.

Built-Up Girders. If built-up girders of solid lumber are used, a minimum of two members are fastened together with three 16d spikes at each end. The other spikes are staggered not farther than 32 inches apart from end to end. Sometimes 1/2-inch bolts are required. Applying glue between the pieces makes the bond stronger. Laminated veneer

Fig. 32-1 A floor frame of platform construction. (*From* Architectural Drafting & Design, *Delmar Publishers*)

Fig. 32-2 A large glulam beam is used for a girder. (*Courtesy of Trus Joist MacMillan*)

lumber may also be built up for use as girders (Fig. 32-3). The end joints of both types are placed directly over supports.

Girder Location. The ends of the girder are usually supported by a *pocket* formed in the foundation wall. The pocket should provide at least a 4-inch bearing for the girder. It should be wide enough to provide 1/2-inch clearance on both sides and the end. This allows any moisture to be evaporated by circulation of air. Thus no moisture will get into the girder, which would cause decay of the timber.

The pocket is formed deep enough to provide for shimming the girder to its designated height. An iron bearing plate is *grouted* in under the girder while it is supported temporarily (Fig. 32-4). Grouting is the process of filling in the small space between the iron plate and the bottom of the pocket

Fig. 32-4 The girder pocket in the foundation wall should be large enough to provide air space around the end of the girder.

ARRANGEMENT AND NAILING FOR
BUILT-UP GIRDERS

BUILT-UP LUMBER

1³/₄" WIDTH PIECES
• MINIMUM OF 2 ROWS 16D
  NAILS @ 12" OC
• MINIMUM OF 3 ROWS OF 16D NAILS
  @ 12" OC FOR 14", 16",
  AMD 18" BEAMS.
  2¹¹/₁₆" OR 3¹/₂"
  WIDTH PIECES
• MINIMUM OF 2 ROWS
  ¹/₂" BOLTS
  @ 24" OC STAGGERED.

BUILT-UP LVL

Fig. 32-3 Spacing of fasteners and joints of built-up girders.

with a thick paste of portland cement. Wood shims are not suitable for use under girders. The weight imposed on them compresses the wood, causing the girder to sink below its designated level.

Installing Girders

Steel girders usually come in one piece and are set in place with a crane. Wood girders are usually built-up and are erected in sections. Build one section. Set one end in the pocket in the foundation wall. Place and fasten the other end on a braced temporary support. Continue building and erecting sections until the girder is completed to the opposite pocket. A solid wood girder is installed in the same manner as a built-up girder. Half lap joints are made directly over posts or columns.

Sight the girder by eye from one end to the other. Place wedges under the temporary supports to straighten the girder. Permanent posts or columns are usually installed after the girder has some weight imposed on it by the *floor joists.* Temporary posts should be strong enough to support the weight imposed on them until permanent ones are installed.

Sills

Sills, also called *mudsills* or *sill plates,* are horizontal members of a floor frame. They lie directly on the foundation wall and provide a bearing for *floor*

joists. Sills may consist of single 2×6, 4×6, or, in many cases, a doubled 2×6 solid lumber (Fig. 32-5).

The sill is attached to the foundation wall with anchor bolts. Their size, type, and spacing are specified on the blueprints. To take up irregularities between the foundation wall and the sill, a *sill sealer* is used.

The sill sealer is a glass wool insulating material. It is used to seal against drafts, dirt, and insects. It comes 6 inches wide and in rolls of 50 feet. Unused, its thickness is one inch. It compresses to as little as 1/32 inch when the weight of the structure is upon it.

Installing Sills

Sills must be installed so they are straight, level, and to the specified dimension of the building. The level of all other framing members depends on the care taken with the installation of the sill.

Sometimes the outside edge of the sill is flush with the outside of the foundation wall. Sometimes it is set in the thickness of the wall sheathing, depending on custom or design. In the case of brick veneered exterior walls, the sill plate may be set back even farther (Fig. 32-6). Remove washers and nuts from the anchor bolts. Snap a chalk line on the top of the foundation wall in line with the inside edge of the sill.

Fig. 32-5 Sill details at the corner.

Fig. 32-6 The setback of the sill plate on the foundation wall may vary.

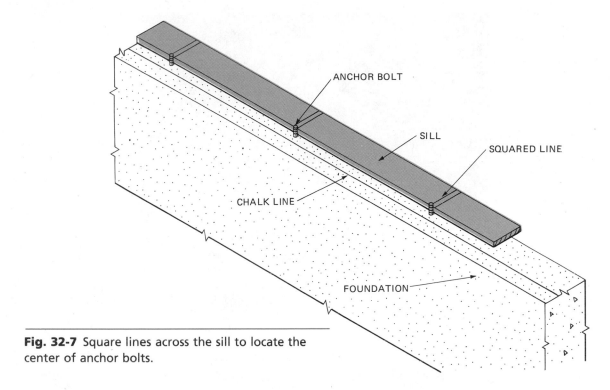

Fig. 32-7 Square lines across the sill to locate the center of anchor bolts.

Locating Anchor Bolts. Cut the sill sections to length. Hold the sill in place against the anchor bolts. Square lines across the sill on each side of the bolts. Measure the distance from the center of each bolt to the chalk line. Transfer this distance at each bolt location to the sill by measuring from the inside edge (Fig. 32-7).

Bore holes in the sill for each anchor bolt. Bore the holes at least 1/8-inch oversize to allow for adjustments. Place the sill sections in position over the anchor bolts after installing the sill sealer, if used. The inside edges of the sill sections should be on the chalk line. Replace the nuts and washers. Be careful not to overtighten the nuts, especially if the concrete wall is still *green* (not thoroughly dry and hard). This may crack the wall.

If the inside edge of the sill plate comes inside the girder pocket, notch the sill plate around the end of the girder (Fig. 32-8). Raise the ends of the girder up so it is flush with the top of the sill plate.

Floor Joists

Floor joists are horizontal members of a frame. They rest on and transfer the load to sills and girders. In residential construction, nominal 2-inch thick lumber placed on edge has been traditionally used. Wood I-beams, with lengths up to 80 feet, are being

Fig. 32-8 Notch the sill around the end of the girder, if necessary.

Fig. 32-9 In addition to solid lumber, wood I-beams are used for floor joists. A girder of parallel strand lumber supports the joists. (*Courtesy of Trus Joist MacMillan*)

specified more often today. In commercial work, steel or a combination of steel and wood trusses is frequently used (Fig. 32-9).

Joists are generally spaced 16 inches OC in conventional framing. They may be spaced 12 or 24 inches OC, depending on the type of construction and the load. The size of floor joists should be determined from the construction drawings.

Joist Framing at the Sill. Joists should rest on at least 1 1/2 inches of bearing on wood and 3 inches on masonry. In platform construction, the ends of floor joists are capped with a *band joist,* also called a *rim joist.* In a balloon frame, joists are cut flush with the outside edge of the sill (Fig. 32-10). The use of wood I-beams requires sill construction as recommended by the manufacturer for satisfactory performance of the frame (Fig. 32-11).

Joist Framing at the Girder. If joists are lapped over the girder, the minimum amount of lap is 4 inches and the maximum overhang is 12 inches. There is no need to lap wood I-beams. They come in lengths long enough to span the building. However, they may need to be supported by girders depending on the span and size of the I-beam.

Sometimes, to gain more headroom, joists may be framed into the side of the girder. There are a number of ways to do this (Fig. 32-12). Joist hangers must be used to support wood I-beams. *Web stiffeners* should be applied to the beam ends if the hanger does not reach the top flange of the beam.

Notching and Boring of Joists

Notches in the bottom or top of sawn lumber floor joists should not exceed 1/6 of the joist depth.

Notches should not be located in the middle 1/3 of the joist span. Notches on the ends should not exceed 1/4 of the joist depth.

Fig. 32-10 Typical framing at the sill with lumber joists.

A1 — BLOCKING PANEL

A2 — 2x4 MIN. "SQUASH" BLOCKS / 3/4" RIM JOIST / 1/16" / SEE DETAIL "A6" FOR ADDITIONAL INFORMATION.

A3 — USE ONLY FOR SINGLE STORY APPLICATIONS OR SECOND STORY OF TWO-STORY APPLICATIONS. / 9-1/2" AND 11-7/8" TJI® JOISTS ONLY / 3/4" RIM JOIST / SEE DETAIL "A6" FOR ADDITIONAL INFORMATION.

A4 — TJI® RIM JOIST / NOTE: MUST HAVE 1-3/4" MINIMUM JOIST BEARING AT ENDS.

A5 — 9-1/2" AND 11-7/8" TJI® JOISTS ONLY WHEN USING DOUBLE 3/4" RIM JOIST. / DOUBLE 3/4" OR MICRO=LAM® LVL RIM JOIST / SEE DETAIL "A6" FOR ADDITIONAL INFORMATION.

A6 — BLOCKING PANELS USED FOR BRACING. / 3/4" RIM JOIST. ALLOWED WITH 9-1/2" AND 11-7/8" JOIST ONLY, UNLESS USED WITH 2x4 MIN. "SQUASH" BLOCKS AS SHOWN IN DETAIL "A2."

WOOD I-BEAM JOISTS

LOAD BEARING OR SHEAR WALL ABOVE (MUST STACK OVER WALL BELOW). / BLOCKING PANEL / GIRDER / NOTE: WEB STIFFENERS MAY BE REQUIRED.

LOAD BEARING WALL ABOVE (MUST STACK OVER WALL BELOW). / 2x4 MIN. "SQUASH" BLOCKS / GIRDER / NOTE: WEB STIFFENERS MAY BE REQUIRED. / SQUASH BLOCK DETAIL / 1/16"

NOTE: WEB STIFFENERS MAY BE REQUIRED.

GIRDER / TOP MOUNT HANGER / FACE MOUNT HANGER / NOTE: WEB STIFFENERS ARE REQUIRED IF THE SIDES OF THE HANGER DO NOT LATER-ALLY SUPPORT THE JOIST TOP FLANGE.

CHECK MANUFACTURERS' APPLICATION RECOMMENDATIONS

Fig. 32-11 Wood I-beam framing details at the sill.

Fig. 32-12 Joist framing details at the girder.

Holes bored in joists for piping or wiring should not be larger than 1/3 of the joist depth. They should not be closer than 2 inches to the top or bottom of the joist (Fig. 32-13).

Some wood I-beams are manufactured with 1 1/2 inch perforated knockouts in the web at approximately 12 inches on center along its length. This allows easy installation of wiring and pipes. To cut other size holes in the web, consult the manufacturer's specifications guide. Do not cut or notch the flanges of wood I-beams.

Laying Out Floor Joists

The locations of floor joists are marked on the sill plate. A squared line marks the side of the joist. An X to one side of the line indicates on which side of the line the joist is to be placed (Fig. 32-14).

Floor joists must be laid out so that the ends of *plywood subfloor* sheets fall directly on the center of floor joists. Start the joist layout by measuring the joist spacing from the end of the sill. Measure back one-half the thickness of the joist. Square a line

NO NOTCHES IN THE MIDDLE THIRD OF THE SPAN

⅓ SPAN ⅓ SPAN ⅓ SPAN

DIAMETER OF HOLES
NOT TO EXCEED ⅓
JOIST DEPTH

SOLID LUMBER
FLOOR JOIST

DEPTH OF NOTCH
NOT TO EXCEED
⅙ JOIST DEPTH

HOLES NOT CLOSER
THAN 2″ TO TOP OR
BOTTOM EDGE OF
JOIST

NOTCHES ON END
NOT TO EXCEED
¼ JOIST DEPTH

A 1/2" HOLE
CAN BE CUT
ANYWHERE
IN THE WEB.

WOOD I-BEAM

DO NOT CUT
HOLES IN
CANTILEVER

FOR HOLES LARGER THAN 1-1/2", AND SQUARE
OR RECTANGULAR CUT-OUTS, CHECK THE
MANUFACTURER'S APPLICATION GUIDE.

WOOD I-BEAM JOIST

DO NOT CUT OR
NOTCH THE FLANGES
OF WOOD I-BEAMS

Fig. 32-13 Allowable notches, holes, and cutouts in floor joists.

across the sill. This line indicates the side of the joist closest to the corner. Place an X on the side of the line on which the joist is to be placed (Fig. 32-15).

From the squared line, measure and mark the spacing of the joists along the length of the building. Place an X on the same side of each line as for the first joist location.

When measuring for the spacing of the joists, use a tape stretched along the length of the build-

ing. Most tapes have prominent markings for 16 inch spacing. Using a tape in this manner is more accurate. Measuring and marking each space individually with a rule or pocket tape generally causes a gain in the spacing. If the spacing is not laid out accurately, the plywood subfloor may not fall in the center of some floor joists. Time will then be lost either cutting the plywood back or adding strips of lumber to the floor joists (Fig. 32-16).

ON CENTER SPACING

16"

16"

JOIST
(KEEP CROWNED
EDGES UP)

SILL PLATE

FOUNDATION
WALL

Fig. 32-14 A line marks the side of a joist. An X tells on which side of the line it is placed.

STEP BY STEP
P R O C E D U R E S

THIS DISTANCE IS 1/2 THE
JOIST THICKNESS. IF THE
JOIST IS 1 1/2" THICK,
THE DISTANCE IS 3/4".

DASHED LINE IS CENTER LINE OF JOIST

16"

STEP 1– MEASURE IN FROM
CORNER THE JOIST SPACING.

16"

STEP 2– MEASURE BACK OR
FORWARD 1/2 THE
JOIST THICKNESS.
IN THIS CASE, THE
MEASUREMENT IS
FORWARD.

STEP 3 – SQUARE A LINE
ACROSS THE SILL PLATE
AND PLACE AN X
ON THE SIDE OF THE
LINE WHERE THE JOIST
IS TO BE PLACED.

SILL PLATE

FOUNDATION

STEP 4 – CONTINUE THE ON CENTER
SPACING ALONG THE
LENGTH OF THE BUILDING.
USE A STEEL TAPE
STRETCHED OVER THE
ENTIRE LENGTH.

16"
16"
16"
16"

SILL PLATE

FOUNDATION

Fig. 32-15 Laying out the location of floor joists on the sill plate.

Laying Out Floor Openings. After marking floor joists for the whole length of the building, study the plans for the location of floor openings. Mark the sill plate, where joists are to be doubled, on each side of large floor openings. Identify the layout marks that are not for full-length floor joists. Shortened floor joists at the ends of floor openings are called *tail joists.* They are usually identified by changing the X to a T. Lay out for partition supports, or wherever doubled floor joists are required (Fig. 32-17).

Lay out the floor joists on the girder and on the opposite wall. If the joists are in-line, Xs are made on the same side of the mark on both the girder and the sill plate on the opposite wall. If the joists are lapped, an X is placed on both sides of the mark at the girder and on the opposite side of the mark on the other wall (Fig. 32-18).

Installing Floor Joists

Stack the necessary number of full-length floor joists at intervals along both walls. Each joist is carefully sighted along its length by eye. Any joist with a severe crook or other warp should not be used

Fig. 32-16

(Fig. 32-19). Joists are installed with the crowned edge up.

Keep the end of the floor joist in from the outside edge of the sill plate by the thickness of the band joist. Toenail the joists to the sill and girder with 10d common nails. Spike the joists together if they lap at the girder (Fig. 32-20). When all floor joists are in position, they are sighted by eye from end to end and straightened. They are held straight by strips of 1×3s tacked to the top of the joists about in the middle of the joist span.

Wood I-beams are installed using standard tools. They can be easily cut to any required length at the job site. A minimum bearing of 1 3/4 inches is required at joist ends and 3 1/2 inches over the girder. The wide, straight wood flanges on the joist make nailing easier, especially with pneumatic framing nailers (Fig. 32-21). Nail joists at each bear-

Fig. 32-17 Joists are doubled on each side of large floor openings and under some partitions.

Fig. 32-18 Lay out lines for the floor joists on the girder and the opposite wall (joists are lapped over the girder).

Fig. 32-19 Stack floor joists against both walls. Sight each joist. Install joists with the crowned edge up.

Fig. 32-20 A floor frame with joists lapped over the girder.

Fig. 32-21 Installing wood I-beam floor joists. Notice how extremely long lengths are easily handled. (*Courtesy of Trus Joist MacMillan*)

ing with one 8d or 10d nail on each side. Keep nails at least 1 1/2 inches from the ends to avoid splitting.

Doubling Floor Joists

For added strength, doubled floor joists must be securely fastened together. Their top edges must be even. In most cases, the top edges do not lie flush with each other. They must be brought even before they can be spiked together.

To bring them flush, toenail down through the top edge of the higher one, at about the center of their length. At the same time squeeze both together tightly by hand. Use as many toenails as necessary, spaced where needed, to bring the top edges flush (Fig. 32-22). Usually no more than two or three nails are needed. Then, fasten the two pieces securely together. Drive nails from both sides, staggered from top to bottom, about 2 feet apart.

Framing Floor Openings

Large openings in floors should be framed before floor joists are installed. This is because room is needed for end nailing. To frame an opening in a floor, first fasten the *trimmer joists* in place. Trimmer joists are full-length joists that run along the inside of the opening. Mark the location of the *header joists* on the trimmers. Header joists are members of the opening that run at right angles to the floor joists.

Cut four *header joists* to length by taking the measurement at the sill between the trimmers. Taking the measurement at the sill where the trimmers are fastened, rather than at the opening, is standard practice. A measurement between trimmers taken at the opening may not be accurate. There may be bow in the trimmer joists (Fig. 32-23).

HELPFUL HINTS

Bring the tops of doubled floor joists flush by first toenailing the edges before spiking them together.

STEP 1. FASTEN BOTH ENDS OF THE JOISTS TO THEIR BEARINGS.

STEP 2. TOENAIL INTO THE TOP EDGE OF THE JOIST WITH THE GREATER CROWN TO BRING TOP EDGES FLUSH BEFORE FASTENING TOGETHER.

STEP 3. SPIKE JOISTS TOGETHER.

DOUBLED FLOOR JOISTS

SECTION

Fig. 32-22

Place two headers, one for each end of the opening, on the sill between the trimmers. Transfer the layout of the tail joists on sill to the headers. Fasten a header on each end of the opening in position by driving spikes through the side of the trimmer into the ends of the headers. Fasten the tail joists in position. Double up the headers. Finally, double up the trimmer joists. Figure 32-24 shows the sequence of operations used to frame a floor opening. This particular sequence allows you to end nail the members rather than toenailing them. End nails are sufficient for headers up to 6 feet. Use joist hangers on headers over 6 feet long and on tail joists over 12 feet long.

Installing the Band Joist. After all the openings have been framed and all floor joists are fastened, install the band joist. This closes in the ends of the floor joists. Band joists may be lumber of the same size as the floor joists. They also may be a single or double layer of 3/4-inch laminated veneer lumber.

Fasten the band joist into the end of each floor joist. If wood I-beams are used as floor joists, drive one nail into the top and bottom flange. The band joist is also toenailed to the sill plate at about 16-inch intervals.

Bridging

Bridging is installed in rows between floor joists at intervals not exceeding 8 feet. For instance, floor

GIRDER

LOCATION OF OPENING HEADERS

FLOOR OPENING

TRIMMER JOIST

DETERMINE HEADER LENGTH HERE

SILL PLATE

Fig. 32-23 The length of the header for the opening should be determined at the sill plate.

STEP BY STEP
P R O C E D U R E S

Fig. 32-24 Sequence of operations to frame a floor opening.

GIRDER

STEP 5
INSTALL OUTSIDE JOISTS

TAIL JOISTS

STEP 4
INSTALL INSIDE HEADERS

STEP 6 - INSTALL
REGULAR SPACED
FULL-LENGTH JOISTS

TAIL JOISTS

SILL PLATE

GIRDER

TRIMMERS MAY NEED TO BE
INSTALLED IF THE SIDES OF THE
OPENING DO NOT FALL ON THE
SPACING

JOIST HANGER
SHOULD BE
USED TO
SUPPORT TRIMMER

REGULAR SPACED
FULL-LENGTH JOISTS

TAIL JOISTS

SILL PLATE

JOIST
HANGER

Fig. 32-24 *(continued)*

WOOD CROSS BRIDGING

JOIST

(A) WOOD CROSS BRIDGING

SOLID BRIDGING

(B) SOLID BRIDGING

(C) METAL CROSS-BRIDGING INSTALLS VERY QUICKLY AFTER SUB-FLOORING IS IN PLACE.

Fig. 32-25 Types of bridging.

joists with spans 8 to 16 feet need one row of bridging in the center of the span. Its purpose is to distribute a concentrated load on the floor over a wider area. Although codes may not require bridging, many builders install it because of customary practice.

Bridging may be solid wood, wood cross-bridging, or metal cross-bridging (Fig. 32-25). Usually solid wood bridging is the same size as the floor joists. It is installed in an offset fashion to permit end nailing.

Wood cross-bridging should be at least nominal 1×3 lumber with two 6d nails at each end. It is placed in double rows that cross each other in the joist space.

Metal cross-bridging is available in different lengths for particular joist size and spacing. It is usually made of 18-gauge steel, and is 3/4 inch wide. It comes in a variety of styles. It is applied in a way similar to that used for wood cross-bridging.

Laying Out and Cutting Wood Cross-Bridging

Wood cross-bridging may be laid out using a framing square. Determine the actual distance between floor joists and the actual depth of the joist. For example, 2×10 floor joists 24 inches OC measure 22 1/2 inches between them. The actual depth of the joist is 9 1/4 inches.

Hold the framing square on the edge of a piece of bridging stock. Make sure the 9 1/4-inch mark of the tongue lines up with the lower edge of the stock. Also make sure the 22 1/2-inch mark of the blade lines up with the upper edge of the stock. Mark lines along the tongue and blade across the stock.

Turn the square. Keeping the same face up, mark along the tongue (Fig. 32-26). The bridging may then be cut using a radial arm saw. Tilt the blade and use a stop set to cut duplicate lengths.

Installing Bridging

Determine the centerline of the bridging. Snap a chalk line across the tops of the floor joist from one end to the other. Square down from the chalk line to the bottom edge of the floor joists on both sides.

HELPFUL HINTS

How to lay out wood cross-bridging with a framing square.

DISTANCE BETWEEN JOISTS

JOIST DEPTH

EDGE OF BRIDGING STOCK

MARK HERE

MARK HERE

STEP 1 – HOLD SQUARE IN FIRST POSITION AS INDICATED BY DOTTED LINE. MARK STOCK ALONG TONGUE & BLADE OF SQUARE.
STEP 2 – ROTATE THE SQUARE TO THE POSITION AS INDICATED BY THE SOLID LINE. MARK ALONG THE TONGUE OF THE SQUARE

Fig. 32-26

CHALK LINE

FIRST INSTALL SOLID BRIDGING IN EVERY OTHER SPACE FOR THE WHOLE LENGTH OF THE BUILDING.

THEN FILL IN THE REMAINING SPACES.

Fig. 32-27

Solid Wood Bridging. To install solid wood bridging, cut the pieces to length. Install pieces in every other joist space on one side of the chalk line. Fasten the pieces by spiking through the joists into their ends. Keep the top edges flush with the floor joists. Install pieces in the remaining spaces on the opposite side of the line in a similar manner (Fig. 32-27).

Wood Cross-Bridging. To install wood cross-bridging, start two 6d nails in one end of the bridging. Fasten it flush with the top of the joist on one side of the line. Nail only the top end. The bottom ends are not nailed until the subfloor is fastened down.

Fasten another piece of bridging in the same space to the other joist. Make sure it is flush with the top and on the opposite side of the line. Continue installing bridging in the other spaces, but alternate so that the top ends in each space are opposite each other (Fig. 32-28).

Metal Cross-Bridging. Metal cross-bridging is fastened in a manner similar to that used for wood cross-bridging. The method of fastening may differ according to the style of the bridging. Usually the bridging is fastened to the top of the joists through predrilled holes in the bridging. Because the metal is thin, nailing to the top of the joists does not interfere with the subfloor.

Some types of metal cross-bridging have steel prongs that are driven into the side of the floor joists. This bridging can be installed from below to layout lines made previous to the installation of the subfloor.

Columns

Girders may be supported by framed walls, wood posts, or steel columns. Usually, columns of steel pipe filled and reinforced with concrete are used. These are commonly called *lally columns* (Fig. 32-29). Metal plates are used at the top and bottom of the columns to distribute the load over a wider area. The plates have predrilled holes so that they may be fastened to the girder. Notched sections prevent the columns from slipping off the plates. Column size should be determined from the blueprints.

Installing Columns

After the floor joists are installed and before any more weight is placed on the floor, the temporary posts supporting the girder are replaced with permanent posts or columns. Straighten the girder by stretching a line from end to end. Measure accurately from the column footing to the bottom of the girder. Hold a strip of lumber on the column footing. Mark it at the bottom of the girder. Transfer this mark to the column. Deduct the thickness of the top and bottom plate.

To mark around the column so it has a square end, wrap a sheet of paper around it. Keeping the

STEP BY STEP
P R O C E D U R E S

STEP 1– SNAP CHALK LINE ACROSS
TOPS OF FLOOR JOISTS
IN CENTER OF BRIDGING ROW.

STEP 3– FASTEN TOP ENDS OF BRIDGING
SO THEY OPPOSE EACH OTHER
ON THE SAME SIDE
OF THE CHALK LINE.

STEP 2– SQUARE LINES
DOWN FROM CHALK LINE
ON BOTH SIDES OF
FLOOR JOISTS.

FLOOR
JOIST

STEP 4– LEAVE BOTTOM ENDS LOOSE
UNTIL SUBFLOOR IS APPLIED.
THEN FASTEN SO EDGE LINES
UP WITH SQUARED LINE.

Fig. 32-28 Installing cross-bridging.

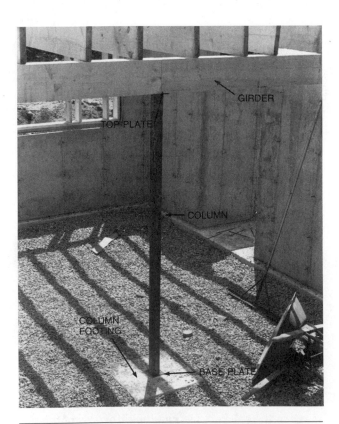

Fig. 32-29 Columns of steel pipe filled and reinforced with concrete are commonly called lally columns.

edges even, mark along the edge of the paper (Fig. 32-30).

Cut through the metal along the line using a hacksaw. Snap the column waste off by hitting the end with a hammer. Trim off any protruding concrete. Install the columns in a plumb position under the girder and centered on the footing. Fasten the top plates to the girder with lag screws. If the girder is steel, then holes must be drilled. The plates are then bolted to the girder, or they may be welded to the girder. The bottom of the columns are held in place when the finish concrete basement floor is placed around them. If there is to be no floor, then the bottom plate must be anchored to the footing.

Wood posts are installed in a similar manner, except their bottoms are placed on a pedestal footing (Fig. 32-31).

Subflooring

Subflooring is used over joists to form a working platform. This is also a base for finish flooring, such as hardwood flooring, or underlayment for carpet or resilient tiles. APA-Rated Sheathing Exposure 1 is generally used for subflooring in a two-layer floor system. Specifications for selecting and fastening panels is shown in Figure 32-32. APA-Rated Sturd-I-Floor panels are used when a single-layer subfloor and underlayment system is desired.

STEP 2 – KEEP EDGES OF PAPER EVEN

STEP 1 – WRAP WIDE SHEET OF PAPER AROUND COLUMN

STEP 3 – MARK AROUND COLUMN ALONG EDGE OF PAPER

LALLY COLUMN

Fig. 32-30

STEEL PIN
MOISTURE BARRIER
CONCRETE PEDESTAL
CONCRETE FOOTING

OPTIONAL NON-SHRINK GROUT

COLUMN BASE

CONCRETE PEDESTAL BASE

ELEVATED COLUMN BASE

SOLID WOOD & PARALLEL STRAND LUMBER POST BASE DETAILS

Fig. 32-31 The bottom of wood posts usually rests on a pedestal type footing to keep them above the grade.

Blocking is required under the joints of these panels unless tongue-and-groove edges are used. Figure 32-33 is a selection and fastening guide for APA Sturd-I-Floor panels.

Applying Plywood Subflooring

Measure in 4 feet from each corner. Snap a line across the tops of the floor joists from one end to the other. Start with a full panel. Fasten the first row to the chalk line (Fig. 32-34). Leave a 1/16-inch space at all panel end joints to allow for expansion.

Start the second row with a half-sheet to stagger the end joints. Continue with full panels to finish the row. Leave a 1/8-inch space between panel edges. All end joints are made over joists.

Continue laying and fastening plywood sheets in this manner until the entire floor is covered (Fig. 32-35, p. 342). Leave out sheets where there are to be openings in the floor. Snap chalk lines across the edges and ends of the building.

Trim overhanging plywood with a portable electric circular saw.

Estimating Materials

Floor Joists. To determine the number of floor joists needed in a floor frame, divide the length of the building by the spacing and add one. Multiply by the number of rows of floor joists. Add the number needed for doubling and for band joists.

Bridging. To determine the total *linear feet* of wood cross-bridging needed, multiply the length of the building by 3 for each row of bridging. Linear feet is a measurement of length.

Panel Subfloor. To determine the number of rated panels of subflooring required, divide the floor area by 32. A rectangular floor area is found by multiplying its width by its length. The area of other shapes can be found by using appropriate formulas.

WOOD STRIP, WOOD BLOCKS, OR
LIGHTWEIGHT CONCRETE FLOORING

STAGGER END JOINTS (OPTIONAL)

1/8" SPACING IS RECOMMENDED AT
ALL EDGE AND END JOINTS UNLESS
OTHERWISE INDICATED BY PANEL
MANUFACTURER

STRENGTH AXIS

2x JOISTS, "I" JOISTS
OR FLOOR TRUSSES

APA RATED SHEATHING

NOTE:
PROVIDE ADEQUATE VENTILATION
AND USE GROUND COVER VAPOR
BARRIER IN CRAWL SPACE. SUBFLOOR
MUST BE DRY BEFORE APPLYING
SUBSEQUENT LAYERS.

APA Panel Subflooring (APA RATED SHEATHING)[a]

Panel Span Rating	Minimum Panel Thickness (in.)	Maximum Span (in.)	Nail Size & Type[e]	Maximum Nail Spacing (in.)	
				Supported Panel Edges	Intermediate Supports
24/16	7/16	16	6d common	6	12
32/16	15/32	16[b]	8d common[c]	6	12
40/20	19/32	20[b][d]	8d common	6	12
48/24	23/32	24	8d common	6	12
60/32	7/8	32	8d common	6	12

(a) For subfloor recommendations under ceramic title, refer to Table 9. For subfloor recommendations under gypsum concrete, contact manufacturer of floor topping.

(b) Span may be 24 inches if 3/4-inch wood strip flooring is installed at right angles to joists.

(c) 6d common nail permitted if panel is 1/2 inch or thinner.

(d) Span may be 24 inches if a minimum 1-1/2 inches of lightweight concrete is applied over panels.

(e) Other code-approved fasteners may be used.

Fig. 32-32 APA Panel Subflooring selection and fastening guide. (*Courtesy of American Plywood Association*)

APA Rated Sturd-I-Floor [a]

Span Rating (Maximum Joist Spacing) (in.)	Minimum Panel Thickness[b] (in.)	Fastening: Glue-Nailed[c]			Fastening: Nailed-Only		
		Nail Size and Type	Maximum Spacing (in.)		Nail Size and Type	Maximum Spacing (in.)	
			Supported Panel Edges	Intermediate Supports		Supported Panel Edges	Intermediate Supports
16	19/32	6d ring- or screw-shank[d]	12	12	6d ring- or screw-shank	6	12
20	19/32	6d ring- or screw-shank[d]	12	12	6d ring- or screw-shank	6	12
24	23/32	6d ring- or screw-shank[d]	12	12	6d ring- or screw-shank	6	12
	7/8	8d ring- or screw-shank[d]	6	12	8d ring- or screw-shank	6	12
32	7/8	8d ring- or screw-shank[d]	6	12	8d ring- or screw-shank	6	12
48	1-3/32	8d ring- or screw-shank[e]	6	(f)	8d ring- or screw-shank[e]	6	(f)

(a) Special conditions may impose heavy traffic and concentrated loads that require construction in excess of the minimums shown. See page 23 for heavy duty floor recommendations.

(b) Panels in a given thickness may be manufactured in more than one Span Rating. Panels with a Span Rating greater than the actual joist spacing may be substituted for panels of the same thickness with a Span Rating matching the actual joist spacing. For example, 19/32-inch-thick Sturd-I-Floor 20 oc may be substituted for 19/32-inch-thick Sturd-I-Floor 16 oc over joists 16 inches on center.

(c) Use only adhesives conforming to APA Specification AFG-01, applied in accordance with the manufacturer's recommendations. If OSB panels with sealed surfaces and edges are to be used, use only solvent-based glues; check with panel manufacturer.

(d) 8d common nails may be substituted if ring- or screw-shank nails are not available.

(e) 10d common nails may be substituted with 1-1/8-inch panels if supports are well seasoned.

(f) Space nails maximum 6 inches for 48-inch spans and 12 inches for 32-inch spans.

Fig. 32-33 APA-Rated Sturd-I-Floor selection and fastening guide. (*Courtesy of American Plywood Association*)

Fig. 32-34 Applying a rated panel subfloor.

STEP BY STEP
P R O C E D U R E S

CHALK LINE

4'–0"

4'–0"

STEP 1– SNAP A LINE ACROSS
THE TOPS OF THE
FLOOR JOISTS 4'–0" IN
FROM THE OUTSIDE EDGE.

STEP 5– SNAP LINES
AND TRIM EDGES AND
ENDS FLUSH WITH FRAME.

STEP 4– CONTINUE
LAYING SHEETS
AND STAGGERING
JOINTS UNTIL FLOOR
IS COVERED.

STEP 3– START 2ND
ROW WITH A HALF SHEET
& CONTINUE WITH FULL
SHEETS.

STEP 2– START 1ST ROW WITH
A FULL SHEET WITH EDGE
TO THE CHALK LINE. CONTINUE
LAYING SHEETS TO OPPOSITE END.

Fig. 32-35 Layout of a panel subfloor.

CHAPTER 33

CONSTRUCTION TO PREVENT TERMITE ATTACK

Of all the destructive wood pests, *termites* are the most common. They cause tremendous economic loss annually. They attack wood throughout most of the country, but are more prevalent in the warmer sections (Fig. 33-1). Buildings should be designed and constructed to minimize termite attack.

Termites

Termites play a beneficial role in their natural habitat. They break down dead or dying plant material to enrich the soil However, when termites feed on wood structures, they become pests.

Kinds of Termites

There are three kinds of termites: *drywood, dampwood,* and *subterranean.*

Drywood Termites. Drywood termites enter a building, usually through attic or foundation vents. They attack sheathing and structural members. They also infest wood door and window frames and furniture. However, the colonies are small. Unless unchecked, they cause less damage to buildings than other kinds of termites. Treatment is usually tent fumigation of the entire building with a toxic gas.

Dampwood Termites. Dampwood termites infest wood kept wet by lawn sprinklers, leaking toilets, showers, pipes, roofs, and other places where wood is kept damp. They are capable of doing great damage to a structure if undetected. Entrance into the building is usually where a continual source of moisture keeps wood wet. Discovery of a leak sometimes reveals a dampwood termite infestation. Once the wood is dry, dampwood termites will leave.

Subterranean Termites. Subterranean termites live in the ground. They are the most destructive species because they have such large underground colonies (Fig. 33-2). For protection against drying out, they must stay in close contact with the soil and its moisture. Above ground, they build earthen *shelter tubes* to protect themselves from the drying effects of the air. These tubes are usually built over a surface for support. Termites can build unsup-

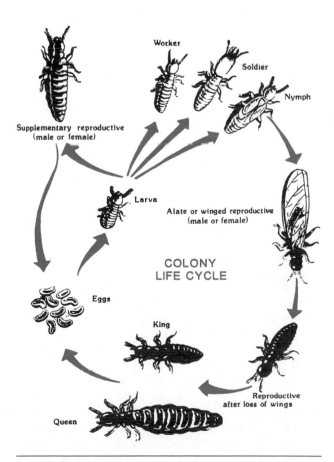

○ Region I Very Heavy
◐ Region II Moderate to Heavy
◔ Region III Slight to Moderate
○ Region IV None to Slight

Fig. 33-1 Degree of subterranean termite hazard in the United States.

Fig. 33-2 Typical subterranean termite life cycle.

Fig. 33-3 Subterranean termites can build unsupported shelter tubes as high as 12 inches. Heat attracts them in their attempt to reach wood. (*Courtesy of* The Termite Report, *Pear Publishing; Don Pearman, photographer*)

ported tubes as high as 12 inches in their effort to reach wood (Fig. 33-3).

Treatment for subterranean termites generally consists of correcting conditions favorable to infestation, installation of ground-to-wood termite barriers, and chemical treatment of the foundation and soil. Chemicals should only be applied by trained technicians.

Techniques to Prevent Termites

Protection against subterranean termites should be considered during planning and construction of a building. Improper design and poor construction practices could lead to termite infestation after completion of the building. Preventative efforts in the planning stage and during construction may save the future owner much anxiety and expense. All the techniques used for the prevention of termite attack are based on keeping the wood in the structure dry (equilibrium moisture content) and making it as difficult as possible for termites to get to the wood. In lumber, a moisture content below 20 percent also

prevents the growth of fungi, which causes wood to rot.

The Site

All tree stumps, roots, branches, and other wood debris should be removed from the building site. Do not bury it on the site. Footing and wall form plank, boards, stakes, spreaders, and scraps of lumber should be removed from the area before backfilling around the foundation. Lumber scraps should not be buried anywhere on the building site. None should be left on the ground beneath or around the building after construction is completed.

The site should be graded to slope away from the building on all sides. The outside finished grade should always be equal to or below the level of the soil in crawl spaces. This ensures that water is not trapped underneath the building (Fig. 33-4).

Chemical treatment of the soil before construction is one of the best methods of preventing termite attack. This should be a supplement to, and not a substitute for proper building practices.

Perforated drain pipe should be placed around the foundation, alongside the footing. This will drain water away from the foundation (Fig. 33-5). The foundation drain pipe should be sloped so water can drain to a lower elevation or a *dry well* some distance from the building. A dry well is a pit in the ground filled with stone to absorb the water from the drain pipe. Gutters and downspouts should be installed to lead roof water away from the foundation. Down-

Fig. 33-4 The finished grade in the crawl space and around the outside of the foundation.

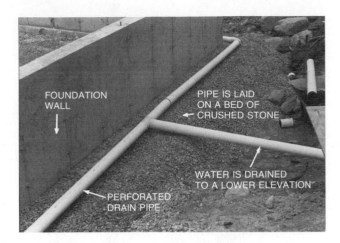

Fig. 33-5 Perforated drain pipe is placed alongside the foundation footing to drain water away from the building.

Fig. 33-6 The top course of concrete block walls should be capped or filled.

spouts may be connected to the foundation drain pipe to facilitate moving the water quickly.

Crawl Spaces

Solid concrete foundation walls should be properly reinforced and cured to prevent the formation of cracks. Cracks as little as 1/32 inch wide permit the passage of termites.

Concrete block walls should be either capped with a minimum of 4 inches of reinforced concrete or the top course filled completely with concrete (Fig. 33-6).

Air should be circulated in crawl spaces by means of ventilators placed to leave no pockets of stagnant air. In general, the total area of ventilation openings should be equal to 1/150th of the ground area of the crawl space. Shrubbery should be kept away from openings to permit free circulation of air. There should be access to the crawl space for inspection of inner wall surfaces for termite tubes.

In crawl spaces and other concealed areas, clearance between the bottom of floor joists and the ground should be at least 18 inches and at least 12 inches for beams and girders (Fig. 33-7).

Keep all plumbing and electrical conduits clear of the ground in crawl spaces. Suspend them from girders or joists. Do not support them with wood blocks or stakes in the ground. The soil around pipes extending from the ground to the wood above should be treated with chemicals.

Slab-on-Grade

Slab-on-grade is one of the most susceptible types of construction to termite attack. Termites gain access to the building over the edge of the slab, or through expansion joints, openings around plumbing, and cracks in the slab. Termite infestations in

this type of construction are difficult to detect and control. For slab-on-grade construction, it is important to have the soil treated with chemicals before placing the concrete slab.

The monolithic slab provides the best protection against termites. The floor and footing are placed in one continous operation, eliminating joints that permit hidden termite entry (Fig. 33-8). Proper curing of the slab helps eliminate the development of cracks through which termites can gain access to the wood above.

One type of independent slab extends completely across the top of the foundation. This prevents hidden termite entry. The lower edge of the slab should be open to view from the outside.

The top of the slab should be at least 8 inches above the grade (Fig. 33-9).

Independent slabs that rest either part-way on or against the side of the foundation wall are the least reliable. Termites may gain hidden access to the wood through expansion joints (Fig. 33-10). Fill the spaces around expansion joints, pipes, conduit, ducts, or steel columns with hot roofing grade coaltar pitch.

Exterior Slabs. Spaces beneath concrete slabs for porch floors, entrance platforms, and similar units against the foundation should not be filled. Leave them open with access doors for inspection. If this cannot be done and spaces must be filled, have the soil treated for termites by a professional.

SECTION

Fig. 33-7 Provide adequate clearance between wood and soil in crawl spaces.

Fig. 33-8 In slab-on-grade construction, the monolithic slab provides the best protection against termites.

Fig. 33-9 An independent slab that extends across the top of the foundation wall prevents hidden termite attack.

(A) EDGE OF SLAB RESTS ON LEDGE OF THE FOUNDATION WALL,
(B) SLAB RESTS ENTIRELY ON THE GROUND (FLOATING).

SECTIONS THRU SLABS AT FOUNDATION WALLS.

Fig. 33-10 Slab construction where termites may obtain access to wood through expansion joints.

Exterior Woodwork

Wall siding should not extend more than 2 inches below the top of foundation walls. It should be at least 6 inches above the finished grade.

Porch supports should be placed not closer than 2 inches from the building to prevent hidden access by termites. Wood steps should rest on a concrete base that extends at least 6 inches above the ground.

Door jambs, posts, and similar wood parts should never extend into or through concrete floors.

Termite Shields

If termite shields are properly designed, constructed, installed, and maintained, they will force termites in the open. This will reveal any tubes constructed around the edge and over the upper surface of the shield (Fig. 33-11). However, research has shown that termite shields have not been effective in preventing termite infestations. Because of improper installation and infrequent inspection, they are not presently recommended by government agencies for detection and prevention of termite attack. However, check local building codes that may mandate their use.

Use of Pressure-Treated Lumber

Lumber in which preservatives are forced into the wood's cells under pressure is commonly called *pressure-treated lumber* (Fig. 33-12). Although other grades are available, two are generally used. *Above ground* is used for sill plates, joists, girders, decks, and similar members. *Ground contact* is suitable for contact with soil or fresh water. Typical grade stamps are shown in Figure 33-13. Special grades are manufactured for saltwater immersion and wood foundations.

Check building codes for requirements concerning the use of pressure-treated lumber. Generally, building codes require the use of pressure-treated lumber for the following structural members:

- Wood joists or the bottom of structural floors without joists that are located closer than 18 inches to exposed soil.

- Wood girders that are closer than 12 inches to exposed soil in crawl spaces or unexcavated areas.

- Sleeper, sills, and foundation plates on a concrete or masonry slab that is in direct contact with the soil.

(A) EXTERIOR WALL WITH WOOD SIDING

TERMITE
SHIELD

SOLID MASONRY
CAP

45° 2"

SOLDER

SOIL COVER
WHEN NEEDED

2"
2"

45°

12" MIN.

(B) EXTERIOR WALL WITH BRICK VENEER

FLASHING

TERMITE
SHIELD

2"

2"

WEEP HOLE

SHEATHING PAPER

8" MIN

DAMP-PROOFED

(C) OVER INTERIOR PIER

FLOOR JOIST

2"

2"

GIRDER

TERMITE
SHIELD

3 1/2"

12" MIN

Fig. 33-11 Installation of termite shields.

Fig. 33-12 Preservatives are forced into lumber under pressure in large cylindrical tanks. (*Courtesy of Southern Pine Marketing Council*)

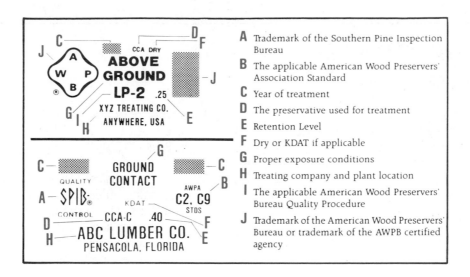

A	Trademark of the Southern Pine Inspection Bureau
B	The applicable American Wood Preservers' Association Standard
C	Year of treatment
D	The preservative used for treatment
E	Retention Level
F	Dry or KDAT if applicable
G	Proper exposure conditions
H	Treating company and plant location
I	The applicable American Wood Preservers' Bureau Quality Procedure
J	Trademark of the American Wood Preservers' Bureau or trademark of the AWPB certified agency

Fig. 33-13 Typical grade stamps for pressure-treated lumber. (*Courtesy of Southern Pine Marketing Council*)

Termites generally will not eat treated lumber. They will tunnel over it to reach untreated wood. Their shelter tubes then may be exposed to view and their presence easily detected upon inspection.

Follow these safety rules when handling pressure-treated lumber.

- Wear eye protection and a dust mask when sawing or machining treated wood (Fig. 33-14).

- When the work is completed, wash areas of skin contact thoroughly before eating or drinking.

- Clothing that accumulates sawdust should be laundered separately from other clothing and before reuse.

- Dispose of treated wood by ordinary trash collection or burial. Do not burn treated wood. The chemical is retained in the ash could pose a health hazard.

Fig. 33-14 Know the safety precautions when handling pressure-treated lumber. (*Courtesy of Southern Pine Marketing Council*)

Review Questions

Select the most appropriate answer.

1. A platform frame is easy to erect because
   a. only one-story buildings are constructed with this type of frame.
   b. each platform may be constructed on the ground.
   c. at each level a flat surface is provided on which to work.
   d. less framing members are required.

2. One of the advantages of the balloon frame is that
   a. the bottom plates act as firestops.
   b. there is little vertical shrinkage in the frame.
   c. the second-floor joists rest on a ribbon instead of a plate.
   d. it is stronger, stiffer, and more resistant to lateral pressures.

3. A heavy beam that supports the inner ends of floor joists is called a
   a. pier.   c. stud.
   b. girder.   d. sill.

4. That member of a floor frame that rests directly on the foundation wall is called a
   a. pier.   c. stud.
   b. girder.   d. sill.

5. To mark a square end on a round column
   a. use a square.
   b. measure down from the existing end.
   c. use a pair of dividers.
   d. wrap a piece of paper around it.

6. Anchor bolts are spaced a maximum of
   a. 6 feet OC.   c. 10 feet OC.
   b. 8 feet OC.   d. 12 feet OC.

7. To protect against termites keep wood in crawl spaces and other concealed areas at least
   a. 8 inches above the ground.
   b. 12 inches above the ground.
   c. 18 inches above the ground.
   d. 24 inches above the ground.

8. When floor joists rest on wood they should have at least
   a. 4 inches of bearing.
   b. 3 1/2 inches of bearing.
   c. 2 1/2 inches of bearing.
   d. 1 1/2 inches of bearing.

9. Floor joists should have a minimum lap of
   a. 2 inches.   c. 6 inches.
   b. 4 inches.   d. 8 inches.

10. It is extremely important when installing floor joists to
    a. fasten them securely with enough nails.
    b. fasten the girder end first.
    c. have the crowned edges up.
    d. notch and size the ends for any differences in width.

■ BUILDING FOR SUCCESS ■

Sound Work Ethics: Obsolete or Essential?

Builders have a tremendous responsibility to maintain integrity in their businesses. Contractors and construction workers are highly visible in the public's eye. Their work rarely goes unnoticed. Few "professional" critics will be capable of doing better work than

the worker. But the worker is the one in the spotlight and under inspection.

A builder's clientele today exhibits the same characteristics as those in the past. They desire quality work at reasonable prices, utilizing quality products. People desiring a new or remodeled home today typically do more homework when selecting a contractor. They are concerned about environmental issues that affect housing, financing, costs, and legalities. But as all of these issues become settled, today's clients still desire to know if they can trust the builder to give them the quality for which they are paying.

How does a builder or construction worker send the message of quality and trust? For the answer, we must return to the basics of long-term success.

Everyone desires to be treated fairly in their transactions with other people. They need to treat others as they themselves wish to be treated. This simple formula has traditionally set in motion the values and ethics necessary to demonstrate honesty and good will.

There are basic rules of conduct that are recognized by every class, group, and culture. These rules or principles are the guidelines for meaningful accomplishments, cooperation, satisfaction, and success. Any deviation from the accepted ethics could very well bring disaster.

Rules of conduct demonstrate the integrity of each person who adopts them. Integrity is defined in part, as "having moral principles, honesty, wholeness, or the quality of being undivided and incorruptible." A person's performance at work and at home should be transparent and require no clarification to show integrity. Integrity requires convictions and standards of ethical performance. In construction, when the specifications for the project are ignored, there can be no assurance of quality. Ignoring the requirements for quality negates integrity. It takes courage to adhere to specs, codes, and sound construction principles. That courage is the mainstay for exhibiting personal integrity.

The construction industry has building codes, state statutes, and local regulations that must be learned and adhered to daily. Every national, state, and local building organization implements a code of ethics for its membership. Individuals who desire to belong to and remain in good standing with an organization must be knowledgeable of the code and willing to apply those ethics on every job. This willingness becomes the basis for voluntary compliance.

Why adhere to a code of ethics? Because integrity must be a visual part of quality workmanship. Without sound work ethics, little credibility can be associated with a person, company, or enterprise. People that do not trust a builder or craftsman will not initiate any transactions with them.

One of the most noticeable characteristics of someone that exhibits sound work ethics is hard, dedicated work. Technology has alleviated many of the extremely difficult, laborious, and dangerous jobs that were once commonplace. Many physically demanding tasks have been minimized by new procedures that allow people to work smarter and more efficiently. Construction workers can learn to properly blend new technology, improved application methods, and honest work to accomplish a task. Although it may not be appealing to many coming into the workforce, meaningful labor and hard work remain a key ingredient to success. We should strive to work smarter, more diligently, and with purpose. Hard work always involves learning along the way. This learning translates into experience and wisdom. In the end, the skilled worker will generate more satisfaction and perhaps more profit. Good work habits are as hard to break as bad ones.

Remember, "the greatest ability may be dependability" to a builder. A successful construction business is based on people who work well on a regular basis. Punctuality on a regular basis is interpreted as proof that the employee values the working relationship with the employer. An employer strongly desires dependable employees.

As the costs of quality construction continue to rise at an accelerated rate, people become even more cautious about spending their savings or making long-term financial commitments. The competition for builders and skilled workers will also continue to increase. In turn, this challenges everyone to demonstrate credibility. The person who desires a career in a construction-related business today and well into the twenty-first century will have to compete for it. A sense of credibility will continue to be a survival technique. By believing in a sound code of ethics, and building a business on fairness and good moral principles, the chances of success are greatly increased.

Sound work ethics will never become obsolete because they are the basis for success. It has been said that "what goes around, comes around." Eventually we will be treated as we treat others. Builders or skilled workers will never have to be ashamed of their work if they perceive their jobs as serving others. Developing and maintaining sound work ethics in the beginning brings about the characteristics of integrity for a lifetime.

The following examples of sound work ethics and conduct reflect the characteristics needed by every worker who desires credibility.

- A day's work for a day's pay.
- Avoid shortcuts that save time but reduce quality.
- Give the customer the quality you yourself desire.
- Clearly state in writing the work to be done by all parties.
- Do not deceive or misrepresent anyone.
- Answer any questions honestly.
- Be straightforward in responding to the concerns of clients.
- Conduct your business in a professional manner.
- Charge your clients fairly.
- Assume your responsibility to your customers, community, and country.
- Pledge your support to the overall success of the project.

With the implementation of sound work ethics and principles, a construction worker will automatically send a message of integrity to the business world. This in turn will bring in a clientele that is comfortable and trusting.

The business world today does not appear to attach less value to sound ethics and principles than in the past. Therefore, each person entering the construction industry should be encouraged to learn and practice good work ethics. This will assure a successful and enjoyable career.

FOCUS QUESTIONS: For individual or group discussion.

1. Describe in your own words what you feel are work ethics.

2. From your personal experiences, have any of these characteristics been observable in skilled workers or builders you have known?

3. What would be some outcomes of not expecting a skilled worker to have sound work ethics?

4. How might you personally go about acquiring or building on your present sound work ethics?

12

Exterior Wall Framing

Exterior walls must be constructed to the correct height, corners braced plumb, walls straightened from corner to corner, and window and door openings framed to specified size. The techniques described in this unit will enable the apprentice carpenter to frame exterior walls with competence.

OBJECTIVES

After completing this unit, the student will be able to:

- identify and describe the function of each part of the wall frame.

- determine the length of exterior wall studs and the size of rough openings, and lay out a story pole.

- build corner posts and partition intersections and describe several methods of forming them.

- lay out the wall plates.

- construct and erect wall sections to form the exterior wall frame.

- plumb, brace, and straighten the exterior wall frame.

- apply wall sheathing.

- describe the construction of wood foundation walls.

CONTENTS

CHAPTER 34

EXTERIOR WALL FRAME PARTS

The wall frame consists of a number of different parts. The student should know the names, function, location, and usual size of each member. Sometimes the names given to certain parts of a structure may differ according to the geographical area. For that reason, some members may be identified with more than one term.

Parts of an Exterior Wall Frame

An exterior wall frame consists of *plates, studs, headers, trimmers, corner posts, partition intersections, ribbons,* and *braces* (Fig. 34-1).

Plates

The top and bottom horizontal members of a wall frame are called *plates*. The bottom member is called a *soleplate*. It is also referred to as the *bottom plate* or *shoe*. The top members are called *top plates*. They usually consist of doubled 2-inch stock. In a balloon frame, the soleplate is not used. Instead, the studs rest directly on the sill plate.

Studs

Studs are vertical members of the wall frame. They run full-length between plates. Shortened studs that do not extend full length because of an opening in the wall are called *jack studs* or *cripple studs*. If one end of the shortened stud butts against either top or bottom plate, it should be called a jack stud. Except in the case of windows and similar openings, where they are called *trimmers*, if neither end touches a plate the stud is said to be a *cripple*. This is consistent with roof framing terms described in later units.

Studs are usually 2×4s, but 2×6s are used when 6-inch insulation is desired in exterior walls. Studs are usually spaced 16 or 24 inches OC.

Headers

Headers run at right angles to studs. They form the top of window, door, and other wall openings, such as fireplaces. Headers must be strong enough to support the load above the opening. The depth of the header depends on the width of the opening. As the width of the opening increases, so must the strength of the header. Check drawings, specifications, codes or manufacturers' literature for header size.

Kinds of Headers. Solid or built-up lumber may be used for headers. For 4 inch walls, two pieces of 2-inch lumber with 1/2-inch plywood or strand board sandwiched in between them gives the head-

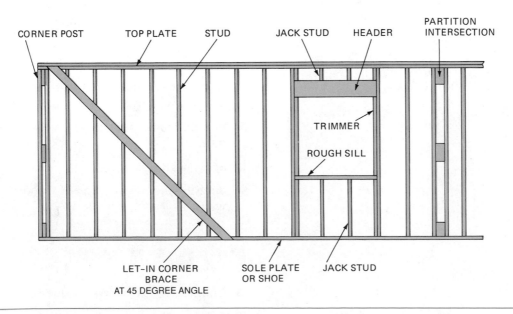

CORNER POST   TOP PLATE   STUD   JACK STUD   HEADER   PARTITION INTERSECTION

TRIMMER

ROUGH SILL

LET-IN CORNER BRACE AT 45 DEGREE ANGLE

SOLE PLATE OR SHOE   JACK STUD

Fig. 34-1 Parts of an exterior wall frame.

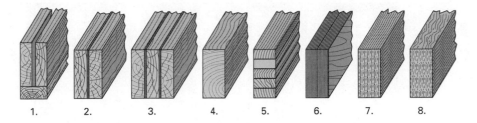

1. A BUILT-UP HEADER WITH A 2 × 4 OR 2 × 6 LAID FLAT ON THE BOTTOM.
2. A BUILT-UP HEADER WITH A 1/2" SPACER SANDWICHED IN BETWEEN.
3. A BUILT-UP HEADER FOR A 6" WALL.
4. A HEADER OF SOLID SAWN LUMBER.
5. GLULAM BEAMS ARE OFTEN USED FOR HEADERS.
6. A BUILT-UP HEADER OF LAMINATED VENEER LUMBER.
7. PARALLEL STRAND LUMBER MAKES EXCELLENT HEADERS.
8. LAMINATED STRAND LUMBER IS USED FOR LIGHT DUTY HEADERS.

Fig. 34-2 Several types of solid and built-up headers are used.

er the full 3 1/2-inch thickness of the wall. In 6-inch walls, three pieces of 2-inch lumber with two pieces of 1/2-inch plywood or strand board in between makes up the 5 1/2-inch wall thickness (Fig. 34-2).

Much engineered lumber is now being used for window and door opening headers. Figures 5-6, 6-1, and 8-1 show the use of laminated veneer lumber, parallel strand lumber, and glulam beams as opening headers. In addition, a header called a *1.2E Light Duty Header* manufactured of laminated strand lumber is now available (Fig. 34-3). The use of engineered lumber permits the spanning of wide openings, such as double garage door openings. Such openings otherwise might need intermediate supports (Fig. 34-4).

In many buildings, when the opening must be supported without increasing the header size, the top of a wall opening may be trussed to provide support (Fig. 34-5). However, when the opening is fairly close to the top plate, the depth of the header is increased. This completely fills the space between the plate and the top of the opening. In this case, the same size header is usually used for all wall openings, regardless of the width of the opening (Fig. 34-6). This eliminates the need to install short jack studs above the header.

Rough Sills

Forming the bottom of a window opening, at right angles to the studs are members called *rough sills*. They usually consist of a single 2-inch thickness. They carry little load. However, many carpenters prefer to use a double 2-inch thickness rough sill. This provides more surface on which to fasten window trim in the later stages of construction.

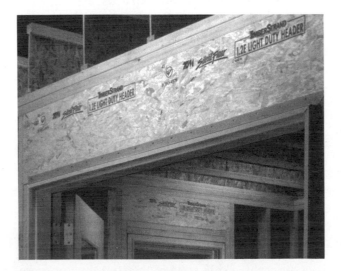

Fig. 34-3 A 1.2E Light Duty Header of laminated strand lumber is used in a window opening. (*Courtesy of Trus Joist MacMillan*)

Fig. 34-4 Laminated veneer lumber is used for a garage door opening header. (*Courtesy of Trus Joist MacMillan*)

Fig. 34-5 Sometimes the top of a wall opening is trussed instead of increasing the header depth. Two methods of trussing an opening are shown.

Fig. 34-6 It is common practice to use the same header depth. Completely fill the space between the opening and the top plate for all openings regardless of the opening width.

Trimmers

Trimmers are shortened studs that support the headers. They are fastened to the studs on each side of the opening. In window openings, they should fit snugly between the header and rough sill (Fig. 34-7). In door openings, the trimmers, sometimes called *door jacks*, fit between the header and the soleplate.

Some carpenters prefer to cut out the soleplate between the studs of a door opening. This lets the bottom end of the trimmers rest on the subfloor (Fig. 34-8). However, cutting out the soleplate may make it difficult to build wall sections on the subfloor and erect them into place. In some cases, door trimmers are left out until the wall section is erected. The soleplate is then cut out and trimmers installed.

Corner Posts

Corner posts are the same length as studs. They are constructed in a manner that provides an outside and an inside corner on which to fasten the exterior

and interior wall coverings. Corner posts are built in a number of ways. Most of the time, three 2×4s are used. Sometimes a solid 4×6 is used along with a single 2×4 (Fig. 34-9). Parallel strand lumber is used frequently for corner posts.

Partition Intersections

Wherever interior partitions meet an exterior wall, extra studs need to be put in the exterior wall. This provides wood for fastening the interior wall covering in the corner. In most cases, the *partition intersection* is made of two studs spiked to the edge of 2×4 blocks about a foot long. One block is placed at the bottom, one at the top, and one about center on the studs.

Another method is to maintain the regular spacing of the studs. Blocking is installed between them wherever partitions occur. The block is set back from the inside edge of the stud the thickness of a board. A 1×6 board is then fastened vertically on the inside of the wall so that it is centered on the partition.

Fig. 34-7 Typical framing for a window opening.

Fig. 34-8 Typical framing for a door opening.

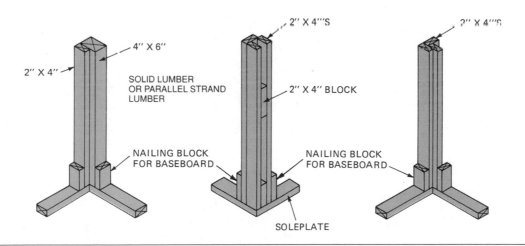

Fig. 34-9 Several methods of making corner posts are used.

Another method is to spike a continuous 2×8 backer to a full length stud. The edges of the backer project an equal distance beyond the edges of the stud (Fig. 34-10).

Ribbons

Ribbons are horizontal members of the exterior wall frame in balloon construction. They are used to sup-

port the second floor joists. The inside edge of the wall studs is notched so that the ribbon lays flush with the edge (Fig. 34-11). Ribbons are usually made of 1×4 stock. Notches in the stud should be made carefully so the ribbon fits snugly in the notch. This prevents the floor joists from settling. If the notch is cut too deep the stud will be unnecessarily weakened.

Fig. 34-10 Partition intersections are constructed in several ways.

Fig. 34-11 Ribbons are used to support floor joists in a balloon frame.

Corner Braces

Generally, no wall bracing is required if rated panel wall sheathing is used. In other cases, such as when insulating board sheathing is used, walls are braced by using nominal 1×4 continuous diagonal strips of lumber. These are set into the face of the studs, top plate, and soleplate at each corner of the building. Or, 2×4s are cut in between the studs (Fig. 34-12).

TOP PLATES

LAP TOP
PLATES
AND NAIL

TEMPORARY
BRACE

WINDOW
HEADER

STUD

LET-IN
CORNER BRACE

ROUGH SILL

SOLEPLATE

RIM JOIST

SUBFLOOR

JOIST

FOUNDATION WALL

ANCHORED SILL PLATE

SPACER BLOCK

1" X 4" LET-IN CORNER BRACES

Fig. 34-12 Studs and plates are notched to receive the corner braces.

CHAPTER 35 FRAMING THE EXTERIOR WALL

Careful construction of the wall frame makes application of the exterior and interior finish easier. It also reduces problems for those who apply it later. This chapter shows how to frame and erect exterior walls in a proficient manner.

Exterior Wall Framing

Plans or blueprints usually indicate the height from finish floor to ceiling. This dimension is found in a wall section. It is needed to determine the length of wall studs.

Determining the Length of Studs

The stud length must be calculated so that, after the wall is framed, the distance from finish floor to ceiling will be as specified in the drawings. To determine the stud length, the thickness of the finish floor and the ceiling thickness below the joist must be known.

To the specified floor-to-ceiling height, add the thickness of the finish floor and the combined thickness of the ceiling below the ceiling joist. Include the thickness of *furring strips* and / or *plaster lath* in the ceiling, if used. Deduct the total thickness of the top plates and the soleplates to find the length of the stud (Fig. 35-1). (This applies only to platform frame construction.)

For example, the finish floor-to-ceiling height is shown to be 7'-9". Add 3/4 inch for the finish floor and 1 1/4 inches for the ceiling thickness below the joist (3/4 inch for furring and 1/2 inch for gypsum board). The total is 7'-11". Deduct the combined thickness of the top plates and soleplates (4 1/2 inches). This leaves a stud length of 7'-6 1/2".

For balloon frame construction, find the height from the sill to the top of the top plates by adding:

- finish floor to ceiling heights of all the stories
- the thickness of all ceilings from the ceiling finish to the joists
- the thickness of all finish floors and subfloors
- and the combined width of all floor joists.

Then deduct the total thickness of the top plates (Fig. 35-2).

The location of the ribbon that supports the second-floor joists in a balloon frame is found by adding:

- the specified first floor to ceiling height
- the thickness of the first-floor ceiling
- finish floor
- subfloor
- and the width of the first-floor joists.

Measure this distance from the bottom end of the stud to the top edge of the ribbon.

Story Pole

The *story pole* is usually a straight strip of 1×2 or 1×3 lumber. Its length is the distance from the subfloor to the bottom of the joist above. Lines are squared across its width, indicating the height of the soleplates and top plates, headers, and rough sills for all openings.

The story pole is used to determine the length of studs, jack studs, and trimmers, even if opening heights differ. The thickness and height of rough

Fig. 35-1 Finding the length of a stud in an exterior wall frame of platform construction.

Fig. 35-2 Finding the length of a stud in a balloon frame.

sills and the location of headers are also indicated and identified on the story pole (Fig. 35-3).

A different story pole is made for each floor of the building. As framing progresses, the story pole is used to test the accurate location of framing members in the wall.

Determining the Size of Rough Openings

A *rough opening* is an opening framed in the wall in which to install doors and windows. The width and height of rough openings are not usually indicated in the plans. It is the carpenter's responsibility to determine the rough opening size for the particular unit from the information given in the door and window schedule. The door and window schedule contains the kind, style, manufacturer's model number, and size of each unit.

Rough Opening Sizes for Exterior Doors

The rough opening for an exterior door must be large enough to accommodate the door, door frame, and space for shimming the frame to a level and plumb position. Usually 1/2 inch is allowed for shimming, at the top and both sides, between the door frame and the rough opening. The amount allowed for the door frame itself depends on the thickness of the door frame beyond the door.

Care must be taken not to make the rough opening oversize. If the opening is made too large, the window or door finish may not cover it.

The sides and top of a door frame are called *jambs*. Jambs may vary in thickness. Sometimes *rab-*

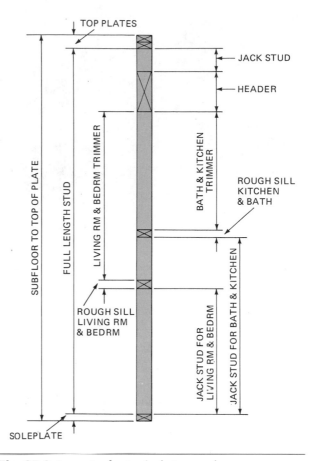

Fig. 35-3 Layout of a typical story pole.

beled wood jambs are used. The rabbet is that part of the jamb that the door stops up against. At other times nominal one- or 2-inch lumber is used for the jamb. A separate *stop* is applied (Fig. 35-4). Steel jambs have the door stop built-in similar to a rabbeted wood jamb.

NOTE: ROUGH OPENING WIDTHS FOR THE SAME SIZE DOOR MAY VARY ACCORDING TO THE THICKNESS OF THE JAMB STOCK.

Fig. 35-4 Determining the rough opening width of a door opening.

RABBET

STOP

RABBETED JAMB

JAMB WITH APPLIED STOP

Fig. 35-4 *(continued)*

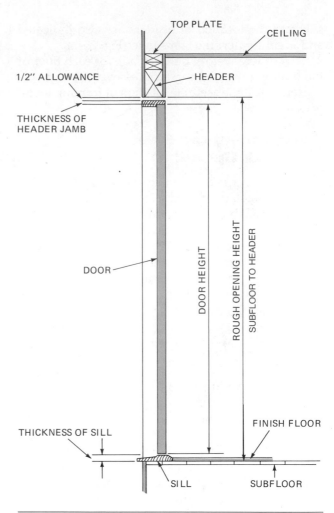

TOP PLATE

CEILING

1/2" ALLOWANCE

HEADER

THICKNESS OF HEADER JAMB

DOOR

DOOR HEIGHT

ROUGH OPENING HEIGHT

SUBFLOOR TO HEADER

FINISH FLOOR

THICKNESS OF SILL

SILL

SUBFLOOR

Fig. 35-5 Determining the rough opening height of an exterior door opening.

The bottom member of the door frame is called a *sill* or *stool*. Sills may be hardwood, metal, or a combination of wood and metal. The type of sill and its thickness must be known in order to figure the rough opening height (Fig. 35-5).

An exterior door may come *prehung* in its frame and the entire unit set in the opening. At other times, only the frame is set in the opening. The door is then hung in the frame at a later stage in the construction. In either case, the rough opening must be calculated in the same manner.

Rough Opening Sizes for Windows

Many kinds of windows are manufactured by a number of firms. Because of the number of styles, sizes, and variety of construction methods, it is best to consult the manufacturer's catalog to obtain the rough opening sizes. These catalogs show the style and size of the window unit. They also give the rough opening (RO) for each unit (Fig. 35-6). Catalogs are available from the lumber company that sells the windows.

Laying Out the Plates

To lay out the plates, measure in at the corners the thickness of the wall. Snap lines on the subfloor between the marks. *Tack* the soleplate in position so its inside edge is to the chalk line. Do not drive nails home (they have to be pulled later). Use only as many as are needed to hold the piece in place. Plan soleplate lengths so that joints between them fall in the center of a full-length stud for the convenient erection of wall sections later.

Wall Openings. From the blueprints, determine the centerline dimension of all the openings in the wall. Lay these out on the soleplate. Measure in each direction from the centerline one-half the width of the rough opening. Square lines at these points across the soleplate. Mark a *T*, for trimmer, on the side of the line away from the center. To distinguish openings, a *T* may be used for window trimmers and a *J* for door jacks. It makes little difference what marks are used as long as the builder understands what they mean.

From the squared lines, measure back the stud thickness. Square lines across. Mark *X*s, on the side of the line away from center, for the full-length studs on each side of the openings (Fig. 35-7).

Partition Intersections

On blueprints, interior partitions are usually dimensioned to their centerline. Mark on the soleplate the centerline of all partitions intersecting the wall. From the centerlines measure in each direction one-

BASIC UNIT SIZES

Fig. 35-6 Sample of a manufacturer's catalog showing rough opening sizes for window units. (*Courtesy of Andersen Corporation*)

half the partition stud thickness. Square lines across the soleplate. Mark Xs, on the side of the lines away from center, for the location of partition intersection studs (Fig. 35-8).

Studs and Jack Studs

After all openings and partitions have been laid out, start laying out all full-length studs and jack studs. Proceed in the same manner and from the same end as laying out floor joists. This keeps studs directly in line with the joists below.

Measure in from the outside corner the regular stud spacing. From this mark, measure in one direction or the other, one-half the stud thickness. Square a line across the soleplate. Place an X on the side of the line where the stud will be located.

Stretch a tape along the length of the soleplate from this first stud location. Square lines across the soleplate at each specified stud spacing. Place Xs on the same side of the line as the first line.

Where openings occur, mark a J on the side of the line to indicate the location of jack studs (Fig. 35-9). All regular and jack studs should line up with the floor joists below. When laying out the opposite wall, start from the same end as the first wall to keep all framing lined up.

Laying Out the Top Plate

Use straight lengths of lumber for the top plate. Place a length alongside the soleplate. With one end flush with the end of the soleplate, mark and cut the other end of the top plate. Make sure it is centered on a stud and lines up with a joint in the soleplate (Fig. 35-10). Cut enough top plates for the length of

Fig. 35-7 Laying out a rough opening width on the soleplate.

Fig. 35-8 Laying out a partition intersection on the soleplate.

STEP BY STEP
P R O C E D U R E S

Fig. 35-9 Laying out the soleplate for regular and jack studs.

the wall. Tack the top plate in position alongside the soleplate. Transfer the marks on the soleplate to the top plate (Fig. 35-11).

Assembling and Erecting Wall Sections

The usual method of framing the exterior wall is to precut the wall frame members, assemble the wall frame on the subfloor, and erect the frame. With a small crew, without special equipment, the walls

are raised section by section. When the frame is erected, the corners are plumbed and braced. Then the walls are straightened between corners. They are also braced securely in position. To prevent problems with the installation of the finish work later, it is important to keep the edges of the frame members flush wherever they join each other.

Precutting Wall Frame Members

Full-length Studs. To assemble a wall section, first cut the needed number of full-length studs to

Fig. 35-10 Joints in the top plate should fall on the center of a stud.

length. A radial arm saw is an effective tool for cutting studs and other framing to length. Set a stop the desired distance from the saw blade to cut duplicate lengths. If a radial arm saw is not available, a jig can be made for a portable electric circular saw, commonly known as a *skilsaw,* to cut duplicate lengths of framing (Fig. 35-12). Reject any studs that are severely warped. They may be cut into shorter lengths for blocking when making corner posts and partition intersections or between studs.

Corner Posts and Partition Intersections. Make up the necessary number of corner posts and partition intersections. Corner posts are usually made by using two full-length studs and spiking them together with short blocks of the same material laid flat between them. One block is placed flush with each end and one block placed about center (Fig. 35-13).

Partition intersections are made in a similar manner except for the way the blocks are placed. They are placed on edge in the same location between the studs. They are spiked to the studs with

their sides flush with the edges of the studs (Fig. 35-14).

Headers, Rough Sills, Jack Studs, and Trimmers. Cut all headers and rough sills. Their length can be determined from the layout on the plates.

Make a story pole. From it determine the length of all trimmers and jack studs. Cut them accordingly. It may be necessary to place identifying marks on headers, rough sills, jacks, and trimmers if rough openings are different sizes.

Assembling Wall Sections

Separate the top plate from the soleplate. Stand them on edge. Place all full-length studs, corner posts, and partition intersections in between them. To avoid a mistake, be careful not to turn one of the plates around. Be certain that the layout lines on top and bottom plates line up.

Very few studs are absolutely straight from end to end. Carefully sight each full-length stud. Place it on the subfloor between the plates with its crowned edge down. It will be difficult for those who apply the interior finish, if no attention is paid to the manner in which studs are installed in the wall. A stud that is installed with its crowned edge out, next to one with its crowned edge in, will certainly present problems later.

It is important to lay all studs on the subfloor with the crowned edges down. When the wall is erected, each stud in the wall will be bowed in. The weight eventually imposed on the wall causes the studs to bow out. Therefore, walls built with an inward bow will have a tendency to straighten when weight is imposed on them.

SQUARE LINES ACROSS TOP PLATE. PUT ALL MARKS ON SOLEPLATE ON TOP PLATE

SOLEPLATE

TOP PLATE

Fig. 35-11 Transferring the layout of a soleplate to the top plate.

HELPFUL HINTS

How to make and use a jig for the portable electric circular saw to cut framing to length.

STOP

FRAMING STOCK

RABBET

BASE OF JIG

SECTION STOP DETAIL

PLANK OR PLYWOOD BASE OF JIG

RABBET END OF STOP

2" × 2"

BUILD JIG TO REQUIRED LENGTH

1. SET STOP IN LOCATION TO CUT DESIRED LENGTH OF FRAMING.

2. SLIDE FRAMING STOCK IN JIG AGAINST STOP AND MAKE CUT.

3. CONTINUE PROCEDURE TO MAKE DUPLICATE LENGTHS.

1" × 2"

3/8" PLYWOOD

FRAMING STOCK

Fig. 35-12

SHORT BLOCK

SHORT BLOCK

STUD

END STUD OF ADJOINING WALL COMPLETES MAKE-UP OF THE CORNER POST.

SHORT BLOCK

SOLE PLATE

SHORT BLOCK

END STUD OF INTERSECTING PARTITION MAKES UP CORNERS.

SHORT BLOCK

STUD

SOLE PLATE OF INTERSECTING PARTITION

SHORT BLOCK

SOLE PLATE OF EXTERIOR WALL

Fig. 35-13 Construction of a corner post.

Fig. 35-14 Construction of a partition intersection.

Fasten each stud, corner post, and partition intersection in the proper position by driving two 16d spikes through the plates into the ends of the members. A pneumatic framing nailer makes the work easier.

Using the story pole, mark the position of all headers and rough sills on the full-length studs on each side of each opening. Fasten the headers and rough sills in position by driving 16d spikes through the studs into their ends. If headers butt the top plate, spike through the plate into the header.

Fasten all jack studs in position between top plates and headers and between soleplates and rough sills. Fasten all trimmers in position between headers and rough sills for window openings and between the header and soleplate for door openings.

It may be necessary to toenail the ends of some members with 8d nails if spikes cannot be driven in their ends. When doubling up studs, toenail their edges flush before spiking them together.

Bracing End Sections

If let-in braces are to be used, it is much easier to notch out the studs and plates and install the braces while the wall section is lying on the subfloor. To make sure the wall section is square before laying out for the brace, measure each section from corner to corner both ways. If the measurements are the same, the section is square and will be plumb when erected. If the measurements are not the same, rack the section by pushing on one corner until the corner-to-corner measurements are equal (Fig. 35-15).

When the section is square, tack the brace in position against the studs and plates. If the brace is to be at a 45 degree angle, measure in from the corner the same distance as the height of the wall.

Mark the studs and plates along each edge of the brace. Remove the brace. Mark the depth of cut on both sides of each member. Use a portable electric circular saw with the blade set for the depth of

the notch. Make multiple saw cuts between the layout lines. Use a wood chisel and trim the remaining waste to form the notch. Fasten the brace in the notches using two 8d common nails in each framing member (Fig. 35-16).

Erecting Wall Sections

To erect the wall, raise one section at a time into position. Spike the soleplate to the chalk line with 16d spikes spaced about 16 inches on center into the floor joists and band joist. Sections are fastened together by nailing the top plates to the projecting end of the studs of the adjoining wall sections. In corners, fasten end studs together to complete the construction of the corner post. A completed corner post provides surfaces to fasten both exterior and interior wall finish.

Brace each section temporarily as erected. Fasten one end of a 1×6 board to a 2×4 block that has been nailed to the subfloor and the other end to the side of a stud (Fig. 35-17). Do not drive nails all the

Fig. 35-16 A let-in corner brace.

Fig. 35-17 Raising an exterior wall section. The sheathing has already been applied.

Fig. 35-15 When the diagonal measurements of a wall section are equal, the section is square.

way home or use duplex head nails in temporary braces so they can be removed later.

Plumbing and Bracing the Corners

If the walls have not been previously braced while being framed on the subfloor, or if no permanent bracing is required, all corners must be plumbed and temporarily braced. Cut 1×6 braces for both sides of each corner with their ends cut at a 45 degree angle. Fasten the top end of the brace to the top plate near the corner post on the inside of the wall.

Temporary braces are fastened to the inside of the wall. They can remain in position until the exterior wall sheathing is applied. Sometimes they remain until it is absolutely necessary to remove them for the application of the interior wall finish. Care must be taken not to let the ends of the braces extend beyond the corner post or top plate. This might interfere with the application of wall sheathing or subsequent ceiling or roof framing.

Method of Plumbing Corner Posts. Plumb the corner post, and fasten the bottom end of the brace. Use duplex head nails or drive nails only partway in for easy removal. For accurate plumbing of the corner posts, use a 6-foot level with accessory aluminum blocks attached to each end. The blocks keep the level from resting against the entire surface of the corner post. This prevents any bow or irregularity in the surface from affecting the accurate reading of the level.

An alternate method is to use the carpenter's hand level in combination with a long straightedge. On the ends of the straightedge, two blocks of equal thickness are fastened to keep the edge of the straightedge away from the surface of the corner post (Fig. 35-18). A plumb bob, transit-level, or laser level may also be used to plumb corner posts.

Doubling the Top Plate

If the complete wall is to be raised in one piece instead of in sections, the top plate can be doubled while the wall is still on the subfloor. Let the plate extend beyond the wall to make a lapped joint at the corner (Fig. 35-19).

Using Wall Jacks. Usually, if the full length of the wall is framed in one piece, the wall sheathing is also applied before the wall is raised. Sometimes even the windows, doors, and exterior wall finish are applied. When this is done, the wall is very heavy. Wall jacks have to be used to raise the wall into position (Fig. 35-20).

If the wall is erected in sections, after all the sections are raised the top plate is doubled. First, toenail into the edges to bring them flush. Then

Fig. 35-18 Using a straightedge and level to plumb corner posts.

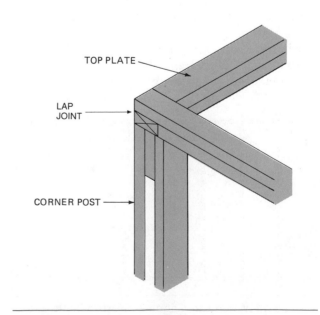

Fig. 35-19 Top plate is lapped at the corners.

1. THE LIFTING BRACKETS ARE PLACED UNDER THE TOP PLATE OR HEADER AND SECURELY NAILED. THE BOTTOM OF THE WALL IS TOE-NAILED TO THE SUBFLOOR, OR OTHERWISE SECURED TO PREVENT IT FROM SLIPPING DURING THE LIFT.

2. THE JACKS ARE PLACED UPRIGHT, FLUSH AGAINST THE LIFTING BRACKET. THE HINGED FOOT IS SECURELY NAILED TO THE SUB-FLOOR NEAR A FLOOR JOIST.

3. THE LIFT BEGINS, BOTH JACKS BEING OPERATED IN UNISON.
   • IF ONE MAN IS RAISING THE WALL, HE SIMPLY MOVES FROM ONE JACK TO THE OTHER AVOIDING EXCESSIVE TWISTING OF THE WALL.

4. THE JACKS REMAIN IN CONTACT WITH THE TOP EDGE OF THE WALL DURING THE ENTIRE LIFT.

5. WHEN THE WALL REACHES THE UPRIGHT POSITION, IT IS STOPPED AND HELD FIRMLY BY THE PRESET WALL STOPS, WHICH ARE ADJUSTABLE.

6. THE WALL JACKS PROVIDE A SUBSTANTIAL BRACE UNTIL THE WALL CAN BE PERMANENTLY BRACED.

Fig. 35-20 A small crew can easily erect a heavy wall using wall jacks. (*Courtesy of Proctor Products Company*)

spike them together with 16d spikes staggered from edge to edge about every 16 inches. Stagger the joints in the top member as far as possible away from those in the bottom member.

Partition Intersections. Where partitions intersect the exterior wall, a space is left for the top plate of the partition to lap the plate of the exterior wall (Fig. 35-21). Lapping the plates of the interior partitions with those of the exterior walls ties them together. This makes a more rigid structure. Usually the only interior partitions with a doubled top plate are *bearing partitions*. Bearing partitions are those that carry a load.

Straightening the Walls

After the corner posts have been plumbed and the top plates doubled, the tops of the walls must be straightened and braced. To straighten each wall, use three blocks of one-inch lumber of exactly the same thickness. To assure equal thickness of the blocks, cut them from the same piece. Fasten two blocks to the inside of the top plate at each end.

Stretch a line tightly over and between the two corner blocks. Use the third block as a test block. Move the wall in or out at about 8- or 10-foot intervals until the test block just clears the line when held against the top plate (Fig. 35-22). It is impor-

Fig. 35-21 The top plate of bearing partitions laps the exterior wall.

Fig. 35-22 Straightening the wall with a stretched line and test block.

tant that the block does not touch the line. It must come as close as possible *without touching*. If allowed to touch, the line is no longer straight.

Combination Wall Aligner and Brace. It may be difficult to hold the wall in alignment while bracing it. Either a manufactured, combination wall aligner and brace or a job-built type is then used. A manufactured type is similar to that used on concrete forms. A few turns on the adjusting screw moves the wall in or out as desired (Fig. 35-23).

A job-built combination wall aligner and brace is called a *spring brace*. A spring brace consists of a 1×8 board about 12 or 14 feet long. The top end is securely fastened, with four or five 8d common nails driven home, to the underside of the top plate. The bottom end is fastened to the subfloor in the same manner so the top of the wall has to be brought in about an inch or less.

Spring the board upward by installing a piece of the same material in a vertical position between it and the subfloor about midway on the brace (Fig.

JUST A TURN OR TWO ON THE FAST THREAD SCREW, AND THE WALL IS ALIGNED AND READY FOR NAILING. WALL BRACES CAN BE USED TO POSITION BEAM SUPPORTS AND PARTITION WALLS, AS WELL AS EXTERIOR WALLS.

Fig. 35-23 A manufactured adjustable wall brace aligns and braces the wall. (*Courtesy of Proctor Products Company*)

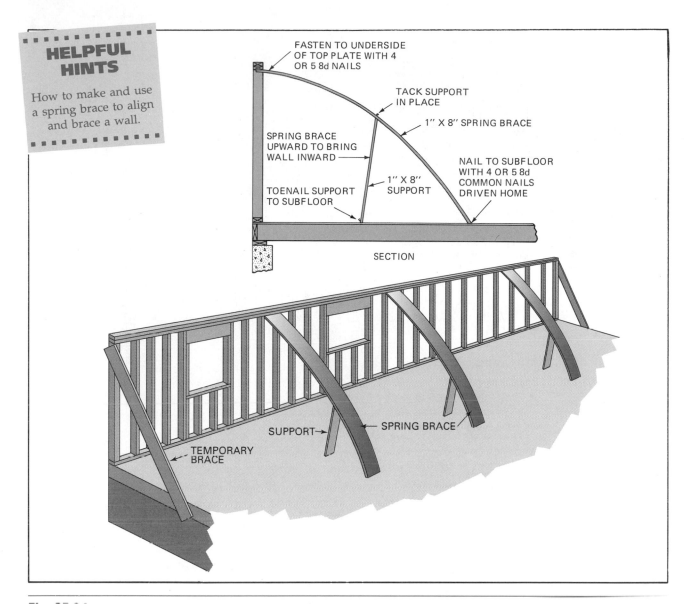

HELPFUL HINTS

How to make and use a spring brace to align and brace a wall.

FASTEN TO UNDERSIDE OF TOP PLATE WITH 4 OR 5 8d NAILS

TACK SUPPORT IN PLACE

1″ X 8″ SPRING BRACE

SPRING BRACE UPWARD TO BRING WALL INWARD

NAIL TO SUBFLOOR WITH 4 OR 5 8d COMMON NAILS DRIVEN HOME

1″ X 8″ SUPPORT

TOENAIL SUPPORT TO SUBFLOOR

SECTION

SUPPORT

SPRING BRACE

TEMPORARY BRACE

Fig. 35-24

35-24). Within limits, the more the brace is sprung upward, the more the wall is brought inward. Bring the wall inward so the test block just fits between the stretched line and the top plate without touching the line. When the wall is straight, fasten the uprights by nailing through the brace and toenailing to the subfloor.

Installing Baseboard Blocking. Use as many braces as necessary to straighten the wall along its entire length. Fasten short 2×4 blocks against both sides of the corner posts and partition intersections and down on the soleplate to provide fastening for the ends of the baseboard (Fig. 35-25).

Wall Sheathing

Wall sheathing covers the exterior walls. It may consist of boards, rated panels, fiberboard, gypsum board, or rigid foam board.

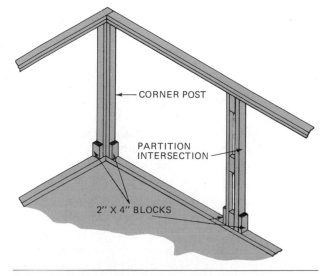

CORNER POST

PARTITION INTERSECTION

2″ X 4″ BLOCKS

Fig. 35-25 Short blocks against the sides of corner posts and partition intersections provide fastening for the ends of the baseboard.

APA Panel Wall Sheathing (a)
(APA RATED SHEATHING panels continuous over two or more spans.)

Panel Span Rating	Maximum Stud Spacing (in.)	Nail size (b) (c)	Maximum Nail spacing (in.)	
			Supported Panel Edges	Intermediate Supports
12/0, 16/0, 20/0 or Wall- 16 oc	16	6d for panels 1/2" thick or less; 8d for thicker panels	6	12
24/0, 24/16, 32/16 or Wall- 24 oc	24			

(a) See requirements for nailable panel sheathing when exterior covering is to be nailed to sheathing.

(b) Common, smooth, annular, spiral-thread, or galvanized box.

(c) Other code-approved fasteners may be used.

Note: Shaded stud spacing meet Code Plus recommendations.

1/8" SPACING IS RECOMENDED AT ALL EDGE AND END JOINTS UNLESS OTHERWISE INDICATED BY PANEL MANUFACTURER

"BLOCK" HORIZONTAL JOINTS IN PANELS USED FOR BRACING

APA RATED SHEATHING APPLIED WITH LONG DIMENSION PARALLEL TO STUDS

APA RATED SHEATHING APPLIED WITH LONG DIMENSION ACROSS STUDS

FILTER STRIP IF REQUIRED

6" MINIMUM CLEARANCE, SIDING TO GRADE

SIDING

Fig. 35-26 Selection and fastening guide for APA-rated panel wall sheathing. (*Courtesy of American Plywood Association*)

Boards. Boards are seldom used because of the high cost of material and labor in comparison to other kinds of wall sheathing. If used, they may be applied diagonally or horizontally. Diagonal sheathed walls require no other bracing. They make the frame stiffer and stronger than boards applied horizontally. Use two 8d common nails at each stud for 6- and 8-inch boards. Use three nails at each stud for 10- and 12-inch boards. End joints must fall over the center of studs. They must be staggered so no two successive end joints fall on the same stud.

Rated Panels. American Plywood Association-rated wall sheathing panels may be applied horizontally or vertically. There is no need for corner braces. A minimum 3/8-inch thickness is recommended when the sheathing is to be covered by exterior siding. Greater thicknesses are recommended when the sheathing also acts as the exterior finish siding.

Use 6d nails spaced 6 inches apart on the edges and 12 inches apart on intermediate studs for panels 1/2 inch thick or less. Use 8d nails for thicker sheathing panels (Fig. 35-26).

Rated sheathing panels are often used in combination with fiberboard, rigid foam, or gypsum board. When panels are applied vertically on both sides of the corner, no other corner bracing is necessary (Fig. 35-27).

Other Sheathing Panels. *Fiberboard sheathing,* commonly called insulation board, is a soft, synthetic sheet material. Common sheet sizes are 2×8 and 4×8. Usually the edges and ends are *matched* (tongue-and-groove). Standard thicknesses are 1/2 inch and 25/32 inch. Most sheets are coated with asphalt for water resistance.

Fiberboard sheathing is fastened with roofing nails spaced 3 inches apart around the edges and 6 inches apart in the center. Fasteners must be kept a

Fig. 35-27 When plywood is used at the corners in combination with fiberboard sheathing, no additional corner bracing is required.

minimum of 3/8 inch from the edges. It may be applied vertically or horizontally.

Gypsum sheathing consists of a treated gypsum filler between sheets of water-resistant paper. Usually 1/2 inch thick, 2 feet wide, and 8 feet long, the sheets have matched edges to provide a tighter wall. Because of the soft material, roofing nails must be used to fasten gypsum sheathing.

Space the nails about 4 inches around the edges and 8 inches in the center. Gypsum board sheathing is used when a more fire-resistant sheathing is required.

Rigid foam sheathing is used when greater insulation is required (Fig. 35-28). A disadvantage of using fiberboard, gypsum board, or rigid foam sheathing is that the exterior wall siding that covers it cannot be nailed to it. Either the siding must be fastened to the studs or special fasteners must be used.

Application of Sheathing Panels. Sheathing panels are applied in a manner similar to applying subfloor. Snap a line 4 feet up across the wall. Fasten the first row of horizontally applied panels so their top edges are to the line. Start the next row with a half-panel to stagger the joints. It may be helpful to drive either some wood shingle tips or spikes at intervals between the foundation and the sill plate. Let them project somewhat so the first row of panels can rest on them (Fig. 35-29). The same technique may be used to support the bottom ends of vertically applied panels.

Estimating Materials for Exterior Walls

To estimate the amount of material needed for exterior walls, first determine the total linear feet of exterior wall. Then, figure one stud for every linear foot of wall, if spaced 16 inches on center. This allows for the extras needed for corner posts, partition intersections, trimmers, door jacks, and blocking openings. For 24-inch spacing, divide the total linear feet of wall by 2. Allow an extra number for openings, corners, and partition intersections.

For plates, multiply the total linear feet of wall by 3 (one soleplate and 2 top plates). Add 5 percent for waste in cutting.

For headers and rough sills, calculate the size for each opening. Add together the material needed for different sizes.

For wall sheathing, first find the total area to be covered. Multiply the total linear feet of wall by the wall height to find the total number of square feet of wall area. Deduct the area of any large openings. Disregard small openings.

For board sheathing, add a percentage for waste to the area to be covered to find the total

Fig. 35-28 Rigid foam insulation is used for exterior wall sheathing. Insulation board sheathing is applied before the walls are erected. (*Courtesy of Dow Chemical Corporation*)

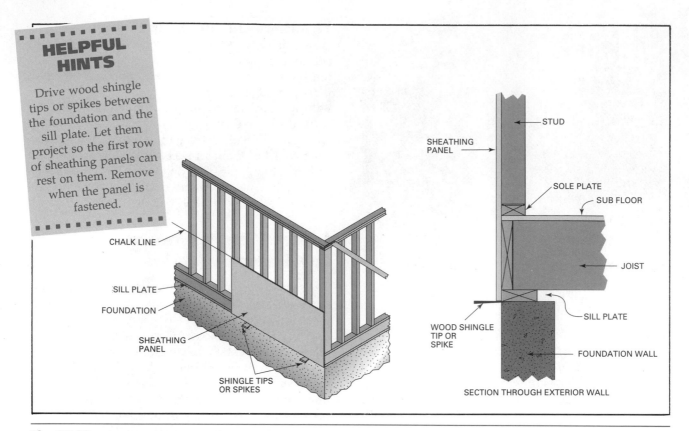

CHALK LINE

SILL PLATE

FOUNDATION

SHEATHING PANEL

SHINGLE TIPS OR SPIKES

STUD

SHEATHING PANEL

SOLE PLATE

SUB FLOOR

JOIST

WOOD SHINGLE TIP OR SPIKE

SILL PLATE

FOUNDATION WALL

SECTION THROUGH EXTERIOR WALL

Fig. 35-29

number of board feet. Add 20 percent for 6-inch boards, 15 percent for 8-inch boards, and 10 percent for 10-inch boards. These large percentages are not actually waste in cutting, but are mostly the difference between the nominal and actual sizes of the boards.

To find the number of sheathing panels, divide the total wall area to be covered by the number of square feet in each sheet. For instance, if a panel measures 4 feet by 8 feet, it contains 32 square feet. Divide the wall area to be covered by 32 to find the number of panels required. Add about 5 percent for waste. If the answer does not come out even, round it up to the next whole panel.

Summary of Ways to Assemble and Erect Exterior Walls

There are a number of ways to assemble and erect exterior walls. The method described in this unit is probably the most common.

Another method is to fasten the soleplate securely. Nail all full-length studs to the top plate. Then, raise the sections. Toenail the studs to the soleplate. In this method, the rough openings are framed after the wall is erected. Then the sheathing is applied. This method is used when the framing crew is small.

Still another method consists of framing and sheathing the entire wall on the subfloor. Sometimes the windows and doors are installed and even the wall siding applied. A disadvantage to this method is that the wall is very heavy to raise into position and may require special raising equipment. Sometimes complete wall sections with doors, windows, and siding installed are prefabricated in the shop, transported to the job site, and erected in position.

All-Weather Wood Foundation

A wood foundation built of pressure-preservative-treated lumber and plywood is called the *All-Weather Wood Foundation (AWWF)* (Fig. 35-30). It can be used to support light-frame buildings, such as houses, apartments, schools, and office buildings. It is accepted by major model building codes and by federal agencies such as the VA and FHA.

AWWF is mentioned in this chapter because much of the system is composed of wood exterior walls. The same techniques are used in their assembly and erection as described in this chapter.

For complete design and construction recommendations on the All-Weather Wood Foundation system, contact the American Plywood Association for literature titled *Permanent Wood Foundation— Guide to Design and Construction*.

PLYWOOD FLOOR

FACE GRAIN DIRECTION

PLYWOOD SIDING

2 × 4 STUD (TYPICAL)

FIELD APPLIED 2 × _ TOP PLATE

2 × _ TOP PLATE(A)

8″ MIN.

PLYWOOD STRIP PROTECTING TOP OF POLYETHYLENE FILM (5½″ MIN.)

30 LB ASPHALT OR 6 MIL POLYETHYLENE FILM STRIPS

PLYWOOD

INSULATION

2 × _ BOTTOM PLATE

2 × _ FOOTING PLATE

SCREED BOARD (OPTIONAL)

POLYETHYLENE FILM

VAPOR BARRIER

PERFORATED 4″ DIAMETER PLASTIC OR FIBER DRAIN PIPE

¾″ OR LARGER GRAVEL OR CRUSHED STONE (MUST NOT PLUG PERFORATED PIPE)

TO STORM SEWER OR DAYLIGHT

(B) A WOOD BASEMENT FLOOR SYSTEM IS UNDER DEVELOPMENT. WRITE APA FOR PRELIMINARY DETAILS.

COVER PLATE

OPTIONAL INTERIOR FINISH

FLOOR JOIST

EXCAVATED SUMP PIT

TONGUE AND GROOVE JOINT

PLYWOOD MAY OVERLAP FIELD APPLIED TOP PLATE FOR SHEAR TRANSFER (FLASHING NOT REQUIRED IF SIDING OVERLAPS)

BLOCKING

CAULK

8″ MIN.

POLYETHYLENE FILM

4″ CONCRETE SLAB(B) OVER VAPOR BARRIER OVER GRAVEL, COARSE SAND OR CRUSHED ROCK

■ PRESSURE TREATED WOOD

(A) NOT REQUIRED TO BE TREATED IF BACKFILL IS MORE THAN 8″ BELOW BOTTOM OF PLATE. TYPICAL FOR ALL FOLLOWING DETAILS

Fig. 35-30 Typical wood foundation. *(Courtesy of American Plywood Association)*

Review Questions

Select the most appropriate answer.

1. The top and bottom horizontal members of a wall frame are called
   a. headers.   c. trimmers.
   b. plates.   d. sills.

2. The horizontal wall member supporting the load over an opening is called a
   a. header.   c. plate.
   b. rough sill.   d. truss.

3. Shortened studs above and below openings are called
   a. shorts.   c. jacks.
   b. lame.   d. stubs.

4. The finish floor to ceiling height in a platform frame is specified to be 7'-10". The finish floor is 3/4 inch thick and the ceiling material below the joist is 1 1/2 inches thick. A single soleplate and a double top plate are used, each of which has an actual thickness of 1 1/2 inches. What is the stud height?
   a. 7'-6 3/4"   c. 7'-9 1/4"
   b. 7'-7 3/4"   d. 7'-11 1/2"

5. A rabbeted jamb is 3/4 inch thick. Allowing 1/2 inch on each side for shimming the frame, what is the rough opening width for a door that is 2'-8" wide?
   a. 2'-9 1/2"   c. 2'-11 1/2"
   b. 2'-10 1/2"   d. 3'-0 1/2"

6. A story pole shows
   a. the length of headers.
   b. the length of rough sills.
   c. the length of trimmers.
   d. the width of the rough opening.

7. Short blocks fastened to the sides of corner posts and partition intersections and against the soleplate provide
   a. added strength to the corner posts and partition intersections.
   b. more secure fastening of the corner posts and partition intersections to the soleplate.
   c. a surface to attach lines when straightening walls.
   d. more surface on which to fasten the ends of baseboard.

8. When assembling wall sections, studs are laid on the floor with the crowned edge
   a. up.   c. in either direction.
   b. down.   d. alternated.

9. Exterior walls are straightened by
   a. sighting by eye.
   b. using a line stretched between two blocks and testing with another block.
   c. using a plumb bob dropped to the soleplate at intervals along the wall.
   d. by sighting along the length of the wall using a builder's level.

■ B U I L D I N G F O R S U C C E S S ■

Job Site Housekeeping

Traditionally, one of the most overlooked components of a successful contracting company is maintaining a clean, functional, and safe work environment. The amount of productive work accomplished at a job site is directly related to the structure and operation of the work activities. Job site housekeeping plays an important part in this structure.

One of the common goals shared by all builders is a safe working environment. This environment is conducive to high production, efficiency, and generally more profit. With more federal, state, and local regulations for construction sites, a builder cannot afford to be ignorant of or neglect those regulations. Insurance companies will continue to have major input into the safety affairs of all businesses. Job site housekeeping will continue to be a focus of attention for insurance compa-

nies, fire marshalls, and many governmental agencies that look after the welfare of workers.

Traditionally skilled workers have shared a sense of concern for the welfare of each person on the crew. They understand the need to maintain a clean and functional work site. Usually each person looks out for any job site hazards that could lead to accidents. Corrections are made once the hazard has been identified. With experience, the skill of spotting potential accidents becomes automatic.

Construction work has not changed drastically over the years. Production creates waste and by-products that must be disposed of regularly. In a very short time work locations can become cluttered and in a state of disarray. Waste and debris, in turn, become the basis for unsafe working conditions. A plan to properly store, remove, or in some manner discard the waste must be established.

The ability to maintain a clean and safe working area has to begin with an understanding of the problem and how it is created. This is the appropriate time to analyze how much of the problem is simply neglect of good housekeeping.

There are many things to consider in the establishment of a safe work environment. One of the first considerations is helping the entire workforce understand what a safe environment is and how it relates to them personally. They must buy into the program or it is doomed to fail. Primarily, the longevity in their job will be affected by job site working conditions. They must see that fewer accidents occur at a clean and well-organized job site. As they ask the question, "What's in it for me?" evidence must be presented to demonstrate how they will be in a "win-win" situation.

Second, each worker must be willing to enforce sound housekeeping principles daily. Building materials must be properly stored so they do not interfere with the construction activities. Appropriate storage locations must be chosen that are out of the way but yet convenient.

Third, time has to be set aside each day for site cleanup. This activity has to be included in the bid or estimate for the construction project. If done on a daily basis, each person can carry out a housekeeping assignment in minimal time.

Fourth, housekeeping efforts must be visible and show the workers proof that the cleanup tasks have been beneficial and perhaps even profitable. An incentive program may help employees, subcontractors, and others see and feel the benefits. One proven benefit has been recognition by the company of individuals who have no or minimal accidents during a certain time period. Another is company recognition for traditionally clean and safe working environments by the community, professional construction organizations, or insurance companies. This, in turn, may have an impact on hiring quality personnel for a particular company. Many potential employees are concerned about the safety record or safe working conditions promoted by a company.

A possible employment situation involving job site housekeeping can be presented by the following example. In typical custom cabinet shop each worker produces an abundance of wood waste daily. Debris left on the floor can cause walking hazards that turn into accidents. A buildup of wood by-products such as sawdust, wood chips, and shavings can become a fire hazard. If the fire marshall or insurance agent stops by unexpectedly, violations may cause production to halt until they are corrected. The cluttered or hazardous work site may be the grounds for insurance premium increases.

Trips, falls, and other accidents can be minimized or eliminated with good housekeeping habits. The time needed to daily clean up the work site will be small, if an organized plan is implemented. The regular elimination of waste can be completed in 10 minutes or less when specific cleanup jobs are scheduled.

Historically, workers are more productive and have a better sense of belonging to a company that has their welfare at heart. People tend to take pride in the fact that they can work for hundreds or thousands of hours without an accident. The vast majority of construction workers who are not involved in accidents in the workplace are working in a situation where good job site housekeeping is practiced. This practice is an integral part of the company safety program for employees. The end results of good housekeeping means many things, but characteristically they will mean a longer and safer construction career.

FOCUS QUESTIONS: For individual or group discussion.

1. How would you explain to your employees the need for and benefits of a good housekeeping program in your place of business?

2. What do you feel might be the biggest obstacles to implementing an efficient, ongoing housekeeping program at a construction site?

3. How would you structure a housekeeping program for a construction company, custom cabinet shop, or factory-built housing organization?

4. What representatives from business and industry might be resources for ideas and support in establishing a good housekeeping structure within your company's safety program?

UNIT
13

Interior Rough Work

OBJECTIVES

After completing this unit, the student will be able to:

- assemble, erect, brace, and straighten bearing partitions.
- determine and make rough openings for interior doors.
- lay out, cut, and install ceiling joists.
- apply furring strips to attach ceiling material.
- lay out and erect nonbearing partitions and install backing in walls for fixtures.
- apply plaster grounds and plaster bases.
- describe various components of light-gauge steel framing.
- lay out and frame light-gauge steel interior partitions.

Interior rough work is constructed in the inside of a structure and later covered by some type of finish work. The interior rough work described in this unit includes the installation of *partitions, ceiling joists, furring strips, backing* and *blocking, plaster bases,* and *plaster grounds.* The term *rough work* does not imply that the work is crude. It is a kind of work that will eventually be covered by other material. Careful construction of the rough frame makes application of the finish work less complicated.

CONTENTS

CHAPTER 36 — INTERIOR PARTITIONS AND CEILING JOISTS

Partitions and *ceiling joists* constitute some of the interior framing. Ceiling joists tie the exterior side walls together and provide a base for the ceiling finish.

Interior partitions supporting a load are called *bearing partitions.* Partitions that merely divide the area into rooms are called *nonbearing* partitions (Fig. 36-1).

Bearing Partitions

Bearing partitions support the inner ends of ceiling joists, or in multistory wood frame structures, the inner ends of joists are used to frame the floor above. They are placed directly over the girder or the bearing partition in the lower level. If several bearing partitions are used on the same floor, supported girders or walls are placed directly under each.

Bearing partitions are erected in a manner similar to exterior walls. A doubled top plate is usually required because of the weight on the wall. Headers over openings must be strong enough to support the load imposed on them. Common practice is to erect only bearing partitions at this stage. The framing of nonbearing partitions is left until later so the roof can be made tight as soon as possible.

If *roof trusses* are used, bearing partitions are not needed on single-story buildings or on the top floor of multistory buildings. The load is carried by the trusses and transmitted to the exterior walls. (Details on roof trusses are given in a later unit.)

Layout and Framing of Bearing Partitions

From the floor plan, determine the location of the bearing partitions. Dimensions are usually given to the centerline. Measure half of the thickness of the partition from the centerline. Snap a chalk line on the subfloor.

Lay the edge of the soleplate to the chalk line, tacking it in position. Lay out the openings, partition intersections, and all regular and jack studs. Lay the top plate next to the soleplate. Transfer the layout of the soleplate to the top plate. All joints in the plates should come in the center of a stud.

Frame, erect, straighten, and temporarily brace the bearing partition in the same manner as for exterior walls. Double the top plate, letting it lap the plate of the exterior wall (Fig. 36-2).

Rough Opening Sizes for Interior Doors

The rough opening width for an interior door opening is found by adding to the door width twice the thickness of the jamb stock and one inch for shimming (1/2 inch for each side). The rough opening height is found by adding the following:

- the thickness of the finish floor
- 1/2 inch for clearance between the finish floor and the bottom of the door
- the height of the door

Fig. 36-1 Load-bearing partitions support the weight of the floor or ceiling above. Nonload-bearing partitions merely divide an area into rooms.

Fig. 36-2 The top plate of the bearing partition laps the plate of the exterior wall.

- the thickness of the head jamb
- 1/2-inch clearance between the head jamb and the rough header.

Usually no threshold is used under interior doors. However, in case a threshold is used, substitute its thickness for the 1/2-inch clearance allowed under the door (Fig. 36-3).

Ceiling Joists

Ceiling joists generally run from the exterior walls to the bearing partition across the width of the building. Construction design varies according to geographical location, traditional practices, and the size and style of the building. The size of ceiling joists is based on the span, spacing, load, and the kind and grade of lumber used. Determine the size and spacing from the plans or from local building codes.

Methods of Installing Joists

At the exterior walls, the ceiling joists are placed so that the roof rafters can be fastened to their sides. In a conventionally framed roof the *rafters* must be fastened to the sides of the joists. This ties the walls together and keeps them from spreading due to the weight of the roof exerting an outward thrust on the exterior walls (Fig. 36-4). The opposing rafters should be in line and directly over the studs in the exterior wall (Fig. 36-5).

Sometimes the ceiling joists are installed in-line. Their ends butt each other at the centerline of the bearing partition. The joint must be *scabbed* to tie the joints together (Fig. 36-6). *Scabs* are short boards fastened to the side of the joist and centered on the joint. They should be a minimum of 24 inches long. The location for the ceiling joists on each exterior wall is on the same side of the rafter location when

Fig. 36-3 Figuring the rough opening size for an interior door.

joists are in-line. Wood I-beams, available in lengths up to 60 feet, may also be used for ceiling joists. Their long length can span most buildings without splicing.

Another method of installing ceiling joists is to place the joists on opposite sides of the rafter on the exterior walls. On the bearing partition, lap the

Fig. 36-4 The weight of the roof exerts pressures that tend to thrust the walls outward. Ceiling joists tie the frame together to resist the outward thrust.

Fig. 36-5 Ceiling joists are located so they can be fastened to the side of rafters.

Fig. 36-6 The joint of in-line ceiling joists must be scabbed at the bearing partition.

joists with a short block of joist stock between them (Fig. 36-7).

Some carpenters install ceiling joists by using the same layout on the exterior walls just mentioned. However, they offset the joists on the bearing partition with the centerline of the lap falling on the centerline of the rafters (Fig. 36-8).

Cutting the Ends of Ceiling Joists

The ends of ceiling joists on the exterior walls usually project above the top of the rafter. This is especially true when the roof has a low slope. These ends must be cut to the slope of the roof, flush with or slightly below the top edge of the rafter.

Lay out the cut, using a framing square. Cut one joist for a pattern. Use the pattern to mark the rest. Make sure when laying out the joists that you sight each for a crown. Make the cut on the crowned edge so that edge is up when the joists are installed. Cut the taper on the ends of all ceiling joists before installation. Make sure the length of the taper cut does not exceed three times the depth of the member. Also make sure that the end of the joist is, at least, half the member's depth (Fig. 36-9).

Stub Joists

Usually, ceiling joists run parallel to the end walls and in the same direction as the roof rafters. When low-pitched hip roofs are used, *stub joists,* which run at right angles to the end wall, must be installed. The use of stub joists allows clearance for the bottom edge of the rafters that run at right angles to the end wall (Fig. 36-10). The stub joists should be laid out so their centerlines coincide with

Fig. 36-7 Short blocks are sometimes used between ceiling joists at the bearing partition.

Fig. 36-8 Ceiling joists may be offset at the bearing partition.

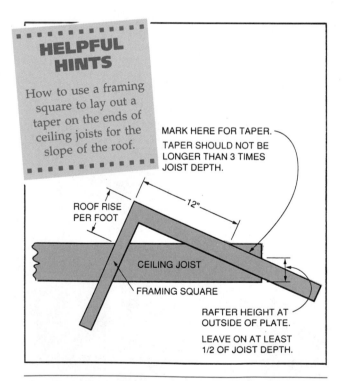

HELPFUL HINTS

How to use a framing square to lay out a taper on the ends of ceiling joists for the slope of the roof.

MARK HERE FOR TAPER.

TAPER SHOULD NOT BE LONGER THAN 3 TIMES JOIST DEPTH.

ROOF RISE PER FOOT

12"

CEILING JOIST

FRAMING SQUARE

RAFTER HEIGHT AT OUTSIDE OF PLATE.

LEAVE ON AT LEAST 1/2 OF JOIST DEPTH.

Fig. 36-9

Fig. 36-10 Stub joints are used for low-pitched hip roofs.

Framing Ceiling Joists to a Beam

In many cases, the bearing partition does not run the length of the building because of large room areas. Some type of beam is then needed to support the inner ends of the ceiling joists in place of the supporting wall. Similar in purpose and design to a girder, the beam may be of built-up solid lumber or engineered lumber.

If the beam is to project below the ceiling, it is installed in the same manner as a header for an opening. The joists are then installed over the beam in the same manner as over the bearing partition (Fig. 36-11).

those of any ceiling *furring strips*, if used. This provides a backing and a nailing surface for the ends of the furring strips. (Description of and application instructions for furring strips are given later in this chapter.)

Fig. 36-11 Framing ceiling joists over a beam.

Fig. 36-12 Framing ceiling joists into a beam for a flush ceiling.

If the beam is to be raised in order to make a flush and continuous ceiling below, then the ends of the beam are set on the top of the bearing partition and the exterior end wall. The joists are butted to the beam. They may be supported by a ledger strip or by metal joist hangers (Fig. 36-12).

Openings

Openings in ceiling joists may need to be made for such things as chimneys, attic access (*scuttle*), or disappearing stairs. Large openings are framed in the same manner as for floor joists. For small openings, there is no need to double the joists or headers (Fig. 36-13).

Fig. 36-13 Joists and headers need not be doubled for small ceiling openings.

Layout and Spacing of Ceiling Joists

Roof rafters rest on the plate directly over the regular spaced studs in the exterior wall. Ceiling joists are installed against the side of the rafters and fastened to them. Spacing of the ceiling joists and rafters should be the same so they can be tied together at the plate line.

■ Square up from the same side of each regular spaced stud or jack stud in the exterior wall and across the top of the plate. Mark an *X* on the side of the line over the stud for the rafter. Place a *J* on the other side of the line for the location of the ceiling joist.

■ On the bearing partition and the opposite exterior wall, use the same layout if the ceiling joists are to be installed in-line.

■ If the joists are to be lapped on the bearing partition with short blocks in between, the joist layout on one exterior wall is made on one side of the rafter layout and on the opposite side of the rafter layout on the other exterior wall. The bearing partition is laid out for short blocks placed in line with the rafters. The joists are installed on both sides of the each block.

■ If no blocks are used, the bearing partition is laid out with lines coinciding with the centerline of the rafters. Joists are fastened on each side of the lines laid out on the bearing partition.

The layouts for various methods of installing ceiling joists are shown in Fig. 36-14.

Installing Ceiling Joists

The ceiling joists on each end of the building are placed so the outside face is flush with the inside of the wall. They are installed in-line with their inner ends butting each other regardless of how the other joists are laid out (Fig. 36-15).

In addition to other functions, these end joists provide fastening for the ends of furring strips or

Fig. 36-15 The end ceiling joist is located with its outside face flush with the inside of the wall.

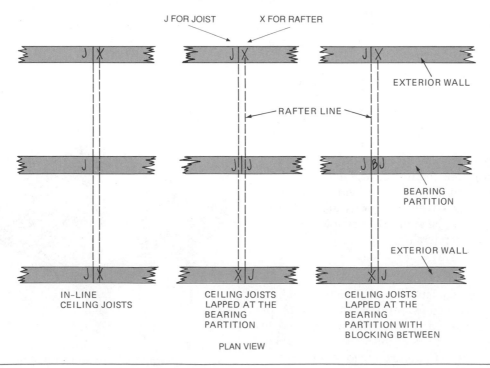

Fig. 36-14 Methods of laying out ceiling joists.

for the ceiling finish, if furring strips are not used. All other joists are fastened in position with their sides to the layout lines. If the outside ends of the joists have not been tapered, sight each joist for a crown. Install each one with the crowned edge up. If the outside ends have been tapered, install the ceiling joists with the cut edge up. Reject any badly warped joists.

Toenail joists with at least two 10d nails into the plates to the layout lines. Remember to keep joists on the correct side of the lines. Spike lapped joists together with at least four 10d nails, two from each side.

Furring Strips

Furring strips for ceilings, sometimes called *strapping* or *stripping,* are usually 1×3 wood strips. They are fastened to the underside of ceiling joists for the attachment of the ceiling finish. They are fastened at right angles to the ceiling joists in rows of 12, 16, or 24 inches on center depending on the type of ceiling finish to be used (Fig. 36-16).

Advantages of Using Furring Strips. Often, furring strips are omitted and the ceiling finish is attached directly to the ceiling joists. However, many builders prefer to install furring strips for a number of reasons. Electricians can run wiring at right angles to the joists, without boring holes in each one, by running cables under the joists between furring strips.

Furring strips can be spaced to accommodate any size ceiling material. *Ceiling tile* and *metal lath,* for instance, cannot be fastened directly to the joists because the joist spacing may be too great.

With the use of furring strips, the base for the ceiling can be straightened wherever necessary by shimming the strips. Also, insulation can be laid between the joists without fastening because the furring strips hold it up. The use of furring strips reduces the transmission of sound through the ceiling to the floor above. The small cost of installing ceiling furring strips is offset by the advantages they provide.

Laying Out Furring Strips

Usually the furring strips are installed before any of the nonbearing partitions are erected if plaster or gypsum board is to be used for ceiling material. If ceiling tiles are to be used, then the nonbearing partitions are erected. Each room is individually laid out for the furring strips. (The layout of furring strips for ceiling tile is discussed in a later unit.)

Layout for Plaster or Gypsum Board Ceiling. For the application of plaster base or gypsum

Fig. 36-16 Furring strips are used for the attachment of ceiling finish.

board, lay out the spacing of furring strips on the bottom edge of the ceiling joist against each end wall. Starting from one side wall, measure in 16 inches to the centerline of a row of furring. Measure back half the width of the furring strip. Square a line representing the edge of the strip across the bottom edge of the joist. Place an X on the side of the line on which the strip is to be fastened. From the squared line, continue the spacing across the end wall. Place Xs on the same side of the line. Make a similar layout on the opposite end wall (Fig. 36-17). Snap chalk lines between the marks on the end walls for the location of the strips.

Installing Furring Strips

Start the installation by constructing a working platform of convenient height. Usually short sawhorses with planks laid on top are used. This platform can be moved as work progresses.

Cut the first strip so that when one end butts against the end wall, the other end falls on the center of a joist. Cut ends using a handsaw while up on the platform. Steady the strip on your knee while cutting. There is no need to square a line across the piece or to make tight joints between the ends of furring strips. Just sight by eye for squareness when cutting. This eliminates getting up and down from the platform whenever a piece needs to be cut, and saves much time and energy.

Apply the first strip with its end against the end wall and its edge against the side wall. Drive two 8d common nails into each joist. Continue applying strips until the opposite end wall is reached.

Fasten the other rows of furring strips with their edges to the chalk line in a similar manner (Fig. 36-18). Toenail the edge of the strip, where necessary, to straighten it and bring it to the chalk line. The final strip is placed with its edge against the bearing partition. If no bearing partition is used,

Fig. 36-17 Laying out the location of furring strips.

Fig. 36-18 Installing furring strips.

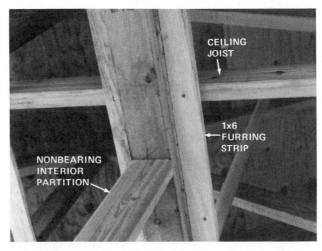

Fig. 36-19 View from floor looking up shows 1×6 furring installed for a partition that runs between, and parallel to, regular furring. Notice that the 1×6 extends about equally on both sides of the partition.

the final strip is placed against the opposite side wall.

Some nonbearing partitions run parallel to the furring strips. Additional furring is required to provide fastening for the ceiling material on each side of the partitions. Install 1×6 furring between regular furring strips so its centerline falls on the centerline of the partitions (Fig. 36-19).

Sight each row of furring by eye. Shim down any places necessary to straighten the rows. Use wood shingles split into about 2-inch widths for shims. Tap the furring down where too high. Insert the shingle tips between the strip and ceiling joist. Then, drive another nail when the strip is shimmed straight (Fig. 36-20).

Nonbearing Partitions

Nonbearing partitions carry no load. They divide the floor area into rooms. Most building codes state that the studs may be spaced up to 28 inches on center and may be set with the long dimension parallel to the wall. Openings may be framed with single studs and headers. However, the general practice is to use 2×4 studs spaced 24 inches on center with their edge parallel to the wall. Because the wall carries no load, headers are usually doubled 2×4s.

Fig. 36-20 Straightening the furring strips by shimming with wood shingles.

Installing Soleplates

The location of the partitions is determined from the floor plans. Dimensions are usually given to their centerline. Locate each partition on the subfloor. Snap chalk lines to indicate one side of the partition. Place Xs on the side of the line the partitions are to be placed. Mark the end of each partition and all openings.

Fasten the soleplates for all partitions to the subfloor, keeping the edges to the chalk line. Straighten the soleplates as necessary by toenailing 8d common nails into their edges. When possible, spike into the floor joists below with one spike into each floor joist staggered along the length of the soleplate. In case partitions run between and parallel to the joists, fasten the soleplates by toenailing through both edges into the subfloor about 16 inches apart. Cut all soleplates plates to length. Leave spaces for door openings (Fig. 36-21).

Fig. 36-21 Soleplates for nonbearing interior partitions are fastened to the subfloor.

Fig. 36-22 Hang a plumb bob from the ceiling to locate the top plate of interior partitions.

Locating Top Plates

The position of the top plate must be marked on the furring strips above. Hang a plumb bob from the furring strip so it lines up with the edge of the soleplate at one end of the partition. Mark the ceiling at this point (Fig. 36-22). Do the same at the other end of the partition. Snap a chalk line between the marks. Locate the top plates for all other partitions in the same manner.

Laying Out the Plates

Cut the top plate to run from one end of the partition to the other. Although the soleplate has been cut out for openings, the top plate runs uncut for the total length of the partition, except if the opening goes all the way to the ceiling.

Tack the top plate on top of the soleplate so that it can be removed later. Lay out the plates in a manner similar to exterior walls. Use the specified stud spacing (Fig. 36-23). Build corner posts and partition intersections as necessary. Find the stud

Fig. 36-23 Laying out the plates for interior partitions.

length. Cut the needed number of full length studs. The stud length for nonbearing partitions is different than in bearing partitions or exterior walls.

To find the stud length for the nonbearing partition, lay two thicknesses of plate material on the subfloor. Measure up to the furring strip (Fig. 36-24). Instead of actually measuring, a pair of slip sticks is useful for determining the length. Or, hold a strip of lumber on top of the bottom plate material. Let the strip run by the top plate. Mark the strip in line with the bottom surface of the top plate.

Constructing Nonbearing Partitions

Separate the top plate from the soleplate. Place all full-length studs, corner posts, and partition intersections on the floor. Make sure their crowned edges run in the same direction, either all up or all down, and their ends are against the fastened soleplate. Spike the top plate to these members to the layout lines.

Raise the section into position. Toenail the studs to the soleplate. Use four 8d common nails in each stud, two on each side. Fasten the top plate into position to the chalk line by toenailing 8d com-

mon nails through the edges of the plate into the furring strips.

Straighten the end stud that butts against another wall by toenailing through its edge into the center block of the partition intersection. Test for straightness by using a long straightedge against its edge (Fig. 36-25).

Construct all nonbearing partitions in a similar manner. For ease of erecting, construct the longer partitions first. Then construct shorter cross-partitions, such as for closets. There is no hard and fast rule for constructing partitions. Some carpenters prefer to erect them in sections by spiking both top and bottom plates to the studs and raising them in position. Others prefer to first fasten both plates in place and then install studs in between. It matters little which method is used as long as the partitions are carefully laid out and constructed.

Bathroom and kitchen walls sometimes must be made thicker to accommodate plumbing. Sometimes 2×6 plates and studs are used or a double 2×4 partition is erected. If the wall thickness needs to be increased only slightly, furring strips may be added to the edges of the studs and plates (Fig. 36-26).

Fig. 36-24 Finding the length of a stud in a nonload-bearing partition.

END STUD
OF PARTITION

CENTER BLOCK
OF PARTITION
INTERSECTION

PARTITION
INTERSECTION

TOENAIL THROUGH
CROWNED EDGE
OF END STUD INTO
CENTER BLOCK
TO STRAIGHTEN

8d COMMON
NAIL

END STUD OF
INTERIOR PARTITION

Fig. 36-25

FURRING
STRIP

FURRING STRIP

STUD

DETAIL

Fig. 36-26 The wall thickness may be increased by fastening furring strips to the edges of the studs.

Openings

Besides door openings, the carpenter must frame openings for heating and air-conditioning ducts, medicine cabinets, electrical panels, and similar items. If the items do not come in a stud space, the stud must be cut and a header installed. When ducts run through the floor, the soleplate and subfloor must be cut out (Fig. 36-27). The reciprocating saw is a useful tool for making these cuts.

OPENING FOR
MEDICINE CABINET

OPENING FOR
HEATING DUCT

Fig. 36-27 Miscellaneous openings in interior partitions.

CHAPTER 37

BACKING, BLOCKING, AND BASES

This chapter deals with *backing, blocking,* and *plaster bases.* A *backing* is, ordinarily, a short block of lumber placed in floor, wall, and ceiling cavities to provide fastening for various parts and fixtures.

Blocking is installed for different purposes, such as providing support for parts of the structure, weathertightness, and firestopping. Sometimes blocking serves as backing.

Several types of *plaster bases* are used to provide a surface for plaster to be spread and on which to adhere. Various accessories are necessary to complete the installation of the plaster base.

Backing and Blocking

There are many places in the structure where blocking and backing should be placed. It would be unusual to find directions for the placement of backing and blocking in a set of plans. It is the wise builder who installs them, much to the delight of those who must install fixtures and finish of all types in later stages of construction.

Backing

Unit 12 recommended installing short blocks with their ends against the sole plate and their sides against the studs on inside corners and on both sides of door openings. This provides more fastening surface for the ends of baseboard (Fig. 37-1).

Much backing is needed in bathrooms. Plumbing rough-in work varies with the make and style of plumbing fixtures. The experienced carpenter

will obtain the rough-in schedule from the plumber. He or she then installs backing in the proper location for such things as bathtub faucets, shower-heads, lavatories, and water closets (Fig. 37-2). Backing should also be installed around the top of the bathtub.

In the kitchen, backing should be provided for the tops and bottoms of wall cabinets and for the tops of base cabinets. If the ceiling is to be built down to form a *soffit* at the tops of wall cabinets, backing should be installed to provide fastening for the soffit (Fig. 37-3).

A homeowner will appreciate the thoughtfulness of the builder who provides backing in appropriate locations in all rooms for the fastening of curtain and drapery hardware (Fig. 37-4).

Blocking

Some types of blocking have already been described in earlier units. See Floor Framing for the installation of solid lumber blocking used for bridging. See Wall Framing for the use of wood blocking for firestops.

When floor panels are used as a combination subfloor and *underlayment* (under carpet and pad), the panel edges must be tongue-and-groove or supported on 2-inch lumber blocking installed between joists (Fig. 37-5).

When ceiling furring strips are not used, ladder type blocking is needed between ceiling joists to support the top ends of partitions that run parallel to and between joists (Fig. 37-6).

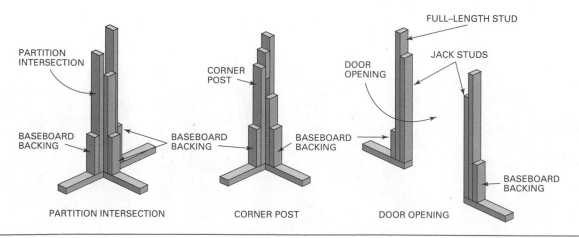

Fig. 37-1 Backing is installed at corners and door openings for baseboard.

Fig. 37-2 Considerable backing is needed in bathrooms.

Fig. 37-3 Location and purpose of backing in kitchens.

Fig. 37-4 An experienced builder will install backing in all rooms for curtain and drapery hardware.

CARPET AND PAD

STRENGTH AXIS

1/8" SPACING IS RECOMMENDED AT ALL EDGE AND END JOINTS UNLESS OTHERWISE INDICATED BY PANEL MANUFACTURER

STAGGER END JOINTS

SITE APPLIED GLUE, BOTH JOISTS AND TONGUE-AND-GROOVE JOINT (OR BETWEEN PANELS AND EDGE BLOCKING)

TONGUE-AND-GROOVE EDGES (OR 2" LUMBER BLOCKING BETWEEN SUPPORTS)

2x JOISTS, PLYWOOD-WEBBED "I" JOISTS OR FLOOR TRUSSES—16", 19.2", 24" OR 32" OC (4x SUPPORTS FOR 48" OC SPACING)

APA RATED STURD-I-FLOOR 16, 20, 24, 32 OR 48 OC

Fig. 37-5 The edges of floor panels used as a combination subfloor and underlayment must be tongue-and-groove or be supported by blocking.

CEILING JOIST

LADDER TYPE BLOCKING

LADDER TYPE BLOCKING

TOP PLATE

STUD

INTERIOR PARTITION RUNNING PARALLEL TO AND IN BETWEEN CEILING JOISTS

Fig. 37-6 Ladder type blocking must be provided for some interior partitions not supported by furring strips.

Blocks are sometimes installed to support bathtubs (Fig. 37-7). There is a negligible amount of shrinkage when the blocks are placed on end.

Wall blocking is required for weathertightness and support at the horizontal edges of wall sheathing panels permanently exposed to the weather. In addition, blocking is required behind all plywood

seams in a wood foundation. When used for these purposes, they must be installed in a straight line (Fig. 37-8).

Blocking between studs is required in walls

STUD

BLOCK

SUBFLOORING

DOUBLE TRIMMER

JOIST

SUPPORT OF BATHTUB AGAINST WALL FRAMING

Fig. 37-7 Supporting the inner edge of a bathtub with blocking.

WALL SHEATHING
PERMANENTLY EXPOSED
TO THE WEATHER

EXTERIOR
WALL BLOCKING

Fig. 37-8 Blocking for wall sheathing must be installed in a straight line.

the studs and strengthen the structure. The required blocking also functions as a firestop in stud spaces. Blocking for these purposes may be installed in staggered fashion (Fig. 37-9).

Installing Backing and Blocking

Install blocking in a staggered row. Fasten by spiking through the studs into each end of each block in

the same manner as staggered solid wood bridging described previously. Installing blocking in a straight line is more difficult. The ends of some pieces must be toenailed (Fig. 37-10). Snap a line across the framing. Square lines in from the chalk line on the sides of the studs.

It may be helpful to fasten short cleats on one side of each stud to support the blocking while the end is being toenailed. If cleats cannot be used or are not desired, start the toenails in the end of the block before positioning it (Fig. 37-11). Backing may

TOP
PLATE

BLOCKING IS
INSTALLED IN
A STAGGERED ROW
FOR EASE OF
FASTENING.

EACH BLOCK MAY BE
FASTENED BY SPIKING
INTO ITS ENDS

INSTALL BLOCKING IN EVERY
OTHER SPACE FIRST. THEN,
INSTALL IN REMAINING SPACES

STAGGERED
WALL BLOCKING

STUD

SOLE
PLATE

Fig. 37-9 Blocking used for stiffening walls or for firestopping may be installed in a staggered fashion.

TOP PLATE

STUD

SOLE PLATE

THE FIRST SET OF IN-LINE BLOCKING, INSTALLED IN EVERY OTHER STUD SPACE, CAN BE FASTENED BY SPIKING INTO THE ENDS OF EACH PIECE

THE SECOND SET OF IN-LINE BLOCKING INSTALLED IN THE REMAINING SPACES MUST BE FASTENED BY TOENAILING THE ENDS

Fig. 37-10 Straight in-line blocking requires that the ends of some pieces be fastened by toenailing.

also be installed in a continuous length by notching the studs and fastening into its edges (Fig. 37-12). If installing blocking between studs, fasten pieces in every other space first. Then go back and fill in. This prevents gaining on the stud space and bowing the studs, especially if the blocking is precut.

Plaster Bases

Plaster bases of some type must be fastened to the wall studs and ceiling joists or furring strips for the application of *plaster*.

Plaster

Plaster has been used for many years for wall and

HELPFUL HINTS

How to toenail wall blocking more easily.

2" x 4" BLOCKING

8d COMMON NAIL

BENCH OR SAWHORSE TOP

1. PLACE BLOCKING AGAINST STOP BLOCK AND START TOENAILS UNTIL POINTS JUST COME THROUGH ENDS. TURN BLOCKING END FOR END AND REPEAT PROCEDURE.

STOP BLOCK

LAYOUT LINE

DRIVE PREVIOUSLY STARTED TOENAIL WHILE HOLDING CHISEL WITH UPWARD PRESSURE

BLOCKING

1" WOOD CHISEL

DRIVE WOOD CHISEL, BEVEL SIDE UP, INTO SIDE OF STUD UNDER BLOCKING

STUD

Fig. 37-11

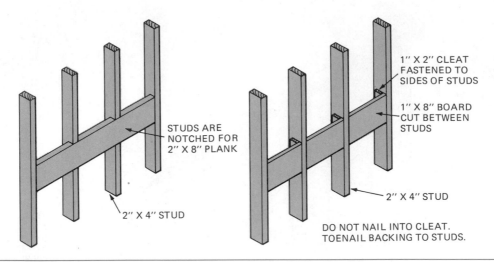

STUDS ARE
NOTCHED FOR
2" X 8" PLANK

1" X 2" CLEAT
FASTENED TO
SIDES OF STUDS

1" X 8" BOARD
CUT BETWEEN
STUDS

2" X 4" STUD

2" X 4" STUD

DO NOT NAIL INTO CLEAT.
TOENAIL BACKING TO STUDS.

Fig. 37-12 Method of installing continuous backing for plumbing and other type fixtures.

ceiling finish. It is made from a white mineral called *gypsum.* Gypsum is mined from the earth in rock form. The rock is crushed and ground to a powder form. It is then mixed with sand and a binder to make plaster.

The plaster powder is mixed with water to form a paste. When spread thinly on a flat surface, it dries to form a hard and smooth surface. The surface on which the plaster is spread is called a *plaster base* (Fig. 37-13).

Lath

In the past, thin wood strips called *wood lath* were used as a plaster base. They are still found in many

Fig. 37-13 Applying plaster to a gypsum plaster base. (*Courtesy of U.S. Gypsum Company*)

older buildings, and can still be purchased in some lumber yards. They were applied with a space between each strip that served to form a *key* or a lock for the plaster. Wood lath is no longer used, except perhaps in remodeling to match the existing construction. At the present time, *gypsum lath,* commonly called *rock lath,* and *metal lath* are used.

Gypsum Lath. Gypsum lath is a thin board consisting of a gypsum core sandwiched between a special paper covering. On the face side, the outer layers of a multilayer paper absorb moisture from the plaster quickly to form a tight bond. The inner layers are treated to resist moisture and prevent softening of the gypsum core.

For most jobs regular rock lath is used. When additional fire resistance is required, Firecode rock lath, whose core contains special mineral materials, may be used.

Gypsum lath is made in several sizes. For framing spaced 16 inches on center, a 3/8-inch thickness may be used. A 1/2-inch thickness is also available. The maximum frame spacing is dependent on the thickness and type of lath used. The standard width is 16 inches. The standard length is 48 inches. A 24x96 gypsum lath is available in a 1/2-inch thickness only.

Metal Lath. Metal lath is formed from sheet steel that has been slit and expanded to form numerous small openings to key the plaster. The metal is expanded in several forms, such as diamond mesh, ribbed, or flat ribbed for various purposes (Fig. 37-14). The diamond mesh design is a general all-purpose type of metal lath. It is available in a *self-furring* type. It has dimple indentations spaced closely each way. It is used as an exterior stucco base, for column fireproofing, and for replastering over old surfaces. Metal lath is available in widths

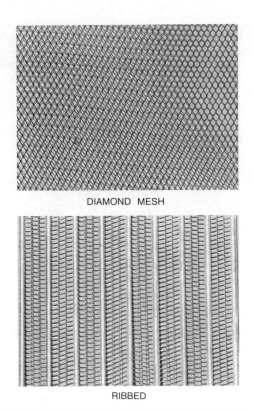

DIAMOND MESH

RIBBED

Fig. 37-14 Types of metal lath.

1" MIN. LAP VERTICAL
1/2" MIN. LAP HORIZONTAL

TIE WIRE

STEEL STUDS →

TRUSS HEAD
SCREWS

METAL LATH →

STEEL RUNNER

FINISH COAT
BROWN COAT
SCRATCH COAT

Fig. 37-15 Gypsum plaster applied to metal lath.

SPECIAL STEEL
CLIPS ARE USED
TO SUPPORT &
ALIGN END JOINTS

Fig. 37-16 Gypsum lath is applied with its end joints staggered. (*Courtesy of U.S. Gypsum Corporation*)

of 27 inches and in lengths of 96 inches. It is coated with black asphaltum paint or is galvanized.

Metal lath, commonly called *wire lath*, is used extensively in commercial construction. On large jobs it is installed by specialists called *lathers*. On smaller jobs, the carpenter may be required to apply either gypsum or metal lath. In residential construction, metal lath is recommended for use on ceilings that have a living area above. In shower and bathtub areas, where excessive moisture is present, only galvanized metal lath is recommended for a plaster base.

Metal lath requires three coats of plaster: a scratch coat, a brown coat, and a finish coat. Gypsum lath requires only two coats: the brown coat and the finish coat. Each coat must be partially dry before the next coat is applied. The combined thickness of the scratch and brown coat is 11/16 inch. The thickness of the finish coat is 1/16 inch (Fig. 37-15). Total thickness of the lath and plaster combined is 3/4 inch.

Applying Gypsum Lath

Gypsum lath is applied with its face out and long dimension across framing members. End joints should fall between framing members. They should be staggered in successive *courses* (rows) with none occurring in line with the sides of openings. Where end joints occur between framing, secure with clips.

Attach gypsum lath to framing with nails, screws, or staples (Fig. 37-16). Drive five fasteners into each framing member across the width of the sheet, not less than 3/8 inch from ends and edges. For 24-inch wide, use six fasteners in each framing member. Be careful not to let the fastener heads cut into the face paper.

Gypsum lath may be cut to length by scoring one side with a utility knife. Bend and break the sheet along the scored line. Then score the back side on the bend. Snap off the waste (Fig. 37-17). A compass saw or wallboard saw is used to make cutouts for pipes or electrical outlets.

Reinforcing Plaster Bases

To prevent plaster cracks, certain sections are reinforced with expanded metal. Fasten small strips diagonally across the upper corners of all window and door openings (Fig. 37-18). Inside corners at the

Fig. 37-17 Cutting gypsum board lath.

Fig. 37-18 Reinforce the tops of windows and door openings with small strips of metal lath.

Fig. 37-19 Interior corners are reinforced with wire lath commonly called cornerite.

intersection of walls and ceilings are reinforced with metal lath called *cornerite*. These are strips bent at a 90 degree angle to fit in the corners. They project about 2 inches out on each side (Fig. 37-19).

Corner beads of expanded metal are applied on all exterior corners. Use a straightedge. Install them level and plumb. Plumb corner beads from the *plaster grounds* (strips of wood used as guides to control thickness of plaster) installed at the base. Where the corner beads meet at the corner of an opening,

Fig. 37-20 Exterior corners are reinforced with corner bead.

make sure the bead of each is exactly flush (Fig. 37-20).

Applying Metal Lath

Metal lath is applied so that its edges and ends lap a minimum of one inch. No extra reinforcing is necessary except using corner beads at the exterior corners. Fasten the expanded metal to metal framing with tie wire and to wood framing with fasteners that catch two strands of the lath and spaced about 6 inches on center into each framing member. Cut expanded metal where necessary using tin snips.

> **CAUTION** Handle the cut sheets carefully. The sharp ends of the metal can cut deeply.

Plaster Grounds

Plaster grounds are strips of wood or metal used as guides by the plasterer to control the thickness of the plaster. The thickness of the ground is the total thickness of the lath and plaster combined. Recommended thickness is 3/4 inch.

Grounds are located at the base of walls and around door and window openings (Fig. 37-21). They are placed in a position so the ground will later be covered by the interior trim. They are also installed wherever it is imperative that a straight and even plaster line be provided for high-quality application of interior trim. Therefore, plaster grounds must be installed in a straight and true line.

Usually 3/4×2 1/2-inch strips are used at the base. Wider strips may be used if the baseboard will cover them. Two narrow strips may be used, one at the floor line and one just below the top of the baseboard. Around door and window openings and at other locations usually 3/4×1-inch strips are used.

In some cases, the inside edges of the window frame act as grounds when the jamb is the proper

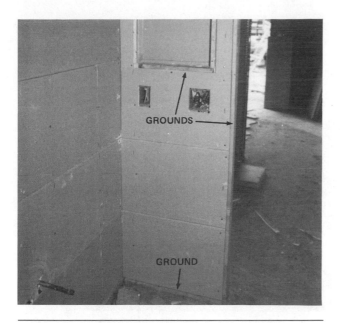

Fig. 37-21 Plaster grounds serve as a guide to control the thickness of plaster.

width. In most cases, the grounds are fastened permanently in position. They are left in place after the plaster is applied. However, they may be installed temporarily, especially around door openings. They are removed when the plastering is completed (Fig. 37-22). The carpenter is responsible for the installation of plaster grounds unless a specialist installs them.

Fig. 37-22 Grounds may be installed temporarily or permanently around openings.

CHAPTER 38 STEEL FRAMING

Light steel framing is used for structural framing and interior nonload-bearing partitions. Specialists in light steel are required when framing is extensive. However, carpenters often frame interior partitions and apply *furring channels* of steel. This chapter, therefore, is limited to their installation.

The strength of steel framing members of the same design and size may vary with the manufacturer. The size and spacing of steel framing members should be determined from the drawings or by those qualified to do so.

Interior Steel Framing

The framing of steel interior partitions is quite similar to the framing of wood partitions. Different kinds of fasteners are used. Some special tools may be helpful.

Steel Framing Components

All steel framing components are coated with material to resist corrosion. The main parts of an interior steel framing system are *studs, runners, channels,* and various *accessories.*

Studs. For interior nonload-bearing applications, *studs* are manufactured from 25-, 22-, and 20-gauge steel. The stud *web* has punchouts at intervals through which to run pipes and conduit. Studs

come in widths of 1 5/8, 2 1/2, 3 5/8, 4, and 6 inches, with 1 and 1 1/4 leg thickness. Studs are available in stock lengths of 8, 9, 10, 12, and 16 feet (Fig. 38-1). Custom lengths up to 28 feet are also available.

Runners. The top and bottom horizontal members of a steel-framed wall are called *runners* or *trucks.* They are installed on floors and ceilings to receive the studs. They are manufactured in gauges, widths, and leg thicknesses to match studs (Fig. 38-2). Runners are available in standard lengths of 10 feet.

Channels. Steel *channels* are formed from 16-gauge steel. They are is available in several widths. They come in lengths of 10, 16, and 20 feet. Channels are used in suspended ceilings and walls. When used for lateral bracing of walls, the channel is inserted through the stud punchouts. It is fastened with welds or clip angles to the studs (Fig. 38-3).

Furring Channels. *Furring channels* are hat-shaped pieces made of 20- and 25-gauge steel. Their overall cross-section size is 7/8 inch by 2 9/16 inches. They are available in lengths of 12 feet (Fig. 38-4). Furring channels are applied to walls and ceilings for the screw attachment of gypsum panels and plaster bases. Framing members may exceed

1 5/8", 2 1/2", 3 5/8", 4", 6"

1 1/4"

WEB PUNCHOUTS
24" O.C.

Fig. 38-1 Steel studs for light framing come in several widths, lengths, and gauges.

AVAILABLE
IN 10 FT. LENGTHS

LEGS ARE
"TOED" INWARD
TO PROVIDE A
FRICTION GRIP ON
STUDS FOR EASE OF
INSTALLATION

WEB

1" & 1 1/4"
LEG HEIGHT

MANUFACTURED TO MATCH STUD WIDTHS

STUD

RUNNER

RUNNERS,
ALSO CALLED
TRACKS

STEEL STUD

STEEL–STUDDED
WALL

Fig. 38-2 Runners are the top and bottom tracks of steel studded walls.

spacing limits for various coverings. Furring can then be installed to meet spacing requirements and provide necessary support for the surfacing material.

Layout and Framing of Steel Partitions

Lay out steel framed partitions as you would wood framed partitions. Snap chalk lines on the floor. Plumb up from partition ends. Snap lines on the ceiling. Make sure that partitions will be plumb. Using a laser level is an efficient way to lay out floor and ceiling lines for partitions. See Figure 25-43.

Lay out the stud spacing and the wall openings on the bottom runner. The top runner is laid out after the first stud away from the wall is plumbed and fastened.

Installing Runners. Fasten runners to floor and

Fig. 38-3 Steel channel is used for several purposes in walls and ceilings.

16 GAUGE STEEL CHANNEL

1/2"

3/4", 1 1/2", 2", 2 1/2"

ANGLE CLIP

STEEL CHANNEL

STEEL STUD

STEEL CHANNEL IS USED TO PROVIDE LATERAL SUPPORT OF WALLS, AMONG OTHER PURPOSES.

Fig. 38-4 Furring channels are used in both ceiling and wall installations.

METAL FURRING CHANNEL

7/8"

2 9/16"

SIZE MAY VARY SLIGHTLY WITH MANUFACTURER

METAL FURRING CHANNEL IS USED FOR THE SCREW ATTACHMENT OF GYPSUM PANELS

ceiling so one edge is to the chalk line. Make sure both floor and ceiling runners are on the same side of the line. Leave openings in floor runners for door frames. Allow for the width of the door and thickness of the door frame. Runners are usually fastened into concrete with power-driven fasteners. Stub concrete nails or masonry screws may also be used. Fasten into wood with 1 1/4-inch oval head screws.

Attach the runners with two fasteners about 2 inches from each end and a maximum of 24 inches on center in between. At corners, extend one runner to the end. Then butt or overlap the other runner (Fig. 38-5). It is not desirable or necessary to make mitered joints.

To cut metal framing to length, tin snips may be used on 25-gauge steel. Using tin snips becomes difficult on thicker metal. A power miter box, com-

RUNNERS

BUTTED   LAPPED

Fig. 38-5 Runners may be butted or overlapped at intersections.

CAUTION The sharp ends of cut metal can cause a serious injury. A pointed end presents an even greater danger. Avoid miter cuts on thin metal. Do not leave short ends of cut metal scattered around the job site. Dispose of them in a container as you cut them.

monly called a *chop saw,* with a metal-cutting saw blade is the preferred tool.

Installing Studs. Cut the necessary number of full-length studs needed. For ease of installation, cut them about 1/4 inch short. Install studs at partition intersections and corners. Fasten to bottom and top runners. Use 3/8-inch self-drilling pan head screws. If moisture may be present where a stud butts an exterior wall, place a strip of asphalt felt between the stud and the wall.

Place the first stud in from the corner between runners. Fasten the bottom in position at the layout line. Plumb the stud. Clamp the top end when plumb, and fasten to the top runner. Lay out the stud spacing on the top runner from this stud (Fig. 38-6.)

Place all full-length studs in position between runners with the open side facing in the same direction. The web punchouts should be aligned vertically. This provides for lateral bracing of the wall and the running of plumbing and wiring (Fig. 38-7). Fasten all studs except those on each side of door openings securely to top and bottom runners.

Wall Openings

The method of framing door openings depends on the type of door frame used. A one-piece metal door frame must be installed before the gypsum board is applied. A three-piece, knocked-down frame is set in place after the wall covering is applied (Fig. 38-8).

Framing a Door Opening for a Three-Piece Frame. First, place full-length studs on each side

of the opening in a plumb position. Fasten securely to the bottom and top plates. Cut a length of runner for use as a header. Cut it 6 inches longer than the width of the opening. Cut the flanges in appropriate places. Bend the web to fit over the studs on each side of the opening. Fasten the fabricated header to the studs at the proper height. Install jack studs over the opening in positions that continue the regular stud spacing (Fig. 38-9). Window openings are framed in the same manner. However, a rough sill is installed at the bottom of the opening. Jack studs are placed both above and below.

Framing a Door Opening for a One-Piece Frame. Place the studs on each side of the opening. Do not fasten to the runners. Set the one-piece door frame in place. Level the door frame header by shimming under a jamb, if necessary. Fasten the bottom ends of the door jambs to the floor in the proper location. Fasten the studs to the door jambs. Then, fasten the studs to the bottom runner. Plumb the door frame. Clamp the stud to the top runner, and fasten with screws. Install header and jack studs in the same manner as described previously (Fig. 38-10).

The steel framing described above is suitable for average weight doors up to 2'-8" wide. For wider and heavier doors, the framing should be strengthened by using 20-gauge steel framing. Also, double the studs on each side of the door opening (Fig. 38-11.)

Chase Walls

A *chase wall* is made by constructing two closely spaced, parallel walls for the running of plumbing, heating and cooling ducts, and similar items. They are constructed in the same manner as described previously. However, the spacing between the outside edges of the wall frame must not exceed 24 inches.

The studs in each wall should be installed with the flanges running in the same direction. They should be directly across from each other. The

STEP BY STEP
P R O C E D U R E S

WALL

2. PLUMB UP FIRST STUD AWAY FROM WALL

3. LAY OUT STUD SPACING ON TOP RUNNER FROM PLUMBED STUD

1. LAY OUT STUD LOCATION ON BOTTOM RUNNER

Fig. 38-6 Lay out the top runner from the plumbed end stud.

Fig. 38-7 Installing steel studs between top and bottom runners. (*Courtesy of U.S. Gypsum Corporation*)

GYPSUM BOARD

3-PIECE KNOCK DOWN STEEL FRAME

GROUT

STEEL STUD

JAMB ANCHOR

STEEL STUD

1-PIECE STEEL DOOR FRAME

INSTALLED AFTER GYPSUM BOARD WALL COVERING

MUST BE INSTALLED BEFORE GYPSUM BOARD WALL COVERING

PLAN OF JAMBS

Fig. 38-8 In a steel studded wall, either one-piece or three-piece, knocked-down metal door frames are used. (*Courtesy of U.S. Gypsum Corporation*)

JAMB STUD

SLIT LEGS OF RUNNER AND BEND WEB TO FIT AGAINST JAMB STUD. FASTEN WITH SCREWS TO STUD

ALTERNATE METHOD–NOTCH RUNNER AND LET LEGS RUN BY FACE OF STUD. FASTEN LEGS TO STUD.

RUNNER USED FOR HEADER

TOP RUNNER

USE RUNNER FOR A HEADER

STUD

JAMB STUDS

BOTTOM RUNNER

ANCHOR RUNNER SECURELY TO FLOOR ON BOTH SIDES OF OPENING

Fig. 38-9 An opening framed for a 3-piece, knock-down, metal door frame.

STEP BY STEP
P R O C E D U R E S

TOP RUNNER

LEAVE JAMB STUDS LOOSE

BOTTOM RUNNER

FASTEN RUNNER SECURELY ON BOTH SIDES OF OPENING

1. FASTEN TOP AND BOTTOM RUNNERS IN POSITION. LEAVE SPACE IN BOTTOM RUNNER FOR OPENING.

2. FASTEN ALL STUDS IN POSITION BUT LEAVE JAMB STUDS LOOSE.

LEVEL HEADER OF DOOR FRAME. SHIM UNDER SIDE JAMB, IF NECESSARY.

3. MOVE JAMB STUDS OUT OF THE WAY.

4. SET DOOR FRAME IN OPENING. LEVEL HEADER AND FASTEN BOTTOM OF FRAME TO FLOOR AT CORRECT WIDTH.

Fig. 38-10 Method of framing the opening for and setting a one-piece door frame.

STEP BY STEP
P R O C E D U R E S

(continued)

8. INSTALL HEADER AND JACK STUDS.

TOP RUNNER

JAMB STUD

DOOR JAMB

STUD

6 FT. LEVEL

JAMB STUD

5. FASTEN JAMB STUDS ON BOTH SIDES TO SIDE JAMBS OF DOOR FRAME.

6. FASTEN JAMB STUDS ON BOTH SIDES TO BOTTOM RUNNER.

7. PLUMB SIDE JAMB OF DOOR FRAME AND FASTEN JAMB STUDS TO TOP RUNNER.

Fig. 38-10 (continued)

■

GROUT

GROUT

JAMB ANCHOR

STEEL STUDS

TWO NESTED STEEL STUDS

JAMB ANCHOR

PLAN OF JAMBS

Fig. 38-11 The rough opening must be strengthened when heavy or wide doors are used. (*Courtesy of U.S. Gypsum Corporation*)

①

FLANGES MANUFACTURED WITH SLIGHT DIFFERENCES IN WIDTHS TO ENABLE "NESTING"

②

③

NESTING STUDS

Fig. 38-12 Chase wall construction details. (*Courtesy of U.S. Gypsum Corporation*)

walls should be tied to each other either with pieces of 12-inch wide gypsum board or short lengths of steel stud. If the wall studs are not opposite each other, install lengths of steel stud horizontally inside both walls. Tie together with shorter lengths of stud material spaced 24 inches on center (Fig. 38-12). Wall ties should be spaced 48 inches on center vertically.

Installing Metal Furring

Metal furring may be used on ceilings applied at right angles to joists. They may be applied vertically or horizontally to masonry walls. Space metal furring channels a maximum of 24 inches on center.

Ceiling Furring. Metal furring channels may be attached directly to structural ceiling members or suspended from them. For direct attachment, saddle tie with double-strand 18-gauge wire to each member. Leave a one inch clearance between ends of furring and walls. Metal furring channels may be spliced. Overlap the ends by at least 8 inches. Tie each end with wire (Fig. 38-13). Steel studs may be used with their open side up for furring when sup-

Fig. 38-13 Method of splicing furring channels. (*Courtesy of U.S. Gypsum Corporation*)

porting framing is widely spaced. Several methods of utilizing metal furring channels or steel studs in suspended ceiling applications are shown in Figure 38-14.

Wall Furring. Vertical application of steel furring channels is preferred. Secure the channels by staggering the fasteners from one side to the other not more than 24 inches on center (Fig. 38-15). For horizontal application on walls, attach furring channels not more than 4 inches from the floor and ceiling. Fasten in the same manner as vertical furring.

Fig. 38-14 Several methods of furring are used in suspended ceiling applications. (*Courtesy of U.S. Gypsum Corporation*)

Fig. 38-15 Furring channels are attached directly to masonry walls. (*Courtesy of U.S. Gypsum Corporation*)

Review Questions

Select the most appropriate answer.

1. Bearing partitions
   a. have a single top plate.
   b. carry no load.
   c. are constructed like exterior walls.
   d. are erected after the roof is tight.

2. The top plate of the bearing partition
   a. laps the plate of the exterior wall.
   b. is a single member.
   c. butts the top plate of the exterior wall.
   d. is applied after the ceiling joists are installed.

3. What is the rough opening height of a door opening for a 6'-8" door if the finish floor is 3/4-inch thick, 1/2-inch clearance is allowed between the door and the finish floor, and the jamb thickness is 3/4 inch?
   a. 6'-9".   c. 6'-10".
   b. 6'-9 1/2".   d. 6'-10 1/2".

4. If ceiling joists are installed in-line,
   a. their end joints lap at the bearing partition.
   b. the joint over the bearing partition must be scabbed.
   c. they are fastened to opposite sides of the rafter.
   d. blocks are placed between them at the bearing partition.

5. The ends of ceiling joists are cut to the pitch of the roof
   a. for easy application of the roof sheathing.
   b. so they will not project above the rafters.
   c. so their crowned edges will be up.
   d. after they are fastened in position.

6. Stub joists
   a. run at right angles to regular ceiling joists.
   b. are used on low pitched common roofs.
   c. are short pieces cut between regular joists.
   d. span the bearing partition and exterior wall.

7. Strapping and stripping are other names given to furring strips fastened to the underside of ceiling joists
   a. to hold the ceiling joists straight from end to end.
   b. for the attachment of the ceiling material.
   c. to make the erection of nonbearing partitions easier.
   d. to provide an air space between the joists and the ceiling.

8. The top plate of nonbearing partitions is located by
   a. measuring from the exterior walls the same distance as the soleplate.
   b. by plumbing with a transit from the soleplate.
   c. by hanging a plumb bob from the ceiling to the soleplate.
   d. any of the above methods.

9. The end stud of partitions that butt against another wall
   a. must be straightened.
   b. must be fastened securely to the wall.
   c. is left out until the wall is erected.
   d. must not be fastened in its center.

10. Gypsum board lath, commonly called rock lath, must be fastened to each framing member with at least
   a. three nails.   c. five nails.
   b. four nails.   d. six nails.

■ BUILDING FOR SUCCESS ■

Job Planning to Avoid Stress

Someday you will have the opportunity to structure projects on the job. Many construction jobs involve tasks beyond the capabilities of one person. It is irrelevant whether the tasks are of a technical or business nature. The planning process is critical in order to es-tablish some control over the outcome of the project. Undoubtedly there will be unexpected interruptions, stress, and people problems to overcome along the way. It is important that we know what we can control or change, and what we can't.

Influences on job planning in construction are many, but include weather, customers, employees, financial lenders, schedules, subcontractors, material suppliers, and families. Identifying which elements can be controlled or at least predicted will help in the planning process.

Job planning involves developing a realistic time frame and sequence of tasks that culminate in the anticipated outcome. Most of the factors listed previously will affect job planning. Some will be more controllable and predictable. Others will remain relatively unpredictable. Weather, for example, can be forecast well at times. With enough notice, schedules can be made to coincide with favorable weather conditions. At other times inclement weather can cause undue hardships during the construction process. The best advice is to pay close attention to the forecasts, and plan the work accordingly.

Customers are sometimes unpredictable. Good communication skills will play an important role in working well with customers. Understand what they want and expect. In return, be clear about what you can and cannot do in relation to their desires. It is essential not to misunderstand or to give false impressions about what is to be done.

Employees will be a key in job planning. Make sure the technical and business skills of employees can deliver the expected results. Subcontractors may be required if the builder's own workforce is small or lacks skills. Subcontractors can be quite varied in their interpretations of what is expected on a particular job. They will often be involved in a variety of projects at the same time. They must clearly understand the schedule for the entire project. They must also understand what is to be done by what dates.

Local lending agencies are closely tied with their own lenders and national influences such as the prime interest rate and the local economy. At times, it may be difficult for you or your customers to borrow the necessary money to finance the construction project.

Material suppliers may cause delays. Usually control or scheduling here is not a major long-term problem. At times backorders cause delays in construction. This may require that a subcontractor return at an inconvenient time days or weeks down the road.

Personal family situations will often affect the job planning process. Time has to be taken for builders and employees to fulfill family obligations. Sometimes family members assist with the business.

Stress in the job planning process can be decreased by understanding what can be controlled or influenced and what can't. Avoiding job-related stress can be accomplished by not worrying about uncontrollable items. Be as flexible as possible. Don't try to control every aspect of a construction project. Breakdowns and delays will happen. People sometimes will not show up for work or fail to give notice of an absence.

Some good ways to reduce the level of stress include the following:

1. Learn when to temporarily walk away from a stressful situation. This gives you time to regenerate, and address it later.
2. Seek others you trust to help solve the problem.
3. Understand what you can realistically accomplish in a day's work. Know your limits.
4. Plan either short or extended vacations. Allow time to concentrate on things you enjoy, especially your family.
5. Reduce large problems into manageable steps that move toward solutions. Complete the job one step at a time.
6. Understand and plan for the things you can personally do, then involve others with the balance of the job tasks.
7. Schedule time for a hobby.

Each person who has worked in construction could add many meaningful suggestions to this list. The important thing to remember is knowing your capabilities, what the job calls for, and how to be flexible in solving problems. Plan some form of task analysis to help reach the expected outcomes.

FOCUS QUESTIONS: For individual or group discussion.

1. What might be some good reasons to plan for a particular job in the construction process?
2. What job-related tasks can lead to stress?
3. How can a worker deal with stress on the job?
4. Why is it important to know what we can control or influence as we make job-related decisions?
5. How would you go about planning a job where people, tasks, schedules, and the weather play major roles? How would a person's flexibility play a part in the planning process?

UNIT

14

Scaffolds, Ladders, and Horses

OBJECTIVES

After completing this unit, the student will be able to:

■ name the parts of, safely erect, and dismantle single-pole and double-pole wood scaffolds.

■ erect and dismantle metal scaffolding in accordance with recommended, safe procedures.

■ build safe staging using wall and roof brackets.

■ safely set up, use, and dismantle pump jack scaffolding.

■ describe the safe use of ladders, ladder jacks, trestles, and horses.

■ build a ladder, sawhorse, and other construction aids.

CONTENTS

Scaffolding or staging are temporary working platforms. They are constructed at convenient heights above the floor or ground. They help carpenters perform their jobs quickly and safely. This unit includes information on the safe erection of various kinds of working platforms and the building of several construction aids.

CHAPTER 39

WOOD, METAL, AND PUMP JACK SCAFFOLDS

Scaffolds must be strong enough to support workers, tools, and materials. They must also provide an extra safety margin. All scaffolds must be capable of supporting at least four times the maximum intended load (Fig. 39-1).

Carpenters usually erect scaffolding. They must be familiar with the different types and methods of construction to provide a safe working platform. The type of scaffold depends on its location, the kind of work being performed, the distance above ground, and the load it is required to support.

Wood Scaffolds

Wood scaffolds are *single-pole* or *double-pole*. They are used when working on walls. The single-pole is used when it can be attached to the wall and does not interfere with the work (Fig. 39-2). The double-pole is used when the scaffolding must be kept clear of the wall for the application of materials or for other reasons (Fig. 39-3). Wood scaffolds are designated as light-, medium-, or heavy-duty, according to the loads they are required to support.

Scaffolding Terms

Poles. The vertical members of a scaffold are called *poles*. All poles should be set plumb. They should bear on a footing of sufficient size and strength to spread the load. This prevents the poles

Fig. 39-1 Scaffolds are temporary working platforms used so work can be performed easily at various heights.

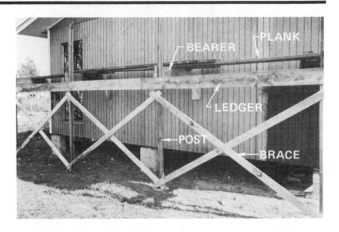

Fig. 39-2 A light-duty single-pole scaffold. Guardrails are required when the scaffolding is over 10 feet in height.

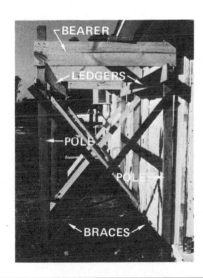

Fig. 39-3 View from the end of a double-pole wood scaffold showing cross braces from inner to outer poles.

from settling (Fig. 39-4). If wood poles need to be spliced for additional height, the ends are squared so the upper pole rests squarely on the lower pole. The joint is scabbed on at least two adjacent sides. Scabs should be at least 4 feet long. The scabs are fastened to the poles so they overlap the butted ends equally (Fig. 39-5).

Bearers. *Bearers* or *putlogs* are horizontal load-carrying members. They run from building to pole in a single-pole staging. Bearers run from pole to

Fig. 39-4 The bottom end of scaffold poles are set on footings or pads to prevent them from sinking into the ground.

pole at right angles to the wall of the building in a double-pole staging. They are set with their greatest dimension vertical. They must be long enough to project a few inches outside the staging pole.

When placed against the side of a building, bearers must be fastened to a notched *wall ledger*. At each end of the wall, bearers are fastened to the corners of the building (Fig. 39-6).

Ledgers. *Ledgers* run horizontally from pole to pole. They are parallel with the building, and support the bearers. Ledgers must be long enough to extend over two pole spaces. They must be overlapped at the pole and not spliced between them (Fig. 39-7).

Braces. *Braces* are diagonal members. They stiffen the scaffolding and prevent the poles from moving or buckling. Full diagonal face bracing is applied across the entire face of the scaffold in both directions. On medium- and heavy-duty double-pole scaffolds, the inner row of poles is braced in the same manner. Cross-bracing is also provided between the inner and outer sets of poles on all double-pole scaffolds. All braces are spliced on the poles (Fig. 39-8).

Plank. *Staging planks* rest on the bearers. They are laid with the edges close together so the platform is tight. There should be no spaces through which tools or materials can fall. All planking should be scaffold grade or its equivalent. Planking should have the ends banded with steel to prevent excessive checking.

Overlapped planks should extend at least 6 inches beyond the bearer. Where the end of planks butt each other to form a flush floor, the butt joint is

Fig. 39-5 Splicing a scaffold pole for additional height.

STEP BY STEP
P R O C E D U R E S

Fig. 39-6 Bearers must be fastened in a notch of a wall ledger for placement against the side of a building.

placed at the centerline of the pole. Each end rests on separate bearers. The planks are secured to prevent movement. End planks should not overhang the bearer by more than 12 inches (Fig. 39-9).

Guardrails. *Guardrails* are installed on all open sides and ends of scaffolds that are more than 10 feet in height. The top rail is usually of 2×4 lumber. It is fastened to the poles about 42 inches above the working platform. A middle rail of 1×6 lumber and

a toeboard with a minimum height of 4 inches are also installed (Fig. 39-10).

Size and Spacing of Scaffold Members

Carpenters must know the minimum sizes of lumber to use and the maximum spacing of the members in order to build a safe scaffold. See Figure

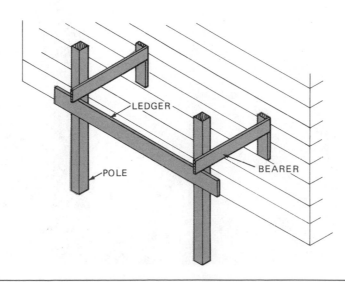

Fig. 39-7 Ledgers run horizontally from pole to pole and support the bearers.

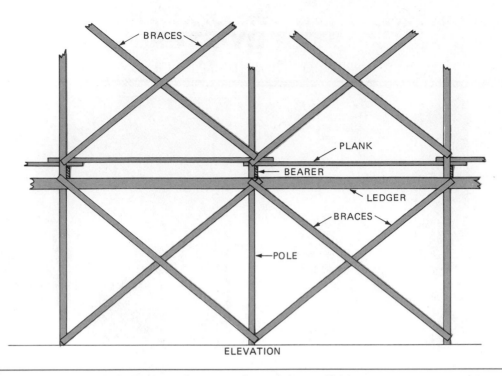

ELEVATION

Fig. 39-8 Diagonal bracing is applied across the entire face of the scaffold.

Fig. 39-9 Recommended placement of scaffold plank.

Fig. 39-10 Recommended guardrail specifications.

WOOD SCAFFOLDING				
MINIMUM SIZES AND MAXIMUM SPACING OF MEMBERS				
	Light Duty		Medium Duty	Heavy Duty
SCAFFOLD HEIGHT	**20 Ft**	**60 Ft**	**60 Ft**	**60 Ft**
Poles	2"x4"	4"x4"	4"x4"	4"x6"
Pole Spacing	10'-0"	10'-0"	8'-0"	6'-0"
Scaffold Width	5'-0"	5'-0"	5'-0"	5'-0"
Bearers to 3'-0"	2"x4" or 1"x8"	2"x4" or 1"x8"		
Bearers to 5'-0"	2"x6"	2"x6"	2"x10"	2"x10"
Ledgers	1"x4"	1 1/4"x9"	2"x10"	2"x10"
Plank	2"x10"	2"x10"	2"x10"	2"x10"
Braces	1"x4"	1"x4"	1"x4"	2"x4"
Vertical Distance Between Stages	7'-0"	9'-0"	7'-0"	6'-6"

Fig. 39-11 Minimum sizes and maximum spacing of wood scaffold members.

39-11 for size and spacing recommendations of scaffold members.

Erecting a Light-Duty Single-Pole Scaffold

Determine the height of the first stage. Mark it on both ends of the building. Snap a chalk line between the marks. Fasten a 1×8 bearer on each end. Make sure it projects out from the side wall the desired distance, and that its top edge is on the chalk line.

The maximum distance from wall to pole for this size bearer is 3'-0". Allow enough extra length for fastening to the wall and pole and to extend beyond the pole a few inches. Use at least five duplex nails through the bearer into the wall.

Fasten intermediate bearers to the side wall with their top edges to the chalk line. Use at least five duplex nails of sufficient length into each wall ledger. Space the bearers no more than 10 feet apart (Fig. 39-12).

Measure and mark each bearer the same distance from the wall for the location of the poles. Lean the necessary number of poles against the wall alongside each bearer. Fasten the end bearer to the

STEP BY STEP

P R O C E D U R E S

1. SNAP CHALK LINE

BEARERS

2. INSTALL BEARERS NOT MORE THAN 10 FT. APART

5. SIGHT THE TOP EDGE OF EACH BEARER FOR LEVEL BEFORE FASTENING TO POLE

6. PLUMB EACH POLE AS LEDGER IS FASTENED

SCAFFOLD POLES

LEDGER

3. MEASURE OUT FROM THE WALL THE SAME DISTANCE ON EACH BEARER FOR POLE LOCATION

4. INSTALL ADEQUATE SILLS

9. INSTALL GUARD RAIL

8. LAY SCAFFOLD PLANK

7. INSTALL CROSS BRACES

Fig. 39-12 Steps in the construction of a light-duty, single-pole wood scaffold.

pole at the mark. Make sure the pole is plumb and that the bearer is level. Again, use at least five duplex nails. Sighting each member by eye is usually sufficient. Sight the edge of each member with plumb and level lines of a building or other object in the distance. Place a footing under the pole if there is danger of the pole sinking into the ground. Fasten the rest of the bearers to the poles in a similar manner. Sight each bearer by eye with the first one so that the top edges are in line. Sight each pole with the first one so that all poles line up.

Install ledgers under the bearers by fastening them to the outside of each pole. Use at least two duplex nails into each pole. Remember that ledgers should not be spliced between poles.

Install 1×4 diagonal cross-braces across the face of the staging. Nail one end of the brace to the pole just above the bearer. Fasten the other end near the bottom of the post. Crisscross the braces between each set of poles. Use two 8d duplex nails in each end of the brace. Drive one nail where the braces cross to tie them together. If the end of the nail protrudes, bend it over.

Lay two staging planks side by side on the bearers in every other section. Let the plank overhang the bearer by the recommended distance. Usually two planks are sufficient for a light-duty scaffold. Plank the remaining sections by placing the plank on those previously laid. Make sure these planks also overhang the bearer by at least 6 inches. The planks should be of equal length so that the ends are flush with each other (Fig. 39-13). Unequal lengths are dangerous. The worker may not expect two drops in level in such a short distance.

In the center of the span fasten a short piece of

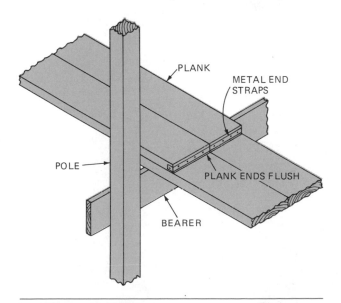

Fig. 39-13 The end of overlapped plank should be flush with each other.

Fig. 39-14 Strengthen the scaffold planks with a short piece of 1×6 across their centers.

1×6 lumber across the planks. This ties them together and strengthens them. Use five 8d common nails driven home into each plank (Fig. 39-14). Install guardrails after the planks are laid. Upper stages of a single-pole scaffold are built in the same manner, except that the poles are already erected.

Always raise scaffolds to the desired height. Do not use ladders, boxes, or other makeshift devices to increase their height. If scaffold poles need to be spliced, do it in the recommended manner.

Erecting a Double-Pole Scaffold

Double-pole scaffolds should be set as close to the wall of a building as possible without hampering the work being done. Their construction is similar to that used for single-pole scaffolds. However, the inner end of the bearers is fastened to an inner set of poles.

If the ground is not level, snap a line across the wall of the building at the height of the first stage. Stand the first inner pole upright against the wall. Mark it at the same height as the chalk line. Fasten a bearer to it so its top edge is on the mark. Stand the outside post up. Fasten the bearer to it so the top edge of the bearer is level. Temporary braces are required to hold the posts upright. At least two workers are needed. Fasten diagonal cross-braces between the bearer and the bottom of the posts.

Continue making sets of inner and outer posts until the desired number is built. Tie the sets together with ledgers on the inside and outside faces. Cross-brace the outside face on light-duty staging. Cross-brace both the outside and inside faces on medium- and heavy-duty staging.

Install the planking and guardrails in the same manner as for single-pole staging (Fig. 39-15). Upper stages are built in the same manner as the first.

STEP BY STEP
P R O C E D U R E S

1. SNAP CHALK LINE ACROSS BUILDING

2. STAND INNER POLES IN POSITION AND MARK FLUSH WITH CHALK LINE. FASTEN BEARERS TO POLES

3. SIGHT BEARERS FOR LEVEL AND FASTEN OUTER POLES EQUAL DISTANCES FROM INNER POLES

BRACE EACH SECTION TEMPORARILY

4. PROVIDE ADEQUATE FOUNDATION FOR SCAFFOLD

5. INSTALL LEDGERS UNDER BEARERS ON INSIDE AND OUTSIDE POLES

6. INSTALL CROSS BRACES BETWEEN INNER AND OUTER POLES

8. LAY SCAFFOLD PLANK

9. INSTALL GUARD RAIL

7. INSTALL CROSS BRACES ON FACE OF SCAFFOLD

Fig. 39-15 Steps in the erection of a light-duty, double-pole wood scaffold on sloping land.

The height can be determined by measuring up from the first stage.

If the height or length of a double-pole scaffold exceeds 25 feet, it must be secured to the building at intervals no greater than 25 feet vertically and horizontally. Ties may be made of 1×4 lumber or reinforcing rod wire.

Metal Scaffolding

Metal scaffolding consists of manufactured tubular welded *end frames* with accessories, such as *folding braces, adjustable screw legs, base plates,* and *guardrail hardware.* They are erected in sections. Each section consists of two end frames and two folding braces (Fig. 39-16). The braces are inserted into pins at the top and bottom of the end frame. They are then locked into position. These hold the end frames securely so the erected scaffold is always plumb, square, and rigid. Braces should not be forced to fit. Level the scaffold until the braces fit easily.

The scaffolding is extended lengthwise by adding braces and end frames until the desired length is reached. Adjusting screws fit in the bottom of the legs of the end frames to level the staging on sloping ground. In addition, scaffolding components of different heights may be used in combination with adjusting screws to level the scaffold. Base plates and adequate sills for scaffold posts must be provided to keep the scaffold from settling.

The height of the scaffold is increased simply by stacking end frames on top of each other. The bottoms of the legs of one end frame slide into the tops of the legs of another. Plumb and level the scaffold as erection progresses. To prevent movement, the scaffold is secured to the structure at intervals not to exceed 26 feet vertically and 30 feet horizontally (Fig. 39-17).

Fig. 39-16 Erecting metal wall scaffolding.

Fig. 39-17 Scaffolding should be securely anchored to walls at least every 26 feet of height and every 30 feet of length.

Scaffold planks are installed on the crossmembers of the end frames in a manner similar to wood scaffolds. Guardrails must be installed on metal staging that is more than 10 feet above the ground.

Mobile Scaffolds

Freestanding mobile scaffold towers may be constructed using metal scaffolding. Simply insert casters in the bottom of the legs. Casters must support four times the maximum intended load. They must also be provided with a positive locking device to hold the scaffold in position.

The height of mobile towers must not exceed four times the smaller base dimension (Fig. 39-18). Platforms must be tightly planked to the full width of the scaffold except for necessary entrance openings.

Move mobile scaffolds only on level floors free of obstructions and openings. To avoid the possibility of tipping the mobile scaffold over, apply the force necessary to move as close to the base as possible.

Pump Jack Scaffolds

Pump jack scaffolds consist of 4×4 poles, a pump jack mechanism, and metal braces for each pole (Fig. 39-19). The braces are attached to the pole at

Fig. 39-18 The height of a mobile scaffold must not exceed four times its smaller base dimension.

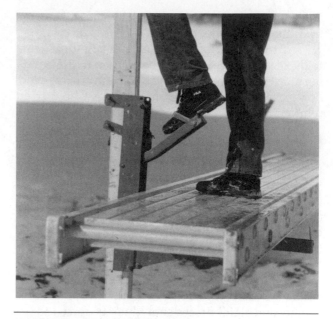

Fig. 39-19 A pump jack scaffold.

intervals and near the top. The arms of the bracket extend from both sides of the pole at 45 degree angles. The arms are attached to the sidewall or roof to hold the pole steady.

The scaffold is raised by pressing on the foot pedal of the pump jack. The mechanism has brackets on which to place the scaffold plank. Other brackets hold a guardrail or platform. Reversing a lever allows the staging to be pumped downward.

Pump jack scaffolds are used widely for wood shingling of sidewalls, where staging must be kept away from the walls, and when a steady working

height is desired. However, pump jack scaffolds have their limitations. They should not be used when the working load exceeds 500 pounds. No more than two persons are permitted at one time between any two supports. Poles must not exceed 30 feet in height. Braces must be installed at a maximum vertical spacing of not more than 10 feet.

In order to pump the scaffold past a brace location, temporary braces are used. The temporary bracing is installed about 4 feet above the original bracing. Once the scaffold is past the location of the original brace, it can be installed. The temporary brace is then removed.

Usually pump jack poles are constructed of two 2×4s spiked together. They must be fastened with the joint parallel to the scaffold plank bracket with 10d common nails. The nails should be no less than 12 inches apart, staggered uniformly from opposite outside edges.

CHAPTER 40 BRACKETS, HORSES, AND LADDERS

Wall Brackets

Scaffold planks may be supported by folding metal *wall brackets*. They are usually made of angle iron. When opened, they form a triangular shape (Fig. 40-1). A bolt attached to the top member passes through a hole in the wall sheathing. A short

block of 2×4 lumber with a hole bored for the bolt is placed against at least two studs on the inside. The bracket is then tightened with a nut against the block (Fig. 40-2). Accessories may be attached to the brackets for installing guardrails for bracket scaffolding over 10 feet in height.

Fig. 40-1 Wall brackets are used frequently on wood frame residential construction. A worker is reinforcing the wall to prevent the leg of the bracket from puncturing the sheathing.

Fig. 40-2 Wall brackets are tightened against a 2×4 that crosses the inside surface of at least two studs.

Installing Wall Brackets

Snap a chalk line across the wall at the desired height of the staging. Bore 5/8-inch holes through the wall sheathing along the chalk line and in stud spaces not more than 8 feet apart. Insert the bolt of one bracket in each of the holes. On the inside, insert a prebored 2×4 over the bolt. The 2×4 must cross over at least two studs. Tighten the nut on the bolt. Secure the handle with a nail driven alongside it.

If the outside wall sheathing is thin, insert a short piece of at least 1×6 stock between the bottom of the bracket and the sheathing. Cross over at least two studs. This prevents the bottom of the bracket from puncturing the sheathing when weight is applied.

Lay plank on the brackets in the same manner as for other scaffolds. Install guardrails if the staging is over 10 feet in height.

Roofing Brackets

Roofing brackets are used when the pitch of the roof is too steep for carpenters to work on without slip-

ping (Fig. 40-3). Usually any roof with more than 5-inch vertical rise per horizontal foot of run requires roof brackets. Roofing brackets are made of wood or metal. They are adjustable for roofs of different pitches.

A metal plate at the top of the bracket has three slots in which to drive nails to fasten the bracket to the roof. The bottom of the slot is round and large enough to slip over the nailhead. This enables removal of the bracket from the fasteners without pulling the nails. The bracket is simply tapped upward from the bottom, and then lifted over the nailheads. The nails that remain are then driven home.

Applying Roof Brackets

Roof brackets are usually used when the roof is being shingled. On steep-pitched roofs, brackets are necessary when shingling. They keep the worker from slipping, and also hold the roofing materials. Apply roof brackets in rows. Space them out so that they can be reached without climbing off the roof bracket staging below.

On asphalt-shingled roofs, place the brackets at about 8-foot horizontal intervals. The top end of the bracket should be just below the next course of shingles. Nail the bracket over a joint or cutout in the tab of the shingle course below. No joint or cutout in the course above should fall in line with the nails holding the bracket. Otherwise, the roof will leak. Use three 8d common nails driven home. Try to get at least one nail into a rafter.

Open the brackets so the top member is approximately level or slightly leaning toward the roof. Place staging plank on the top of the brackets. Overlap them as in wall scaffolds. Keep the inner edges against the roof for greater support. A toe-board made of 1×6 or 1×8 lumber is usually placed flat on the roof with its bottom edge on top of the

Fig. 40-3 Roof brackets are used when the roof pitch is too steep for carpenters to work on without slipping.

Fig. 40-4 The placement of a toeboard and plank used on roof brackets.

brackets. This protects the new roofing from the workers' toes when the roofing has progressed that far (Fig. 40-4).

Trestles

Trestles are used when a low scaffold on a level surface is desired. A trestle is a low working platform supported by a bearer with spreading legs at each end. This type of scaffold is used frequently in the building interior by lathers, plasterers, and drywall applicators working on ceilings.

Trestle Jacks

Manufactured metal trestle supports, called *trestle jacks*, are available for frequent and prolonged use. Trestle jacks are adjustable in height at about 3-inch intervals. They are clamped to a 2-inch wood support on which the scaffold planks are placed. The

Fig. 40-5 Adjustable metal trestle jacks are manufactured for prolonged and repetitive use.

size of the support depends on how far apart the trestle jacks are placed. Metal braces hold the trestle rigid (Fig. 40-5).

Sawhorses are used for trestle supports when occasional use is required and their height meets the requirements of the job (Fig. 40-6). For light-duty work, horses and trestle jacks should not be spaced more than 8 feet apart. Do not use horses that have become weak or defective. The horizontal bearing member of horses used for trestle support should be at least a 3×4. The horse legs should be made of nominal 1 1/4×5 stock.

If a horse scaffold is arranged in tiers, no more than two tiers should be used. The legs of the

Fig. 40-6 Sawhorses are sometimes used for supports for a trestle scaffold.

horses in the upper tier should be nailed to the planks. Each tier should be cross-braced.

Ladders

Carpenters must often use *ladders* to work from or to reach working platforms above the ground. Most commonly used ladders are the *stepladder* and the *extension ladder*. They are usually made of wood, aluminum, or fiberglass. Make sure all ladders are in good condition before using them.

Extension Ladders

To raise an extension ladder, place its feet against a solid object. Pick up the other end. Walk forward under the ladder, pushing upward on each rung until the ladder is upright (Fig. 40-7). With the ladder vertical and close to a wall, extend the ladder by pulling on the rope with one hand while holding the ladder upright with the other. Raise the ladder to the desired height. Make sure the spring-loaded hooks are over the rungs on both sides.

Lean the top of the ladder against the wall. Move the base out until the distance from the wall is about 1/4 the vertical height. This will give the proper angle to the ladder. The proper angle for climbing the ladder can also be determined, as shown in Figure 40-8.

Fig. 40-7 Raising an extension ladder.

If the ladder is used to reach a roof or working platform, it must extend above the top support by at least 3 feet.

HELPFUL HINTS

A method of setting a ladder at a proper angle for climbing.

LADDER

GRIP RUNG OF LADDER WITH ARM OUT STRAIGHT

STAND WITH BACK STRAIGHT

PROPER ANGLE FOR CLIMBING

FOOT CENTERED ON LADDER

Fig. 40-8

When the ladder is in position, shim one leg, if necessary, to prevent wobbling. Face the ladder when climbing. Grasp the side rails or rungs with both hands (Fig. 40-9).

Stepladders

When using a stepladder, open the legs fully so the brackets are straight and locked. Make sure the ladder does not wobble. If necessary, place a shim under the leg to steady the ladder. Never work above the second step from the top. Do not use the ledge in back of the ladder as a step. The ledge is used to hold tools and materials only. Move the ladder as necessary to avoid overreaching. Make sure all materials and tools are removed from the ladder before moving it.

Ladder Jacks

Ladder jacks are metal brackets installed on ladders to hold scaffold plank. At least two ladders and two jacks are necessary for a section. Ladders should be heavy-duty, free from defects, and placed no more

than 8 feet apart. They should have devices to keep them from slipping.

The ladder jack should bear on the side rails in addition to the ladder rungs. If bearing on the rungs only, the bearing area should be at least 10 inches on each rung. No more than two persons should occupy any 8 feet of ladder jack scaffold at any one time. The platform width must not be less than 18 inches. Planks must overlap the bearing surface by at least 10 inches (Fig. 40-10).

Construction Aids

Sawhorses, work stools, ladders, and other construction aids are sometimes custom-built by the carpenter on the job or in the shop.

Sawhorses

Sawhorses are used on practically every construction job. They support material that is being laid out or cut to size. Unless they are being used as supports for a trestle scaffold, sawhorses are usually made with a 2×4 or 2×6 top, 1×6 legs, and 3/8 or 1/2-inch plywood leg braces.

Sawhorses are constructed in a number of ways according to the preference of the individual. However, they should be of sufficient width, a comfortable working height, and light enough to be moved easily from one place to another. A typical sawhorse is 36 inches wide with 24 inch legs (Fig. 40-11). A tall person may wish to make the leg 26 inches long.

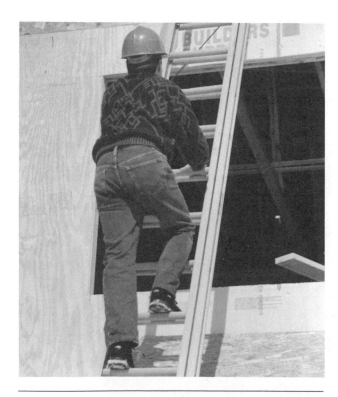

Fig. 40-9 Face the ladder when climbing. Hold on with both hands.

Fig. 40-10 Ladder jacks are used to support scaffold plank for short-term, light repair work.

STEP BY STEP
P R O C E D U R E S

2"x6" TOP

TOP VIEW

36"

1"x6" LEGS

SIDE VIEW

24" OR AS DESIRED

10"

3/8" PLYWOOD BRACE

SIDE VIEW

SAWHORSE - MULTI-VIEW

1. CUT 2" × 6" SAWHORSE TOP 36" LONG AND BEVEL BOTH EDGES OF EACH END AS SHOWN. IF DESIRED, BEVEL MAY BE RIPPED ON BOTH EDGES FOR ENTIRE LENGTH OF TOP.

4. HOLD PLYWOOD BRACE AS SHOWN AND MARK ITS LENGTH AT THE TOP EDGE.

FROM EACH MARK, LAY OUT SAME ANGLE AS TOP AND LEGS. CUT, MAKE DUPLICATE, AND FASTEN ONE ON EACH END OF HORSE FLUSH WITH OUTSIDE FACE OF LEGS.

6"

3/8"

DETAIL OF END

3/8" PLY.

END VIEW

3. FASTEN ALL FOUR LEGS TO SAWHORSE TOP

PERSPECTIVE

2. CUT FOUR LEGS TO 24" LENGTH OR AS DESIRED, WITH BEVEL ON EACH END AT SAME ANGLE AS TOP

TOP

LEG

Fig. 40-11 Construction of a typical sawhorse.

Work Stools

Work stools or *crickets* are used, especially by finish carpenters. They support work for cutting. They serve as step stools to reach objects high on the wall and ceiling. A tray is usually installed near the bottom to hold a few tools needed for repetitive work. A slot is cut in the top to carry the stool (Fig. 40-12).

STEP BY STEP
P R O C E D U R E S

CUT ALL PARTS TO OVERALL SIZE

					CUTTING LIST
AMT.	PART	T"	W"	L"	MATERIAL
1	A	3/4	14	24	PLYWOOD
2	B	3/4	13	13 1/4	PLYWOOD
1	C	3/4	11 1/2	21 1/2	PLYWOOD
2	D	3/4	3 1/2	23	PINE
2	E	3/4	1 1/2	23	PINE

PLAN

MAY VARY ACCORDING TO HEIGHT OF WORKER

ELEVATION

SECTION

② MACHINE TWO LEGS AS SHOWN BELOW.

③ ASSEMBLE LEGS, RAILS AND SHELF.

TOOL TRAY

④ BORE 1" HOLES AT EACH END OF HAND SLOT. RIP BETWEEN HOLES AND ROUND OVER EDGES.

FASTEN TOP TO FRAME OF STOOL.

RIP

1" HOLES

Fig. 40-12 Plan and instructions for building a work stool.

Job-Made Ladders

At times it is necessary to build a ladder on the job. These are usually short, straight ladders no more than 24 feet in length. The side rails are made of clear, straight-grained 2×4 stock spaced 15 to 20 inches apart. *Cleats* or *rungs* are cut from 1×4 stock and inset into the edges of the side rails not more than 3/8 inch. Filler blocks are sometimes used on the rails between the cleats. Cleats are uniformly spaced at 12 inches top to top (Fig. 40-13).

Scaffold Safety

The safety of those working at a height depends partly on properly constructed scaffolds. Those who have the responsibility of constructing scaffolds must be thoroughly familiar with the sizes, spacing, and fastening of scaffold members and other scaffold construction techniques.

STEP BY STEP
P R O C E D U R E S

1. TEMPORARILY FASTEN RAILS OF LADDER SIDE BY SIDE AND LAYOUT DADOES 12" O.C. AS SHOWN BELOW.

2. CUT DADOES. IF AVAILABLE, USE A DADO HEAD INSTALLED ON A RADIAL ARM SAW, AND MAKE CUTS WITH THE RAILS STILL FASTENED TOGETHER. CUT ONLY 3/8" DEEP. ANY DEEPER WILL WEAKEN THE RAIL.

3. CUT RUNGS 20" LONG. SEPARATE RAILS AND FASTEN RUNGS IN DADOES. KEEP ENDS OF RUNGS FLUSH WITH OUTSIDE FACE OF RAILS.

CUT DADO 3/8" DEEP

WIDTH OF RUNG

RAIL OF LADDER

20" WIDE

Fig. 40-13 Constructing a job-built ladder.

Review Questions

Select the most appropriate answer.

1. The vertical members of a scaffold are called
   a. columns.
   b. piers.
   c. poles.
   d. uprights.

2. Bearers support
   a. ledgers.
   b. plank.
   c. rails.
   d. braces.

3. Scaffold planks should be at least
   a. 2×6.
   b. 2×8.
   c. 2×10.
   d. 2×12.

4. Overlapped scaffold planks should extend beyond the bearer at least
   a. 3 inches.
   b. 6 inches.
   c. 8 inches.
   d. 12 inches.

5. End planks should not overhang the bearer by more than
   a. 6 inches.
   b. 8 inches.
   c. 10 inches.
   d. 12 inches.

6. On a light-duty wood scaffold, bearers should not be spaced horizontally more than
   a. 8 feet.
   b. 10 feet.
   c. 12 feet.
   d. 14 feet.

7. The minimum size of cross braces in a wood scaffold is
   a. 1×2.
   b. 1×3.
   c. 1×4.
   d. 2×4.

8. A double pole wood scaffold must be secured to the building if its length or height exceeds
   a. 20 feet.
   b. 25 feet.
   c. 30 feet.
   d. 35 feet.

9. The height of a mobile scaffold must not exceed the minimum base dimension by
   a. three times.
   b. four times.
   c. five times.
   d. six times.

10. Guardrails must be installed on all scaffolds more than
   a. 10 feet in height.
   b. 16 feet in height.
   c. 20 feet in height.
   d. 24 feet in height.

■ BUILDING FOR SUCCESS ■

The Construction Team Networking Process

The concept of teamwork has always been at the heart of the construction process. A network implies the use of interconnected or interrelated groups to accomplish tasks. A team network will have common goals. These shared goals become the desired outcomes that formulate and direct the team's activities. Successful teamwork outcomes are reflected in a project that is completed on or ahead of schedule, within budget, and with no or minimal worker injuries. Satisfaction for the client and all interested parties will also be achieved with the proper use of a team-centered structure.

Effective teamwork has many characteristics, but some of the most important are coordination, cooperation, scheduling, problem solving, and quality assurance. If these are implemented successfully, the goals are most generally met.

Coordination is the process of harmonizing the various people and jobs on a construction project. This action is directed through a team facilitator who might be an architect, general contractor, or field superintendent. A good coordinator is someone who has a clear picture of the entire project. He or she must understand the skilled trades involved, the building materials / processes, and job scheduling.

Cooperation is necessary for a team to accomplish its goals. Good interaction and communication will insure cooperation. People find it very difficult to cooperate without trust. Flexibility is a part of cooperation, and must also be possessed by team members. This helps insure that job tasks are completed, even by alternate routes.

Good scheduling is vital if deadlines are to be met. The project coordinator has the responsibility of implementing the best delivery times for materials and the optimal time to move people on and off the work site. Planning these sequential moves will demand experience and flexibility.

Problem solving becomes the responsibility of everyone involved in the construction process. Even the best planning by experienced people will be plagued at times by the unexpected. Typical examples of the unexpected are inclement weather, delivery delays, accidents, and disagreements. As problems arise, teamwork will be implemented to solve them. Each person is called upon to share his or her expertise and ideas in finding solutions. A problem solving process should be agreed upon by all parties in advance. Usually the general contractor has implemented a procedure based on experience.

Quality assurance is a desired outcome with the construction team. Ongoing monitoring is required by everyone to maintain product quality. The client will become more aware of job and material quality as the construction progresses. At each level of construction, the skilled workers, technicians, and laborers must be committed to achieving the highest possible standards. The key to a successful team network concept is a well-prepared plan. The need for such a plan must be integrated into everyone's thinking from the beginning. People who do not buy into the team network process will cause breakdowns. These will ultimately cost the construction process time and money. Losses of this type are first felt in the profit margin. The worst possible consequence could be the lack of coordination and scheduling that might cause personal injuries to workers.

In order to assure the welfare of the construction workers and the general public, the team network has to be accepted by everyone at the job site. Efficiency will be an outcome when full acceptance of teamwork is gained. Negative factors such as hostilities, down time, and accidents will decrease when the team network concept is accepted and implemented by everyone involved in the project. Many times the common denominator is the profit margin. All builders and construction workers will understand that teamwork will assure greater chances for a profit to be maintained on each project.

FOCUS QUESTIONS: For individual or group discussion.

1. The team network process in construction has its base established in working together to reach goals. What might be some good reasons to consider communicating or networking well with others in the construction industry?

2. What do you feel are necessary components in a network system approach to construction? What understandings have to be present for this system to be functional?

3. Can other components become part of a network system when considering all of the varied types of trades represented in the world of construction? What are they?

4. By concentrating on any one of the team network components, how can you defend its need?

UNIT

15

Roof Framing

OBJECTIVES

After completing this unit, the student will be able to:

- describe several roof types and define roof framing terms.

- describe the members of gable, gambrel, hip, intersecting, and shed roofs.

- lay out a common rafter and erect gable, gambrel, and shed roofs.

- lay out and install gable studs.

- lay out the members of and frame hip and equal-pitch intersecting roofs.

- erect a trussed roof.

- apply roof sheathing.

CONTENTS

The ability to lay out rafters and frame all types of roofs is an indication of an experienced carpenter. On most jobs, the boss lays out the different rafters, while workers make duplicates of them. Those persons aspiring to supervisory positions in the construction field must know how to frame various kinds of roofs.

CHAPTER 41 ROOF TYPES AND TERMS

Knowledge of diverse roof types and the terms used in their construction are essential to framing roofs.

Roof Types

Several roof styles are in common use. These roofs are described in the following material and are illustrated in Figure 41-1.

- The most common roof style is the *gable roof.* Two sloping roof surfaces meet at the top. They form triangular shapes at each end of the building called *gables.*

- The *shed roof* slopes in one direction. It is sometimes referred to as a *lean-to.* It is commonly used on additions to existing structures. It is also used extensively on contemporary homes.

- The *hip roof* slopes upward from all walls of the building to the top. This style is used when the same overhang is desired all around the building. The hip roof diminishes the height of a building. It also eliminates maintenance of gable ends.

- An *intersecting roof* is required on buildings that have wings. Where two roofs intersect, valleys are formed. This requires several types of rafters.

- The *gambrel roof* is a variation of the gable roof. It has two slopes on each side instead of one. The lower slope is much steeper than the upper slope. It is framed somewhat like two separate gable roofs.

- The *mansard roof* is a variation of the hip roof. It has two slopes on each of the four sides. It is framed somewhat like two separate hip roofs.

- The *butterfly roof* is an inverted gable roof. It is used on many modern homes. It resembles two shed roofs with their low ends placed against each other.

- Other roof styles are a combination of the styles just mentioned. The shape of the roof can be one of the most distinctive features of a building.

Roof Framing Terms

It is important for the carpenter who wants to become proficient in roof framing to be familiar with

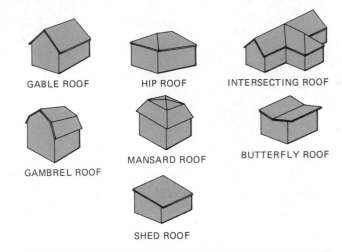

GABLE ROOF HIP ROOF INTERSECTING ROOF

GAMBREL ROOF MANSARD ROOF BUTTERFLY ROOF

SHED ROOF

Fig. 41-1 Several roof styles are used for residential buildings.

roof framing terms. The terms are defined in the following material and illustrated in Figure 41-2.

- The *span* of a roof is the horizontal distance covered by the roof. This is usually the width of the building measured from the outer faces of the frame.

- A *rafter* is one of the sloping members of a roof frame. It supports the roof covering.

- The *total run* of a rafter is the horizontal distance over which the rafter rises. This is 1/2 the span, except in roofs of unequal pitch.

- The *unit of run* is one foot for all rafters that run at right angles to the plate. The unit of run for hip and valley rafters that run at a 45 degree angle to the plate line is 17 inches. Rafters that run at odd angles to the plate line have various units of run.

- The *ridge* is the uppermost horizontal line of the roof.

- The *total rise* is the total vertical distance that the roof rises.

- The *line length* of a rafter is the *hypotenuse* (longest side) of a right triangle formed with its base as the total run and its other leg as the total rise. The line length gives no consideration to the thickness or width of the framing stock.

- The *pitch* is the amount of slope of a roof. This is usually expressed as a fraction. The pitch is

Fig. 41-2 Roof framing terms.

found by dividing the total rise by the span. For example, if the span of the building is 32 feet and total rise is 8 feet, then 8 divided by 32 equals 1/4. The roof is said to be 1/4 pitch.

■ The slope of a roof is also expressed as the *rise per foot*. This means the number of inches it rises vertically for every foot it runs horizontally. In drawings, the slope or *cut* of a roof is indicated by a symbol. This symbol shows the rise in inches per foot of run (4 and 12, 6 and 12, or 8 and 12, for example) as shown in Figure 22-8.

■ A *plumb cut* is any cut on the rafter that is vertical when the rafter is in position. Always mark along the tongue of the framing square when laying out plumb cuts on rafters.

■ A *level cut* is any cut on the rafter that is horizontal when the rafter is in position. Always mark along the blade of the framing square when laying out level cuts on rafters (Fig. 41-3).

Fig. 41-3 Plumb cuts are laid out on the tongue of the square. Level cuts are laid out on the blade.

Calculating Rough Rafter Lengths

It is necessary to determine the length of lumber to order for use as rafters. The rough length of a rafter from plate to ridge can be calculated using the line length method. Let the blade of the square represent the total run. Let the tongue of the square represent the total rise. Use a scale of one inch to the foot. Measure the distance from blade to tongue from a given total rise and total run to find the length of the rafter (Fig. 41-4). Add extra if the rafter overhangs the wall of the building.

Fig. 41-4 Scaling the rough length of a common rafter.

EXAMPLE Assume the total rise of a rafter is 9 feet. Assume the total run is 12 feet. Locate 9 inches on the tongue of the square and 12 inches on the blade. Measure the distance between. This will be 15 inches. Using a scale of one inch equals one foot, the rough rafter length from plate to ridge is 15 feet. Standard stock length of 16 feet would have to be used for the rafter with no overhang.

CHAPTER 42 GABLE AND GAMBREL ROOFS

The *gable roof* is the style most commonly used. A *gambrel roof* is like a gable roof with two different slopes.

Framing the Gable Roof

An equal-pitched gable roof is a roof that has an equal slope on both sides intersecting the ridge in the center of the span (Fig. 42-1). This roof is the simplest to frame. Only one type of rafter, the *common rafter*, needs to be laid out.

Gable roofs may have different slopes on each side, such as the *saltbox* style roof (Fig. 42-2). The principles for framing both equal or unequal-pitched gable roofs are the same.

Laying Out a Common Rafter

The *common rafter* extends at right angles from the plate upward to the ridge. It is called a common rafter because it is common to all types of roofs. It is used as a basis for laying out other kinds of rafters.

The *ridge,* although not absolutely necessary, simplifies the erection of the roof. It provides a means of tying the rafters together before the roof is sheathed. Erecting a roof frame is called *raising* the roof.

Cuts on the Common Rafter

The common rafter requires layouts for making several cuts. The cut at the top is called the *plumb cut* or *ridge cut.* It fits against the ridge board. The cut at the plate, called the *bird's mouth* or *seat cut,* consists of a plumb line and a level line layout. It fits against the top and outside edge of the wall plate. The distance between the ridge cut and the seat cut, called the *line length* or *theoretical length,* must be determined. At the bottom end of the rafter, a *tail cut* is made on the rafter *tail* or *overhang* that extends beyond the building (Fig. 42-3).

Common Rafter Pattern

Lay a piece of rafter stock across two sawhorses. Sight the stock along the edge for straightness. Select the straightest piece possible, because it will be used as a pattern to mark the required number of rafters for the roof. If there is any slight crook in the edge of the pattern, the carpenter stands on the side

RIDGE BOARD

COLLAR TIE

COMMON RAFTER

Fig. 42-1 The gable roof is framed with common rafters, ridgeboard, collar ties, and gable studs.

GABLE STUD

Fig. 42-2 The ridge of the saltbox roof is off-center.

RIDGE

PLUMB CUT AT RIDGE

COMMON RAFTER

SEAT CUT

TOP PLATE OF WALL

LEVEL CUT OF SEAT

PLUMB CUT OF SEAT

TAIL CUT

Fig. 42-3 The cuts of the common rafter.

that has the crowned edge. This will be the top edge of the rafter when it is installed.

Laying Out the Ridge Cut

Place the square down on the side of the stock at its left end. Hold the tongue of the square with the left hand and the blade with the right hand with the heel of the square pointing away. Move the square until the outside edge of the tongue and the near edge of the stock line up with the specified rise in inches. Make sure the blade of the square and the edge of the stock line up with 12 inches. Mark along the outside edge of the tongue for the plumb cut at the ridge (Fig. 42-4).

EXAMPLE Assume the cut of the roof is 6 and 12. Hold the square so the 6-inch mark on the tongue and the 12-inch mark on the blade line up with the top edge of the rafter stock. Mark along the outside edge of the tongue of the square.

Finding the Length of a Common Rafter

The distance between the plumb cut at the ridge and the seat cut can be determined by the *step-off method* or by using the *rafter tables.*

The Step-Off Method. The step-off method is based on the unit of run (12 inches for the common rafter). The rafter stock is stepped-off for each unit of run until the desired number of units or parts of units are stepped-off.

First, lay out the ridge cut. Mark where the blade intersects with the top edge of the rafter. Hold the square at the same angle. Shift it until the tongue lines up with this mark. Move the square in

Fig. 42-4 Laying out the plumb cut of the common rafter at the ridge.

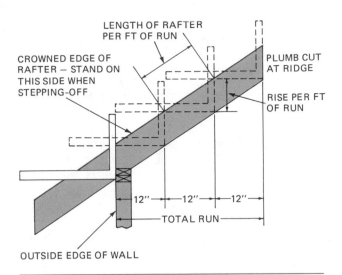

Fig. 42-5 Stepping-off the common rafter.

a similar manner until the total run of the rafter is laid out (Fig. 42-5). Mark a plumb line along the tongue of the square at the last step. This line will be parallel to the ridge cut.

EXAMPLE If the rafter has a total run of 16 feet, the square is moved 16 times.

Stepping-Off a Fractional Part of a Unit of Run. First, step-off for the total whole units of run. Move the square as if to step-off one more time. Instead of marking at the blade edge of the square, measure the fractional part of the unit of run along the blade. Mark it on the rafter. Holding the square in the same position, move it so the outside of the tongue lines up with the mark. Lay out a plumb line along the tongue of the square (Fig. 42-6).

EXAMPLE If the rafter has a total run of 16 feet, 8 inches, step off 16 times. Hold the square for the 17th step. Mark along the blade at the 8-inch mark. With the square held at the same angle, move it to the mark. Lay out the plumb cut of the bird's mouth by marking along the tongue.

Fig. 42-6 Laying out a fractional part of the total run of a common rafter.

Layout Aids for the Framing Square

Several aids may be used to keep the square at the same angle for each step during the step-off process.

Framing Square Gauges. *Framing square* gauges may be attached to the square. They are used as stops against the top edge of the rafter. These gauges are attached to the tongue of the square for the desired rise per foot of run and to the blade at the unit of run (Fig. 42-7).

Straightedge. A small straightedge clamped across the tongue and blade of the square at the desired location will insure that the square is held at the same angle for every step. The square is moved by sliding the straightedge against the top edge of the rafter stock (Fig. 42-8).

Pitch Board. A *pitch board* may be used in place of the framing square (Fig. 42-9). It is made by lay-

Fig. 42-7 Framing square gauges are attached to the square to hold it in the same position for every step-off.

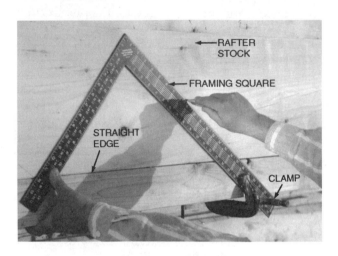

Fig. 42-8 A straightedge clamped to the square helps the step-off procedure.

Fig. 42-9 A pitch board may be used instead of the framing square for stepping-off.

ing out the rise and run at right angles to each other on a scrap piece of plywood or similar material. A short piece of 1×2 stock is fastened to the long edge of the pitch board to serve as a guide against the top edge of the rafter.

Calculating the Common Rafter Length

Rafter tables give information for finding the length and cuts on various kinds of rafters. They come in booklet form. They are also stamped on one side of most framing squares (Fig. 42-10). On the square, the inch marks indicate the rise of the rafter per foot of run. Directly below inch marks 2 through 18 are the length of several kinds of rafters in inches per foot of run.

Using Rafter Tables. To find the line length of a common rafter, first find the length per foot of run in the top line of the rafter table. This number will be under the inch mark that corresponds to pitch of the roof being framed. Multiply the length per foot of run found in the table by the number of units of run.

EXAMPLE Find the line length of a common rafter with a rise of 8 inches per foot of run for a building 28 feet wide. Reading below the 8-inch mark on the square, the length of the common rafter per foot of run is found to be 14.42 inches. Multiplying 14.42 inches by 14, which is the run of the rafter, gives 201.88 inches. Dividing by 12, to change inches to feet gives 16.82 feet. Changing 0.82 foot to inches by multiplying by 12 gives 9.84 inches. Changing 0.84 inch to 16ths of an inch by multiplying by 16, gives slightly over 13, making the total length of the rafter 16'-9 13/16" long. A small calculator makes the mathematics easy.

Finding Rafter Length per Unit of Run with a Calculator. A calculator can be used to find the length of a rafter of any run and rise. Multiply the rise by itself, and multiply the run by itself. Add these numbers together, and find the square root of the total.

EXAMPLE For the same rafter in the example above, multiplying 8, which is the rise of the rafter per foot of run, by itself gives 64. Multiplying 12, which is the number of inches in the unit of run, by itself gives 144. Added together, the sum is 208. Pressing the square root button on the calculator reveals an answer of 14.422205 which, when rounded off to the nearest hundredth, equals 14.42.

Laying Out the Common Rafter Length

Mark a plumb cut at the ridge. From the ridge cut, lay out, along the top edge of the rafter, the length of the rafter as determined by calculations. Mark the length. At this point, and make another plumb cut at the seat (Fig. 42-11).

Laying Out the Seat Cut of the Rafter

The *seat cut* of the rafter is a combination of a level cut and a plumb cut. The level cut rests on top of the wall plate. The plumb cut fits snugly against the outside edge of the wall.

On the plumb line of the seat cut, which has already been laid out, measure down from the top edge of the rafter a distance that will leave sufficient stock to insure enough strength to support any overhang. Mark on the plumb line. (This is usually a minimum of 2/3 the width of the stock.) Hold the

Fig. 42-10 Rafter tables are found on the framing square.

Fig. 42-11 Laying out the line length of the common rafter.

square in the same position as for stepping-off. Slide it along the stock until the blade lines up with the mark. Mark along the blade for the level cut of the seat (Fig. 42-12). On roofs with moderate slopes, the length of the level cut of the seat is usually the width of the wall plate. For steep roofs, the level cut is shorter. Otherwise too much stock would be cut out of the rafter, weakening it.

Overhang of the Common Rafter

In most plans, the rafter overhang is given in terms of a level measurement. To lay out the overhang, simply step off the necessary amount of run from the plumb line of the seat cut (Fig. 42-13). The *tail cut* is the cut at the end of the rafter tail. It may be a plumb cut, level cut, a combination of cuts, or a square cut (Fig. 42-14). It is common practice, however, to let the rafter tails run *wild*. This means they are slightly longer than needed. They are cut off in a

Fig. 42-12 Laying out the seat cut of the common rafter.

Fig. 42-13 Laying out the overhang of the common rafter.

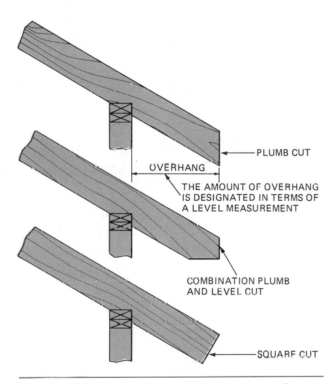

Fig. 42-14 Various tail cuts of the common rafter.

straight line after the roof frame is erected. In some cases, when the rafter tails are left exposed, any decorative, curved cuts are made on the tail ends before the roof is raised.

Shortening the Common Rafter

The length of the rafter at this point is from the outside of the wall to the centerline of the ridge (its line length). Therefore, the rafter must be shortened to allow for the thickness of a ridge board, if one is used (Fig. 42-15). If no ridge board is used, no shortening is necessary.

To shorten the rafter, measure back, at a right angle, from the plumb line at the ridge, a distance equal to 1/2 the thickness of the ridge board. Lay

Fig. 42-15 Shortening the common rafter.

out another plumb line at this point for the actual length of the rafter.

NOTE Shortening is always measured at right angles to the ridge cut, regardless of the pitch of the roof.

Cutting the Required Number of Rafters

To find the number of rafters required for a gable roof, divide the length of the building by the spacing of the rafters. Add one, and then multiply the total by two.

EXAMPLE A building is 42 feet long. The rafter spacing is 16 inches OC. Divide 42 by 1 1/3 (16 inches divided by 12 equals 1 1/3 feet) to get 31 1/2 spaces. Change 31 1/2 to the next whole number to get 32. Add 1 to make 33. Multiply by 2. Show that sixty-six rafters are needed.

Pile the required number of rafter stock with the crowned edges all facing in the same direction. Lay the pattern on each piece so the crowned edge of the stock will be the top edge of the rafter. Mark and make the necessary cuts. Usually a skilsaw or radial arm saw is used to make these cuts.

CAUTION On some cuts that are at a sharp angle with the edge of the stock (i.e., the level cut of the seat), the guard of the skilsaw may not retract. In this case, retract the guard by hand until the cut is made a few inches into the stock. Never wedge the guard in an open position to overcome this difficulty.

Fig. 42-16 Wood I-beams are frequently used for roof rafters. (*Courtesy of Trus Joist MacMillan*)

Wood I-Beam Roof Details

In addition to solid lumber, wood I-beams may be used for rafters (Fig. 42-16). Some roof framing details for wood I-beams are shown in Figure 42-17. To determine sizes for various spans and more spe-

Fig. 42-17 Wood I-beam roof framing details. (*Courtesy of Trus Joist MacMillan*)

BEVELED WEB STIFFENER EACH SIDE OF TJI® JOIST WEB.

BIRDSMOUTH CUT MUST NOT OVERHANG INSIDE FACE OF PLATE, SO THAT TJI® JOIST FLANGE BEARS FULLY ON PLATE.

BIRDSMOUTH CUT

MICRO-LAM® LVL OR PARALLAM® PSL BEAM.

Fig. 42-17 *(continued)*

cific information, consult the manufacturer's literature.

Laying Out the Ridge Board

Lay a sufficient number of boards to be used for a ridge alongside the wall plate. Transfer the rafter layout from the plate to the ridge board. Joints in the ridge board are centered on a rafter. The total length of the ridge board should be the same as the length of the building, unless there is an overhang at the gable ends. If there is an overhang, add the necessary amount on both ends.

Erecting the Gable Roof Frame

Construct a double-pole scaffold, if necessary, in the center of the building at a convenient height from the ridge (usually about 5 feet below the ridge). Lay the ridge board on top of the scaffold in the same direction it was laid out. Be careful not to turn the pieces around, end for end.

Lean the cut rafters against the building. Space them along each side, with the ridge cut up. Select four straight pieces for the end rafters, commonly called the *rake* rafters. One carpenter on each side of the building and one at the ridge are needed to raise the roof with efficiency.

Fasten a rafter to each end of the first section of ridge board. Raise the ridge board and two rafters into position. Fasten the rafters at the seat. Install two opposing rafters on the other side. Fasten the opposing rafters to the ridge board by nailing through the board at an angle into the cut end of the rafter. Drive the nail home with a nail set if necessary. Keep the top edges of the rafters and ridge board flush with each other.

Plumb and brace this section. Plumb the section by dropping a plumb bob from the end of the ridge board to the outside edge of the plate at the end of the building. Brace the section temporarily from attic floor to ridge. Raise all other sections in a similar manner.

Install the rest of the rafters. Fasten one rafter and then the opposing one. Do not install a number of rafters on one side only. Sight along the top edge of the ridge board for straightness as framing progresses. If not straight, make adjustments when installing the rafters. In addition to fastening rafters to the wall plate, fasten them to the sides of ceiling joists.

Collar Ties. Install *collar ties* 48 inches on center, or as required by codes or drawings. Collar ties are horizontal members fastened to pairs of rafters. They help strengthen the roof (Fig. 42-18).

Plumbing and Bracing. Check the roof for plumb again. Brace the roof frame securely with braces fastened to the under edges of the rafters running from plate to ridge. Applying the brace to the underside of the roof frame allows the roof to be sheathed without disturbing the braces.

Rafter Tails. If the rafter tails have not previously been cut, measure and mark the end rafters for the amount of overhang. This is usually a level measurement. Plumb the marks up to the top edge of the rafters. Stretch a line between these two marks across the top edges of all the rafters. Using a level, plumb down on the side of each rafter from the chalk line. Using a skilsaw, cut each rafter. Start the cut from the top edge, and follow the line carefully. If the tail cut is a combination of plumb and level cuts, make the plumb cuts first. Snap a line

STEP BY STEP
P R O C E D U R E S

RIDGEBOARD IS FASTENED TO
CENTERLINE OF RAFTER TO
PROVIDE FASTENING FOR
NEXT SECTION OF RIDGE.

RIDGEBOARD

RAFTER

COMMON
RAFTER

RIDGEBOARD

WALL PLATE

SEAT CUT
OF RAFTER

1. FASTEN ENDS OF A SECTION OF
RIDGEBOARD TO TWO RAFTERS.

RAISE & HOLD UP WHILE RAFTERS
ARE FASTENED AT THE SEAT.

NOTE: CEILING JOISTS ARE NOT
SHOWN FOR REASONS OF CLARITY.

3. ADD SECTIONS
OF RIDGEBOARD SUPPORTED
BY RAFTERS UNTIL
END OF BUILDING
IS REACHED.

2. INSTALL
OPPOSING RAFTERS.

Fig. 42-18 Steps in the erection of a gable roof.

STEP BY STEP
P R O C E D U R E S

(*continued*)

(4.) INSTALL REMAINING RAFTERS IN PAIRS OPPOSING EACH OTHER.

(5.) INSTALL COLLAR TIES, IF REQUIRED.

INSTALLING RAFTERS IN A GABLE ROOF

RIDGE

RAFTER

(6.) PLUMB THE GABLE END BY HANGING A PLUMB BOB FROM THE RIDGE

PLUMB BOB

PLATE

BRACE

(7.) WHEN PLUMB, BRACE THE ROOF.

RIDGE

PLATE

Fig. 42-18 (*continued*)

STEP BY STEP

P R O C E D U R E S

RAFTER

GABLE STUD →

WALL PLATE

WALL
BRACE

CORNER
POST

STUD →

ELEVATION

②ON BOTH ENDS, FROM THE
BOTTOM MARK, LAY OUT A
PLUMB LINE USING A
CARPENTER'S HAND LEVEL.

①. ON BOTH ENDS OF BUILDING,
LAY OUT A LEVEL MEASUREMENT
FROM THE WALL, AND MARK THE
BOTTOM EDGE OF THE
RAFTER.

PLUMB
LAYOUT
LINE

RAFTER

CEILING
JOIST

WALL PLATE

③.STRETCH CHALK
LINE FROM END TO END.
DO NOT SNAP LINE.
USE LINE AS A GUIDE.

IF TOP EDGES OF RAFTERS ARE NOT
EXACTLY THE SAME HEIGHT, A SNAPPED LINE
WILL PRODUCE END CUTS NOT IN LINE.

④.USING CHALK LINE AS A
GUIDE, PLUMB DOWN ALONG ONE
SIDE OF ALL RAFTERS USING A
HAND LEVEL.

HAND LEVEL

Fig. 42-19 Laying out the tail cut of common rafters.

across the cut ends where specified. Level in from the chalk line. Make the level cuts with a skilsaw. Work from the outside toward the wall (Fig. 42-19).

Constructing a Rake Overhang

If an overhang is specified at the rakes, horizontal structural members called *lookouts* must be installed. They support the rake rafter as shown in Figure 42-20.

Installing Gable Studs

The triangular areas formed by the rake rafters and the wall plate at the ends of the building are called *gables*. They must be framed with studs. These studs are called *gable studs*. The bottom ends are cut square. They fit against the top of the wall plate.

The top ends fit snugly against the bottom edge and inside face of the end rafter. They are cut off flush with the top edge of the rafter (Fig. 42-21).

Laying Out Gable Studs

The end wall plate is laid out for the location of the gable studs. Plumb down from the center of the ridge board to the wall plate. Square a line across the plate. Measure half a stud space on each side of the centerline for the studs on each side of center. From these marks, lay out the stud spacing in both directions to the outside of the building. With a straightedge and level, plumb these marks up to the inside face of the end rafters to lay out the top end of the gable studs.

LOOKOUTS

GABLE
END

LOOKOUTS NOTCHED
OVER GABLE END.

LOOKOUTS

GABLE END

GABLE END LOWERED
TO SUPPORT LOOKOUTS.

Fig. 42-20 Lookouts support the rake overhang.

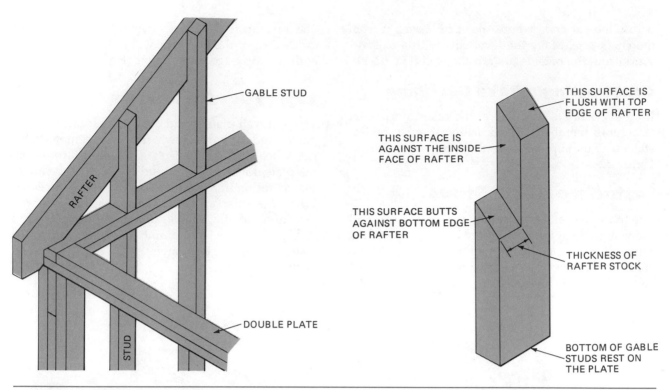

Fig. 42-21 The cut at the top end of the gable stud.

Finding the Length of Gable Studs

One method of finding the length and cut of the stud is to stand it up plumb on the mark at the plate and on the inside face of the rafter. Mark it along the bottom and top edge of the rafter (Fig. 42-22). Remove the stud. Use a scrap piece of rafter stock to mark the depth of cut on the stud. Mark and cut all studs in a similar manner.

The studs are fastened by toenailing to the plate and by nailing through the rafter into the edge of the stud. Care must be taken when installing gable studs not to force the end rafters up. This creates a crown in them. Sight the top edge of the end rafters for straightness as gable studs are installed. After all gable studs are installed, the end ceiling joist is spiked to the inside edges of the studs. The ceiling joist is straightened as it is fastened.

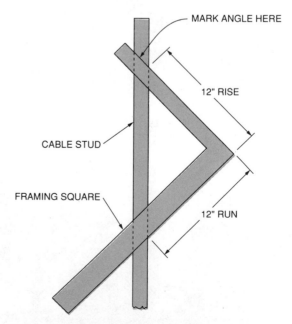

Fig. 42-22 Finding the length and cut of the gable stud.

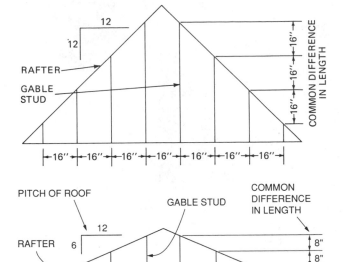

Fig. 42-23 Gable studs spaced equally have a common difference in length.

Laying Out Gable Studs by a Common Difference

Gable studs that are spaced equally have a common difference in length. Each stud is shorter than the next one by the same amount (Fig. 42-23). Once the length of the first stud and the common difference is known, gable studs can be easily laid out and cut on the ground.

To find the common difference in the length of gable studs, multiply the spacing in feet by the rise per foot of the roof.

The quickest method of finding the length of the first stud is to place it in position. Mark it as described previously. Find the length of the longest

EXAMPLE Gable studs are spaced 16 inches OC. The roof rises 6 inches per foot of run. Change 16 inches to 1 1/3 feet. Multiply 1 1/3 by 6 to get 8. The common difference is 8 inches.

gable stud. Mark the common difference successively along its length for a pattern to lay out all gable studs.

Framing a Gambrel Roof

The two slopes of a *real* gambrel roof are chords of a semicircle whose diameter is the width of the building. In order to frame a real gambrel roof, a full-size layout must be made on a large flat surface, such as the subfloor. From the layout, rafter lengths and angles of the cuts can be found (Fig. 42-24).

Usually the pitch of each slope of a gambrel roof is given in the drawings. The slopes, then, may not actually be chords of a semicircle (Fig. 42-25). In

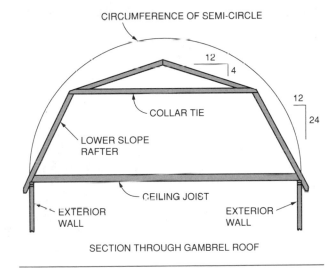

Fig. 42-25 In most cases, the pitch of each slope of the gambrel roof is given in the drawings.

Fig. 42-24 The chords of a semicircle are formed by the slopes of a true gambrel roof.

UPPER SLOPE RAFTER

PURLIN

30°

IF A RAFTER IS WITHIN 30° OR LESS OF VERTICAL IT MAY BE SIZED AS A WALL MEMBER

CEILING JOIST

LOWER SLOPE RAFTERS

SECTION THROUGH PURLIN

UPPER SLOPE RAFTER

DOUBLE PLATE

CEILING JOIST

KNEE WALL

SECTION THROUGH KNEE WALL

Fig. 42-26 Gambrel roof rafters intersect each other at a purlin or knee wall.

this case, the rafter lengths and cuts can also be found with a full-size layout. They are usually laid out with a framing square, however, similar to the layout of gable roof rafters.

Laying Out Gambrel Roof Rafters

Determine the construction at the intersection of the two slopes. Usually gambrel roof rafters meet at a continuous member, similar to the ridge, called a *purlin.* Any structural member that runs at right angles to and supports rafters is called a purlin. Sometimes, the rafters of a gambrel roof meet at the top of a *knee wall.* In the lower and steeper slope, the rafters may be sized as wall members rather than roof members if the rafter is within 30 degrees of vertical (Fig. 42-26).

Determine the run and rise of the rafters for both slopes. Find their line length in the same way as for rafters in a gable roof. Lay out plumb lines at the top and bottom of rafters for each slope. Make the seat cut on the lower rafter. Notch all rafters where they intersect at the purlin (Fig. 42-27). Because of the steep slope of the lower roof, the level cut of the seat cannot be made the full width of the plate. At least 2/3 of the width of the rafter stock must remain.

Erecting the Gambrel Roof

If a purlin is used, fasten a rafter to each end of a section of purlin. Raise the assembly. Fasten the lower end of the rafters against ceiling joists and on the top plate of the wall. Brace the section under

• •

STEP BY STEP
P R O C E D U R E S

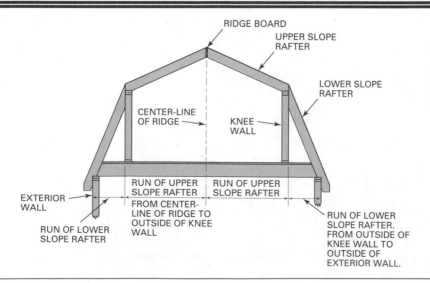

RIDGE BOARD

UPPER SLOPE RAFTER

LOWER SLOPE RAFTER

CENTER-LINE OF RIDGE

KNEE WALL

EXTERIOR WALL

RUN OF LOWER SLOPE RAFTER

RUN OF UPPER SLOPE RAFTER

RUN OF UPPER SLOPE RAFTER FROM CENTER-LINE OF RIDGE TO OUTSIDE OF KNEE WALL

RUN OF LOWER SLOPE RAFTER. FROM OUTSIDE OF KNEE WALL TO OUTSIDE OF EXTERIOR WALL.

Fig. 42-27 Layout of the gambrel roof rafters for the lower and upper slopes.

STEP BY STEP

P R O C E D U R E S

(continued)

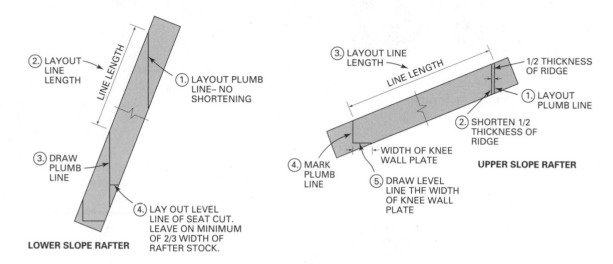

②. LAYOUT LINE LENGTH

①. LAYOUT PLUMB LINE– NO SHORTENING

LINE LENGTH

③. DRAW PLUMB LINE

④. LAY OUT LEVEL LINE OF SEAT CUT. LEAVE ON MINIMUM OF 2/3 WIDTH OF RAFTER STOCK.

LOWER SLOPE RAFTER

③. LAYOUT LINE LENGTH

LINE LENGTH

1/2 THICKNESS OF RIDGE

①. LAYOUT PLUMB LINE

②. SHORTEN 1/2 THICKNESS OF RIDGE

④. MARK PLUMB LINE

WIDTH OF KNEE WALL PLATE

⑤. DRAW LEVEL LINE THE WIDTH OF KNEE WALL PLATE

UPPER SLOPE RAFTER

Fig. 42-27 (continued)

each end of the purlin. Continue framing sections until the other end of the building is reached. Place temporary and adequate bracing under the purlin where needed. If a knee wall is used, build, straighten, and brace the wall.

Frame the roof by installing both lower and upper slope rafters opposite each other. This maintains equal stress on the frame. Plumb and brace the roof after a few rafters have been installed. Sight along the ridge and purlin or knee wall for straightness as framing progresses. After all rafters have been erected, install collar ties with their bottom edge resting on the purlin.

The open ends formed by the gambrel roof are framed with studs in the same manner as installing studs in gable ends.

CHAPTER 43 HIP ROOFS

The hip roof is a little more complicated to frame than the gable roof. Two additional kinds of rafters need to be laid out.

Framing the Hip Roof

To frame the hip roof it is necessary to lay out, not only common rafters and a ridge, but also *hip* rafters and *hip jack* rafters (Fig. 43-1). Hip rafters are required where the slopes of the hip roof meet. The hip jack rafter runs at right angles from the plate to the hip rafter. Several lengths of hip jacks are required.

Run of the Hip Rafter

Because the hip rafter runs at a 45 degree angle from the plate, the amount of horizontal distance it covers (total run) is greater than that of the common rafter. The hip rafter must rise the same amount as the common rafter with the same number of steps (units of run) to meet at the ridge. Therefore, the unit of run for the hip rafter is increased.

The total run of the hip is the diagonal of a square formed by the total run of adjacent common rafters and the exterior walls. Because the unit of run of the common rafter is 12 inches, the unit of

Fig. 43-1 Members of a hip roof frame.

run of the hip rafter becomes the diagonal of a 12-inch square, which is 16.97 inches or, 17 inches, for all practical purposes (Fig. 43-2).

FOR EXAMPLE, if the pitch of a hip roof is an 8-inch rise per foot of run, the common rafter is stepped-off by holding the square at 8 and 12. The hip rafter would be laid out by holding the square at 8 and 17. The same number of steps are taken as for the common.

STEP BY STEP
P R O C E D U R E S

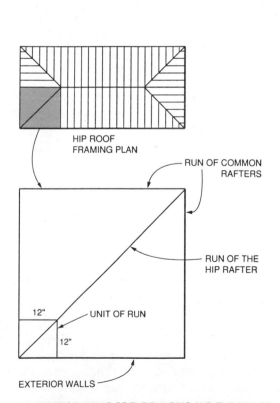

① THE EXTERIOR WALLS OF THE BUILDING AND THE RUN OF COMMON RAFTERS FORM A SQUARE. THE RUN OF THE HIP RAFTER IS THE DIAGONAL OF THE SQUARE.

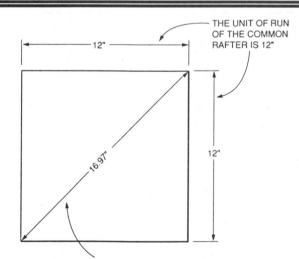

② THE UNIT OF RUN OF THE HIP RAFTER IS THE DIAGONAL OF A 12" SQUARE, WHICH IS 16.97" OR 17" FOR STEPPING OFF PURPOSES.

③ WHEN STEPPING OFF, HOLD THE FRAMING SQUARE AT 17" FOR THE RUN OF THE HIP RAFTER.

Fig. 43-2 The unit of run of the hip rafter is 17 inches.

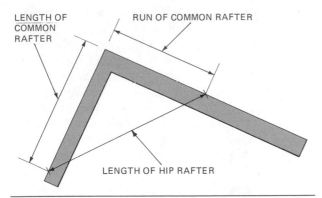

Fig. 43-3 Scaling across the framing square to determine the rough length of the hip rafter.

Estimating the Rough Length of the Hip Rafter

The rough length of the hip rafter can be found in the manner previously described for the common rafter. Let the tongue of the square represent the length of common rafter. Let the blade of the square represent its total run. Measure across the square the distance between the two points. Using a scale of one inch equals one foot gives the length of the hip rafter from plate to ridge (Fig. 43-3).

EXAMPLE The common rafter length is 15 feet. Its total run is 12 feet. A measurement across these points scales off to 19'-2 1/2". It will be necessary to order 20-foot lengths for the hip rafters unless extra is needed for overhang.

Laying Out the Hip Rafter

To lay out the hip rafter, make the cut at the ridge. Lay out its length. Make the seat cut. Lay out the length of the tail. Make the tail cut.

Ridge Cut of the Hip. The ridge cut of a hip rafter is a compound angle called a *cheek cut* or *side cut*. A *single cheek* cut or a *double cheek* cut may be made on the hip rafter according to the way it is framed at the ridge (Fig. 43-4).

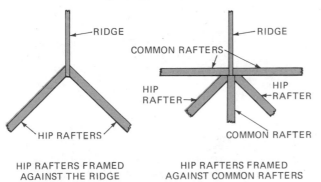

HIP RAFTERS FRAMED AGAINST THE RIDGE   HIP RAFTERS FRAMED AGAINST COMMON RAFTERS

Shortening the Hip. The rafter should be shortened before the cheek cuts are laid out. The amount of shortening depends on how it is framed at the ridge. When framed against the ridge, the hip rafter is shortened by half the 45 degree thickness of the ridge board. When framed against common rafters, the hip is shortened half the 45 degree thickness of the common rafter (Fig. 43-5).

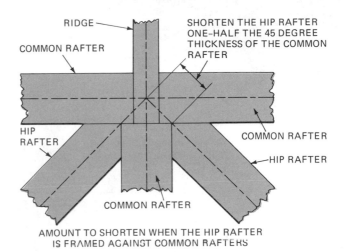

AMOUNT TO SHORTEN WHEN THE HIP RAFTER IS FRAMED AGAINST COMMON RAFTERS

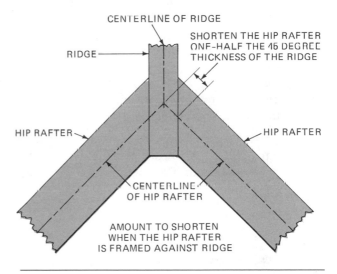

AMOUNT TO SHORTEN WHEN THE HIP RAFTER IS FRAMED AGAINST RIDGE

Fig. 43-5 Amount to shorten the hip rafter for both methods of framing.

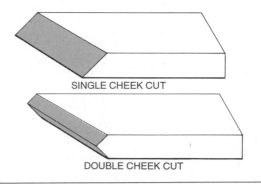

SINGLE CHEEK CUT

DOUBLE CHEEK CUT

Fig. 43-4 Single or double cheek cuts are used depending on the method of framing the hip rafter at the ridge.

HELPFUL HINT –
SHORTEN RAFTERS BEFORE
LAYING OUT CHEEK CUTS

③ DRAW SECOND
PLUMB LINE

HIP RAFTER STOCK

① DRAW
PLUMB
LINE

② MEASURE AT A RIGHT
ANGLE TO PLUMB LINE
THE AMOUNT TO SHORTEN,
DETERMINED BY THE METHOD
OF FRAMING.

Fig. 43-6

To lay out the ridge cut, select a straight length of stock for a pattern. Lay it across two sawhorses. Stand on the side that will be the top edge when in position. Mark a plumb line near the left end. Hold the tongue of the square at the rise and the blade of the square at 17 inches, the unit of run for the hip rafter. Shorten the rafter by measuring at right angles from the plumb line. Lay out another plumb line at that point (Fig. 43-6).

Laying Out Cheek Cuts. To complete the layout of the ridge cut, mark lines for a single or double cheek cut as required. The method of laying out cheek cuts shown in Figure 43-7 gives accurate results regardless of the pitch of the roof.

The two bottom lines of the rafter tables on the framing square can be also be used to find the angle of side cuts of hip and jack rafters. A measurement is given under the inch marks for various pitches. The square is held with the measurement at the tongue and 12 inches at the blade across the top edge of the rafter. The angle is laid out by marking along the tongue (Fig. 43-8).

Laying Out the Hip Rafter Length

The length of the hip rafter is laid out in a manner similar to that used for the common rafter. It may be found by the step-off method, rafter tables, or calculating. Remember to always start any layout for length from the original plumb line before any shortening.

Stepping-Off the Hip. In the step-off method, the number of steps for the hip rafter is the same as for the common rafter in the same roof. The same rise is used, but the unit of run for the hip is 17. For example, for a roof with a rise of 6 inches per foot of run, the square is held at 6 and 12 for the common rafter and 6 and 17 for the hip rafter of the same roof (Fig. 43–9).

If the total run of the rafter contains a fractional part, the same fractional part of 17 must be used to determine the extra length of the hip rafter.

Using the Rafter Tables. Finding the length of the hip using the tables found on the framing square is similar to finding the length of the common rafter. However, the figures from the second line are used. These figures give the length of the

STEP BY STEP
P R O C E D U R E S

5. DRAW DIAGONAL THROUGH CENTERLINE

2. SQUARE PLUMB LINE ACROSS TOP EDGE

3. LAY OUT CENTERLINE ON TOP EDGE

1. LAY OUT PLUMB LINE

4. MEASURE AT RIGHT ANGLES TO PLUMB LINE ONE-HALF THICKNESS OF RAFTER STOCK AND LAY OUT ANOTHER PLUMB LINE

SINGLE CHEEK CUT LAYOUT

5. SQUARE 2ND PLUMB LINE ACROSS TOP

2. SQUARE FIRST PLUMB LINE ACROSS TOP EDGE

6. DRAW DIAGONALS IN BOTH DIRECTIONS

3. LAY OUT CENTERLINE ON TOP EDGE

1. LAY OUT PLUMBLINE

4. MEASURE AT RIGHT ANGLES TO PLUMB LINE ONE-HALF THICKNESS OF RAFTER STOCK AND LAY OUT ANOTHER PLUMB LINE

DOUBLE CHEEK CUT LAYOUT

Fig. 43-7 Laying out single and double cheek cuts.

(continued)

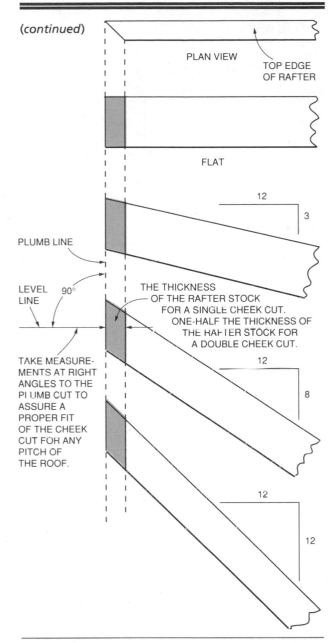

Fig. 43-7 (continued)

EXAMPLE If the total run of the rafter is 15'-6", step-off 15 times. Use 17 as the unit of run. The 6 inches is 1/2 foot. One-half of 17 is 8 1/2 inches. Measure and mark this dimension along the blade of the square when held at the last step. Move the square until the tongue lines up with the mark. Mark along the tongue for the plumb line of the seat cut, which is the outside edge of the wall plate (Fig. 43-10).

hip rafter in inches for every foot of run. It is only necessary to multiply this figure by the number of feet of run. The figure found in the tables may also be found by calculating as described previously for common rafters.

EXAMPLE Find the length of a hip rafter for a roof with a rise of 8 inches per foot of run with a total run of 14 feet. On the square below the 8-inch mark, it is found that the length of the hip rafter per foot of run is 18.76 inches. This multiplied by 14 equals 262.64 inches. Dividing by 12 gives 21.87 feet. Multiplying 0.87 foot by 12 gives 10.44 inches. Change 0.44 inch to 7/16 inch by multiplying by 16. The total length of the hip rafter is 21'-10 7/16".

If the total run of the rafter contains a fractional part of a foot, multiply the figure in the tables by the whole number of feet plus the fractional part changed to a decimal. For instance, if the total run of the rafter is 15'-6", multiply the figure in the tables by 15 1/2 changed to 15.5.

Lay out the length obtained by either method along the top edge of the hip rafter stock from the original plumb line at the ridge before any shortening.

Laying Out the Seat Cut of the Hip Rafter

Like the common rafter, the seat cut of the hip rafter is a combination of plumb and level cuts. The last plumb line made when laying out the rafter length is the plumb cut for the seat of the rafter. This cut is usually made square across the bottom edge of the hip rafter. No consideration needs to be given to fitting it around the corner of the wall. When making the level cut of the seat, consideration must be given to *dropping* the hip rafter.

Backing and Dropping the Hip. Because the hip rafter is at the intersection of the side and end slopes of the roof, the top outside corners of the rafter project above the plane of the roof. When the corners of the rafter are beveled flush with the roof, it is called *backing* the hip. When the level line of the seat is cut higher, it is called *dropping* the hip (Fig. 43-11).

Dropping the hip is much easier and is more frequently used. The procedure for determining the amount of drop is shown in Figure 43-12. If you want to back the hip, make the seat cut of the hip rafter so the same amount of stock is left on above it as the common rafter. The amount of backing, then, is the same as the amount of drop.

USE THE DIMENSION GIVEN IN THE TABLE IN COMBINATION WITH 12" TO LAY OUT SIDE CUTS OF JACKS. ALWAYS MARK ON THE 12" SIDE.

ALWAYS HOLD THIS SIDE AT 12".
ALWAYS MARK ALONG THIS SIDE.

12"

11-3/8" DIMENSION FROM THE RAFTER TABLE

FRAMING SQUARE

TOP EDGE OF JACK RAFTER STOCK

EXAMPLE: LAYING OUT THE SIDE CUT, ALSO CALLED CHEEK CUT, OF A JACK RAFTER FOR A ROOF WITH A PITCH OF 4" RISE PER FOOT OF RUN.

THE SAME PROCEDURE IS USED FOR LAYING OUT SIDE CUTS OF HIP & VALLEY RAFTERS, EXCEPT THE BOTTOM LINE OF THE RAFTER TABLE IS USED.

Fig. 43-8 The rafter tables can be used to lay out the side cut of hip and jack rafters.

STEP BY STEP

P R O C E D U R E S

① HOLD SQUARE AT RISE ON TONGUE AND 17" ON BLADE AND LAYOUT PLUMB LINE.

② LAY OUT REQUIRED NUMBER OF STEPS AND FRACTIONS THEREOF.

③ MOVE SQUARE ONE MORE STEP AND LAYOUT PLUMB LINE. DISTANCE BETWEEN PLUMB LINES IS LINE LENGTH OF RAFTER.

17"

RISE IN INCHES PER FOOT OF RUN

ALTHOUGH 16.97" IS THE LENGTH OF THE HIP RAFTER PER UNIT OF RUN, 17" IS USED FOR STEPPING-OFF PURPOSES.

Fig. 43-9 Stepping-off the hip rafter.

Laying Out the Tail of the Hip Rafter

If you wish to lay out the tail of the hip rafter, multiply its length per unit of run by the number of units of run of the common rafter overhang to find its total length. Lay the length out along its top edge

from the plumb line at the seat. Draw another plumb line. The tail cut of the hip rafter is usually a double cheek cut (Fig. 43-13).

Laying Out the Hip Jack Rafter

The *hip jack rafter* runs at right angles from the plate

LAY OUT FRACTIONAL UNIT OF
RUN TIMES 17.
EXAMPLE: 6" RUN = 1/2 UNIT OF RUN
1/2 TIMES 17 = 8-1/2"
LAY OUT 8-1/2"

RISE PER
FT OF RUN

17"

17"

MARK PLUMB LINE OF
BIRD'S MOUTH HERE

Fig. 43-10 Stepping-off a fractional part of a foot of run on the hip rafter.

ROOF SHEATHING

BACKING

HIP RAFTER

Fig. 43-11 "Backing" the hip.

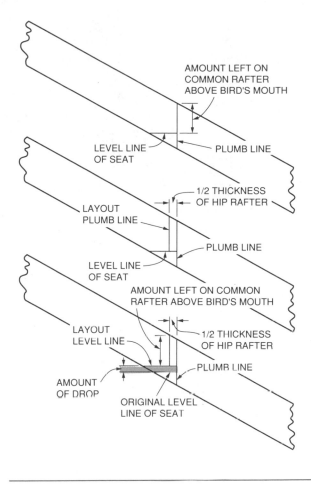

AMOUNT LEFT ON
COMMON RAFTER
ABOVE BIRD'S MOUTH

LEVEL LINE
OF SEAT

PLUMB LINE

1/2 THICKNESS
OF HIP RAFTER

LAYOUT
PLUMB LINE

PLUMB LINE

LEVEL LINE
OF SEAT

AMOUNT LEFT ON COMMON
RAFTER ABOVE BIRD'S MOUTH

LAYOUT
LEVEL LINE

1/2 THICKNESS
OF HIP RAFTER

PLUMB LINE

AMOUNT
OF DROP

ORIGINAL LEVEL
LINE OF SEAT

Fig. 43-12 Lay out to determine the amount to drop the hip rafter.

STEP BY STEP
P R O C E D U R E S

THE LINE LENGTH OF
THE HIP RAFTER TAIL

5.) LAY OUT
DIAGONALS

4.) LAY OUT
CENTER-LINE

PLUMB LINE OF HIP
RAFTER SEAT CUT

3.) DRAW PLUMB
LINE AND SQUARE
ACROSS TOP EDGE

2.) MEASURE TOWARD
SEAT CUT, 1/2 THICKNESS
OF THE HIP RAFTER STOCK

1.) LAY OUT PLUMB
LINE AND SQUARE
ACROSS TOP EDGE

NOTE: THE OVERHANG OF THE ROOF IS GIVEN IN THE DRAWINGS
AS A LEVEL DIMENSION OF THE OVERHANG OF THE COMMON
RAFTER, OR RUN OF THE COMMON RAFTER TAIL. USE THIS
DIMENSION TO STEP-OFF OR CALCULATE THE LINE LENGTH OF
THE HIP RAFTER TAIL.

Fig. 43-13 Laying out the tail of the hip rafter.

Fig. 43-14 Hip jack rafters are shortened common rafters.

to the hip rafter. Its seat cut and overhang are the same as that of the common rafter. It is actually a shortened common rafter. It has the same unit of run, 12 inches (Fig. 43-14).

Finding the Length of Hip Jack Rafters

Hip jack rafters are framed in pairs against the side of the hip rafter. Each set is shorter than the next set by the same amount. This is called the *common difference* (Fig. 43-15). The common difference is found in the rafter tables on the framing square for jacks 16 and 24 inches OC.

Once the length of the longest jack is determined, the length of all others can be found by making each set shorter by the common difference.

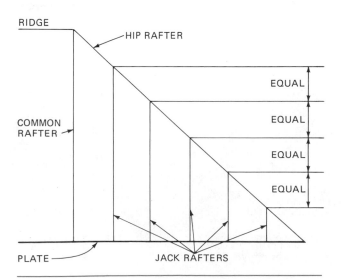

Fig. 43-15 Common difference in the length of hip jack rafters.

In order to find the length of the longest jack, its total run must be known.

Finding the Total Run of the Longest Jack Rafter. The total run of any pair of jack rafters and the outside edge of the wall plate form a square. Since all sides of a square are equal, the total run of any hip jack rafter is equal to its distance from the corner of the building. At the plate, measure from the corner of the building to the centerline of the longest jack to find its total run (Fig. 43-16).

Find the length of the longest hip jack rafter by stepping-off. Or, use the rafter tables in the same way as for common rafters except that the total run of the jack as found above is used to determine the number of steps. Lay out the seat cut and tail exactly the same as for the common rafter.

Shortening the Hip Jack Rafter. Since the hip jack rafter meets the hip rafter at a 45 degree angle, it must be shortened half the 45 degree angle thickness of the hip rafter stock. Measure the distance at right angles to the original plumb line. Draw another plumb line.

Laying Out the Cheek Cut of the Hip Jack Rafter. The hip jack rafter has a single cheek cut. Square the shortened plumb line across the top of the rafter. Draw an intersecting centerline along the top edge. Measure back at right angles from the last plumb line a distance equal to half the jack rafter stock. Lay out another plumb line. On the top edge, draw a diagonal from the last plumb line through the intersection of the center and the second plumb line (Fig. 43-17). The direction of the diagonal depends on which side of the hip the jack rafter is framed.

Determine the common difference of hip jack rafters from the rafter tables under the inch mark that coincides with the pitch of the roof being framed. The common difference may also be calculated. For rafters spaced 24 inches OC, the common

Fig. 43-16 The total run of any hip jack rafter is equal to its distance from the corner of the building.

STEP BY STEP
P R O C E D U R E S

1. LAY OUT PLUMB LINE

2. MEASURE AT RIGHT ANGLE TO PLUMB LINE ONE-HALF 45 DEGREE THICKNESS OF HIP RAFTER AND DRAW SHORTENED PLUMB LINE

3. SQUARE SHORTENED PLUMB LINE ACROSS TOP EDGE OF RAFTER STOCK

4. MEASURE AT RIGHT ANGLE FROM SHORTENED PLUMB LINE ONE-HALF THE THICKNESS OF THE JACK RAFTER STOCK AND DRAW ANOTHER PLUMB LINE.

5. DRAW CENTERLINE ALONG TOP EDGE OF RAFTER

6. DRAW DIAGONAL FROM LAST PLUMB LINE THROUGH CENTERLINE.

THE CUTTING LINES ARE THE DIAGONAL AND THE LAST PLUMB LINE.

Fig. 43-17 Steps in the layout of the cheek cut of the hip jack rafter.

difference is equal to the common rafter length for two units of run. For rafters spaced 16 inches OC, the common difference is equal to the common rafter length for 1 1/3 units of run (Fig. 43-18).

Once the common difference is determined, measure the distance successively along the top edge of the pattern for the longest jack rafter. This pattern is then used to cut all the jack rafters necessary to frame the roof. Cut *pairs* with cheek cuts going in opposite directions.

Finding the Length of the Hip Roof Ridge

The total run of the hip rafter is the diagonal of a square whose sides are equal to the run of the common rafter. Since the run of the common rafter is equal to half the width of the building, the *line length* of the ridge is found by subtracting the width of the building from the length of the building (Fig. 43-19).

The *actual length* of the hip roof ridge must be cut longer than the line length, however. The amount of increase depends on the construction. If the hip rafters are framed against the ridge, increase the length at *each* end by half the 45 degree thickness of the hip rafter plus half the ridge board. If the hip rafters are framed against the common rafters, the line length of the ridge is increased at *each* end half the thickness of the common rafter stock (Fig. 43-20).

Raising the Hip Roof

Transfer the common rafter layout at the wall plate to the ridge board in a manner similar to the layout of the gable roof ridge. Erect the common roof section in the same manner as described previously. If

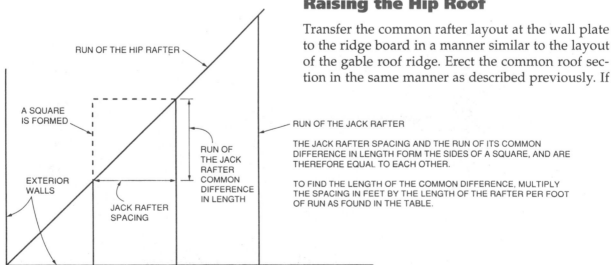

RUN OF THE HIP RAFTER

A SQUARE IS FORMED

EXTERIOR WALLS

JACK RAFTER SPACING

RUN OF THE JACK RAFTER COMMON DIFFERENCE IN LENGTH

RUN OF THE JACK RAFTER

THE JACK RAFTER SPACING AND THE RUN OF ITS COMMON DIFFERENCE IN LENGTH FORM THE SIDES OF A SQUARE, AND ARE THEREFORE EQUAL TO EACH OTHER.

TO FIND THE LENGTH OF THE COMMON DIFFERENCE, MULTIPLY THE SPACING IN FEET BY THE LENGTH OF THE RAFTER PER FOOT OF RUN AS FOUND IN THE TABLE.

Fig. 43-18 Determining the common difference in hip jack rafters by calculating.

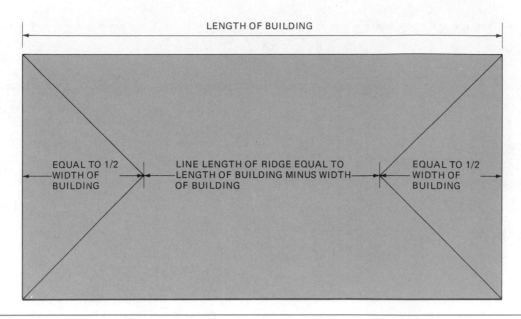

Fig. 43-19 Determining the line length of the hip roof ridge.

TO FIND ACTUAL LENGTH OF RIDGE, INCREASE ITS LINE LENGTH BY THE 45° THICKNESS OF THE HIP RAFTER PLUS THE THICKNESS OF THE RIDGE BOARD.

PLAN OF HIP RAFTERS FRAMED TO RIDGE

TO FIND THE ACTUAL LENGTH OF THE RIDGE, INCREASE ITS LINE LENGTH BY THE THICKNESS OF THE COMMON RAFTER.

PLAN OF HIP RAFTERS FRAMED AGAINST COMMON RAFTERS

Fig. 43-20 Actual length of the hip roof ridge for both methods of framing the hip rafter.

the hip rafters are framed against the ridge, install them next. If they are framed against the common rafters, install the commons against the end of the ridge. Then install the hip rafters.

Fasten jack rafters to the plate and to hip raf-

ters in pairs. As each pair of jacks is fastened, sight the hip rafter along its length. Keep it straight. Any *bow* in the hip is straightened by driving the jack a little tighter against the bowed out side of the hip as the roof is framed (Fig. 43-21).

. . . .

STEP BY STEP

P R O C E D U R E S

ROOF FRAMING PLAN

1. ERECT THE RIDGE WITH ONLY AS MANY COMMON RAFTERS AS NEEDED. THEN INSTALL THE HIP RAFTERS. THE HIP RAFTERS EFFECTIVELY BRACE THE ASSEMBLY.

   NOTE: THIS PROCEDURE SHOWS THE HIPS FRAMED TO THE RIDGE. IF HIPS ARE FRAMED TO THE COMMON RAFTERS, INSTALL THE COMMONS TO THE END OF THE RIDGE BEFORE THE HIPS.

ROOF FRAMING PLAN

2. INSTALL THE REMAINING COMMON RAFTERS IN PAIRS OPPOSING EACH OTHER. SIGHT THE TOP EDGE OF THE RIDGE FOR STRAIGHTNESS AS FRAMING PROGRESSES.

ROOF FRAMING PLAN

3. INSTALL THE HIP JACK RAFTERS IN PAIRS OPPOSING EACH OTHER. SIGHT THE HIP FOR STRAIGHTNESS AS JACKS ARE INSTALLED. IT IS BEST TO INSTALL A PAIR OF JACKS ABOUT HALF-WAY UP THE HIP TO STRAIGHTEN IT. THE HIP CAN BE STRAIGHTENED BY DRIVING THE JACK, ON THE CROWN SIDE OF BOW, DOWN TIGHTER.

Fig. 43-21 Steps in the erection of the hip roof frame.

■

CHAPTER 44 THE INTERSECTING ROOF

Buildings of irregular shape, such as L-, H-, or T-shaped, require a roof for each section. These roofs meet at an intersection, where *valleys* are formed. They are called *intersecting roofs*. The roof may be a gable, a hip, or a combination of types.

The Intersecting Roof

The intersecting roof requires the layout and installation of several kinds of rafters not previously described.

■ *Valley rafters* form the intersection of the

slopes of two roofs. If the heights of both roofs are different, two kinds of valley rafters are required.

■ The *supporting valley rafter* runs from the plate to the ridge of the main roof.

■ The *shortened valley rafter* runs from the plate to the supporting valley rafter.

■ *Valley jack rafters* run from the ridge to the valley rafter.

■ The *valley cripple jack rafter* runs between the supporting and shortened valley rafter.

■ The *hip-valley cripple jack rafter* runs between a hip rafter and a valley rafter (Fig. 44-1).

Confusion concerning the layout of so many different kinds of rafters can be eliminated by remembering:

■ The length of any kind of rafter can be found by determining its run.

■ The amount of shortening is always measured at right angles to the plumb cut.

■ The method previously described for laying out cheek cuts on rafters for any pitch.

Layout of Supporting Valley Rafters

The layout of valley rafters is similar to that of hip rafters. The unit of run for both is 17 inches. The total run of the supporting valley is the run of the common rafter of the main roof, called the *major span*. Its line length is found either by the step-off method, or by the rafter tables in the same manner

as for hip rafters. A single cheek cut is made at the ridge. The rafter is shortened by half the 45 degree thickness of the ridge board (Fig. 44-2).

Seat Cut of the Valley Rafter

The seat cut of the valley rafter must be extended to clear the inside corner of the wall plate. From the plumb line at the plate, measure toward the tail end of the rafter a distance equal to half the thickness of the rafter stock. Lay out another plumb line. Measure down, along the last plumb line from the top edge of the rafter, a distance equal to the amount left on the common rafter. Draw a level line at this point (Fig. 44-3).

The valley rafter is not dropped like the hip. Its top corners do not project above the slope of the roof. However, one edge, from the shortened valley rafter to the ridge, needs backing.

Overhang and Tail Cut of the Valley Rafter

The length of the valley rafter overhang is found in the same manner as that used for the hip rafter. Step-off the same number of times, from the plumb line of the seat cut, as stepped off for the common rafter. However, a unit run of 17 is used instead of 12.

The tail cut of the valley rafter is a double cheek cut. It is similar to that of the hip rafter, but angles inward instead of outward. From the last plumb line, measure toward the tail a distance that is half the thickness of the rafter stock. Lay out another plumb line. Square both plumb lines across

Fig. 44-1 Members of the intersecting roof frame.

STEP BY STEP
P R O C E D U R E S

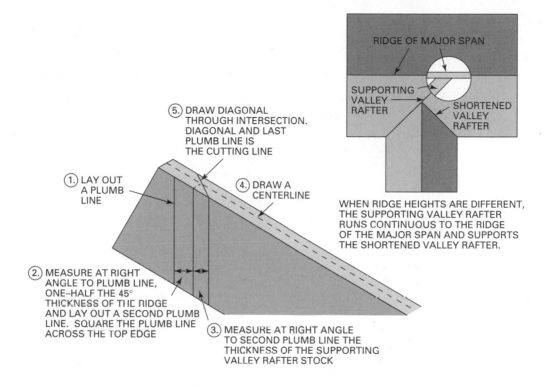

(5.) DRAW DIAGONAL THROUGH INTERSECTION. DIAGONAL AND LAST PLUMB LINE IS THE CUTTING LINE

(1.) LAY OUT A PLUMB LINE

(4.) DRAW A CENTERLINE

(2.) MEASURE AT RIGHT ANGLE TO PLUMB LINE, ONE–HALF THE 45° THICKNESS OF THE RIDGE AND LAY OUT A SECOND PLUMB LINE. SQUARE THE PLUMB LINE ACROSS THE TOP EDGE

(3.) MEASURE AT RIGHT ANGLE TO SECOND PLUMB LINE THE THICKNESS OF THE SUPPORTING VALLEY RAFTER STOCK

RIDGE OF MAJOR SPAN

SUPPORTING VALLEY RAFTER

SHORTENED VALLEY RAFTER

WHEN RIDGE HEIGHTS ARE DIFFERENT, THE SUPPORTING VALLEY RAFTER RUNS CONTINUOUS TO THE RIDGE OF THE MAJOR SPAN AND SUPPORTS THE SHORTENED VALLEY RAFTER.

(5.) DRAW DIAGONALS AS SHOWN

(1.) LAY OUT PLUMB LINE.

(4.) DRAW A CENTERLINE

(2.) MEASURE AT RIGHT ANGLE TO PLUMB LINE ONE–HALF 45° THICKNESS OF RIDGE AND LAY OUT A SECOND PLUMB LINE. SQUARE THE PLUMB LINE ACROSS THE TOP EDGE.

(3.) MEASURE AT RIGHT ANGLE TO SECOND PLUMB LINE ONE–HALF THE THICKNESS OF THE VALLEY RAFTER STOCK. SQUARE THE PLUMB LINE ACROSS THE TOP EDGE.

RIDGE

VALLEY RAFTER

VALLEY RAFTER

RIDGE

RIDGE HEIGHTS ARE THE SAME WHEN BOTH SPANS WITH THE SAME ROOF PITCH ARE EQUAL.

BOTH VALLEY RAFTERS MEET AT THE RIDGE

Fig. 44-2 The ridge cut of the supporting valley rafter.

STEP BY STEP
P R O C E D U R E S

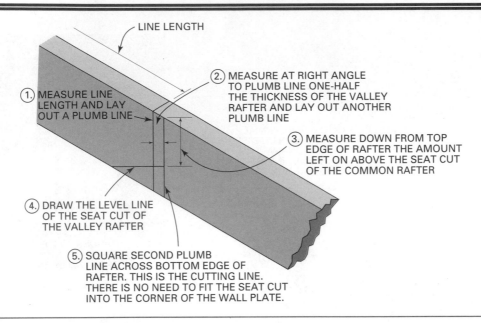

— LINE LENGTH

① MEASURE LINE LENGTH AND LAY OUT A PLUMB LINE

② MEASURE AT RIGHT ANGLE TO PLUMB LINE ONE-HALF THE THICKNESS OF THE VALLEY RAFTER AND LAY OUT ANOTHER PLUMB LINE

③ MEASURE DOWN FROM TOP EDGE OF RAFTER THE AMOUNT LEFT ON ABOVE THE SEAT CUT OF THE COMMON RAFTER

④ DRAW THE LEVEL LINE OF THE SEAT CUT OF THE VALLEY RAFTER

⑤ SQUARE SECOND PLUMB LINE ACROSS BOTTOM EDGE OF RAFTER. THIS IS THE CUTTING LINE. THERE IS NO NEED TO FIT THE SEAT CUT INTO THE CORNER OF THE WALL PLATE.

Fig. 44-3 The layout of the valley rafter seat cut.

the top of the rafter. Draw diagonals from the center (Fig. 44-4).

Layout of the Shortened Valley

The length of the shortened valley is found by using the run of the common rafter of the smaller roof, called the *minor span*. Its seat and tail are laid out the same as for the supporting valley.

However, the plumb cut at the top end is different from that of the supporting valley. Because the two valley rafters meet at right angles, the cheek cut of the shortened valley is a square cut. To shorten this rafter, measure back at right angles from the plumb line at the top end a distance that is half the thickness of the supporting valley rafter stock. Lay out another plumb line. (Fig. 44-5).

Layout of the Valley Jack

The length of the valley jack can be found if its total run is known. The total run of any valley jack rafter is equal to the run of the common rafter minus the horizontal distance that it is located in from the corner of the building (Fig. 44-6). Remember that all jack rafters are shortened common rafters and that the unit of run is 12.

The ridge cut of the valley jack is the same as a common rafter. It is shortened in the same way. The cheek cut against the valley rafter is a single cheek cut. The valley jack is shortened at this end by half

the 45 degree angle thickness of the valley rafter stock. The layout of all other valley jack rafters is made by making each set shorter by the common difference found in the rafter tables.

STEP BY STEP
P R O C E D U R E S

— FINISHED CUT

③ CENTERLINE

④ DRAW DIAGONALS

① LINE LENGTH TO HERE

② 1/2 THICKNESS OF VALLEY RAFTER

Fig. 44-4 The layout of the valley rafter tail cut.

Fig. 44-5 Plumb cut layout on the top end of the shortened valley rafter.

Layout of the Valley Cripple Jack

As stated before, the length of any rafter can be found if its total run is known. The run of the valley cripple jack is always twice its horizontal distance from the intersection of the valley rafters (Fig. 44-7). Use the common rafter tables, or step-off in a manner similar to that used for common rafters to find its length. On *each* end, shorten by half the 45 degree thickness of the valley rafter stock (Fig. 44-8). Make a single cheek cut.

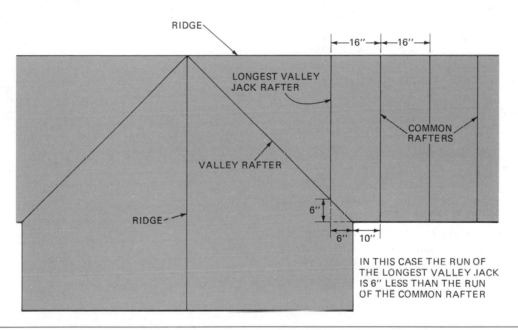

Fig. 44-6 How to determine the run of the longest valley jack rafter.

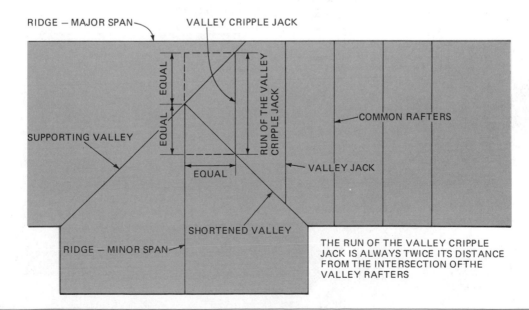

Fig. 44-7 Determining the run of the valley cripple jack rafter.

Fig. 44-8 The layout of the valley cripple jack rafter.

Hip-Valley Cripple Jack Rafter Layout

Determine the length of the rafter by first finding its total run. The run of a hip-valley cripple jack rafter is equal to the plate line distance between the seat cuts of the hip and valley rafters (Fig. 44-9). Lay out the length using the cut of the common rafter. All hip-valley cripple jacks cut between the same hip and valley rafters are the same length. On each end, shorten by half the 45 degree thickness of the hip and valley rafters. Make single cheek cuts (Fig. 44-10).

Ridge Length of the Minor Span

When the ridge heights of both roofs are the same, the line length of the ridge of the minor span is

Fig. 44-9 Finding the total run of the hip-valley cripple jack rafter.

Fig. 44-10 Layout of the hip-valley cripple jack rafter.

equal to the building projection plus half the width of the major span (Fig. 44-11). When ridge heights are different, the line length of the ridge is equal to the building projection plus half the width of the minor span (Fig. 44-12). The actual length of the ridge in both cases is shorter than the line length. However, it is customary to make the intersecting ridge longer than needed. The end rafters are located by plumbing up from the plate. The surplus end of the ridge is cut off after the roof is framed.

The end of the ridge at the intersection of the valleys is simply a square cut. It is not necessary to make the ridge fit perfectly into the intersection.

Erecting the Intersecting Roof

The intersecting roof is raised by framing opposing members of the main span first. Then, the valley rafters are installed. To prevent bowing the ridge of the main span, install rafters to oppose the valley rafters. Install common and jack rafters in sets opposing each other. Sight members of the roof as framing progresses, keeping all members in a straight line.

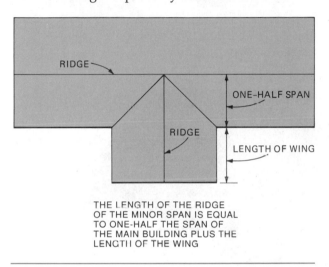

Fig. 44-11 Finding the line length of the ridge of the minor span when ridge heights are the same.

Fig. 44-12 Finding the line length of the ridge of the minor span when ridge heights are different.

CHAPTER 45

SHED ROOFS, DORMERS, AND OTHER ROOF FRAMING

The shed roof slopes in only one direction. It is relatively easy to frame. A shed roof may be freestanding or one edge may rest on an exterior wall while the other edge butts against an existing wall (Fig. 45-1). Shed and other type roofs are also used on *dormers*. A dormer is a framed projection above the plane of the roof. It contains one or more windows for the purpose of providing light and ventilation, or for enhancement of the exterior design (Fig. 45-2).

Framing a Shed Roof

A *shed roof rafter* runs at a right angle to the plate line, similar to a common rafter. However, it has two seat cuts instead of one. The unit of run is 12

inches, the same as the common rafter. The total run of the rafter, if both ends rest on exterior walls, is the width of the building minus the thickness of one of the walls (Fig. 45-3). For shed roofs that butt against an existing wall, the rafters are laid out the same as common rafters.

Layout of the Shed Roof Rafter

To lay out the shed roof rafter, mark a plumb line and a level line for a seat cut at one end. From the plumb line, step-off the rafter. Or, find its length using the common rafter tables. Lay out another seat cut. Leave enough beyond the seat cut on both ends for overhangs at top and bottom. The overhang may be trimmed after the rafters have been

Fig. 45-1 A building with shed roofs butting each other.

Fig. 45-2 A roof with gable and shed dormers.

Fig. 45-3 The shed roof rafter has two seat cuts.

installed, as described previously for common rafters. If preferred, the overhangs can be laid out and cut before installation (Fig. 45-4).

Shed roofs are framed by toenailing the rafters into the plate at the designated spacing. It is important that the plumb cut of the seat be kept snug up against the walls. In addition to nailing the rafters, metal *framing anchors* or *ties* may be required, not only for shed roofs, but for all types of framed roofs.

Dormers

Usually dormers have either gable or shed roofs. A gable dormer roof is framed similar to an intersecting gable roof (Fig. 45-5). In most cases, the shed dormer roof extends to the ridge of the main roof to gain enough pitch (Fig. 45-6).

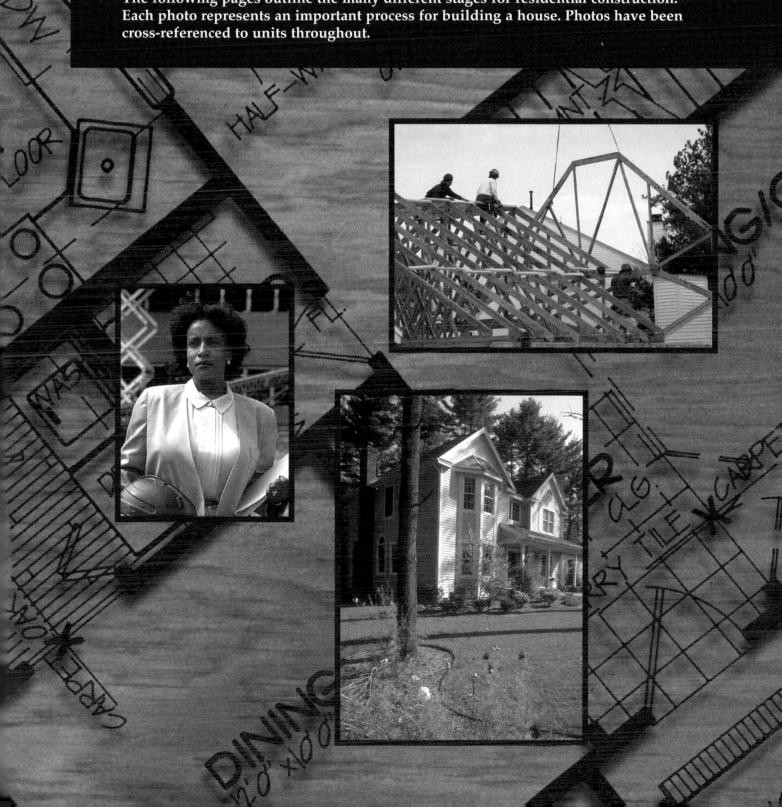

From rough work to a *smooth finish.*

The following pages outline the many different stages for residential construction. Each photo represents an important process for building a house. Photos have been cross-referenced to units throughout.

The prospective buyers choose the site with the assistance of a general contractor, real estate agent, architect, or engineer.

refer to
UNIT
8

The architectural team draws up plans to meet state and local regulations.

refer to
UNIT
8

The architect reviews plans and schedules with the clients.

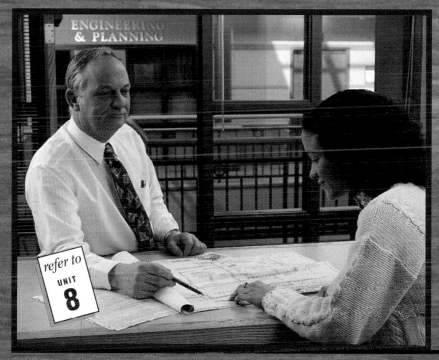

The architect applies for the necessary permits.

The excavator uses heavy equipment to clear the land.

A general contractor, or engineer, lays out the project.

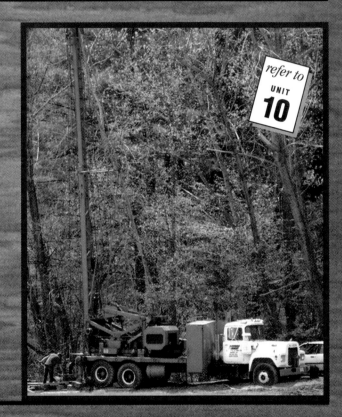

refer to
UNIT
9

refer to
UNIT
10

Installation of septic, well, and other utilities will begin.

refer to
UNIT
10

Heavy equipment begins the excavation process.

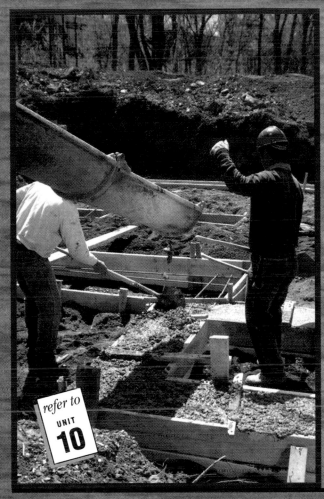

refer to
UNIT
10

The general contractor, or mason contractor, pours the footings.

refer to
UNIT
10

These photos show a block foundation and a poured concrete foundation.

The excavator backfills after the foundation is poured.

refer to
UNIT
10

Carpenters, or framers, apply decking material to floor joists.

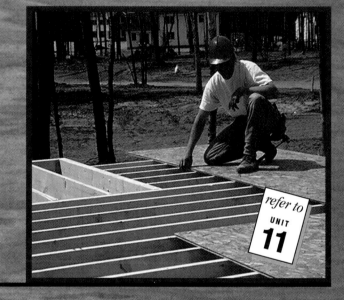

refer to
UNIT
11

This photo shows the carpenter, or framer, raising an exterior wall.

refer to
UNIT
12

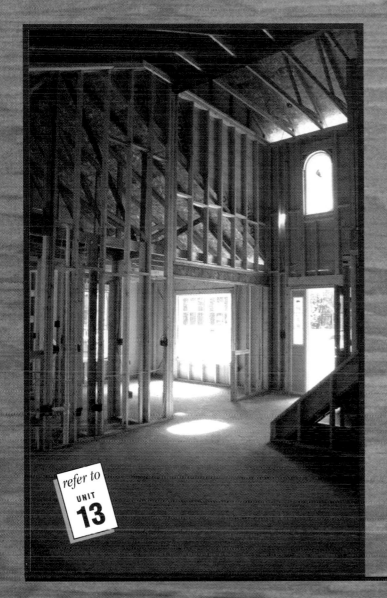

refer to
UNIT
13

This photo shows interior partitions, headers, rafters, staircase, etc.

refer to
UNIT
18

At this stage, the carpenter applied the sheathing to exterior walls and roof, and completed the rakes and fascias.

The building inspector inspects all phases of construction.

refer to
UNIT
8

When the building is ready for roofing, the general contractor, or roofing contractor, applies the shingles (cedar, slate, asphalt, etc.).

refer to
UNIT
19

The builder applied exterior (rigid) insulation to increase the R-value, and to allow for different building applications.

refer to
UNIT
17

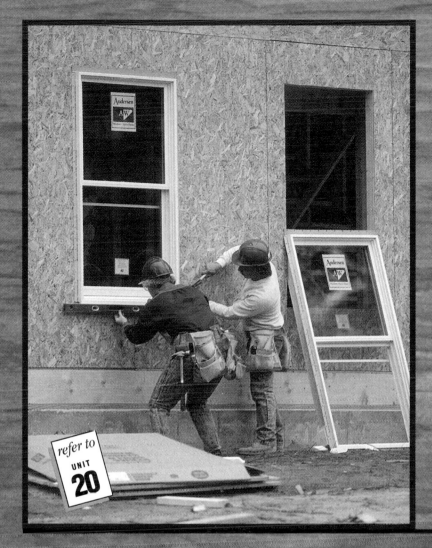

refer to
UNIT
20

Carpenters install
double hung windows.

refer to
UNIT
22

The builder applies vinyl,
aluminum, or cedar siding.

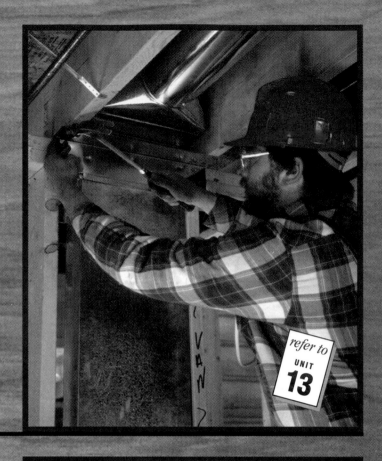

refer to
UNIT
13

Plumbers, or HVAC contractors, install HVAC (Heating, Ventilation and Air Conditioning). In this picture, the plumber is installing heating and cooling ducts.

refer to
UNIT
13

At this point in the construction process, the plumber installs rough plumbing.

refer to UNIT **13**

The electrician installs the rough electrical work.

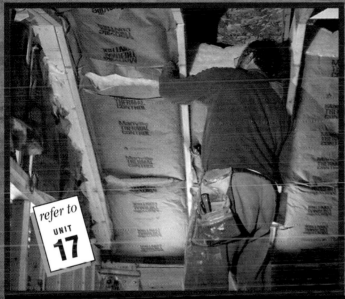

refer to UNIT **17**

The insulation contractor, or general contractor, insulates the ceiling.

refer to UNIT **24**

A drywall contractor, or general contractor, installs drywall, which is commonly referred to as sheetrock.

The OSHA inspector examines the building site to ensure the safety of the work force.

refer to
UNIT
8

The worker tapes joints after drywall is installed.

refer to
UNIT
24

The finish carpenter installs the cabinets.

refer to
UNIT
31

refer to UNIT **30**

The hardwood flooring installer sands the flooring to prepare the wood for finishing.

refer to UNIT **30**

This photo shows a variety of finished floors.

This photo shows a plumber setting a sink, which is part of the finish plumbing process.

refer to
UNIT
28

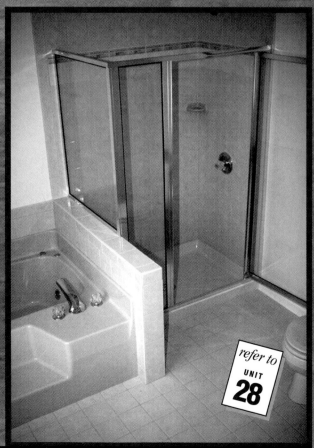

Bath fixtures, showers, tubs, and toilets are all part of finish plumbing.

refer to
UNIT
28

refer to
UNIT
28

The electrician completes the finish electrical work at this time.

refer to
UNIT
28

The trimmer installs the baseboard.

refer to
UNIT
23

The landscapers plant shrubs and trees to add a finishing touch.

STEP BY STEP
P R O C E D U R E S

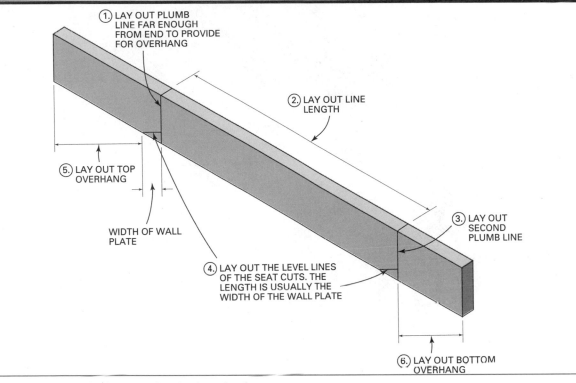

(1.) LAY OUT PLUMB LINE FAR ENOUGH FROM END TO PROVIDE FOR OVERHANG

(2.) LAY OUT LINE LENGTH

(5.) LAY OUT TOP OVERHANG

WIDTH OF WALL PLATE

(3.) LAY OUT SECOND PLUMB LINE

(4.) LAY OUT THE LEVEL LINES OF THE SEAT CUTS. THE LENGTH IS USUALLY THE WIDTH OF THE WALL PLATE

(6.) LAY OUT BOTTOM OVERHANG

Fig. 45-4 Steps in the layout of a shed roof rafter.

VALLEY JACK RAFTER

DORMER RIDGE

RIDGE

GABLE STUD

COLLAR TIE

VALLEY RAFTER

COMMON RAFTER

DOUBLED RAFTER ON EACH SIDE OF DORMER OPENING

Fig. 45-5 Framing a gable dormer is similar to framing intersecting gable roofs.

Fig. 45-6 Framing details of a shed dormer. Usually the rafters of the dormer extend to the ridge of the main roof to gain enough pitch.

When framing openings for dormers, the rafters on both sides of the opening are doubled. Some dormers have their front walls directly over the exterior wall below. When dormers are framed with their front wall partway up the main roof, top and bottom headers of sufficient strength must be installed.

Other Roof Framing

There are a number of other roof framing problems related in some way to the framing of roofs previously described. Solutions to some of the most commonly encountered problems are given.

Fitting a Valley Jack Rafter to Roof Sheathing

Some intersecting roofs are built after the main roof has been framed and sheathed. In this type of construction, the valley rafters are eliminated and common rafters are used full-length, making the main roof easier to frame. However, an extensive repair job will be required if the sheathing in the valley deteriorates because of a leak in the roof. Steps to lay out the seat cut of the valley jack rafter to fit against roofs of the same and different pitches are shown in Figure 45-7.

Layout of the Ridge Cut against Roof Sheathing

The layout of the ridge of the intersecting gable roof that fits against the roof sheathing is shown in Figure 45-8.

Cut of the Shed-Roof Rafter against a Roof

The top ends of shed-roof rafters are occasionally fitted against a roof of a different and steeper pitch. The cuts can be laid out by using a framing square as outlined in Figure 45-9.

STEP BY STEP
P R O C E D U R E S

RAFTER STOCK

④ DRAW DIAGONAL

③ LAY OUT PLUMB LINE AND SQUARE ACROSS TOP EDGE OF RAFTER

① LAY OUT LEVEL LINE

② MEASURE BACK ON LEVEL LINE THE THICKNESS OF THE RAFTER STOCK

SEAT CUT FOR ROOF OF SAME PITCH

① SQUARE ACROSS TOP EDGE AND LAY OUT A PLUMB LINE

④ DRAW DIAGONAL

② MEASURE DOWN ON THE PLUMB LINE A DISTANCE EQUAL TO RISE IN INCHES OF THE MAIN ROOF OVER A RUN OF THE RAFTER THICKNESS

③ LAY OUT A LEVEL LINE

SEAT CUT FOR ROOF OF DIFFERENT PITCH

Fig. 45-7 Laying out the seats of valley jack rafters that fit against roofs of the same and different pitches.

FRAMING SQUARE

RISE OF THE ROOF IN INCHES PER FT. OF RUN

12"

HOLD SQUARE AS SHOWN AND MARK ALONG THE BLADE

RIDGE STOCK

Fig. 45-8 Layout of a ridge that fits against roof sheathing.

STEP BY STEP
P R O C E D U R E S

② HOLD THE SQUARE ON THE
LEVEL LINE TO THE PITCH
OF THE MAIN ROOF

12"

RISE OF
THE MAIN
ROOF

③ DRAW CUTTING LINE ALONG
THE TONGUE OF THE SQUARE.
EXTEND THE LINE ACROSS THE
SIDE OF THE RAFTER STOCK

① DRAW A LEVEL LINE
TO THE PITCH OF
THE SHED DORMER RAFTER

Fig. 45-9 Steps in laying out a shed roof rafter that fits against a roof of a different pitch.

CHAPTER **46** TRUSSED ROOFS

Roof Trusses

Roof trusses are extensively used in residential and commercial construction (Fig. 46-1). Because of their design, they can support the roof over a wide span. In residential construction, the use of roof trusses

Fig. 46-1 Trusses are used extensively for roof framing.

eliminates the need for load-bearing partitions below. The roof is also framed in much less time. However, because of their design, much usable attic space is lost.

Truss Construction

A roof truss consists of upper and lower *chords* and diagonals called *web members*. The upper chords act as rafters. The lower chords serve as ceiling joists. Joints are fastened securely with metal or wood gusset plates (Fig. 46-2).

Truss Design

Most trusses are made in fabricating plants. They are transported to the job site. Trusses are designed by engineers to support prescribed loads. Trusses may also be built on the job, but approved designs must be used. The carpenter should not attempt to design a truss. Approved designs and instructions for job-built trusses are available from the American Plywood Association and the Truss Plate Institute.

Fig. 46-2 The members of roof trusses are securely fastened with metal gussets.

The most common truss design for residential construction is the *Fink* truss (Fig. 46-3). Other truss shapes are designed to meet special requirements (Fig. 46-4).

Erecting a Trussed Roof

Carpenters are more involved in the erection than the construction of trusses. The erection and bracing of a trussed roof is a critical stage of construction. Failure to observe recommendations for erection and bracing could cause a collapse of the structure. This could result in loss of life or serious injury, not to mention the loss of time and material.

The recommendations contained herein are technically sound. However, they are not the only method of bracing a roof system. They serve only as a guide. The builder must take necessary precautions during handling and erection to insure trusses are not damaged, which might reduce their strength.

Small trusses, which can be handled by hand, are placed upside down, hanging on the wall plates,

toward one end of the building. Trusses for wide spans require the use of a crane to lift them into position. One at a time is lifted, fastened, and braced in place. The end truss is installed first by swinging it up into place and bracing it securely in a plumb position.

Bracing the First Truss

All other trusses are tied to the first truss. Therefore, the bracing system depends, to a great extent, on how well the first truss is braced. One satisfactory method is for the first truss to be braced to stakes driven and securely anchored in the ground. The ground braces should be located directly in line with all rows of top chord continuous lateral bracing (Fig. 46-5). The first truss is fastened to the plate and temporarily to the end bracing. Do not nail scabs to the end of the building to brace the first truss. These scabs can break off or pull out, causing a collapse of the roof.

Temporary Bracing of Trusses

As trusses are set in place, they are nailed to the plate. Metal framing ties are usually applied (Fig. 46-6). Sufficient temporary bracing must be applied. Temporary bracing should be no less than 2×4 lumber as long as practical, with a minimum length of 8 feet. The 2×4s should be fastened with two 16d duplex nails at every intersection. Exact spacing should be maintained as bracing is installed. Adjusting the spacing later can cause trusses to collapse if a key brace is unfastened at the wrong time.

Temporary bracing must be applied to three planes of the truss assembly: the top chord or sheathing plane, the bottom chord or ceiling plane, and vertical web plane at right angles to the trusses.

Bracing the Plane of the Top Chord. Continuous lateral bracing should be installed within 6

Fig. 46-3 The Fink truss is widely used in residential construction.

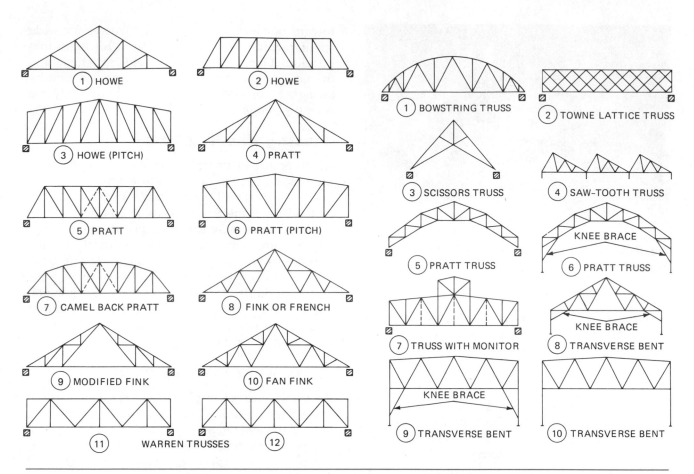

Fig. 46-4 Truss shapes are designed for special requirements.

Fig. 46-5 The end truss must be well braced before the erection of other trusses.

inches of the ridge and at about 8 to 10 feet intervals between the ridge and wall plate. Diagonal bracing should be set at approximately 45 degree angles between the rows of lateral bracing. It forms triangles that provide stability to the plane of the top chord (Fig. 46-7). If possible, the bracing should be placed on the underside of the top chord. Then, it will not have to be removed as the roof sheathing is applied.

Bracing the Web Member Plane. Temporary bracing in the plane of the web members are diagonals placed at right angles to the trusses from top to bottom chords (Fig. 46-8). They usually become permanent braces of the web member plane.

Bracing the Plane of the Bottom Chord. In order to maintain the proper spacing on the bottom chord, continuous lateral bracing for the full length

Fig. 46-6 Metal framing ties are used to fasten trusses to the wall plate.

Fig. 46-7 Temporary bracing of the plane of the top chord.

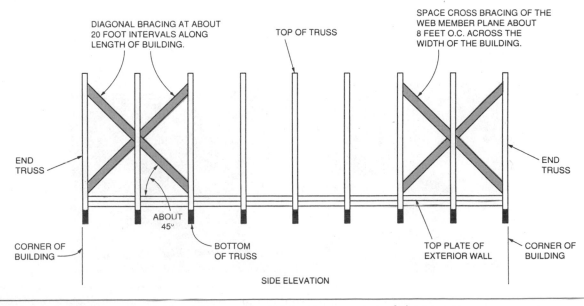

Fig. 46-8 Bracing of the web member plane prevents lateral movement of the trusses.

DIAGONAL BRACING
AT EACH END AND AT
20 FOOT INTERVALS
ALONG LENGTH OF
BUILDING

END
TRUSS

LATERAL
BRACES

BOTTOM
CHORDS

CORNER OF
BUILDING

CORNER OF
BUILDING

LATERAL BRACING
8 TO 10 FEET O.C.

TOP CHORDS
OF TRUSSES

CORNER OF
BUILDING

NOTE: PORTIONS OF THE
TRUSSES ARE CUT AWAY
SO THE PLANE OF THE BOTTOM
CHORD MAY BE SEEN MORE
CLEARLY.

Fig. 46-9 Bracing the plane of the bottom chord.

of the building must be applied. The bracing should be nailed to the top of the bottom chord at intervals no greater than 8 to 10 feet along the width of the building.

Diagonal bracing should be installed, at least, at each end of the building (Fig. 46-9). In most cases, temporary bracing of the plane of the bottom chord is left in place as permanent bracing.

Permanent Bracing

Permanent bracing is designed by the architect or engineer for the structural safety of a building. It is installed at the appropriate time as specified.

Roof Sheathing

Roof sheathing is applied after the roof frame is complete. Sheathing gives rigidity to the roof frame. It also provides a nailing base for the roof covering. Rated panels of plywood and strand board are com-

monly used to sheath roofs. Solid lumber sheathing boards are used only occasionally.

Lumber Sheathing

Lumber sheathing, if used, is usually square-edge or matched 1×8 or 1×10 boards. They are laid close together when a solid and continuous base is needed, such as for asphalt shingles. The boards may be spaced when the roof covering is wood shingles, shakes, or other stiff roofing. Spaced sheathing is usually 1×4 strips with the on center spacing the same as the shingle exposure. Joints between boards must be made on a rafter and staggered. Each board must bear on at least two rafters.

Plank decking is used in post and beam construction where the roof supports are spaced farther apart. Plank roof sheathing may be of 2-inch nominal thickness or greater, depending on the span. Usually both edges and ends are matched. The plank roof often serves as the finish ceiling for the rooms below (Fig. 46-10).

Fig. 46-10 Matched plank being used to cover a flat roof of post and beam construction. (*Courtesy of Georgia Pacific*)

Panel Sheathing

Plywood and other rated panel roof sheathing is laid with the face grain running across the rafters for greater strength. End joints are made on the rafters and staggered. Nails are spaced 6 inches apart on the ends and 12 inches apart on intermediate supports (Fig. 46-11).

Adequate blocking, tongue and grooved edges, or other suitable edge support such as *panel clips* must be used when spans exceed the indicated value of the plywood roof sheathing. Panel clips are small metal pieces shaped like a capital I. They are used between adjoining edges of the plywood between rafters (Fig. 46-12). Two panel clips are used for 48-inch or greater spans. One clip is used for lesser spans.

The ends of sheathing are allowed to slightly and randomly overhang at the *rakes* until sheathing is completed. A chalk line is snapped between the ridge and plate at each end of the roof. The sheathing ends are trimmed to the line with a skilsaw.

Fig. 46-11 Sheathing a trussed roof with plywood.

APA PANEL ROOF SHEATHING

APA RATED SHEATHING

1/8" SPACING IS RECOMMENDED AT ALL EDGE AND END JOINTS UNLESS OTHERWISE INDICATED BY PANEL MANUFACTURER.

PANEL CLIP

PANEL CLIP OR TONGUE-AND-GROOVE EDGES IF REQUIRED.

STAGGER END JOINTS (OPTIONAL)

ASPHALT OR WOOD SHINGLES OR STAKES. FOLLOW ROOFING MANUFACTURER'S RECOMMENDATIONS FOR ROOFING FELT.

PROTECT EDGES OF EXPOSURE 1 OR 2 PANELS AGAINST EXPOSURE TO WEATHER, OR USE EXTERIOR PANEL STARTER STRIP.

NOTE: COVER SHEATHING AS SOON AS POSSIBLE WITH ROOFING FELT FOR EXTRA PROTECTION AGAINST EXCESSIVE MOISTURE PRIOR TO ROOFING APPLICATION.

NOTE: FOR PITCHED ROOFS, PLACE SCREENED SURFACE OR SIDE WITH SKID-RESISTANT COATING UP IF OSB PANELS ARE USED. WEAR SKID-RESISTANT SHOES WHEN INSTALLING ROOF SHEATHING.

Recommended Minimum Fastening Schedule for APA Panel Roof Sheathing (Increased nail schedules may be required in high wind zones.)

Panel Thickness [b] (in.)	Nailing [c] [d]		
		Maximum Spacing (in.)	
	Size	Panel Edges	Intermediate
5/16 - 1	8d	6	12 [a]
1 - 1/8	8d or 10d	6	12 [a]

(a) For spans 48 inches or greater, space nails 6 inches at all supports.
(b) For stapling asphalt shingles to 5/16-inch and thicker panels, use staples with a 15/16-inch minimum crown width and a 1-inch leg length. Space according to shingle manufacturer's recommendations.
(c) Use common smooth or deformed shank nails with panels to 1 inch thick. For 1-1/8-inch panels, use 8d ring- or screw-shank or 10d common smooth-shank nails.
(d) Other code-approved fasteners may be used.

Fig. 46-12 Recommendations for the application of APA panel roof sheathing. (*Courtesy of American Plywood Association*)

Review Questions

Select the most appropriate answer.

1. The minimum amount of stock left above the seat cut of the common rafter to insure enough strength to support the overhang is usually
   a. one-quarter of the rafter stock width.
   b. one-half of the rafter stock width.
   c. two-thirds of the rafter stock width.
   d. three-quarters of the rafter stock width.

2. What is the line length of a common rafter from the centerline of the ridge to the plate with a rise of 5 inches per foot of run, if the building is 28'-0" wide?
   a. 14'-8 1/2".
   c. 15'-6".
   b. 15'-2".
   d. 16'-2".

3. The common difference in the length of gable studs spaced 24 inches OC for a roof with a pitch of 8 inches rise per foot of run is
   a. 8 inches.
   c. 16 inches.
   b. 12 inches.
   d. 20 inches.

4. How much is laid out on the blade of the square for 6 inches of run when stepping-off a hip rafter?
   a. 6 inches.
   c. 8 1/2 inches.
   b. 6/17 foot.
   d. 17 inches minus 6 inches.

5. Using the rafter tables, the line length of a hip rafter with a rise of 6 inches per foot of run and a total run of 12 feet is
   a. 16'-6".
   c. 18'-0".
   b. 17'-4".
   d. 18'-8".

6. The unit of run of any jack rafter is
   a. 12 inches.
   c. 17 inches.
   b. 16 inches.
   d. the same as hip and valley rafters.

7. The length of any rafter in a roof of specified pitch can be found if
   a. its total rise is known.
   b. its total run is known.
   c. the rise per foot of run is known.
   d. its length per foot of run is known.

8. The total run of any hip jack rafter is equal to
   a. its distance from the corner.
   b. one-third of its total rise.
   c. its length minus the total rise.
   d. its common difference in length.

9. The total run of the shortened valley rafter is equal to
   a. the total run of the minor span.
   b. the total run of the major span.
   c. the total run of the common rafter.
   d. the total run of the hip rafter.

10. The total run of the hip-valley cripple jack rafter is equal to
    a. one-half the run of the hip jack rafter.
    b. one-half the run of the longest valley jack rafter.
    c. the distance between bird's mouths of the hip and valley rafters.
    d. the difference in run between the supporting and shortened valley rafters.

■ BUILDING FOR SUCCESS ■

Communication Skills in Construction

The world of construction has always relied on strong oral and written communication skills to relay information. Today we communicate electronically more than ever before. Yet the most effective exchange of information is still based on good speaking, writing, and listening skills.

Along with the personal computer, fax machines, fiber optics, and telecommunications, there is still a need to develop strong oral and written communication skills. These skills form the basis for using high-technology achievements successfully.

Oral communication has always been important in providing the desired "personal touch" for customers. The vast majority of construction projects require good oral and written interaction both in the office and at the job site. As we become a more service-oriented nation, we must continue to value the power of the spoken and written word in conducting business. To enhance the success of the project, the builder, subcontractor, or technician on the job must maintain exactness and clarity when dealing with clients.

The diverse populations that make up the nation's workforce today bring new dimensions to effective communication. Construction personnel represent more cultures, education levels, and diversities than ever before. This alone calls for a greater need to promote clear and concise oral and written communication at the workplace. These new dimensions call for increased abilities in reading, writing, and listening.

Technicians today are more involved in the planning, coordinating, and marketing as well as the construction of a project. This will continue to require more verbal and writing skills. Legalities, warranties, quality assurances, and guarantees have taken on expanded roles. Interpreting codes, specifications, statutes, and other stipulations demands better reading and listening skills. Mistakes due to errors in reading or interpretation can be costly and even endanger lives. All workers must develop a proficiency in affective communication.

One of the biggest challenges to young people entering the industry today appears to be in the area of using appropriate language. The use of inappropriate or abusive language in the business world can seriously restrict the success of the project. Associated with this problem is the concern for forming appropriate business attitudes. To illustrate this concern, we need to look at the issue from the client's viewpoint.

Let's take, for example, the family that is having a new house built in a community. Excluding a business venture, this will be the largest investment they make in their lifetime. They are envisioning a new abode where they will be rearing their children, sharing family experiences, and possibly spending their retirement years. They take it seriously. Beginning with the planning phase, their expectation level is high.

Now enter the builders and workers representing the various trades. These representatives from the construction industry have the responsibility of being professional. They should have the trust of the customer as part of the construction package. An agreement has been reached. The builder and his or her team are obligated to provide the best quality for the price being paid. This must be communicated.

The builder must demonstrate professional attitudes and use professional language. If the builder or construction representative uses abusive or profane language on the job site, it sends a negative message to the client. The message will say, "This project is not important; it's just another routine job." That, in turn, will probably tell the client that the quality is less than what should be expected.

A worker with the proper attitude and command of the language will not spit chewing tobacco on the subfloor or present himself or herself vocally as crude or uncaring. We must remember that clients will advertise for the builder whether the builder likes it or not. They will share with others how they perceived the builder's professionalism and work ethics in addition to his or her quality of work. Not all communication is conducted orally or in writing. We also communicate who we are by our performance. People read each other continually.

We must ask ourselves, "How can I improve my communication skills?" Each person needs to take a personal inventory to see how well equipped he or she is. Some questions and suggestions follow that may help students determine their own strengths or weaknesses in personal communication skills.

- Can I list, in writing, my career goals?
- How much do these goals involve oral and written communications?
- What is my current skill level in communicating?
- Do I listen, write, and speak well? Is my vocabulary acceptable?
- What does my past communication performance in school indicate?
- Do I understand what personal communication skills will be demanded of me in a construction career?
- Do I understand how communication

skills will describe my professional image, either positively or negatively?

- Do I know someone I can talk to that knows what communication skills will help me reach my goals?

- Do I know how these skills can be obtained?

- Once I understand what good communication skills will be required of me to be successful, will I desire to obtain those skills?

An important recommendation to any student pursuing a career in construction is to talk to knowledgeable people who can provide accurate information about the necessary communication requirements for the desired job. Trade representatives, counselors, parents, teachers, and the CEOs of construction businesses will be the people with that knowledge.

Someone once said, "It is not so much what we know as how well we use what we know." A personal vocabulary does not have to equal the volume of a dictionary. But words must be selected carefully. A good policy may be one that says, "If you wouldn't write it and sign it, don't say it."

In conclusion, the element of effective communication must never be diminished or disregarded. It is a vital part of good business ethics. It has become even more important as we strive for success in the workplace.

FOCUS QUESTIONS: For individual or group discussion.

1. Address each question presented in the preceding study.

2. What would be the consequences of not developing good communication skills as I pursue a career in construction?

Stair Framing

A great degree of skill is necessary for the design and construction of staircases. The carpenter must know how to lay out and build well-designed staircases with first-class workmanship. Staircases must be comfortable to use and safe. This unit deals with stair framing. A later unit describes the application of stair finish.

OBJECTIVES

After completing this unit, the student will be able to:

■ describe several stairway designs.

■ define terms used in stair framing.

■ determine the rise and tread run of a stairway.

■ determine the length of and frame a stairwell.

■ lay out a stair carriage and frame a straight stairway.

■ lay out and frame a stairway with a landing.

■ lay out and frame a stairway with winders.

■ lay out and frame basement stairs.

CONTENTS

In addition to the staircase, consideration must be given to the layout and framing of the *stairwell*. The stairwell is the opening in the floor through which a person must pass when climbing and going down the stairs (Fig. 47-1). The stairwell is framed at the same time as the floor.

Stairway Design

Stairs, stairway, and *staircase* are terms used to designate one or more flights of steps leading from one level of a structure to another. Stairs are further defined as *finish* or *service* stairs. Finish stairs extend from one habitable level of a house to another. Service stairs extend from a habitable to a nonhabitable level. Stairways in residential construction should be at least 36 inches wide—preferably even wider. This allows the passage of two persons at a time and for the moving of furniture (Fig. 47-2).

Types of Stairways

A *straight* stairway is continuous from one floor to another. There are no turns or landings. *Platform* stairs have intermediate landings between floors. Platform-type stairs sometimes change direction at the landing. An L-type platform stairway changes direction 90 degrees. A U-type platform stairway changes direction 180 degrees.

Platform stairs are installed in buildings in which there is a higher floor to floor level. They also provide a temporary resting place. They are a safety feature in case of falls. The landing is usually constructed at about the middle of the staircase.

A *winding* staircase gradually changes direction as it ascends from one floor to another. In many cases, only a part of the staircase winds. Winding stairs' may solve the problem of a shorter straight horizontal run. However, their use is not recommended. They are relatively difficult to construct. They pose a danger because of their tapered treads, which are very narrow on one end (Fig. 47-3).

Stairways constructed between walls are called *closed* stairways. Closed stairways are more economical to build. However, they add little charm or beauty to a building. Stairways that have one or both sides open to a room are called *open* stairways. One side of the staircase may be closed while the other side is open for all or part of the flight. Usually, a staircase is the outstanding feature of an entrance. It provides beauty and grace to a room. A staircase generally affects the character of the entire interior.

THE STAIRWELL IS THE OPENING IN A FLOOR FOR A STAIRWAY.

STAIR CARRIAGES ARE THE FRAMING MEMBERS OF A STAIRCASE.

Fig. 47-1 Frame for stairs and stairwell.

STAIR WIDTH-3'-0" MINIMUM
3'-2" TO 3'-4" PREFERRED

Fig. 47-2 Recommended stair widths.

Stair Framing Terms

The terms used in stair framing are defined in the following material. Figure 47-4 illustrates the relationship of the various terms to each other and to the total staircase.

- The *total rise* of a stairway is the vertical distance between finish floors.

- The *total run* is the total horizontal distance that the stairway covers.

- A *rise* is the vertical distance from one step to another. A *riser* is the finish material that covers this distance.

- The *tread run* is the horizontal distance between the faces of the risers. A *tread* is the horizontal member on which the feet are placed when climbing or descending the stairs. The *nosing* is that part of the tread that extends beyond the face of the riser.

- A *stair carriage* is sometimes called a *stair horse* or *rough stringer*. It is usually a nominal 2×10 or 2×12 framing member that supports the treads and risers.

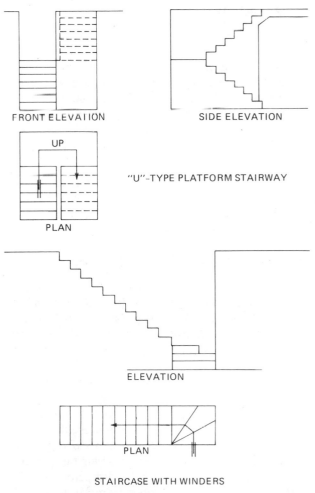

Fig. 47-3 Various types of stairways.

Fig. 47-4 Stair framing terms.

■ A *stairwell* is an opening in the floor for the stairway to pass through. It provides adequate headroom for persons using the stairs.

Methods of Stair Construction

There are two principal methods of stair construction. The *housed finished stringer* is usually fabricated in a shop. Framing stairs on the job usually requires using *stair carriages.*

The Housed Stringer Staircase

For the housed stringer staircase, the framing crew frames only the stairwell. The staircase is installed when the house is ready for finishing. Dadoes are routed into the sides of the finish stringer. They *house* (make a place for) and support the risers and treads (Fig. 47-5).

Occasionally, the finish carpenter builds a housed stringer staircase on the job site. A router and jig are used to dado the stringers. Then, the treads, risers, and other stair parts are cut to size. The staircase is then assembled. Stair carriages are not required when the housed finish stringer method of construction is used. (Building housed finished stringer stairs is described in more detail in a later unit on stair finish.)

HOUSED FINISHED STRINGER METHOD

Fig. 47-5 A housed finish stringer method of stair construction.

The Job-Built Staircase

The *job-built* staircase uses stair carriages that are installed when the structure is being framed. The carriage is laid out for cutouts. Risers and treads are fastened to the cutouts (Fig. 47-6). Temporary rough treads are installed for easy access to upper levels until the stairs are ready for the finish treads and other stair trim. The layout for both housed stringers and stair carriages is made in a similar manner.

TREAD

RISER

STAIR CARRIAGE

FINISH STRINGER

STAIR CARRIAGE METHOD

Fig. 47-6 Job-built stairs are usually framed using stair carriages.

Determining the Rise and Tread Run

Staircases must be constructed at a proper angle for maximum ease in climbing and for safe descent. The relationship of the rise and run determines this angle (Fig. 47-7). The preferred angle is between 30 and 35 degrees. Stairs with a slope of less than 20 degrees waste a lot of valuable space. Stairs with a slope that is excessively steep (50 degrees or over) are difficult to climb and dangerous to descend.

Some building codes specify that the height of a riser shall not exceed 8 1/4 inches and that the width of a tread not be less than 9 inches. However, a rise of between 7 and 7 1/2 inches and a tread run of between 10 and 12 inches is recommended. The rise should be determined first.

Determining the Riser Height

To determine the individual rise, the total rise of the stairway must be known. Divide the total rise by 8 inches. (This measure is more than recommended for each rise.) The result is a usually a whole number and a fraction. The next largest whole number may be the total number of risers required.

For example, a total rise of 9'-0" (108 inches) is divided by 8 inches. The result is 13 1/2. The next largest whole number is 14. This is the number of risers that may be desired.

Dividing the total rise by the number of risers gives the height of each rise. For example, dividing the total rise (108 inches) by the number of risers (14) gives a unit rise of 7 5/8 inches.

To decrease the riser height, increase the number of risers by one. Divide the total rise by that number. For example, increasing 14 by one and dividing 108 inches by 15 equals slightly over 7 3/16 inches.

Another method of determining the riser height does not involve mathematics. Stand a *story rod* (usually a 1×2 strip of lumber) vertically, with one end on the floor. Mark on it the finish floor to finish floor height (total rise of the stairs). Set the dividers at a comfortable rise. Divide the rod into equal spaces, adjusting the dividers slightly as necessary (Fig. 47-8).

Determining the Tread Run

The tread run (width) is measured from the face of one riser to the next. It does not include the nosing (Fig. 47-9).

To find the tread run, apply the following rule:

■ The sum of one rise and one tread should equal between 17 and 18. For example, if the riser height is 7 3/8 inches, then the minimum tread width may be 17 inches minus 7 3/8 inches. This equals 9 5/8 inches. The maximum tread width may be 18 inches minus 7 3/8 inches. This equals 10 5/8 inches.

LADDERS

PREFERRED 75°

STAIRS

CRITICAL ANGLE 50°

PREFERRED – 30° TO 35°

MIN. ANGLE 20°

CRITICAL ANGLE 15°

RAMPS

Fig. 47-7 Recommended angles for stairs.

9"

8-1/4"

42°

10"

7-1/2"

35°

THESE STAIRS HAVE THE MAXIMUM RISE AND MINIMUM TREAD RUN ALLOWED BY CODES. THE ANGLE AT WHICH THEY RUN IS STEEPER THAN PREFERRED.

STAIRS WITH THE RISE AND TREAD RUN SHOWN ABOVE RUN AT A PREFERRED ANGLE FOR SAFER AND MORE COMFORTABLE USE.

STEP BY STEP
P R O C E D U R E S

Fig. 47-8 The riser can be laid out on a story pole using dividers.

Fig. 47-9 The tread run does not include the nosing.

HELPFUL HINTS

Formulas for determining the tread run.

7-1/2" + 10" = 17-1/2"

7-1/2"

10"

FORMULA #1 - ONE RISE PLUS ONE TREAD RUN SHOULD EQUAL BETWEEN 17 AND 18 INCHES.

7-1/2" + 7-1/2" + 9-1/2" = 24-1/2"

RISE

7-1/2"

TREAD RUN

9-1/2"

FORMULA #2 - THE SUM OF TWO RISES AND ONE TREAD RUN SHOULD EQUAL BETWEEN 24 & 25 INCHES.

NOTE: MOST BUILDING CODES SPECIFY A MAXIMUM RISE OF 8-1/4 INCHES AND A MINIMUM TREAD RUN OF 9 INCHES.

Fig. 47-10

Another formula to find the tread width is found in many building codes. It states that the sum of two risers and one tread shall not be less than 24 inches nor more than 25 inches (Fig. 47-10). With this formula, a rise of 7 3/8 inches calls for a minimum tread width of 9 1/4 inches and maximum of 10 1/4 inches.

Decreasing the rise increases the run of the stairs. This uses up more space. Increasing the riser height decreases the run. This makes the stairs steeper and more difficult to climb. The carpenter must use good judgment and adapt the rise and run dimensions to the space in which the stairway is to be constructed in conformance with the building code. A riser height of 7 1/2 inches and a tread run of 10 inches makes a safe, comfortable stairway.

Framing a Stairwell

A stairwell is framed in the same manner as any large floor opening. Unit 11, Floor Framing, describes methods of framing floor openings. Several methods of framing stairwells are illustrated in Figure 47-11.

Determining the Size of the Stairwell

Stairwell Width. The width of the stairwell depends on the width of the staircase. The drawings show the finish width of the staircase. However, the stairwell must be made wider than the staircase to allow for wall and stair finish (Fig. 47-12). Extra width will be required for a handrail and other finish parts of an open staircase that makes a U-turn on the landing above. The carpenter must be able to determine the width of the stairwell by studying the size, type, and placement of the stair finish before framing the stairs.

Length of the Stairwell. The length of the stairwell depends on the slope of the stairway. Stairs with a low angle require a longer stairwell to provide adequate *headroom*. Headroom is the vertical distance measured from the outside corner of a tread and riser to the ceiling above (Fig. 47-13). Most building codes require a minimum of 6'-8" for headroom. However, 7'-0" headroom or even more is preferred.

To find the length of the stairwell, add the thickness of the floor assembly above (including subfloor, floor joists, furring, and ceiling finish) to the desired headroom. Divide by the riser height. The result will probably be a whole number and a fraction. Use the next largest whole number. Multiply by the tread width to find the length of the stairwell.

Fig. 47-11 Methods of framing stairwells.

For example, a stairway has a riser height of 7 1/2 inches and a tread width of 10 inches. The total thickness of the floor above is 11 3/4 inches. When added to the desired headroom of 84 inches, this equals 95 3/4 inches. Dividing this total by 7 1/2 equals 12.77. Changing 12.77 to 13 and multiplying by 10 inches equals 130 inches or 10'-10", the length of the stairwell (Fig. 47-14).

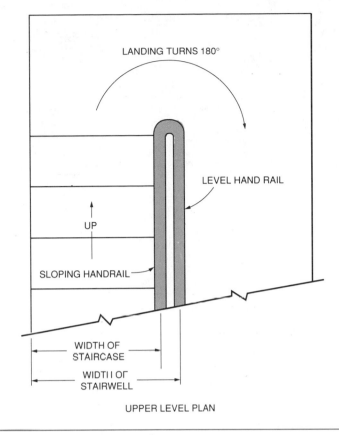

UPPER LEVEL PLAN

Fig. 47-12 The stairwell must be made wider than the staircase. This allows for wall and stair finish.

Fig. 47-13 Low-angle stairs require longer stairwells to provide adequate headroom.

This length is correct if the header of the stairwell acts as the top riser. If the carriage is framed flush with the top of the opening, add another tread width to the length of the stairwell. A longer stairwell will provide more headroom. Additional headroom can be obtained by framing the header above the low end of the staircase at the same angle as the stairs (Fig. 47-15).

IF THE STAIR CARRIAGE IS FRAMED FLUSH WITH THE FLOOR ABOVE, ADD ANOTHER TREAD RUN TO THE LENGTH OF THE STAIRWELL.

LENGTH OF STAIRWELL

HEADROOM

THICKNESS OF FLOOR

DESIRED HEADROOM

1. DIVIDE SUM OF FLOOR THICKNESS & DESIRED HEADROOM BY THE RISER HEIGHT. CHANGE ANSWER TO NEXT WHOLE NUMBER.

2. MULTIPLY ANSWER FOUND IN 1 BY THE TREAD RUN TO FIND LENGTH OF STAIRWELL.

Fig. 47-14 How to calculate the length of a stairwell.

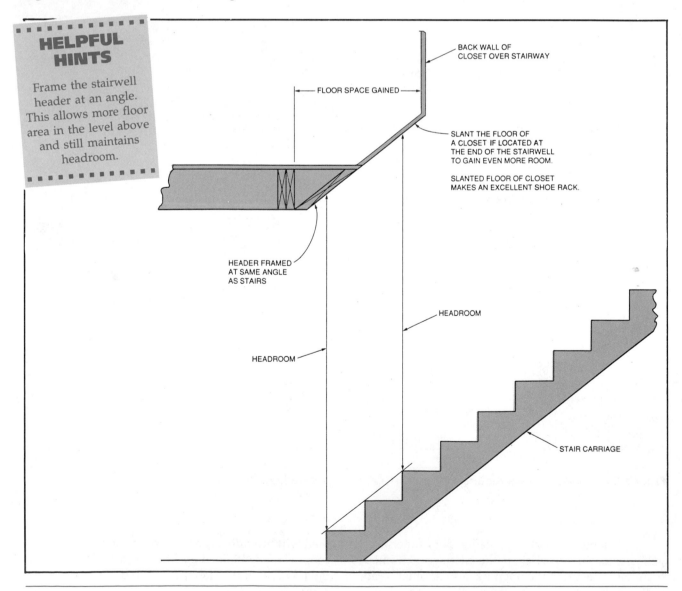

HELPFUL HINTS

Frame the stairwell header at an angle. This allows more floor area in the level above and still maintains headroom.

BACK WALL OF CLOSET OVER STAIRWAY

FLOOR SPACE GAINED

SLANT THE FLOOR OF A CLOSET IF LOCATED AT THE END OF THE STAIRWELL TO GAIN EVEN MORE ROOM.

SLANTED FLOOR OF CLOSET MAKES AN EXCELLENT SHOE RACK.

HEADER FRAMED AT SAME ANGLE AS STAIRS

HEADROOM

HEADROOM

STAIR CARRIAGE

Fig. 47-15

CHAPTER 48 STAIR LAYOUT AND CONSTRUCTION

All stairs are laid out in approximately the same way. The use, location, and cost of stairs determine the way they are built. Regardless of the kind of stairs, where they are located, or how much they cost, care should be taken in their layout and construction.

Laying Out a Stair Carriage

When laying out stair carriages, make sure that every riser height will be the same. Also make sure all tread widths will be equal when the staircase is finished. It is very dangerous to descend a flight of stairs and unexpectedly step down a longer or shorter distance than the previous step. It is also dangerous for a person accustomed to stepping on treads of a certain width to suddenly step on a narrower or wider one when descending stairs. Stairs that are not layed out and constructed properly could cause a fatal accident.

Determining the Rough Length of a Stair Carriage

The length of lumber needed for a stair carriage can be determined by scaling, across the framing square. Use the edge of the square that is graduated in 12ths of an inch. Mark the total rise on the tongue. Then mark the total run on the blade. Scale off in between the marks (Fig. 48-1).

For example, a stairway has a total rise of 8'-9" and a total run of 12'-3". Scaling across the square between 8–9/12ths and 12–3/12ths at a scale of 1" = 1'-0" reads a little over 15. A 16-foot length of lumber is needed for the stair carriage.

Fig. 48-1 Scale across the framing square to find the rough length of a stair carriage.

Stepping-Off the Stair Carriage

Place the stair carriage stock on a pair of sawhorses. Sight the stock for a crowned edge. Stand on the side of the crowned edge when laying out. This will be the top edge of the carriage. Set gauges on the framing square with the rise on the tongue and the tread run on the blade. Or, use a pitch block. Lines laid out along the tongue will be plumb lines. Those laid out along the blade will be level lines when the stair carriage is in its final position. Step-off the necessary number of times, marking along the outside of the tongue and blade. These lines are the back sides of the finish risers and undersides of the finish treads (Fig. 48-2). Lay out a level line at the top of the carriage for the tread. Lay out a level line on the bottom where the carriage sits on the floor.

Equalizing the Bottom Riser

A certain amount may have to be cut off the bottom end of the stair carriage. This will make the bottom riser equal in height to all the other risers when the staircase is finished. The thickness of the finish floor and the finish stair treads must be known.

■ If the tread stock is thicker than the finish floor, and the carriage rests on the subfloor, then the first riser is made less by the difference.

■ If the carriage rests on the finish floor, the first riser is made less by the thickness of the tread stock.

■ If the bottom of the carriage rests on the subfloor, and the finish floor and tread stock are the same thickness, the height of the first rise is the same as all the rest. Nothing more is cut off the bottom end of the stair carriage.

Lay out a level line where the carriage will rest on the floor, according to the conditions. The first riser should be the same as all the rest when the staircase is complete. Cutting the bottom end off to equalize the bottom riser is called *dropping* the stair carriage (Fig. 48-3).

Equalizing the Top Riser

No amount is cut from the level line at the top of the carriage. The top riser is equalized by fastening the carriage a certain distance below the subfloor of

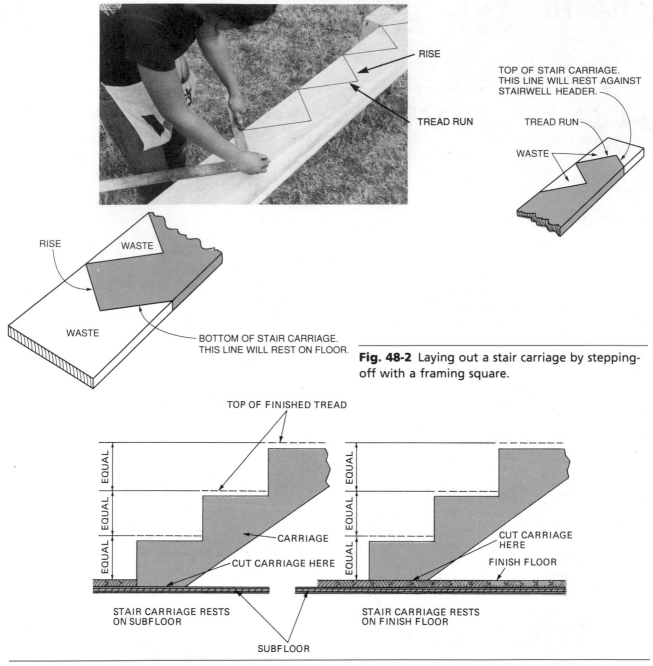

Fig. 48-2 Laying out a stair carriage by stepping-off with a framing square.

Fig. 48-3 Dropping the stair carriage to equalize the first riser.

the top landing. This distance depends on the thickness of the stair tread and the finish floor above.

■ With the header of the stairwell acting as the top riser, if the stair tread thickness is the same as the finish floor, the carriage is fastened a riser height below the subfloor.

■ If the stair tread is thicker than the finish floor, fasten the carriage a distance of a riser, plus the difference, below the subfloor of the top landing.

For example, if the tread stock is 1 1/16 inches and the finish floor above is 3/4 inch, the difference

is 5/16 inch more than the riser height below the top of the subfloor above (Fig. 48-4). Whatever conditions may be encountered, take steps to equalize the top and bottom risers for the construction of a safe staircase.

Width of the Top Tread

The width of the top finished tread must be the same as all others in the staircase. If the carriage is framed against the stairwell *header,* and the same thickness of finish is applied to the header as the thickness of the riser stock, then the tread cut at the

Fig. 48-4 Equalizing the top riser.

top of the carriage is made the same as all others. Measure the tread run from the top riser. Lay out a plumb line.

   If the carriage is framed to the stairwell in a different manner, allowances must be made so that the top finished tread width will be the same as all others in the staircase (Fig. 48-5).

Cutting the Stair Carriages

After the first carriage is laid out, cut it. Use a skilsaw. Follow the layout lines carefully because this will be a pattern for others.

Fig. 48-5 Methods of framing the stair carriage to the stairwell header.

CAUTION When making a cut at a sharp angle to the edge, the guard of the saw may not retract. Retract the guard by hand until the cut is made a few inches into the stock. Then release the guard and continue the cut. Never wedge the guard in an open position.

Finish the cuts at the intersection of the riser and tread run with a handsaw. Use the first carriage as a pattern. Lay out and cut as many other carriages as needed. Usually, three carriages are used for residential staircases of average width. For wider stairs, the number of carriages depends on such factors as whether or not risers are used and the thickness of the tread stock. Check the drawings or building code for the spacing of carriages for wider staircases.

Built-Up Stair Carriage

In order to conserve materials, the triangular blocks cut out of the stair carriage may be fastened to the edge of a fairly clear, straight-grained 2×4 to form a *built-up stair carriage* (Fig. 48-6). The blocks are fastened by using the first stair carriage as a pattern. Built-up stair carriages are usually used as intermediate carriages in the stairway.

Framing a Straight Stairway

If the stairway is either completely closed or closed on one side, the walls must be prepared before the stair carriages are fastened in position.

Preparing the Walls of the Staircase

If a wall is to be plastered, a *ground* must be installed as a guide for the thickness of the plaster. To find the location of the ground, lay the stair carriage

against the wall in its proper position. Check several tread runs for level. If they are not level, determine the reason and make the necessary adjustments.

Mark the studs in the wall near the top and bottom of the carriage about one inch above the outside corner of the tread and riser cut. Remove the carriage. Snap a chalk line between the marks. Fasten a nominal 1×10 board so its top edge is to the line. Use three 8d common nails into each stud and also into the bottom plate of the wall. Apply backing for lath between the studs and above the plaster ground (Fig. 48-7).

Gypsum board (drywall) is sometimes applied to walls before the stair carriages are installed against them. This procedure saves time. It eliminates the need to cut the drywall around the cutouts of the stair carriage. This method also requires no blocking between the studs to fasten the ends of the drywall panels. However, the time saved may be offset by some disadvantages.

Blocking between the studs in back of the stair carriage provides backing for fastening the stair trim. Lack of it may cause difficulty for those who apply the finish. Also, any repair or remodeling work will be more difficult if gypsum board is applied before the stair carriages. In addition, drywall is usually not applied until after the framing is complete. Workers are deprived of the stair frame for access to upper levels until that time.

If the drywall is to be applied after the stairs are framed, blocking is required between studs in back of the stair carriage to fasten the ends of the gypsum board. Snap a line in the same manner as for plaster grounds. Install 2×6 or 2×8 blocking, on edge, between and flush with the edges of the

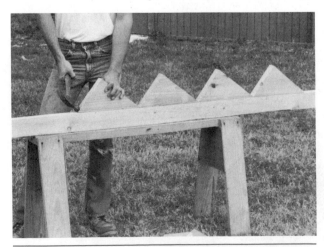

Fig. 48-6 Making a built-up stair carriage.

1″ X 10″ PLASTER GROUND. THE TOP EDGE OF THE GROUND MUST BE BELOW THE TOP EDGE OF THE FINISH STRINGER.

STUD

CARRIAGE

BLOCKING FOR FASTENING END OF LATH

PARTITION SOLEPLATE

SUBFLOOR

Fig. 48-7 Preparing the wall of a staircase for plastering by applying ground and backing for a plaster base.

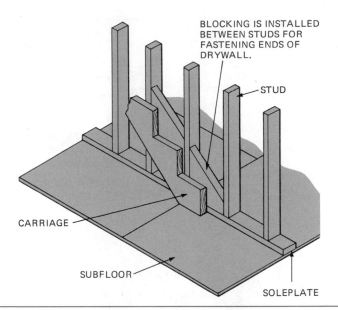

Fig. 48-8 Preparation of the wall for application of gypsum board before stair carriages are installed.

studs. Their top edge should be to the chalk line (Fig. 48-8).

Housed staircases may be installed after the walls are finished. Sometimes they are installed before the walls are finished. Then, they must be furred out away from the studs. This allows the wall finish to extend below the top edge of the finished stringer.

Installing the Stair Carriage

When installing the stair carriage, fasten the first carriage in position on one side of the stairway. Attach it at the top to the stairwell header. Make sure the distance from the subfloor above to the top tread run is correct.

Fasten the bottom end of the stair carriage to the soleplate of the wall and with intermediate fastenings into the studs. Drive spikes near the bottom edge of the carriage. This prevents splitting the triangular sections.

Fasten a second carriage on the other wall in the same manner as the first. If the stairway is to be open on one side, fasten the carriage at the top and bottom of the staircase. The location of the stair carriage on the open end of a stairway is in relation to the position of the *handrail*. First, determine the location of the centerline of the handrail. Then, position the stair carriage on the open side of a staircase. Make sure its outside face will be in a line plumb with the centerline of the handrail when it is installed (Fig. 48-9).

Fasten intermediate carriages at the top into the stairwell header and at the bottom into the subfloor. Test the tread run and riser cuts with a straightedge placed across the outside carriages (Fig. 48-10). About halfway up the flight, or where necessary, fasten a temporary riser board. This straightens and maintains the spacing of the carriages (Fig. 48-11).

If a wall is to be framed under the stair carriage at the open side, fasten a bottom plate to the subfloor plumb with the outside face of the carriage. Lay out the studs on the plate. Cut and install studs under the carriage in a manner similar to that used to install gable studs. Be careful to keep the carriage straight. Do not crown it up in the center (Fig. 48-12). Install rough lumber treads on the carriages until the stairway is ready for the finish treads.

Laying Out and Framing a Stairway with a Landing

A stair landing is an intermediate platform between two flights of stairs. The landing is usually floored with the same materials as the main floors of the structure. Some codes require that the minimum length of a landing be not less than 2'-6". Other codes require the minimum dimension to be the width of the stairway. L-type stairs may have the landing near the bottom or the top. U-type stairs usually have the landing about midway on the flight. Many codes state that no flight of stairs shall have a vertical rise of more than 12 feet. Therefore, any staircase running between floors with a vertical distance of more than 12 feet must have at least one intermediate landing or platform.

Platform stairs are built by first erecting the platform. The finish floor of the platform should be the same height as if it were a finish tread in the

HELPFUL HINTS

How to position a stair carriage on the open side of a staircase.

CENTERLINE OF THE FUTURE HANDRAIL

FUTURE HANDRAIL

OUTSIDE FACE OF THE OPEN STAIR CARRIAGE

POSITION THE OUTER FACE OF THE OUTSIDE STAIR CARRIAGE SO IT IS PLUMB WITH THE CENTERLINE OF THE HANDRAIL WHEN INSTALLED

Fig. 48-9

STRAIGHTEDGE AGAINST RISER CUT

STRAIGHTEDGE AGAINST TREAD CUT

STAIR CARRIAGE

INTERMEDIATE STAIR CARRIAGE

Fig. 48-10 Check the position of tread and riser cuts on intermediate carriages with a straightedge.

Fig. 48-11 Fasten a temporary riser about halfway up the flight. This straightens and maintains the carriage spacing.

staircase. This allows an equal riser height for the total flight (Fig. 48-13). The stairs are then framed to the platform as if two straight flights were being framed. Either the stair carriage or the housed

Fig. 48-13 The platform is located so that its top is in line with one of the tread runs.

stringer method of construction may be used (Fig. 48-14).

Laying Out Winding Stairs

Winders are not recommended for safety reasons. However, some codes state that they may be used

Fig. 48-12 Install studs under the stair carriage on the open side of the stairway.

Fig. 48-14 Stair carriages are framed to the platform in a way similar to that used for two straight flights.

in individual dwelling units, if the required tread width is provided along an arc. This is called the *line of travel*. It is a certain distance from the side of the stairway where the treads are narrower. The same code also states that in no case shall any width of tread be less than 6 inches at any point (Fig. 48-15). Some other codes may permit a narrower tread. Check the building code for the area in which the work is being performed.

To lay out a winding turn of 90 degrees, draw a full-size winder layout on the floor directly below where the winders are to be installed. On a closed staircase, the walls at the floor line represent its sides. For stairs open on one side, lay out lines on the floor representing the outside of the staircase. The wall represents the inside. Swing an arc, showing the line of travel. Use the corner of the wall or layout lines as center. The radius of the arc may be 12 to 18 inches, as the codes permit. Fifteen inches is usually used.

From the same center, lay out the width of the narrow end of the treads in both directions. Square lines from the end points to the opposite side of the staircase. Divide the arc into equal parts. Project lines from the narrow end of the tread, through the intersections at the arc, to the wide end at the wall. These lines represent the faces of the risers. Draw

lines parallel to these to indicate the riser thickness. Plumb these lines up the wall to intersect with the tread run for each winder (Fig. 48-16).

The cuts on the stair carriage for the winding steps are obtained from the full-size layout. Lay out and cut the carriage. Fasten it to the wall. Install rough treads until the stairs are ready for finishing.

If one side of the staircase is to be open, a *newel post* is installed. Then, the risers are mitered to or *mortised* into the post (Fig. 48-17). A mortise is a rectangular cavity in which the riser is inserted. Newel posts are part of the stair finish. They are described in more detail in a following chapter.

Framing Basement Stairs

Basement stairs are sometimes built without riser boards. Two carriages are used. Nominal 2×10 treads are cut in between them. The carriages are not cut out, like those previously described. They may be dadoed to receive the treads. An alternative method is to fasten *cleats* to the carriages to support the treads (Fig. 48-18).

Lay out the carriages in the usual manner. Cut the bottoms, on a level line, to fit the floor. Cut the tops, along a plumb line, to fit against the header of the stairwell. "Drop" the carriages as necessary to

Fig. 48-15 Specifications for winding stairs.

STEP BY STEP
P R O C E D U R E S

Fig. 48-16 Draw a full-size layout of a set of winders on the floor to find the rise and tread run of the stair carriage.

Fig. 48-17 On the open side, winder risers are mitered against or mortised into a newel post.

provide a starting riser with a height equal to the rest of the risers.

Dadoed Carriages

If the treads are to be dadoed into the carriages, lay out the thickness of the tread on the stringer below the layout line. The top of the tread is to the original layout line. Mark the depth of the dado on both edges of the carriage. Set the skilsaw to cut the depth of the dado. Make cuts along the layout lines for the top and bottom of each tread. Then, make a series of saw cuts between those just made.

Chisel from both edges toward the center. Remove the excess to make the dado. Spike all treads into the dadoes. Assemble the staircase. Fasten the assembled staircase in position. Locate the top tread at a height to obtain an equal riser at the top. The lower end of a basement stairway is sometimes anchored by installing a kicker plate, which is fastened to the floor (Fig. 48-19).

Fig. 48-18 Basement and service stair carriages may be dadoed or cleated to support the treads.

Fig. 48-19 A kicker plate may be installed to fasten the bottom end of basement stairs.

Cleated Carriages

If the treads are to be supported by cleats, measure down from the top of each tread a distance equal to its thickness. Draw another level line. Fasten 1×3 cleats to the carriages. Make sure their top edges are to the bottom line. Fasten the carriages in position. Cut the treads to length. Install the treads between the carriages so the treads rest on the cleats. Fasten the treads by spiking the stair carriage into their ends.

Review Questions

Select the most appropriate answer.

1. Stairways in residential construction should have a minimum width of
a. 30 inches.   c. 36 inches.
b. 32 inches.   d. 40 inches.

2. A nosing is
a. a horizontal member of a step.
b. a vertical member of a step.
c. the top molded edge of a finish stringer.
d. the overhang of the tread beyond the riser.

3. The maximum riser height stated in this unit is
a. 8 1/4 inches.   c. 7 1/2 inches.
b. 7 3/4 inches.   d. 7 1/4 inches.

4. The riser height for a stairway with a total vertical rise of 8'-9" is
a. 8 inches.   c. 7 5/8 inches.
b. 7 3/4 inches.   d. 7 1/2 inches.

5. Applying the rules stated in this unit, what is the tread run if the riser height in a stairway is 7 1/4 inches?
a. 8 3/4 to 9 3/4 inches
b. 9 to 10 inches
c. 9 3/4 to 10 3/4 inches
d. 10 to 11 inches

6. Many building codes specify a minimum tread width of
a. 8 1/2 inches.   c. 9 1/2 inches.
b. 9 inches.   d. 10 inches.

7. A flight of stairs has a riser height of 7 1/2 inches and a tread run of 10 inches. The total thickness of the upper floor is 10 inches. What is the length of the stairwell if the stairwell header acts as the top riser and the desired minimum headroom is 6'-8"?
a. 9'-0".   c. 10'-0".
b. 9'-6".   d. 10'-2".

8. Most building codes specify a minimum headroom clearance of

a. 6'-6".   c. 7'-0".

b. 6'-8".   d. 7'-6".

9. The stair carriage of a stairway with a riser height of 7 1/2 inches rests on the finish floor. What is the riser height of the first step if the tread thickness is 3/4 inch?

a. 6 3/4 inches.   c. 7 3/4 inches.

b. 7 1/2 inches.   d. 8 1/4 inches.

10. A stairway has a riser height of 7 1/2 inches. The tread stock thickness is 1 1/16 inches. The finish floor thickness of the upper floor is 3/4 inch. What distance down from the top of the upper subfloor must the rough carriage be fastened if the stairwell header acts as the top riser?

a. 7 3/16 inches.   c. 7 1/2 inches.

b. 7 3/8 inches.   d. 7 3/16 inches.

■ BUILDING FOR SUCCESS ■

Construction Associations and the College Student

How can a professional organization benefit college students who desire careers in construction?

This question is asked annually by many colleges and universities across the nation. The following true account is presented as a possible answer to this question. Although it is only one community college's response to involvement beyond the classroom, it may motivate other colleges as well.

In 1988 Larry Kness, the program chair for the building construction program at Southeast Community College, Milford, Nebraska, was looking for ways to provide more career information to students. Classrooms, textbooks, and labs were providing the essential opportunities and experiences for students in the construction program. The educational experiences were geared to prepare competent, skilled technicians to work in residential and light commercial construction. The program had proven to be exceptional in its capacity to send well-trained carpenters, masons, drafters, and estimators into the construction world. However, there was a need to make students more aware of the business aspect of construction.

General education courses were providing the basics in business, communications, accounting, finance, and math, but direct exposure to builders was lacking. The need for more student exposure to the local, state, and national construction world loomed as the missing link. After the need was identified,

the program chair responded to an invitation from the Lincoln Home Builders Association (HEAL) to form a National Association of Home Builders (NAHB) student chapter on the Milford campus. HBAL offered to become the sponsors of the college student chapter. Through the combined efforts of builders, educators, and students, a new relationship was formed to promote quality construction education.

The idea was presented to the building construction students, who overwhelmingly accepted the proposal. Chuck Brazie, executive vice president of HBAL, initially proposed the formation of the chapter. Nadine Condello, the new CEO of HBAL, became the coordinator. She proved to be the catalyst who finalized the chapter formation at the national level. Condello demonstrated the necessary vision needed to make the student outreach a success.

Many local association members had been graduates of SCC's construction program, were employers of graduates, or were continuing supporters of the program. This membership and the board of directors were instrumental in the entire effort to create a local student chapter.

This new educational outreach opened the door to more student involvement in the construction industry. New opportunities to visit construction sites, listen to guest speakers, attend local HBAL meetings, and participate in student chapter competitions at the national level were introduced or expanded.

Students were provided an opportunity to look inside a national organization that has a goal of promoting quality construction businesses nationally. Avenues for meeting local, state, and national construction representatives became more readily available. The student chapter was introduced to new challenges and rewards through student involvement.

Participation in local activities with HBAL provided a means of generating finances to fund trips to the national NAHB convention. (This convention participation will be presented in the case study centering on the local HBAL activities.)

As we studied the opportunities available for chapter involvement at the national level, we saw dozens of other two-year and four-year colleges and universities actively participating with the NAHB. We saw what other schools were doing to supplement their construction programs. We were excited to belong to an organization that included many highly respected construction programs from colleges and universities such as Texas A&M, Purdue, Penn State, BYU, Snow College (Utah), Hinds Community College (Mississippi), and Ohio State. These schools represented a vast array of institutions that radiated quality. They became examples of excellence as we pursued more involvement.

Annually the NAHB student chapter community participates in national competitions such as construction marathons, essays on affordable housing, and "Outstanding Student Chapter." Through open competition, any chapter may compete with others within their classification (two-year or four-year schools) from across the country for national honors. The question we had to answer was, Can we realistically compete with the proven leaders across the nation?

After looking at the competition process, our student chapter and sponsoring association felt we should enter the "race" and learn how to compete at the national level. Our application for consideration as "Outstanding Student Chapter" was evaluated as a worth-

while competitor. We were notified that we had placed as one of the three finalists. We were invited to attend the national convention in Atlanta for the announcement of the winners. Our chapter set a goal to attend the national convention held in February 1991. At that convention we were personally awarded the honor of Outstanding Student Chapter for 1990. Second place was awarded to Texas A&M. Third place went to Purdue University. This was the first time that a two-year construction program had won the top honor.

What began as an experimentation to assist students in becoming better informed about their chosen industry has become an ongoing educational experience for our student chapter members. The potential financial awards, competitions, and attendance at national conventions are just a few of the benefits for students belonging to a college student chapter. The challenge for students to become more knowledgeable about the construction industry is an important part of belonging to a student chapter. This example of Southeast Community College and the NAHB is offered as proof to all that many opportunities exist for colleges to provide meaningful education beyond the classroom by way of a construction-related student chapter. ·

FOCUS QUESTIONS: For individual or group discussion.

1. Considering your own educational institution, what might be some benefits of establishing a construction-related student chapter?

2. How could you initiate a charter chapter? Whom could you contact?

3. Do you feel that there would be institutional, faculty, industry, and student support to maintain an ongoing chapter?

Insulation and Ventilation

Thermal insulation prevents the loss of heat in buildings during cold seasons. It also resists the passage of heat into air-conditioned buildings in hot seasons. *Acoustical* insulation reduces the passage of sound from one area to another. Adequate *ventilation* must be provided in a building. This evaporates moisture that may be formed in the space between thermal insulation and the cold surface.

OBJECTIVES

After completing this unit, the student will be able to:

- describe how insulation works and define insulating terms and requirements.
- describe the commonly used insulating materials and state where insulation is placed.
- properly install various kinds of insulation.
- state the purpose of and install vapor barriers.
- describe various methods of construction to reduce the transmission of sound.
- explain the need for ventilating a structure, describe types of ventilators, and state the minimum recommended sizes.

CONTENTS

CHAPTER 49 THERMAL AND ACOUSTICAL INSULATION

Most materials used in construction have some insulating value (Fig. 49-1). Even the air spaces between studs resist the passage of heat. After insulating material is installed, however, the stud space has many times the resistance to the passage of heat and sound than the air space alone had.

How Insulation Works

Air is an excellent insulator if confined to a small space and kept still. The insulation becomes more effective as the air spaces become smaller and greater in number. Millions of tiny air cells, trapped in its unique cellular structure make wood a natural and effective insulating material. Most insulating materials are manufactured by trapping dead air in tiny spaces to provide resistance to the flow of heat and sound.

Thermal Insulation

Among the materials used for insulating are glass fibers, mineral fibers (rock), organic fibers (paper), and plastic foam. Aluminum foil is also used. It works by reflecting heat instead of blocking its passage.

Resistance Value of Insulation

The *R-value* of insulation is a number that indicates its measured resistance to the flow of heat. The higher the R-value, the more efficient the material is in retarding the passage of heat. Insulation is rated

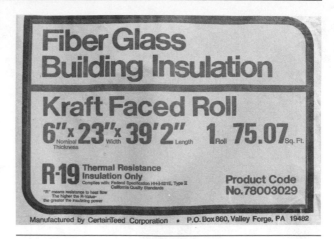

Fig. 49-2 Insulation is rated according to its thermal resistance. It is stamped with an R-value.

according to its efficiency. Its R-value is clearly printed on the wrapper (Fig. 49-2).

Insulation Requirements

The rising costs of energy coupled with the need to conserve it have resulted in higher R-value recommendations for new construction than in previous years. Average winter low-temperature zones of the United States are shown in Figure 49-3. This information is used to determine the R-value of insulation installed in walls, ceilings, and floors. Insulation requirements vary according to the average low temperature (Fig. 49-4).

In warmer climates, less insulation is needed to conserve energy and provide comfort in the cold season. However, air-conditioned homes should receive more insulation in walls, ceilings, and floors. This assures economy in the operation of air-conditioning equipment during the hot season.

Comfort and operating economy are dual benefits. Insulating for maximum comfort automatically provides maximum economy of heating and cooling operations. It also reduces the initial costs of heating and cooling equipment because smaller units may be adequate.

Where Heat is Lost

The amount of heat lost from the average house varies with the type of construction. The principal areas and approximate amount of heat loss for a typical house with moderate insulation are shown in Figure 49-5.

THERMAL PROPERTIES OF VARIOUS BUILDING MATERIALS, PER INCH OF THICKNESS	
Material	Thermal Resistance R
Wood	1.25
Air Space	0.97
Cinder Block	0.28
Common Brick	0.20
Face Brick	0.11
Concrete (Sand and Gravel)	0.08
Stone (Lime or Sand)	0.08
Steel	0.0032
Aluminum	0.00070

Fig. 49-1 Insulating values of various building materials per inch of thickness.

Fig. 49-3 Average low-temperature zones of the United States. (*Courtesy U.S. Department of Agriculture, Forest Service*)

RECOMMENDED R-VALUES OF INSULATION			
AVERAGE LOW TEMPERATURE Degrees Fahrenheit	CEILINGS	WALLS	FLOORS
0 & BELOW	38	17	19
0 TO +10	30	17	19
+10 TO +40	19	11	0

Fig. 49-4 Insulation requirements for various climates.

5% THROUGH CEILINGS

16% THROUGH WINDOWS

17% THROUGH FRAME WALLS

3% THROUGH DOORS

38% AIR LEAKAGE THROUGH CRACKS IN WALLS, WINDOWS, AND DOORS

20% THROUGH BASEMENT WALLS

1% THROUGH BASEMENT FLOOR

Fig. 49-5 Areas and amounts of heat loss for a typical house with moderate insulation. (*Courtesy of Dow Chemical Company*)

Houses of different architectural styles vary in their heat loss characteristics. A single-story house, for example, contains less exterior wall area than a two-story house. It also has a proportionately greater ceiling area. Greater heat loss through floors is experienced in homes erected over concrete slabs or unheated crawl spaces unless these areas are well insulated.

The transfer of heat through uninsulated ceilings, walls, or floors can be reduced almost any desired amount by installing insulation. Maximum quantities in these areas can cut heat losses by up to 90 percent. The use of 2×6 studs in exterior walls permits installation of 6-inch-thick insulation. This achieves an R-19 value, or, with improved insulation, an R-21 value.

Windows and doors are generally sources of great heat loss. By reducing the heat transfer through other sections of the house, discomfort caused by these sources can be offset. Weatherstripping around windows and doors reduces heat loss. Heat loss through glass surfaces can be reduced 50 percent or more by installing double glazed windows or by adding storm sash and storm doors.

Fig. 49-6 Blanket insulation comes in rolls.

Types of Insulation

Insulation is manufactured in a variety of forms and types. These are commonly grouped as *flexible, loose-fill, rigid, reflective,* and *miscellaneous.*

Flexible Insulation

Flexible insulation is manufactured in *blanket* and *batt* form. Blanket insulation (Fig. 49-6) comes in rolls. Widths are suited to 16- and 24-inch stud and joist spacing. The usual thickness is from one inch to 6 1/4 inches. The body of the blanket is made of fluffy rock or glass wool, wood fiber, or cotton. Organic insulation materials are treated to make them resistant to fire, decay, insects, and vermin. Most blanket insulation is faced with asphalt-laminated Kraft paper or aluminum foil with flanges on both edges for fastening to studs or joists. The facing of the blanket serves as a vapor barrier. It resists the movement of water vapor. It should always face the warm side of the wall.

Batt insulation (Fig. 49-7) is made of the same material as blanket insulation. It comes in thicknesses up to 12 inches. Widths are for standard stud

Fig. 49-7 Batt insulation is made up to 12 inches thick.

and joist spacing. Lengths are either 48 or 93 inches. Batts come faced with Kraft paper or unfaced.

Loose-Fill Insulation

Loose-fill insulation is usually composed of materials in bulk form. It is supplied in bags or bales. It is placed by pouring, blowing, or packing by hand. Materials include rock or glass wool, wood fibers, shredded redwood bark, cork, wood pulp products, and vermiculite.

Loose-fill insulation is suited for use between ceiling joists in unheated attics. It is also used in the sidewalls of existing houses that were not insulated during construction.

Rigid Insulation

Rigid insulation is usually a fiber or foamed plastic material in sheet or board forms (Fig. 49-8). The material is available in a wide variety of sizes, with widths up to 4 feet and lengths up to 12 feet. The most common types are made from polystyrene, polyurethane, glass fibers, processed wood, and sugar cane or other fibrous vegetable materials.

Structural insulating boards come in densities ranging from 15 to 31 pounds per cubic foot. They are used as building boards, roof decking, sheathing, and wallboard. Their primary purpose is structural. They also serve as insulation. In house construction, perhaps the most common forms are sheathing and decorative coverings in sheet or tile forms.

Reflective Insulation

Reflective insulation usually consists of outer layers of aluminum foil bonded to inner layers of various materials for added strength to resist heat flow. Reflective insulation should be installed facing an air space with a depth of 3/4 inch or more.

Fig. 49-8 Rigid insulation being installed between floor joists over a crawl space.

Miscellaneous Insulation

Foamed-in-place insulation is sometimes used. A urethane foam is produced by mixing two chemicals together. It is blown into place. It expands on contact with the surface.

Sprayed insulation is usually inorganic fibrous material. It is blown against a surface that has been primed with an adhesive coating. It is often left exposed for acoustical as well as insulating properties.

Other types of insulating material, such as lightweight *vermiculite* and *perlite aggregates,* are sometimes used in plaster to reduce heat flow.

Where and How to Insulate

To reduce heat loss all walls, ceilings, roofs, and floors that separate heated from unheated spaces should be insulated. Insulation should be placed in all outside walls and in the ceiling. In houses with unheated crawl spaces, it should be placed between the floor joists.

A ground cover of roll roofing or plastic film such as polyethylene should be placed on the soil of crawl spaces. This decreases the moisture content of the space as well as of the wood members. Provision should also be made for ventilation of unheated areas. (The use of vapor barriers, the cause of moisture condensation, and the need for ventilation are discussed in greater detail later in this unit.)

In houses with flat or low-pitched roofs, insulation should be used in the ceiling area only if sufficient space is left above the insulation for ventilation. Insulation is used along the perimeter of houses built on slabs when required (Fig. 49-9).

Installing Flexible Insulation

Cut the batts or blankets with a knife. Make lengths slightly oversize so stapling flanges can be formed at the top and bottom. Measure out from one wall a distance equal to the desired lengths of insulation. Draw a line on the floor. Roll out the material from the wall. Compress the insulation with a straightedge on the line. Cut it with a sharp knife (Fig. 49-10). Cut the necessary number of lengths.

Fig. 49-10 Compress flexible insulation with a straightedge. Cut it with a sharp knife.

Fig. 49-9 Placement of insulation in various types of structures.

IN (A), THE INSULATION FLANGE IS STAPLED TO THE INSIDE EDGE
OF THE STUD. IN(B), THE INSULATION IS RECESSED.

Fig. 49-11 Methods of installing wall insulation.

Place the batts or blankets between the studs. Staple the flanges of the vapor barrier either to the sides or to the inside edges of the studs as well as the top and bottom plates (Fig. 49-11). Use a hand stapler to fasten the insulation in place (Fig. 49-12).

Fill any spaces around doors and windows. Do not pack the insulation tightly. Squeezing or compressing it reduces its effectiveness. Use narrow strips of vapor barrier material along the top, bottom, and sides of wall openings. This protects these areas (Fig. 49-13). Ordinarily these are not adequately covered by the barrier on the blanket or batt. If the insulation has no covering or stapling tabs, it is friction-fitted between the framing members.

Ceiling Insulation

Ceiling insulation is installed by stapling it to the ceiling joists or by friction-fitting it between them. If furring strips have been applied to the ceiling joists, the insulation is simply laid on top of the furring strips. In most cases, the use of unfaced insulation is recommended for ceilings. This allows moisture into the attic, where it can be evaporated by the circulation of air.

Extend the insulation across the top plate. However, if the eaves are vented, care must be

Fig. 49-13 Fill spaces around doors and windows with insulation. Install vapor barrier material over them.

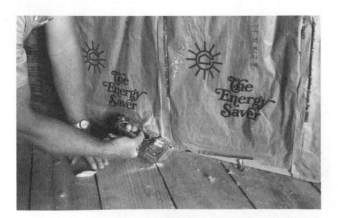

Fig. 49-12 Use a hand stapler to fasten wall insulation in place.

RAFTER

ALLOW FOR AIRWAY
ABOVE INSULATION

BAFFLE

NAILING BLOCK

LOOKOUT

SCREEN

FASCIA

SOFFIT

INLET VENTILATOR

Fig. 49-14 Compress insulation at the top of the wall plate, if necessary, to permit a free flow of air from the eaves.

taken not to block the flow of air into the attic. It may be necessary to compress the insulation against the top of the wall plate to permit the free flow of air (Fig. 49-14).

Insulating Floors over Crawl Spaces

Flexible insulation installed between floor joists over crawl spaces may be held in place by wire mesh or pieces of heavy-gauge wire wedged between the joists. The vapor barrier on the insulation faces the ground (Fig. 49-15) instead of the heated side. This prevents moisture from entering from below.

Insulating Masonry Walls

Masonry walls require furring strips for one method of insulation. Fasten 1×2 furring strips 16 or 24 inches OC. Insulate as for wood frame walls (Fig. 49-16).

Installing Loose-Fill Insulation

Loose-fill insulation is poured by hand or blown into place with special equipment. Level the surface of the material between ceiling joists to the desired depth (Fig. 49-17).

CAUTION Use a dust mask to prevent inhaling fine dust particles.

(A)   (B)

BUILDING PAPER

FINISH FLOOR

SUBFLOOR

INSULATION

WIRE MESH

INSULATION SUPPORTS

SIMPLY PUSH UP
IN PLACE

In (a), wire mesh is stapled to the edges of the joists.
In (b), pieces of heavy gauge wire, pointed at each end, are wedged between the joists to support the insulation.

Fig. 49-15 Methods of installing insulation between floor joists in a crawl space.

Fig. 49-16 Method of insulating masonry walls.

Fig. 49-17 Level loose-fill insulation to the desired depth.

Installing Rigid Insulation

Rigid insulation is easily cut with a knife or saw. It is usually applied by friction-fitting between the framing members. However, it may be applied to masonry walls with special adhesives (Fig. 49-18). Polystyrene foam plastic, commonly called Styrofoam, has excellent insulating qualities. It is also waterproof. Therefore, it can also act as a vapor barrier. It is frequently used to insulate existing structures before new siding is applied (Fig. 49-19). It is also used extensively on roofs either below or above the roof sheathing (Fig. 49-20).

Fig. 49-18 Rigid insulation is applied to basement walls before backfilling.

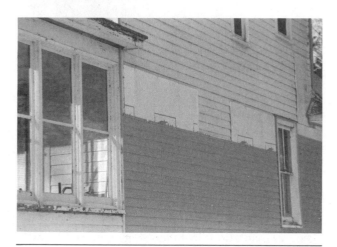

Fig. 49-19 Rigid insulation applied over old existing siding.

Installing Reflective Insulation

Reflective insulation is usually installed between framing members in the same manner as blanket insulation. It is attached to either the face or side of the studs. However, an air space of at least 3/4 inch must be maintained between its surface and the inside edge of the stud.

Properly Installed Insulation

Insulation that is not properly installed can cause serious problems. It could render the material useless. It could cause wood to decay and paint to blister. The carpenter must be aware of the consequences of improper application and not take the installation of insulation lightly. The carpenter must know how to install insulation and vapor barriers in the recommended fashion and how to provide proper and adequate ventilation. Follow the manufacturer's directions carefully when installing any type of insulation.

Fig. 49-20 Method of installing rigid insulation on roofs.

Acoustical Insulation

Acoustical or *sound* insulation resists the passage of noise through a building section. In the past, the reduction of sound transfer between rooms was important in offices, apartments, and motels. Now, more attention is being paid to insulating for sound in homes. Excessive noise is not only annoying but harmful. It not only causes fatigue and irritability. It can damage the sensitive hearing nerves of the inner ear.

Sound insulation between active areas and quiet areas of the home is desirable. Sound insulation between the bedroom area and the living area is important. Bathrooms should also be insulated. Sound control is an important part of residential and commercial construction.

Sound Transmission

Noises such as a loud radio, barking dog, or shouting create sound waves. These waves radiate outward from the source until they strike a wall, floor, or ceiling. These surfaces then begin to vibrate by the pressure of the sound waves in the air. Because the wall vibrates, it conducts sound to the other side in varying degrees, depending on the wall construction.

Sound Transmission Class. The resistance of a building section, such as a wall, to the passage of sound is rated by its *Sound Transmission Class* (STC). The higher the number, the better the sound barrier. The approximate effectiveness of walls with varying STC numbers is shown in Figure 49-21.

Sound travels readily through the air and through some materials. When airborne sound strikes a wall, the studs act as conductors unless they are separated in some way from the covering material. Electrical outlet boxes placed back to back in a wall easily transmit sound. Faulty construction, such as poorly fitted doors, often allows sound to pass through. Therefore, good construction practices aid in controlling sound.

Wall Construction

A wall that provides sufficient resistance to airborne sound should have an STC rating of 45 or greater. At one time, the resistance was usually provided only by double walls, which resulted in increased costs. However, a system of using sound-deadening *insulating board* with a gypsum board outer covering has been developed. This system provides good sound resistance. Resilient steel channels placed at right angles to the studs isolate the gypsum board from the stud. Figure 49-22 shows various types of wall construction and their STC rating.

25	Normal speech can be understood quite easily
30	Loud speech can be understood fairly well
35	Loud speech audible but not intelligible
42	Loud speech audible as a murmur
45	Must strain to hear loud speech
48	Some loud speech barely audible
50	Loud speech not audible

This chart from the Acoustical and Insulating Materials Association illustrates the degree of noise control achieved with barriers having different STC numbers.

Fig. 49-21 Approximate effectiveness of walls with varying STC ratings.

WALL DETAIL	DESCRIPTION	STC RATING
(A) 16″ — 2 X 4	1/2″ GYPSUM WALLBOARD	32
	5/8″ GYPSUM WALLBOARD	37
(B) 16″ — 2 X 4	3/8″ GYPSUM LATH (NAILED) PLUS 1/2″ GYPSUM PLASTER WITH WHITECOAT FINISH (EACH SIDE)	39
(C)	8″ CONCRETE BLOCK	45
(D) 16″ — 2 X 4	1/2″ SOUND DEADENING BOARD (NAILED) 1/2″ GYPSUM WALLBOARD (LAMINATED) (EACH SIDE)	46
(E) 16″ — 2 X 4	RESILIENT CLIPS TO 3/8″ GYPSUM BACKER BOARD 1/2″ FIBERBOARD (LAMINATED) (EACH SIDE)	52

SOUND INSULATION OF SINGLE WALLS

WALL DETAIL	DESCRIPTION	STC RATING
(A) 16″ — 2 X 4	1/2″ GYPSUM WALLBOARD	
(B) 2 X 4	5/8″ GYPSUM WALLBOARD (DOUBLE LAYER EACH SIDE)	45
(C) 2 X 4 BETWEEN OR "WOVEN"	1/2″ GYPSUM WALLBOARD 1-1/2″ FIBROUS INSULATION	49
(D) 2 X 4	1/2″ SOUND DEADENING BOARD (NAILED) 1/2″ GYPSUM WALLBOARD (LAMINATED)	50

SOUND INSULATION OF DOUBLE WALLS

Fig. 49-22 Sound transmission rating of various types of wall construction.

Floor and Ceiling Construction

Sound insulation between an upper floor and the ceiling below involves not only the resistance of airborne sounds, but also that of *impact noise*. Impact noise is caused by such things as dropped objects, footsteps, or moving furniture. The floor is vibrated by the impact. Sound is then radiated from both sides of the floor. Impact noise control must be considered as well as airborne sounds when constructing floor sections for sound insulation.

An *Impact Noise Rating* shows the resistance of various types of floor-ceiling construction to impact noises. The higher the positive value of the INR, the more resistant to impact noise transfer. Figure 49-23 shows various types of floor-ceiling construction with their STC and INR ratings.

Sound Absorption

The amount of noise in a room can be minimized by the use of *sound absorbing materials*. Perhaps the most commonly used material is *acoustical tile.* These are most often used in the ceiling where they are not subjected to damage. The tiles are soft and are made of wood fiber or similar materials. The tile surface consists of small holes or fissures or a combination of both (Fig. 49-24). These holes or fissures act as sound traps. The sound waves enter, bounce back and forth, and finally die out. (A more complete description and method of installing ceiling tile can be found in a later unit.)

DETAIL	DESCRIPTION	ESTIMATED VALUE	
		STC RATING	APPROX. INR
(A) 16" 2 X 8	FLOOR 7/8" T. & G. FLOORING CEILING 3/8" GYPSUM BOARD	30	−18
(B) 2 X 8	FLOOR 3/4" SUBFLOOR 3/4" FINISH FLOOR CEILING 3/4" FIBERBOARD	42	−12
(C) 2 X 8	FLOOR 3/4" SUBFLOOR 3/ 3/4" FINISH FLOOR CEILING 1/2" FIBERBOARD LATH 1/2" GYPSUM PLASTER 3/4" FIBERBOARD	45	−4

RELATIVE IMPACT AND SOUND TRANSFER IN FLOOR–CEILING COMBINATIONS
(2- BY 8-IN. JOISTS)

Fig. 49-23 Floor and ceiling construction rated for STC and INR.

DETAIL	DESCRIPTION	ESTIMATED VALUE	
		STC RATING	APPROX. INR

(A) 16″ 2 X 10	FLOOR
	  3/4″ SUBFLOOR
	  (BUILDING PAPER)
	  3/4″ FINISH FLOOR
	CEILING
	  GPYSUM LATH AND
	  SPRING CLIPS
	  1/2″ GYPSUM PLASTER
		52	−2

(B) 2 X 10	FLOOR
	  5/8″ PLYWOOD
	  SUBFLOOR
	  1/2″ PLYWOOD
	   UNDERLAYMENT
	  1/8″ VINYL–ASBESTOS
	   TILE
	CEILING
	  1/2″ GYPSUM
	  WALLBOARD
		31	−17

(C) 2 X 10	FLOOR
	  5/8″ PLYWOOD
	  SUBFLOOR
	  1/2″ PLYWOOD
	   UNDERLAYMENT
	  FOAM RUBBER PAD
	  3/8″ NYLON CARPET
	CEILING
	  1/2″ GYPSUM
	  WALLBOARD
		45	+5

RELATIVE IMPACT AND SOUND TRANSFER IN FLOOR-CEILING COMBINATIONS
(2- BY 10-IN. JOISTS)

Fig. 49-23 (*continued*)

Fig. 49-24 Sound-absorbing ceiling tile.

CHAPTER 50 CONDENSATION AND VENTILATION

Energy use and costs are reduced when insulation is installed in buildings. Problems of moisture condensation must be solved with vapor barriers and ventilation.

Condensation

Warm air holds more moisture than cold air. The warmer the air, the more moisture it can hold. As moisture-laden air is cooled, it reaches the *dew point.* At this point it is completely saturated with moisture. Fog is air that has reached its dew point. Any further lowering of the air temperature causes it to lose some of its moisture as *condensation* on the cold surface. Evidence of this can be seen as dew on grass in the morning when warm humid air has cooled off at night. The moisture it has lost is condensed on the ground.

Moisture in Buildings

Moisture in the form of water vapor comes from many activities. It is produced by mainly cooking, taking showers and baths, washing and drying clothes, and using a humidifier. The vapor in the inside air penetrates through materials in its effort to mix and equalize with colder and drier outside air.

In an insulated building, water vapor strives to move through the walls, ceilings, and floors. It may condense under certain conditions when it comes in contact with a cold surface (Fig. 50-1). In a wall, this contact is made on the inside surface of the exterior wall sheathing. In a crawl space, condensation can form on the floor frame and subfloor. In attics, the ceiling joists and roof frame can become saturated with moisture. Condensation of water vapor in walls, attics, roofs, and floors leads to serious problems.

Problems Caused by Condensation

- If insulation absorbs condensed water vapor, the dead air spaces in the insulation may become filled with water. This renders the insulation useless.

- Condensed moisture absorbed by the wood frame raises the moisture content of the wood. This improves the conditions for fungi to grow and decay the wood.

- If moisture is absorbed by the sheathing, it is eventually absorbed by the exterior trim and siding. This causes the exterior paint to blister and peel. Excessive condensation of water vapor may even damage the interior finish.

- A warm attic that is inadequately ventilated may cause the formation of ice dams at the eaves. After a heavy snowfall, heat causes the snow next to the roof to melt. Water running down the roof freezes on the cold roof overhang, forming an ice dam. This causes the water to back up on the roof, under the shingles, and into the walls and ceiling (Fig. 50-2).

Fig. 50-1 Moisture in warm air condenses when it comes in contact with a cold surface.

TRAPPED WATER

ICE DAM

WATER ENTRY

INSUFFICIENT
INSULATION

GUTTER

NO SOFFIT
VENTILATION

(A)

FLASHING
(ROLL ROOFING)

RAFTER

AIRWAY

SUFFICIENT
INSULATION

INSTALL METAL
BAFFLES BETWEEN
RAFTERS TO PROVIDE
AN AIRWAY

GUTTER

SOFFIT VENTILATION

(B)

Fig. 50-2 A warm attic may cause the formation of ice dams on the roof during cold weather.

Prevention of Vapor Condensation

Logically, to prevent the condensation of moisture in a building, steps should be taken to reduce the amount of moisture content of the inside warm air. This can be accomplished by the installation of exhaust fans in kitchens and bathrooms that vent the heavily moisture-laden air to the outside. Check the exhaust piping of the clothes dryer. Make sure it is not constricted and is working properly.

Ventilate the attic. Install no barrier to the passage of vapor to the attic where it can flow to the outside.

Install a barrier to the passage of water vapor to wall cavities and other areas that cannot be ventilated for the evaporation of condensation (Fig. 50-3).

Ventilation

Ventilation helps combat condensation of moisture. In a well-ventilated area, any condensed moisture is removed by evaporation. It is important that areas on the cold side of the insulation be well ventilated, if possible. One of the areas where this can be effectively accomplished is in the attic.

With a well-insulated ceiling and adequate ventilation, attic temperatures are low. Melting of snow on the roof over the attic space is greatly reduced.

In crawl spaces under floors, ventilation can be easily provided to evaporate any condensation of moisture.

In walls, ventilation of the air space on the

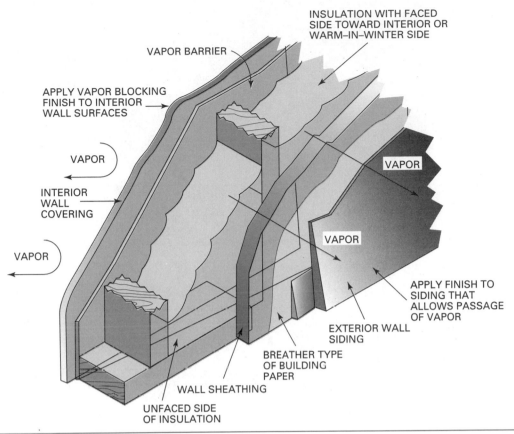

Fig. 50-3 A wall should be constructed to prevent the entrance of vapor from the inside and to allow its passage to the outside.

cold side of the insulation is often difficult. This is especially true if the wall is full of insulation. This is why some contractors object to completely filling the stud space with insulation. They insist on leaving at least 3/4 inch of air space between the insulation and the inside of the exterior wall sheathing. However, using a properly installed vapor barrier on the room side of the insulation should stop most of the vapor from passing through into the stud space.

Any vapor that passes the barrier should be allowed to leave the stud space easily. This can be done by using a building paper on wall sheathing that is not a vapor barrier and allows its passage. Apply a *breather* type of exterior finish to the wall siding that allows the passage of vapor through it. Use siding vents, if practical, to help move vapor to the outside. When the wall is constructed in this manner, filling the stud space completely with insulation should pose no problem.

On flat roofs, or on other roofs where the ceiling finish is attached to the roof rafters, insulation is usually installed between the rafters. An adequate air space of at least 1 1/2 inches must be maintained between the insulation and the roof sheathing. The air space must be amply vented with air inlets in the soffit and outlets at the ridge (Fig. 50-4).

Failure to do so is likely to result in decay of the roof frame, causing expensive, major repairs. Application of a vapor barrier on the heated-in-winter side of the insulation helps prevent any condensation of

Using either method, it is important to provide space between the insulation and roof sheathing to permit air circulation.

Fig. 50-4 Methods of installing flexible insulation between rafters.

Ventilating areas of gable roofs: A, Louvers in end walls; B, louvers in end walls with additional openings in soffit area; C, louvers at end walls with additional openings at eaves and dormers. Cross section of C shows free opening for air movement between roof boards and ceiling insulation of attic room.

Fig. 50-5 Methods of ventilating gable roofs.

moisture in the rafter space by blocking the passage of vapor.

Types of Ventilators

There are many types of ventilators. Their location and size are factors in providing adequate ventilation.

Ventilating Gable Roofs. *Louvers* are generally provided in both end walls of gable roofs. They should be as close to the ridge as possible. Louvers are manufactured millwork. They come in various shapes with slats at an angle and spaced to keep out the weather. A wire mesh screen is applied to prevent insects and small animals from getting into the building. Better movement of air can be obtained if additional openings are provided in the *soffit* (roof overhang) area (Fig. 50-5).

Ventilating Hip Roofs. Hip roofs should have air inlets in the soffit and outlets at or near the peak of the roof. The most efficient type of inlet opening is a continuous screened opening in the soffit (Fig. 50-6). With a hip roof, the soffit vent is excellent. It can extend completely around the building. Round or rectangular metal soffit vents are manufactured in various sizes. They may also be used in each

Fig. 50-6 A continuous screened vent in the soffit is a most efficient type of air inlet.

rafter space for air inlets to the attic. In addition, corrugated and perforated prefinished metal is widely used for soffit material. It provides ventilation through the total soffit width and length.

Continuous *ridge* and *hip* vents (Fig. 50-7) are widely used. They provide ample attic air outlets. Soffit vents combined with continuous ridge and

RIDGE VENT   HIP VENT

Fig. 50-7 Continuous ridge and hip vents are efficent attic air outlets. (*Courtesy of CorAvent, Inc.*)

hip vents provide a very effective way of ventilating an attic.

The air outlet openings near the peak can also be provided by installing globe type metal *turbine ventilators* (Fig. 50-8). They can be located below the peak on the rear slope of the roof. Then they are not visible from the front of the house.

A variation of a hip roof, called a *Dutch hip*, has gabled extensions designed specifically to house large triangular louvers on both ends. This allows a breeze to blow across the attic, providing an excellent method of ventilation (Fig. 50-9).

Ventilating Crawl Spaces. For crawl spaces, usually rectangular metal or plastic louvered ven-

Fig. 50-8 Globe-shaped turbine roof ventilators are sometimes used to vent hip roofs.

Ventilating areas of hip roofs: A, Inlet openings beneath eaves and outlet vent near peak; B, inlet openings beneath eaves and ridge outlets.

Fig. 50-9 Methods of ventilating hip roofs.

tilators are used. Some are made to fit into a space made by leaving out a concrete block in the foundation. Vents should be placed near each corner of the crawl space foundation and in intermediate equally spaced locations (Fig. 50-10). The use of a ground cover moisture barrier is recommended under most conditions. It protects wood framing members from ground moisture. It also allows the use of a smaller number of ventilators.

Determining Ventilator Size

The location of ventilators is a factor in their efficiency. Ventilators must be of adequate size.

Size of Attic Ventilators. The minimum free-air area for attic ventilators is based on the ceiling area of the rooms below. The free-air area for the openings should be 1/300ths of the ceiling area. For example, if the ceiling area is 1,200 square feet, the minimum total free-air area of the ventilators should be 4 square feet.

Crawl Space Ventilators. The total free-air area of crawl space ventilators should be 1/1600ths of the ground area when a vapor barrier ground cover is used. For example, if a crawl space measures 50 feet by 100 feet, its ground area is 5,000 square feet. The free-air area of ventilators must be a little over 3 square feet.

The free-air area of a ventilator is the area through which air is able to pass. The actual area must be increased to allow for any restrictions such as louvers or wire screen mesh. Use as coarse a screen as conditions permit. Lint and dirt tend to clog fine-mesh screens. When painting, be careful not to contact the screen and clog the mesh with paint. Ventilators should be designed to keep the weather out. They should not be closed. In cold climates, it is especially important to leave them open in the winter as well as during the summer.

Description of Vapor Barriers

The most commonly used material for a vapor barrier is *polyethylene film*. It is commonly called *Visqueen*. It is a transparent plastic sheet. It comes in rolls of usually 100 feet in length and 10, 12, 14, 16, and 20 feet in width. The most commmonly used thicknesses are 4 mil and 6 mil.

Other materials used as vapor barriers are *asphalt-laminated papers* and *aluminum foil*. Most blanket and batt insulations are manufactured with a vapor barrier on one side.

Installing a Polyethylene Vapor Barrier

Unroll a length long enough to cover the wall. Make sure to allow a few extra inches in length. Unfold the section. Staple it along the top plate, letting the rest drape down to the floor. Smooth out wrinkles. Staple about 6 to 12 inches apart to every stud (Fig. 50-11). Cut the material. Carefully fit it around all openings. Lap all joints. Repair any tears with tape. Cut off the excess at the floor line.

Fig. 50-10 Crawl space ventilator and vapor barriers.

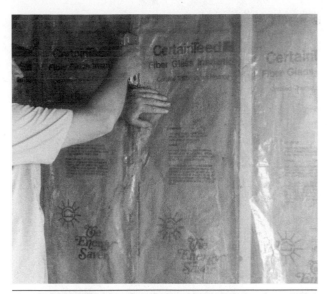

Fig. 50-11 Installing polyethylene film vapor barrier.

Some contractors apply the film over all openings. They then cut the film after the finish is applied. This assures a more positive seal. The barrier is fitted tightly around outlet boxes. A ribbon of sealing compound around outlets or switch boxes minimizes vapor penetration.

Polyethylene is applied over insulation having no vapor barrier facing. It also insures excellent protection over any type of insulation.

Vapor barriers are not usually installed on ceilings. This allows water vapor to pass through the ceilings where it can be easily evaporated by movement of air in the attic. By allowing the vapor to pass through ceilings, its entrance into wall cavities, where the movement of air is not as great, is reduced. However, where temperatures commonly fall below −20°F or where roof slopes are less than 3 in 12, vapor barriers are recommended in ceilings.

No vapor barrier can be considered 100 percent impervious. Some vapor leakage into the wall is to be expected. Thus, the flow of vapor to the outside should not be impeded by materials of high vapor resistance on the cold side of the wall. For example, sheathing paper should not be highly vapor resistant. This also applies to exterior siding and exterior finishes. Bathroom and kitchen vents and fans should be used to expel generated moisture to the outside.

Review Questions

Select the most appropriate answer.

1. Of the common building materials listed below the most efficient insulator is
   a. cinder block.   c. stone.
   b. concrete.   d. wood.

2. The insulating term *R-value* is defined as the measure of
   a. resistance of a material to the flow of heat.
   b. the heat loss through a building section.
   c. the conductivity of a material.
   d. the total heat transfer through a building section.

3. If insulation is covered on both sides, only one side acts as a vapor barrier. It should face the
   a. warm side.   c. inside.
   b. cold side.   d. outside.

4. Reflective insulation usually consists of
   a. foamed plastic.   c. aluminum foil.
   b. glass wool.   d. vermiculite.

5. Reflective insulation is effective only when
   a. it is placed on the warm side.
   b. it comes in contact with the wall covering.
   c. it faces an air space of at least 3/4 inch deep.
   d. it is used under the exterior finish.

6. For a more effective seal, the tabs of vapor barrier covering flexible insulation are stapled to
   a. the sides of the framing members.
   b. the inside edges of the framing members.
   c. the inside surface of the wall sheathing.
   d. the top and bottom plates.

7. Squeezing or compressing flexible insulation tightly into spaces
   a. reduces its effectiveness.
   b. increases its efficiency since more insulation can be installed.
   c. is necessary to hold it in place.
   d. helps prevent air leakage by sealing cracks.

8. When insulation is placed between roof framing members, there should be an air space
between the insulation and the roof sheathing
of at least

a. 3/4 inch.   c. 1 1/2 inches.
b. 1 inch.   d. 2 inches.

9. Ventilators

a. are closed in winter to conserve heat.
b. are left open year round.
c. are closed in stormy weather.
d. are closed when the house is to be left
vacant for long periods of time.

10. Vapor barriers

a. should be installed in exterior walls and
ceilings.
b. should be installed on ceilings only.
c. should not be installed on ceilings unless the
roof pitch is less than 3 in 12.
d. "envelop" the house and are installed on all
walls, floors, and ceilings.

■ BUILDING FOR SUCCESS ■

Nontraditional Workers in the Workplace

Across the country today, the traditional white male "turf" is being shared by more women and other minority groups. This demands fresh perspectives and increased flexibility. There will be a need for more understanding, demonstrations of respect, and cooperation by everyone.

All ethnic and diversity groups in the workplace must be accepted. Acceptance must be authentic and demonstrated if workers are to be successful. This acceptance is based on the understanding of a person's qualifications to do the job. These include, but are not limited to, technical skills, talents, abilities, knowledge, and experience.

Competition for jobs must be clearly centered on identifiable criteria that avoid discrimination and/or harassment. The criteria must focus on what is needed to do the job satisfactorily. Performance on the job should be evaluated by a person's ability to perform, and not whether the worker is male, female, white, black, of color, or handicapped.

Nontraditional workers in the construction workforce today represent a variety of cultures and ethnic groups. We must all continue to learn more about each co-worker. Common goals must be shared by all workers regardless of their culture. There will always be a need for good company orientation programs that help introduce each worker to a compatible work environment. Supervisors who work with a specific ethnic group other than their own may need to learn another lan-

guage for effective communication. There will continue to be a need to gain a better understanding of other customs and cultural traditions that can cause misunderstandings on the job. With the workforce gradually incorporating more females into its ranks, there may be a need to entertain short or extended maternity leaves. Flexibility has always played an important part in adjusting to new developments in the work world.

Increased ethnic and cultural diversity groups entering the workforce will influence the number of skilled and unskilled workers. The underlying criteria for determining who will obtain and maintain quality jobs in construction must center on the skills and abilities of the workers. Obtaining and retaining those jobs should be the responsibility and goal of every construction student, apprentice, and trainee nationwide. Good education and skill development will be the center of attention for those offering training programs. The person who is best prepared with the necessary technical skills, business skills, and abilities will generally gain an edge when seeking employment. Additionally, the person who has a good understanding of how to be productive as a team member that includes nontraditional workers will be more successful then others.

The avenue for obtaining the necessary skills and abilities may not be as important as learning the right technology and developing the required skills that accompany that technology. There are many educational opportunities available to people who desire construction careers. Universities, colleges, trade unions, professional construction organizations, and construction companies offer quality preparation and training. Competence in applying technical skills will be a must. So will competence in communicating, reading, teamwork, social skills, and self management.

The best job offer may take people out of their familiar environment and place them in one that has a major population that represents another culture or diverse group. This will take a good understanding of non traditional workers and flexibility. We must be prepared to work alongside those representing other ethnic groups. The nontraditional worker will continue to be an important part of the skilled workforce. All successful carpenters, plumbers, electricians, and masons that represent a vast array of cultures and ethnic groups will possess many similar technical skills. They will also need the capability to establish common goals and work alongside many nontraditional workers.

FOCUS QUESTIONS: For individual or group discussion.

1. What is your current understanding of the nation's workforce in reference to nontraditional workers?

2. What impact is a diversified workforce having in your area of desired employment?

3. What elements of the population do you think will be represented in your technical specialty during the next decade? Will this population representation show a contrast from the present-day workforce?

4. How can you plan to be part of a workforce that promotes teamwork in problem solving? What would be some examples of new skills and understandings that would help bring success to your job?

STUDENT ACTIVITIES for

Section 2: Rough Carpentry

Unit 9 Building Layout

1. By using the two site plans, lay out the buildings, erect batter boards, and identify and mark the building line, footing line, and digging line. Set up the builder's level in the site. Take elevation readings at each corner of the lot and building.

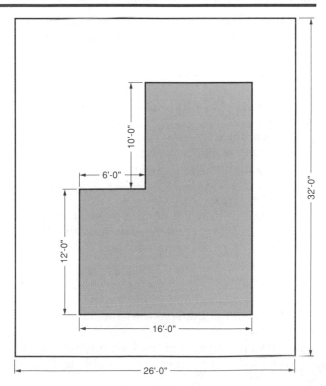

SITE LAYOUT #1
SCALE 1/4"=1'—0"

SITE LAYOUT #2
SCALE 1/4"=4'—0"

Unit 10 Concrete and Concrete Forms

1. Lay out and form a 4-inch patio slab, 12' × 16', with a slope of one inch across the 12-foot dimension. Use 1" × 4" form lumber.

Check the slope with a straightedge and level or a builder's level.

Unit 11 Floor Framing

Using the drawings, complete the following framing activities by selecting all the needed tools and materials. Have the instructor check your work for accuracy. Then make any necessary corrections.

1. Using the drawing of the box header, measure and lay out the header to receive floor joists. Materials required: 1 ea. 2" × 4" × 12' (box header).

Layout Criteria:
1. Floor joists to be at 16 inches OC.
2. Mark end of the box header at 12' × 0" for cutting.
3. Make all changes in the joist positions to allow for stair trimmers and double joists under the walls.

4. Draw the accepted symbols or markings on the lumber.
   a. on center,   c. trimmer,
   b. floor joist,   d. wall.

BOX HEADER

Unit 12 Wall Framing.

Materials required: 1 ea. 2″ × 4″ × 12′, 1 ea. 2″ × 4″ × 8′.

Layout Criteria (platform frame):

1. Studs to be at 16inches OC.

2. Window rough openings located as shown.

3. Single trimmers at opening sides.

4. Locate intersecting walls as shown.

5. Calculate header lengths.

6. Mark end of the wall plate at 12 feet for cutting.

7. Mark all changes in the stud positions to allow for openings, trimmers, and walls.

WALL PLATES
SCALE 1/2″=1′—0″

Unit 15 Roof Framing

Solve the following problems by finding the dimensions and calculations needed in the roof frame layout.

1. Given: 6/12 roof slope, 16 inches OC rafter spacing. Find the theoretical lengths of entire roof frame, with rafter lengths to plate and to the tail end.

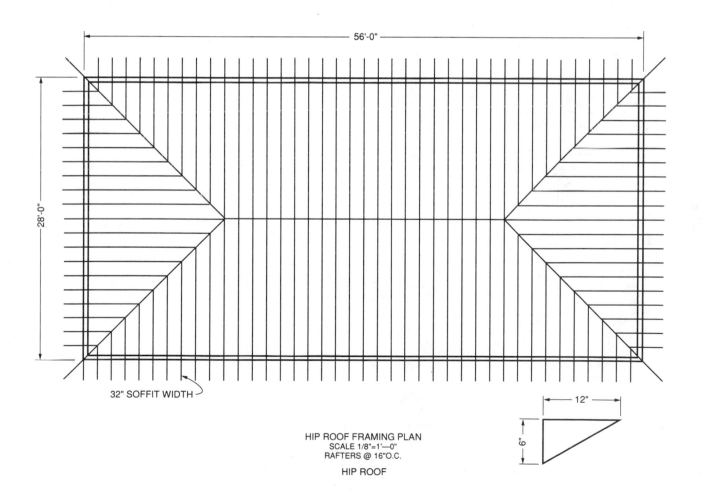

HIP ROOF FRAMING PLAN
SCALE 1/8″=1′—0″
RAFTERS @ 16″O.C.

HIP ROOF

2. Given: 5/12 roof slope, 16 inches OC rafter
spacing. Find the theoretical lengths of entire
roof frame of roofs A and B, to plate and tail

end. Roof B may be shortened 4 feet to 6 feet
to create cripple jacks.

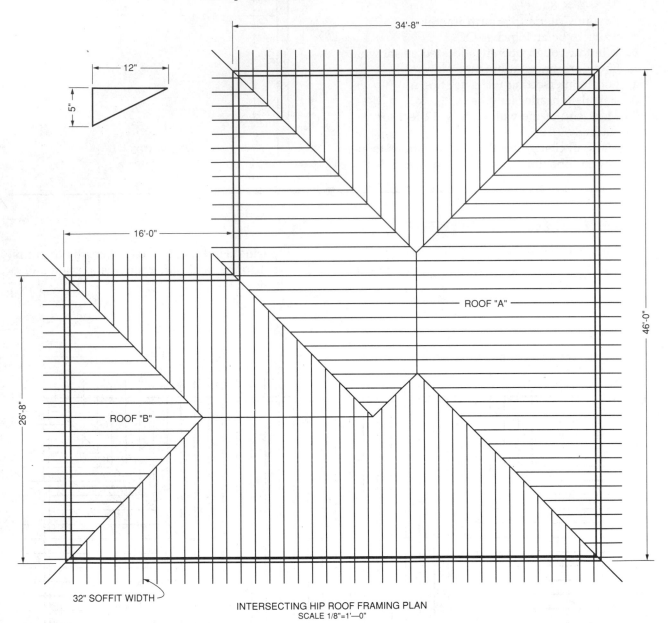

INTERSECTING HIP ROOF FRAMING PLAN
SCALE 1/8"=1'—0"

INTERSECTING HIP ROOF

3. Substitute new dimensions and slope in roof
plan #2 for additional practice.

4. Lay out the end cut for a 2″ × 6″ ceiling joist

having a 7/12 slope cut at 4 inches above the
plate.

Unit 16 Stair Framing

1. Solve for the missing dimensions in the stair problem. Total rise, unit rise, total run, unit run, and headroom.

STAIR SECTION
SCALE 3/8"=1'—0"

2. Lay out and cut the stair stringer in the drawing.

STAIR STRINGER
SCALE 3/8"=1'—0"

Unit 17 Insulation and Ventilation

1. Make a display board of the insulation used in residential and light commercial construction. List the manufacturer, product name, and costs.

2. Make a display of selected roof and soffit venting systems. List the manufacturer, product name, and costs.

Exterior Finish

UNIT

18

Cornice Construction

OBJECTIVES

After completing this unit, the student will be able to:

- describe various types of cornices and name their parts.
- build box, snub, and open cornices.
- apply rake trim and build a cornice return.
- install gutters and downspouts.

CONTENTS

The exterior finish work where the lower portion of the roof overhangs the wall is called the *cornice*. It is also commonly known as the *eaves*. In some parts of the country, this section is called the *coving*. (In this unit, it will be called the cornice.)

When constructing a building, one of the objectives is to make the roof tight as soon as practical. This permits interior work to progress without getting wet. The roof covering cannot be completed unless the cornice is installed. The cornice is a very visible and major part of the architectural design of a building. Therefore, much care should be taken in its construction. Because the cornice is so prominent, it is important that all cornice parts be installed straight and true with well-fitted joints.

CHAPTER 51

CORNICE TERMS AND DESIGN

Cornice terms from earlier times still remain in use. However, cornice design has changed considerably. In earlier times, cornices were very elaborate and required much time to build. Now, cornice design, in most cases, is much more simplified. Only occasionally is a building designed with an ornate cornice similar to those built in years gone by.

Parts of a Cornice

Several finish parts are used to build the cornice. In some cases, additional framing members are required.

Subfascia

The *subfascia* is sometimes called the *false fascia* or *rough fascia* (Fig. 51-1). It is a horizontal framing member fastened to the rafter tails. It provides an even, solid, and continuous surface for the attachment of other cornice members. When used, the subfascia is generally a nominal one- or 2-inch-thick piece. Its width depends on the slope of the roof, the tail cut of the rafters, or the type of cornice construction.

Lookouts

Lookouts are framing members, usually of 2×4 stock. They provide a fastening surface for the soffit of a wide cornice. They run from the wall to the end of the rafter. Lookouts may be installed on every rafter. Or, they may be spaced up to 48 inches OC, depending on the material used for the soffit (Fig. 51-2).

Soffit

The *soffit* is sometimes called the *planchir*. It is the finish member on the underside of a box cornice. Soffit material includes solid lumber, matched boards, plywood, strand board, or corrugated aluminum (usually with perforations for ventilation). Fiberboard, made especially for soffit use, is also available. It has precut screened openings for attic ventilation. Soffits may be installed horizontally or fastened to the bottom edge of the rafter tails to the slope of the roof. The soffit is an ideal location for the placement of attic ventilation (Fig. 51-3).

Fascia

The *fascia* is fastened to the subfacia or to the ends of the rafter tails. It is sometimes grooved to receive the soffit. It provides a surface for the attachment of a *gutter*. The fascia may be built up of one or more members to enhance the beauty of the cornice.

The bottom edge of the fascia usually extends below the soffit 1/4 to 3/8 inch. The portion of the fascia that extends below the soffit is called the *drip*. The drip is necessary when the fascia is grooved to support the soffit. Even if the fascia is not grooved,

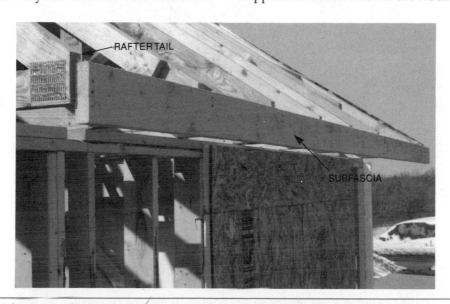

Fig. 51-1 The subfascia is a framing member that is fastened to the ends of the rafter tails.

RAFTER

ROOF SHEATHING

LOOKOUTS

WALL SHEATHING

Fig. 51-2 Lookouts are necessary framing members on wide box cornices. They provide backing for the soffit.

ROOF SHEATHING

RAFTER

PLATE

AIR
FLOW

SOFFIT
VENT

LOOKOUT

SOFFIT

WALL SHEATHING

A LEVEL SOFFIT IS FASTENED TO LOOKOUTS

Fig. 51-3 The soffit is the bottom finish member of the cornice. It may be fastened in a level or sloping position. Through it ventilation is provided to the attic.

A SLOPING SOFFIT IS FASTENED
TO THE BOTTOM EDGES OF RAFTER TAILS

Fig. 51-3 (*continued*)

providing a drip prevents rainwater from being swept back against the walls of the building by the wind. In addition, a drip makes the cornice more attractive. It is also easier than making a flush joint at that intersection (Fig. 51-4).

Fig. 51-4 The fascia is a member of the cornice finish that is attached to the ends of the rafters. It may be grooved to receive the soffit.

Frieze

The *frieze* is fastened to the sidewall with its top edge against the soffit. Its bottom edge is sometimes *rabbeted* to receive the sidewall finish. In other cases, the frieze may be furred away from the sidewall to allow the exterior wall *siding* to extend above and behind its bottom edge. However, the frieze is not always used. The sidewall finish may be allowed to come up to the soffit. The joint between the siding and the soffit is then covered by a molding.

Cornice Molding

The *cornice molding* is used to cover the joint between the frieze and the soffit. If the frieze is not used, the cornice molding covers the joint between the siding and the soffit (Fig. 51-5).

Cornice Design

Cornices are generally classified into three main types: *box, snub,* and *open.*

The Box Cornice

The *box cornice* is probably most common. It gives a finished appearance to this section of the exterior. Because of its overhang, it helps protect the side-walls from the weather. It also provides shade for windows.

Box cornices may be designated as narrow or wide. They may be constructed with level or sloping soffits. A *narrow box cornice* (Fig. 51-6) is one in which the cuts on the rafters serve as nailing surfaces for the cornice members. A *wide box cornice* may be constructed with a level or sloping soffit. A wide, level soffit requires the installation of lookouts.

The Snub Cornice

The *snub cornice* (Fig. 51-7) does not provide as attractive an appearance. Nor does it give as much protection to the sidewalls of a building because of its small overhang. There is no rafter projection beyond the wall. The snub cornice is chosen primarily to cut down the cost of material and labor. A snub cornice is frequently used on the *rakes* of a gable end in combination with a boxed cornice on the sides of a building.

The Open Cornice

The *open cornice* (Fig. 51-8) has no soffit. It is used when it is desired to expose the rafter tails. It is often used when the rafters are large, laminated or

RAFTER

ROOF SHEATHING

PLATE

FASCIA

SOFFIT

LOOKOUT

THE FRIEZE MAY BE RABBETED

FRIEZE

CORNICE MOULDING

CORNICE MOULDING

THE FRIEZE MAY BE FURRED

FRIEZE

Fig. 51-5 The frieze is attached to the sidewall just below the soffit. The cornice molding covers the joint between the frieze and the soffit.

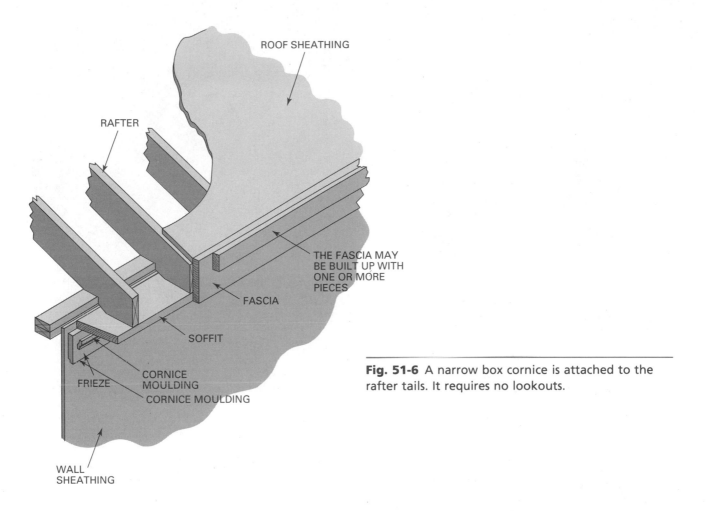

ROOF SHEATHING

RAFTER

THE FASCIA MAY
BE BUILT UP WITH
ONE OR MORE
PIECES

FASCIA

SOFFIT

CORNICE
MOULDING

FRIEZE

CORNICE MOULDING

WALL
SHEATHING

Fig. 51-6 A narrow box cornice is attached to the rafter tails. It requires no lookouts.

ROOF SHEATHING

RAFTER

FASCIA

THE FASCIA MAY
BE RABBETED AS
SHOWN OR FURRED
AWAY FROM THE
WALL TO RECEIVE
THE WALL SIDING.

WALL PLATE

WALL
SHEATHING

Fig. 51-7 The snub cornice projects only a small amount beyond the sidewalls of a building.

Fig. 51-8 The rafter tails are exposed when an open cornice is used.

solid beams with a wide overhang that exposes the roof decking on the bottom side. Open cornices give contemporary or rustic design to post-and-beam framing. They provide protection to sidewalls at low cost. This cornice might also be used for conventionally framed buildings for reasons of design and to reduce costs.

The bottom side of the roof sheathing on open cornices must present an attractive appearance. Nails or staples must not be driven completely through the exposed sheathing. To present a more attractive appearance, *tongue-and-grooved* stock, either *V-grooved* or *beaded*, may be used. It is used face down as *roof sheathing* on the part that overhangs the wall. If the thickness of the sheathing covering the roof is not the same as the sheathing covering

the overhang, the difference is feathered out by using wood shingles as wedges on top of the rafter (Fig. 51-9).

Rake Cornices

The *main cornice* is constructed on the rafter tails where they meet the walls of a building. On buildings with hip or mansard roofs, the main cornice, regardless of the type, extends around the entire building.

On buildings with gable roofs, a boxed main cornice with a sloping soffit, attached to the bottom edge of the rafter tails, may be returned up the rakes to the ridge. The cornice that runs up the rakes is called a *rake cornice* (Fig. 51-10).

SECTION THROUGH OPEN CORNICE

Fig. 51-9 The underside of exposed roof sheathing in an open cornice must present an attractive appearance.

Fig. 51-10 A boxed main cornice with a sloping soffit may be returned up the rakes of a gable roof.

Cornice Returns. A main cornice with a horizontal soffit attached to level lookouts may, at times, be terminated at each end wall against a snub rake cornice (Fig. 51-11). At other times, a *cornice return* must be constructed to change the level box cornice to the angle of the roof.

A main cornice with a level soffit may also be *returned upon itself.* That is, the main cornice is mitered at each end. It is turned 90 degrees toward and beyond the corner as much as it overhangs on the side of the building. This short section on each end of a gable roof is called a *cornice return.* The cornice return provides a *stop* for the *rake cornice.* It adds to the design of the building at this point (Fig. 51-12). However, cornice returns of this type are rarely built today. A large amount of labor is involved in their construction. The main cornice is returned on the rakes of the gable end in a much more simple fashion as described.

Fig. 51-11 A boxed main cornice with a level soffit may be terminated at the gable ends against a snub rake cornice.

Fig. 51-12A A boxed main cornice with a level soffit may be changed to the angle of the roof with a simple cornice return.

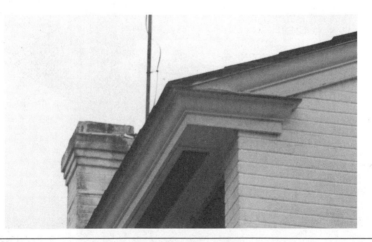

Fig. 51-12B A main cornice with a level soffit may be returned upon itself at the gable ends.

CHAPTER 52 BUILDING CORNICES

Whenever possible, the *ends* of each piece of finish are *butted* or *stopped* against other pieces. This is so that either all, or most of the end, is hidden from view. It is not good practice to leave the ends of any finish exposed. The end grain of lumber takes a finish differently than the rest of the lumber. This detracts from the appearance of the finish.

All exposed *edges* and *sides* of finish work, although not necessarily sanded, should be planed smooth, unless specified to be rough. All exposed sharp corners should be slightly rounded over. A finish, whether paint, stain, or some other kind, will shrink away from a sharp corner. Sharp corners can cut the skin. A slightly rounded corner is more attractive than a sharp one. It is another indication of a professional's work (Fig. 52-1).

Building a Narrow Box Cornice

A narrow box cornice requires no lookouts for support of the soffit. The first member of the narrow box cornice to be installed is the soffit. The level cut on the tail of the rafter to which the soffit is to be attached has already been made.

Installing the Soffit

Assuming that the narrow box cornice soffit material is nominal one-inch solid lumber and not to be fitted into a groove in the fascia, prepare the soffit pieces. Slightly back-bevel the outside edges with a jack plane. This assures a tight fit between the fascia and the soffit on the bottom, where it is seen from the ground (Fig. 52-2). When back-beveling, make sure the outside edges are straight and have no

crook. If necessary, straighten any crook while planing. Sight the edges to make sure each is straight from one end to the other.

Square and cut one end of the soffit material if the cornice is not returned around the corner. If the cornice is to be returned, miter the end at a 45 de-

Fig. 52-2 The outside edge of the soffit is slightly back-beveled. This assures a tight fit with the fascia.

Fig. 52-1

PLYWOOD ROOF
SHEATHING

RAFTER

STRAIGHTEDGE

LOOKOUT

STUD

USE A DADOED
STRAIGHTEDGE
IF THE SOFFIT
PROJECTS BEYOND
THE RAFTER.

SOFFIT

Fig. 52-3 Use a short straightedge to position the outside edge of the soffit.

gree angle. Measure and cut the other end square so it falls on the center of one of the rafter tails. Tack this piece in position against the level cut of the rafter tails. Keep the outside edge even with the plumb cut on the rafter tail. Use a short straightedge on the plumb cut of the rafter tail to determine this, if necessary (Fig. 52-3).

The position of the first piece of soffit depends on the type of rake cornice. Position the piece so its end will be compatible with the rake cornice or the cornice return. In the case of a hip or mansard roof, position it so the soffit material can be joined by the finish on the other side. Square the ends and tack successive lengths of soffit material in position until the end of the cornice is reached. Make the last end-cut so it can be returned or stopped as the case may be.

Undercutting the Ends of Cornice Finish.

Make a slight *undercut* on ends of finish that are to be joined when only the face side shows. If one or both edges are exposed, make a square cut for a short distance from the edge. Then, undercut slightly. Undercutting ends allows a tight joint to be made on the face side more easily. Undercutting also assures that a *standing cut* is not made on the end. A *standing cut* is the opposite of an undercut. The ends on the back side butt up before the face side comes up tight (Fig. 52-4).

Sighting the Outside Edge of the Soffit.

After all lengths of soffit material are tacked in place, sight the outside edge of the entire length for straightness in and out (horizontally). Remove nails, as necessary. Move the soffit boards in or out until the outside edges of all boards are in a straight line from one end of the building to the other. Sight by eye from both ends. Make sure the outside edges are in a perfectly straight line.

Once the soffit is straight horizontally, sight it for straightness up and down (vertically). Shim any high spots with wood shingles between the soffit and the rafter.

HELPFUL HINTS

Slightly undercut end joints on exterior trim. This assures a tight joint on the face side.

UNDERCUT (EXAGGERATED)

FACE SIDE

Fig. 52-4

When all boards are straight, both horizontally and vertically, nail them into position. Boards that are 8 inches wide or over require three nails. Those that are 6 inches wide or less require only two nails into each rafter tail. Use galvanized common or casing nails that are at least three times longer than the thickness of the soffit material. Set each nail below the surface. Install soffits in the same way wherever they are required on the rest of the building.

Installing the Fascia

Cut a square or mitered end of the fascia material as may be necessary. Cut the other end square so it rests on the center of a rafter tail. Tack the first piece in position. Tack succeeding pieces in position in a way similar to that used in installing the soffit. When all pieces are tacked into position, sight the

Fig. 52-5 The fascia drips below the soffit by at least 1/4 inch.

fascia for straightness up and down. There is no need to sight for straightness horizontally. Nailing the fascia against the edge of the soffit will straighten it in this plane. The fascia should extend below the soffit at least 1/4 inch (Fig. 52-5). Move the fascia boards as necessary by pulling the nails and retacking to straighten them.

When all of the fascia material is in a straight line, nail it to the rafter tails (Fig. 52-6). Use two or three nails in each rafter tail as required. Use galvanized nails and set them. Nail through the fascia at about 16-inch intervals into the edge of the soffit.

Use additional fasteners where required to assure a tight joint between the fascia and soffit.

A strip of 1×2 or 1×3 is usually added at the top edge of the fascia. This provides more overhang and improves its appearance. Snap a line on the fascia. Nail the strip to the fascia so its bottom edge is to the line. Use galvanized nails. Set them below the surface. Plane the top edge of the strip to the slope of the roof, if necessary.

Installing the Frieze

Before installing the frieze, first apply a narrow strip of building paper on the wall with its top edge against the soffit. Its bottom edge should project below the frieze about 2 to 3 inches (Fig. 52-7). This assures a more weatherproof joint between the bottom edge of the frieze and the exterior siding. When the rest of the building paper is applied prior to the exterior siding, the top edge is tucked under this strip.

Apply *furring strips* if the frieze is to be *furred* away from the wall sheathing. Or, rabbet the bottom edge of the frieze stock to receive the exterior siding (Fig. 52-8). Fasten the frieze to the sidewall. Keep its top edge against the soffit. Use two or three galvanized common or casing nails into each stud according to the width of the frieze. Set all nails below the surface.

The ends of the frieze are mitered if they are to be returned around the corner of the building. The ends are cut square if they are to be butted against

Fig. 52-6 The fascia is fastened to rafter tails.

Fig. 52-7 Fasten sheathing paper in back of and extending below the frieze.

Fig. 52-8 Methods of preparing the frieze to receive the wall finish.

corner boards. If butted against corner boards, the corner boards are installed before the frieze. The installation of corner boards is described in a later unit.

Installing Cornice Molding

The cornice molding covers the joint between the frieze and the soffit. The molding is cut square on the ends if butted against the rake trim. It is mitered if it is to be returned around the corner. Fasten the molding by driving galvanized finish, casing or common nails through its center at about a 45 degree angle, approximately 16 inches OC, or as necessary. Make a tight joint between the frieze and soffit. Sometimes no frieze is used. The cornice molding is installed after the exterior siding has been applied (Fig. 52-9).

Building a Wide Box Cornice

A wide box cornice differs from a narrow box cornice. Lookouts must be installed to support the soffit. To install lookouts, first level from the end of the rafter tail to the wall sheathing. Do this on both ends of the wall on which the cornice is to be built. Snap a line against the wall sheathing between the two marks. Apply a piece of 1×3 lumber to the wall sheathing so its bottom edge is to the line.

Fasten the inside end of the lookouts by toenailing into the 1×3. Keep the bottom edges flush. Fasten the outside end of the lookouts to the rafter tails. Keep the bottom edge flush with the level cut on the tail. See Figure 51-2.

Installation of other members of the cornice is then completed in a manner similar to that used for building a narrow box cornice. If the outside edge of the soffit fits into a groove in the fascia, the soffit

edge is extended beyond the tail cut of the rafter a little less than the depth of the fascia groove. A short straightedge with a dado can be used to position the soffit (Fig. 52-3).

Building a Sloping Box Cornice

In a sloping box cornice (Fig. 52-10), the soffit is applied to the bottom edges of the rafter overhang. No lookouts are required. If the fascia is to be in-

CORNICE SECTION

Fig. 52-9 The cornice molding covers the joint between the frieze and the soffit.

Fig. 52-10 The sloping box cornice eliminates the need of lookouts.

stalled plumb, the outside edge of the soffit must be beveled to the slope of the roof. If the fascia is installed square to the roof slope, the outside edge of the soffit is cut square. The soffit is installed and straightened in the same way as horizontal soffits. The fascia and other members of the cornice are then applied in a manner similar to that described previously.

Building a Snub Cornice

In a snub cornice (Fig. 52-11) the rafter tail cut is made flush with the outside edge of the wall framing. The wall sheathing is applied up to the top edge of the rafter. The roof sheathing is then applied over the top edge of the wall sheathing.

In a snub cornice, no soffit is applied. The fascia acts also as the frieze. The bottom edge may be rabbeted to receive the exterior wall siding. The fascia may also be furred away from the wall. This allows the siding to be applied under it.

To install the fascia, first apply a narrow strip of sheathing paper to the top of the wall. Its bottom edge should extend a few inches below the fascia when installed. Mark the bottom edge of the fascia on both ends of the building. Snap a chalk line between the two marks. Install the fascia so its bottom edge is to the line.

For ventilation, cut holes in the fascia. Center them on the rafter spaces to receive circular aluminum louvers. These louvers are available in sizes from 1 to 3 inches in diameter. They press fit into the holes. The holes are usually cut using a *hole saw* of the desired diameter.

Fig. 52-11 The snub cornice has no soffit. It provides little protection for the sidewall.

Use two or three galvanized nails, as necessary, through the fascia into the ends of each rafter. Set all nails below the surface. Make end-cuts that are compatible with the rake finish or for returning around the corner of the building.

Make square cuts between fascia pieces. Block plane the joints as necessary to make a tight fit. Make a slight undercut on end joints. This facilitates a tight joint on the face of the boards. Some carpenters prefer to make mitered end joints between sections of fascia (Fig. 52-12). The type of joint is just a matter of preference. If necessary, plane the top edge of the fascia flush with the slope of the roof.

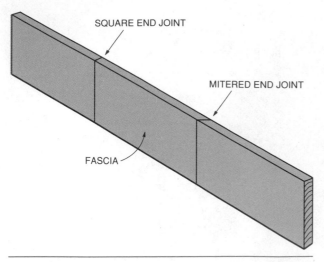

Fig. 52-12 End joints on fascia members may be square or mitered as preferred.

Building an Open Cornice

The open cornice is somewhat simpler to build. No lookouts or soffit are required. However, the frieze must be cut between the rafters (Fig. 52-13). Plumb cuts on rafter tails must be made in an absolutely straight line. Any shimming of the fascia to straighten it will be exposed to view.

Test the straightness of rafter tail cuts with a straightedge. Or, stretch a line over small blocks from one end rafter to the other. Trim any rafter tails that do not lie in a straight line.

Installing the Frieze

The frieze is usually installed first. Cut and apply some building paper. It should lie behind the frieze when installed, and extend below its bottom edge by about 2 or 3 inches. The frieze is sometimes installed so its bottom edge is flush with the bottom edge of the rafters. In this case, separate pieces of building paper are cut between rafters.

In other cases, the bottom of the frieze may extend a few inches below the bottom edge of the rafters. The frieze is installed in long lengths. Notches in the pieces fit around each rafter.

In both methods, the frieze must be marked and cut to fit snugly against the sides of the rafters. These cuts are not necessarily square to the edge of the frieze. The rafters may be tilted slightly one way or the other.

To assure tight fits, tack a section of frieze stock to the wall with its top edge, previously beveled to the roof slope, against the bottom edge of the rafters. Hold a straightedge that extends across the face of the frieze against the side of the rafter. Mark a line across the frieze. Lay out for the next rafter in the same manner.

Square lines from a point where the top of the frieze meets the side of the rafters to the bottom edge of the frieze. From a point where the squared line meets the bottom edge of the frieze, draw lines parallel to the first lines laid out. These are the cutting lines. Make the cuts with a slight undercut (Fig. 52-14).

If the frieze is to be notched around the rafters, the squared line is brought down to the bottom of the notch. The depth of the notch is determined by measuring from a chalk line snapped on the exterior wall for the location of the bottom edge of the frieze to the bottom edge of the rafter.

STEP BY STEP
P R O C E D U R E S

Fig. 52-14 Laying out the frieze of an open cornice to fit between the rafters.

Fig. 52-13 In an open cornice, the frieze is cut between the rafter tails. It extends up to the bottom side of the roof sheathing.

The frieze is fitted between rafters in this manner along the entire length of the wall. It is fastened in position. Nailing surfaces for the frieze are provided by nailing cleats to the side of each rafter or by installing blocking between the rafters. The top edge of the frieze, when properly installed, should fit snugly against the bottom side of the roof sheathing.

The frieze may be vented in the same manner as the snub cornice. Small metal circular louvers may be installed between rafters. Bore holes of the specified size. Simply tap the louvers into the bored holes, as described previously for a snub cornice.

Installing the Fascia

Installing the fascia on an open cornice is done in much the same manner as in a closed cornice. Since no soffit is used, the fascia usually drips below the bottom edge of the rafter tail cut by at least 1/4 inch. However, designs differ. In some cases, the bottom part of the tail cut of the rafter is left exposed. In any case, the fascia must be installed in a straight, true line with well-fitted joints between the sections.

Rake Cornices

The simplest type of rake trim is the *snub rake cornice*. This usually consists of a 1×6 or 1×8 board, furred away from the wall sheathing by the thickness of the exterior siding. Usually a 1×2 or 1×3 is added flush with the top edge. This gives it added dimension and provides more overhang (Fig. 52-15). This cornice may be used in combination with any kind of main cornice. If the main cornice is boxed, a piece, called a *dutchman* is added to the bottom edge of the rake fascia. This closes in the box cornice (Fig. 52-16). The term *dutchman* is given to any odd piece applied to fill or cover an opening or a defect.

Installing a Snub Cornice on the Rakes

Determine the thickness of the gable end exterior siding. Apply a furring strip of the same thickness to the gable end. Its top edge should be flush with the top side of the roof sheathing.

Tack the rake fascia to the furring strip. Keep the top edges flush with each other. Let the top end of the fascia extend above the roof on the opposite side. Let the lower end extend by the fascia of the main cornice by an amount sufficient to stop the *gutter,* plus a few inches. Mark the intersection at the peak of the roof. Hold a level at the centerline of the ridge. Mark a plumb line on the rake fascia.

Remove the fascia. Cut to the line. Fasten in position so that the plumb cut is in the centerline of

Fig. 52-15 A snub cornice is frequently used on the rakes of a gable roof in combination with other main cornice types.

Fig. 52-16 A "dutchman" is added to the rake cornice to close in the end of the boxed main cornice.

the ridge. To make sure the fascia is fastened in the same position as when marked, drive nails into the same holes made when the fascia was originally tacked in position. Fasten the fascia along the top edge only about every 16 inches. The bottom edge is fastened after the wall siding is applied. Mark, cut, and install the furring strip and rake fascia on the opposite side in the same manner. Build up the fascia with 1×2 or 1×3 strips as may be specified.

The lower end of the rake fascia is usually trimmed after the gutter is installed. Usually this end is trimmed with a combination of plumb and level cuts. The plumb cut is made about an inch outside the gutter. The level cut is usually made flush with the bottom edge of the fascia of the main cornice. The dutchman is usually installed after the wall siding has been applied.

Box Cornices on the Rakes

Box cornices on the rakes are built in about the same manner as they are on sidewalls. However, they are built to the slope of the roof. When the rake extension is only 6 to 8 inches, the members of the rake cornice can be nailed to a series of short lookout blocks. The blocks are toenailed to the gable end sheathing and also fastened to the projecting roof sheathing (Fig. 52-17).

In a moderate overhang of up to 20 inches, both the extending sheathing and a fly rafter aid in supporting the rake cornice. The *fly rafter* or *barge rafter* extends from the extended ridge board to the subfascia that connects the ends of the rafters. The roof sheathing should extend from inner rafters to the end of the rake projection. This provides rigidity and strength (Fig. 52-18). An extrawide overhang at the rake requires a ladder type frame for support of the rake cornice (Fig. 52-19).

Fig. 52-17 A narrow box cornice on the rakes.

Fig. 52-18 A moderate overhang at the rakes requires a fly rafter.

Fig. 52-19 An extrawide rake cornice requires a ladder type frame for support.

A *gutter* is a shallow trough or conduit set below the edge of the roof along the fascia. It catches and carries off rainwater from the roof. A *downspout,* also called a *conductor,* is a rectangular or round pipe. It carries water from the gutter downward and away from the foundation (Fig. 53-1).

Gutters

Gutters may be made of wood, galvanized iron, aluminum, copper, or vinyl. Copper gutters require no finishing. Vinyl and aluminun gutters are prefinished and ready to install. Wood and galvanized metal gutters need several protective coats of finish after installation.

The size of a gutter is determined by area of the roof from which it handles the water runoff. Under ordinary conditions, one square inch of gutter cross-section is required for every 100 square feet of roof area. For instance, a 4×5 inch gutter has a cross-section area of 20 square inches. It is, therefore, capable of handling the runoff from 2,000 square feet of roof surface.

Wood Gutters

Wood gutters are usually made of Douglas fir, Western red cedar, or California redwood. They come in sizes of 3×4, 4×5, 4×6, and 5x7, and lengths of over 30 feet. Although wood gutters are not used as extensively as in the past, they enhance the cornice design. When properly installed and maintained, they will last as long as the building itself.

Metal Gutters

Metal gutters are made in rectangular, beveled, ogee, or semicircular shapes (Fig. 53-2). They come in a variety of sizes, from 2 1/2 inches to 6 inches in height and from 3 inches to 8 inches in width. Stock lengths run from 10 to 40 feet. *Forming machines* are often brought to the job site to form the gutters to practically any desired length.

Besides straight lengths, metal gutter systems have components comprised of *inside* and *outside corners, joint connectors, outlets, end caps,* and others. *Metal brackets* or *spikes and ferrules* are used to support the gutter sections on the fascia (Fig. 53-3).

Laying Out the Gutter Position

Gutters should be installed with a slight pitch of about 1/4 inch to 10 feet. This allows water to drain toward the downspout. A gutter of 20 feet or less in length may be installed to slope in one direction with a downspout on one end. On longer buildings, the gutter is usually crowned in the center. This allows water to drain to both ends.

On both ends of the fascia, mark the location of the bottom side of the gutter. The top outside edge of the gutter should be in relation to a straight line projected from the top surface of the roof. The height of the gutter depends on the pitch of the roof (Fig. 53-4).

Stretch a chalk line between the two marks. Move the center of the chalk line up enough to give

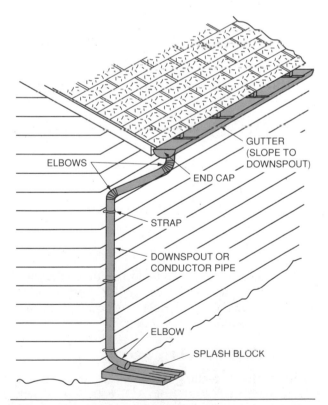

Fig. 53-1 Gutters and downspouts are an important system for conducting water away from the building. Otherwise, it might accumulate in crawl spaces or seep into basements.

OTHER SHAPES ARE AVAILABLE

Fig. 53-2 Metal gutters are available in several shapes.

PIECE NEEDED	DESCRIPTION
	GUTTER COMES IN 10' LENGTH
	SLIP JOINT CONNECTOR USED TO CONNECT JOINTS OF GUTTER
	END CAPS — WITH OUTLET USED AT ENDS OF GUTTER RUNS
	END PIECE — WITH OUTLET USED WHERE DOWNSPOUT CONNECTS
	OUTSIDE MITRE USED FOR OUTSIDE TURN IN GUTTER
	INSIDE MITRE USED FOR INSIDE TURN IN GUTTER
	FASCIA BRACKET USED TO HOLD GUTTER TO FASCIA ON WALL
	STRAP HANGER CONNECTS TO EAVE OF ROOF TO HOLD GUTTER
	STRAINER CAP SLIPS OVER OUTLET IN END PIECE AS A STRAINER
	DOWNSPOUT COMES IN 10' LENGTHS
	ELBOW — STYLE B FOR DIVERTING DOWNSPOUT IN OR OUT FROM WALL
	ELBOW — STYLE A FOR DIVERTING DOWNSPOUT TO LEFT OR RIGHT
	CONNECTOR PIPE BAND OR CLINCHER USED TO HOLD DOWNSPOUT TO LEFT OR RIGHT
	SHOE USED TO LEAD WATER TO SPLASHER BLOCK
	MASTIC USED TO SEAL ALUMINUM GUTTERS AT JOINTS
	SPIKE & FERRULE USED TO HOLD GUTTER TO EAVE OF ROOF

STRAP HANGER

FASCIA BRACKET

END PIECE

SPIKE & FERRULE

GUTTER

END CAP   SLIP CONNECTOR

DOWNSPOUT

CONDUCTOR PIPE
BAND (CLINCHER)

GUTTERS AND DOWNSPOUTS ARE MADE UP OF MANY
SEPARATE PARTS

Fig. 53-3 Components of a metal gutter system.

the gutter the proper pitch. Thumb the line down. Snap it on both sides of the center. For a slope in one direction only, snap a straight line lower on one end than the other to obtain the proper pitch. It is important to install gutters on the chalk line. This avoids any dips that may prevent complete draining of the gutter.

Installing Wood Gutters

The gutter should be *furred* away from the fascia. This allows air to circulate behind the gutter. It prevents moisture from accumulating there with the resulting growth of fungi and rotting of the gutter, fascia, and other cornice trim. Use short lengths of *lattice*, as furring. Apply them vertically to the fascia opposite the end of each rafter.

Applying Furring Strips

Thin strips of rectangular molding approximately 5/16 inch thick and from 1 1/8 to 1 3/4 inches wide are called lattice. Ordinarily the 1 3/4-inch width is used to fur a wood gutter. Cut the strips so they will extend from the top of the gutter to the bottom of the *gutter molding* when fastened to the

Fig. 53-4 The height of the gutter on the fascia is in relation to the pitch of the roof.

fascia. The gutter molding is applied below the gutter and against the furring strips (Fig. 53-5).

Methods of Stopping the Gutter

The gutter is mitered around the cornice return. Or, it is butted against the inside of a snubbed rake cornice (Fig. 53-6). Turn the gutter upside down in the miter box to make the cut. Make the cut from the bottom side. It is the bottom side that is viewed from the ground. Also, it is easier to hold in that position in a miter box. Cut both ends of the first section of gutter.

Installing Goosenecks

Bore holes and install *goosenecks* in the gutter wherever downspouts are to be located. Goosenecks are short sections of lead pipe with a flange on one end used for gutter outlets. The inside of the gutter is rabbeted around the hole, so the flange of the gooseneck will be flush when installed. A bed of caulking is placed in the rabbet. The gooseneck is fastened in place with copper tacks through the flange (Fig. 53-7).

Fastening the Gutter

Drive 8d common nails partway in at about 2-foot intervals along the chalk line. This supports the gutter while marking, fitting, and fastening it. Support the first section of gutter on the previously driven nails. Fasten it by toenailing through the top and bottom into the furring strip, fascia, and rafter tail (Fig. 53-8).

SNUB RAKE
CORNICE

LATTICE
FURRING STRIPS

GUTTER

GUTTER MOULDING

Fig. 53-5 The gutter is furred away from the fascia. This allows for circulation of air and to prevent the accumulation of moisture.

THE GUTTER RETURN IS LATER COVERED WITH A "WATER TABLE" ON WHICH THE RAKE CORNICE IS STOPPED.

PLAN

PICTORIAL

PLAN

PICTORIAL

Fig. 53-6 A wood gutter is either mitered around the cornice return or butted against the snub rake cornice.

COPPER TACKS

CAULK UNDER FLANGE

WOOD GUTTER

GOOSENECK

SECTION THROUGH GOOSENECK

Fig. 53-7 A gooseneck is a water outlet installed in a wood gutter.

Treating Joints between Gutter Sections

Mark, cut, and fit other sections, if necessary, in a similar manner until returned or butted on the other end. Usually wood gutters are long enough to be installed in one piece. However, if a joint between sections is necessary, a short length of sheet lead is installed over the joint. Bed it in caulking. Fasten with copper tacks flush with the inside of the gutter (Fig. 53-9).

Waterproofing End Joints

To prevent water from laying in the end of the gutter beyond the gooseneck, a length of sheet lead is

Fig. 53-8 Method of fastening a wood gutter to the fascia.

Fig. 53-9 Sealing the joint between sections of wood gutter with sheet lead.

formed to the inside of the gutter, up on the roof and rake trim or cornice return. Trim to size after forming. Caulk and fasten with copper tacks (Fig. 53-10). There is no need to recess it flush with the inside of the gutter.

Installing Metal and Vinyl Gutters

Fasten the brackets to the chalk line on the fascia with screws. The screws should be of the same met-

al as the brackets to prevent electrolysis. Aluminum brackets may be spaced up to 30 inches OC. Steel brackets may be spaced up to 48 inches OC. Install the gutter sections in the brackets. Use slip-joint connectors to join the sections. Apply the recommended gutter sealant to connectors before joining.

Install outlet sections at each end or where required. Join with a connector. Add the end cap. Use either inside or outside corners where gutters make a turn.

Vinyl gutters and components are installed in a manner similar to metal ones.

SHEET LEAD

BED EDGES IN CAULKING AND FASTEN WITH COPPER TACKS

GOOSENECK

FASCIA

SNUB RAKE CORNICE

GUTTER

GUTTER MOULDING

FURRING STRIP

Fig. 53-10 Method of preventing water from accumulating in the ends of wood gutters beyond the gooseneck.

Installing Downspouts

Metal or vinyl downspouts are fastened to the wall of the building in specified locations with aluminum straps. Galvanized iron rings, called *darts*, are used to fasten round corrugated galvanized iron downspouts. A dart consists of a hinged ring with a spike-like projection that can be driven into the wall. The dart is opened. It is then driven in the wall where desired. The downspout is placed in the open end. The hinged half is closed on it and fastened with copper wire. Downspouts should be fastened at the top and bottom and every 6 feet in between for long lengths.

The connection between the downspout and the gutter is made with 45 degree elbows and short straight lengths of downspout. The connection will depend on the offset of the gutter from the downspout. An elbow, called a *shoe*, should be used with a splash block at the bottom of the downspout. This leads water away from the foundation. An alternate method is to connect the downspout with piping to the foundation drain. This carries water away from the building into storm drains or drywells. Strainer caps should be placed over gutter outlets if water is

conducted by this alternate method. Leaves and other debris that fall into gutters flow into the foundation drainage system and can cause serious problems (Fig. 53-11).

Practical Tips for Installing Cornices

Install cornice members in a straight and true line with well-fitting, tight joints. Do not dismiss the use of hand tools for cutting, fitting, and fastening exterior trim.

Many times a hand tool is a better choice than a power tool. Because of its cutting action, a power saw splinters out the face side. A sharp handsaw can do the job just as fast and makes a neater job than a power saw. Joints may be fitted by block planing or by running a few handsaw cuts through the joint when possible.

All fasteners should be noncorrosive. Hot-dipped galvanized fasteners provide better protection against rusting than plated ones. Fasteners should be well set. They should then be puttied to conceal the fasteners. This also further prevents the heads from rusting or corroding. To set and conceal fasteners in exterior trim is a mark of quality.

Paint or otherwise protect all exterior trim as soon as possible after installation. Properly installed and protected wood exterior trim will last indefinitely.

Fig. 53-11 Downspouts are fastened in place with pipe bands or darts. Elbows are used at the top to connect downspouts to the gutter and at the bottom to lead water away from the foundation.

Review Questions

Select the most appropriate answer.

1. That part of the exterior finish where the walls of a building meet the roof is called the
a. cornice. c. ridge.
b. fascia. d. soffit.

2. The exterior trim that extends up the slope of the roof on the gable ends is called the
a. box finish. c. return finish.
b. rake finish. d. snub finish.

3. A member of the cornice generally fastened to the rafter tails is called the
a. drip. c. planchir.
b. fascia. d. soffit.

4. A member of the cornice usually applied under the soffit and against the sidewall is the
a. fascia. c. planchir.
b. frieze. d. rake.

5. The member of a box cornice usually installed first is the
a. fascia. c. soffit.
b. frieze. d. gutter.

6. The amount the fascia extends below the soffit is called the
a. drip. c. projection.
b. nosing. d. return.

7. The top edge of the fascia is flush with
a. the top edge of the rafter.
b. the top side of the roof sheathing.
c. the bottom side of the roof sheathing.
d. the top edge of the lookout.

8. A cornice that has no soffit is called a
a. box cornice. c. rake cornice.
b. closed cornice. d. snub cornice.

9. A decorative piece added to the bottom edge
of the rake finish to close in the end of a main
box cornice is sometimes called a
a. dutchman.   c. scab.
b. furring.   d. splice.

10. A device that carries water downward from the
roof and away from the building is called a
a. drip.   c. downspout.
b. gooseneck.   d. gutter.

■ BUILDING FOR SUCCESS ■

Construction-Related Student Chapters

A reasonable question each construction education program should ask itself is, Could our program be enhanced by implementing a construction-related student chapter? Every secondary and postsecondary education program that has some type of construction curriculum as its core needs to look closely at this question.

First, and probably foremost, the administration and faculty has to determine how an outreach of this type will benefit the program curriculum. They also have to find out if one or more people are willing to commit themselves to making it work. At least one faculty member has to be completely sold on the idea. Without this commitment, the chances of success will be minimal. The most success will come if two or three faculty members are convinced and willing to invest time in a student chapter.

The administration will need a clear picture of what the expected outcomes will be before support will be given. That picture can be clarified by the vision of the instructor, students, and local construction organizations. The buy-in plan has to be well presented in order for everyone involved to see how the students will benefit. Everyone who supports the formation of a student chapter will need to see how they will benefit from their time commitment and possible financial contribution. All parties will be somewhat tuned into to radio station WIIFM. (Refer to the Building for Success "Implementing A Safety Program.")

One of the most appropriate ways to research the need or desire for a student chapter is to meet with the local, district, or state trade associations that employ the program's graduates. The local advisory teams that advise the construction-related programs will often be the most interested in strengthening the curriculum. They usually see better employees as a benefit of student chapters. Many times this team will have a trade representative that will be willing to sponsor a student chapter. The representatives of the state or local home builders association or general contractors association, generally will be on the advisory team. They may be willing to sponsor a student chapter. Dozens of organizations, such as the National Association of Home Builders (NAHB), the Associated General Contractors (AGC), the Associated Builders and Contractors (ABC), and the American Society of Heating, Refrigeration and Air Conditioning Engineers (ASHRAE) support student chapters at high schools and colleges. In researching the need or desire to form a trade-related student chapter, a plan should be developed to accurately look at the benefits for everyone, the time involvement, the financial investment, the work, the commitment, the expected outcomes, and continuance of the chapter.

Essentially, the plan has to be accurate. It must reflect the interests of each particular school and construction program. The support from industry has to be there from the beginning. Few, if any, student chapters are effective if no industrial support exists.

The plan should begin with one or more meetings involving all interested parties. Initiation of the concept can begin at the regular advisory meeting. Or, it can be researched by a special team in preparation for the regular meeting. The designated local construction association will provide a leadership role along with the faculty member who initiates the chapter formation.

The benefits for all interested parties will need to be identified. Student benefits may center on more involvement with the local sponsoring association, scholarship funds for chapter members, community service opportunities, field trips to construction projects, learning more about the business of construction, or attending the local association meetings or national conventions. Institutional benefits could include more exposure to industry, state, and national recognition of a quality construction program and scholarship funds. Employer benefits could be better-prepared employees, workers who are more in tune with desired business or communication skills, and employees with a better understanding of how the various trades interrelate. Instructors will be working with students who desire to excel in their studies, learn more in-depth about the construction industry, and seek opportunities where they will meet other construction students across the nation.

The time involvement will be flexible depending how active the chapter becomes. Traditionally a chapter will become involved with fund raising activities, community service work, field trips, and local, state, or national functions. The faculty advisor and chapter officers may meet on a frequent basis. The chapter should meet once or twice a month on a regular basis. The local sponsoring association will be involved occasionally.

The financial involvement for the students will be minimal. Membership dues may be required by the national organization. Fund raising usually generates more finances to support chapter activities. Sponsoring local associations may assist in obtaining scholarship funds for student chapter members.

There will need to be time set aside for work and social activities. The faculty advisor will spend extra time each week planning chapter activities with the executive committee. Everyone who participates will have to make a commitment to reach the desired outcomes or goals. These goals must be identified and stated in the constitution and bylaws.

Perhaps the desire for the continuance of the student chapter will be the greatest challenge. Perseverance will become a great ally as the student chapter gets underway. For success, the focus on purpose must be fixed.

FOCUS QUESTIONS: For individual or group discussion.

1. What might be the purpose of participating in a construction-centered student chapter outreach?

2. How important would be the support of the local sponsoring construction association?

3. What might be some sound ideas for generating funds to support chapter activities?

4. What topics could become components of the student chapter constitution and bylaws?

U N I T

19

Roofing

O BJECTIVES

After completing this unit, the student will be able to:

- define roofing terms.
- describe and apply roofing felt, organic or fiber glass asphalt shingles, and roll roofing.
- describe various grades and sizes of wood shingles and shakes and apply them.
- flash valleys, sidewalls, chimneys, and other roof obstructions.
- estimate needed roofing materials.

Materials used to cover a roof and make it tight are part of the exterior finish called *roofing*. Roofing adds beauty to the exterior and protects the interior. Before roofing is applied, the roof deck must be securely fastened. There must be no loose or protruding nails. All cornice trim should be in place and primed. Properly applied roofing gives years of dependable service.

C ONTENTS

CHAPTER 54 ASPHALT SHINGLES

*A*sphalt shingles are the most commonly used roof covering for residential and light commercial construction. They are designed to provide protection from the weather for a period ranging from twenty to thirty years. They are available in many colors and styles.

Roofing Terms

An understanding of the terms most commonly used in connection with roofing is essential for efficient application of roofing material.

- A *square* is the amount of roofing required to cover 100 square feet of roof surface.

- *Deck* is the wood roof surface to which roofing materials are applied.

- *Coverage* is the number of overlapping layers of roofing and the degree of weather protection offered by roofing material. Roofing may be called single, double, or triple coverage, for example.

- A *shingle butt* is the bottom exposed edge of a shingle.

- *Courses* are horizontal rows of shingles or roofing.

- *Exposure* is the distance between courses of roofing measured from butt to butt. It is the amount of roofing in each course exposed to the weather (Fig. 54-1).

- The *top lap* is the height of the shingle or other roofing minus the exposure. In roll roofing this is also known as the *selvage*.

- The *head lap* is the distance from the butt of an overlapping shingle to the tip of a shingle two courses below measured up the slope.

- *End lap* is the horizontal distance that the ends of roofing in the same course overlap each other.

- *Flashing* are strips of thin sheet metal. They are usually made of lead, zinc, copper, or alumi-

num. They may also be strips of roofing material used to make watertight joints on a roof. Metal flashing comes in rolls of various widths. They are cut to the desired length.

- *Asphalt cements* and *coatings* are manufactured to various consistencies depending on the purpose for which they are to be used. *Cements* are classified as *plastic, lap,* and *quick-setting*. They will not flow at summer temperatures. They are used as adhesives to bond asphalt roofing products and flashings. They are usually troweled on the surface. *Coatings* are usually thin enough to be applied with a brush. They are used to resurface old roofing or metal that has become weathered.

Electrolysis is a reaction that occurs when unlike metals come in contact with water. This contact causes one of the metals to erode. A simple way to prevent the disintegration caused by electrolysis is to secure metal roofing material with fasteners of the same metal.

Preparing the Deck

A metal *drip edge* or a single course of wood shingles is installed along the roof edges. The metal drip edge is usually made of aluminum or galvanized iron. It is applied along the lower edge of the roof (Fig. 54-2). The drip edge or wood shingle course is used to support the asphalt shingle overhang. Otherwise the shingles would droop from the heat of the sun.

Install the metal drip edge by using roofing nails of the same metal spaced 8 to 10 inches along its inner edge. Lap end joints by about 2 inches.

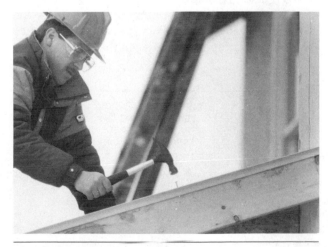

Fig. 54-2 Applying a metal drip edge.

Fig. 54-1 Asphalt strip shingle exposure and lap.

Fig. 54-3 Installing a wood shingle starter course.

A wood shingle starter course serves the same purpose as the metal drip edge. Apply a shingle with the desired overhang at each end of the roof. Stretch a line between them. Install temporary shingles at intermediate points. Fasten the line to their butts to provide a straight line from one end of the building to the other (Fig. 54-3).

Apply wood shingles along the edge of the roof. Their butts should come as close to the line as possible without touching it. Sight each shingle for a bow. Fasten it with the crown of the bow up. Use two 3d galvanized nails halfway up the shingle about 1/2 inch in from each edge. Also fasten two nails through each shingle into the top edge of the fascia. Continue applying shingles until the course is completed.

Underlayment

The deck should next be covered with an asphalt shingle *underlayment.* The underlayment protects the sheathing from moisture until the roofing is applied. It also gives additional protection to the roof afterward. Use an underlayment that allows the passage of water vapor. This prevents moisture or frost from accumulating between the underlayment and the deck.

Fig. 54-5 Application of asphalt shingle underlayment felt. (*Courtesy of Asphalt Roofing Manufacturers' Association*)

Asphalt Felts

Asphalt-saturated felts consist of heavy felt paper saturated with asphalt or coal tar. They are usually made in various weights of pounds per square (Fig. 54-4). Asphalt felt comes in 36-inch wide rolls. The rolls are 72 or 144 feet long covering 2 or 4 square. Usually the lightest weight felt is used as an underlayment for asphalt shingles.

Apply a layer of asphalt felt underlayment over the deck. Lay each course of felt over the lower course at least 2 inches. Make any end laps at least 4 inches. Lap the felt 6 inches from both sides over all hips and ridges (Fig. 54-5).

Nail or staple through each lap and through the center of each layer about 16 inches apart. Roofing nails driven through the center of metal discs or specially designed, large head felt fasteners hold the underlayment securely in strong winds until shingles are applied (Fig. 54-6). A metal drip edge should be installed along the rakes after the application of the underlayment.

	APPROX. WEIGHT PER ROLL	APPROX. WEIGHT PER SQUARE	SQUARES PER ROLL	ROLL LENGTH	ROLL WIDTH	SIDE OR END LAPS	TOP LAP	EXPOSURE
	60 #	15 #	4	144'	36"	4"	2"	34"
	60 #	30 #	2	72'	36"	TO		
SATURATED FELT	60 #	60 #	1	36'	36"	6"		

Fig. 54-4 Sizes and weights of asphalt-saturated felts.

Fig. 54-6 An asphalt felt underlayment is applied to the roof deck before shingling.

Low Slope Underlayment Application

If asphalt shingles are to be used on slopes less than 4 inches per foot, down to 2 inches per foot, special application of the underlayment is required. Lay a 19-inch course of felt along the eaves. This course should completely cover the drip edge or starter course. Then lay full-width sheets, each overlapping the preceding course by 19 inches and exposing 17 inches of the underlying sheet (Fig. 54-7). Use only enough fasteners to hold the felt securely to the deck until the shingles are applied.

The underlayment plies are then cemented together to form an *eaves flashing*. The eaves flashing should extend up the roof at least 24 inches beyond the inside wall line of the building. The layers are cemented together using a continuous layer of plastic asphalt cement. The cement should be applied uniformly with a notched trowel at a rate of two gallons per 100 square feet.

Kinds of Asphalt Shingles

Two types of *asphalt shingles* are manufactured. *Organic* shingles have a base made of heavy asphalt-saturated paper felt coated with additional asphalt. *Fiber glass* shingles have a base mat of glass fibers. The mat does not require the saturation process and only requires an asphalt coating. Both kinds of shingles are surfaced with selected mineral granules. The asphalt coating provides weatherproofing qualities. The granules protect the shingle from the sun, provide color, and protect against fire.

Fiber glass-based asphalt shingles have an Underwriters Laboratories Class A fire resistance rating. The Class A rating is the highest standard for resistance to fire. The Class C rating for organic shingles, while not as high, will meet most residential building codes.

Organic and fiber glass shingles come in a wide variety of colors, shapes, and weights (Fig. 54-8). They are applied in the same manner. Shingle quality is generally determined by the weight per square. Most asphalt shingles are manufactured with factory-applied adhesive. This increases their resistance to the wind.

Applying Asphalt Shingles

Before applying strip shingles, make sure that the roof deck is properly prepared. The underlayment and drip edge should be applied. Asphalt roofing products become soft in hot weather. Be careful not to damage them by digging in with heavy shoes during application or by unnecessary walking on the surface after application.

Roof Slope

The pitch of a roof should not be less than 4 inches per foot when conventional methods of asphalt

METAL DRIP EDGE AT RAKE APPLIED OVER UNDERLAYMENT

TIGHT WOOD DECK

METAL DRIP EDGE DIRECTLY ON DECK

FELT PLIES OF UNDERLAYMENT CEMENTED UP FROM EAVES FAR ENOUGH TO OVERLIE A POINT 24" INSIDE THE INSIDE WALL LINE OF THE BUILDING TO PROVIDE AN EAVE FLASHING

19"
36"
19"
19"
19"
17"

NO. 15 SATURATED FELT STARTER COURSE COVERED WITH ASPHALT CEMENT

FIRST AND SUCCEEDING COURSES OF UNDERLAY TO BE 36" WIDE AND LAPPED 19"

Fig. 54-7 Method of applying underlayment prior to shingling roofs with low slopes. (*Courtesy of Asphalt Roofing Manufactures' Association*)

| PRODUCT | Configuration | Per Square | | | Size | | Exposure | Under-writers' Listing |
		Approx. Shipping Weight	Shingles	Bundles	Width	Length		
Wood Appearance Strip Shingle More Than One Thickness Per Strip Laminated or Job Applied	Various Edge, Surface Texture & Application Treatments	285 # to 390 #	67 to 90	4 or 5	11-1/2" to 15"	36" or 40"	4" to 6"	A or C — Many Wind Resistant
Wood Appearance Strip Shingle Single Thickness Per Strip	Various Edge, Surface Texture & Application Treatments	Various 250 # to 350 #	78 to 90	3 or 4	12" or 12-1/4"	36" or 40"	4" to 5-1/8"	A or C — Many Wind Resistant
Self-Sealing Strip Shingle	Conventional 3 Tab	205 #– 240 #	78 or 80	3	12" or 12-1/4"	36"	5" or 5-1/8"	A or C — All Wind Resistant
	2 or 4 Tab	Various 215 # to 325 #	78 or 80	3 or 4	12" or 12-1/4"	36"	5" or 5-1/8"	
Self-Sealing Strip Shingle No Cut Out	Various Edge and Texture Treatments	Various 215 # to 290 #	78 to 81	3 or 4	12" or 12-1/4"	36" or 36-1/4"	5"	A or C — All Wind Resistant
Individual Lock Down Basic Design	Several Design Variations	180 # to 250 #	72 to 120	3 or 4	18" to 22-1/4"	20" to 22-1/2"	—	C — Many Wind Resistant

Fig. 54-8 Asphalt shingles are available in a wide variety of sizes, shapes, and weights. (*Courtesy of Asphalt Roofing Manufacturers; Association*)

shingle application are used. On lower roof slopes, down to 2 inches rise per foot of run, *self-sealing* asphalt shingles are used. If "free tab" strip shingles are used, cement each tab down with a spot of quick-setting asphalt cement about 1 1/2 inches in diameter (Fig. 54-9). A double coverage felt underlayment for low slopes must also be provided as previously described.

Asphalt Shingle Layout

On small roofs, strip shingles are applied by starting from either rake. On long buildings, a more accurate vertical alignment is insured by starting at the center and working both ways (Fig. 54-10). Mark the center of the roof at the eaves and at the ridge. Snap a chalk line between the marks. Snap a series of chalk lines 6 inches apart on each side of

the centerline if the shingle tab cutouts are to break on the halves. Snap lines 4 inches apart if the cutouts are to break on the thirds. When applying the shingles, start the course with the end of the shingle to the vertical chalk line. Start succeeding courses in the same manner. Break the joints as necessary. Pyramid the shingles up in the center. Work both ways toward the rakes.

If it is decided to start at the rakes and cutouts are to break on the halves, start the first course with a whole tab. The second course is started with a shingle from which 6 inches have been cut; the third course, with a strip from which the entire first tab is removed; the fourth, with one and one-half tabs removed, and so on (Fig. 54-11). These starting strips are precut for faster application. Waste from these strips is used on the opposite rake.

LOCATION OF CEMENT UNDER
SQUARE BUTT TABS

SPOT OF ASPHALT
CEMENT UNDER CENTER
OF EACH TAB

LOCATION OF CEMENT UNDER
UNDER HEX TABS

Fig. 54-9 On low slopes, use self-sealing shingles or spot free tabs with asphalt cement. (*Courtesy of Asphalt Roofing Manufacturers; Association*)

Fig. 54-10 On long roofs, start shingling in the center. Work toward the rakes.

If the cutouts are to break on the thirds, cut the starting strip for the second course by removing 4 inches. Remove 8 inches from the strip for the third course, and so on (Fig. 54-12).

Cut the shingles by scoring them on the back side with a utility knife. Use a square as a guide for the knife. Bend the shingle. It will break on the scored line.

The layouts may have to be adjusted so that tabs on opposite rakes will be of approximatly equal widths. No rake tab should be less than 3 inches in width.

Starter Course of Asphalt Shingles

The *starter course* backs up and fills in the spaces between tabs of the first regular course of shingles. Cut a minimum of 3 inches off the butt of the shingles needed for the starter course. Apply them with the factory-applied adhesive along the roof edge. Start the course so that no end joint will fall in line with an end joint or tab cutout of the regular first

NAILING

2" HEAD LAP

UNDERLAYMENT

1"

5-5/8"

4" END LAP

WOOD DECK

EAVES FLASHING
STRIP

METAL DRIP
EDGE APPLIED
OVER FELT
ALONG RAKE

SELF-SEALING STRIP

START 3RD COURSE
WITH FULL STRIP
MINUS FIRST TAB

START SECOND COURSE
WITH FULL STRIP MINUS
1/2 TAB

START FIRST COURSE
WITH FULL STRIP

METAL DRIP
EDGE

STARTER – FULL
SHINGLE MINUS 3"
WITH TABS CUT OFF

Fig. 54-11 Layout to stagger cutouts on the halves starting from the rake.

Fig. 54-12 Layout to stagger cutouts on the thirds starting from the rake.

course of shingles. If the rake trim projects beyond the fascia, a shingle tab about 4 or 5 inches wide is installed to cover the top edge of the rake trim (Fig. 54-13). A wood shingle should first be placed under the tab to give it support.

Fastening Asphalt Shingles

Selecting suitable fasteners, using the recommended number, and putting them in the right places are important steps in the application of asphalt shingles. Lay the first regular course of shingles on top of the starter course. Keep their bottom edges flush with each other. Use a minimum of four fasteners in each strip shingle. Do not nail into or above the factory-applied adhesive (Fig. 54-14).

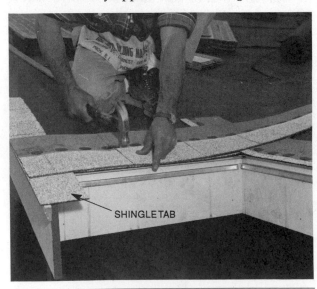

Fig. 54-13 A shingle tab is cut to cover the projecting edge of the rake trim.

NAIL LOCATIONS FOR THREE-TAB STRIP SHINGLE

NAIL LOCATIONS FOR TWO-TAB STRIP SHINGLE

NAIL LOCATIONS FOR NO-CUTOUT STRIP SHINGLE

Fig. 54-14 Recommended fastener locations for asphalt strip shingles. (*Courtesy of Asphalt Roofing Manufacturers' Association*)

The fastener length should be sufficient to penetrate the sheathing at least 3/4 inch, or through approved panel sheathing. Roofing nails should be 11- or 12-gauge galvanized or aluminum with barbed shanks. They should have 3/8 to 7/16 inch heads. Roofing nails may be driven by hand or with power nailers. Power-driven staples may be used in place of nails. However, their use is limited to shingles with factory-applied adhesive (Fig. 54-15). Staples must be at least 16-gauge. They should have a minimum crown width of 15/16 inch.

Align each shingle of the first course carefully. Fasten each shingle from the end nearest the shingle just laid. This prevents buckling. Drive fasteners straight so that the nail heads or staple crowns will not cut into the shingles. The entire crown of the staple or nail head should bear tightly against the shingle. It should not penetrate its surface (Fig. 54-16). Continue applying shingles until the first course is complete. Both ends of the course should be flush with the rake drip edge.

Shingle Exposure

The maximum *exposure* of asphalt shingles to the weather depends on the type of shingle. Recommended maximum exposures range from 4 to 6 inches. Most commonly used asphalt shingles have a maximum exposure of 5 inches. Less than the maximum recommended exposure may be used, if desired.

Laying Out Shingle Courses

When laying out shingle courses, space the desired exposure up each rake from the top edge of the first course of shingles. Snap lines across the roof for five or six courses. When snapping a long line, it may be necessary to thumb the line down against the roof at about center. Then snap the lines on both sides of the thumb.

Fig. 54-15 Pneumatic staplers and nailers are often used to fasten asphalt shingles.

Lay succeeding courses so their top edges are to the chalk line (Fig. 54-17). Start each course so the cutouts are staggered in the desired manner. Continue snapping lines and applying courses until a point 3 or 4 feet below the ridge is reached. Some carpenters snap a line to straighten out the course after using the top of the cutout for the shingle exposure for a number of courses.

Fig. 54-16 It is important to fasten asphalt shingles correctly. (*Courtesy of Asphalt Roofing Manufacturers' Association*)

Fig. 54-17 Lay shingles so their top edge is to the chalk line.

Spacing Shingle Courses to the Ridge

The last course of shingles should be exposed below the *ridge cap* by about the same amount as all the other shingle courses. Cut a full tab from an asphalt strip. Center it on the ridge. Then bend it over the ridge. Do this at both ends of the building. Mark the bottom edges on the roof. It may be necessary, when reroofing older buildings, to mark and space up near the center also, if the ridge is sagged.

Measure up 2 inches. Snap a line between the marks. This line will be the top edge of the next to last course of shingles. It should be about 3 1/2 inches down from the ridge. Divide the distance between this line and the top of the last course of shingles applied into spaces as close as possible to the exposure of previous courses. Do not exceed the maximum exposure (Fig. 54-18). Snap lines across the roof and shingle up to the ridge.

The line for the last course of shingles is snapped on the face of the course below. Lay out the exposure from the bottom edge of the course on both ends of the roof. Snap a line across the roof. Fasten the last course of shingles by placing their butts to the line. Bend the top edges over the roof. Fasten their top edges to the opposite slope.

Applying the Ridge Cap

Cut hip and ridge shingles from shingle strips to make approximately 12×12-inch squares. Cut shingles from the top of the cutout to the top edge on a slight taper. The top edge should be narrower than the bottom edge (Fig. 54-19). Cutting the shingles in this manner keeps the top half of the shingle from protruding when it is bent over the ridge.

The ridge cap is applied after both sides of the roof have been shingled. At each end of the roof, center a shingle on the ridge. Bend it over the ridge. Mark its bottom edge on the front slope or the one most visible. Snap a line between the marks.

Beginning at the bottom of a hip or at one end of the ridge, apply the shingles over the hip or ridge. Expose each 5 inches. In cold weather, ridge cap shingles may have to be warmed in order to prevent cracking when bending them over the ridge. On the ridge, shingles are started from the end away from prevailing winds. The wind should blow over the shingle butts, not against them. Keep one edge, from the butt to the start of the tapered cut, to the chalk line. Secure each shingle with one fastener on each side, 5 1/2 inches from the butt and one inch up from each edge (Fig. 54-20). Apply the cap across the ridge until 3 or 4 feet from the end. Then space the cap to the end in the same manner as spacing the shingle course to the ridge. The last ridge shingle is cut to size. It is applied with

2. MEASURE DOWN FROM RIDGE ABOUT 3-1/2" TO MARK TOP OF NEXT TO LAST COURSE OF SHINGLES.

3. DIVIDE THIS DISTANCE INTO SPACES AS CLOSE TO, BUT NOT MORE THAN, THE SHINGLE EXPOSURE USED BELOW. — 3-1/2"

1. STOP LAST COURSE 3 OR 4 FEET FROM RIDGE.

Fig. 54-18 Space shingle courses evenly to the ridge.

CUT ALONG DOTTED LINE, TAPERING TOP PORTION SLIGHTLY.

12"

12"   12"   12"

Fig. 54-19 Hip and ridge shingles are cut from strip shingles. (*Courtesy of Asphalt Roofing Manufacturers' Association*)

NAIL LOCATION FOR HIP AND RIDGE SHINGLES

DIRECTION OF PREVAILING WIND

5" EXPOSURE RIDGE

START HERE

HIP

START HERE 5" EXPOSURE

Fig. 54-20 Applying ridge shingles. (*Courtesy of Asphalt Roofing Manufacturers' Association*)

one fastener on each side of the ridge. The two fasteners are covered with asphalt cement to prevent leakage.

Steep Slope Application

Steep roof slopes, like those on lower *mansard roof* slopes, reduce the effectiveness of factory-applied adhesive. This is especially true in cold or shaded areas. The maximum slope recommended for normal asphalt shingle application is 60 degrees or 21 inches rise per foot of run.

On steeper slopes, apply spots of quick-setting asphalt plastic cement about one inch in diameter under all shingle tabs immediately after installation. Apply one spot of cement under each tab of shingles with three or more tabs. Use two spots of cement under tabs of shingles with two tabs. For shingles with no cutouts, use three spots of cement.

The steep slope roof must be provided with through ventilation to evaporate moisture in the air behind the roof sheathing.

CHAPTER 55 ROLL ROOFING

Roll roofing can be used on roof slopes as little as one inch rise per foot of run. On steeper roofs, roll roofing is used when economy is the major factor and appearance is not so important.

Types of Roll Roofing

Roll roofing is made of the same materials as asphalt shingles. Various types are made with a base sheet of organic felt or glass fibers in a number of weights (Fig. 55-1). Some types are applied with exposed or concealed fasteners. They have a top lap of 2 to 4 inches. A concealed nail type, called *double coverage* roll roofing, has a top lap of 19 inches. All kinds come in rolls that are 36 inches wide.

Roll roofing is recommended for use on roofs with slopes less than 4 inches rise per foot. How-

TYPICAL ASPHALT ROLLS

1	2		3	4		5		6	7
PRODUCT	Approximate Shipping Weight		Sqs. Per Package	Length	Width	Side or End Lap	Top Lap	Exposure	Underwriters' Listing
	Per Roll	Per Sq.							
MINERAL SURFACE ROLL	75# to 90#	75# to 90#	One	36' 38'	36'' 36''	6''	2'' 4''	34'' 32''	C
MINERAL SURFACE ROLL DOUBLE COVERAGE	55# to 70#	55# to 70#	One Half	36'	36''	6''	19''	17''	C

Fig. 55-1 Types of roll roofing.

ever, the exposed nail type should not be used on pitches less than 2 inches rise per foot. Roll roofing applied with concealed nails and having a top lap of at least 3 inches may be used on pitches as low as one inch per foot. The exposed fastener type is rarely used. Only the concealed fastener type is recommended for use and described in this chapter. Use the same type and length of nails as for asphalt shingles.

General Application Methods

Apply all roll roofing when the temperature is above 45 degrees Fahrenheit. This prevents cracking the coating. Cut the roll into 12- to 18-foot lengths. Spread in a pile on a smooth surface to allow them to flatten out.

Use only the lap or quick-setting cement recommended by the manufacturer. Store cement in a warm place until ready for use. To warm it rapidly, place the unopened container in hot water.

CAUTION These materials are flammable. Never warm them over an open fire or place them in direct contact with a hot surface.

Apply roll roofing only on a solid, smooth, well-seasoned deck. Make sure the area below has

sufficient ventilation to prevent the deck from absorbing condensation. This would cause the roofing to warp and buckle. A felt underlayment is not used with roll roofing.

Applying Roll Roofing with Concealed Fasteners

Apply 9-inch wide strips of the roofing along the eaves and rakes overhanging about 3/8 inch. Fasten the strips with two rows of nails one inch from each edge. Space them about 4 inches apart.

Apply the first course with its edges and ends flush with the strips. Secure the upper edge with nails staggered about 4 inches apart. Do not fasten within 18 inches of the rake edge. Apply cement only to that part of the edge strips covered by the first course. Press the lower edge and rake edges of the first course firmly in place over the edge strips. Finish nailing the upper edge out to the rakes. Apply succeeding courses in like manner. Make all end laps 6 inches wide. Apply cement the full width of the lap.

After all courses are in place, lift the lower edge of each course. Apply the cement in a continuous layer over the full width and length of the lap. Press the lower edges of the upper courses firmly

NAILS 4" APART IN ROW 1" IN FROM EDGE OF STRIP

ROOF DECK

UPPER EDGE NAILED-NAILS 4" APART AND STAGGERED

9" EDGE STRIP TO OVERHANG EAVES AND RAKES 1/4" TO 3/8"

CEMENT OVER ENTIRE LAP AREAS

6"

TOP LAP NOT LESS THAN 3"

CEMENT OVER ENTIRE LAP AREAS

ALL EXPOSED EDGES TO BE FIRMLY EMBEDDED IN CEMENT

Fig. 55-2 Recommended procedure for applying concealed-nail type roll roofing.

into the cement. A small bead should appear along the entire edge of the sheet (Fig. 55-2). Care must be taken to apply the correct amount of cement.

Covering Hips and Ridges

Cut strips of 12×36 roofing. Bend the pieces lengthwise through their centers. Snap a chalk line on both sides down 5 1/2 inches from the hip or ridge. Apply cement from one line over the top to the other line.

Fit the first strip over the hip or ridge. Press it firmly into place. Start at the lower end of a hip and at either end of a ridge. Lap each strip 6 inches over the preceding one. Nail each strip only on the end that is to be covered by the overlapping piece. Spread cement on the end of each strip that is lapped before the next one is applied. Continue in this manner until the end is reached (Fig. 55-3).

Applying Double Coverage Roll Roofing

Cut the 19-inch *selvage* portion from enough double coverage roll roofing to cover the length of the roof. Save the surfaced portion for the last course. Apply the selvage portion parallel to the eaves. It should overhang the drip edge by 3/8 inch. Secure it to the roof deck with three rows of nails. Place the top row 4 1/2 inches below the upper edge. Put the bottom row one inch above the bottom edge. Place the other row halfway between. Place the nails in the upper and middle rows slightly staggered about 12 inches apart. Place the nails in the lower row about 6 inches apart and slightly staggered. Nail along rakes in the same manner.

Apply the first course. Secure it with two rows

NAIL HERE ONLY

6"

QUICK SETTING CEMENT OVER RIDGE

Fig. 55-3 Finishing hips and ridges using concealed-nail type roll roofing.

of nails in the selvage portion. Place one row about 4 3/4 inches below the upper edge. Put the second row about 8 1/2 inches below the first. Space the nails about 12 inches apart in each row and stagger them (Fig. 55-4).

Apply succeeding courses in the same manner. Lap the full width of the 19-inch selvage each time. Make all end laps 6 inches wide. End laps are made in the manner shown in Figure 55-5. Stagger end laps in succeeding courses.

Lift and roll back the surface portion of each course. Starting at the bottom, apply cement to the entire selvage portion of each course. Apply it to within 1/4 inch of the surfaced portion. Press the overlying sheet firmly into the cement. Apply pressure over the entire area using a light roller to insure adhesion between the sheets at all points.

Apply the remaining surfaced portion left from the first course as the last course. This type roofing may also be applied in like manner parallel to the rakes (Fig. 55-6). Hips and ridges are covered in the same manner shown in Figure 55-7.

ROOF DECK

NAIL LOCATIONS

12"

8-1/2"

4-3/4"

19" SELVAGE LAP

17" EXPOSURE

6" END LAP

METAL DRIP EDGE

ASPHALT CEMENT AS SPECIFIED BY THE MANUFACTURER

FIRST COURSE

STARTER STRIP OF 19" SELVAGE CUT FROM A FULL SHEET LAID TO OVERHANG EAVE AND RAKE 1/4" TO 3/8"

Fig. 55-4 Method of applying double coverage roll roofing.

STEP BY STEP
P R O C E D U R E S

STEP 1 APPLY NAILS ON 4" CENTERS

ROOF DECK

STEP 3 — APPLY NAILS ON 4" CENTERS 1" FROM END OF SHEET

6"

1"

STEP 2 — APPLY CEMENT TO ENTIRE LAPPED AREA

Fig. 55-5 Method of making end laps on double coverage roll roofing.

ROOF DECK

4-3/4"

12"

12"

2" TO 4"

6" END LAP

SPECIAL ASPHALT CEMENT

19" SELVAGE

METAL DRIP EDGE ON EAVES AND RAKES

17" EXPOSURE

19" SELVAGE STARTER STRIP

Fig. 55-6 Vertical application of double coverage roll roofing.

Fig. 55-7 Method of covering hips and ridges with double coverage roofing.

It is important to follow specific application instructions because of differences in the manufacture of roll roofing. Some instructions call for hot asphalt. Others call for cold cement. Others give the option of either. Specific requirements for quantities and types of adhesive must be followed.

CHAPTER 56 WOOD SHINGLES AND SHAKES

Wood *shingles* and *shakes* are used extensively to cover roofs (Fig. 56-1). Most shingles and shakes are produced from western red cedar. Cedar logs are first cut into desired lengths. They are then split into sections from which shingles and shakes are sawn or split. All shingles are sawed. Most shakes are split. Shingles, therefore, have a relatively smooth surface. Most shakes have at least one highly textured, natural grain. surface. Most shakes also have thicker butts than shingles (Fig. 56-2).

Most wood shingles and shakes are produced by mills that are members of the Cedar Shake and Shingle Bureau. Their product label (Fig. 56-3) is assurance that the products meet quality standards.

Fig. 56-2 All shingles are sawed from the log. Most shakes are split and have a rough surface. (*Courtesy of Cedar Shake and Shingle Bureau*)

Fig. 56-1 Wood shingles and shakes are used extensively for roof covering.

Fig. 56-3 The product label assures that quality requirements are met. (*Courtesy of Cedar Shake and Shingle Bureau*)

Description of Wood Shingles and Shakes

Wood shingles are available for use on roofs in four standard grades. Shakes are manufactured by different methods to produce four types. Both shingles and shakes may be treated to resist fire or premature decay in areas of high humidity (Fig. 56-4).

SHINGLES

NO. 1 BLUE LABEL®

NO. 2 RED LABEL

NO. 3 BLACK LABEL

NO. 4 UNDERCOURSING

SHAKES

NO. 1 HANDSPLIT & RESAWN

NO. 1 CERTI-SAWN ®

NO. 1 TAPERSPLIT

NO. 1 STRAIGHTSPLIT

Fig. 56-4 Description of grades and kinds of wood shingles and shakes. (*Courtesy of Cedar Shake and Shingle Bureau*)

Sizes and Coverage

Shingles come in lengths of 16, 18, and 24 inches. The butt thickness increases with the length. Shakes are available in lengths of 15, 18, and 24 inches. Their butt thicknesses are from 3/8 to 3/4 inch. The 15-inch length is used for starter and finish courses. Figure 56-5 shows the sizes and the amount of roof area that one square, laid at various exposures, will cover.

Maximum Exposures

The area covered by one square of shingles or shakes depends on the amount they are exposed to the weather. The maximum amount of shingle exposure depends upon the length and grade of the shingle or shake and the pitch of the roof. Shakes are not generally applied to roofs with slopes of less than 4 inches rise per foot. Shingles, with reduced exposures, may be used on slopes down to 3 inches rise per foot. Figure 56-6 shows the maximum recommended roof exposure for wood shingles and shakes.

Sheathing and Underlayment

Shingles and shakes may be applied over spaced or solid roof sheathing. *Spaced sheathing* is usually 1×4 or 1×6 boards. *Solid sheathing* is usually APA-rated panels of plywood or strand board. It may be required in regions subject to frequent earthquakes or under treated shingles and shakes. It is also recommended for use with shakes in areas where wind-driven snow is common.

Spaced Sheathing

Solid wood sheathing is applied from the eaves up to a point that is plumb with a line 12 to 24 inches inside the wall line. An eaves flashing is installed, if required. Spaced sheathing may then be used above the solid sheathing to the ridge.

For shingles, either 1×4 or 1×6 spaced sheathing may be used. Space 4-inch boards the same amount as the shingles are exposed to the weather. In this method of application, each course of shingles is nailed to the center of the board. If 6-inch boards are used, they are spaced two exposures.

COVERAGE AND EXPOSURE TABLES

Shingle Coverage Table

LENGTH AND THICKNESS	Approximate coverage of one square (4 bundles) of shingles based on following weather exposures								
	3½"	4"	4½"	5"	5½"	6"	6½"	7"	7½"
16" × 5/2"	70	80	90	100*	—	—	—	—	—
18" × 5/21/4"	—	72½	81½	90½	100*	—	—	—	—
24" × 4/2"	—	—	—	—	73½	80	86½	93	100*
NOTE: *Maximum exposure recommended for roofs.									

Shake Coverage Table

SHAKE TYPE, LENGTH AND THICKNESS	Approximate coverage (in sq. ft.) of one square, when shakes are applied with an average ½" spacing, at following weather exposures, in inches (d):				
	5	5½	7½	8½	10
18" × ½" Handsplit-and-Resawn Mediums (a)	—	55(b)	75(c)	—	—
18" × ¾" Handsplit-and-Resawn Heavies (a)	—	55(b)	75(c)	—	—
18" × 5/8" Tapersawn	—	55(b)	75(c)	—	
24" × 3/8" Handsplit	50(e)	—	75(b)	—	—
24" × ½" Handsplit-and-Resawn Mediums	—	—	75(b)	85	100(c)
24" × ¾" Handsplit-and-Resawn Heavies	—	—	75(b)	85	100(c)
24" × 5/8" Tapersawn	—	—	75(b)	85	100(c)
24" × ½" Tapersplit	—	—	75(b)	85	100(c)
18" × 3/8" Straight-Split	—	65(b)	90(c)	—	—
24" × 3/8" Straight-Split	—	—	75(b)	85	100(c)
15" Starter-Finish course	Use supplementary with shakes applied not over 10" weather exposure.				
(a) 5 bundles will cover 100 sq. ft. roof area when used as starter-finish course at 10" weather exposure; 7 bundles will cover 100 sq. ft. roof area at 7½" weather exposure; see footnote (d). (b) Maximum recommended weather exposure for 3-ply roof construction. (c) Maximum recommended weather exposure for 2-ply roof construction. (d) All coverage based on an average ½" spacing between shakes. (e) Maximum recommended weather exposure.					

Fig. 56-5 Tables show the sizes and coverage of wood shingles and shakes. (*Courtesy of Cedar Shingle and Shake Bureau*)

Shingle Exposure Table

PITCH	Maximum exposure recommended for roofs								
	Length								
	No. 1 Blue Label			No. 2 Red Label			No. 3 Black Label		
	16″	18″	24″	16″	18″	24″	16″	18″	24″
3/12 to 4/12	3¾″	4¼″	5¾″	3½″	4″	5½″	3″	3½″	5″
4/12 and steeper	5″	5½″	7½″	4″	4½″	6½″	3½″	4″	5½″

Shake Exposure Table

PITCH	Maximum exposure recommended for roofs	
	Length	
	18″	24″
4/12 and steeper	7½″	10″(a)
(a) 24″ × ⅜″ handsplit shakes limited to 5″ maximum weather exposure, per UBC.		

Formula for calculating material at reduced exposures:

square footage ÷ reduced coverage = total material required

e.g. You are estimating a roof that measures 3200 square feet (32 squares). You have decided to put 16″ shingles at 4″ exposure. A 4-bundle square at 4″ exposure covers 80 square feet.

3200 ÷ 80 = 40 squares of material

Fig. 56-6 Maximum exposures of wood shingles and shakes for various roof pitches. (*Courtesy of Cedar Shake and Shingle Bureau*)

Two courses of shingles are nailed to the same board when courses are exposed up to, but not exceeding, 5 1/2 inches. For shingles with greater exposures, the sheathing is spaced a distance of one exposure (Fig. 56-7).

In shake application, spaced sheathing is usually 1×6 boards spaced the same distance, on center, as the shake exposure (Fig. 56-8). The spacing should never be more than 7 1/2 inches for 18-inch shakes and 10 inches for 24-inch shakes installed on roofs.

Underlayment

No underlayment is required under wood shingles.

A breather-type roofing felt may be used over solid or spaced sheathing.

Application Tools and Fasteners

Shingles and shakes are usually applied with a shingling hatchet. Recommendations for the type and size of fasteners should be closely followed.

Shingling Hatchet

A *shingling hatchet* (Fig. 56-9), should be lightweight. It should have both a sharp *blade* and a *heel*. A *sliding gauge* is sometimes used for fast and accu-

Shingles Applied Over 1″ × 4″ Boards

Shingles Applied Over 1″ × 6″ Boards

Fig. 56-7 Application of wood shingles on spaced sheathing. (*Courtesy of Cedar Shake and Shingle Bureau*)

Shakes Applied Over 1" × 6" Boards

Fig. 56-8 Method of applying shakes on spaced sheathing. (*Courtesy of Cedar Shake and Shingle Bureau*)

Fig. 56-9 A shingling hatchet is commonly used to apply wood shingles and shakes.

Type of Shingle or Shake	Nail Type and Minimum Length	
Shingles—New Roof	Type	(in)
16" and 18" Shingles	3d Box	1¼
24" Shingles	4d Box	1½
Shakes—New Roof	Type	(in)
18" Straight-Split	5d Box	1¾
18" and 24" Handsplit-and-Resawn	6d Box	2
24" Tapersplit	5d Box	1¾
18" and 24" Taper-sawn	6d Box	2

Fig. 56-10 Recommended nail types and sizes for wood shingles and shakes. (*Courtesy of Cedar Shake and Shingle Bureau*)

rate checking of shingle exposure. The gauge permits laying several shingle courses at a time without snapping a chalk line. A power nailer may be used. However, shingles and shakes often need to be trimmed or split with the hatchet. More time may be lost than gained by using a power nailer.

Fasteners

Apply each shingle with only two corrosion-resistant nails, such as stainless steel, hot-dipped galvanized, or aluminum. Minimum nail lengths for shingles and shakes are shown in Figure 56-10. Staples should be 16-gauge aluminum or stainless steel with a minimum crown of 7/16 inch. The staple should be driven with its crown across the grain of the shingle or shake. Staples should be long enough to penetrate the sheathing at least 1/2 inch.

Applying Wood Shingles

Install the first layer of a starter course of wood shingles as described in a previous chapter. If a gutter is used, overhang the shingles plumb with the center of the gutter. If no gutter is installed, overhang the starter course 1 1/2 inches beyond the fascia. Space adjacent shingles between 1/4 and 3/8 inch apart. Place each fastener about 3/4 inch in from the edge of the shingle and not more than one inch above the exposure line. Drive fasteners flush with the surface. Make sure that the head does not crush the surface.

Apply another layer of shingles on top of the first layer of the starter course. The starter course may be tripled, if desired, for appearance. This procedure is recommended in regions of heavy snowfall.

Fig. 56-11 Details of wood shingle application. (*Courtesy of Cedar Shake and Shingle Bureau*)

Lay succeeding courses across the roof. Apply several courses at a time. Stagger the joints in adjacent courses at least 1 1/2 inches. There should be no joint in any three adjacent courses in alignment. Joints should not line up with the centerline of the heart of wood or any knots or defects. *Flat* or *slash grain* shingles wider than 8 inches should be split in two before fastening. Trim shingle edges with the hatchet to keep their butts in line (Fig. 56-11).

After laying several courses, snap a chalk line to straighten the next course. Proceed shingling up the roof. When 3 or 4 feet from the ridge, check the distance on both ends of the roof. Divide the distance as close as possible to the exposure used. A full course should show below the ridge cap. Shingle tips are cut flush with the ridge. Tips can be easily cut across the grain with a hatchet if cut on an angle to the side of the shingle.

On intersecting roofs, stop shingling the roof a few feet away from the valley. Select and cut a shingle at a proper taper. Apply it to the valley. Do not break joints in the valley. Do not lay shingles with their grain parallel to the valley centerline. Work back out. Fit a shingle to complete the course (Fig. 56-12).

STEP BY STEP
P R O C E D U R E S

Order of applying shingles or shakes at valley

1. Stop course line here

2. Place pre-cut piece so that cut angle is positioned on chalk line with butt on course line

3. Select a shingle or shake of the required width to complete the course

Fig. 56-12 Method of applying shingles or shakes along an open valley. (*Courtesy of Cedar Shake and Shingle Bureau*)

Hips and Ridges

After the roof is shingled, 4- to 5-inch wide hip and ridge caps are applied. Measure down on both sides of the hip or ridge at each end for a distance equal to the exposure. Snap a line between the marks.

Lay a shingle so its bottom edge is to the line. Fasten with two nails. Use longer nails so that they penetrate at least 1/2 inch into or through the sheathing. With the hatchet, trim the top edge flush with the opposite slope.

Lay another shingle on the other side with the butt even and its upper edge overlapping the first shingle laid. Trim its top edge at a bevel and flush with the side of the first shingle. Double this first set of shingles, alternating the joint. Apply succeeding layers of cap, with the same exposure as used on the roof. Alternate the overlap of each layer (Fig. 56-13). Space the exposure when nearing the end so all caps are about equally exposed.

Applying Wood Shakes

Shakes are applied in much the same manner as shingles (Fig. 56-14). Mark the handle of the shingling hatchet at 7 1/2 and 10 inches from the top of the head. These are the exposures that are used most of the time when applying wood shakes.

An underlayment of felt is applied over a starter course of one or two layers of shingles or shakes that are overlaid with a course of shakes to

Fig. 56-14 Three or four shake courses at a time are carried across the roof. (*Courtesy of Cedar Shake and Shingle Bureau*)

be exposed. Butts of the starter course should project 1 1/2 inches beyond the fascia.

Interlayment

Next, lay an 18-inch wide strip of #30 roofing felt over the starter course. Its bottom edge should be positioned at a distance equal to twice the shake exposure above the butt line of the first course of shakes. For example, 24-inch shakes, laid 10 inches to the weather, would have felt applied 20 inches above the butts of the shakes. The felt will cover the top 4 inches of the shakes. It will extend up 14 inches on the sheathing. The top edge must rest on the spaced sheathing, if used.

Nail only the top edge of the felt. Fasten successive strips on their top edge only. Their bottom edges should be one shake exposure from the bottom of the previous strip. It is important to lay the felt straight. It serves as a guide for applying shakes. After the roof is felted, the tips of the shakes are tucked under the felt. The bottom should be exposed by the distance of twice the exposure (Fig. 56-15).

Apply the second and successive courses with joints staggered and fasteners placed the same as for shingles. The spacing between shakes should be at least 3/8 inch, but not more than 5/8 inch. Lay straight-split shakes with their smooth end toward the ridge (Fig. 56-16).

Maintaining Shake Exposure

There is a tendency to angle toward the ground. Therefore, check the exposure regularly with the hatchet handle. An easy way to be sure of correct exposure is to look through the joint between the edges of the shakes in the course below the one being nailed. The bottom edge of the felt will be

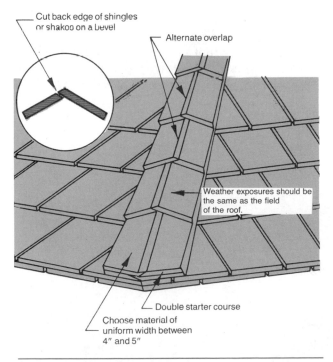

Cut back edge of shingles or shakes on a level

Alternate overlap

Weather exposures should be the same as the field of the roof.

Double starter course

Choose material of uniform width between 4" and 5"

Fig. 56-13 When applying hip and ridge shingles, alternate the overlap. (*Courtesy of Cedar Shake and Shingle Bureau*)

DOUBLE STARTER COURSE

18-INCH, 30-LB, FELT LAID OVER TOP PORTION OF EACH COURSE

Fig. 56-15 An interlayment of felt is required when laying shakes.

Fig. 56-17 The tip of the shake is inserted between the layers of interlayment. Its butt is lined up with the bottom edge of another layer sighted between the edge joints of the course below. (*Courtesy of Cedar Shake and Shingle Bureau*)

visible. The butt of the shake being nailed is positioned directly above it (Fig. 56-17).

Ridges and Hips

Adjust the exposure so that tips of shakes in the next-to-last course just come to the ridge. Use economical 15-inch starter-finish shakes for the last course. They save time by eliminating the need to trim shake tips at the ridge. Cap ridges and hips with shakes in the same manner as that used with wood shingles.

Estimating Roofing Materials

Find the area of the roof in square feet. Divide the total by 100 to determine the number of squares needed. Add about 5 to 10 percent for waste. A simple roof with no dormers, valleys, or other obstructions requires less allowance for waste to be added. A complicated roof requires more.

For wood shingles, add one square for every 240 linear feet of starter course. Add one square of shakes for 120 linear feet of starter course.

Add one extra square of shingles for every 100 linear feet of valleys and about two squares for shakes.

Add an extra bundle of shakes or shingles for every 16 feet of hip and ridge to be covered.

Figure 2 pounds of nails per square at standard exposure.

Remember that a square of roofing will cover 100 square feet of roof surface only when applied at standard exposures. Allow proportionally more material when these exposures are reduced.

Nail with 2 nails per shake approx. 3/4" to 1" from edge and 1 1/2" to 2" above butt line

Felt laid over top portion of each course

Space shakes 3/8" to 5/8" apart

Exposure

Spaced sheathing should be 1" × 6" or wider

Solid wood sheathing

Adjacent courses should be offset 1 1/2" minimum

Eave protection

Gutter

15" starter course

Fascia

Rafter

Rafter header

Drip edge 1 1/2"

Fig. 56-16 Shake application details. (*Courtesy of Cedar Shake and Shingle Bureau*)

CHAPTER 57 FLASHING

Flashing is a material used in various susceptible locations. It prevents water from entering a building (Fig. 57-1). The words *flash, flashed,* and *flashing* are also used as verbs to describe the installation of the material. Various kinds of flashing are applied at the eaves, valleys, chimneys, vents, and other roof projections. They prevent leakage at the intersections.

Kinds of Flashing

Flashing material may be sheet copper, zinc, aluminum, galvanized iron, or mineral-surfaced asphalt roll roofing. Copper and zinc are high-quality flashing material, but are more expensive. Roll roofing is less expensive. Colors that match or contrast with the roof covering can be used. If properly applied, roll roofing used as a valley flashing will outlast the main roof covering. Sheet metal, especially copper, may last longer. However, it is good practice to replace all flashing when reroofing.

Eaves Flashing

Whenever there is a possibility of ice dams forming along the eaves and causing a backup of water, an *eaves flashing* is needed. For a slope of at least 4 inches rise per foot of run, install a course of 36-inch wide, smooth or mineral surface roll roofing, 50

pounds per square or heavier over the underlayment and drip edge. On lower slopes protection against water leakage from ice dams is gained by applying the underlayment as described previously and shown in Figure 54-7.

A course of a special 36-inch wide, self-adhering eaves flashing, called an *ice and water shield,* may also be used. Apply the flashing. Let it overhang the drip edge by 1/4 to 3/8 inch. The flashing should extend up the roof far enough to cover a point at least 12 inches inside the wall line of the building. If the overhang of the eaves requires that the flashing be wider than 36 inches, the necessary horizontal lap joint is cemented and located on the portion of the roof that extends outside the wall line (Fig. 57-2).

Flashing Valleys

Roof valleys are especially vulnerable to leakage. This is because of the great volume of water that flows down through them. Valleys must be carefully flashed according to recommended procedures. Valleys are flashed in two ways: *open* or *closed.*

Open Valley Flashing

Apply a 36-inch wide strip of asphalt felt centered in the valley. Fasten it with only enough nails along its edges to hold it in place. Let the courses of felt underlayment applied to the roof overlap the valley underlayment by not less than 6 inches. Seat the felt

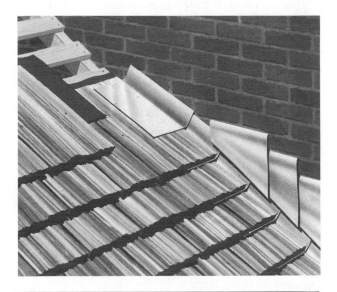

Fig. 57-1 Flashings are used to seal against leakage where the roof butts against adjoining walls and at other intersections. (*Courtesy of Cedar Shake and Shingle Bureau*)

EAVE FLASHING STRIP

LAP JOINT

OVERHANG 1/4" TO 3/8"

METAL DRIP EDGE

12" MIN

Fig. 57-2 An eaves flashing is installed if there is danger of ice dams forming along the eaves.

Fig. 57-3 A felt underlay is applied in the valley before flashing.

well into the valley. Be careful not to cause any break in the felt. The eave flashing, if required, is then applied (Fig. 57-3).

Using Roll Roofing Flashing

Lay an 18-inch wide layer of mineral-surfaced roll roofing centered in the valley. Its mineral-surfaced side should be down. Use only enough nails spaced one inch in from each edge to hold the strip smoothly in place. Press the roofing firmly in the bottom of the valley when nailing the opposite edge. On top of the first strip, lay a 36-inch wide strip with its surfaced side up. Center it in the valley. Fasten it in the same manner as the first strip.

Snap a chalk line on each side of the valley. Use them as guides for trimming the ends of the shingle courses. These lines are spaced 6 inches apart at the ridge. They are spread 1/8 inch per foot as they approach the roof edge. Thus, a valley 16 feet long will be 8 inches wide at the eaves.

The upper corner of each end asphalt shingle is clipped. This helps keep water from entering between the courses. The roof shingles are cemented to the valley flashing with plastic asphalt cement (Fig. 57-4).

Using Metal Flashing

Prepare the valley with underlayment in the same manner as described previously. Lay a strip of sheet metal flashing centered in the valley. The metal should extend at least 10 inches on each side of the valley centerline for slopes with a 6-inch rise or less and 7 inches on each side for a steeper slope. Carefully press and form it into the valley. Fasten the metal with nails of similar material spaced close to its outside edges. Use only enough fasteners to hold it smoothly in place.

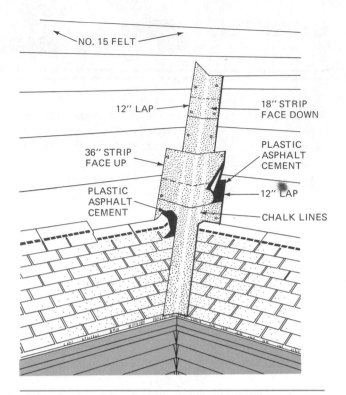

Fig. 57-4 Method of applying a roll roofing open valley flashing.

Snap lines on each side of the valley, as described previously. Use them as guides for cutting the ends of the shingle courses. Trim the last shingle of each course to fit on the chalk line. Clip one inch from its upper corner at a 45 degree angle. To form a tight seal, cement the shingle to the metal flashing with a 3-inch width of asphalt plastic cement (Fig. 57-5).

If a valley is formed by the intersection of a low-pitched roof and a much steeper one, a one-inch high, crimped standing seam should be made in the center of the metal flashing of an open valley. The seam will keep heavy rain water flowing down the steeper roof from overrunning the valley and possibly being forced under the shingles of the lower slope.

Closed Valley Flashing

Closed valleys are those where the roof shingles meet in the center of the valley and completely cover the valley flashing. Using a closed valley protects the valley flashing. It thus adds to the weather resistance at very vulnerable points. Several methods are used to flash closed valleys.

The first step for any method is to apply the asphalt felt underlayment as previously described for open valleys. Then, center a 36-inch width of smooth or mineral surface roll roofing, 50-pound per square, or heavier, in the valley over the under-

36" WIDE FELT CENTERED
IN VALLEY AND OVERLAPPING
ROOF UNDERLAYMENT

METAL FLASHING APPLIED
10" ON EACH SIDE OF
CENTER ON ROOFS WITH
6" RISE OR LESS. APPLY
7" EACH SIDE OF CENTER
ON STEEPER ROOFS.

END SHINGLES
ARE BEDDED IN
ASPHALT CEMENT

CHALK LINES TO
GUIDE SHINGLES

Fig. 57-5 Procedure for flashing an open valley with metal.

layment. Form it smoothly in the valley. Secure it with only as many nails as necessary.

Woven Valley Method

Valleys are commonly flashed by applying asphalt shingles on both sides of the valley and alternately weaving each course over and across the valley. This is called the *woven valley* (Fig. 57-6).

Lay the first course of shingles along the edge of one roof up to and over the valley for a distance of at least 12 inches. Lay the first course along the edge of the adjacent roof. Extend the shingles over the valley on top of the previously applied shingles.

Succeeding courses are then applied. Weave the valley shingles alternately, first on one roof and then on the other. When weaving the shingles, make sure they are pressed tightly into the valley. Also make sure no nail is closer than 6 inches to the valley centerline. Fasten the end of the woven shingle with two nails (Fig. 57-7). Most carpenters prefer to cover each roof area with shingles to a point approximately 3 feet from the valley. They weave the valley shingles in place later.

No end joints should occur in or near the center of the valley. Therefore, it may be necessary to occasionally cut a strip short that would otherwise

Fig. 57-6 Valleys are often flashed by weaving shingles.

36" ROLL ROOFING
50# OR HEAVIER

EACH STRIP
TO EXTEND
AT LEAST 12"
BEYOND
CENTER
OF VALLEY

6" MIN

EXTRA NAIL IN
END OF STRIP

Fig. 57-7 Specifications for a woven valley flashing.

end near the center. Continue from this cut end with a full-length strip over the valley.

Closed Cut Valley Method

Apply the shingles to one roof area. Let the end shingle of every course overlap the valley by at least 12 inches. Make sure no end joints occur at or near the center of the valley. Place fasteners no closer than 6 inches from the center of the valley. Form the end shingle of each course snugly in the valley. Secure its end with two fasteners.

Snap a chalk line along the center of the valley on top of the overlapping shingles. Apply shingles to the adjacent roof area. Cut the end shingle to the chalk line. Clip the upper corner of each shingle as described previously for open valleys. Bed the end that lies in the valley in about a 3-inch wide strip of asphalt cement. Make sure that no fasteners are located closer than 6 inches to the valley centerline (Fig. 57-8).

Step Flashing Method

Step flashings are individual metal pieces tucked between courses of shingles (see Fig. 57–1). When applying step flashings in valleys, first estimate the number of shingle courses required to reach the ridge. Using tin snips, cut a piece of metal flashing for each course of shingles. Each piece should be at least 18 inches wide for slopes with a 6-inch rise or greater and 24 inches wide for slopes with less pitch. The height of each piece should be at least 3 inches more than the shingle exposure.

36" ROLL ROOFING 50# OR HEAVIER

EACH STRIP TO EXTEND AT LEAST 12" BEYOND CENTER OF VALLEY

EXTRA NAIL IN END OF STRIP

6" MIN

PLASTIC ASPHALT CEMENT

Fig. 57-8 Recommendations for applying a closed cut valley flashing.

Prepare the valley with underlayment as described previously. Snap a chalk line in the center of the valley. Apply the starter course on both roofs. Trim the end shingle of each course to the chalk line. Fit and form the first piece of flashing to the valley. Trim its bottom edge flush with the drip edge. Fasten it in the valley over the first layer of the starter course. Use fasteners of like material to prevent electrolysis. Fasten the upper corners of the flashing only.

Apply the first regular course of shingles to both roofs on each side of the valley. Trim the valley shingles so their ends lay on the chalk line. Bed them in plastic asphalt cement. Do not drive nails through the metal flashing.

Apply the next piece of flashing in the valley over the first course of shingles. Keep its bottom edge about 1/2 inch above the butts of the next course of shingles. Apply the second course of shingles in the same manner as the first. Secure a flashing over the second course. Apply succeeding courses and flashings in this manner (Fig. 57-9). Remember, a flashing is placed over each course of shingles. Do not leave any flashings out. When the valley is completely flashed, no metal flashing surface is exposed. If the valley does not extend all the way to the ridge of the main roof, a *lead saddle* is applied over the ridge of the minor roof (Fig. 57-10).

Flashing against a Wall

When roof shingles butt up against a vertical wall, the joint must be made watertight. The usual method of making the joint tight is with the use of metal step flashings.

The flashings are cut about 8 inches in width. They are bent at right angles in the center so they will lay about 4 inches on the roof and extend about 4 inches up the sidewall. The length of the flashings is about 3 inches more than the exposure of the shingles. When used with shingles exposed 5 inches to the weather, they are made 8 inches in length. Cut and bend the necessary number of metal flashings.

The roofing is applied and flashed before the siding is applied to the vertical wall. First, apply an underlayment of asphalt felt to the roof deck. Turn the ends up on the vertical wall by about 3 to 4 inches.

Apply the first layer of the starter course, working toward the vertical wall. Fasten a metal flashing, on top of the first layer of shingles. Its bottom edge should be flush with the drip edge. Use one fastener in each top corner. Lay the first regular course with its end shingle over the flashing and against the sheathing of the sidewall. Do not drive any fasteners through the flashings. It is usu-

ASPHALT FELT
UNDERLAYMENT

36 INCH WIDE FELT
CENTERED IN VALLEY
OVERLAPPING ROOF UNDERLAYMENT

PLACE A FLASHING ON TOP
OF EACH COURSE OF SHINGLES
SO ITS BOTTOM EDGE IS JUST
ABOVE THE BUTT OF THE
SHINGLE TO BE PLACED
OVER IT. FASTEN
IN THE UPPER AND
OUTER CORNERS

INDIVIDUAL STEP FLASHING–
18" WIDE FOR ROOFS
WITH 6" RISE OR GREATER–
24" WIDE FOR ROOFS
LESS THAN 6" RISE–
FLASHING HEIGHT IS
AT LEAST 3" MORE THAN
SHINGLE EXPOSURE.

ASPHALT
STRIP SHINGLES

Fig. 57-9 Specifications for applying metal step flashings in a valley.

MAIN ROOF

LEAD SADDLE OVER JOINT OF
TWO VALLEYS BENT TO FIT

RIDGE OF
MINOR ROOF

TYPICAL SADDLE FLASHING

Fig. 57-10 A lead saddle is installed over the ridge of a minor roof where it intersects with the main roof.

ally not necessary to bed the shingles to the flashings with asphalt cement. The step flashing holds down the end of the shingle below it.

Apply a flashing over the first course and against the wall. Keep its bottom edge at a point that will be about 1/2 inch above the butt of the next course of shingles. Continue applying shingles and flashings in this manner until the ridge is reached (Fig. 57-11). Some carpenters prefer to cut the shingles back if a tab cutout occurs over a flashing. This prevents metal from being exposed to view.

Flashing a Chimney

In many cases, especially on steep pitch roofs, a *cricket* or *saddle* is built between the upper side of the chimney and roof deck. The saddle, although not a flashing in itself, prevents accumulation of water behind the chimney (Fig. 57-12).

On chimneys, the flashings are usually made of lead. They are installed by *brick masons* who build the chimney. The upper ends of the flashing are bent around and *mortared* in between the courses of brick as the chimney is built. The flashings are long enough to be bent at and over the roof sheathing for tucking between shingles. These flashings are usually in place before the carpenter applies the roof covering.

The underlayment is applied and tucked under the existing flashings. The shingle courses are brought up to the chimney. They are applied under the flashing on the lower side of the chimney. This is called the *apron flashing*. Shingles are tucked under the apron. The top edge of the shingles is cut as

PLACE A FLASHING ON TOP OF
EACH COURSE OF SHINGLES SO
ITS BOTTOM END IS JUST ABOVE
THE BUTT OF THE SHINGLE THAT
WILL BE PLACED OVER IT.
FASTEN THE FLASHING WITH ONE
FASTENER IN EACH UPPER AND
OUTER CORNER.

3" LAP

ROOFING FELT

3" LAP

4" MINIMUM

STEP FLASHING

SIDING

SHINGLES

2" CLEARANCE
FOR SIDING

SIDING

Fig. 57-11 Using metal step flashings where a roof
joins a wall.

STEP FLASHING DETAIL

CHIMNEY

COUNTERFLASHING

METAL COVER

SADDLE

BOARDS OR
PLYWOOD

CALK EDGES

FLASHING

SHINGLES

Fig. 57-12 A saddle is built to prevent the
accumulation of water behind the chimney.

necessary, until the shingle exposure shows below the apron. The apron is then pressed into place on top of the shingles in a bed of plastic cement. Its projecting ends are carefully and gently formed up around the sides of the chimney and under the lowest side flashings.

Along the sides of the chimney, the flashings are tucked in between the shingles in the same manner as in flashing against a wall. No nails are used in the flashings. The standing portions of the *side flashings* are bedded to the chimney with asphalt cement. The roof portion is bedded to the shingle. The projecting edges of the lowest side flashings are carefully formed around the corner. They are folded against the low side on the chimney. The top edges of the highest side flashings are also folded around the corner and under the *head flashing* on the upper side of the chimney.

The head flashing is cemented to the roof. Shingles are applied over it. They are bedded to it with asphalt cement. Its projecting ends are also carefully formed around the corner on top of the side flashings.

The projecting ends of chimney flashings are carefully formed and folded around the corners of the chimney. Gently and carefully tap the metal with a hammer handle. Care must be taken not to break through the flashings. Chimneys may be flashed by other methods and materials, other than described above, depending on the custom of certain geographical areas (Fig. 57-13).

Other rectangular roof obstructions, such as skylights, are flashed in a similar manner. The carpenter usually applies the flashings to these obstructions.

Flashing Vents

Flashings for round pipes, such as *stack vents* for plumbing systems and *roof ventilators*, usually come as *flashing collars* made for various roof pitches. They fit around the stack. They have a wide flange on the bottom that rests on the roof deck. All joints in the flashing are soldered to make them watertight. The top end sometimes is bent and fitted inside the top end of the stack. Or, it may be sealed to

Fig. 57-13 Chimney flashings are installed by masons for use by those who shingle the roof.

the side with a lead ring and asphalt cement. The flashing is installed over the stack vent, with its flange on the roof sheathing. It is fastened in place with one fastener in each upper corner.

Shingle up to the lower end of the stack vent flashing. Lift the lower part of the flange. Apply shingle courses under it. Cut the top edge of the shingles, where necessary, until the shingle exposure, or less, shows below the lower edge of the flashing. Apply asphalt cement under the lower end of the flashing. Press it into place on top of the shingle courses.

Apply shingles around the stack and over the flange. Do not drive nails through the flashing. Bed shingles to the flashing with asphalt cement, where necessary (Fig. 57-14).

Fig. 57-14 Method of shingling around a stack vent flashing.

REVIEW QUESTIONS

Select the most appropriate answer.

1. A square is the amount of roofing required to cover
a. 50 square feet.   c. 150 square feet.
b. 100 square feet.   d. 200 square feet.

2. One roll of #15 asphalt felt will cover
a. one square.   c. 3 square.
b. 2 square.   d. 4 square.

3. When applying asphalt felt on a roof deck as underlayment, lap each course over the lower course by at least
a. 2 inches.   c. 4 inches.
b. 3 inches.   d. 6 inches.

4. Under no circumstances should asphalt strip shingles be used on slopes lower than
a. one inch rise per foot of run.
b. 2 inches rise per foot of run.
c. 3 inches rise per foot of run.
d. 4 inches rise per foot of run.

5. When applying asphalt shingles, it is recommended that no rake tab be less than
a. 2 inches in width.   c. 4 inches in width.
b. 3 inches in width.   d. 5 inches in width.

6. For slopes with a 6-inch rise per foot of run or less, metal valley flashings should extend on each side of the valley centerline by at least
a. 6 inches.   c. 10 inches.
b. 8 inches.   d. 12 inches.

7. When flashing a valley by weaving shingles, do not locate any nails closer to the valley centerline than
a. 6 inches.   c. 10 inches.
b. 8 inches.   d. 12 inches.

8. Flashings about 8 inches wide are used when flashing a roof that butts against a vertical wall. They are bent so that
a. 3 inches lays on the wall and 5 inches lays on the roof.
b. 4 inches lays on the wall and 4 inches lays on the roof.
c. 5 inches lays on the wall and 3 inches lays on the roof.
d. 2 inches lays on the wall and 6 inches lays on the roof.

9. A built-up section between the roof and the upper side of a chimney is called a
a. saddle.   c. furring.
b. dutchman.   d. breast.

10. Concealed nail roll roofing may be used on roofs with slopes as low as
a. one inch rise per foot of run.
b. 2 inches rise per foot of run.
c. 3 inches rise per foot of run.
d. 4 inches rise per foot of run.

11. Most wood shingles and shakes are made from
a. cypress.
b. redwood.
c. eastern white cedar.
d. western red cedar.

12. The longest available length of wood shingles and shakes is
a. 16 inches.
b. 18 inches.
c. 24 inches.
d. 28 inches.

13. The maximum exposure for No. 1, 16-inch wood shingles laid on roofs at least as steep as 4 in 12 is
a. 4 1/2 inches.
b. 5 inches.
c. 5 1/2 inches.
d. 6 inches.

14. Wood shingles normally overhang the fascia by
a. 3/8 inch.
b. one inch.
c. 1 1/2 inches.
d. 2 inches.

15. It is important to lay interlayment straight prior to applying shakes
a. so there are no wrinkles in the felt.
b. to obtain the proper lap.
c. for ease in nailing.
d. because the felt serves as a guide.

■ BUILDING FOR SUCCESS ■

What Employers are Looking For

It has been said, "There are three kinds of people: those who make things happen, those who watch things happen, and those who wonder what happened." More likely than not, we all have been in each of these categories at one time or another. In preparing for immediate and long-term employment, each person should desire to make positive things happen.

Study and careful preparations will make a difference in a person's employment opportunities. We have the responsibility to make positive things happen to ourselves and to people around us. Contractors and construction related companies are always looking for people who take the initiative to make a company or business progressive. That will not change as we continue into the twenty-first century.

Everyone entering a construction-related career must clearly understand the demands placed on technically skilled workers. As in the past, various construction trades call for workers to perform technical tasks on a daily basis. Additional business skills are increasingly important. Skilled workers, such as carpenters, will be expected to possess many social, personal management, and problem-solving skills. They have to be able to work with minimal supervision. More and more industries are calling for both new and current

skilled workers to be more adaptable, autonomous, and willing to learn on the job.

Workplace basics are changing and affecting larger groups of our nation's population. Many employers desire each worker to respond well to new technology, assure quality control, be involved in participative management, and acquire new skills quickly. Due to intensive competition, the time to adjust and continue producing quality products has been shortened. Therefore, all industries must respond to new demands with a workforce that is flexible, well educated, and progressive in its thinking. The most successful companies understand the need to assist in the training and development of their employees. More contractors today look for both local and national training opportunities. They want their employees to obtain new skills that will support the company as they attain their goals. This is one of the many criteria the new construction graduates must look for in the job search process. A construction-related business that desires to be on the cutting edge in competition will have some element of employee development in place. These companies should be a high priority in the job selection process.

Learning on the job will continue to be a requirement for all skilled workers well into the next century. Employees at all levels must become proficient in observing, listening, and solving problems. Innovative ideas, reinforced with strong interpersonal communication skills, will play a larger role in the progress of a construction business. Teamwork will continue to increase as problems are addressed. Learning to learn begins in a person's early years and must continue throughout the employment years. These new skills will become the foundation on which more refined skills will be developed on the job.

Many construction leaders, managers, and supervisors have diversified educational backgrounds. Some arrived at their leadership positions through on-the-job training, apprenticeship programs, trade schools, community colleges, or a four-year construction management programs. They possessed many of the basic reading, writing, and computation skills needed for the job. They also have accumulated an abundance of additional necessary technical and business skills while employed. Good communication skills, involving observing, listening, writing, and speaking, are essentials for all skilled workers who desire to excel. Students who do not place great importance on communication and social skills will not readily advance in their employment.

Economics in the workplace will continue to call for creative thinking and problem-solving skills. Competition for a share of the market, quality products, and cost efficiency will be a driving forces for each construction-related business. Both management and skilled workers will have excellent opportunities for input into a company's creative efforts. Through constant observations of trends and developments, each employee must be ready to offer new ideas that will help make his or her employer a pace-setter or strong competitor.

As each student prepares for a career in construction, he or she must set goals that are challenging, realistic, and obtainable. Most goals will involve education, training, and work experience. The selected program of study should contain a curriculum that will, as a minimum, prepare each person for the entry-level job. This becomes the first educational step that can be combined with other training opportunities and work experiences that lead to the next step. The entire sequence may take years to complete. All things of quality take investment of time, planning, and commitment to reach completion.

Each person has the opportunity to make things happen with each situation. No two people will have the same circumstances, backgrounds, or resources. What is done with abilities and desire is the key to success. Reasonable questions that must be asked are: Where do you want to go? How can you get there? Do you want to watch things happen around you? Do you want to forever wonder what happened? Or do you want to make positive things happen in your life? Ask those around you who have insight into the construction industry or career counseling to assist you in making these challenging decisions. That may possibly be the first step to a rewarding career.

FOCUS QUESTIONS: For individual or group discussion.

1. What are some skills you feel an employer in construction will desire of you?

2. What ongoing changes in business and industry will affect your job? How well you can advance? How will new technology, trends, economic factors, and societal developments affect your opportunities for obtaining a job in construction?

3. How do you think being diversified in technical and business skills could help assure you of being considered for the job you desire?

Windows

indows are normally installed prior to the application of exterior *siding*. Care must be taken to provide easy-operating, weathertight, attractive units. Quality workmanship results in a more comfortable interior, saves energy by reducing fuel costs, minimizes maintenance, gives longer life to the units, and makes application of the exterior siding easier.

OBJECTIVES

After completing this unit, the student will be able to:

- describe the most popular styles of windows and name their parts.

- select and specify desired sizes and styles of windows from manufacturers' catalogs.

- install various types of windows in an approved manner.

- cut glass and glaze a sash.

CONTENTS

CHAPTER 58

WINDOW TERMS AND TYPES

Wood *windows* are one of many types of *mill-work* (Fig. 58-1). Millwork is a term used to describe products, such as windows, doors, and cabinets, fabricated in woodworking plants that are used in the construction of a building. Windows are usually fully assembled and ready for installation when delivered to the construction site. Windows are also made of aluminum and steel. Windows made with exposed wood parts encased in vinyl are called *vinyl-clad* windows. The names given to various parts of a window are the same, in most cases, regardless of the window type.

Parts of a Window

When shipped from the factory, the window is complete except for the interior trim. It is important that the installer know the names, location, and functions of the parts of a window in order to understand, or to give, instructions concerning them.

The Sash

The *sash* is a frame in a window that holds the glass. The type of window is generally determined by the way the sash operates. The sash may be installed in a fixed position, move vertically or horizontally, or swing outward or inward.

Sash Parts. Vertical edge members of the sash are called *stiles.* Top and bottom horizontal members are called *rails.* The pieces of glass in a sash are

called *lights.* There may be more than one light of glass in a sash. Small strips of wood that divide the glass into smaller lights are called *muntins.* Muntins may divide the glass into rectangular, diamond, or other shapes (Fig. 58-2).

Many windows come with false muntins called *grilles.* Grilles do not actually separate or support the glass. They are applied as an overlay to simulate small lights of glass. They are made of wood or plastic. They snap in and out of the sash for easy cleaning of the lights (Fig. 58-3).

Window Glass

Several qualities and thicknesses of sheet glass are manufactured for *glazing* and other purposes. The

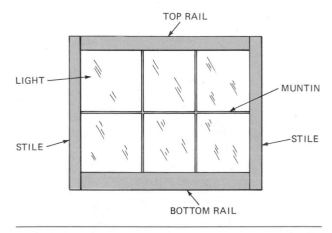

Fig. 58-2 A sash and its parts.

Fig. 58-1 Windows of many types and sizes are fully assembled in millwork plants and ready for installation. (*Courtesy of Andersen Windows, Inc.*)

Fig. 58-3 Removable grilles simulate true divided light muntins. (*Courtesy of Andersen Windows, Inc.*)

Sheet		Range				Approx. Oz./Sq. ft.
Picture		.043–.053	AA, A, B		50[1]	9–11
		.054–.069 and .070–.080	AA, A		70[1]	12–14 and 15–17
			B		80[1]	
Window	SS	.085–.101	AA, A, B		120[1]	19.5
	DS	.115–.134			140[1]	26.0
Heavy Sheet	3/16	.182–.205	AA, A, B		84×120	40.0

[1] United inches (sum of width plus length.)

Fig. 58-4 Kinds and sizes of glass commonly used in residential construction.

installation of glass in a window sash is called glazing. *Single strength* (SS) glass is about 3/32-inch thick. It is used for small lights of glass. For larger lights, *double strength* (DS) glass about 1/8-inch thick may be used. *Heavy sheet* glass about 3/16-inch thick is also manufactured (Fig. 58-4). Many other kinds of glass are made for use in construction.

Safety Glass

Most windows are not glazed with safety glass. If broken, they could fragment and cause injury. Care must be taken to handle windows in a manner to prevent breaking the glass. Some codes require a type of *safety glass* in windows with low sill heights or located near doors. Skylights and roof windows are generally required to be glazed with safety glass.

Safety glass is constructed, treated, or combined with other materials to minimize the possibility of injuries resulting from contact with it. Several types of safety glass are manufactured.

Laminated glass consists of two or more layers of glass with inner layers of transparent plastic bonded together. *Tempered glass* is treated with heat or chemicals. When broken at any point, the entire piece immediately disintegrates into a multitude of small granular pieces. *Wired glass* consists of a single sheet with wire mesh imbedded in the glass. *Transparent plastic* is also used for safety glazing material.

Insulating Glass

To help prevent heat loss, and to avoid condensation of moisture on glass surfaces, *insulating glass* is used frequently in sash in place of a single thickness.

Insulating glass consists of two or three (generally two) layers of glass separated by a sealed air space 3/16 to one inch in thickness (Fig. 58-5). Moisture is removed from the air between the lay-

Fig. 58-5 Cutaway of insulating glass used to increase the R-value in a window.

ers before the edges are sealed. To raise the R-value of insulating glass, the space between the layers is filled with *argon* gas. Argon conducts heat at a lower rate than air. Additional window insulation may be provided with the use of *removable glass panels* or *combination storm sash.*

Solar Control Glass. The R-value of windows may also be increased by using special *solar control* insulating glass, called *high performance* or *LoE* glass. LoE is an abbreviation for *low emissivity.* It is used to designate a type of glazing that reflects heat back into the room in winter and blocks heat from entering in the summer (Fig. 58-6). An invisible, thin,

BOTH HIGH-PERFORMANCE AND HIGH-PERFORMANCE SUN WINDOWS ARE PROGRAMMED
TO DETECT HEAT AND KEEP IT WHERE YOU WANT IT—REGARDLESS OF WHERE THE WINDOWS
ARE, THE ANGLE OF THE SUN OR THE SIZE OF THE HOME.

Fig. 58-6 Special solar control glass is used in windows to help keep heat in during cold weather and out during hot weather. (*Courtesy of Andersen Windows, Inc.*)

metallic coating is bonded to the air space side of the inner glass. This lets light through, but reflects heat.

The Window Frame

The sash is hinged to, slides, or is fixed in a *window frame.* The frame usually comes with the exterior trim applied. It consists of several distinct parts (Fig. 58-7).

The Sill

The bottom horizontal member of the window frame is called a *sill* or *stool.* It is set or shaped at an angle to shed water. Its bottom side is usually grooved so a weathertight joint can be made with the wall siding.

Jambs

The vertical sides of the window frame are called *side jambs.* The top horizontal member is called a *header jamb.*

Extension Jambs. The inside edge of the jamb should be flush with the finished interior wall surface when the window is installed. In some cases, windows can be ordered with jamb widths for standard wall thicknesses. In other cases, jambs are made narrow. *Extension jambs* are then provided with the window unit. The extensions are cut to width to accomodate various wall thicknesses. They are applied to the inside edge of the jambs of the window frame (Fig. 58-8). The extension jambs are installed at a later stage of construction when the interior trim is applied. They should be stored for safekeeping until needed.

Fig. 58-7 A window frame consists of parts with specific terms.

VERTICAL DETAIL SCALE 3" = 1'0"

2-7/8"

INTERIOR TRIM

HEAD

PINE EXTENSION JAMB

EXTENSION JAMS FIT BETWEEN THE WINDOW JAMB AND THE INTERIOR WALL, AND ARE PRE-DRILLED FOR NAILING DIRECTLY INTO THE EDGE OF THE WINDOW JAMB.

ROUGH OPENING HEIGHT

UNIT DIMENSION HEIGHT

BOLD LINES INDICATE VINYL CLAD PARTS & VINYL FLANGES

1-1/8"

SILL

4-1/2"

SECTION

HORIZONTAL DETAIL 3" = 1'0"

1-3/8"

PINE EXTENSION JAMBS

2-7/8"

1-5/16"

CAULK

JAMB

JAMB

1/4"

UNIT DIMENSION WIDTH

1/4"

ROUGH OPENING WIDTH

PLAN

Fig. 58-8 To compensate for varying wall thicknesses, extension jambs are provided with some window units.

Blind Stops

Blind stops are sometimes applied to certain types of window frames. They are strips of wood attached to the outside edges of the jambs. Their inside edges project about 1/2 inch inside the frame. They provide a weathertight joint between the outside cas-

ings and the frame. They also act as stops for screens and storm sash. They make the outer edge of the channel for top sash of *double-hung* windows.

Casings

Window units usually come with *exterior casings* applied. The side members are called *side casings*. In most windows, their lower ends are cut at a bevel and rest on the sill. The top member is called the *header casing*. On flat casings, a weathertight *rabbeted* or *tongue-and-grooved* joint is made between them. When molded casings are used, the mitered joints at the head are usually bedded in compound (Fig. 58-9).

When windows are installed or manufactured, side by side, in multiple units, a *mullion* is formed where the two side jambs are joined together. The casing covering the joint is called a *mullion casing*.

The Drip Cap

A *drip cap* comes with some windows. It is applied on the top edge of the header casing. It is shaped with a sloping top to carry rain water out over the window unit. Its inside edge is rabbeted. The wall siding is applied over it to make a weathertight joint.

Window Flashing

In some cases, a *window flashing* is also provided. This is a piece of metal as long as the header casing. It is bent to fit over the header casing and against the exterior wall (Fig. 58-10). The flashing prevents the entrance of water at this point. Flashings are usually made of aluminum or zinc. The vinyl flanges of vinyl-clad wood windows are usually formed as an integral part of the window. No additional head flashings are required.

Making a Flashing on the Job. If window flashings are not provided with the unit, they can be made on the job site. Cut a length of sheet zinc or aluminum from a roll of the desired width. The metal should be wide enough to extend up the wall about 2 inches, over the top edge of the header casing, and about 1/4 inch down over its face side.

Fig. 58-10 A window flashing covers the top edge of the header casing. It extends up the sidewall above the window.

Tack the metal along its top edge to a plank about 2 inches in from the edge of the stock. Bend the metal by tapping the overhanging edge with a short 2×4 block (Fig. 58-11). The other small bend is made when the flashing is installed.

Protective Coatings

Most window units are *primed* with a first coat of paint applied at the factory. Vinyl-clad wood windows are designed to eliminate painting. However, they are available in only a few colors. In case units need priming, this should be done before installation. Store the units under cover and protected from

Fig. 58-9 A weathertight joint is made between side and header casings.

Fig. 58-11 Bending a window flashing on the job site.

the weather until installed. Additional protective coats should be applied as soon as practical.

Screens

Most manufacturers provide screens as optional accessories for all kinds of windows. On outswinging and sliding windows the screens are attached to the inside of the frame. On double-hung windows they are mounted on the outside of the frame.

The screen mesh is usually plastic or aluminum. Bronze is used in more expensive screens. Screens may be manufactured specifically for various types of windows. They are mounted in place with hardware already installed at the factory.

Wood screens are also available in various sizes. They may be installed full or half size. Half screens slide in tracks installed on the inside edges of the window's side casings. Figure 58-12 shows how a window screen with a wood frame is made.

STEP BY STEP
P R O C E D U R E S

1. TO MAKE THE SCREEN SHOWN ABOVE
   a.) CUT 2 STILES 22″ LONG
   b.) CUT RAILS 24″ LONG
2. MAKE THROUGH MORTISE & TENON JOINTS BY:
   a.) MORTISE BOTH ENDS OF THE STILES
   b.) TENON BOTH ENDS OF THE RAILS
3. ASSEMBLE THE FRAME WITH GLUE & SCREWS IN EACH JOINT.
   a.) SQUARE THE FRAME BY MAKING DIAGONAL DISTANCES EQUAL

Fig. 58-12 Instructions for making a wood framed window screen.

Fig. 58-13 Fixed windows are often used in combination with other types of windows. Here, fixed half-round windows are used above casement windows. (*Courtesy of Andersen Windows, Inc.*)

Types of Windows

Common types of windows are *fixed, single-* or *double-hung, casement, sliding, awning, hopper,* and *jalousie* windows.

Fixed Windows

Fixed windows consist of a frame in which a sash is fitted in a fixed position. They are manufactured in many shapes (Fig. 58-13).

Oval and *circular* windows are usually installed as individual units. *Elliptical, half rounds,* and *quarter rounds* are widely used in combination with other types (Fig. 58-14). In addition, fixed windows are manufactured in *geometric* shapes (*squares, rectangles, triangles, parallelograms, diamonds, trapezoids, pentagons, hexagons* and *octagons*). They may be assembled or combined with other types of windows in a great variety of shapes (Fig. 58-15). In addition to factory-assembled units, lengths of the frame stock may be purchased for cutting and assembling of odd shape or size units on the job.

Arch windows have a curved top or head that make them well-suited to be joined in combination with a number of other types of windows or doors (Fig. 58-16). All of the windows mentioned come in a variety of sizes. With so many shapes and sizes, hundreds of interesting and pleasing combinations can be made.

Single- and Double-Hung Windows

The *double-hung window* consists of an upper and a lower sash that slide vertically by each other in separate channels of the side jambs (Fig. 58-17). The *single-hung* window is similar except the upper sash is fixed. A strip separating the sash is called a *parting bead.*

In most units, sash slide in metal channels that are installed in the frames. Each sash is provided

Fig. 58-14 Elliptical, half round, and quarter round windows. (*Courtesy of Andersen Windows, Inc.*)

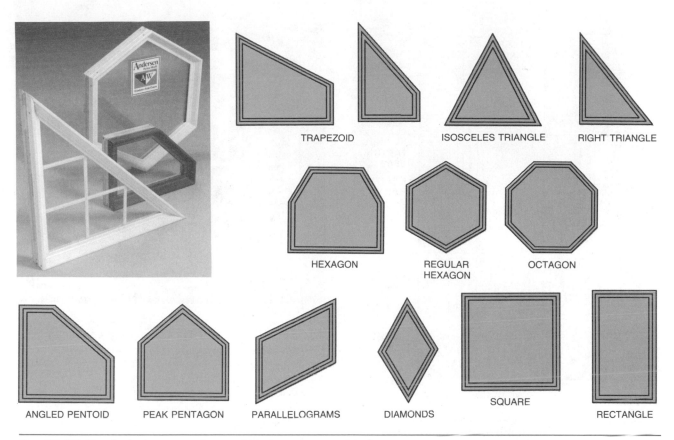

TRAPEZOID   ISOSCELES TRIANGLE   RIGHT TRIANGLE

HEXAGON   REGULAR HEXAGON   OCTAGON

ANGLED PENTOID   PEAK PENTAGON   PARALLELOGRAMS   DIAMONDS   SQUARE   RECTANGLE

Fig. 58-15 Fixed windows are also made in many geometric forms.

with *springs, balances,* or *compression weatherstripping* to hold it in place in any position. Compression weatherstripping prevents air infiltration, provides tension, and acts as a counterbalance. Some types provide for easy removal of the sash for painting and repair.

When the sash are closed, specially shaped *meeting rails* come together to make a weathertight joint. *Sash locks* located at this point not only lock the window, but draw the rails tightly together. Other hardware consists of *sash lifts* that are fastened to the bottom rail of the bottom sash. They are

ARCH WINDOWS USED ABOVE DOUBLE-HUNG WINDOWS

Fig. 58-16 Arch windows have curved head jambs and casings. (*Courtesy of Andersen Windows, Inc.*)

ELEVATION　　　　　SECTION

Fig. 58-17 The double-hung window and its parts.

sometimes eliminated by providing finger grooves in the rail.

Double-hung windows can be arranged in a number of ways. They can be installed side by side in multiple units or in combination with other types. Figure 58-18 shows them used in a *bay window* unit. Detachable vinyl grille simulates muntins.

Casement Windows

The *casement window* consists of sash hinged at the side. It swings outward by means of a crank or lever. Most casements swing outward. The inswinging type is very difficult to make weathertight. An advantage of the casement type is that the entire sash can be opened for maximum ventilation. Figure 58-19 shows the use of casement windows in a *bow window* unit.

Sliding Windows

Sliding windows have sash that slide horizontally in separate tracks located on the header jamb and sill (Fig. 58-20). When a window-wall effect is desired, many units can be placed side by side. Most units come with all necessary hardware applied.

Awning and Hopper Windows

An *awning window* unit consists of a frame in which a sash hinged at the top swings outward by means

Fig. 58-18 Double-hung windows are combined in this bay window unit. (*Courtesy of Andersen Windows, Inc.*)

of a crank or lever. A similar type, called the *hopper window,* is hinged at the bottom and swings inward.

Each sash is provided with an individual frame so that many combinations of width and height can be used. These windows are often used in combination with other types (Fig. 58-21).

Jalousie Windows

A *jalousie window* consists of a metal frame that holds a series of horizontal glass slats. The glass slats open and close together when operated by a single crank (Fig. 58-22). The use of jalousie windows is limited to porches and breezeways or to warm climates. They are not very weathertight.

Fig. 58-19 Casement windows are combined in this bow window unit. (*Courtesy of Andersen Windows, Inc.*)

Fig. 58-20 The sash in sliding windows move horizontally by each other. (*Courtesy of Andersen Windows, Inc.*)

Fig. 58-21 Awning windows are often used in stacks or in combination with other types of windows. (*Courtesy of Andersen Windows, Inc.*)

Fig. 58-22 A jalousie window has many horizontal glass slats that operate similar to a venetian blind.

Skylights and Roof Windows

Skylights provide light only. *Roof windows* contain operating sash to provide light and ventilation (Fig. 58-23). One type of roof window comes with a tilting sash that allows access to the outside surface for cleaning. Special flashings are used when multiple skylights or roof windows are ganged together.

Fig. 58-23 Skylights and roof windows are made in a number of styles and sizes. (*Courtesy of Andersen Windows, Inc.*)

CHAPTER 59 WINDOW INSTALLATION AND GLAZING

There are numerous window manufacturers that produce hundreds of kinds, shapes, and sizes of windows. Because of the tremendous variety and design differences, follow the manufacturer's instructions closely to insure a correct installation. Directions in this unit are basic to most window installations. They are intended as a guide to be supplemented by procedures recommended by the manufacturer.

Selecting and Ordering Windows

The builder must study the plans to find the type and location of the windows to be installed. The floor plan shows the location of each unit. Each unit is usually identified by a number or a letter next to the window symbol.

Those responsible for designing and drawing plans for building or selecting windows must be aware of, and comply with, building codes that set certain standards in regard to windows. Most codes require minimum areas of natural light. Codes also require minimum ventilation by windows unless provided by other means. Some codes stipulate minimum window sizes in certain rooms for use as emergency egress.

Outswinging windows, such as awning and casement windows, should not swing out over decks, patios, and similar areas unless they are high enough to permit persons to travel under them. When lower, the projecting sash could cause serious injury.

Window Schedule

The numbers or letters found in the floor plan identify the window in more detail in the *window schedule.* This is, usually, part of a set of plans (Fig. 59-1). This schedule normally includes window style, size, manufacturer's name, and unit number. Rough opening sizes may or may not be shown.

Manufacturers' Catalogs

Sometimes a window schedule is not included. Units are identified only by the manufacturer's name and number on the floor plan. In order to gain more information, the builder must refer to the window *manufacturer's catalog.*

The catalog usually includes a complete description of the manufactured units and optional accessories, such as insect screens, glazing panels, and grilles. For a particular window style, the catalog typically shows overall unit dimensions, rough opening widths and heights, and glass sizes of manufactured units. Large-scale cross-section details of the window unit are also usually included so the builder can more clearly understand its construction (Fig. 59-2).

Order window units giving the type and identification letters and/or numbers found in the window schedule or manufacturer's catalog. The size of all existing rough openings should be checked to make sure they correspond to the size given in the catalog before windows are ordered.

		WINDOW	SCHEDULE	
IDENT.	QUAN.	MANUFACTURER	SIZE	REMARKS
A	6	MORGAN	2'-8" X 3'-10"	D.H. SINGLE
B	3	MORGAN	2'-8" X 3'-10"	D.H. TRIPLE
C	2	MORGAN	3'-4" X 3'-10"	D.H. SINGLE
D	1	ANDERSEN	CW24	CASEMENT SINGLE
E	1	ANDERSEN	C34	CASEMENT TRIPLE
F	1	ANDERSEN	C23	CASEMENT DOUBLE

Fig. 59-1 Typical window schedule found in a set of plans.

BASIC UNIT SIZES

VERTICAL DETAIL Scale 3" = 1'0"

NOTE: Hinging of sash that is shown is standard. Single sash indicates hinging left, right or stationary. Specify left or right, as viewed from the outside. For other hinging of multiple units contact your local Andersen supplier.

*CW series units (except CW2 and CW3 height) open to 20" clear opening width using sill hinge control bracket. Bracket can be pivoted allowing for cleaning position.

CW series units are also available with a 22" clear opening width. Please contact distributor for availability.

When ordering be sure to specify color desired.

HORIZONTAL DETAIL Scale 3" = 1'0"

Fig. 59-2 Typical page from a window manufacturer's catalog. (*Courtesy of Andersen Windows, Inc.*)

Installing Wood Windows

All rough window openings should be prepared to insure weathertight window installations. Two methods of preparation are generally used.

Applying Felt Strips

Prepare the openings in wood frame walls by applying strips of #15 asphalt felt about 8 inches wide against the wall sheathing on each side of the window openings. Let each end of the strip project above and below the opening about 4 inches. Fasten the strips along each edge about 12 inches apart. Do not fasten the lower end below the window opening (Fig. 59-3).

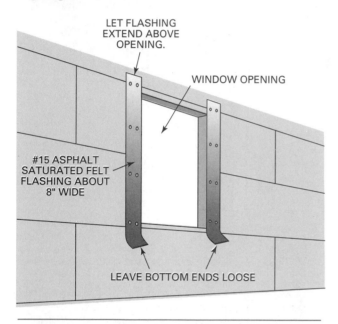

LET FLASHING
EXTEND ABOVE
OPENING.

WINDOW OPENING

#15 ASPHALT
SATURATED FELT
FLASHING ABOUT
8" WIDE

LEAVE BOTTOM ENDS LOOSE

Fig. 59-3 Flash each side of the window opening with felt strips prior to installing the window.

Housewraps and Building Paper

Exterior walls are sometimes covered with a building paper prior to the application of *siding*. This prevents the infiltration of air into the structure. Yet, it also allows the passage of water vapor to the outside. In place of building paper, exterior walls may be covered with a type of air infiltration barrier commonly called *housewrap*.

Housewrap is a very thin, tough, plastic material. It is used to cover the sheathing on exterior walls for the same purpose as building paper (Fig. 59-4). Housewraps are commonly known by the brand names of Typar and Tyvek. They are more resistant than building paper to air leakage and are virtually tearproof. Yet they are also breatheable to allow water vapor to escape. This prevents moisture build-up inside wall cavities, which could cause fungus growth and rot. Building paper only comes in 36-inch wide rolls. Housewrap rolls are 1.5, 3, 4.5, 5, 9, and 10 feet wide.

Building paper must be applied after windows and doors are installed. If installed before, there may be a prolonged period before it is covered by siding. During this time it could get wet and deteriorate or be torn by high winds. Therefore rough openings are prepared with strips of felt, because a flashing must extend under the window casing.

Housewrap is designed to survive prolonged periods of exposure to the weather. It can and usually is applied immediately after framing is completed, but before doors and windows have been installed. The wrap, then, serves also as a flashing for the sides of windows and doors. It is not necessary to apply flashing strips to the sides of rough openings. Housewrap get its name because it is completely wrapped around the building. It covers

Fig. 59-4 Housewrap is widely used as an air infiltration barrier on sidewalls. (*Courtesy of Reemay, Inc.*)

Fig. 59-5 Housewrap being applied to a sidewall. (*Courtesy of Reemay, Inc.*)

corners, window and door openings, plates, and sills.

Applying Housewrap. Begin at the corner, holding the roll vertically on the wall. Unroll it a short distance. Make sure the roll is plumb. Secure the sheet to the corner, leaving about a foot extending beyond the corner to overlap later. Continue to unroll. Make sure the sheet is straight, with no buckles. Fasten every 12 to 18 inches (Fig. 59-5).

Unroll directly over window and door openings and around the entire perimeter of the building. Overlap all joints by at least 3 inches. Secure them with a special housewrap tape. On horizontal joints, the upper layer should overlap the lower layer.

Make a cut in the housewrap from corner to corner of rough openings. Fold the triangular flaps in and around the opening (Fig. 59-6). Secure the flaps on the inside with fasteners about every 6 inches apart.

> **CAUTION** Housewraps are slippery. They should not be used in any application where they will be walked on.

Establishing Window Height

Most of the windows in each story of a building are set so their header casings lie in a straight line at the same level even if some of the windows are shorter than others. Snap a chalk line all around the building to the height of the header casing. Windows are then set by lining the top edges of the header casing to the chalk line.

Another way of establishing this height is by the use of a story pole. Lay out the height of the header jamb from the subfloor on the story pole. When setting windows, hold the story pole with one end against the subfloor. Line up the header jamb of the window with the layout line on the story pole (Fig. 59-7).

Installing Windows

Remove all protection blocks from the window unit. Cut off any *horns* left on the unit. Horns are exten-

Fig. 59-6 Housewrap is cut and folded in wall openings. (*Courtesy of Reemay, Inc.*)

Fig. 59-7 Window heights are sometimes established by using a story pole.

sions of the side jambs above the header jamb and below the sill. The extensions or horns protect the units during storage, shipment to the dealer, and delivery to the construction site.

Do not remove any diagonal braces applied at the factory. Close and lock the sash. If windows are stored inside, they can easily be moved through the openings and set in place (Fig. 59-8). Center the unit in the opening on the rough sill with the window casing overlapping the wall sheathing.

> **CAUTION** Have sufficient help when setting large units. Handle them carefully to avoid damaging the unit or breaking the glass. Broken glass can cut through protective clothing and cause serious injury.

Place a level on the window sill. If not level, determine which side of the window is the highest. Check to see if the high side of the unit is at the desired height. If not, bring it up by shimming with a wood shingle tip between the rough sill and the bottom end of the window's side jamb (Fig. 59-9). Tack through the lower end of the casing into the sheathing on the high side. Shim the other side of the window, in the same manner, so the window sill is level. Tack the lower end of the casing on that side.

Fig. 59-8 Windows can be easily installed, through the opening, from the inside.

On wide windows with long sills, shim at intermediate points so that the sill is perfectly straight and level with no sag. Use either a long level or a shorter level in combination with a straightedge. Also, sight the sill by eye from end to end to make sure it is straight.

Plumb the ends of the side jambs with a level. Tack the top ends of the side casings. Straighten the side jambs between sill and header jamb. Tack through the side casings at intermediate points. Straighten and tack the header casing.

Check the joint between sash and jamb. Make sure the sash operates properly. If not, make necessary adjustments. Then fasten the window permanently in place. Use galvanized casing or common

Fig. 59-9 Shim under the side jambs to level the window unit. (*Courtesy of Andersen Windows, Inc.*)

Fig. 59-10 Roofing nails are used to fasten the flanges of vinyl-clad windows. (*Courtesy of Andersen Windows, Inc.*)

nails spaced about 16 inches apart. Keep nails about 2 inches back from the ends of the casings to avoid splitting. Nail length depends on the thickness of the casing. Nails should be long enough to penetrate the sheathing and into the framing members. Set the nails so they can be puttied over later.

Vinyl-clad windows have a vinyl nailing flange. Large head roofing nails are driven through the flange instead of through the casing (Fig. 59-10).

The installation of wood windows in masonry and brick veneer walls is similar to that of frame walls. Adequate clearance should be left for caulking around the entire perimeter between the window and masonry (Fig. 59-11).

Flashing the Header Casing

To flash the header casing, cut the flashing with tin snips. Its length should be equal to the overall width of the window. Do not let the ends project beyond the side casings. This will make the application of siding difficult. If the flashing must be ap-

Fig. 59-11 Windows are installed in masonry openings against wood bucks. (*Courtesy of Andersen Windows, Inc.*)

Fig. 59-12 A window flashing is used over the top of the header casing.

plied in more than one piece, lap the joint about 3 inches. Place the flashing firmly on top of the header casing. Secure with fasteners along its top edge and into the wall sheathing (Fig. 59-12).

Flashing the Sill

If the sides of the window openings have been previously flashed with felt strips, cut a strip of felt about 8 inches wide. It should be long enough to extend to the outside edges of the side flashings. Tuck the edge of the strip under the loose ends of the felt flashings on each side and well into the groove in the bottom of the window sill. Fasten the strip. Make sure to leave a few inches of the bottom edge loose so that building paper can be applied under it later (Fig. 59-13).

If housewrap is used, apply a strip of it in the same manner and well up in the siding groove in the bottom of window sill. It must, however, be applied on top of housewrap already installed. Seal all overlapping edges and ends with housewrap tape.

Fig. 59-13 Flashing the sill of a wood window with asphalt felt.

Fig. 59-14 Installing a metal awning window.

Fig. 59-15 Method of setting a metal window in a masonry opening.

Metal Windows

Metal windows are available in the same styles as wood windows. The shape and sizes of the parts vary with the manufacturer and the intended use.

In frame construction, if metal windows are used, they may be set in a wood frame. The frame is then installed in the same manner as for wood windows. They may also be set in the opening with their flanges overlapping the siding or sheathing. Caulking is applied under the flanges. The unit is then screwed to the wall (Fig. 59-14).

In masonry construction, wood *bucks* are fastened to the sides of the opening. Metal windows are installed against them. The bottom usually rests against a rabbet of a preformed concrete sill. The flanges on the two sides and top are bedded in caulking. They are then screwed to the bucks. Mortar is applied to cover the flanges and beveled to the outside corner of the masonry opening (Fig. 59-15).

Carefully follow the installation directions provided with the units, whether wood or metal.

Glazing

The art of cutting and installing lights of glass in sash and doors is called *glazing*. Those who do such work are called *glaziers*. Sometimes the carpenter may have to replace a light of glass.

Glazing Materials

Sash are made so that the glass is usually held in place with *glazing points* and *glazing compound* (Fig. 59-16). Glazing points are small triangular or diamond-shaped pieces of thin metal. They are driven into the sash parts to hold the glass in place.

Fig. 59-16 Lights of glass are secured with glazing points and glazing compound.

Glazing compound is commonly called *putty.* It is used to cover and seal around the edges of the light.

A light of glass is installed with its convex side, or crown of its bow, up in a thin bed of compound against the rabbet of the opening. Glass set in this manner is not as apt to break when installed.

Installing Glass

To replace a light of glass, first remove the broken glass.

CAUTION Use heavy gloves to handle the broken glass. Broken glass edges are sharp and can cut easily.

Clean all compound and glazier points from the rabbeted section of the sash. Apply a thin bed of glazing compound to the surface of the rabbet on which the glass will lay. Lay the glass in the sash with its crowned side up. Carefully seat the glass in the bed by moving it back and forth slightly.

Fasten the glass in place with glazier points. Slide the driver along the glass. Do not lift the driver off the glass. Special glazier point driving tools prevent glass breakage. If a driving tool is not available, drive the points with the side of a chisel or a putty knife.

Lay a bead of compound on the glass along the rabbet on top of the light of glass. Trim the compound at a bevel by drawing the putty knife along. One edge should be flush with the inside edge of the glass opening. The other edge of the compound should feather to the outside edge of the opening. Prime the compound as soon as possible after glazing. Lap the paint about 1/16th inch onto the glass to make a seal against the weather.

Cutting Glass

Sometimes it may be necessary to cut a light of glass to size so it will fit in the opening. Lay the glass on a clean, smooth surface. Brush some mineral spirits along the line of cut. Hold a straightedge on the line. Draw a glass cutter with firm pressure along the straightedge to make a clean, uniformly scored line (Fig. 59-17).

Do not go over the scored line. This will dull the glass cutter. The line must be scored along the whole length the first time with no skips. Otherwise, the glass may not break where desired. Practice on scrap pieces to become proficient in making clean breaks.

Move the glass so the scored line is even with the edge of the workbench. Apply downward pressure on the overhanging glass. If the glass is properly scored, it will break along the scored line (Fig. 59-18).

Fig. 59-17 The glass must be scored with a single stroke of the glass cutter.

Fig. 59-18 If the glass is scored correctly, it will break easily and cleanly along the scored line.

Review Questions

Select the most appropriate answer.

1. A frame in a window that holds the glass is called a
   a. light.   c. sash.
   b. mullion.   d. stile.

2. Small wood strips that divide the glass into smaller lights are called
   a. mantels.   c. mullions.
   b. margins.   d. muntins.

3. When windows are installed in multiple units, the joining of the side jambs forms a
   a. mantel.   c. mullion.
   b. margin.   d. muntin.

4. A window that consists of an upper and a lower sash that slide vertically is called a
   a. casement window.
   b. double-hung window.
   c. hopper window.
   d. sliding window.

5. A window that has a sash hinged on one side and swings outward is called
   a. an awning window.
   b. a casement window.
   c. a double-hung window.
   d. a hopper window.

6. The difference between a hopper and an awning window is that the hopper window
   a. swings inward instead of outward.
   b. swings outward instead of inward.
   c. is hinged at the top rather than at the bottom.
   d. is hinged on the side rather than on the bottom.

7. Before setting a window in an opening
   a. flash all sides of the opening.
   b. flash the bottom.
   c. flash two sides.
   d. flash the top.

8. Horns on windows are
   a. placed on the header casing.
   b. extensions of the side jambs.
   c. moldings applied to a flat casing.
   d. extensions of the header jamb.

9. The art of cutting and installing glass is called
   a. gauging.   c. gouging.
   b. glazing.   d. grouting.

10. Glass is installed
   a. with its crowned side up.
   b. with its concave side up.
   c. with either side up.
   d. on the inside face of the sash.

■ BUILDING FOR SUCCESS ■

The Building Inspector: Friend or Foe?

A career in construction will introduce you to a variety of skilled and professional trade representatives. One of these representatives will be the building inspector. In most cases, the inspector will be engaged by a municipality. He or she serves the community as a quality control agent who assures public safety during the construction of buildings. A primary function of the local building inspector is to examine and evaluate construction. He or she determines if it is in compliance with the ap-

proved standards. Codes become the criteria by which the plans, specifications, and entire construction process is gauged.

The building inspector has the responsibility of determining if a building or structure is safe for occupancy, meets minimum code requirements, and is safe from fire.

The working relationship among contractors, skilled workers, and building inspectors must be one of mutual respect, courtesy, and confidence. It is imperative that building contractors and subcontractors clearly understand the building inspector's role and how the inspector can assist in the building process. Everyone involved in construction should strive to establish good working relations with each other.

Trust can be built between the inspector and builder when both understand the inspector's function and purpose. This trust is based on how well the inspector knows the adopted codes, evaluates construction projects, and relates his or her assessments to the builder. Equally important is the builder's history of performing quality work. Together the two professionals can set the pace for meaningful and safe construction projects.

The local building standards should relate to the building's structural integrity, life safety, and environmental health capabilities. The pertinent community ordinances are reflected along with the local building codes. These codes will have many parts, possibly consisting of building, electrical, fire prevention, health, housing, life safety, mechanical, and plumbing requirements. Additionally, the local zoning ordinances will play a major role in the inspector's job. Workable codes should be performance codes rather than specification codes. As long as the materials and workmanship perform to code standards, many systems can be used. New technology, new products, climatic conditions, and environmental issues will continue to influence local codes and regulations.

Inspections will generally occur in a specific order. For example, site inspection may be followed by the inspections of footings and foundation forms, reinforcing steel, embedded electrical, mechanical, and plumbing, or steel reinforcing before the concrete is placed. It is the builder's responsibility to properly plan for and schedule the appropriate inspection throughout the construction process. Assisting the inspector as needed during the inspection process will establish the best working relationships. Partial inspections may be needed so the construction assembly process can remain on schedule.

Inspection records will be required at all construction sites to properly document each inspection until the project is completed. Occasionally corrections will have to be made when violations or omissions have taken place. Adequate time will be given to make the necessary adjustments. Quick attention by the contractor will generally result in remaining on schedule and keep unforeseen expenses to a minimum.

This may be a time when all parties will need to implement their best communication and business skills to avoid frustrations and misunderstandings. The outcomes of these visits by the building inspector will form the basis for future cooperation. The inspector must remain impartial in interpreting the relevant codes and ordinances. The builder needs to recognize the infraction, make the necessary changes, and maintain the integrity of the relationship.

The building inspector has an extremely important role in the construction process. There will always be a need for someone to assure adequate and safe construction methods for the general populace. Occupants of buildings have great concern for a safe and well-built structure.

The building inspector is generally looked upon in a favorable manner by quality builders. They have learned to respect and trust the integrity of the position. Contractors can assure their customers a well-built home, office building, or place of business. The inspector's documentation verifies that the community's public safety requirements have been met throughout the construction process. In the vast majority of cases, the building inspector is a builder's friend and advisor in the construction process. The working relationships involving the inspector should remain favorable for a lifetime.

FOCUS QUESTIONS: For individual or group discussion.

1. How would you define the role and purpose of the local building inspector?

2. How do you think a building inspector could help you establish and maintain a reputable business?

3. What might be some reasons to clearly understand the role and mission of the building inspector?

4. What could be some possible consequences of not cooperating with the inspector and the agency he or she represents?

5. What do you envision as the best working relationship between the contractor and the building inspection official?

UNIT
21

Exterior Doors

Exterior doors, like windows, are manufactured in millwork plants in a wide range of styles and sizes. Many entrance doors come prehung in frames, complete with exterior casings applied and ready for installation. In other cases, the door frame must be constructed and set in the rough opening. The door is then fitted and hinged to it.

Careful construction and installation of the door frame makes it easier to fit and hang the door afterward. In addition, it contributes to smooth operation of the door and protection against the weather for many years.

Parts of an Exterior Door Frame

Terms given to members of an exterior door frame are the same as similar members of a window frame. The bottom member is called a *sill* or *stool*. The vertical side members are called *side jambs*. The top horizontal part is a *header jamb*. The exterior door trim may consist of many parts to make an elaborate and eye-appealing entrance or a few parts for a more simple doorway. The *door casings*, if not too complex, are usually applied to the door frame before it is set (Fig. 60-1). When more intricate trim is specified, it is usually applied after the frame is set (Fig. 60-2).

Sills

In residential construction, door frames are usually designed and constructed for entrance doors that swing inward. Codes require that doors swing outward in buildings used by the general public. The shape of a wood door sill for an inswinging door is different from that for an outswinging door (Fig. 60-3).

In addition to wood, extruded aluminum sills of many styles are manufactured for both inswinging and outswinging doors. They usually come with vinyl inserts to weatherstrip the bottom of the door. Some are adjustable for exact fitting between the sill and door (Fig. 60-4).

Jambs

Side and header jambs are the same shape. Jambs may be square edge pieces of stock to which *door stops* are later applied. Or, they may be *rabbeted jambs*, with *single* or *double* rabbets. On double-rabbeted jambs, one rabbet is used as a stop for the main door. The other is used as a stop for storm and screen doors (Fig. 60-5). Several jamb widths are available for different wall thicknesses. For walls of

Fig. 60-1 Parts of an exterior door frame.

Fig. 60-2 Elaborate entrance trim is available in knocked-down form, ready for assembly. (*Courtesy of Morgan Manufacturing*)

SILLS FOR INSWINGING DOORS

SILLS FOR OUTSWINGING DOORS

Fig. 60-3 Wood sill shapes vary according to the swing of the door.

Fig. 60-4 Some metal sills are adjustable for exact fitting at the bottom of the door.

odd thicknesses, jambs, except double-rabbeted, may be ripped to any desired width.

Exterior Casings

Exterior casings may be plain square edge stock. Moldings are sometimes applied around the outside edges of the casings. This is done to improve the appearance of the entrance. Because the main entrance is such a distinctive feature of a building, the exterior casing may be enhanced on the job with *fluted*, or otherwise shaped, pieces and appropriate caps and moldings applied (Fig. 60-6). Flutes are narrow, closely spaced, concave grooves that run parallel to the edge of the trim. In addition, ornate main entrance trim may be purchased in knocked-down form. It is then assembled at the job site (Fig. 60-7).

Making a Door Frame

One of the most common types of door frames consists of single-rabbeted jambs with an oak sill for an inswinging door. (Instructions for building and setting this frame are given in following paragraphs. The specifics of the instructions may not apply to other types of door frames, but the basics are similar.) If the construction and setting of the door

Fig. 60-5 Door jamb cross-sections may be square-edge, single-rabbeted, or double-rabbeted.

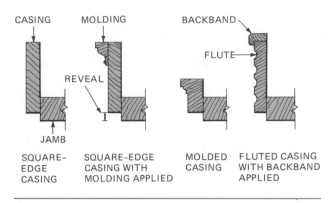

Fig. 60-6 Exterior door casings may be enhanced by applying moldings and by shaping.

Fig. 60-7 A few samples of manufactured entrance door trim. Many other styles are available. (*Courtesy of Morgan Manufacturing*)

frame described herein are mastered, competence can be easily gained with other types of door frames.

A more accurately sized door frame can be constructed if the door itself is available for measurement. Many doors are manufactured slightly over the nominal size. Some are manufactured slightly undersize. Making the frame too small will result in excessive fitting of the door. Unacceptable wide joints between door and frame is the result of making the frame too big.

Determining Jamb Width

The first step in making any door frame is to determine the jamb width. The jamb width is equal to the overall wall thickness from the outside surface of the wall sheathing to the inside surface of interior wall covering. Because the interior wall covering probably has not been applied at this point, its planned thickness must be known.

Rip the jamb stock so its width corresponds to the wall thickness. Rip the edge opposite the rabbet. Smooth and slightly back-bevel the ripped edge with a jack plane (Fig. 60-8). Back-bevel the rabbeted edge, also, if this has not already done at the mill. The reason for back-beveling is to assure a tight fit between the edges of the jamb and door casings when they are applied. Slightly round over the sharp outside corners that will be left exposed.

Dadoing the Side Jambs

The side jambs are usually *dadoed* to receive the ends of the header jamb and the sill. Lay the side jambs on a pair of sawhorses, side by side. Make sure their outside edges are against each other and their ends flush.

Dadoing for the Header Jamb. Square a line across both pieces about one inch down from the top end. To lay out the width of the dado, hold a piece of jamb stock with its end on the squared line. Mark its overall thickness toward the bottom end of the side jambs (Fig. 60-9). Mark the depth of the dado on the edge of the jamb opposite the rabbet.

To make the dado, first cut along both layout lines with a fine-toothed hand crosscut saw to the depth of the dado. Be careful when making the bottom cut not to score the face of the rabbet. If scored, it will be exposed when the frame is assembled. Chisel in from both edges to make the dado. The bottom surface of the dado should lie in a straight line from edge to edge with no crown. A slight

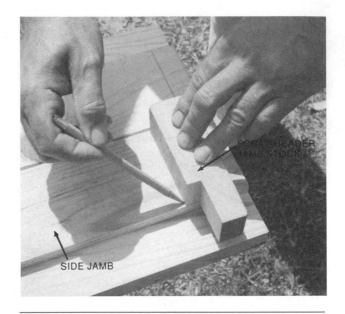

Fig. 60-9 A scrap piece of jamb stock is used to lay out the width of the dado for the header jamb.

Fig. 60-8

Fig. 60-10 A completed dado to receive the header jamb of the door frame.

hollow in the surface is not objectionable (Fig. 60-10).

Dadoing for the Sill. Place the scrap piece of jamb stock, previously used in the dado, with its rabbeted side toward the bottom end of the side jamb. From the rabbet of the scrap piece, lay out a distance equal to the door height. Allow an addition 3/16 inch for top and bottom joints, along the side jamb. Square a line across the face of the rabbet.

Hold a piece of scrap sill stock so the short straight surface, on its face side, lines up with the mark. The top of the *chamfer* on the inside edge of the sill should line up with the inside edge of the side jamb. Mark the side jamb along both sides of the scrap sill piece when held in this position (Fig. 60-11).

Place the side jambs back together again. Transfer the layout for the sill dado from one side jamb to the other. Dado the side jambs for the sill in

a manner similar to that for the header jamb (Fig. 60-12). Again, be careful not to score the face of the rabbet when making the upper saw cut.

Sill and Header Jamb

Cut the header jamb so it is square on both ends. Its length should be the width of the door plus 3/16 inch. The extra 3/16 inch allows for a 3/32-inch joint between door and frame on each side.

Lay out the same length on the sill stock length. Square lines across the face of the sill. From the squared lines, measure toward each end a distance of 5/16 inch more. Square across only on the chamfer. Notch both ends of the sill as shown in Figure 60-13. The projecting ends of the sill eventually butt against edges of the interior casing when applied.

Assembling the Door Frame

Although some have different-shaped jambs and sills, all door frames are assembled in much the same manner. Sand any layout lines from the side jambs. Stand the header and sill on end. Place the side jamb on them with their ends inserted in the dadoes. Nail through the side jambs into the ends of the sill and header jamb. To assure a tight fit on the face side, drive a chisel between the back sides of both sill and header jamb and the shoulders of the dado before nailing. This will drive the sill and header jamb tight against the dado shoulders on the exposed side. Drive three 8d common nails through the side jambs into the end of both pieces. Turn the assembly over. Install the other side jamb in the

Fig. 60-11 Mark both sides of the scrap sill stock to lay out the dado for the sill.

Fig. 60-12 Completed dado to receive the sill of the door frame.

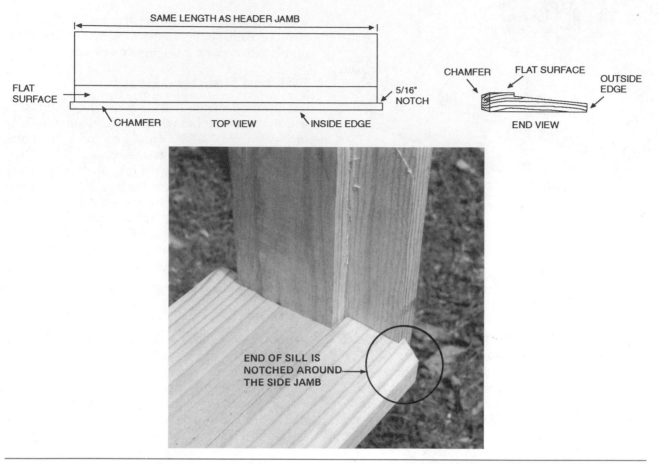

Fig. 60-13 The ends of the sill are notched so the inside edge of interior casings will later butt against them.

same manner. Keep the edges of the side and header jamb flush with each other. The notches in the sill should fit tightly against the inside edges of the side jambs (Fig. 60-14).

Applying the Exterior Casing

The exterior casing is applied so its inside edge is set back 1/2 inch from the inside face of the jambs. This setback is called a *reveal*. It is used as a stop for outswinging screen or storm doors (Fig. 60-6).

Joints between side and header door casings are made similar to those on windows. This prevents wind and rain from entering the wall cavity. If square-edged casings are used, a *blind rabbet* makes a weathertight joint between them (Fig. 60-15).

Applying Square Edge Casings. Rabbet the top end of both side casings. Apply both side casings with noncorroding fasteners spaced about 12 inches apart into the edge of the side jamb. The casings should be positioned so the top ends and inside edges will allow a 1/2-inch reveal on the jambs. Let the bottom ends project below the sill to the bottom edge of the wall sheathing or any desired amount.

Cut the header casing to a length equal to the distance between the outside edges of the side casings. Make a blind rabbet on both ends to fit over the rabbeted ends of the side casings.

Applying Molded Casings. Molded casings are applied in the same manner as square edge casings. However, a *plain miter* or a *splined miter* joint is made at the head. Plain mitered joints are usually caulked with compound and fastened together. A splined miter joint is used when exterior door casings are subjected to severe weather. It is made by inserting a thin strip of wood or metal across the joint into the ends of the miters (Fig. 60-16).

Setting the Door Frame

When setting a door frame, cut off the horns that project beyond the sill and header jamb. Flash both sides of the rough door opening with saturated felt in the same manner as for windows, except if housewrap has been used. Set the door frame in the opening with the sill on the subfloor (Fig. 60-17). Center the bottom of the frame in the opening. Level the sill by shimming under the side jambs, if necessary. Tack the lower end of both side casings to the wall.

Fig. 60-14 The assembled frame before exterior casings are applied.

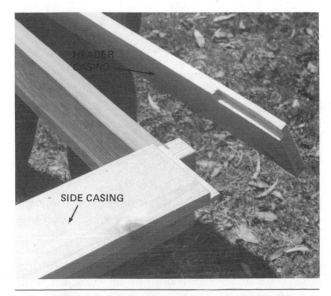

Fig. 60-15 A blind rabbet joint is used between exterior side and header door casings.

Fig. 60-16 A splined miter is sometimes used to join molded exterior door casings.

Fig. 60-17 Setting an exterior door frame.

② SUSPEND PLUMB BOB TO SILL FROM HEADER JAMB A CONVENIENT DISTANCE FROM SIDE JAMB.

HEADER CASING

③ PUSH TOP OF FRAME ONE WAY OR OTHER & FASTEN WHEN DISTANCE AT BOTTOM EQUALS THAT AT TOP.

HEADER JAMB

SIDE CASING

SIDE JAMB

FASTENER

SILL

④ THE FRAME IS PLUMB WHEN THIS DISTANCE EQUALS DISTANCE AT TOP.

① SHIM & LEVEL SILL & FASTEN THROUGH CASING ON BOTH SIDES OPPOSITE SILL.

Fig. 60-18

The side jambs must be plumbed next. A 26- or 28-inch carpenter's level may not be accurate when plumbing the sides because of any bow that may be in the jambs. Use either a 6-foot level with blocks attached to each end, a 6-foot straightedge in combination with a carpenter's level, or a plumb bob.

To use the plumb bob, measure from the side jamb a short distance on the header jamb. Hang the plumb bob from the mark to the sill. Move the top of the frame sideways, one way or the other, until the same distance is obtained at the sill (Fig. 60-18). When the side jambs are plumb, tack the top of the side casings to the wall.

Straighten the jambs between the top and bottom using a long straightedge placed against their sides. Tack the side casings to the wall at intermediate points when they are straight. Make sure the header jamb is straight. Tack the header casing to the wall.

Flash under the sill with felt before driving nails home. Felt flashing can then be installed under both the side casings and the side felt flashings. Check all parts again for level and plumb. Then drive and set all nails. Apply any decorative molding or other trim, if specified. Install the door flashing on the top edge of the header casing.

Setting Door Frames in Masonry Walls

In commercial construction, exterior wood or metal door frames are sometimes set in place before masonry walls are built. The frames must be set and firmly braced in a level and plumb position. The header jamb is checked for level. The bottom ends of the side jambs are secured in place. It may be necessary to shim one or the other side jamb in order to level the header jamb. The side jambs are then plumbed in a sideways direction. They are braced in position. Then, the frame is plumbed and braced at a right angle to the wall (Fig. 60-19).

Finally, the frame is checked to see if it has a *wind*. A wind is a twist in the door frame caused when the side jambs do not line up vertically with each other. No matter how carefully the side jambs

ADJUSTABLE WIRE ANCHORS
ARE INSTALLED AS WALL IS BUILT

LEVEL HEADER JAMB

PLUMB EACH SIDE
JAMB IN OR OUT
AT RIGHT ANGLE
TO WALL

BRACE DOOR
FRAME WHEN
PLUMB

PLUMB
DOOR
FRAME
SIDEWAYS

ANCHOR BOTTOM ENDS
TO FLOOR. SHIM, IF NECESSARY, TO LEVEL HEADER JAMB

Fig. 60-19 Installation of an exterior door frame in a masonry wall.

of a door frame are plumbed, it is always best to check the frame to see if it has a wind.

One method of checking the door frame for a wind is to stand to one side. Sight through the door frame to see if the outer edge of one side jamb lines up with the inner edge of the other side jamb. If they line up, the frame is in wind. If they do not line up, make adjustments until they do. One way of making the adjustment is to plumb and brace one side at a right angle to the wall. Then sight, line up, and brace the other side jamb.

Some workers check for a wind by stretching two lines diagonally from the corners of the frame. If both lines meet accurately at their intersections, the frame does not have a wind (Fig. 60-20).

HELPFUL HINTS

How to check for a wind in a door frame.

INNER EDGE OF
OPPOSITE SIDE
JAMB

SIGHT THROUGH DOOR
FRAME TO SEE IF OUTER
EDGE OF THIS JAMB
LINES UP WITH INNER
EDGE OF OPPOSITE SIDE
JAMB

SIGHT

FRAME WITH A WIND

FRAME WITH NO WIND

METHOD #1– SIGHT THROUGH DOOR FRAME

Fig. 60-20

DOOR FRAME

STRETCHED LINES

IF THERE IS A SPACE BETWEEN LINES AT INTERSECTION, OR LINES ARE PRESSED TIGHTLY TOGETHER, AN ADJUSTMENT NEEDS TO BE MADE.

PLUMB BOTH SIDES OF DOOR FRAME. STRETCH LINES AND SEE IF THEY MEET ACCURATELY AT INTERSECTION

METHOD #2– STRETCH LINES DIAGONALLY

Fig. 60-20 *(continued)*

CHAPTER 61   DOOR FITTING AND HANGING

Fitting and hanging of wood doors is still an important part of the carpentry trade despite the increasing use of prehung door units. There are many situations in new construction and in remodeling work that require fitting and hanging doors on the job. It is helpful to be able to identify door styles, know their sizes, and understand their construction.

Door Styles and Sizes

Exterior *flush* and *panel* doors are available in many styles. Architects have many choices when designing entrances.

Flush Doors

An exterior flush door has a smooth, flat surface of wood veneer or metal. It has a framed, *solid core* of staggered wood blocks or composition board. Wood *core blocks* are inserted in appropriate locations in composition cores. They serve as *backing* for door locks (Fig. 61-1). Openings may be cut in flush doors either in the factory or on the job. Lights of various kinds and shapes are installed in them. Molding of various shapes may be applied in many designs to make the door more attractive.

Panel Doors

Panel doors are classified by one large manufacturer as *high-style, panel, sash, fire, insulated, French, Dutch, and ventilating* doors. *Sidelights,* although not actually doors, constitute part of some entrances. They are fixed in the door frame on one or both sides of the door (Fig. 61-2). *Transoms* are similar to sidelights. When used, they are installed above the door.

Panel Door Styles

High-style doors, as the name implies, are highly crafted designer doors. They may have a variety of cut-glass designs. Panel doors are made with raised panels of various shapes. Sash doors have panels of tempered or insulated glass that allow the passage of light. Fire doors are used where required by codes. These doors prevent the spread of fire for a certain period of time. Insulated doors have thicker panels with LoE or argon-filled insulated glass. French doors may contain one, five, ten, twelve, fifteen, or eighteen lights of glass for the total width and length inside the frame. A Dutch door consists of top and bottom units, hinged independently of

MAT-FORMED COMPOSITION BOARD CORE
7 PLY CONSTRUCTION ILLUSTRATED

FRAMED BLOCK NON-GLUED CORE
5 PLY CONSTRUCTION ILLUSTRATED

TOP AND BOTTOM RAILS
MINIMUM 1 INCH

WIDTH OF STILES
MINIMUM 1 INCH

CORE OF MAT-FORMED
COMPOSITION BOARD

FACE VENEER

TOP AND BOTTOM RAILS
MINIMUM 1 INCH

WIDTH OF STILES
MINIMUM 1 INCH

WOOD CORE BLOCKS

FACE VENEER

Fig. 61-1 Solid wood or composition cores are required in exterior flush doors. (*Courtesy of National Wood Window and Door Association*)

Fig. 61-2 Sidelights are installed on one or both sides of the main entrance door.

each other. Ventilating doors are made with operating sash and screen combinations (Fig. 61-3).

Parts of a Panel Door

A panel door consists of a frame that surrounds panels of solid wood and glass, or louvers. Some door parts are given the same terms as a window sash.

The outside vertical members are called *stiles.* Horizontal frame members are called *rails.* The *top rail* is generally the same width as the stiles. The *bottom rail* is the widest of all rails. A rail situated at lockset height, usually 38 inches from the finish floor to its center, is called the *lock rail.* Almost all other rails are called *intermediate rails. Mullions* are vertical members between rails dividing panels in a door.

Fig. 61-3 Several kinds of exterior doors are made in many designs. (*Courtesy of Morgan Manufacturing*)

The molded shape on the edges of stiles, rails, mullions, and bars, adjacent to panels, is called the *sticking*. The name is derived from the molding machine, commonly called a *sticker*, used to shape the parts. Several shapes are used to *stick* frame members.

Bars are narrow horizontal or vertical rabbeted members. They extend the total length or width of a glass opening from rail to rail or from stile to stile. Door *muntins* are short members, similar to and extending from bars to a stile, rail, or another bar. Bars and muntins divide the overall length and width of the glass area into smaller lights.

Panels fit between and are usually thinner than the stiles, rails, and mullions. They may be *raised* on one side or on both sides for improved appearance (Fig. 61-4).

DOORS ARE STUCK WITH VARIOUS SHAPES

Fig. 61-4 The parts of an exterior paneled door.

Exterior Door Sizes

Practically all exterior entrance doors are manufactured in a thickness of 1 3/4 inches, in widths of 2'-8" and 3'-0", and in heights of 6'-8" and 7'-0". Doors that are less than 6'-8" in height and less than 2'-8" in width are not generally used in entrances. A few styles are produced in widths of 2'-6" and 2'-10" and in heights of 6'-6", 6'-10", and 8'-0". Some styles, such as French, panel, and sash doors, are available in narrower widths, such as 1'-6", 2'-0", and 2'-4", when double doors are used, for instance. Sidelights are made in both 1 3/8- and 1 3/4-inch thickness, and 1/2 inch greater in height than the entrance door. This allows for fitting the bottom against the sloping sill of the door frame. Auxiliary doors, such as combination storm and screen doors are made in 1 1/8-inch thickness.

There may be styles and sizes of stock doors other than those described above. Door manufacturers range from those that employ a few workers and serve only a local market to large manufacturers that ship their doors all over the world. It is not possible or desirable to describe all their products here. Study door manufacturers' literature to become better informed about door styles and sizes.

Check local building codes for entrance door requirements. Many codes specify minimum sizes, number, kind, and location of entrance doors.

Prehung Doors

Some doors come already fitted, and hinged in the door frame with the *lockset* and outside casings installed. A *prehung* door unit is set in the rough opening similar to setting door frames.

Installing a Prehung Door Unit

Working from the outside, center the unit in the opening with the casings removed. The door should be in a closed position with factory-installed spacers between door and frame still in place. Level the sill. Shim between the side jambs and the jack stud at the bottom. Tack through the side jambs and shims into the stud at the bottom using finishing nails. Plumb the side jambs. Shim at the top and tack. Shim at intermediate points along the side jamb. Fasten through the shims (Fig. 61-5).

Fig. 61-5 The door remains in the frame while a prehung unit is being set.

Make sure the spacers between the door and the jamb are in place to maintain the proper joint between the door and jamb. Also make sure that the outside edge of the jamb is flush with the wall sheathing. Open the door to make sure is operates properly. Make any necessary adjustments. Drive additional nails as required. Set them all. Do not make any hammer marks on the finish. Drive the nail until it is almost flush. Then use a nail set to drive it the rest of the way (Fig. 61-6). Some installers prefer to leave the exterior casing on. They fasten the unit through the casing first. Then, they go inside to install the shims and additional fastening.

Fitting the Door

The first step in fitting a door is to determine the side that will close against the stops on the door frame. If the design of the door permits that either side may be used toward the stops, sight along the door stiles from top to bottom to see if the door is bowed. Hardly any doors are perfectly straight. Most are bowed to some extent.

Determining Stop Side of Door

The door should be fitted to the opening so that the hollow side of the bow will be against the door stops. Hanging a door in this manner allows the top and bottom of the closed door to come up tight against stops. The center comes up tight when the door is latched. Also, the door will not rattle.

If no attention is paid to which side the door stops against, then the reverse may happen. The door will come up against the stop at the center and away from the stop at the bottom and top (Fig. 61-7).

Determining Exposed Side of Sash Doors

It is important to hang exterior doors containing lights of glass with the proper side exposed to the weather. This prevents wind-driven rainwater from seeping through joints. Manufacturers clearly indicate, with warning labels glued to the door, which side should face the exterior. Any door warranty is voided if the door is improperly hung. Do not hang exterior doors with the removable *glass bead* facing outward. Glass bead is small molding used to hold lights of glass in the opening. The bead can be iden-

STEP BY STEP

P R O C E D U R E S

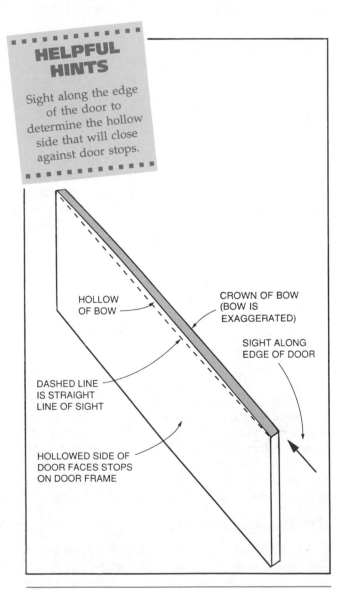

1. DRIVE NAIL ALMOST HOME
2. THEN, USE NAIL SET

NAIL SET

3. SET NAIL AT LEAST 1/8" DEEP

Fig. 61-6 To avoid marring finish work, drive the nail until almost home. Then use a nail set.

HELPFUL HINTS

Sight along the edge of the door to determine the hollow side that will close against door stops.

HOLLOW OF BOW

CROWN OF BOW (BOW IS EXAGGERATED)

SIGHT ALONG EDGE OF DOOR

DASHED LINE IS STRAIGHT LINE OF SIGHT

HOLLOWED SIDE OF DOOR FACES STOPS ON DOOR FRAME

Fig. 61-7

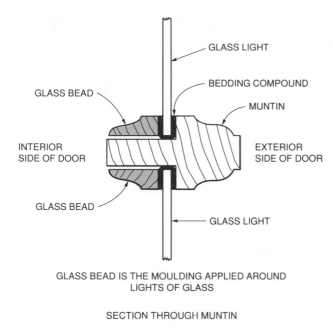

GLASS LIGHT

BEDDING COMPOUND

GLASS BEAD

MUNTIN

INTERIOR
SIDE OF DOOR

EXTERIOR
SIDE OF DOOR

GLASS BEAD

GLASS LIGHT

GLASS BEAD IS THE MOULDING APPLIED AROUND
LIGHTS OF GLASS

SECTION THROUGH MUNTIN

Fig. 61-8 It is important that the side of doors containing glass lights face in the direction recommended by the manufacturer.

tified by holes made when fasteners of the bead were set (Fig. 61-8).

Some doors are manufactured with *compression glazing.* This virtually eliminates the possibility of any water seeping through the joints. When LoE insulating glass is used in a door, it is especially important to have the door facing in the direction indicated by the manufacturer.

Determining the Swing of the Door

Determine the swing of the door from the plans. A door is designated as being *right-hand* or *left-hand*, depending on the direction it swings. The designation is determined by standing on the side of the door that swings away from the viewer. It is a right-hand door if the hinges are on the viewer's right. It is a left-hand door if the hinges are on the left (Fig. 61-9). Lightly mark the door to identify the hinge edge and the hinge pin side.

Place the door on sawhorses with its face side up. Cut any horns from the stiles. Measure carefully the width and height of the door frame. The frame should be level and plumb but, this may not be the case and should not be taken for granted.

The process of fitting a door into a frame is called *jointing. The door must be carefully jointed. An even joint* of approximately 3/32 inch must be made between the door and the frame on all sides. A wider joint of approximately 1/8 inch must be made to allow for swelling of the door and frame in extremely damp weather.

Use a *door jack* to hold the door steady. A manufactured or job-made jack may be used (Fig. 61-10). Use a *jointer plane*, either hand or power. Plane the edges and ends of the door so that it fits snugly in the door frame.

Fit the top end against the header jamb. Then fit the bottom against the sill so that a proper joint is obtained. Careful jointing may require moving the door in and out of the frame several times.

Next, joint the hinge edge against the side jamb. Finally, plane the lock edge so the desired joint is obtained on both sides. The lock edge must also be planed on a *bevel*. This is so the back corner does not strike the jamb when the door is opened. The bevel is determined by making the same joint between the back corner of the door and the side jamb when the door is slightly open, as when the door is closed (Fig. 61-11).

Extreme care must be taken when fitting doors not to get them undersize. The door can always be jointed a little more. However, it cannot be made wider or longer. Do not cut more than 1/2 inch total from the width of a door. Cut no more

LEFT-HAND SWING

RIGHT-HAND SWING

VIEWING DIRECTION

VIEWING DIRECTION

IF HINGES ARE ON THE LEFT
SIDE WHEN VIEWED WITH THE
DOOR SWINGING AWAY, IT IS
A LEFT-HAND DOOR

IF HINGES ARE ON THE RIGHT
SIDE WHEN VIEWED WITH THE
DOOR SWINGING AWAY, IT IS
A RIGHT-HAND DOOR

Fig. 61-9 Determining the swing of a door.

WEIGHT OF DOOR BOWS
PLYWOOD BASE CAUSING
BRACKETS TO CLAMP
DOOR.

DOOR

3/4" OR THICKER
BRACKETS

3/8" PLYWOOD

SPACE BETWEEN BRACKETS
SLIGHTLY MORE THAN
DOOR THICKNESS.

A DOOR JACK CAN BE MADE ON THE JOB
FROM SCRAP LUMBER.

2" × 4" BLOCKS

Fig. 61-10 When fitting a door, hold it in a door jack.

JAMB

JAMB

DOOR

LOCK EDGE OF DOOR
PLANED AT BEVEL

PLAN

AMOUNT OF BEVEL —
1/8" IN 2"

1-3/8" DOOR — APPROX 1/16"
1-3/4" DOOR — APPROX 3/32"

AMOUNT OF
DOOR BEVEL

JAMB

DOOR

3/32" JOINT WHEN
DOOR IS CLOSED

JAMB

3/32" JOINT
WHEN DOOR IS
PARTLY OPEN

DOOR

Fig. 61-11 The lock edge of a door must be planed at a bevel to clear the jamb when opened.

than 2 inches from its height. Cut equal amounts from ends and edges when approaching maximum amounts. Check the fit frequently by placing the door in the opening, even if this takes a little extra time. Most entrance doors are very expensive. Care should be taken not to ruin one. Speed will come with practice. Handle the door carefully to avoid marring it or other finish. After the door is fitted, *ease* all sharp corners slightly with a block plane and sandpaper. To ease sharp corners means to round them over slightly.

Hanging the Door

On swinging doors, the *loose-pin* type *butt hinge* is ordinarily used. The pin is removed. Each *leaf* of the hinge is applied separately to the door and frame. The door is hung by placing it in the opening. The pins are inserted to rejoin the separated hinge leaves. Extreme care must be taken so that the hinge leaves line up exactly on the door and frame. Use three 4×4 hinges on 1 3/4-inch doors 7'-0", or less in height. Use four hinges on doors over 7'-0" in height.

The hinge leaves are recessed in flush with the door edge and only partway across. The recess for the hinge is called a *hinge gain* or *hinge mortise*. Hinge gains are only made partway across the edge of the door. This is so that the edge of the hinge is not exposed when the door is opened. The remaining distance, from the edge of the hinge to the side of the door, is called the *backset* of the hinge. Butt hinges must be wide enough so that the pin is lo-

cated far enough beyond the door face to allow the door to clear the door trim when fully opened (Fig. 61-12).

Location and Size of Door Hinges

On paneled doors, the top hinge is usually placed with its upper end in line with the bottom edge of the top rail. The bottom hinge is placed with its lower end in line with the top edge of the bottom rail. The middle hinge is centered between them.

On flush doors, the recommended placement is not more than 9 3/4 inches down from the top nor more than 10 3/8 inches up from the bottom. The middle hinge is centered between the two (Fig. 61-13).

Laying Out Hinge Locations

Place the door in the frame. Shim the top and bottom so the proper joint is obtained. Place shims between the lock edge of the door and side jamb of the frame. The hinge edge should be tightly against the door jamb.

Use a sharp knife. Mark across the door and jamb at the desired location for one end of each hinge. A knife is used because it makes a finer line than a pencil. The marks on the door and jamb should not be any longer than the hinge thickness. Place a small X, with a pencil, on both the door and the jamb. This indicates which side of the knife mark to cut the gain for the hinge (Fig. 61-14). Care must be taken to cut hinge gains on the same side of

Fig. 61-12 Hinges are backset from the side of the door and the edge of the door stop.

Fig. 61-13 Recommended placement of hinges on flush doors.

the layout line on both the door and the door frame. Remove the door from the frame. Place it in the door jack with its hinge edge up in order to lay out and cut the hinge gains.

Laying Out the Hinge Gain

The first step in laying out a hinge gain is to mark its ends. Place a hinge leaf on the door edge with its end on the knife mark previously made. With the *barrel* of the leaf against the side of the door, hold

Fig. 61-14 Mark the location of hinges on the door and frame with a sharp knife.

Fig. 61-15 The ends of hinge gains are laid out by scoring with a knife along a hinge leaf.

the leaf firmly. Score a line along one end with a sharp knife. Then, tap the other end, until the leaf just covers the line. Score a line along the other end (Fig. 61-15). Score only partway across the door edge.

Using a Butt Gauge

The *butt gauge* contains two sliding rods that can be secured with set screws. One rod is adjusted to lay out the depth of the hinge gain. The other rod is used to lay out the backset of the hinge on both the door and door frame. Adjust this rod for the desired backset and the other rod for the depth of the hinge gain. Lay out the backset between the two end marks previously made (Fig. 61-16). Use the other rod and lay out the gain depth. Scoring lines between and not beyond the ends of the gain is a sign of a professional. Scored layout lines beyond the gain will remain exposed on the door after the gain is cut.

Fig. 61-16 Laying out the backset of the hinge gain with a butt gauge.

HELPFUL HINTS

Use a knife to deepen the scored backset line when mortising for hinges.

Fig. 61-17

Cutting Hinge Gains

Use a sharp knife to deepen the scored backset line. It may be necessary to draw the knife along the line several times to score to the bottom of the gain. Do not use a chisel for scoring. Using a chisel will likely split the edge of the door (Fig. 61-17).

With the bevel of the chisel down, cut a small chip from each end of the gain. The chips will break off at the scored end marks (see Fig. 12–22). With the flat of the chisel against the shoulders of the gain, cut down to the bottom of the gain.

Make a series of small chisel cuts along the length of the gain. Brush off the chips. Then, with the flat of the chisel down, pare and smooth the excess down to the depth of the gain. Be careful not to slip and cut off the backset (Fig. 61-18).

After the gain is made, the hinge leaf should press-fit into it. It should be flush with the door edge. If the hinge leaf is above the surface, deepen the gain until the leaf lies flush. If the leaf is below the surface, it may be shimmed flush with thin pieces of cardboard from the hinge carton.

Applying Hinges

Press the hinge leaf in the gain. Drill pilot holes for the screws. Center the pilot holes carefully on the countersunk holes of the hinge leaf. A *self-centering punch* is often used for this purpose. Drilling off-center will cause the hinge to move from its position when the screw is driven. Fasten the hinge leaf with screws provided with the hinges. Cut gains and apply all hinge leafs on the door and frame in the same manner.

Fig. 61-18 Chiseling out the hinge gain.

Shimming Hinges

Hang the door in the frame by inserting a pin in the barrels of the top hinge leaves first. Insert the other pins. Try the swing of the door. If the door binds against or is too close to the jamb on the hinged side, shim between the hinge leaves and the gain. Use a narrow strip of cardboard on the side of the screws nearest the pin. This will move the door toward the lock side of the door frame. If the door binds against or is too close to the jamb on the lock side, apply shims in the same manner. However, apply them on the opposite side of the hinge screws (Fig. 61-19). Check the bevel on the lock edge of the door. Plane to the proper bevel, if necessary. Ease all sharp, exposed corners.

Other Methods of Hanging Doors

Other methods are used to lay out and cut hinge gains. They may be faster and more efficient if the necessary tools and equipment are available.

Using Butt Hinge Markers

Instead of laying out hinge gains with a butt gauge and knife, *butt hinge markers* of several sizes are often used (see Fig. 11–27). With markers, location of the hinge leaves must still be marked on the door and jamb, as described. The width and length of the hinge gain are outlined by simply placing the marker in the proper location on the door edge. Tap it with a hammer (Fig. 61-20). However, the depth of the gain must be scored with a butt gauge or some other gauging method. The gain is then chiseled out in the manner previously described.

Using a Butt Hinge Template and Router

A *butt hinge template* fixture or jig and portable electric router are usually used when many doors need

Fig. 61-20 Sometimes butt hinge markers are used to lay out hinge gains.

to be hung. Most hinge routing jigs contain three adjustable templates secured on a long rod. The templates are positioned and tacked in place with pins. The jig is used on both the door and frame as a guide to rout hinge gains. The templates are adjustable for different size hinges, hinge locations, and door thicknesses. An attachment positions the hinge template jig to provide the required joint between the top of the door and the frame.

The gains are cut using a router with a special hinge mortising bit. The bit is set to cut the gain to the required depth. A template guide is attached to the base of the router. It rides against the jig template when routing the hinge gain (Fig. 61-21).

PLACING THE SHIM TOWARD THE OUTSIDE OF THE HINGE MOVES THE DOOR TOWARD THE LOCK EDGE

PLACING THE SHIM TOWARD THE INSIDE OF THE HINGE MOVES THE DOOR AWAY FROM THE LOCK EDGE

Fig. 61-19 Shimming the hinges will move the door toward or away from the lock edge.

Fig. 61-21 A butt hinge template is used on both door and frame when routing hinge gains.

Butt hinges with rounded corners are used in routed gains. Hinges with square corners require that the rounded corner of a routed gain be chiseled to a square corner.

Other Exterior Doors

Other exterior doors, such as swinging *double doors*, are fitted and hung in a similar manner to single doors. Allowance must be made, when fitting swinging double doors, for an *astragal* between them for weathertightness (Fig. 61-22). An astragal is a molding that is rabbeted on both edges. It is designed to cover the joint between double doors. One edge has a square rabbet. The other has a beveled rabbet to allow for the swing of one of the doors.

Patio Doors

Patio door units normally consist of two or three sliding or swinging glass doors completely assembled in the frame (Fig. 61-23). Instructions for as-

Fig. 61-22 An astragal is required between double doors for weathertightness. (*Courtesy of Andersen Corporation*)

Fig. 61-23 Two or three doors are usually used in sliding or swinging type patio door units. (*Courtesy of Andersen Corporation*)

sembly are included with the unit. They should be followed carefully. Installation of patio door frames is similar to setting frames for swinging doors. After the frame is set, the doors are installed using special hardware supplied with the unit.

In sliding patio door units, one door is usually stationary. In two-door swinging units, either one or the other door is the swinging door. In three-door units, the center door is usually the swinging door.

Garage Doors

Overhead garage doors come in many styles, kinds, and sizes. Two popular kinds used in residential construction are the *one-piece* and the *sectional* door. The rigid one-piece unit swings out at the bottom. It then slides up and overhead. The sectional type has hinged sections. These sections move upward on rollers and turn to a horizontal position overhead. A *rolling steel door*, used mostly in commercial construction, consists of narrow metal slats that roll up on a drum installed above the opening.

Special hardware, required for all types, is supplied with the door. Equipment is available for power operation of garage doors, including remote control. Also supplied are the manufacturers' directions for installation. These should be followed carefully. There are differences in the door design and hardware of many manufacturers.

CHAPTER 62 INSTALLING EXTERIOR DOOR LOCKSETS

After the door has been fitted and hung in the frame, the *lockset* and other door hardware is installed. A large variety of locks are available from several large manufacturers in numerous styles and qualities, providing a wide range of choices.

Kinds of Locksets

Locksets can be grouped into two broad categories: *cylindrical* and *mortise* types. A distinguishing difference is the basic shape. Consequently, the preparation of the door to receive the lockset varies (Fig. 62-1).

Cylindrical Locksets

Cylindrical locksets are often called *key-in-the-knob locksets* (Fig. 62-2). They are the most commonly used type in both residential and commercial construction. This is primarily because of the ease and speed of installing them. They may be obtained in several groups, from light-duty residential to extra-heavy-duty commercial and industrial applications.

In place of knobs, *lever handles* are provided on locksets likely to be used by handicapped persons or in other situations where a lever is more suited than a knob (Fig. 62-3). *Deadbolt locks* are used for

CYLINDRICAL LOCK

MORTISE LOCK

Fig. 62-1 The method of preparing the door for cylindrical locks differs from that for mortise locks.

Fig. 62-2 Cylindrical locksets are the most commonly used type in both residential and commercial construction. (*Courtesy of Schlage Lock Co.*)

Fig. 62-3 Locks with lever handles are used when difficulty in turning knobs is expected. (*Courtesy of Schlage Lock Co.*)

Fig. 62-4 Deadbolt locks are used primarily as auxiliary locks for added security. (*Courtesy of Schlage Lock Co.*)

Fig. 62-5 A grip-handle lockset combines well with a deadbolt lock. (*Courtesy of Schlage Lock Co.*)

both primary and auxiliary locking of doors in residential and commercial buildings (Fig. 62-4). They provide additional security. They also make an attractive design in combination with *grip handle* locksets or latches (Fig. 62-5).

An *interconnected* lock combines deadbolt locking with a standard key-in-knob set (Fig. 62-6). By turning the knob on the inside both the locks are retracted for emergency exiting.

Mortise Locksets

The *mortise* lockset (Fig. 62-7) is more commonly used in commercial construction, such as hospitals, schools, office and similar buildings, where extra-heavy-duty locks of high quality are required. Mortise locks are generally not used in residential construction. This is because of the high cost of the lock and the time required to install it.

Lock Trim and Finish

The kind of trim and finish are factors, in addition to the kind and style, which determine the quality of a lockset. *Escutcheons* are decorative plates of various shapes that are installed between the door and the lock handle or knob. Locksets and escutcheons are available in various metals and finishes. More expensive locksets and trim are made of brass, bronze, or stainless steel. Less expensive ones may be of steel that is plated or coated with a finish. It is important that conditions and usage be taken into

Fig. 62-6 The interconnected lock combines a deadbolt with a standard latch lockset. (*Courtesy of Schlage Lock Co.*)

Fig. 62-7 Mortise locks are generally used in commercial construction. (*Courtesy of Schlage Lock Co.*)

consideration when selecting locksets. This is especially the case in areas that have a humid climate or are near salt water.

Installing Cylindrical Locksets

To install a cylindrical lockset, first check the contents and read the manufacturer's directions carefully (Fig. 62-8). There are so many kinds of locks manufactured that the mechanisms vary greatly. The directions included with the lockset must be followed carefully.

However, there are certain basic procedures. Open the door to a convenient position. Wedge the bottom to hold it in place (Fig. 62-9). Measure up, from the floor, the recommended distance to the centerline of the lock. This is usually 38 inches. At this height, square a light line across the edge and stile of the door.

Marking and Boring Holes

Position the center of the paper template supplied with the lock on the squared lines. Lay out the centers of the holes that need to be bored (Fig. 62-10). It is important that the template be folded over the high corner of the beveled door edge. The distance from the door edge to the center of the hole through the side of the door is called the *backset* of the lock. Usual backsets are 2 3/8 and 2 3/4 inches. Make sure the backset is marked correctly before boring the hole. One hole must be bored through the side and one into the edge of the door. The manufacturers' directions specify the hole sizes.

The hole through the side of the door should be bored first. Stock for the center of the boring bit is lost if the hole in the edge of the door is bored first. The hole is usually 2 1/8 inch in diameter. It can be bored with hand tools, using an expansion bit in a bit brace. However, it is a difficult job. If using hand tools, bore from one side until only the point of the bit comes through. Then bore from the other side to avoid splintering the door.

Using a Boring Jig. A *boring jig* is frequently used. It is clamped to the door to guide power-driven *multispur* bits. With a boring jig, holes can be bored completely through the door from one side. The clamping action of the jig prevents splintering (Fig. 62-11).

After the holes are bored, insert the latch in the hole bored in the door edge. Hold it firmly, and score around its faceplate with a sharp knife. Remove the latch unit. Deepen the vertical lines with the knife in the same manner as with hinges. Do not use a chisel along these lines. This may split out the edge of the door. Then, chisel out the recess so that the faceplate of the latch lays flush with the door edge.

Faceplate markers, if available, may be used to lay out the mortise for the latch faceplate. A marker of the appropriate size is held in the bored latch hole and tapped with a hammer (Fig. 62-12). Complete the installation of the lockset by following specific manufacturers' directions.

SCHLAGE®

HOW TO INSTALL THIS A SERIES LOCK

MANUFACTURED BY SCHLAGE LOCK COMPANY

Part of worldwide Ingersoll-Rand

1. MARK DOOR

Mark height line (center line of latchbolt) on edge of door. Suggested height from floor 36". Mark center point of door thickness. Position center line of template on height line. Hold in place and mark center point for 2 1/8" hole.

Mark For 2⅛" Hole

Height Line (Center Line of Latchbolt).

Note: "A" series locks are normally extended for doors as thick as 1⅞". For doors over 1⅞" thick, mark hole for latch ¾" from outside face of door.

2. BORE TWO HOLES

Bore a 2⅛" hole from both sides of door (to avoid splintering or otherwise damaging door).

2⅛" Hole

⅞" Hole (Standard)

Bore a ⅞" or 1" hole, depending on latch housing diameter, straight into door edge to intersect with center of 2⅛" hole.

Use latchfront faceplate to pattern for cutout. Front should fit flush with door surface. Install latch with screws provided.

2a. CIRCULAR LATCH INSTALLATION

For circular latch installation. Drill 1" diameter latch-bolt hole.

Insert latch partially into latchbolt hole. Line up beveled face of latchbolt with edge of jamb. Push latch into hole as far as it will go.

Place wooden block against bolt. Apply enough force to depress bolt. Tap block with hammer or mallet to drive latch into hole. Surface of latch faceplate should be flush with edge of door.

3. INSTALL STRIKE

Mark vertical line and height line on jamb exactly opposite center point of latch hole. Clean out hole and install strike.

Step 1. FOR T STRIKE bore two ⅞" holes, ¹¹⁄₁₆" deep in frame on vertical line ⅜" above and below height line.

Clear Out

Height Line

⅜"

⅜"

Strike

Strike Box

1⅛"

2¾"

⅛" Deep

Step 2. FOR FULL LIP STRIKE mark screw holes for strike so that screws lie on same vertical center line as latch screws. Cut out frame providing for clearance of latch bolt and strike tongue and install strike.

Height Line

Strike Screws Are Located On Vertical Center Line

4. REMOVE INSIDE TRIM

Depress knob catch, slide knob off spindle and remove appropriate rose design as described.

Snap On Rose

Mounting Plate

Threaded Rose

Fig. 62-8 Directions for installation are provided with the lockset. (*Courtesy of Schlage Lock Co.*)

5. ADJUST ROSE

Rotate Rose 1/16" Short of Housing for 1 3/8" Thick Door.

Outside Rose   Housing

Rotate Out To 3/16" For 1 7/8" Door. This Is The Maximum Adjustment.

6. INTERLOCK UNITS

Latch unit must be in place before installing lock. Be sure lock housing engages with latch prongs and retractor interlocks with latch bar.

Lock Housing

Retractor Engages Latch Bar.

Latch Prongs Engage Lock Housing.

For proper installation, deadlocking plunger on latchbolt must stop against strike, preventing forcing when door is closed.

DEADLOCKING PLUNGER

STRIKE

Caution—Do Not Attempt to Mount Lock Unit with Door Closed.

7. ATTACH TRIM & ROSE

Step 1: SNAP ON ROSE. Slip mounting plate over spindle and fasten securely with two machine screws. Snap rose over spring clip on mounting plate.

Notch for Spring Clip

Snap rose over spring clip slot to remove rose

Depress Knob Catch Slide knob all the way onto spindle so catch engages into slot.

Step 2: THREADED ROSE Slip rose over spindle and screw onto threaded hub. Turn clockwise and tighten with spanner wrench

8. TO CHANGE LOCK HAND

Pin tumbler cylinders are factory assembled in knobs for right or left hand doors as ordered. If necessary to change the hand of a lock so that cylinder is right side up, see following instructions.

RIGHT WAY

WRONG WAY

8a. STANDARD CYLINDER

Insert pointed end of spanner wrench, or small nail, in the outside knob sleeve on side facing latchbolt. Exert pressure and turn key slowly until knob catch depresses; then pull off knob. Turn knob over. With key partially removed from cylinder, replace knob onto spindle. Slide knob up to knob catch. Insert key and turn key one-quarter turn to right, depress knob catch and push knob into position.

8b. INTERCHANGEABLE CORE CYLINDER

Follow steps in 8A for removing knob. CORE MUST NOW BE REMOVED FROM KNOB BEFORE REPLACING ON LOCK. Insert Control key and turn 15° to the right until action stops. Pulling on key will now extract core from knob.

15°

Turn knob over and replace knob on spindle. Slide knob up to knob catch. Insert long end of installation tool into knob until it engages driver. Turn tool one-quarter turn to right, depress knob catch and push knob into position. With control key fully inserted, push core into knob housing, rotate key 15° to the right and insert core completely. Rotate key back to upright and remove control key.

SCHLAGE

Part of worldwide Ingersoll-Rand

©1985 Schlage Lock Company
Printed in USA
P509-403   Rev. 11/90

Fig. 62-8 (continued)

**HELPFUL
HINTS**

How to hold the door
steady while installing
a lockset.

SIDE JAMBS

DOOR

SILL

SCRAP 1 × 3
OR 1 × 4 STOCK
ABOUT 1 FT. LONG

JAM BOTTOM PIECE
IN TIGHTLY

SUBFLOOR

Fig. 62-9

Installing the Striker Plate

The *striker plate* is installed on the door jamb so when the door is closed it latches tightly with no play. If the plate is installed too far out, the door will not close tightly against the stop. It will then rattle. If the plate is installed too far in, the door will not latch.

To locate the striker plate in the correct position, place it over the latch in the door. Close the door snugly against the stops. Push the striker plate in against the latch. Draw a vertical line on the face of the plate flush with the outside face of the door (Fig. 62-13).

Open the door. Place the striker plate on the jamb. The vertical line, previously drawn on it, should be in line with the edge of the jamb. Center the plate on the latch. Hold it firmly while scoring a line around the plate with a sharp knife. Chisel out the mortise so the plate lies flush with the jamb. Screw the plate in place. Chisel out the center to receive the latch.

Mortise Lock

Mortise locks are so-called because the edge of the door must be mortised to the depth of the lock. The mortise is made by boring a series of holes along the centerline of the door edge. The mortise is then shaped with a chisel to receive the lock (Fig. 62-14). A *mortising jig* can be used to guide bits when boring holes for mortises (Fig. 62-15).

After the mortise is made for the body of the lock, a gain is chiseled in the edge of the door for the faceplate. When mortise locks are installed, some of the holes in the side of the door are not bored completely through. Care must be taken to read the manufacturers' directions carefully. A door installation would be unacceptable if a hole was bored through when it was supposed to be bored only partway through.

After the door is fitted, hung, and locked, remove all hardware. Prime the door and all exposed parts of the door frame. Replace the door and hardware after the prime coat is dry.

Fig. 62-10 Using a template to lay out the centers of the holes for a lockset.

Fig. 62-11 Boring jigs are frequently used to guide bits when boring holes for locksets.

Fig. 62-12 Using a faceplate marker.

MARKING THE STRIKER PLATE TO
POSITION IT ON THE DOOR JAMB

EDGE OF JAMB

JAMB

STRIKER PLATE

CENTER LINE
OF DOOR LOCK
HEIGHT

LINE UP MARK
PREVIOUSLY MADE WITH
EDGE OF JAMB

POSITION THE STRIKER PLATE ON THE JAMB
AND SCORE AROUND ITS EDGES AND ENDS WITH
A KNIFE

Fig. 62-13 Installing the striker plate.

Fig. 62-14 A series of holes is bored in the door edge to install a mortise lockset.

FACEPLATE OUTLINE

Fig. 62-15 A mortising jig can be used to bore holes when making mortises for locks. (*Courtesy of Porter Cable Co.*)

Review Questions

Select the most appropriate answer.

1. The standard thickness of exterior wood doors in residential construction is
   a. 1 3/8 inches.   c. 1 3/4 inches.
   b. 1 1/2 inches.   d. 2 1/4 inches.

2. The recommended minimum width of exterior entrance doors is
   a. 3'-0".   c. 2'-6".
   b. 2'-8".   d. 2'-4".

3. The height of exterior entrance doors in residential construction is generally not less than
   a. 7'-0".   c. 6'-8".
   b. 6'-10".   d. 6'-6".

4. The thickness of combination storm and screen auxillary doors is usually
   a. 3/4 inch.   c. 1 3/8 inches.
   b. 1 1/8 inches.   d. 1 1/2 inches.

5. A narrow member dividing the glass in a door into smaller lights is called a
   a. bar.   c. rail.
   b. mullion.   d. stile.

6. The setback of the door casing on the edge of the door jamb is called a
   a. gain.   c. rabbet.
   b. backset.   d. reveal.

7. Projections of the stiles of a door beyond the top and bottom rails are called
   a. horns.   c. casings.
   b. ears.   d. rabbets.

8. Before hanging a door, sight along its length for a bow. The hollow side of the bow should
   a. face the outside.
   b. face the door stops.
   c. face the inside.
   d. be straightened with a plane.

9. The joint between the door and door frame should be close to
  a. 3/32 inch.
  b. 3/64 inch.
  c. 1/4 inch.
  d. 3/16 inch.

10. On paneled doors, the top end of the top hinge is usually placed
  a. in line with the bottom edge of the top rail of the door.
  b. 7 inches down from the top end of the door.
  c. not more than 10 3/4 inches down from the top end of the door.
  d. in line with the top edge of an intermediate rail.

■ BUILDING FOR SUCCESS ■

Preparing for the Job Search

During the final months of your education, your attention will gradually be drawn toward finding gainful employment. Securing employment in an area related to your studies will most likely be where you concentrate your job-seeking efforts. But there may be a need to remain open to other job opportunities that might appear to be somewhat removed from your original employment plans.

For the apprentice carpenter, employment normally accompanies the education and training experience. Apprenticeship programs are one of the more secure avenues to gainful employment in the construction industry. Regardless of the avenue through which you obtained your training and education, you should feel competent and prepared for full-time construction work.

In addition to technical skills, what is needed to help a student in his or her quest for a specific job? Generally, most competitors will possess the necessary technical skills called for in the job. But what other skills, attitudes, and abilities will employers be looking for? How can a person be sure he or she has the additional prerequisites an employer may be looking for?

Before the actual job search is under way, each student or graduate must do some self-assessment. Once people understand more about themselves and compare that data to the desired job, better job decisions may be made. If the position of self-employment in a construction-related business is the ultimate career goal for you, a quick review of your skills, abilities, financial status, and experience will indicate how soon that may become real-

ity. At the very least, you will identify certain things that will have to be accomplished in a given time frame in order for you to become the owner of your own business. Personal work experience must be obtained first, before self-employment can be considered. The amount of experience and the time it takes depend on the individual.

An assessment of each job with which you have an interest must be conducted. This will begin a process of elimination. The job's work conditions, specific technical tasks, climatic conditions, associated stresses, potential dangers, time demands, travel obligations, and potential wages will all help you target a desired job.

Included in the assessment of the job should be the satisfaction element. Usually jobs that bring minimal satisfaction, even if accompanied by larger salaries, will cause people to reconsider their decisions in time. It is commonplace today to see people change jobs for reasons such as advancement opportunities or more money. Other reasons center on increased stress and dissatisfaction.

In much of the business world today, there is an increased emphasis on developing a personal portfolio. This portfolio enables a person to market his or her skills to a variety of businesses. This makes a person more mobile in nature and able to work with more than one employer or client. This is more apt to be an avenue for the four-year college graduate. It may also be a consideration for the technician who has more management or business desires in mind. Many technically skilled workers are continuing to make themselves available to employers in the traditional way by contracting with only one employer.

As each student prepares for graduation,

he or she should further assess personal skills and abilities. Important questions include the following.

1. Do I like to lead others or follow others' instructions? Why?

2. Do I like to make decisions regarding the planning of a project? Why?

3. Do I feel confident that I can perform the technical tasks that the job demands? Why?

4. Do I work well with others as a team player?

5. Do I feel more confident with maximum or minimal supervision and assistance?

6. Will I enjoy performing the same job tasks on a continual basis?

7. Do I understand the physical demands of the job?

8. Will I be satisfied with the starting wage, knowing my present financial obligations?

9. Can I demonstrate to my employer my capabilities and understand a good performance will be necessary for pay raise considerations?

10. Am I prepared to assume the job responsibilities?

The construction education program you are currently experiencing should use these and other questions as part of your job search orientation. Each job, once identified, can be looked at in greater depth with the assistance of the instructor or others with trade experience. The learning outcomes from personal and job assessment discussions will help point each person in the best direction for employment.

The following is an illustration of how an in-depth job assessment helped one student select a job that was related to construction but not involved in the actual building process. This young man had a desire to work with residential construction. He did not desire to be a frame or trim carpenter. He was completing a program of study in building construction at a community college. However, he was open to other ideas for employment. His interests were greater in associating with people, drafting, estimating, cabinet making, and sales. After looking closely at the retail building materials business, he decided to pursue that line of work. His construction background and education gave him the skills needed to work in the construction materials sales business. Prior to graduation he interviewed with a major building materials company. He was offered a job in the location of his choice. In his later years as an educator in construction, he helped form a building materials program of study at a community college.

Preparing for the job search must precede the actual job search. It should be an ongoing process beginning at least, with the technical training and education that follow high school. An earlier preliminary evaluation of employment opportunities in high school would be even more beneficial for responsible decisions. Field trips throughout elementary and secondary school years have exposed many people to a variety of job opportunities. Individuals that can offer assistance in preparing for a job search are parents, teachers, trade representatives, counselors, clergy, and members of technical advisory teams. By seeking the assistance of people who understand various types of construction-related jobs, each student can better decide which job will be appealing. The avenue of education and training pursued by each person can be a result of spending time with one or all of those people.

FOCUS QUESTIONS: For individual or group discussion.

1. Conduct a self-analysis by answering the questions stated in this study.

2. Record your responses to the self-analysis and the job analysis for guidance purposes.

3. Make a list of the potential job opportunities that appear to be worthy of your further research. Narrow this list down to a few that surface as the best for you.

4. With the assistance of your instructor, begin the job search.

UNIT

22

Siding

OBJECTIVES

After completing this unit, the student will be able to:

■ describe the shapes, sizes, and grades of various siding products.

■ install corner boards and prepare sidewalls for siding.

■ apply horizontal and vertical lumber siding.

■ apply plywood and hardboard panel and lap siding.

■ apply wood shingles and shakes to sidewalls.

■ apply aluminum and vinyl siding.

■ estimate required amounts of siding.

CONTENTS

Finish that covers large areas of exterior walls is called *siding*. Siding does not include masonry covering, such as stucco or brick veneer. Siding is used extensively in both residential and commercial construction.

CHAPTER 63 WOOD SIDING TYPES AND SIZES

Fig. 63-1 Wood siding is used extensively in both residential and commercial construction. (*Courtesy of California Redwood Association*)

Siding is manufactured from *solid lumber, plywood, hardboard, aluminum,* and *vinyl.* It comes in many different patterns. Prefinished types eliminate the need to refinish for many years, if at all. Siding may be applied horizontally, vertically, or in other directions, to make many interesting designs (Fig. 63-1).

Wood Siding

The natural beauty and durability of solid wood have long made it an ideal material for siding. The *Western Wood Products Association (WWPA)* and the *California Redwood Association (CRA)* are two major organizations whose member mills manufacture siding and other wood products. They have to meet standards supervised by their associations. Grade stamps of the WWPA and CRA and other associations of lumber manufacturers are placed on siding produced by member mills. Grade stamps assure the consumer of a quality product that complies with established standards (Fig. 63-2). WWPA member mills produce wood siding from species such as *fir, larch, hemlock, pine, spruce,* and *cedar.* Most *redwood* siding is produced by mills that belong to the CRA.

Wood Siding Grades

Grain. Some siding is available in *vertical grain, flat grain,* or *mixed grain.* In vertical grain siding the annual growth rings, viewed in cross-section, must form an angle of 45 degrees or more with the surface of the piece. All other lumber is classified as flat grain (Fig. 63-3). Vertical grain siding is the highest

quality. It warps less, takes and holds finishes better, has less defects, and is easier to work.

Surface Texture. Sidings are manufactured with *smooth, rough,* or *saw-textured* surfaces. Saw-textured

WESTERN WOODS PRODUCTS ASSOCIATION

CALIFORNIA REDWOOD ASSOCIATION &
REDWOOD INSPECTION SERVICE

Fig. 63-2 Association trademarks assure that the product has met established standards of quality. (*Courtesy of Western Wood Products Association; Courtesy of California Redwood Association*)

Fig. 63-3 In some species, siding is available in vertical, flat, or mixed grain. (*Courtesy of California Redwood Association*)

surface finishes are obtained by resawing in the mill. They generally hold finishes longer than smooth surfaces.

WWPA Grades. Grades published by the WWPA for siding products are shown in Figure 63-4. Siding graded as *premium* has fewer defects such as knots and pitch pockets. The highest premium grade is produced from clear, all-heart lumber. *Knotty* grade siding is divided into #1, #2, and #3 *common*. The grade depends on the type and number of knots and other defects.

CRA Grades. There are over thirty grades of redwood lumber. The best grades are grouped in a category called *architectural*. They are used for high-quality exterior and interior uses, including siding (Fig. 63-5).

Siding Patterns and Sizes

The names, descriptions, and sizes of siding patterns are shown in Figure 63-6. Some patterns can only be used for either horizontal or vertical applications. Others can be used for both. *Drop* and *tongue-and-grooved* sidings are manufactured in a variety of patterns other than shown. *Bevel* siding, more commonly known as *clapboards*, is a widely used kind (Fig. 63-7).

GRADE	DESCRIPTION
CLEAR ALL HEART	A SUPERIOR GRADE FOR FINE SIDINGS AND ARCHITECTURAL USES. IT IS ALL HEARTWOOD AND THE GRADED FACE OF EACH PIECE IS FREE OF KNOTS.
CLEAR	SIMILAR IN QUALITY TO CLEAR ALL HEART, EXCEPT THAT IT INCLUDES SAPWOOD IN VARYING AMOUNTS. SOME BOARDS MAY HAVE ONE OR TWO SMALL, TIGHT KNOTS ON THE GRADED FACE.
HEART B	AN ECONOMICAL ALL-HEART-WOOD GRADE CONTAINING A LIMITED NUMBER OF TIGHT KNOTS AND CHARACTERISTICS NOT PERMITTED IN CLEAR OR CLEAR ALL HEART. IT IS GRADED ON ONE FACE ON ONE EDGE.
B GRADE	AN ECONOMICAL GRADE CONTAINING A LIMITED NUMBER IF TIGHT KNOTS WITH SAPWOOD ACCENTING THE HEARTWOOD.

REDWOOD GRADES ARE ESTABLISHED BY THE REDWOOD INSPECTION SERVICE

Fig. 63-5 Redwood siding grades. (*Courtesy of California Redwood Association*)

General Categories (Note that there are additional grades for bevel pattern)		GRADES		
		Western Species		**Cedar**
		Selects	**Finish**	**Western & Canadian**
ALL PATTERNS	**PREMIUM GRADES**	C Select D Select	Superior Prime	Clear Heart A Grade B Grade
Additional Grades for Bevel Patterns	Premium		Superior Bevel	Clear VG Heart A Bevel B Bevel Rustic C Bevel
	Knotty		Prime Bevel	Select Knotty Quality Knotty
ALL PATTERNS	**KNOTTY GRADES**	Commons	Alternate Boards	
		#2 Common	Select Merch.	Select Knotty
		#3 Common	Construction	Quality Knotty
			Standard	

Fig. 63-4 Western Wood Products Association grade rules for siding products. (*Courtesy of Western Wood Products Association*)

SIDING PATTERNS		NOMINAL SIZES
		Thickness & Width
	TRIM **BOARD-ON-BOARD** **BOARD-AND-BATTEN** Boards are surfaced smooth, rough or saw-textured. Rustic ranch-style appearance. Provide horizontal nailing members. Do not nail through overlapping pieces. Vertical applications only.	1 x 2 1 x 4 1 x 6 1 x 8 1 x 10 1 x 12 1¼ x 6 1¼ x 8 1¼ x 10 1¼ x 12
³/₁₆ ³/₁₆ ¹⁵/₃₂ ³/₄	**BEVEL OR BUNGALOW** Bungalow ("Colonial") is slightly thicker than Bevel. Either can be used with the smooth or saw-faced surface exposed. Patterns provide a traditional-style appearance. Recommend a 1" overlap. Do not nail through overlapping pieces. Horizontal applications only. Cedar Bevel is also available in ⁷/₈ x 10, 12.	½ x 4 ½ x 5 ½ x 6 ⅝ x 8 ⅝ x 10 ¾ x 6 ¾ x 8 ¾ x 10
⁵/₁₆ ¹³/₃₂ ¹¹/₁₆ ¹³/₁₆	**DOLLY VARDEN** Dolly Varden is thicker than bevel and has a rabbeted edge. Surfaced smooth or saw textured. Provides traditional-style appearance. Allows for ½" overlap, including an approximate ⅛" gap. Do not nail through overlapping pieces. Horizontal applications only. Cedar Dolly Varden is also available ⁷/₈" x 10, 12.	Standard Dolly Varden ¾ x 6 ¾ x 8 ¾ x 10 Thick Dolly Varden 1 x 6 1 x 8 1 x 10 1 x 12
	DROP Drop siding is available in 13 patterns, in smooth, rough and saw textured surfaces. Some are T&G (as shown), others are shiplapped. A variety of looks can be achieved with the different patterns. Do not nail through overlapping pieces. Horizontal or vertical applications.	¾ x 6 ¾ x 8 ¾ x 10

SIDING PATTERNS		NOMINAL SIZES
		Thickness & Width
	TONGUE & GROOVE Tongue & groove siding is available in a variety of patterns. T&G lends itself to different effects aesthetically. Sizes given here are for Plain Tongue & Groove. Do not nail through overlapping pieces. Vertical or horizontal applications.	1 x 4 1 x 6 1 x 8 1 x 10 Note: T&G patterns may be ordered with ¹/₄, ³/₈ or ⁷/₁₆" tongues. For wider widths, specify the longer tongue and pattern.
	CHANNEL RUSTIC Channel Rustic has ½" overlap (including an approximate ⅛" gap) and a 1" to 1¼" channel when installed. The profile allows for maximum dimensional change without adversely affecting appearance in climates of highly variable moisture levels between seasons. Available smooth, rough or saw textured. Do not nail through overlapping pieces. Horizontal or vertical applications.	¾ x 6 ¾ x 8 ¾ x 10
	LOG CABIN Log Cabin siding is 1½" thick at the thickest point. Ideally suited to informal buildings in rustic settings. The pattern may be milled from appearance grades (Commons) or dimension grades (2x material). Allows for ¼" overlap, including an approximately ⅛" gap. Do not nail through overlapping pieces. Horizontal or vertical applications.	1½ x 6 1½ x 8 1½ x 10 1½ x 12

Fig. 63-6 Names, descriptions, and sizes of natural wood siding patterns. (*Courtesy of Western Wood Products Association*)

Fig. 63-7 Bevel siding is commonly known as clapboards. (*Courtesy of California Redwood Association*)

Panel and Lap Siding

Most *panel* and *lap* siding is manufactured from plywood and hardboard. It comes in a variety of sizes, patterns, and surface textures. Some hardboard siding comes prefinished in a variety of colors. Or, it is primed for the later application of finish coats of any color.

Panel Siding

Plywood panel siding, manufactured by APA member mills, is known as *APA 303* siding. It is produced in a number of surface textures and patterns (Fig. 63-8). It comes in several thicknesses, 4-foot widths, and lengths of 8, 9, and 10 feet. It is usually

BRUSHED

Brushed or relief-grain surfaces accent the natural grain pattern to create striking textured surfaces. Generally available in 11/32", 3/8", 1/2", 19/32" and 5/8" thicknesses. Available in redwood, Douglas fir, cedar, and other species.

KERFED ROUGH-SAWN

Rough-sawn surface with narrow grooves providing a distinctive effect. Long edges shiplapped for continuous pattern. Grooves are typically 4" o.c. Also available with grooves in multiples of 2" o.c. Generally available in 11/32", 3/8", 1/2", 19/32" and 5/8" thicknesses. Depth of kerfgroove varies with panel thickness.

APA TEXTURE 1-11

Special 303 Siding panel with shiplapped edges and parallel grooves 1/4" deep, 3/8" wide; grooves 4" or 8" o.c. are standard. Other spacings sometimes available are 2", 6" and 12" o.c., check local availability. T 1-11 is generally available in 19/32" and 5/8" thicknesses. Also available with scratch-sanded, overlaid, rough-sawn, brushed and other surfaces. Available in Douglas fir, cedar, redwood, southern pine, other species.

ROUGH-SAWN

Manufactured with a slight, rough-sawn texture running across panel. Available without grooves, or with grooves of various styles; in lap sidings, as well as in panel form. Generally available in 11/32", 3/8", 1/2", 19/32" and 5/8" thicknesses. Rough-sawn also available in Texture 1-11, reverse board-and-batten (5/8" thick), channel groove (3/8" thick), and V-groove (1/2" or 5/8" thick). Available in Douglas fir, redwood, cedar, southern pine, other species.

CHANNEL GROOVE

Shallow grooves typically 1/16" deep, 3/8" wide, cut into faces of 3/8" thick panels, 4" or 8" o.c. Other groove spacings available. Shiplapped for continuous patterns. Generally available in surface patterns and textures similar to Texture 1-11 and in 11/32", 3/8" and 1/2" thicknesses. Available in redwood, Douglas fir, cedar, southern pine and other species.

REVERSE BOARD-AND-BATTEN

Deep, wide grooves cut into brushed, roughsawn, coarse sanded or other textured surfaces. Grooves about 1/4" deep, 1" to 1-1/2" wide, spaced 8", 12" or 16" o.c. with panel thickness of 19/32" and 5/8". Provides deep, sharp shadow lines. Long edges shiplapped for continuous pattern. Available in redwood, cedar, Douglas fir, southern pine and other species.

Fig. 63-8 APA 303 plywood panel siding is produced in a wide variety of sizes, surface textures, and patterns. (*Courtesy of American Plywood Association*)

Fig. 63-9 Shapes of horizontally applied hardboard panel siding.

Fig. 63-10 Shapes and sizes of lap siding.

applied vertically. It may also be applied horizontally.

Hardboard panel siding is available with similar surface treatments and sizes. In addition, some hardboard siding, for horizontal application only, is manufactured 12 inches wide and 16 feet long. It comes in a *drop lap* design of two 6-inch courses or three 4-inch courses (Fig. 63-9). Most panel siding is shaped with *shiplapped* edges for weathertight joints.

Lap Siding

Lap siding is manufactured with rough-sawn, weathered wood grain, or other embossed surface textures. Some surfaces are *grooved* or *beaded* with square or beveled edges. It comes in thicknesses from 7/16 to 19/32 inch, widths of 6, 8, and 12 inches, and lengths of 16 feet. Its long length makes installation go more quickly. One type of hardboard lap siding comes with a *self-aligning spline*, inserted in the back side, for faster installation (Fig. 63-10).

Estimating Siding

First, calculate the wall area by multiplying its length by its height. To calculate the area of a gable end, multiply the length by the height. Then divide the result by two. Add the square foot areas together with the areas of other parts that will be sided, such as dormers, bays, and porches.

Since windows and doors will not be covered,

HOUSE AREA		
FRONT WALL	40' X 9' =	360 sq. ft.
BACK WALL	40' X 9' =	360 sq. ft.
LEFT SIDE WALL	22' X 9' =	198 sq. ft.
RIGHT SIDE WALL	22' X 9' =	198 sq. ft.
LEFT GABLE	$\frac{22' \times 6'}{2}$ =	66 sq. ft.
RIGHT GABLE	$\frac{22' \times 6'}{2}$ =	66 sq. ft.
TOTAL HOUSE AREA		1248 sq. ft.

OPENING AREA	
10 WINDOWS 3' X 4'	= 120 sq. ft.
2 WINDOWS 2' X 4'	= 16 sq. ft.
2 DOORS 3' X 7'	= 42 sq. ft.
TOTAL OPENING AREA	= 178 sq. ft.

Fig. 63-11 Estimating the area to be covered by siding.

their total surface areas must be deducted. Multiply the width by the height (large openings only). Subtract the results from the total area to be covered by siding (Fig. 63-11).

The amount of siding to order depends on the kind of siding used. When calculating, use the *factor* for either linear feet or board feet, according to the way it is sold (Fig. 63-12). (Coverage factors for other kinds of siding, such as wood shingles, can be found in the appropriate chapters.)

Pattern	Nominal Width	Width		Factor for Lineal Feet	Factor for Board Feet
		Dressed	Exposed Face		
Bevel & Bungalow	4	3½	2½	4.8	1.60
	6	5½	4½	2.67	1.33
	8	7¼	6¼	1.92	1.28
	10	9¼	8¼	1.45	1.21
Dolly Varden	4	3½	3	4.0	1.33
	6	5½	5	2.4	1.2
	8	7¼	6¾	1.78	1.19
	10	9¼	8¾	1.37	1.14
	12	11¼	10¾	1.12	1.12
Drop, T&G & Channel Rustic	4	3⅜	3⅛	3.84	1.28
	6	5⅜	5⅛	2.34	1.17
	8	7⅛	6⅞	1.75	1.16
	10	9⅛	8⅞	1.35	1.13
Log Cabin	6	5⁷⁄₁₆	4¹⁵⁄₁₆	2.43	2.43
	8	7⅛	6⅝	1.81	2.42
	10	9⅛	8⅝	1.39	2.32
Boards	2	1½	The exposed face width will vary depending on size selected and on how the boards-and-battens or boards-on-boards are applied.		
	4	3½			
	6	5½			
	8	7¼			
	10	9¼			

COVERAGE ESTIMATOR

Fig. 63-12 Multiply the area to be covered by either the linear feet or board feet factor to find the total footage required. (*Courtesy of Western Wood Products Association*)

CHAPTER 64 APPLYING VERTICAL AND HORIZONTAL WOOD SIDING

The method of siding application varies with the type. This chapter describes the application procedures for the most commonly used kinds of *solid* and *engineered wood* siding (Fig. 64-1).

Preparation for Siding Application

Before siding is applied, it must be determined how it will be *stopped* or treated at the foundation, eaves, and corners. The installation of various kinds of exterior wall trim may first be required.

Foundation Trim

In most cases, no additional trim is applied at the

foundation. The siding is started so that it extends slightly below the top of the foundation. However, a *water table* may be installed for appearance. It sheds water a little farther away from the foundation. The water table usually consists of a board and a drip cap installed around the perimeter. Its bottom edge is slightly below the top of the foundation. The siding is started on top of the water table.

Eaves Treatment

At the eaves the siding may end against the bottom edge of the frieze. The width of the frieze may vary. It is necessary to know its width when laying out horizontal siding courses. The siding may also terminate against the soffit, if no frieze is used. The joint between is then covered by a cornice molding. The size of the molding must be known to plan exposures on courses of horizontal siding.

Rake Trim

At the rakes, the siding may be applied under a furred-out rake fascia. When the rake overhangs the sidewall, the siding may be fitted against the rake frieze. When fitted against the rake soffit the joint is covered with a molding (Fig. 64-2).

Gable End Treatment

Sometimes a different kind of siding is used on the gable ends than on the sidewalls below the plate. The joint between the two types must be weathertight. One method of making the joint is to use a drip cap and flashing between the two types of material.

In another method, the plate and studs of the gable end are extended out from the wall a short distance. This allows the gable end siding to overlap the siding below (Fig. 64-3). Furring strips may also be used on the gable end framing in place of extending the gable plate and studs.

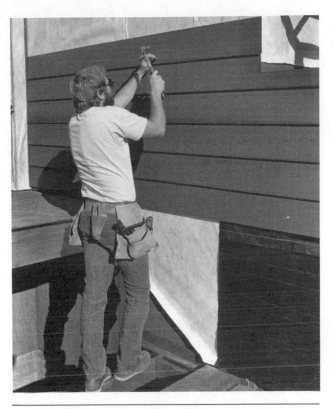

Fig. 64-1 Applying solid redwood bevel siding. (*Courtesy of California Redwood Association*)

Fig. 64-2 Methods of stopping siding at foundation, eaves, and rake areas.

Fig. 64-3 The gable end may be constructed to extend over the siding on the wall below.

Treating Corners

One method of treating corners is with the use of *corner boards.* Horizontal siding may be mitered around exterior corners. Or, metal corners may be used on each course of siding. In interior corners, siding courses may butt against a square corner board or against each other (Fig. 64-4). The thickness of corner boards depends on the type of siding used. The corner boards should be thick enough so that the siding does not project beyond the face of the corner board.

The width of the corner boards depends on the effect desired. However, one of the two pieces, making up an outside corner, should be narrower than the other by the thickness of the stock. Then,

after the wider piece is fastened to the narrower one, the same width is exposed on both sides of the corner. The joint between the two pieces should be on the side of the building that is least viewed.

Installing Corner Boards

Before installing corner boards, flash both sides of the corner. Install a strip of #15 felt vertically on each side. One edge should extend beyond the edge of the corner board at least 2 inches. Tuck the top ends under any previously applied felt.

With a sharp plane, slightly back-bevel one edge of the narrower of the two pieces that make up the outside corner board. This assures a tight fit between the two boards. Cut, fit, and fasten the narrow piece. Start at one end and work toward the bottom. Make sure to keep the beveled edge flush with the corner. Fasten with galvanized or other noncorroding nails spaced about 16 inches apart along its inside edge.

Cut, fit, and fasten the wider piece to the corner in a similar manner. Make sure its outside edge is flush with the surface of the narrower piece. The outside row of nails is driven into the edge of the narrower piece. Plane the outside edge of the wide piece wherever necessary to make it come flush. Slightly round over all sharp exposed corners by planing a small chamfer and sanding. Set all nails so they can be filled over later. Make sure a tight joint is obtained between the two pieces (Fig. 64-5).

Corner boards may also be applied by fastening the two pieces together first. Then install the assembly on the corner.

Applying Building Paper

A description of the kinds and purposes of building paper has been previously given in Chapter 59. A *breatheable* type building paper should be applied to sidewalls if housewrap has not already been applied. Building paper is not designed for prolonged

Fig. 64-4 Methods of returning and stopping courses of horizontal siding.

SATURATED FELT FLASHING

BACK–BEVEL
OUTSIDE EDGE

FIRST PIECE IS
PLACED ON SIDE
OF BUILDING LEAST
VIEWED

BACK–BEVEL
EXAGGERATED

SECTION

THE EXPOSED FACES SHOULD
BE CLOSE TO EQUAL WIDTH
ON BOTH SIDES OF THE CORNER.

Fig. 64-5 Procedure for installing corner boards.

exposure to the weather. It is not generally applied to the total wall area at one time, but only as the application of siding progresses.

Apply the paper horizontally. Start at the bottom of the wall. Make sure the paper lies flat with no wrinkles. Fasten it in position. If nailing, use large head roofing nails. Fasten in rows near the bottom, the center, and the top about 16 inches apart. Each succeeding layer should lap the lower layer by about 4 inches.

The sheathing paper should lap over any flashing applied at the sides and tops of windows and doors and at corner boards. It should be tucked under any flashings applied under the bottoms of windows or frieze boards. In any case, all laps should be over the paper below.

Installing Horizontal Wood Siding

One of the important differences between *bevel siding* and other types with *tongue-and-groove, shiplap,* or *rabbeted* edges is that exposure of courses of bevel siding can be varied somewhat. With other types, the amount exposed to the weather is constant with every course and cannot vary.

The ability to vary the exposure is a decided advantage. It is desirable from the standpoint of appearance, weathertightness, and ease of application to have a full course of horizontal siding exposed above and below windows and over the tops of doors (Fig. 64-6).

Determining Siding Exposure

To determine the siding exposure so that it is about equal both above and below the window sill, divide the overall height of the window frame by the amount of exposure.

Dividing the distance by the recommended exposure results in the number of courses between the top and the bottom of the window. For example, if the overall height of a window is 64 inches and 8-inch siding is used, divide the height by 7, the maximum allowed exposure. You will need slightly over nine courses. To obtain the exposure, divide 64 by 9. The result is 7 1/9 inches or a *slack* 7 1/8 inches.

The next step is to determine the exposure distance from the bottom of the window sill to about one inch below the bottom of the foundation wall. If this distance is 36 inches, then five courses are required at approximately 7 3/16 inches *strong*.

The term *strong* is used to mean slightly more, about 1/32 inch, than the measurement indicated. The term *slack* means slightly less. Thus, the exposure above and below the window sill is almost the same. The difference is not noticeable to the eye. The same procedure is used to lay out courses above the window. The layout lines are then transferred to a story pole. The story pole is used to lay out the courses all around the building (Fig. 64-7). Because of varying window heights, it may not always be possible to lay out full siding courses above and below every window or door.

Instead of calculating the number and height of siding courses mathematically, *dividers* may be used to space off the distances.

Starting Bevel Siding from the Top

Another advantage of using bevel siding is that application may be made starting at the top and working toward the bottom, if more convenient. With this method, a number of chalk lines may be snapped without being covered by a previous course. This saves time. Any scaffolding already erected may be used and then dismantled as work progresses toward the bottom.

Starting Siding from the Bottom

Most horizontal siding, however, is usually started at the bottom. For bevel siding, a *furring strip* of the same thickness and width of the siding *headlap* is

BEVEL SIDING

Fig. 64-6 A full course of horizontal siding should be exposed over and below windows and over doors, if possible.

Fig. 64-7 Laying out a story pole for courses of horizontal siding.

SECTION THROUGH SILL

Fig. 64-8 For bevel siding, a strip of wood of the same thickness and width of the headlap is used as a starter strip.

first fastened along the bottom edge of the sheathing (Fig. 64-8). For the first course, a line is snapped on the wall at a height that is in line with the *top edge* of the first course of siding.

For siding with constant exposure, the only other lines snapped are across the tops of wide entrances, windows, and similar objects to keep the courses in alignment. For lap siding, with exposures that may vary, lines are snapped for each successive course in line with their bottom edge. Stagger joints in adjacent courses as far apart as possible.

Fitting Siding

When applying a course of siding, start from one end and work toward the other end. With this procedure, only the last piece will have to be fitted. Tight fitting butt joints must be made between pieces. If an end joint must be fitted, use a block plane to trim the end as needed. When a piece has to be fitted between other pieces, measure carefully. Cut it just a trifle strong. Place one end in position. Bow the piece outward, position the other end, and snap into place (Fig. 64-9).

A *preacher* is often used for accurate layout of siding where it butts against corner boards, casings, and similar trim. The siding is allowed to overlap the trim. The preacher is held against the trim. A line is then marked on the siding along the face of the preacher (Fig. 64-10).

When fitting siding under windows, make sure the siding fits snugly in the groove on the underside of the window sill for weathertightness (Fig. 64-11).

Fastening Siding

Siding is fastened to each bearing stud or about every 16 inches. On bevel siding, fasten through the butt edge just above the top edge of the course below. Do not fasten through the lap. This prevents splitting of the siding that might be caused by slight swelling or shrinking due to moisture changes. Care must be taken to fasten as low as possible to avoid splitting the siding in the center. The location and number of fasteners recommended for siding are shown in Figure 64-12.

Fig. 64-9 Measure and cut individual pieces of bevel siding carefully for tight fits. (*Courtesy of California Redwood Association*)

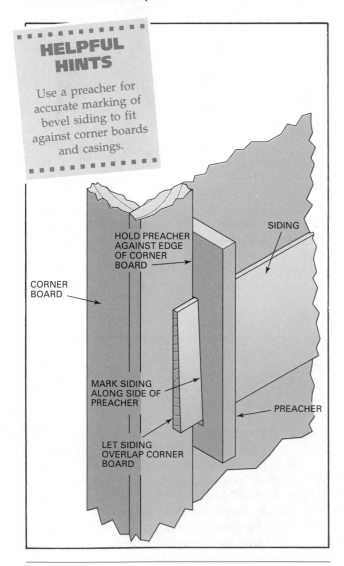

HELPFUL HINTS

Use a preacher for accurate marking of bevel siding to fit against corner boards and casings.

HOLD PREACHER AGAINST EDGE OF CORNER BOARD

SIDING

CORNER BOARD

MARK SIDING ALONG SIDE OF PREACHER

PREACHER

LET SIDING OVERLAP CORNER BOARD

Fig. 64-10

WINDOW SILL

WEATHER GROOVE IN UNDERSIDE OF SILL

SIDING

Fig. 64-11 Fit siding snugly in the weathergroove on the underside of the window sill.

Vertical Application of Wood Siding

Bevel sidings are designed for horizontal applications only. *Board on board* and *board and batten* are applied only vertically. Almost all other patterns may be applied in either direction. The following paragraphs describe the vertical application of *tongue-and-groove* siding. Other edge-shaped siding is installed in a comparable way.

HAIRLINE OVERLAP

BOWED INTO POSITION

SNAPPED TIGHT

6" & NARROWER	8" & WIDER
Plain	Plain
Recommend 1" overlap. One siding or box nail per bearing, just above the 1" overlap.	Recommend 1" overlap. One siding or box nail per bearing, just above the 1" overlap.
Rabbeted Edge	Rabbeted Edge — Approximate 1/8" gap for dry material 8" and wider. 1/2"=full depth of rabbet
Allows for 1/2" overlap. One siding or box nail per bearing, 1" up from bottom edge.	Allows for 1/2" overlap. One siding or box nail per bearing, 1" up from bottom edge.
Use siding or box nail to face nail once per bearing, 1-1/2" up from bottom edge.	Approximate 1/8" gap for dry material 8" and wider. 1/2"=full depth of rabbet. Use two siding or box nails, 3-4" apart, per bearing, to face nail.
T&G Pattern / Shiplap Pattern. Use casing nails to blind nail T&G patterns, one nail per bearing. Use siding or box nails to face nail shiplap patterns,1" up from bottom edge.	T&G Pattern / Shiplap Pattern. Approximate 1/8" gap for dry material 8" and wider. 1/2"=full depth of rabbet. Use two siding or box nails, 3-4" apart, to face nail, 1" up from bottom edge.

SIDING USUALLY APPLIED HORIZONTALLY

Fig. 64-12 Location and number of fasteners recommended for wood siding. (*Courtesy of Western Wood Products Association*)

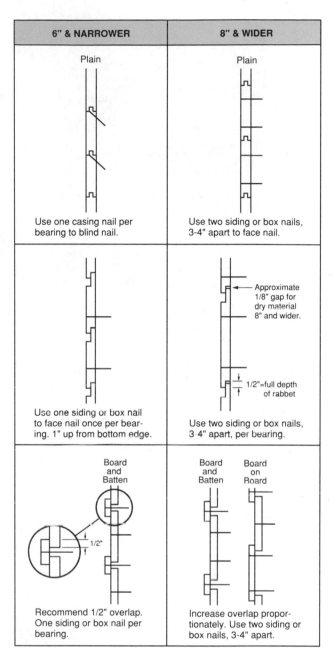

6" & NARROWER	8" & WIDER
Plain	Plain
Use one casing nail per bearing to blind nail.	Use two siding or box nails, 3-4" apart to face nail.
	Approximate 1/8" gap for dry material 8" and wider. 1/2"=full depth of rabbet
Use one siding or box nail to face nail once per bearing. 1" up from bottom edge.	Use two siding or box nails, 3-4" apart, per bearing.
Board and Batten. Recommend 1/2" overlap. One siding or box nail per bearing. 1/2"	Board and Batten / Board on Board. Increase overlap proportionately. Use two siding or box nails, 3-4" apart.

SIDING USUALLY APPLIED VERTICALLY

Installing Vertical Tongue-and-Groove Siding

Corner boards are not usually used when wood siding is applied vertically. The siding boards are fitted around the corner (Fig. 64-13). Rip the grooved edge from the starting piece. Slightly back-bevel the ripped edge. Place it vertically on the wall with the beveled edge flush with the corner similar to making a corner board.

The tongue edge should be plumb, the bottom end should be about one inch below the sheathing. The top end should butt or be tucked under any trim above. *Face nail* the edge nearest the corner. *Blind nail* into the tongue edge. Nails should be

Fig. 64-13 Vertical tongue-and-groove siding needs little accessory trim, such as corner boards. (*Courtesy of California Redwood Association*)

placed from 16 to 24 inches apart. *Blocking* must be provided between studs if siding is applied directly to the frame.

Fasten a temporary piece on the other end of the wall projecting below the sheathing by the same amount. Stretch a line to keep the bottom ends of other pieces in a straight line (Fig. 64-14). Apply succeeding pieces by toenailing into the tongue edge of each piece.

Make sure the edges between boards come up tight. If they do not come up tight by nailing alone,

drive a chisel, with its beveled edge side against the tongue, into the sheathing. Use it as a pry to force the board up tight. When the edge comes up tight, fasten it close to the chisel. If this method is not successful, toenail a short block of the siding with its grooved edge into the tongue of the board (Fig. 64-15). Drive the nail home until it forces the board up tight. Drive nails into the siding on both sides of the scrap block. Remove the scrap block.

Continue applying pieces in the same manner. Make sure to keep the bottom ends in a straight

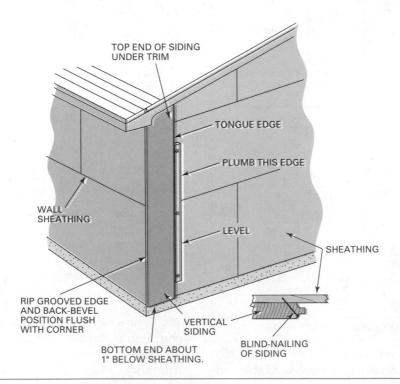

TOP END OF SIDING
UNDER TRIM

TONGUE EDGE

PLUMB THIS EDGE

WALL
SHEATHING

LEVEL

SHEATHING

RIP GROOVED EDGE
AND BACK-BEVEL
POSITION FLUSH
WITH CORNER

VERTICAL
SIDING

BOTTOM END ABOUT
1" BELOW SHEATHING.

BLIND-NAILING
OF SIDING

Fig. 64-14 Starting the application of vertical board siding.

Fig. 64-15

line. Avoid making horizontal joints between lengths. If joints are necessary, use a mitered or rabbeted joint for weathertightness (Fig. 64-16).

Fitting around Doors and Windows

Vertical siding is fitted tightly around window and door casings with different methods than those used for horizontal siding.

Fig. 64-16 Use mitered or rabbeted end joints between lengths of vertical board siding.

Approaching a Wall Opening. When approaching a door or window, cut and fit the piece just before the one to be fitted against the casing. Then remove it. Set it aside for the time being.

Cut, fit, and tack the piece to be fitted against the casing in place of the next to last piece. Level from the top of the window casing and the bottom of the sill. Mark the piece.

To lay out the piece so it will fit snugly against the side casing, first cut a scrap block of the siding material, about 6 inches long. Remove the tongue from one edge. Be careful to remove all of the tongue, but no more. Hold the block so its grooved edge is against the side casing and the other edge is on top of the siding to be fitted. Mark the piece by holding a pencil against the outer edge of the block while moving the block along the length of the side casing (Fig. 64-17).

Cut the piece, following the layout lines carefully. When laying out to fit against the bottom of the sill, make allowance to rabbet the siding to fit in

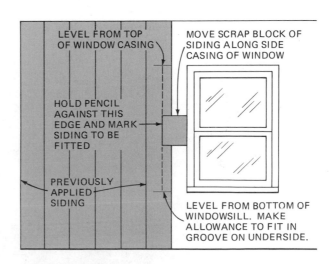

Fig. 64-17 Method of fitting vertical board siding when approaching a window casing.

the weather groove on the bottom side of the window sill. Place and fasten the pieces in position.

Continue to apply the siding with short lengths across the top and bottom of the window. Each length under a window must be rabbeted at the top end to fit in the weather groove at the sill.

Leaving a Wall Opening. A full length must also be fitted to the casing on the other side of the window. To mark the piece, first tack a short length of scrap siding above and below the window and against the last pieces of siding installed. Tack the length of siding to be fitted against these blocks. Level from the top and bottom of the window. Mark the piece for the horizontal cuts.

To lay out the piece for the vertical cut that fits against the side casing, use the same block with the tongue removed, as used previously. Hold the grooved edge against the side casing. With a pencil against the other edge, ride the block along the side casing while marking the piece to be fitted (Fig. 64-18).

Remove the piece and the scrap blocks from the wall. Carefully cut the piece to the layout lines. Then fasten in position. Continue applying the rest of the siding until you are almost to the other end of the wall.

Method of Ending Vertical Siding

The last piece of vertical siding should be close to the same width of previously installed pieces. If siding is installed in *random* widths, plan the application. The width of the last piece should be equal, at least, to the width of the narrowest piece. It is not good practice to allow vertical siding to end with a narrow sliver.

Stop several feet from the end. Space off, and determine the width of the last piece. If it will not be a satisfactory width, install narrower or wider pieces for the remainder, as required. It may be necessary to rip available siding to narrower widths and reshape the grooves (Fig. 64-19).

When the corner is reached, the board is ripped to width along its tongue edge. It is slightly back-beveled for the first piece on the next wall to butt against. When the last corner is reached, the board is ripped in a similar manner. However, it is smoothed to a square edge to fit flush with the surface of the first piece installed. All exposed sharp corners should be eased.

Installing Panel Siding

Plywood, hardboard, and other panel siding is usually installed vertically. It can be installed horizontally, if desired. Lap siding panels are ordinarily applied horizontally.

Installing Vertical Panel Siding

Start a vertical panel so it is plumb, with one edge squared and flush with the starting corner. The inner edge should fall on the center of a stud. Fasten panels, of 1/2-inch thickness or less, with 6d *siding nails*. Use 8d siding nails for thicker panels. Fasten panel edges about every 6 inches and about every 12 inches along intermediate studs.

Apply successive sheets. Leave a 1/8-inch space between panels. Panels must be installed with their bottom ends in a straight line. There should be a minimum of 6 inches above the finished grade line.

Installing Horizontal Panel Siding

Mark the height of the first course of horizontal panel siding on both ends of the wall. Snap a chalk

Fig. 64-18 Laying out vertical board siding to fit against a window casing, when leaving it.

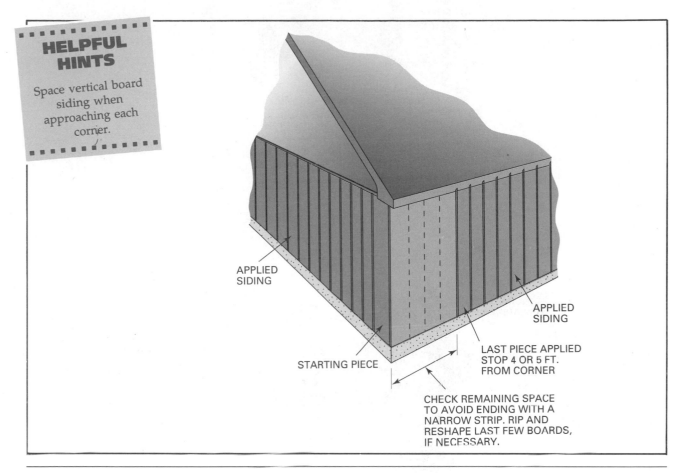

HELPFUL HINTS

Space vertical board siding when approaching each corner.

APPLIED SIDING

APPLIED SIDING

STARTING PIECE

LAST PIECE APPLIED STOP 4 OR 5 FT. FROM CORNER

CHECK REMAINING SPACE TO AVOID ENDING WITH A NARROW STRIP. RIP AND RESHAPE LAST FEW BOARDS, IF NECESSARY.

Fig. 64-19

line between marks. Fasten a full-length panel with its top edge to the line, its inner end on the center of a stud, and its outer end flush with the corner. Fasten in the same way as for vertical panels.

Apply the remaining sheets in the first course in like manner. Trim the end of the last sheet flush with the corner. Start the next course so joints will line up with those in the course below.

Both vertical and horizontal panels may be applied to sheathing or directly to studs if backing is provided for all joints (Fig. 64-20).

Carefully fit and caulk around doors and windows. It is important that horizontal butt joints be either *offset and lapped*, *rabbeted*, or *flashed* (Fig. 64-21). Vertical joints are either shiplapped or covered with *battens*.

Applying Lap Siding Panels

Panels of lap siding are applied in much the same manner as wood bevel siding with some exceptions. First, install a strip, of the same thickness and width of the siding headlap, along the bottom of the wall. Determine the height of the top edge of the first course. Snap a chalk line across the wall. Apply

the first course with its top edge to the snapped line.

When applied over nailable sheathing, space nails 8 inches apart in a line about 3/4 inch above the bottom edge of the siding. When applied directly to framing, fasten at each stud location. A 1/8-inch, caulked joint between the ends of siding and trim is recommended. Joints between ends of siding may also be flashed with a narrow strip of #15 felt centered behind the joint and backed with a wood shingle wedge (Fig. 64-22). In some types of lap panel siding, a 1/8-inch gap is allowed between the ends. A special joint molding is inserted in the gap (Fig. 64-23).

Corners. Individual metal or plastic outside corners may be installed as work progresses. They can also be installed after the entire wall is sided. Applied corners are used because composition material does not miter well. Somewhat better success may be had when mitering plywood at outside corners. Slide the corners up into place with the top slipping under the course above. When the corner is in position, fasten it through the prepunched holes (Fig. 64-24). Lap siding may also be finished with corner boards.

VERTICAL APPLICATION OF TEXTURE 1-11 SIDING

INSULATION AS REQUIRED AND WARM SIDE VAPOR BARRIER

NO DIAGONAL WALL BRACING REQUIRED WITH PLYWOOD PANEL SIDING

PLYWOOD PANEL SIDING

LEAVE 1/16" SPACE AT END AND EDGE JOINTS

JOINTS — NO CAULKING REQUIRED WHERE EDGES ARE SHIPLAPPED, BACKED WITH BUILDING PAPER, OR BATTEN. IF CAULKING IS NOT USED WITH UNBATTENED SQUARE BUTT JOINTS, TREAT PLYWOOD EDGES WITH WATER REPELLENT.

6" MINIMUM CLEARANCE, SIDING TO GRADE

303 SIDING VERTICAL APPLICATION

NO DIAGONAL WALL BRACING NEEDED

CAULK VERTICAL JOINTS OR BACK WITH BUILDING PAPER. IF CAULKING IS NOT USED, TREAT PLYWOOD EDGES WITH WATER REPELLENT.

INSULATION AS REQUIRED AND WARM SIDE VAPOR BARRIER

2 X 4 BLOCKING AT HORIZONTAL JOINTS

STUDS 16" OR 24" o.c. WHERE PLYWOOD PANEL SIDING IS INSTALLED HORIZONTALLY

BATTENS AT 4' OR 8' o.c. TO CONCEAL BUTT JOINTS AT PANEL ENDS. NAIL THROUGH BATTENS MUST PENETRATE STUDS AT LEAST 1"

PLYWOOD PANEL SIDING INSTALLED HORIZONTALLY.

6" MINIMUM CLEARANCE, SIDING TO GRADE

303 SIDING HORIZONTAL APPLICATION

Fig. 64-20 Panel siding may be applied vertically or horizontally to sheathing or directly to studs. (*Courtesy of American Plywood Association*)

VERTICAL WALL JOINTS

Butt & Caulk

Plywood

Caulk or back with building paper

Shiplap

Grooved plywood (reverse board & batten shown), same joint detail for T 1-11 and channel groove

Vertical Batten

Batten

Use ring-shank nails for the battens, applied near the edges in two staggered rows

VERTICAL INSIDE & OUTSIDE CORNER JOINTS

Butt & Caulk

Plywood

Caulk

Rabbet & Caulk

Plywood

Rabbet one piece plywood, caulk and butt

Corner Board Lap Joints

Plywood

Corner boards

HORIZONTAL WALL JOINTS

Butt & Flash

Plywood

Flashing (Galv. or aluminum)

Lap Plywood

Plywood

Lap top plywood over bottom plywood

Shiplap

Plywood

Shiplap joint

Fig. 64-21 Panel siding joint details. (*Courtesy of American Plywood Association*)

WEDGE & FLASH UNDER VERTICAL JOINTS

STAGGER JOINTS IN ADJACENT COURSES

EXPOSURE AS RECOMMENDED BY MANUFACTURER

NAILING STRIP

6" MINIMUM CLEARANCE, SIDING TO GRADE

Fig. 64-22 Lap siding application details.

Fig. 64-23 A special molding is sometimes used on end joints of lap siding.

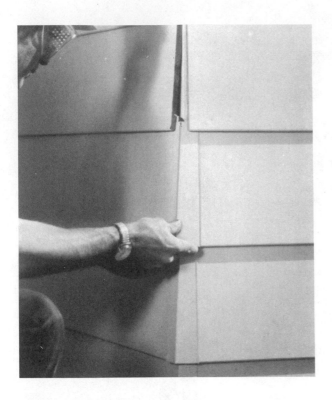

Fig. 64–24 Individual metal trim being applied to corners of lap siding.

CHAPTER **65** WOOD SHINGLE AND SHAKE SIDING

Wood *shingles* and *shakes* are frequently used for siding, as well as roofing (Fig. 65-1). Those previously described in Chapter 56 for roofing may also be applied to sidewalls.

Sidewall Shingles and Shakes

Some kinds of shingles and shakes are designed for sidewall use only (Fig. 65-2). *Rebutted* and *rejointed* ones are machine trimmed with parallel edges and square butts for sidewall application. Rebutted and rejointed machine-grooved, sidewall shakes have *striated faces*.

Special *fancy butt* shingles are available in a variety of designs. They provide interesting patterns in combination with square butts or other types of siding. Fancy butt shingles were widely used in the nineteenth century on Victorian style

Fig. 65-1 Wood shingles and shakes are frequently used as siding. (*Courtesy of Western Wood Products Association*)

ARROW   SQUARE   FISH-SCALE   DIAGONAL   HALF COVE   DIAMOND   ROUND   HEXAGONAL   OCTAGONAL

FANCY-BUTT RED CEDAR SHINGLES. NINE OF THE MOST POPULAR DESIGNS ARE SHOWN. FANCY-BUTT SHINGLES CAN BE CUSTOM PRODUCED TO INDIVIDUAL ORDERS.

REBUTTED AND REJOINTED. MACHINE TRIMMED FOR PARALLEL EDGES WITH BUTTS SAWN AT RIGHT ANGLES. FOR SIDEWALL APPLICATION WHERE TIGHTLY FITTING JOINTS ARE DESIRED.

PANELS. WESTERN RED CEDAR SHINGLES ARE AVAILABLE IN 4- AND 8-FOOT PANEL-IZED FORM.

MACHINE GROOVED. MACHINE-GROOVED SHAKES ARE MANUFACTURED FROM SHINGLES AND HAVE STRIATED FACES AND PARALLEL EDGES. USED DOUBLE-COURSED ON EXTERIOR SIDEWALLS.

Fig. 65-2 Some wood shingles and shakes are made for sidewall application only. (*Courtesy of Cedar Shake and Shingle Bureau*)

buildings. They are once again becoming popular (Fig. 65-3).

Red cedar shingles, factory-applied on 4- and 8-foot panels, are available in several styles and exposures. Panels come with square butt shingles applied in straight or staggered lines and with a number of fancy butt designs.

Applying Wood Shingles and Shakes

Wood shingles and shakes may be applied to sidewalls in either single-layer or double-layer courses. In *single coursing*, shingles are applied to walls with a single layer in each course, in a way similar to roof

application. However, greater exposures are allowed on sidewalls than on roofs (Fig. 65-4).

In *double coursing*, two layers are applied in one course. Consequently, even greater weather exposures are allowed. Double coursing is used when wide courses with deep, bold shadow lines are desired (Fig. 65-5).

Applying the Starter Course

The *starter course* of sidewall shingles and shakes is applied in much the same way as the starter course on roofs. A double layer is used for single-course applications. A triple layer is used for triple coursing. Less expensive *undercourse* shingles are used for underlayers.

ROUND FANCY BUTT SHINGLES

Fig. 65-3 Fancy butt shingles are still used to accent sidewalls with distinctive designs. (*Courtesy of Andersen Corporation*)

SINGLE COURSING

DOUBLE COURSING

Max. Weather Exposure		
Shingle Length	Single Course	Double Course
16"	7 1/2"	12"
18"	8 1/2"	14"
24"	11 1/2"	16"

Max. Weather Exposure		
Shake Length & Type	Single course	Double course
16" Centigroove	7 1/2"	12"
18" Centigroove	8 1/2"	14"
18" resawn	8 1/2"	14"
24" resawn	11 1/2"	18"
18" taper-sawn	8 1/2"	14"
24" tapersplit	11 1/2"	18"
24" taper-sawn	11 1/2"	18"
18" straight-split	8 1/2"	16"

Fig. 65-5 Double-coursed shingles, with two layers in each row, permit even greater exposures when wide courses are desired. (*Courtesy of Cedar Shake and Shingle Bureau*)

Fig. 65-4 Single-coursed shingle wall application is similar to roof application. However, greater weather exposures are allowed. (*Courtesy of Cedar Shake and Shingle Bureau*)

Fasten a shingle, on both ends of the wall, with its butt about one inch below the top of the foundation. Stretch a line between them at the butts. Sight the line for straightness. Fasten additional shingles at necessary intervals. Attach the line to

their butts to straighten it (Fig. 65-6). Even a tightly stretched line will sag in the center over a long distance.

Apply a single course of shingles so the butts are as close to the chalk line as possible without touching it. Remove the line. Apply another course on top of the first course. Offset the joints in the outer layer at least 1 1/2 inches from those in the bottom layer. Untreated shingles should be spaced

Fig. 65-6 Stretch a straight line as a guide for the butts of the starter course.

1/8 to 1/4 inch apart to allow for swelling and to prevent buckling. Shingles can be applied close together, if factory-primed or if treated soon after application (Fig. 65-7).

Single Coursing

A story pole may be used to lay out shingle courses in the same manner as with horizontal wood siding. Snap a chalk line across the wall, at the shingle butt line, to apply the first course. Using only as many finish nails as necessary, tack 1×3 straightedges to the wall with their top edges to the line. Lay individual shingles with their butts on the straightedge (Fig. 65-8). Use a shingling hatchet to trim and fit the edges, if necessary. Butt ends are not trimmed. If rebutted and rejointed shingles are used, no trimming should be necessary.

At times it may be necessary to fit a shingle between others in the same course. Tack the next to

SHINGLES SPACED 1/8" TO 1/4" APART. THESE JOINTS ALLOW FOR EXPANSION AND PREVENT POSSIBLE "BUCKING." FACTORY PRODUCTS MAY BE CLOSER.

LEAVE A SIDE LAP OF AT LEAST 1-1/2" BETWEEN JOINTS IN SUCCESSIVE COURSES.

Fig. 65-7 Stagger joints between shingle courses. Space untreated shingles. (*Courtesy of Cedar Shake and Shingle Bureau*)

Fig. 65-8 Rest shingle butts on a straightedge when single-coursing on sidewalls.

Fig. 65-9

last shingle in place with one nail. Slip the last shingle under it. Score along the overlapping edge with the hatchet. Cut along the scored line. Fasten both shingles in place (Fig. 65-9).

Fasteners. Each shingle, up to 8 inches wide, is fastened with two nails or staples about 3/4 inch in from each edge. On shingles wider than 8 inches, drive two additional nails about one inch apart near the center. Fasteners should be hot-dipped galvanized, stainless steel, or aluminum. They should be driven about one inch above the butt line of the next course. Fasteners must be long enough to penetrate the sheathing by at least 1/2 inch.

Staggered and Ribbon Coursing. An alternative to straight line courses is *staggered* and *ribbon coursing*. In staggered coursing, the butt lines of the shingles are alternately offset below, but not above,

the horizontal line. Maximum offsets are one inch for 16- and 18-inch shingles and 1 1/2 inches for 24-inch shingles.

In ribbon coursing, both layers are applied in straight lines. The outer course is raised about one inch above the inner course (Fig. 65-10).

Shingle Panels. Shingles in *panelized* form are applied in the same manner as horizontal panel siding, described in the previous chapter (Fig. 65-11). The end shingles of each course in the panel are offset for staggered joints. They match up with adjacent panels.

Corners

Shingles may be butted to corner boards like any horizontal wood siding. On outside corners, they may also be applied by alternately overlapping

Fig. 65-10 Staggered and ribbon courses are an alternative to straight-line courses.

Fig. 65-11 Wood shingles are made in panel form for faster application. (*Courtesy of Cedar Shake and Shingle Bureau*)

each course in the same manner as in applying a wood shingle ridge. Inside corners may be woven by alternating the corner shingle first on one wall and then the other (Fig. 65-12).

Double Coursing

When double coursing, the starter course is tripled. The outer layer of the course is applied 1/2 inch lower than the inner layer. For ease in application, use a rabbeted straightedge or one composed of two pieces with offset edges (Fig. 65-13).

Fastening. Each inner layer shingle is applied

A) SHINGLES BUTTED AGAINST CORNER BOARDS

B) SHINGLES BUTTED AGAINST SQUARE WOOD STRIP ON INSIDE CORNER, FLASHING BEHIND

C) LACED OUTSIDE CORNER

D) LACED INSIDE CORNER WITH FLASHING BEHIND

E) MITERED CORNER

Fig. 65-12 Wood shingle corner details. (*Courtesy of Cedar Shake and Shingle Bureau*)

with one fastener at the top center. Each outer course shingle is face-nailed with two 5d galvanized box or special 14-gauge shingle nails. The fasteners are driven about 2 inches above the butts, and about 3/4 inch in from each edge.

Estimating Shingle Siding

The number of squares of shingles needed to cover a certain area depends on how much they are exposed to the weather. One square of shingles will

Fig. 65-13 Use a straightedge with rabbeted edges, or one made of two pieces with offset edges, for double-coursed application.

cover 100 square feet when 16-inch shingles are exposed 5 inches, 18-inch shingles exposed 5 1/2 inches, and 24-inch shingles exposed 7 1/2 inches.

A square of shingles will cover more area with greater exposures and less area with smaller exposures. After calculating the wall area as shown in Figure 63-11, determine the amount of shingles needed by using the coverage tables shown in Figure 65-14.

LENGTH AND THICKNESS	Approximate coverage of one square (4-bundle roof-pack) of shingles at indicated weather exposures:													
	3½″	4″	4½″	5″	5½″	6″	6½″	7″	7½″	8″	8½″	9″	9½″	10″
16″ × 5/2″	70	80	90	100	110	120	130	140	150	160	170	180	190	200
18″ × 5/2¹/⁴″	—	72½	81½	90½	100	109	118	127	136	145½	154½	163½	172½	181½
24″ × 4/2″	—	—	—	—	73½	80	86½	93	100	106½	113	120	126½	133

LENGTH AND THICKNESS	10½″	11″	11½″	12″	12½″	13″	13½″	14″	14½″	15″	15½″	16″
16″ × 5/2″	210	220	230	240	—	—	—	—	—	—	—	—
18″ × 5/2¹/⁴″	191	200	209	218	227	236	245½	254½	—	—	—	—
24″ × 4/2″	140	146½	153	160	166½	173	180	186½	193	200	206½	213

Fig. 65-14 Sidewall shingle coverage at various exposures. (*Courtesy of Cedar Shake and Shingle Bureau*)

CHAPTER 66   ALUMINUM AND VINYL SIDING

Except for the material, *aluminum* and *vinyl* siding systems are similar. Aluminum siding is finished with a baked-on enamel. In vinyl siding, the color is in the material itself. Both kinds are manufactured with interlocking edges, for horizontal and vertical applications. Descriptions and instructions are given here for vinyl siding systems, much of which can be applied to aluminum systems.

Siding Panels and Accessories

Siding systems are composed of *siding panels* and several specially shaped *moldings*. Moldings are used on various parts of the building, to trim the installation. In addition, the system includes shapes for use on *soffits*.

Siding Panels

Siding panels, for horizontal application, are made in 8- and 12-inch widths. They come in configurations to simulate one, two, or three courses of bevel or drop siding. Panels designed for vertical application come in 12-inch widths. They are shaped to resemble boards. They can be used in combination with horizontal siding. Vertical siding panels with solid surfaces may also be used for soffits. For ventilation, *perforated* soffit panels of the same configuration are used (Fig. 66-1).

Siding System Accessories

Siding systems require the use of several specially shaped accessories. *Inside* and *outside corner posts* are used to provide a weather-resistant joint to corners. Corner posts are available with channels of appropriate widths to accommodate various configurations of siding.

Some other accessories include *J-channels*, *starter strips*, and *undersill/finish trim*. *J-channels* are made with several opening sizes. They are used in a number of places such as around doors and windows, at transition of materials, against soffits, and many other places (Fig. 66-2). The majority of vinyl siding panels and accessories are manufactured in 12′-6″ lengths.

Applying Horizontal Siding

The siding may expand and contract as much as 1/4 inch over a 12′-6″ length with changes in temperature. For this reason, it is important to center fasteners in the slots. Do not drive them too tightly. There should be about 1/32 inch between the head of the fastener, when driven, and the siding (Fig. 66-3). Space fasteners 16 inches apart for horizontal

HORIZONTAL-(1) DOUBLE 4"; (2) SINGLE 8"; (3) DOUBLE 5"; (4) TRIPLE 3"; (5) DUTCH LAP.

VERTICAL-CAN BE USED IN COMBINATION WITH HORIZONTAL SIDING. VERTICAL IS AVAILABLE IN DOUBLE 5"; TRIPLE 3"; AND QUAD 4".

PERFORATED FOR SOFFIT USE

Fig. 66-1 Some of the most commonly used configurations of vinyl siding for horizontal and vertical application. (*Courtesy of Vinyl Siding Institute*)

INSIDE CORNER POST

OUTSIDE CORNER POST

Fig. 66-2 Various accessories are used to trim a vinyl siding installation. (*Courtesy of Georgia-Pacific Corporation*)

Undersill Finish Trim

Starter Strip

Frieze Moulding

Drip Cap Vertical Edge

J-Channel

TRIM & MOLDINGS

PERFORATED
SOFFIT

OUTSIDE CORNER POST

HORIZONTAL SIDING

"F" TRIM

FASCIA

"J" CHANNEL

UNDERSILL
TRIM

VERTICAL SIDING
OR SOFFIT

"J" CHANNEL

INSIDE CORNER POST

HORIZONTAL SIDING
STARTER STRIP

UNDERSILL/FINISH TRIM

Fig. 66-2 *(continued)*

1 32"

YES   NO

YES   NO

Fig. 66-3 Fasten siding to allow for expansion and contraction.

siding and 6 to 12 inches apart for accessories, unless otherwise specified by the manufacturer.

Applying the Starter Strip

Snap a level line to the height of the starter strip all around the bottom of the building. Fasten the strips to the wall with their top edges to the chalk line. Leave a 1/4-inch space between them and other accessories to allow for expansion (Fig. 66-4). Make sure the starter strip is applied as straight as possible. It controls the straightness of entire installation.

Installing Corner Posts

Corner posts are installed in corners 1/4 inch below the starting strip and 1/4 inch from the top. Attach the posts by fastening in the top of the upper slot on each side. The posts will hang on these fasteners. The rest of the fasteners should be centered on the slots. Make sure the posts are straight and true from top to bottom (Fig. 66-5).

Installing J-Channel

Install J-channel across the top and along the sides of window and door casings. It may also be installed under windows or doors with the *undersill* nailed inside of the channel. To miter the corners, cut all pieces to extend, on both ends, beyond the casings and sills a distance equal to the width of the channel face. On both ends of the side J-channels,

CHALK LINE   WALL SHEATHING

1 1/2"   1 1/2"

FOUNDATION   STARTER STRIP

WALL SHEATHING   LEAVE SPACE BETWEEN ENDS

1/4"

STARTER STRIP

FOUNDATION

Fig. 66-4 Installation of the starter strip.

Fig. 66-5 Inside and outside corner posts are installed in a similar way.

cut a 3/4-inch notch out of the bottom. Fasten in place. On both ends of the top and bottom channels, make 3/4-inch cuts. Bend down the tabs and miter the faces. Install them so the mitered faces are in front of the faces of the side channels (Fig. 66-6).

Installing Siding Panels

Snap the bottom of the first panel into the starter strip. Fasten it to the wall. Start from a back corner, leaving a 1/4-inch space in the corner post channel. Work toward the front with other panels. Overlap each panel about one inch. The exposed ends should face the direction from which they are least viewed (Fig. 66-7).

Install successive courses by interlocking them with the course below and staggering the joints between courses. Use tin snips, hacksaw, utility knife, or power saw with an abrasive wheel or fine-tooth circular blade. Reverse the blade if a power saw is used, for smooth cutting through the vinyl.

Fitting around Windows. Plan so there will be no joints under a window. Hold the siding panel under the sill. Mark the width of the cutout, allow-

Fig. 66-6 J-channel installation around windows and doors.

Fig. 66-7 Installing the first course of horizontal siding panels.

ing 1/4-inch clearance on each side. Mark the height of the cutout, allowing 1/4-inch clearance below the sill.

Make vertical cuts with tin snips. Score the horizontal layout lines with a utility knife or scoring tool. Bend the section to be removed back and forth until it separates. Using a special *snaplock punch,*

punch the panel 1/4 inch below the cut edge at 6-inch intervals to produce *raised lugs* facing outward. Install the panel under the window and up in the undersill trim. The raised lugs cause the panel to fit snugly in the trim (Fig. 66-8).

Panels are cut and fit over windows in the same manner as under them. However, the lower

Fig. 66-8 Method of fitting a panel under a window.

portion is cut instead of the top. Install the panel by placing it into the J-channel that runs across the top of the window (Fig. 66-9).

Installing the Last Course under the Soffit

The last course of siding panels under the soffit is installed in a manner similar to fitting under a window. An *undersill trim* is applied on the wall and up against the soffit. Panels in the last course are cut to width. Lugs are punched along the cut edges. The panels are then snapped firmly into place into the undersill trim (Fig. 66-10).

Gable End Installation

The rakes of a gable end are first trimmed with J-channels. The panel ends are inserted into the channel with 1/4-inch expansion gap. Make a pattern for cutting gable end panels at an angle where they intersect with the rake. Use two scrap pieces of siding to make the pattern. Interlock one piece with an installed siding panel below. Hold the other piece on top of it and against the rake. Mark along the bottom edge of the slanted piece on the face of the level piece (Fig. 66-11).

Applying Vertical Siding

The installation of vertical siding is similar to that for horizontal siding with a few exceptions. The method of fastening is the same. However, space fasteners about 12 inches apart for vertical siding panels. The starter strip is different. It may be *1/2-inch J-channel* or *drip cap* flush with and fitted into the corner posts (Fig. 66-12). Around windows and doors, under soffits, against rakes, and other locations, 1/2-inch J-channel is used. One of the major

Fig. 66-9 Fitting a panel over a window.

differences is that a vertical layout should be planned so that the same and widest panel is exposed at both ends of the wall.

Installing the First Panel

Install the first vertical panel plumb on the starter strip with one edge into the corner post. Allow 1/4 inch at top and bottom. Place the first nails in the uppermost end of the top nail slots to hold it in position.

The edge of the panel may need to be cut in order for equal widths to be exposed on both ends of the wall. If the panel is cut on the flat surface, place a piece of undersill trim backed by furring into the channel of the corner post. Punch lugs along the cut edge of the panel at 6-inch intervals. Snap the panel into the undersill trim. Edges of vertical panels cut to fit in J-channels around win-

MEASURE FOR LAST COURSE OF SIDING

Fig. 66-10 Fitting the last course of horizontal siding under the soffit.

LAYOUT A PATTERN FOR THE RAKE ANGLE
ON A SCRAP PIECE OF SIDING.

USE THE PATTERN TO MARK THE
RAKE ANGLE ON SIDING PANELS

Fig. 66-11 Fitting horizontal siding panels to the rakes.

dows and doors are treated in the same way (Fig. 66-13).

Estimating Aluminum and Vinyl Siding

Aluminum and vinyl siding panels are sold by the square. Determine the wall area to be covered. Add 10 percent of the area for waste. Divide by 100. This gives you the number of squares needed.

Become familiar with accessories and how they are used. Measure the total linear feet required for each item.

Fig. 66-12 The starter strip shape and its intersection with corner posts is different for vertical application of vinyl siding.

Fig. 66-13 Undersill trim and furring are required when vertical siding is cut to fit into corner posts and J-channel on the sides of windows and doors. (*Courtesy of Vinyl Siding Institute*)

Review Questions

Select the most appropriate answer.

1. Bevel siding is applied
   a. horizontally.
   b. vertically.
   c. horizontally or vertically.
   d. horizontally, vertically, or diagonally.

2. A type of bevel siding, whose exposure cannot be changed, has
   a. tongue-and-grooved edges.
   b. rabbeted butt edges.
   c. beveled back sides to lie flat against studs.
   d. a convex shape.

3. A particular advantage of bevel siding over other types of horizontal siding is that
   a. it comes in a variety of prefinished colors.
   b. it has a constant weather exposure.
   c. the weather exposure can be varied slightly.
   d. application can be made in any direction.

4. Building paper used under siding should
   a. be a vapor barrier.
   b. allow vapor to pass through.
   c. be applied only when no sheathing is used.
   d. be polyethylene film.

5. When applying horizontal siding, it is desirable to
   a. maintain exactly the same exposure with every course.
   b. apply full courses above and below windows.
   c. work from the top down.
   d. use a water table.

6. The height of a window frame is 73 1/2 inches. Dividing this height into spaces as close to 6 inches as possible results in spaces of
   a. 5 7/8 inches.   c. 6 1/16 inches.
   b. 5 15/16 inches.   d. 6 1/8 inches.

7. In order to lay out a story pole for courses of horizontal siding, which of the following must be known?
   a. the width of windows and doors
   b. the kind and size of finish at the eaves and foundation
   c. the location of windows, doors, and other openings
   d. the length of the wall to which siding is to be applied
   e. the size of corner boards

8. Maximum exposure for single-coursed 18-inch wood shakes is
   a. 8 inches.   c. 9 inches.
   b. 8 1/2 inches.   d. 9 1/2 inches.

9. When installing aluminum or vinyl siding, drive nails
   a. tightly against the flange.
   b. loosely against the flange.
   c. with large heads.
   d. colored the same as the siding.

10. To allow for expansion when installing solid vinyl starter strips, leave a space between the ends of, at least,
    a. 1/8 inch.   c. 3/8 inch.
    b. 1/4 inch.   d. 1/2 inch.

■ BUILDING FOR SUCCESS ■

Employer/Employee Relationships

Each person pursuing employment in the construction industry should be aware of the need to establish good working relationships with his or her employer. Goals will have to be set as each skilled worker becomes involved with an employer, customers, and peers. Careful thought and preparation will help make each worker more successful.

The reasons for forming good working relationships with others in the construction field need to be identified. Each day as tasks are carried out, meaningful relationships help the flow and progress of the work. The cooperation that is established in the work process will build friendships. In many cases, people work alongside their peers and employers for hours. What happens in these situations affects the amount of job stress a person has to deal with from day to day. This underscores the importance of team effort in solving problems on the job.

Other reasons for developing good working relationships on the job are centered on job longevity, advancement, customer service, production, and enjoyment. A career of framing wood or steel buildings on a crew will be more enjoyable when people like each other and are compatible. How and to what extent people advance depend on their working relationships. Customers see each worker as representing the business through personal contacts. They are either pleased or displeased by that one-on-one relationship. Production on the job is affected by relationships. Low production may mean that there are problems among workers on the job. Friction in personalities or work habits will create situations that distract from quality work. One of the primary reasons a person takes a job is for enjoyment. That criterion is one of the most important in selecting the job. Without job enjoyment, work is not rewarding and fulfilling. It will be only a matter of time before someone will leave a job if there is no work enjoyment.

How does one go about establishing good working relationships on the job? This is a good time to look at yourself and see how you work and what you expect on the job. Personalities will be one of the first things people notice when placed alongside someone in an employment situation. Compatibility will be established or rejected almost immediately. Usually people will expose their skills, talents, personality traits, and work habits in a very short time on the job. Many times the job itself calls for certain types of actions, skills, and abilities and sets the pace to which other people can react. How you present yourself and react to others will set the stage for working relationships.

Generally, people who are knowledgeable, friendly, productive, and use their skills well develop good working relationships with employers and peers. Being honest with

people will be one good way of telling others what you are like. How can each person benefit from establishing good relationships with others? The dividends are there from the beginning in the form of a productive and safe working environment. Accidents on the job can often be traced to poor working relationships with others.

Some of the continuing benefits that come from having good working relationships are seen as characteristics of trustworthiness, dependability, concern for the progress of the company, and less stress in the workplace. These basic characteristics will be reflected in the attitudes and performance of each skilled worker on the job.

In looking at a way to hear why employers are concerned with good working relationships, try asking them. Contact one or more employers in construction-related businesses. Ask them to share their opinions with you. The information gained from this type of conversation will provide invaluable assistance in making a good job choice.

FOCUS QUESTIONS: For individual or group discussion.

1. What might be some reasons for establishing good work relations with an employer?

2. How could good working relationships benefit you as a carpenter, drafter, estimator, or any other skilled worker?

3. What would you expect from a person who will not make an effort to work well with others?

4. Do you have a good idea of how to establish good working relationships with an employer and peers in a job situation? What would be your approach?

5. How would you help someone establish a good working relationship with you?

Decks, Porches, and Fences

A mong the final steps in finishing the exterior is the building of *porches*, *decks*, *fences*, and other accessory structures. Plans may not always show specific construction details. Therefore, it is important to know some of the techniques used to build these structures.

O BJ EC TIV E S

After completing this unit, the student will be able to:

- describe the construction of and kinds of materials used in decks and porches.

- lay out and construct footings for decks and porches.

- install supporting posts, girders, and joists.

- apply decking in the recommended manner and install flashing, for an exposed deck, against a wall.

- construct deck stairs and railings.

- describe several basic fence styles.

- design and build a straight and sturdy fence.

C O N T EN T S

CHAPTER 67 DECK AND PORCH CONSTRUCTION

Wood porches and decks are built to provide outdoor living areas for various reasons, in both residential and commercial construction (Fig. 67-1). The construction of both is similar. However, a porch is covered by a roof. Its walls may be enclosed with wire mesh screens for protection against insects. With screen and storm window combinations, glass replaces screen to keep the porch comfortably warm in the cold months.

Deck Materials

If not specified, the kind, grade, and sizes of lumber must be selected before building a deck. Also, the size and kind of fasteners, connectors, anchors, and other hardware must be determined.

Lumber

All lumber used to build decks should be *pressure-treated* with preservatives or be from a decay-resistant species, such as *redwood* or *cedar*. Remember, it is the *heartwood* of these species that is resistant to decay, not the *sapwood*. Either *all-heart* or *pressure-treated* lumber should be used wherever there is a potential for decay. This is essential for posts that are close to the ground and other parts subject to constant moisture. (A description of pressure-treated lumber and its uses can be found in Chapter 33.)

Lumber Grades. For pressure-treated southern pine and western cedar, #2 grade is structurally adequate for most applications. Appearance can be a deciding factor when choosing a grade. If a better appearance is desired, higher grades should be considered.

A grade called *Construction Heart* is the most suitable and most economical grade of California redwood for deck posts, beams, and joists. For decking and rails, a grade called *Construction Common* is recommended. Better appearing grades are available. However, they are more expensive.

Two grades of redwood, called *Redwood Deck Heart* and *Redwood Deck Common,* are manufactured especially for exterior walking surfaces. Two grades of decking, *Standard* and *Premium,* are also available in pressure-treated southern pine. Special decking grades of western cedar may also be obtained. The lumber grade or special purpose is shown in the grade stamp (Fig. 67-2).

Lumber Sizes. Specific sizes of supporting posts, girders, and joists depend on the spacing and height of supporting posts and the spacing of girders and joists. In addition, the sizes of structural members depend on the type of wood used and the weight imposed on the members. Too many factors prohibit generalization about sizes of structural members. Check with local building officials or with a

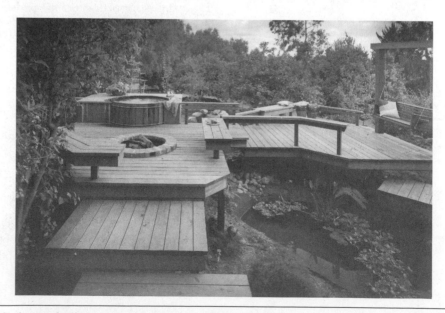

Fig. 67-1 Wood decks are built in many styles. This multilevel deck blends well with the landscape. (*Courtesy of California Redwood Association*)

Fig. 67-2 Grade stamps show the grade and special purpose of lumber used in deck construction. (*Courtesy of California Redwood Association; Courtesy of Southern Forest Products Association*)

professional to determine the sizes of structural members for specific deck construction. Determining sizes with incomplete information may result in failure of undersize members. Unnecessary expense is incurred with the use of oversize members.

Fasteners

All nails, fasteners, and hardware should be stain-less steel, aluminum, or hot-dipped galvanized. Electroplated galvanizing is not acceptable because the coating is too thin. In addition to corroding and eventual failure, poor quality fasteners will react with substances in decay-resistant woods and cause unsightly stains.

Building a Deck

Most wood decks consist of *posts*, set on *footings*, supporting a platform of *girders* and *joists* covered with *deck boards. Posts, rails, balusters,* and other special parts make up the railing (Fig. 67-3). Other parts, such as shading devices, privacy screens, benches, and planters, lend finishing touches to the area.

Installing a Ledger

If the deck is to be built against a building, a *ledger,* usually the same size as the joists, is nailed or bolted to the wall for the entire length of the deck (Fig. 67-4). The ledger acts as a beam to support joists that run at right angles to the wall. It is installed to a level line. Its top edge is located to provide a comfortable step down from the building after the decking is applied. The ledger height may be used as a benchmark for establishing the elevations of supporting posts and girders.

After the deck is applied, a flashing is installed under the siding and on top of the deck board. Caulking is applied between the deck and

Fig. 67-3 The components of a deck.

BUILDING PAPER
OR HOUSEWRAP

WALL SHEATHING

SIDING

SPACE DECKING
ABOUT 1/4"

METAL
FLASHING IS
INSTALLED UNDER
SIDING

FLASHING IS INSTALLED ON
TOP OF DECK BOARDS
WITH CAULKING UNDER AND
FASTENED CLOSE TOGETHER ALONG
THE OUTSIDE EDGE

DECK BOARDS
BARK SIDE UP

EXPANSION ANCHORS
IN MASONRY. BOLTS
OR LAG SCREWS
IN FRAMING

DECK
JOISTS

LEDGER

Fig. 67-4 A ledger is used where the deck joins the building. The joint is made weathertight with a flashing.

the flashing. The flashing is then fastened, close to and along its outside edge, with nails spaced closely together. The outside edge of the flashing should extend beyond the ledger.

Footing Layout and Construction

Footings for the supporting posts must be accurately located. To determine their location, erect batter boards and stretch lines in a manner previously described in Chapter 26. All footings require digging a hole and filling it with concrete (Fig. 67-5).

Footing Size and Style

In stable soil and temperate climate, the footing width is usually made twice the width of the post it is supporting. The footing depth reaches undisturbed soil, at least 12 inches below grade. In cold climates the footing should extend below the frost line.

There are several footing styles that are commonly used. One method is to partially fill the footing hole with concrete to within a few inches from the top. Set a precast pier 2 inches into the wet

INTERIOR
FOOTING

PERIMETER FOOTING

PLUMB BOB

CORNER BATTER BOARD

CORNER FOOTING

FOOTING AND POST
LAYOUT AND EXCAVATION

Fig. 67-5 Footing and post layout and excavation. (*Courtesy of Simpson Strong-Tie Company*)

concrete. After the concrete has set, attach a post anchor to the top of the precast pier (Fig. 67-6). Align piers and anchors with the layout lines.

The top of the footing may be brought above grade with the use of a wood box or fiber-tube form. Place concrete in the footing hole. Bring it to the top of the form. Set post anchors while the concrete is still wet (Fig. 67-7).

Erecting Supporting Posts

All supporting posts are set on footings. They are then plumbed and braced. Cut posts, for each footing, a few inches longer than their final length. Tack the bottom of each post to the anchor. Brace them in a plumb position in both directions (Fig. 67-8).

When all posts are plumbed and braced, the tops must be cut level to the proper height. From the height of the deck, deduct the deck thickness and the depth of the girder. Mark on a corner post. Mark the other posts by leveling from the first post marked. Mark each post completely around using a square. Cut the tops with a portable circular saw (Fig. 67-9).

Installing Girders

Install the girders on the posts using post and beam metal connectors. The size of the connector will depend on the size of the posts and girders. Install

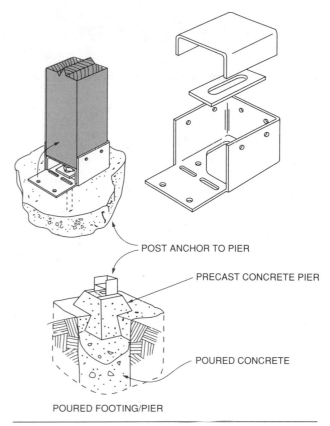

POST ANCHOR TO PIER

PRECAST CONCRETE PIER

POURED CONCRETE

POURED FOOTING/PIER

Fig. 67-6 Precast concrete pier embedded in the footing. (*Courtesy of Simpson Strong-Tie Company*)

SET PB POST BASE JUST AFTER POUR

GRADE

FIBER-TUBE FORM (TEAR AWAY AFTER CONCRETE IS SET)

POURED CONCRETE

SET CB OR CD BASE JUST AFTER CONCRETE POUR

GRADE

WOOD BOX FORM (REMOVE AFTER CONCRETE IS SET)

POURED CONCRETE IN HOLE DUG WITH POST HOLE DIGGER OR SHOVEL

Fig. 67-7 The top of the footing can be brought above grade with the use of a wood or fiber-tube form. (*Courtesy of Simpson Strong-Tie Company*)

③ SECURE POSTS
WITH 1x BOARDS
"STAKE-OUT"

① FASTEN POST TO
ANCHOR WITH NAILS
OR BOLTS

② PLUMB POSTS
EACH WAY

Fig. 67-8 Supporting posts are cut longer than required and braced plumb on footings. (*Courtesy of Simpson Strong-Tie Company*)

■

girders with the crowned edge up. Any splice joints should fall over the center of the post (Fig. 67-10).

Installing Joists

Joists may be placed over the top or between the girders. When joists are hung between the girders, the overall depth of the deck is reduced. This provides more clearance between the frame and the ground. For decking run at right angles, joists may be spaced 24 inches on center. Joists should be spaced 16 inches on center for diagonal decking.

Lay out and install the joists in the same manner as previously described in earlier chapters. Use

appropriate hangers if joists are installed between girders (Fig. 67-11). When joists are installed over girders, use recommended framing anchors. Make sure all joists are installed with their crowned edges facing upward.

Bracing Supporting Posts

If the deck is 4 feet or more above the ground, the supporting posts should be braced in a manner similar to that shown in Figure 67-12. Use minimum 1×6 braces for heights up to 8 feet. Use minimum 1×8 braces for higher decks applied continuously around the perimeter.

Applying Deck Boards

Specially shaped *radius edge decking* is available in both pressure-treated and natural decay-resistant lumber. It is usually used to provide the surface and walking area of the deck. Dimension lumber of 2-inch thickness and widths of 4 and 6 inches is also widely used. Lay boards with the bark side up to minimize cupping (Fig. 67-13).

Boards are usually laid parallel with the long dimension of the deck. Deck boards usually do not come longer than 16 feet. It may be desirable to lay the boards parallel to the short dimension, if their length will span it, to eliminate end joints in the decking. Boards may also be laid in a variety of patterns including diagonal, herringbone, and parquet. Make sure the supporting structure has been designed and built to accommodate the design.

Much care should be taken with the application of deck boards. Snap a straight line as a guide to apply the starting row. Start at the outside edge if the deck is built against a building. A ripped and narrower ending row of decking is not so noticeable

CUT POSTS TOPS WITH CIRCULAR SAW

CARPENTER'S SQUARE
ON 2x4 MAY BE USED
TO MARK HEIGHT, ALSO

MARK POST COMPLETELY AROUND WITH SQUARE

Fig. 67-9 The post height is determined and the tops cut with square ends level with each other. (*Courtesy of Simpson Strong-Tie Company*)

Fig. 67-10 Girders are installed with their crowned edges up on top of supporting posts. (*Courtesy of Simpson Strong-Tie Company*)

Fig. 67-11 Deck joists are installed between girders with joist hangers. (*Courtesy of Simpson Strong-Tie Company*)

Fig. 67-12 Supporting posts must be braced if the deck is 4 feet or more above the ground. (*Courtesy of Simpson Strong-Tie Company*)

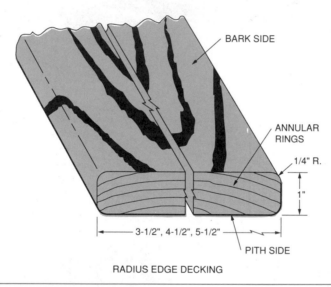

RADIUS EDGE DECKING

Fig. 67-13 Special radius edge decking is used and applied with the bark side up to resist cupping. (*Courtesy of Southern Pine Marketing Council*)

against the building as it is on the outside edge of the deck.

Straighten boards as they are installed. Maintain about a 1/4-inch space between them. If the deck boards do not span the entire width of the deck, cut the boards so their ends are centered over joists. Make tight-fitting end joints. Stagger them between adjacent rows. Predrill holes for fasteners to prevent splitting the ends. Let the end of each row overhang the deck. Use two nails in each joist. Drive nails at an angle. Set the heads below the surface. This will keep the nails from working loose and the heads from staining the surface. When approaching the end, it is better to increase or decrease the spacing of the last six or seven rows of decking. This way you will end with a row that is nearly equal in width to all the rest, rather than

with a narrow strip. When all the deck boards are laid, snap lines. Cut the overhanging ends flush with the framing. Apply a preservative to them (Fig. 67-14).

Applying Trim

A fascia board may be fastened around the perimeter of the deck. Its top edge should be flush with the deck surface. The fascia board conceals the cut ends of the decking and the supporting members below. The fascia board is optional.

Stairs and Railings

Stairs. Most decks require at least one or two steps leading to the ground. To protect the bottom ends of the stair carriage, they should be treated

PART 1

PAINT EXPOSED ENDS WITH PRESERVATIVE

STAGGER JOINTS

MARK EDGE WITH LINE ON DECK BOARDS AND CUT WITH CIRCULAR SAW

SQUARE DECK BOARDS AT THIS END

ANGLE NAILS

STRAIGHTEN EACH BOARD AND MAINTAIN 1/4" SPACE

TRIM WILD ENDS

ALTERNATE LONG AND SHORT BOARDS

PART 2

JOIST   SPACER

DIAGONAL SQUARE

HERRINGBONE

PARQUET

DECK BOARD PATTERNS

PART 3

Fig. 67-14 Much care should be taken when installing deck boards because the surface is so visible. (*Part 1, Courtesy of California Redwood Association; Part 2, Courtesy of Simpson Strong-Tie Company; Part 3, Courtesy of Southern Pine Marketing Council*)

with preservative and supported by an above-grade concrete pad (Fig. 67-15). Stair layout and construction are described in Chapter 48. Stairs with more than four risers are generally required to have at least one handrail. The design and construction of the stair handrail should conform to that of the deck railing.

Rails. There are numerous designs for deck railings (Fig. 67-16). All designs must conform to certain code requirements. Most codes require a 36-inch-

high railing around the exposed sides, if the deck is more than 30 inches above the ground. In addition, codes specify that no railing should have openings greater than 8 inches. Each linear foot of railing must be strong enough to resist a pressure of 20 lbs./sq. ft. applied horizontally at a right angle against the top rail. Check local building codes for deck stair and railing requirements.

Railings may consist of *posts*, also called *stanchions*, top, bottom, and intermediate *rails*, and *balusters*. Sometimes *lattice work* is used to fill in the

Fig. 67-15 Stairs for decks are usually constructed with a simple basement or utility design. (*Courtesy of Simpson Strong-Tie Company*)

Fig. 67-16 The design of this deck railing meets code requirements of height and open spaces. (*Courtesy of California Redwood Association*)

rail spaces above the deck. It is frequently used to close the space between the deck and the ground. Posts, rails, balusters, and other deck parts are manufactured in several shapes especially for use on decks (Fig. 67-17).

Stanchions or posts are sometimes notched on their bottom ends to fit over the edge of the deck. They are usually spaced about 4 feet apart. They are fastened with lag screws or bolts. The top rail may go over the tops or be cut between the posts. The bottom rail is cut between the posts. It is kept a few inches (no more than 8) above the deck. The remaining space may be filled with intermediate rails, balusters, lattice work, or other parts in designs as desired or as specified (Fig. 67-18).

Deck Accessories

There are many details that can turn a plain deck into an attractive and more comfortable living area. *Shading structures* are built in many different designs. They may be completely closed in or spaced to provide filtered light and air circulation. *Benches* partially or entirely around the deck may double as a place to sit and act as a railing (Fig. 67-19). Bench

seats should be 18 inches from the deck. Make allowance for cushion thickness, if used. The width of the seat should be from 18 to 24 inches.

Porches

The porch deck is constructed in a similar manner to an open deck. Members of the supporting structure may need to be increased in size and the spans decreased to support the weight of the walls and roof above. Work from plans drawn by professionals or check with building officials before starting. Porch walls and roofs are framed and finished as described in previous chapters (Fig. 67-20).

Summary

Decks and porches are designed in many different ways. There are other ways to construct them besides the procedures described in this chapter. However, making the layout, building the supporting structure, applying the deck, constructing the railing and stairs, and, in the case of a porch, building the walls and roof are basic steps that can be applied to the construction of practically any deck or porch.

A. COLONIAL STUB POST
B. V-GROOVE STUB POST
C. GOTHIC STUB POST
D. 8' GOTHIC POST

A. 36" BALUSTER
B. 42" BALUSTER
C. 48" BALUSTER
D. 42" ANGLE CUT BALUSTER
E. 36" EARLY AMERICAN BALUSTER
F. 36" COLONIAL BALUSTER

A. 2" x 4" x 12' FLAT TOP RAIL
B. 2" x 4" x 12' RAIL

A. LATTICE
B. 8' LATTICE DIVIDER
C. 8' LATTICE CAP
D. 8' HEAVY DUTY LATTICE CAP

LATTICE WORK

POST & RAIL

SQUARE EDGE POSTS & BALUSTERS

TURNED POSTS & BALUSTERS

Fig. 67-17 Railing parts are manufactured in several shapes.

Fig. 67-18 The deck railing may be constructed with various kinds of parts in a number of designs.

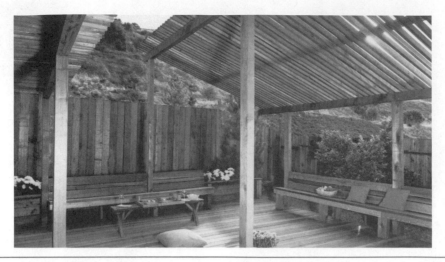

Fig. 67-19 A plain deck can be made more attractive and comfortable with benches and shading devices. (*Courtesy of California Redwood Association*)

Fig. 67-20 A porch is composed of a deck enclosed by walls and a roof. (*Courtesy of Southern Pine Marketing Council*)

CHAPTER 68 FENCE DESIGN AND ERECTION

A superior fence combines utility and beauty (Fig. 68-1). It may define space, create privacy, provide shade or shelter, screen areas from view, or form required barriers around swimming pools and other areas for the protection of small children.

The design of a fence is often the responsibility of the builder. With creativity and imagination, he or she can construct an object of beauty and elegance that also fulfills its function. The objective of this chapter is to create an awareness of the importance of design by showing several styles of fences and the methods used to erect them.

Fence Material

Fences are not supporting structures. Lower, knotty grades of lumber may be used to build them. This is not the case if appearance is a factor and you want to show the natural grain of the wood. Many kinds of softwood may be used as long as they are shielded from the weather by protective coatings of paint, stains, or preservatives. However, for wood posts that are set into the ground and other parts that may be subjected to constant moisture, pressure-treated or all-heart, decay-resistant lumber must be used.

The same type of fasteners that are used on exposed decks should be used to build fences. Inferior hardware and fasteners will corrode and cause unsightly stains when in contact with moisture.

Fence Design

Fences consist of posts, rails, and fence boards. Fences may be constructed in almost limitless designs. Zoning regulations sometimes restrict their height or placement. Often the site will affect the design. For example, the fence may have to be stepped like a staircase on a steep slope (Fig. 68-2). Fences on property lines can be designed to look attractive from either side. Fence designs may block wind and sunlight. Fence boards may be spaced in many attractive patterns. Most fences are constructed in several basic styles, each of which can be designed in numerous ways. Provisions should be made in the fence design to drain water from any area where it may otherwise be trapped.

Picket Fence

The *picket* fence is commonly used on boundary lines or as barriers for pets and small children. Usually not more than 4 feet high, the pickets are spaced to provide plenty of air and also to conserve material. The tops of the pickets may be shaped in various styles. Or, they may be cut square, with ends exposed or capped with a molding. The pickets may be applied with their tops in a straight line or in curves between posts. When pickets are applied with their edges tightly together, the assembly is called a *stockade* fence. Stockade fences are usu-

Fig. 68-1 A fence can serve its intended purpose and also enhance the surroundings. (*Courtesy of California Redwood Association*)

Fig. 68-2 Sometimes the site may affect the design of a fence, such as on this steep slope. (*Courtesy of California Redwood Association*)

Fig. 68-3 The picket fence is constructed in many styles. (*Courtesy of California Redwood Association*)

ally higher and are used when privacy is desired (Fig. 68-3).

Board on Board Fence

The *board on board* fence is similar to a picket fence. However, the boards are alternated from side to side so that the fence looks the same from both sides. The boards may vary in height and spacing according to the degree of privacy and protection from wind and sun desired. The tops or edges of the boards may be shaped in many different designs (Fig. 68-4).

Lattice Fence

This fence gets its name from narrow strips of wood called *lattice.* Strips are spaced by their own width. Two layers are applied at right angles to each other, either diagonally or in a horizontal and vertical fashion, to form a lattice work panel. Panels of various sizes can be prefabricated and installed between posts and rails, similar to a lattice work deck railing as shown in Figure 67-18.

Panel Fence

The *panel* fence creates a solid barrier with boards or panels fitted between top and bottom rails (Fig. 68-5). Fence boards may be installed diagonally or in other appealing designs. Alternating the panel design provides variety and adds to the visual appeal of the fence. A small space should be left between panel boards to allow for swelling in periods of high humidity. Figure 68-2 is a good example of a panel fence that has been made more attractive with the addition of a few members. In many cases, two or more basic styles may be combined to enhance

Fig. 68-4 A board on board fence is similar to a picket fence. However, it looks the same on both sides.

Fig. 68-5 The panel fence restricts air. However, it provides shade and privacy. (*Courtesy of California Redwood Association*)

Fig. 68-6 Solid panels are combined with lattice work panels for an attractive design. (*Courtesy of California Redwood Association*)

Fig. 68-7 The louvered fence provides privacy and lets air through. (*Courtesy of California Redwood Association*)

the design. Figure 68-6 shows lattice panels combined with board panels.

Louvered Fence

The *louvered* fence is a panel fence with vertical boards set at an angle (Fig. 68-7). The fence permits the flow of air through it and yet provides privacy. This fence is usually used around patios and pools. It is not used as a barrier along property lines.

Post and Rail Fence

The *post and rail* fence is a basic and inexpensive style normally used for long boundaries (Fig. 68-8). Designs include two or more square edge or round rails, of various thicknesses, widths, and diameters, cut between the posts or fastened to their edges. Most post and rail fence designs have large openings. They are not intended as barriers.

Fence Post Design

Fence posts are usually wood or iron. Wood posts are usually 4×4. Larger sizes may be used, depending on the design. The post tops may be shaped in various ways to enhance the design of the fence (Fig. 68-9). In order to conserve material and reduce expenses, a 4×4 post may be made to appear larger by applying furring and then boxing it in with one-inch boards. The top may then be capped with shaped members and molding in various designs to make an attractive fence post (Fig. 68-10).

Iron posts may be pipe or solid rod of from one inch to 1 1/4 inches in diameter for fences 3 to 4 feet tall. Larger diameters should be used for higher fences. Iron fence posts may be boxed in to simulate large wood posts. The tops are then usually capped

Fig. 68-8 The post and rail fence is an inexpensive design for long boundaries.

Fig. 68-9 Fence post tops are shaped in various ways to enhance the design of the fence.

with various shaped members similar to boxed wood posts (Fig. 68-11). Iron posts should be galvanized or otherwise coated to resist corrosion.

Building a Fence

The first step in building a fence is to set the fence posts. Locate and stretch a line between the end

MOULDING UNDER CAP

BOX

FURRING

4" × 4" POST

FURRING

BOX

Fig. 68-10 Method of boxing and capping a wood post.

CAP

BOX

SQUARE BLOCK FITS ON PIPE

IRON PIPE OR ROD

BLOCK

Fig. 68-11 Method of boxing and capping an iron fence post.

posts. If it is not possible to stretch a line because of steep sloping land, set up a transit-level to lay out a straight line. (See Chapter 26, Laying Out Foundation Lines.) If the fence is to be built on a property line, make sure that the exact locations of the boundary markers are known.

Setting Posts

Posts are generally placed about 8 feet apart. Mark the post locations with stakes along the fence line. Dig holes about 10 inches in diameter with a post hole digger. The depth of the hole depends on the height of the fence. Higher fences require that posts be set deeper (Fig. 68-12). The bottom of the hole should be filled with gravel or stone. This provides drainage and helps eliminate moisture to extend the life of the post.

Posts may be set in the earth. For the strongest fences, however, set the posts in concrete. Set the end posts first. Use a level to ensure that the posts are plumb in both directions. Brace them securely. The height of the post is determined by measuring from the ground, especially when the ground slopes. If top fence rails run across the tops of the posts, then the posts are left long. The tops are cut later in a straight or contoured line, as necessary. Make sure the face edges of the posts are aligned with the length of the fence. After the posts are braced, stretch lines between the end posts at the top and at the bottom.

WOOD POSTS

METAL ANCHOR

IRON POST

Fig. 68-12 A 3-foot hole is needed for an 8-foot fence. For shorter fences, post holes should be at least 2 feet deep.

Set and brace intermediate posts. The face edge of the posts are kept to the stretched line at the top and bottom. They are plumbed in the other direction with a level. Lag screws or spikes partially driven into the post strengthen the set of fence posts in concrete.

Place concrete around the posts. Tamp it well into the holes. Form the top so the surface pitches down from the posts. Instead of setting the posts in concrete, metal anchors may be embedded. The posts are installed on them after the concrete has hardened (Fig. 68-13).

Installing Fence Rails

Usually two or three horizontal rails are used on most fences. However, the height or the design may require more. Rails may run across the face or be cut in between the fence posts. If rails are cut between posts, they are installed to a snapped chalk line across the edges of the posts. They are secured by toenailing or with metal framing connectors. The bottom rail should be kept at least 6 inches above the ground.

Very often, the faces of wood posts are not in the same line as the rails. Therefore, rail ends must

be cut with other than square ends to fit between posts. Figure 68-14 shows a method of laying out rail ends to fit between twisted posts.

When iron fence posts are used, holes are bored in the rails. The rails are then installed over the previously set iron posts. Splices are made on the ends to continue the rail in a straight, unbroken line. Special metal pipe grips are made to fasten fence rails to iron posts (Fig. 68-15). If iron posts are boxed with wood, then the rails are installed in the same manner as for wood posts.

On some rustic style post and rail fences, the rails are doweled into the posts. In this case, post and rails must be installed together, one section at a time.

Applying Spaced Fenceboards and Pickets

Fasten pickets in plumb positions with their tops to the correct height at the starting and ending points. Stretch a line tightly between the tops of the two pickets. If the fence is long, temporarily install intermediate pickets to support the line from sagging. Sight the line by eye to see if it is straight. If not, make adjustments and add more support pickets if

STEP BY STEP
P R O C E D U R E S

STRETCHED LINE

END MARKER

INTERMEDIATE STAKES

END MARKER

1. LOCATE AND STRETCH LINE BETWEEN END MARKERS

2. DRIVE INTERMEDIATE STAKES ABOUT 8 FEET APART

END POST

POST HOLES

END POST

3. SET AND BRACE END POSTS PLUMB

4. EXCAVATE FOR INTERMEDIATE POSTS

5. STRETCH LINE NEAR TOP OF POSTS

END POST

END POST

POST HOLE

6. STRETCH LINE NEAR BOTTOM OF POSTS

8. PLUMB EDGES OF INTERMEDIATE POSTS AND BRACE

7. SET AND BRACE INTERMEDIATE POSTS TO STRETCHED LINES TOP AND BOTTOM.

Fig. 68-13 Procedure for setting fence posts.

Fig. 68-14

necessary. Use a picket for a spacer, and fasten pickets to the rails. If the spacing is different, rip a piece of lumber for use as a spacer.

Cut only the bottom end of the pickets when trimming their height. The bottom of pickets should not touch the ground. Place a 2-inch block on the ground. Turn the picket upside down with its top end on the block. Mark it at the chalk line. Fasten the picket with its top end to the stretched line (Fig. 68-16).

Continue cutting and fastening pickets using the spacer. Keep their tops to the line. Check the pickets for plumb frequently. If not plumb, bring back into plumb gradually with the installation of three or four pickets.

Stop 3 or 4 feet from the end. Check to see if the spacing will come out even. Usually the spacing has to be either increased or decreased slightly. Set the dividers for the width of a picket plus a space, increased or decreased slightly, whichever is appropriate. Space off the remaining distance. Adjust the dividers until the spacing comes out even. Any slight difference in a few spaces is not noticeable. This is much better than ending up with one narrow, conspicuous space (Fig. 68-17).

Installing Pickets in Concave Curves

In some cases, the pickets or other fenceboards are installed with their tops in concave curved lines between fence posts. If the fenceboards have shaped tops that cannot be cut, erect the fence using the following procedure.

PIPE GRIP TIE

METAL GRIP TIES ARE MANUFACTURED
TO ATTACH FENCE RAILS TO PIPE FENCE POSTS

TOP FENCE RAIL

SPLICES IN RAILS

IRON FENCE POST

BOTTOM FENCE RAIL

Fig. 68-15 Fence rails are installed on iron fence posts in several ways. (*Courtesy of Simpson Strong-Tie Company*)

STEP BY STEP
P R O C E D U R E S

③. STRETCH LINE TIGHTLY BETWEEN STARTING AND ENDING PICKETS

TOP RAIL

①. INSTALL STARTING PICKET IN A PLUMB POSITION

④. TEMPORARILY INSTALL INTERMEDIATE PICKETS TO SUPPORT LINE

②. INSTALL ENDING PICKET IN A PLUMB POSITION

BOTTOM RAIL

SPACER

TURN PICKET UPSIDE DOWN AND MARK AT STRETCHED LINE

STRETCHED LINE

TOP RAIL

INTERMEDIATE SUPPORTING PICKETS

ENDING PICKET

BOTTOM RAIL

WOOD BLOCK

⑤. INSTALL PICKETS USING SPACER

Fig. 68-16 Procedure for installing spaced pickets.

TIGHTLY STRETCHED LINE

SET DIVIDERS TO SLIGHTLY
MORE OR LESS THAN THE
WIDTH OF A PICKET PLUS A
SPACE AS NEEDED TO COME
OUT EVEN AT THE END.

ENDING
PICKET

LAST PICKET APPLIED 4 OR 5
FEET FROM THE END

Fig. 68-17 Divide the remaining distance into equal spaces when ending a picket fence installation.

Install a picket on each end at the high point of the curve. In the center, temporarily install a picket with its top to the low point of the curve. Fasten a flexible strip of wood in a curve to the top of the three pickets. Start from the center. Work both ways to install the remainder of the pickets with their tops to the curved strip. Space the pickets to each end (Fig. 68-18). Other fenceboards, such as board

on board and louvers, are installed in a similar manner.

If fenceboard tops are to be cut in the shape of the curve, tack them in place with their tops above the curve. Bend the flexible strip to the curve. Mark all the fenceboard tops. Remove the fenceboards, if necessary. Cut the tops and replace them.

FASTEN FLEXIBLE STRIP
TO TOP OF STARTING PICKET

TURN PICKET UPSIDE
DOWN AND MARK AT
CURVED STRIP

FASTEN FLEXIBLE STRIP
TO CENTER PICKET

FASTEN FLEXIBLE STRIP
TO TOP OF END PICKET

FLEXIBLE STRIP
BENT TO CURVE

FENCE POST

TOP FENCE
RAIL

SPACER

BOTTOM FENCE RAIL

WOOD BLOCK

FENCE POST

Fig. 68-18 A method to install spaced fenceboards with their tops in a concave curve.

Review Questions

Select the most appropriate answer.

1. No fasteners and hardware used on exposed decks and fences should be
   a. aluminum.
   b. hot-dipped galvanized.
   c. electroplated.
   d. stainless steel.

2. A ledger is a beam
   a. attached to the side of a building.
   b. supported by a girder.
   c. used to support girders.
   d. installed on supporting posts.

3. A footing for supporting posts must extend
   a. at least 12 inches below grade.
   b. 36 inches in depth.
   c. below the frost line.
   d. to stable soil.

4. Deck joists must be installed
   a. between girders.
   b. over girders.
   c. crowned edge up.
   d. using joist hangers.

5. A railing is required on deck stairs with more than
   a. 30 inches total rise.
   b. 2 risers.
   c. 3 feet total rise.
   d. 4 risers.

6. A railing is required on decks
   a. more than 3 stair risers above the ground.
   b. 30 or more inches above the ground.
   c. 4 feet or more above the ground.
   d. of any height.

7. The usual height of a bench seat without a cushion is
   a. 14 inches.   c. 18 inches.
   b. 16 inches.   d. 20 inches.

8. A high fence with picket edges applied tightly together is called a
   a. board on board fence.
   b. panel fence.
   c. post and rail fence.
   d. stockade fence

9. The bottom rail of any type fence is installed above the ground by, at least,
   a. the thickness of a 2×4 block.
   b. the width of a 2×4 block.
   c. 6 inches.
   d. 8 inches.

10. When applying fence pickets to rails
    a. plumb each one before fastening.
    b. plumb them frequently.
    c. cut the top ends to the line.
    d. fasten bottom ends flush with bottom rail.

■ B U I L D I N G   F O R   S U C C E S S ■

Customer Relations in Construction

An important element affecting transactions that occur between the general public and a business is customer relations. Assuming that a primary goal of good customer relations is to gain customer satisfaction, some key considerations must be identified.

To set the stage for establishing good customer relations, businesspersons should follow certain practices. They must exhibit knowledge and credibility, demonstrate skill in their areas of expertise, and inspire customers. They must be aware of the customer's needs while possessing a good attitude. The information that goes out from the business must be accurate and verified.

Any mental image people have about a company or business will relate directly to that company's success. The target population of a business could be large or quite small, depending on the product and cost. The volume of sales will be reflected in the company's customer relations policies.

Another primary goal of customer relations is to create and maintain a positive image using sound business principles. This must be a high priority. As a builder conducts daily business with the public, his or her company is involved with maintaining a good image. The builder has the responsibility to conduct business in an ethical and proper manner if trust is to be established.

Without regular customers, a business will be forced to close its doors. The need for good customer relations must be recognized by every successful businessperson. This will involve developing good personal relations skills, sound business skills, and an awareness of customer satisfaction.

The customer is any person who purchases a commodity or service from another person or place of business. Any lasting business relationship that is created is established with mutual goals. The foundation of this relationship begins when the customer desires a quality service or product from a vendor for an agreed price. Once the transaction is initiated, a business relationship is formed. From that initial agreement, the process of "doing business" will involve the need to honor that agreement. In the process it is important how each party is treated. Basic responsibilities will be for the vendor to supply the service or product as agreed. The customer is to respond with payment once the goods are delivered.

In the more tangible terms of construction, we can use the home builder as an example. A potential client contacts a builder with a desire to build a new home. The builder sees this as an opportunity to "sell" his or her product and service to the client. The customer may have heard of the builder previously or may have seen one of the builder's products. This alone may have set in motion the foundation for any customer relations affecting the project. The customer will bring along any previous image of the builder based on personal observations and/or what others have said. This becomes the starting point of communications and relationships that will affect the success of any business that is conducted.

Good customer relations will have a major impact on the success of all business transactions. A builder's public image and reputation will be affected by the manner in which his or her employees view public relations. Mutual trust and respect must be established in the beginning and continue throughout the duration of the transaction. Sound business principles must be readily seen by the public and customers. Customer relation policies must be consistent in order to assure honesty and fairness. All future business activities between the builder and the clientele will be based on the previously established relationships. By establishing clear oral and written communications at the start of any business discussions, the future working relationships will almost always guarantee success. As a builder, subcontractor, or other technician interacts with the clientele, he or she will need to establish good customer relations. The success of doing business with the public will depend on how well those relations are established.

FOCUS QUESTIONS: For individual or group discussion.

1. How do you define customer relations?

2. What are five important considerations for establishing positive customer relations between a builder and his or her customer?

3. How can effective customer relations assure a framing subcontractor that his or her perceived image is credible?

4. As you prepare to enter the construction world, how would you structure a plan to establish good customer relations for your personal business?

STUDENT ACTIVITIES for

Section 3: Exterior Finish

Unit 18 Cornice Construction

1. Using a typical wall section, explain the construction process and major parts of a box, snub, and open cornice. Emphasize the need for accuracy in building the cornice. Build a model of each type.

2. Using metal soffit and fascia wrap products, explain the installation process for the box cornice. Also demonstrate the tools used in this process.

3. Explain how to determine the proper location for the plough in wood fascia to receive the soffit board.

Unit 19 Roofing

1. Demonstrate and explain the application of asphalt shingles, including felting, layout, nailing, starter course, cutting and spacing on the gable, hip, and valley. Include the installations of metal edge, gutter apron, valley flashing, roof vents, asphalt cement, step flashing, hip and ridge caps.

2. Demonstrate and explain the application of wood shingles and shakes using the same criteria as in activity #1.

3. Demonstrate the installation process of roll roofing with concealed nails and double coverage. Include felting and asphalt cement use.

4. Explain the available grades of asphalt shingles and wood roofing products, their advantages and disadvantages, costs, and how to calculate the quantities needed for the job.

5. Explain in detail any two typical roofing problems and how to prevent or properly repair those problems.

Unit 20 Windows

1. Using a major wood window manufacturer's catalog, explain how to read the catalog and obtain the following information: window type, sizes available, features, accessories, costs, and technical information for calculating Heating, Ventilating, & Air Conditioning data.

2. Using a blueprint and the window manufacturer's catalog, explain the process of selecting windows that will meet local building code requirements. List the information on a window schedule.

3. Duplicate activities #1 and #2 for metal windows.

4. Explain and demonstrate the manufacturer's installation process for a casement, double hung, or gliding window. Include a description of the specific features, such as cladding, hardware, weatherstripping, and sash venting action.

Unit 21 Exterior Doors

1. Describe the styles, sizes, features, and parts of an exterior door.
2. Assemble and set an exterior door frame. Hang a door in the frame. Install the lockset.
3. Rout hinge gains using a butt hinge template and router.
4. Explain the process of installing a sectional overhead garage door.

Unit 22 Exterior Wall Siding

1. Make a product display board showing many samples of sidings, corner boards, accessories, and finishes available. Explain how the siding products are estimated, priced, and sold.
2. Demonstrate the application procedures for horizontal lap and vertical sidings, with the necessary trims.
3. Demonstrate the application procedures for the sheet type plywood, composite, and hardboard sidings, with the necessary trims.
4. Explain the installation process of metal and plastic sidings. Include the necessary accessories for each siding product.
5. Using a residential blueprint, estimate the quantity of siding needed to cover the side walls of the house with a 4' × 8' hardboard product. Make no deductions for cutouts. Figure no waste.
6. Explain the theory of a story pole and demonstrate how to make one for horizontal lap siding.
7. Explain the use of a water table, where it is used, and how it is installed.

Unit 23 Porches, Decks, and Fences

1. Using a set of blueprints of a residence with no front porch, design a porch that adds character and aesthetic value to the house. Design a shed, gable, or hip roof into the plan. The open or enclosed wall design may be used. Include a complete list of construction materials and costs.
2. Using a blueprint, design a wood deck and backyard fence to provide an outside eating and recreation area for the home. Incorporate into the plan the use of pressure-treated and redwood or western red cedar products. The deck may have a sun roof or be open. Prepare a complete list of the building materials and costs.
3. Make a presentation of your design to others, stating how you arrived at the design, the reasons certain materials were selected, a time schedule, and how the project will be built. Exhibit examples of the materials and plans.

Interior Finish

24

Drywall Construction

O B J E C T I V E S

After completing this unit, the student will be able to:

■ describe various kinds, sizes, and uses of gypsum panels.

■ describe the kinds and sizes of nails, screws, and adhesives used to attach gypsum panels.

■ make single-ply and multi-ply gypsum board applications to interior walls and ceilings.

■ conceal gypsum board fasteners and corner beads.

■ reinforce and conceal joints with tape and compound.

The term *drywall construction* generally means the application of gypsum board to interior walls and ceilings. In contrast it does not refer to the application of wet gypsum plaster to a plaster base of gypsum or wire lath.

C O N T E N T S

CHAPTER 69   GYPSUM BOARD

Gypsum board is sometimes called *wallboard, plasterboard, drywall,* or *Sheetrock*®. It is used extensively in construction (Fig. 69-1). The term Sheetrock® is a brand name for gypsum panels made by the U.S. Gypsum Company. However, the brand name is in such popular use, it has become a generic name for gypsum panels. Gypsum board makes a strong, high-quality, fire-resistant wall and ceiling covering. It is readily available, easy to apply, decorate, or repair, and relatively inexpensive.

Gypsum Board

Many types of gypsum board are available for a variety of applications. The board or panel is composed of a gypsum core encased in a strong, smooth-finish paper on the face side and a natural-finish paper on the back side. The face paper is folded around the long edges. This reinforces and protects the core. The long edges are usually tapered. This allows the joints to be concealed with compound without any noticeable *crown joint* (Fig. 69-2). A crown joint is a buildup of the compound above the surface.

Types of Gypsum Panels

Most gypsum panels can be purchased, if desired, with an aluminum foil backing. The backing functions as a vapor retarder. It helps prevent the passage of interior water vapor into wall and ceiling spaces.

Fig. 69-1 The application of gypsum board to interior walls and ceilings is called drywall construction. (*Courtesy of U.S. Gypsum Company*)

Fig. 69-2 The long edges of gypsum panels are usually tapered for effective joint concealment.

Regular. *Regular* gypsum panels are most commonly used for single- or multilayer application. They are applied to interior walls and ceilings in new construction and remodeling.

Eased Edge. An *eased edge* gypsum board has a special tapered, rounded edge. This produces a much stronger concealed joint than a tapered, square edge (Fig. 69-3).

Fig. 69-3 An eased edge panel has a rounded corner that produces a stronger concealed joint.

Type X. *Type X* gypsum board is typically known as *Firecode* board. It has greater resistance to fire because of special additives in the core. Type X gypsum board is manufactured in several degrees of resistance to fire. Type X looks the same as regular gypsum board. However, it is labeled Type X on the edge or on the back.

Water-Resistant. *Water-resistant* gypsum board consists of a special moisture-resistant core and paper cover that is chemically treated to repel moisture. It is frequently used as a base for application of wall tile in bath, showers, and other areas subjected to considerable moisture. It is easily recognized by its distinctive green face. It is frequently called *green board* by workers in the field. Water-resistant panels are available with a Type X core for increased fire resistance.

Special Purpose. *Backing board* is designed to be used as a base layer in multilayer systems. It is available with regular or Type X cores. *Coreboard* is available in one-inch thickness. It is used for various applications, including the core of solid gypsum partitions. It comes in 24-inch widths with a variety of edge shapes. *Predecorated* panels have coated, printed, or overlay surfaces that require no further treatment. *Liner board* has a special fire-resistant core encased in a moisture-repellent paper. It is used to cover shaft walls, stairwells, chaseways, and similar areas.

Veneer Plaster Base. *Veneer plaster bases* are commonly called blue board. They are large, 4-foot wide gypsum board panels faced with a specially treated blue paper. This paper is designed to receive applications of *veneer plaster.* Conventional plaster is applied about 3/8 inch thick and takes considerable time to dry. In contrast, specially formulated veneer plaster is applied in one coat of about 1/16 inch, or two coats totaling about 1/8 inch. It takes only about forty-eight hours to dry.

Gypsum lath is used as a base to receive conventional plaster. (It is more fully described in Chapter 37.) Other gypsum panels, such as soffit board and sheathing, are manufactured for exterior use.

Sizes of Gypsum Panels

Widths and Lengths. Coreboards and liner boards come in 2-foot widths and from 8 to 12 feet long. Other gypsum panels are manufactured 4 feet wide and in lengths of 8, 9, 10, 12, 14, or 16 feet. Gypsum board is made in a number of thicknesses. Not all lengths are available in every thickness.

Thicknesses

- A 1/4-inch thickness is used as a base layer in multilayer applications. It is also used to cover existing walls and ceilings in remodeling work. It can be applied in several layers for forming curved surfaces with short radii.

- For walls and ceilings in manufactured housing, a 5/16-inch thickness is used.

- A 3/8-inch thickness is usually applied as a face layer in repair and remodeling work over existing surfaces. It is also used in multilayer applications in new construction.

- Both 1/2-inch and 5/8-inch thicknesses are the most commonly used gypsum panels for single-layer wall and ceiling application in residential and commercial construction. The 5/8-inch thick panel is more rigid. It has greater resistance to impacts and fire.

- Coreboards and liner boards come in thicknesses of 3/4 and one inch.

Cement Board

Like gypsum board, *cement board* is a panel product. However, it has a core of portland cement reinforced with a glass fiber mesh embedded in both sides (Fig. 69-4). The core resists water penetration and will not deteriorate when wet. It is designed for use in areas that may be subjected to high-moisture conditions. It is used extensively in bathtub, shower, kitchen, and laundry areas as a base for ceramic

Fig. 69-4 Composition of cement board. (*Courtesy of U.S. Gypsum Company*)

tile. In fact, some building codes require its use in these areas.

Panels are manufactured in sizes designed for easy installation in tub and shower areas with a minimum of cutting. Standard cement board panels come in a thickness of 1/2 inch, in widths of 32 or 36 inches, and in 5-foot lengths. Custom panels are available in a thickness of 5/8 inch, widths of 32 or 48 inches, and lengths from 32 to 96 inches.

Cement board is also manufactured in a 5/16-inch thickness. It is used as an underlayment for floors and countertops. Exterior cement board is used primarily as a base for various finishes on building exteriors.

Drywall Fasteners

Specially designed nails and screws are used to fasten drywall panels. Ordinary nails or screws are not recommended. The heads of common nails are too small in relation to the shank. They are likely to cut the paper surface when driven. Staples may only be used to fasten the base layer in multilayer applications. They must penetrate at least 5/8 inch into supports. Using the correct fastener is extremely important for proper performance of the application. Fasteners with corrosion-resistant coatings must be used when applying water-resistant gypsum board or cement board.

Nails

Gypsum board nails should have flat or concave heads that taper to thin edges at the rim. Nails should have relatively small diameter shanks with heads at least 1/4 inch, but no more than 5/16 inch in diameter. For greater holding power, nails with annular ring shanks are used (Fig. 69-5).

Smooth shank nails should penetrate at least 7/8 inch into framing members. Only 3/4-inch penetration is required when ring shank nails are used. Greater nail penetrations are required for fire-rated applications.

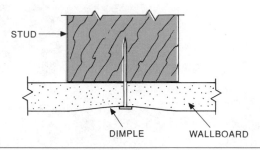

Fig. 69-6 Nails are driven with a convex-faced drywall hammer. This forms a dimple in the board. The dimple is filled with compound to conceal the fastener.

Nails should be driven with a drywall hammer that has a convex face. This hammer is designed to compress the gypsum panel face to form a dimple not more than 1/16 inch when the nail is driven home (Fig. 69-6). The dimple is made so the nail head can later be covered with compound. Care should be taken to drive the nail at a right angle to the board to prevent breaking the face paper.

Screws

Special drywall screws are used to fasten gypsum panels to steel or wood framing or to other panels. They are made with Phillips heads designed to be driven with a drywall power screwdriver (Fig. 69-7). A proper setting of the nosepiece on the power screwdriver assures correct countersinking of the screwhead. When driven correctly, the specially contoured bugle head makes a uniform depression in the panel surface without breaking the paper.

Different kinds of drywall screws are used for fastening into wood, metal, and gypsum panels. *Type W* screws are used for wood, *Type S* and *Type S-12* for metal framing, and *Type G* for fastening to gypsum backing boards (Fig. 69-8).

Type W screws have sharp points. They have specially designed threads for easy penetration and excellent holding power. The screw should pene-

THIN 1/4"
DIAMETER
HEADS

RING SHANK

SMOOTH SHANK

SMALL-HEADED
NAIL FOR USE
WITH PREDECORATED
WALLBOARD

Fig. 69-5 Special nails are required to fasten gypsum board.

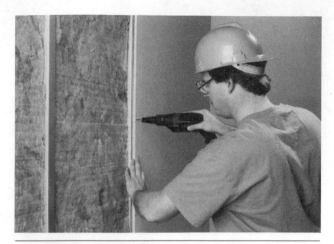

Fig. 69-7 Drywall screws are driven with a power screwdriver to the desired depth. (*Courtesy of U.S. Gypsum Company*)

TYPE S

FOR LIGHT GAUGE METAL FRAMING

TYPE S-12

FOR 20 GAUGE OR HEAVIER METAL FRAMING

TYPE G

FOR FASTENING INTO BASE LAYERS OF GYPSUM BOARD

TYPE W

FOR WOOD FRAMING

Fig. 69-8 Several types of screws are used to fasten gypsum panels. Selection of the proper type is important.

trate the supporting wood frame by at least 5/8 inch.

Type S screws are self-drilling and self-tapping. The point is designed to penetrate sheet metal with little pressure. This is an important feature because thin metal studs have a tendency to bend away when driving screws. Type S-12 screws have a different drill point designed for heavier gauge metal framing.

Type G screws have a deeper, special thread design for effectively fastening gypsum panels together. These screws must penetrate into the supporting board by at least 1/2 inch. If the supporting board is not thick enough, longer fasteners should be used. Make sure there is sufficient penetration into framing members.

Adhesives

Drywall adhesives are used to bond single layers directly to supports or to laminate gypsum board to base layers. Adhesives used to apply gypsum board are classified as *stud adhesives* and *laminating adhesives*.

For bonding gypsum board directly to supports, special *drywall stud adhesive* or approved *construction adhesive* is used. Supplemental fasteners must be used with stud adhesives. Stud adhesives are available in large cartridges. They are applied to framing members with hand or powered adhesive guns (Fig. 69-9).

CAUTION Some types of drywall adhesives may contain a flammable solvent. Do not use these types near an open flame or in poorly ventilated areas.

For laminating gypsum boards to each other, *joint compound adhesives* and *contact adhesives* are used. Joint compound adhesives are applied over the entire board with a suitable spreader prior to lamination. Boards laminated with joint compound adhesive require supplemental fasteners.

When contact adhesives are used, no supplemental fasteners are necessary. However, the board cannot be moved after contact has been made. The adhesive is applied to both surfaces by brush, roller, or spray gun. It is allowed to dry before laminating. A *modified contact* adhesive is also used. It permits an open time of up to thirty minutes during which the board can be repositioned, if necessary.

Fig. 69-9 Applying drywall adhesive to stud edges. (*Courtesy of U.S. Gypsum Company*)

CHAPTER 70

SINGLE-LAYER AND MULTILAYER DRYWALL APPLICATION

Single-layer gypsum board applications are widely used in light commercial and residential construction. They adequately meet building code requirements. *Multilayer* applications are more often used in commercial construction. They have increased resistance to fire and sound transmission. Both systems provide a smooth, unbroken, quality surface if recommended application procedures are followed.

Single-Layer Application

Drywall should be not be delivered to the job site until shortly before installation begins. Boards stored on the job for long periods are subject to damage. The boards must be stored under cover and stacked flat on supports. Supports should be at least 4 inches wide and placed fairly close together (Fig. 70-1). Leaning boards against framing for long periods may cause the boards to warp. This makes application more difficult. To avoid damaging the edges, carry the boards. Do not drag them. Then, set the boards down gently. Be careful not to drop them.

Cutting and Fitting Gypsum Board

Take measurements accurately. Cut the board by first scoring the face side through the paper to the core. Use a utility knife. Guide it with a *drywall T-square*, if cutting a square end (Fig. 70-2). The board is then broken along the scored face. The back paper is scored along the fold. The sheet is then broken by snapping the board in the reverse direction (Fig. 70-3).

GYPSUM PANELS

STICKING SHOULD BE AT LEAST 4" WIDE

Fig. 70-1 Correct method of stacking gypsum board.

Fig. 70-2 Using a drywall T-square as a guide when scoring across the width of a board. (*Courtesy of U.S. Gypsum Company*)

Fig. 70-3 Method of snapping and breaking the board along the scored line. (*Courtesy of U.S. Gypsum Company*)

Fig. 70-4 Lines parallel to edges are scored by gauging with a tape. A tape guide and tape tip are useful accessories for gauging. (*Courtesy of U.S. Gypsum Company*)

Fig. 70-5 A gypsum board stripper scores both sides of parallel cuts close to edges. (*Courtesy of U.S. Gypsum Company*)

To make cuts parallel to the long edges, the board is often gauged with a tape and scored with a utility knife. A *tape guide* and *tape tip* are sometimes used to aid the procedure. The tape guide permits more accurate gauging and protects the fingers. The tape tip contains a slot into which the knife is inserted. This prevents slipping off the end of the tape (Fig. 70-4).

When making parallel cuts close to long edges, it is usually necessary to score both sides of the board to obtain a clean break. A *gypsum board stripper* is a useful tool. It scores both sides of the board with one stroke. The tool is adjustable to make cuts up to 4 1/2 inches wide (Fig. 70-5).

Smooth ragged edges with a drywall rasp, coarse sanding block, or knife. A job-made drywall rasp can be made by fastening a piece of metal lath to a wood block. Cut panels should fit easily into place without being forced. Forcing the panel may cause it to break.

Aligning Framing Members

Before applying the gypsum board, check the framing members for alignment. Stud edges should not be out of alignment more than 1/8 inch with adjacent studs. A wood stud that is out of alignment can be straightened by the procedure shown in Figure 70-6. Ceiling joists are sometimes brought into alignment with the installation of a *strongback* across the tops of the joists at about the center of the span (Fig. 70-7). Furring strips are straightened by shim-

Fig. 70-6

Fig. 70-7 A strongback is sometimes used to align ceiling joists or the bottom chord of roof trusses.

ming with wood shingles between the furring and the joists (see Chapter 36).

Fastening Gypsum Panels

Drywall is fastened to framing members with nails or screws. Hand pressure should be applied on the panel next to the fastener being driven. This ensures that the panel is in tight contact with the framing member. The use of adhesives reduces the number of nails or screws required. A single or double method of nailing may be used.

Single Nailing Method. With this method, nails are spaced a maximum of 7 inches OC on ceilings and 8 inches OC on walls into frame members. Nails should be first driven in the center of the board and then outward toward the edges. Perimeter fasteners should be at least 3/8 inch, but not more than one inch from the edge.

Double Nailing Method. In double nailing, the perimeter fasteners are spaced as for single nailing. In the field of the panel, space a first set of nails 12 inches OC. Space a second set 2 to 2 1/2 inches from the first set. The first nail driven is reseated after driving the second nail of each set. This assures solid contact with framing members (Fig. 70-8).

Screw Attachment. Screws are spaced 12 inches OC on ceilings and 16 inches OC on walls when framing members are spaced 16 inches OC. If framing members are spaced 24 inches OC, then screws are spaced a maximum of 12 inches OC on both walls and ceilings.

Using Adhesives. Apply a straight bead about 1/4 inch in diameter to the centerline of the stud edge. On studs where panels are joined, two parallel beads of adhesive are applied, one on each side of the centerline. Zigzag beads should be avoided to prevent the adhesive from squeezing out at the joint.

On wall application, supplemental fasteners are used around the perimeter. Space them about 16 inches apart. On ceilings, in addition to perimeter fastening, the field is fastened at about 24-inch intervals (Fig. 70-9).

Fig. 70-8 Gypsum panels may be single-nailed or double-nailed.

Gypsum panels may be prebowed. This reduces the number of supplemental fasteners required. Prebow the panels by one of the methods shown in Figure 70-10. Make sure the finish side of the panel faces in the correct direction. Allow them to remain overnight or until the boards have a 2-inch permanent bow. Apply adhesive to the studs. Fasten the panel at top and bottom plates. The bow keeps the center of the board in tight contact with the adhesive until bonded.

Ceiling Application

Gypsum panels are applied first to ceilings and then to the walls. Panels may be applied parallel, or

Fig. 70-9 Beads of stud adhesive are applied (left) straight under the field of a board, and (right) parallel beads under joints.

Fig. 70-10 Prebowing keeps the board in tight contact with the adhesive until bonded and reduces the number of fasteners required.

at right angles, to joists or furring. If applied parallel, edges and ends must bear completely on framing. If applied at right angles, the edges are fastened where they cross over each framing member. Ends must be fastened completely to joists or furring strips.

Carefully measure and cut the first board to width and length. Cut edges should be against the wall. Lay out lines on the panel face indicating the location of the framing in order to place fasteners accurately.

Gypsum board panels are heavy. At least two or more people are needed for ceiling application unless special equipment is available. Lift the panel overhead and place it in position (Fig. 70-11). Install two *deadmen* under the panel to hold it in position.

Deadmen are supports made in the form of a "T." They are easily made on the job using 1×3 lumber with short braces from the vertical member to the horizontal member. The leg of the support is made about 1/4 inch longer than the floor-to-ceiling height. The deadmen are wedged between the floor and the panel. They hold the panel in position while it is being fastened (Fig. 70-12). Using deadmen is much easier than trying to hold the sheet in position and fasten it at the same time.

Fasten the sheet in one of the recommended manners. Hold the board firmly against framing to avoid nail pops or protrusions. Drive fasteners straight into the member. Fasteners that miss supports should be removed. The nail hole should be dimpled so that later it can be covered with joint compound (Fig. 70-13).

Once the sheet is fastened, remove the deadmen for use with the next sheet to be applied. Con-

Fig. 70-11 Two or more persons are needed for ceiling application of gypsum board unless special lifting equipment is used.

tinue applying sheets in this manner, staggering end joints, until the ceiling is covered.

To cut a corner out of a panel in case of a protrusion in the wall, make the shortest cut with a *drywall saw.* Then, score and snap the sheet in the other direction (Fig. 70-14). To cut a circular hole, mark the circle with pencil dividers, twist and push the drywall saw through the board. Cut out the hole, following the circular line.

Fig. 70-12 Deadmen are wedged between floor and ceiling. They hold the board in place while it is being fastened.

Horizontal Application on Walls

When walls are less than 8'-1" high, wallboard is usually installed horizontally, at right angles to the studs. If possible, use a board of sufficient length to go from corner to corner. Otherwise, use as long a board as possible to minimize end joints because they are difficult to conceal. Stagger end joints or center them over and below windows and over

doors, if possible. That way not so much of the joint is visible. End joints should not fall on the same stud as those on the opposite side of the partition.

The top panel is installed first. Cut the board to length to fit easily into place without forcing it. Stand the board on edge against the wall. Start fasteners along the top edge opposite each stud. Raise the sheet so the top edge is firmly against the ceil-

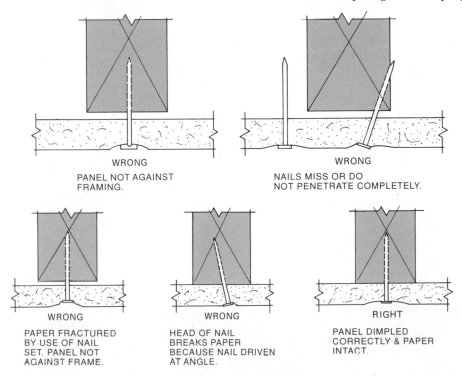

WRONG
PANEL NOT AGAINST FRAMING.

WRONG
NAILS MISS OR DO NOT PENETRATE COMPLETELY.

WRONG
PAPER FRACTURED BY USE OF NAIL SET. PANEL NOT AGAINST FRAME.

WRONG
HEAD OF NAIL BREAKS PAPER BECAUSE NAIL DRIVEN AT ANGLE.

RIGHT
PANEL DIMPLED CORRECTLY & PAPER INTACT.

Fig. 70-13 It is important to drive fasteners correctly for secure attachment of gypsum panels.

Fig. 70-14 Use a knife and drywall saw to cut the corner out of a gypsum panel.

Fig. 70-16 A foot lifter is used to lift gypsum panels in the bottom course up against the top panels. (*Courtesy of U.S. Gypsum Company*)

ing. Drive the fasteners home. Fasten the rest of the sheet in the recommended manner (Fig. 70-15).

Measure and cut the bottom panel to width and length. Cut the width about 1/4 inch narrower than the distance measured. Lay the panel against the wall. Raise it with a *drywall foot lifter* against the bottom edge of the previously installed top panel. A drywall foot lifter is a tool especially designed for this purpose. However, one can be made on the job by tapering the end of a short piece of 1×3 or 1×4 lumber (Fig. 70-16). Fasten the sheet as recommended. Install all others in a similar manner. Stagger any necessary end joints. Locate them as far from the center of the wall as possible so they will be less conspicuous.

Where end joints occur on steel studs, attach the end of the first panel to the open or unsupported edge of the stud. This holds the stud flange in a rigid position for the attachment of the end of the adjoining panel. Making end joints in the opposite manner usually causes the stud edge to deflect. This results in an uneven surface at the joint (Fig. 70-17).

Making Cutouts in Wall Panels. There are several ways of making cutouts in wall panels for electrical outlet boxes, ducts, and similar objects. Care must be taken not to make the cutout much larger than the outlet. Most cover plates do not cover by much. If cut too large, much extra time has to be taken to patch up around the outlet, replace the panel, or install oversize outlet cover plates.

■   One method of making a cutout is to first rub the edges of the outlet box with *carpenter's cake*

Fig. 70-15 Horizontal application of gypsum panels to walls. (*Part 1, Courtesy of U.S. Gypsum Company*)

STUD

NAILING BLOCK OPTIONAL

TAPERED EDGE

SOLEPLATE

DIRECTION OF GYPSUM BOARD INSTALLATION

DIRECTION OF GYPSUM BOARD INSTALLATION

DEFLECTION POINT—BOTH PANELS

CORRECT APPLICATION

DEFLECTION POINT—FIRST PANEL

DEFLECTION POINT—SECOND PANEL

EDGES BIND

INCORRECT APPLICATION

Fig. 70-17 Sequence of making end joints when attaching gypsum panels to steel studs. (*Courtesy of U.S. Gypsum Company*)

chalk or similar chalk. Place the gypsum panel to be cut out in position over the outlet box. Tap the face of the panel several times opposite the box. The outline of the box should be transferred to the back of the panel. The cutout can then be made with a drywall saw.

■ In a second method, the sides of the outlet box are plumbed down to the floor, or up to the previously installed top panel, whichever is is more convenient. The panel is placed in position. Lines for the sides of the box are plumbed on it from the marks on the floor or on the panel. The top and bottom of the box are laid out by measuring down from the bottom edge of the top panel. With a utility knife, score the outline of the box. Then score its diagonals. Punch the intersection of the diagonals with a hammer to fold in the four triangles formed by the scored layout. Cut along the fold on the back side to make the cutout (Fig. 70-18).

■ A fast, easy, and accurate way of making cutouts is with the use of a portable electric *drywall cutout tool.* The approximate location of the center of the outlet box is determined and marked on the panel. The panel is then installed over the box. Using the cutout tool, a hole is plunged through the panel in the approximate center of the outlet box. Care must be taken not to make contact with wiring. The tool is not recommended for use around live wires. The tool is moved in any direction until the bit hits a side of the box. It is then withdrawn slightly to ride over the edge to the outside of the box. The tool is then moved so the bit rides around the outside of the box to make the cutout (Fig. 70-19). Usually cutouts are made for outlets after all the panels in a room have been installed.

To make cutouts around door openings, either mark and cut out the panel before it is applied, or make the cutout after it is applied. To make the cutout after the panel is applied, use a saw to cut in one direction. Then score it flush with the opening on the back side in the other direction. Bend and score it on the face to make the cutout.

Vertical Application on Walls

Vertical application of gypsum panels on walls, with long edges parallel to studs, is more practical if the ceiling height is more than 8'-1" or the wall is 4'-0" wide or less. Cut the first board, in the corner, to length and width. Its length should be about 1/4 inch shorter than the height from floor to ceiling. It should be cut to width so the edge away from the

AFTER SCORING, PUNCH AT INTERSECTION

Fig. 70-18 To make a cutout in the panel, score the outline of the cutout and its diagonals. Then, punch with a hammer and cut out. (*Courtesy of Gold Bond Building Products*)

DRYWALL CUTOUT BIT HAS FLAT SURFACES ON THE TWIST THAT RIDE AGAINST SURFACE OF BOX

BIT RIDES AGAINST OUTSIDE OF OUTLET BOX

Fig. 70-19 A portable electric drywall cutout tool is often used to make cutouts for outlet boxes and similar objects. (*Courtesy of Porter-Cable*)

corner falls on the center of a stud. All cut edges must be in the corners. None should butt edges of adjacent panels.

With a foot lifter, raise the sheet so it is snug against the ceiling. The tapered edge should be plumb and centered on the stud. Fasten it in the specified manner. Continue applying sheets around the room with tapered uncut edges against each other. There should be no horizontal joints between floor and ceiling (Fig. 70-20). Make any necessary cutouts as previously described.

Floating Interior Angle Construction

To help prevent nail popping and cracking due to structural stresses where walls and ceilings meet, the *floating angle* method of drywall application may be used. When joists or furring strips are at right angles to walls, fasteners in ceiling panels are located 7 inches from the wall for single nailing, and 11 to 12 inches for double nailing or screw attachment. When joists or furring are parallel to the wall, nailing is started at the intersection.

Gypsum panels applied on walls are fitted tightly against ceiling panels. This provides a firm support for the floating edges of the ceiling panels. The top fastener into each stud is located 8 inches down from the ceiling for single nailing and 11 to 12 inches down for double nailing or screw attachment.

At interior wall corners, the underlying wallboard is not fastened. The overlapping board is fitted snugly against the underlying board. This brings it in firm contact with the face of the stud. The overlapping panel is nailed or screwed into the interior corner stud (Fig. 70-21).

TAPERED EDGE

GYPSUM BOARD

Fig. 70-20 Applying gypsum panels parallel to wall studs. (*Photo, Courtesy of U.S. Gypsum Company*)

7" TO FIRST FASTENER
ON CEILING

8" TO FIRST
FASTENER
ON WALL

FASTEN THIS
SIDE ONLY

Fig. 70-21 Floating angle method of applying drywall. (*Courtesy of U.S. Gypsum Company*)

Applying Gypsum Panels to Curved Surfaces

Gypsum panels may be applied to curved surfaces. However, closer spacing of the frame members may be required to prevent flat areas from occurring on the face of the panel. If the paper and core of gypsum panels are moistened, they may be bent to curves with shorter radii than when dry. After the boards are thoroughly moistened, they should be stacked on a flat surface. They should be allowed to stand for at least one hour before bending. Moistened boards must be handled very carefully. They will regain their original hardness after drying. The minimum bending radii for dry and wet gypsum panels are shown in Figure 70-22.

To apply panels to a *convex* surface, fasten one end to the framing. Gradually work to the other end by gently pushing and fastening the panel progressively to each framing member. When applying panels to a *concave* curve, fasten a stop at one end. Carefully push on the other end to force the center of the panel against the framing. Work from the end against the stop. Fasten the panel successively to each framing member.

Gypsum board may be applied to the curved inner surfaces of arched openings. If the dry board cannot be bent to the desired curve, it may be moistened or parallel knife scores made about one inch apart across its width.

Drywall Application to Bath and Shower Areas

Water-resistant gypsum board and cement board panels are used in bath and shower areas as bases for the application of ceramic tile. Framing should be 16 inches OC. Steel framing should be at least 20-gauge thickness.

Apply panels horizontally with the bottom edge not less than 1/4 inch above the lip of the shower pan or tub. The bottom edges of gypsum panels should be uncut and paper-covered. Do not apply panels directly over a vapor barrier.

Check the alignment of the framing. If necessary, apply furring strips to bring the face of the board flush with the lip of the tub or shower pan (Fig. 70-23).

Provide blocking between studs about one inch above the top of the tub or shower pan. Install additional blocking between studs behind the horizontal joint of the panels above the tub or shower pan.

Cement board panels are cut in the same manner as gysum board panels. Before attaching panels, apply thinned ceramic tile adhesive to all cut edges around holes and other locations.

Attach panels with corrosion-resistant screws or nails spaced not more than 8 inches apart. When ceramic tile more than 3/8 inch thick is to be applied, the nail or screw spacing should not exceed 4 inches OC.

Multilayer Application

A multilayer application has one or more layers of gypsum board applied over a base layer. It provides

BOARD THICKNESS IN INCHES	DRY	WET
1/4	5 FT.	2 TO 2½ FT.
3/8	7½ FT.	3 TO 3½ FT.
1/2	20 FT.	4 TO 4½ FT.

Fig. 70-22 Minimum bending radii of gypsum panels.

½" GYPSUM PANEL

½" INTERIOR CEMENT BOARD

½" INTERIOR CEMENT BOARD

SUPPORT FRAMING FOR ATTACHMENT OF FIXTURES

SCREWS 8" O.C.

CERAMIC TILE

TAPE

SETTING MATERIAL

2 x 4 WOOD STUDS 16" O.C.

TAPE

SETTING MATERIAL

¼" SPACE

TUB

WOOD PLATE

TYPICAL BATHTUB INSTALLATION

WATER RESISTANT GYPSUM BOARD

ADHESIVE

TILE

PAPER EDGE

CAULK

TUB SUPPORT

CEMENT BOARD OR WATER RESISTANT GYPSUM BOARD

¼"

TUB

FLOOR LINE

¼"

WATER RESISTANT GYPSUM BOARD OR CEMENT BOARD

ADHESIVE

TILE

PAPER EDGE

WATERDAM

CAULK

SHOWER BASE

¼" (6.4 mm)

Fig. 70-23 Installation details around bathtubs and showers.

greater strength, higher fire resistance, and better sound control. The base layer may be gypsum backing board, regular gypsum board, or other gypsum base material.

Base Layer

The base layer is fastened in the same manner as single-layer panels. However, double nailing is not necessary and staples may be used in wood fram-

ing. On ceilings, panels are applied with the long edges either at right angles or parallel to framing members. On walls, the panels are applied with the long edges parallel to the studs.

Face Layer

Joints in the face layer are offset at least 10 inches from joints in the base layer. The face layer is applied either parallel to or at right angles to framing, whichever minimizes end joints and results in the least amount of waste.

The face layer may be attached with nails, screws, or adhesives. If nails or screws are used without adhesive, the maximum spacing and minimum penetration into framing should be the same as for single-layer application.

Adhesive Attachment of the Face Layer

Panels of the face layer may be attached to the base layer by *sheet lamination, strip lamination,* or *spot lamination.* In sheet lamination, the entire back of the face layer is covered with laminating adhesive, using a notched spreader (Fig. 70-24).

In strip lamination, the adhesive is applied in ribbons with a special spreader, spaced 16 to 24 inches OC (Fig. 70-25). In *spot lamination,* plum-sized daubs of adhesive are applied to the panel in a regular pattern.

After the adhesive is applied, the face panel is held firmly against the base layer with supplemental fasteners or braces while the adhesive sets. The panel may be also be secured by temporarily nailing with duplex nails driven through wood or gypsum board scraps. The nail should pentrate into the framing by at least 3/4 inch. Type G screws may be used, but the base layer must be at least 1/2 inch thick.

Fig. 70-24 In sheet lamination, the entire back of the face layer is covered with adhesive. (*Courtesy of U.S. Gypsum Company*)

Fig. 70-25 In strip lamination, ribbons of the adhesive are spaced. (*Courtesy of U.S.Gypsum Company*)

CHAPTER 71 CONCEALING FASTENERS AND JOINTS

After the gypsum board is installed, it is necessary to conceal the fasteners and to reinforce and conceal the joints. One of several levels of finish may be specified for a gypsum board surface. The lowest level of finish may simply require the taping of wallboard joints and *spotting* of fastener heads on surfaces. This is done in warehouses and other areas where appearance is normally not critical. The level of finish depends, among other things, on the number of coats of compound applied to joints and fasteners (Fig. 71-1).

Description of Materials

Fasteners are concealed with *joint compound.* Joints are reinforced with *joint tape* and covered with *joint compound.* Exterior corners are reinforced with *corner bead.* Other kinds of drywall trim may be used around doors, windows, and other openings.

TAPERED EDGES OF
GYPSUM BOARD

TAPE

FINISHING
(THIRD) COAT
LEVEL 4

FINISHING
(SECOND) COAT
LEVEL 3

EMBEDDING
(FIRST) COATS

LEVELS 1 AND 2

LEVEL 5 FINISHING REQUIRES A THIN SKIM COAT
OF JOINT COMPOUND APPLIED TO ENTIRE
SURFACE IN ADDITION TO LEVEL 4 FINISHING

Fig. 71-1 The level of finish varies with the type of final decoration to be applied to drywall panels.

Joint Compounds

Drying type joint compounds for joint finishing and fastener spotting are made in both a dry powder form and a ready-mixed form in three general types. Drying type compounds provide smooth application and ample working time. A *taping compound* is used to embed and adhere tape to the board over the joint. A *topping compound* is used for second and third coats over taped joints. An *all-purpose compound* is used for both bedding the tape and finishing the joint. All-purpose compounds do not possess the strength or workability of two-step taping and topping compound systems.

Setting type joint compounds are used when a faster setting time than that of drying types is desired. Drying type compounds harden through the loss of water by evaporation. They usually cannot be recoated until the next day. Setting type compounds harden through a chemical reaction when water is added to the dry powder. Therefore, they come only in a dry powder form and not ready-mixed. They are formulated in five different setting times. The fastest setting type will harden in as little as twenty to thirty minutes. The slowest type takes four to six hours to set up. Setting type joint compounds permit finishing of drywall interiors in a single day.

Joint Reinforcing Tape

Joint reinforcing tape is used to cover, strengthen, and provide crack resistance to drywall joints. One type is made of *high-strength fiber paper*. It is designed for use with joint compounds on gypsum panels. It is

Fig. 71-2 Applying joint tape to an interior corner.

creased along its center to simplify folding for application in corners (Fig. 71-2).

Another type is made of *glass fiber mesh*. It is designed to reinforce joints on veneer plaster gypsum panels. It is not recommended for use with conventional compounds for general drywall joint finishing. It may be used with special high-strength setting compounds. Glass fiber mesh tape is available with a plain back or with an adhesive backing for quick application (Fig. 71-3). Joint tape is normally available 2 and 2 1/2 inches wide in 300-foot rolls.

Fig. 71-3 An adhesive-backed glass fiber mesh tape is quickly applied to drywall joints. (*Courtesy of U.S. Gypsum Company*)

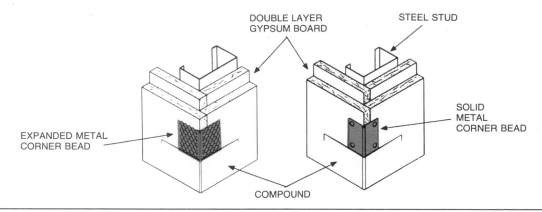

DOUBLE LAYER GYPSUM BOARD

STEEL STUD

EXPANDED METAL CORNER BEAD

SOLID METAL CORNER BEAD

COMPOUND

Fig. 71-4 Corner beads are used to finish and protect exterior corners of drywall panels. (*Courtesy of U.S. Gypsum Company*)

Corner Bead and Other Drywall Trim

Corner beads are applied to protect exterior corners of drywall construction from damage by impact. One type with solid metal flanges is widely used. Another type has flanges of expanded metal with a fine mesh. This provides excellent keying of the compound (Fig. 71-4).

Corner bead is fastened through the drywall panel into the framing with nails or staples. Instead of using fasteners, a *clinching tool* is sometimes used. It crimps the solid flanges and locks the bead to the corner (Fig. 71-5).

Metal corner tape is applied by embedding in joint compound. It is used for corner protection on arches, windows, and other locations (Fig. 71-6).

A variety of *metal trim* is used to provide protection and finished edges to drywall panels. They

Fig. 71-5 A clinching tool is sometimes used to fasten corner beads to exterior corners. (*Courtesy of U.S. Gypsum Company*)

CUT TAPE WITH SNIPS

EMBED IN JOINT COMPOUND

Fig. 71-6 Flexible metal corner tape is applied to exterior corners by embedding in compound. (*Courtesy of U.S. Gypsum Company*)

are used at windows, doors, inside corners, and intersections. They are fastened through their flanges into the framing (Fig. 71-7).

Control joints are metal strips with flanges on both sides of a 1/4-inch V-shaped slot. Control joints are placed in large drywall areas. They relieve stresses induced by expansion or contraction. They are used from floor to ceiling in long partitions and from wall to wall in large ceiling areas (Fig. 71-8). The flanges are concealed with compound in a manner similar to corner beads and other trim.

Applying Joint Compound and Tape

In cold weather, care should be taken to maintain the interior temperature at a minimum of 50 degrees Fahrenheit for twenty-four hours before and during application of joint compound, and for at least four days after application has been completed.

Care should also be taken to use clean tools. Avoid contamination of the compound by foreign material, such as sawdust, hardened, or different type compounds.

Fig. 71-7 Various metal trim is used to provide finished edges to gypsum panels. (*Courtesy of U.S. Gypsum Company*)

Fig. 71-8 Control joints are used in large wall and ceiling areas subject to movement by expansion and contraction. (*Courtesy of U.S. Gypsum Company*)

Fig. 71-9 Taping compound is first applied to the channel formed by the tapered edges between panels.

Fig. 71-10 The tape is embedded into the compound.

Prefilling Joints

Before applying compound to drywall panels, check the surface for fasteners that have not been sufficiently recessed. Also look for other conditions that may affect the finishing. Prefill any joints between panels of 1/4 inch or more and all V-groove joints between *eased-edged* panels with compound. A twenty-four-hour drying period can be eliminated with the use of setting compounds for prefilling operations.

Embedding Joint Tape

Fill the recess formed by the tapered edges of the sheets with the specified type of joint compound. Use a joint knife (Fig. 71-9). Center the tape on the joint. Lightly press it into the compound. Draw the knife along the joint with sufficient pressure to *embed* the tape and remove excess compound (Fig. 71-10).

There should be enough compound under the tape for a proper bond, but not over 1/32 inch under the edges. Make sure there are no air bubbles under the tape. The tape edges should be well ad-

hered to the compound. If not satisfactory, lift the portion. Add compound and embed the tape again. A *taping tool* is sometimes used. It applies the compound and embeds the tape at the same time (Fig. 71-11).

Fig. 71-11 A taping tool applies tape and compound at the same time. (*Courtesy of U.S. Gypsum Company*)

Immediately after embedding, apply a thin coat of joint compound over the tape. This helps prevent the edges from wrinkling. It also makes easier concealment of the tape with following coats. Draw the knife to bring the coat to feather edges on both sides of the joint. Make sure there is no excess compound left on the surface. After the compound has set up, but not completely hardened, wipe the surface with a *damp sponge*. This eliminates the need for sanding any excess after the compound has hardened.

Spotting Fasteners

Fasteners should be *spotted* immediately before or after embedding joint tape. *Spotting* is the application of compound to conceal fastener heads. Apply enough pressure on the taping knife to fill only the depression. Level the compound with the panel surface. Spotting is repeated each time additional coats of compound are applied to joints.

Applying Compound to Corner Beads and Other Trim

The first coat of compound is applied to corner beads and other metal trim when first coats are given to joints and fasteners. The nose of the bead or trim serves as a guide for applying the compound. The compound is applied about 6 inches wide from the nose of the bead to a feather edge on the wall. Each subsequent finishing coat is applied about 2 inches wider than the previous one.

Fill and Finishing Coats

Allow the first coat to dry thoroughly. This may take twenty-four hours, or more, depending on temperature and humidity unless a setting type compound has been used. It is common to use setting compounds for first coats and slower setting types for finishing coats. Feel the entire surface to see if any excess compound has hardened on the surface. Sand any excess, if necessary, to avoid interfering with the next coat of compound.

The second coat is sometimes called a *fill* coat. It is feathered out about 2 inches beyond the edges of all first coats, approximately 7 to 10 inches wide. Care must always be taken to remove all excess compound so that it does not harden on the surface. Professional drywall finishers rarely have to sand any excess in preparation for following coats. Remember, a damp sponge rubbed over the joint after the compound starts to set up will remove any small particles of excess. It will also help bring edges to a feather edge.

If the level of finish requires it, apply a third and *finishing* coat of compound over all fill coats. The edges of the finishing coat should be feathered out about 2 inches beyond the edges of the second coat (Fig. 71-12).

Interior Corners

Interior corners are finished in a similar way. However, the tape is folded in the center to fit in the corner. After the tape is embedded, drywall fin-

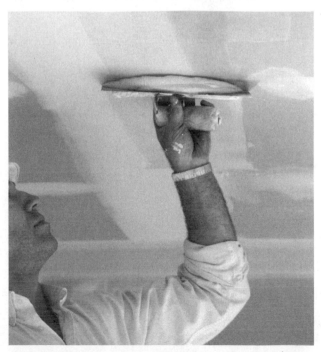

Fig. 71-12 Applying a finishing coat of compound to a taped drywall joint.

ishers usually apply a setting compound to one side only of each interior corner. By the time they have finished all interior corners in a room, the compound has set enough to finish the other side of the corners.

Estimating Drywall

To estimate the amount of drywall material needed, determine the area of the walls and ceilings to be covered. To find the ceiling area, multiply the length of the room by its width. To find the wall area, multiply the perimeter of the room by the height. Subtract all large wall openings. Combine all areas to find the total number of square feet of drywall required. Add about 5 percent of the total for waste. The number of drywall panels can then be determined by dividing the total area to be covered by the area of one panel.

About 1,000 screws are needed for each 1,000 square foot of drywall when applied to framing 16 inches OC and 850 screws for 24 inch OC framing. About 5 pounds of nails are required to fasten each 1,000 square foot of drywall.

Approximately 370 feet of joint tape and 135 pounds of conventional, ready-mixed joint compound are needed for every 1,000 square foot of drywall area.

Review Questions

Select the most appropriate answer.

1. Standard gypsum board width is
   a. 36 inches.   c. 54 inches.
   b. 48 inches.   d. 60 inches.

2. Standard gypsum board lengths are
   a. 8, 10, and 12 feet.
   b. 8, 10, 12, and 14 feet.
   c. 8, 9, 10, 12, and 14 feet.
   d. 8, 9, 10, 12, 14, and 16 feet.

3. When fastening drywall, minimum penetration of ring-shanked nails into the framing member is
   a. 1/2 inch.   c. 7/8 inch.
   b. 3/4 inch.   d. one inch.

4. Gypsum board is usually installed vertically on walls when the wall height is greater than
   a. 8'-0".   c. 8'-4".
   b. 8'-1".   d. 8'-6".

5. Ceiling joists are sometimes aligned by the use of a
   a. deadman.   c. strongback.
   b. dutchman.   d. straightedge.

6. In the single nailing method, nails are spaced a maximum of
   a. 8 inches OC on walls; 7 inches OC on ceilings.
   b. 10 inches OC on walls; 8 inches OC on ceilings.
   c. 12 inches OC on walls; 10 inches OC on ceilings.
   d. 12 inches OC on walls and ceilings.

7. Screws are spaced
   a. 12 inches OC on walls; 10 inches OC on ceilings.
   b. 12 inches OC on walls and ceilings.
   c. 16 inches OC on walls and ceilings.
   d. 16 inches OC on walls; 12 inches OC on ceilings.

8. Joints in the face layer of a multilayer application are offset from joints in the base layer by at least
   a. 6 inches.   c. 10 inches.
   b. 8 inches.   d. 12 inches.

9. The paper-covered edge of water-resistant gypsum board is applied above the lip of the tub or shower pan not less than
   a. 1/4 inch.   c. 1/2 inch.
   b. 3/8 inch.   d. 3/4 inch.

10. When ceramic tile more than 3/8-inch thick is to be applied over water-resistant gypsum board, fasten the board with screws or nails spaced not more than
   a. 4 inches OC.   c. 8 inches OC.
   b. 6 inches OC.   d. 10 inches OC.

■ BUILDING FOR SUCCESS ■

A Formula for Success

The following information comes from an interview with Greg Shinaut, the owner of *Pride Homes*, a custom home builder in Lincoln, Nebraska. We will look at the events that led up to self-employment in the home building industry. Of primary importance are the basics that sets the entire process of owning a business into motion. The initial dream, the goal setting, the plan to reach those goals, the experiences along the way, the skill building, the relationships formed, and the outcomes are presented. This example may appear to be unique, but it actually represents thousands of other success stories throughout the building industry today.

Shinaut's initial interest in building began during his elementary school years in Fremont, Nebraska. When challenged by the question, "What do you want to do with your life," he knew that he desired to work as a carpenter and build houses. His interest was spurred by outside influences since he was not reared in a construction-oriented family. With the assistance of his high school counselor and industrial technology instructor, Shinaut began to formulate ideas and goals for his career. He desired to learn the necessary carpentry skills that would allow him to enter the building field and experience a variety of jobs related to home building.

He sat up a ladder with steps to obtain along the way. Knowing that he would eventually need a good working knowledge of construction systems and procedures, good carpentry skills, business skills, and people skills, he entered the building construction program of study at Southeast Community College. Upon graduation, he was employed by a builder in central Nebraska. He was given a chance to experience a variety of construction jobs. This led to later work as a framer for a builder in the Omaha area. Eventually he worked as a deck builder and siding applicator in the city of Lincoln. While working for a builder in Lincoln, Shinaut served as the "call-back" representative. A lot of the work involved customer relations. The total skills and attributes he gained during those years provided the field and business experiences he needed to reach another stage in the predetermined sequence of events that lead to self-employment.

In focusing on the necessary attributes and abilities needed for being a successful builder and businessperson, Shinaut identifies the following as important. Having an open mind, with the ability to look and listen to everything affecting the construction world, is the pace setter. "Doing what it takes to get it done" and meeting the client's demands are two ingredients that are necessary for success. Being open to learning, using quality people and their strengths, possessing a good attitude, and being able to take constructive criticism are extremely important along the way. Having a conscience, being honest with people in dealings, sharing information with others, and being dedicated to people and work must not be ignored. Shinaut feels that all these attributes and abilities are required to build the necessary trust in people needed for a successful business.

As he continued working toward the goal of long-term self-employment in the home building business, some typical unknowns were carried with him. Not knowing exactly what tomorrow would bring, whether there would be enough business to keep moving, or if success would displace failure, were daily questions.

When asked to share a few observations with others who may be considering a career in construction, Shinaut offered some good advice. He cautions interested persons to diligently work their way into their careers. They need to decide how they might meet their goals and build the necessary skills and attitudes that undergird success. Experience the field, pace yourself, and remain open to learning to allow for confidence building. Being careful not to act too quickly in considering going out on your own before all of your homework is completed is important. Disaster can set in when actions are not backed by experience.

Shinaut believes that education, both formal and informal, is the basis for learning the construction business. In reference to young people entering the construction business, he feels that each person needs to find the correct balance. Those reared in a construction environment need different experiences than those coming from other backgrounds. Eventually the majority of business personnel in the construction world will gain a composite of similar experiences to help focus on successful careers.

The day-to-day challenges center on spending quality time with his family, doing what it takes to get the job, and including the family in the business operations. Without a complete commitment from the family, the frustration and stress are increased. The benefits coming from being self-employed are centered on personal satisfaction. The gratification that comes from seeing a family enter a new house and convert it into a home is extremely satisfying. It is a good feeling to see what hard work and dedication can produce in the lives of people.

The opportunity to work with many community organizations produces other satisfactions. The home builder has the opportunity to provide an important service to a community that affects the lives of many people. Usually the contribution to the community comes back in the way of benefits. These benefits range from providing assistance in solving local problems to helping local children, elderly, and disadvantaged people reach a better standard of living. Local home builder associations are strong in community service due to the caring attitude of the associated members.

When asked if he felt successful and why, Shinaut responded that he did not feel established, but needed that feeling to stay competitive. The feeling of complacency or being overly assured of continual work without competition will bring failure. The quality home builder needs to be challenged each year in the business.

Shinaut advises those who are considering a career in the construction business to "do it." Begin with a dream, and start to set goals. Follow your heart and reinforce that feeling with a willingness to work and learn. Use your imagination. Go as far as you are willing to be challenged. Don't wait for something to happen, because it usually won't. You have to initiate the action. This listing of activities and considerations culminates in an exemplary "formula for success."

FOCUS QUESTIONS: For individual or group discussion.

1. According to Shinaut, what appears to be some of the key considerations a person must think about when anticipating future self-employment? List six or more of the considerations.

2. As you look at the considerations made by Shinaut when concerning self-employment do you agree with his identified "formula for success"? Why or why not? How could you add to this formula to make it better?

3. How important is the planning process to successfully reaching your goals? What would be your recommendations to a person when goals seem out of reach?

4. How important is it to a person entering the business world to understand need for both technical and business skill building? Identify at least three important skills that you feel are necessary for eventual self-employment. Explain why they are necessary and how they can be obtained.

Wall Paneling and Wall Tile

Plans and specifications often call for the installation of *wall paneling* in certain rooms of both residential and commercial construction. *Ceramic wall tile* is widely used in rest rooms, baths, showers, kitchens, and similar areas.

OBJECTIVES

After completing this unit, the student will be able to:

- describe and apply several kinds of sheet wall paneling.

- describe and apply various patterns of solid lumber wall paneling.

- describe and install ceramic wall tile to bathroom walls.

- estimate quantities of wall paneling and ceramic wall tile.

CONTENTS

CHAPTER 72 — TYPES OF WALL PANELING

Two basic kinds of *wall paneling* are *sheets* of various prefinished material and solid *wood boards*. Many compositions, colors, textures, and patterns are available in *sheet* form. Solid *wood boards* of many species and shapes are used for both rustic and elegant interiors (Fig. 72-1).

Description of Sheet Paneling

Sheets of *prefinished plywood, hardboard, particleboard, plastic laminate*, and other material are used to panel walls.

Plywood

Prefinished plywood is probably the most widely used sheet paneling. A tremendous variety is available in both hardwoods and softwoods. The more expensive types have a face veneer of real wood. The less expensive kinds of plywood paneling are prefinished with a printed wood grain or other design on a thin vinyl covering. Care must be taken not to scratch or scrape the surface when handling these types. Unfinished plywood panels are also available.

Some sheets are scored lengthwise at random intervals to imitate solid wood paneling. There is always a score 16, 24, and 32 inches from the edge. This facilitates fastening of the sheets and in case the sheet has to be ripped lengthwise to fit stud spacing.

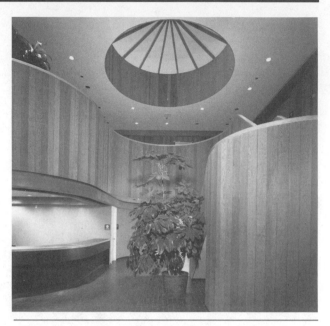

Fig. 72-1 Solid wood board paneling provides warmth and beauty to interiors of residential and commercial buildings.

Most commonly used panel thicknesses are 3/16 and 1/4 inch. Sheets are normally 4 feet wide and 7 to 10 feet long. An 8-foot length is most commonly used. Panels may be shaped with square, beveled, or shiplapped edges (Fig. 72-2). Matching molding is available to cover edges, corners, and joints. Thin ring-shanked nails, called *color pins*, are

Fig. 72-2 Sheet paneling comes with various edge shapes.

available in colors to match panels. They are used when exposed fastening is necessary.

Hardboard

Hardboard is available in many man-made surface colors, textures, and designs. Some of these designs simulate stone, brick, stucco, leather, weathered or smooth wood, and other materials. Unfinished hardboard is also used that has a smooth, dark brown surface suitable for painting and other decorating. Unfinished hardboard may be solid or perforated in a number of designs (Fig. 72-3).

Tileboard is a hardboard panel with a baked-on plastic finish. It is embossed to simulate ceramic wall tile. The sheets come in a variety of solid colors, marble, floral, and other patterns. Tileboard is designed for use in bathrooms, kitchens, and similar areas.

Hardboard paneling comes in widths of 4 feet and in lengths of from 8 to 12 feet. Commonly used thicknesses are from 1/8 to 1/4 inch. Color-coordinated molding and trim are available for use with hardboard paneling.

Particleboard

Panels of *particleboard* come with wood grain or other designs applied to the surface, similar to plywood and hardboard. Sheets are usually 1/4 inch thick, 4 feet wide, and 8 feet long. Prefinished particleboard is used when an inexpensive wall covering is desired. Because the sheets are brittle and break easily, care must be taken when handling. They must be applied only on a solid wall backing.

Unfinished particleboard is not usually used as an interior wall finish. One exception, made from aromatic cedar chips, is used to cover walls in closets to repel moths.

Plastic Laminates

Plastic laminates are commonly called *mica*. They are widely used for surfacing kitchen cabinets and countertops. They are also used to cover walls or parts of walls in kitchens, bathrooms, rest rooms, and similar areas where a durable, easy-to-clean surface is desired. Laminates can be scorched by an open flame. However, they resist heat, alcohol, acids, and stains. They clean easily with a mild detergent.

Laminates are known by such brand names as Formica, Pionite, Wilson Art, and others. They are manufactured in many colors and designs, including wood grain patterns. Surfaces are available in gloss, satin, and textured finishes, among others.

Laminates are ordinarily used in two thicknesses. *Vertical type* laminate is relatively thin (about 1/32 inch). It is used for vertical surfaces, such as walls and cabinet sides. Vertical type laminate is available only in widths of 4 feet and lengths of 8 feet. *Regular* or *standard* laminate is about 1/16 inch thick. It comes in widths of 24, 36, 48, and 60 inches and in lengths of 5, 6, 8, 10, and 12 feet. It is generally used on horizontal surfaces, such as countertops. It can be used on walls, if desired, or if the size required is not available in vertical type. Sheets are usually manufactured one inch wider and longer than the nominal size.

Laminates are difficult to apply to wall surfaces because they are so thin and brittle. Also, because a *contact bond* adhesive is used, the sheet cannot be moved once it makes contact with the surface. Thus, prefabricated panels, with sheets of laminate already bonded to a backer, are normally used to panel walls.

Description of Board Paneling

Board paneling is used on interior walls when the warmth and beauty of solid wood is desired. Wood paneling is available in softwoods and hardwoods of many species. Each has its own distinctive appearance, unique grain, and knot pattern.

Fig. 72-3 Hardboard is also available in a number of perforated designs.

Wood Species

Woods may be described as light, medium, and dark-toned. Light tones include birch, maple, spruce, and white pine. Some medium tones are cherry, cypress, hemlock, oak, ponderosa pine, and fir. Among the darker-toned woods are cedar, mahogany, redwood, and walnut. For special effects, knotty pine, wormy chestnut, pecky cypress, and white-pocketed Douglas fir board paneling may be used.

Surface Textures and Patterns

Wood paneling is available in many shapes. It is either *planed* for smooth finishing, or *rough-sawn* for a rustic, informal effect. *Square-edge* boards may be joined edge to edge, spaced on a dark background, or applied in *board-and-batten* or *board-on-board* patterns. *Tongue-and-grooved* or *shiplapped* paneling comes in patterns, a few of which are illustrated in Figure 72-4.

Sizes

Most wood paneling comes in a 3/4-inch thickness and in nominal widths of 4, 6, 8, 10, and 12 inches. A few patterns are manufactured in a 9/16-inch thickness. *Aromatic cedar* paneling is used in clothes closets. It runs from 3/8 to 5/16 inch thick, depending on the mill. It is usually *edge* and *end matched* (tongue-and-grooved) for application to a backing surface.

Moisture Content

To avoid shrinkage, wood paneling, like all interior finish, should be dried to a *moisture equilibrium* content. That is, its moisture content should be about the same as the area in which it is to be used. Except for arid, desert areas and some coastal regions, the average moisture content of the air in the United States is about 8 percent (Fig. 72-5). Interior finish applied with an excessive moisture content will eventually shrink, causing open joints, warping, loose fasteners, and many other problems.

Other Paneling

The types of paneling described in this chapter are those that are most commonly used. In order to become acquainted with other types and methods of application, a study of manufacturers' catalogs dealing with sheet and board paneling, found in Sweet's Architectural File, is suggested.

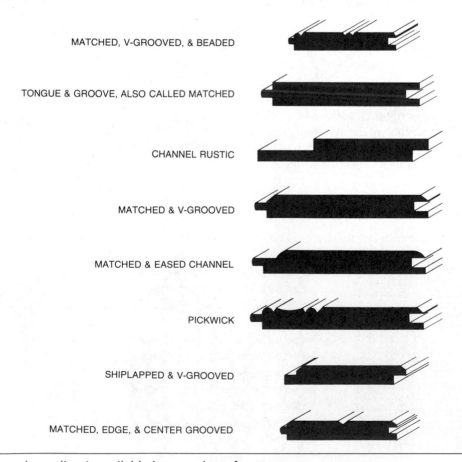

MATCHED, V-GROOVED, & BEADED

TONGUE & GROOVE, ALSO CALLED MATCHED

CHANNEL RUSTIC

MATCHED & V-GROOVED

MATCHED & EASED CHANNEL

PICKWICK

SHIPLAPPED & V-GROOVED

MATCHED, EDGE, & CENTER GROOVED

Fig. 72-4 Solid wood paneling is available in a number of patterns.

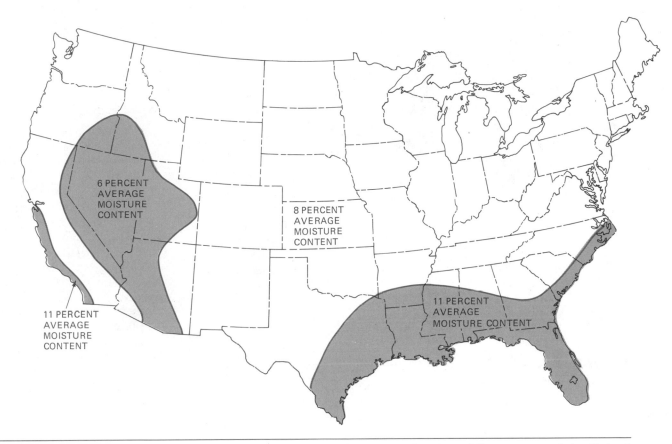

Fig. 72-5 Recommended average moisture content for interior finish woodwork in different parts of the United States.

CHAPTER **73** APPLICATION OF WALL PANELING

Sheet paneling, such as prefinished plywood and hardboard, is usually applied to walls with the long edges vertical. *Board* paneling may be installed vertically, horizontally, diagonally, or in many interesting patterns (Fig. 73-1).

Installation of Sheet Paneling

Some building codes require a base layer of gypsum board applied to studs or furring strips for the installation of sheet paneling. Even if not required, a backer board layer, at least 3/8 inch thick, should be installed on walls prior to the application of sheet paneling. The backing makes a stronger and more fire-resistant wall, helps block sound transmission, tends to bring studs in alignment, and provides a rigid finished surface for application of paneling. Without a base layer, depressions in the

Fig. 73-1 Wood paneling can be applied in many directions. (*Courtesy of California Redwood Association*)

Fig. 73-2 Apply sheet paneling over a gypsum wallboard base.

Labels in figure: SHEET PANELING, GYPSUM BACKER BOARD, PARTITION STUD

Fig. 73-3 Wainscoting is a wall finish, usually paneling, applied to the lower portion of the wall that is different from the upper portion.

Labels in figure: WALLBOARD, CHAIR RAIL, PANELING, BASEBOARD

thin sheets usually can be seen in the surface between the studs (Fig. 73-2).

Furring strips must be applied to masonry walls before the installation of paneling. The strips are usually applied by driving hardened nails with a pneumatic nailer. Instead of using furring strips, a freestanding wood wall close to the masonry wall can be built, if enough space is available.

Preparation for Application

Mark the location of each stud in the wall on the floor and ceiling. Paneling edges must fall on stud centers, even if applied with adhesive over a backer board, in case supplemental nailing of the edges is necessary. Panels are usually fastened with a combination of color pins and adhesive.

Apply narrow strips of black paint on the wall from floor to ceiling. Center them where joints between paneling will occur. If joints between sheets open slightly because of shrinkage during extended dry periods or heating seasons, it is not so noticeable with a black surface behind the joint.

Sometimes paneling does not extend to the ceiling but covers only the lower portion of the wall. This partial paneling is called *wainscoting*. It is usually installed about 3 feet above the floor (Fig. 73-3). If the wall is to be wainscoted, snap a horizontal line across the wall to indicate its height.

Stand panels on their long edge against the wall for at least forty-eight hours before installation. This allows them to adjust to room temperature and humidity (Fig. 73-4). Otherwise sheets may buckle

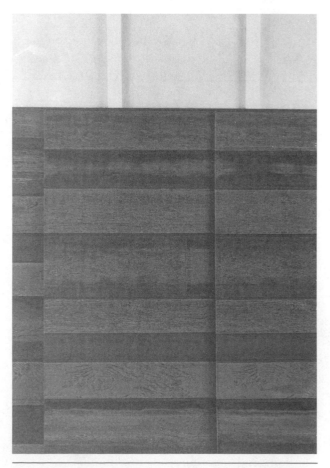

Fig. 73-4 Allow panels to stand in a room for forty-eight hours to adjust to temperature and humidity.

after installation, especially if they were very dry when installed.

Just prior to application, stand panels on end, side by side, around the room. Arrange them by matching the grain and color to obtain the most pleasing appearance.

Starting the Application

Starting in the corner, cut the first sheet to length about 1/4 inch less than the wall height. Place the sheet in the corner. Plumb the outside edge, and tack it in the plumb position. Set the distance between the points of the dividers the same as the amount the sheet overlaps the center of the stud (Fig. 73-5). Scribe this amount on the edge of the sheet butting the corner (Fig. 73-6).

Remove the sheet from the wall. Place it on saw horses on which a sheet of plywood, or two 8-foot lengths of 2×4s, have been placed for support. Cut to the scribed lines with a sharp, fine-toothed, hand crosscut saw. A handsaw is recommended for cutting sheets of prefinished paneling. Power saws usually produce an undesirable splintered edge along the layout line. In addition, the irregularity of some scribed edges makes them difficult to cut with a portable circular saw.

Fig. 73-5 The first sheet must be set plumb in the corner.

Fig. 73-6 Set dividers and scribe the edge of the first sheet to the corner.

Replace the sheet with the cut edge fitting snugly in the corner. The joint at the ceiling need not be fitted if a molding is to be used. If a tight fit between the panel and ceiling is desired, set the dividers and scribe a small amount at the ceiling line. Remove the sheet again. Cut to the scribed line. Replace the sheet, and raise it snugly against the ceiling. The space at the bottom will be covered later by a baseboard.

Fastening

If only nails are used, fasten about 6 inches apart along edges and about 12 inches apart on intermediate studs for 1/4-inch thick paneling. Nails may be spaced farther apart on thicker paneling. Drive nails at a slight angle for better holding power (Fig. 73-7).

If adhesives are used, apply a 1/8-inch continuous bead where panel edges and ends make contact. Apply beads 3 inches long and about 6 inches apart on all intermediate studs (Fig. 73-8). Put the panel in place. Tack it at the top. Be sure the panel is properly placed in position. Press on the panel surface to make contact with the adhesive. Use firm, uniform pressure to spread the adhesive beads evenly between the wall and the panel. Then, grasp the panel and slowly pull the bottom of the sheet a few inches away from the wall (Fig. 73-9). Press the sheet back into position after about two minutes. After about twenty minutes recheck the panel. Apply pressure to assure thorough adhesion and to smooth the panel surface. Apply successive sheets in the same manner. Do not force panels in position. Panels should touch very lightly at joints.

Fig. 73-8 Applying adhesive to a wall for the installation of sheet paneling.

Fig. 73-7 Drive nails at a slight angle for better holding power. Color pins are used on prefinished sheet paneling.

Fig. 73-9 The sheet is pulled a short distance away from the wall to allow the adhesive to dry slightly.

Wall Outlets

To lay out for wall outlets, plumb and mark both sides of the outlet to the floor. If the opening is close to the ceiling, plumb upward and mark lightly on the ceiling. Level the top and bottom of the outlet on the wall beyond the edge of the sheet to be installed. Or, level on the adjacent sheet, if closer. Cut, fit, and tack the sheet in position. Level and plumb marks from the wall and floor onto the sheet for the location of the opening (Fig. 73-10).

Another method is to rub a cake of carpenter's chalk on the edges of the outlet box. Fit and tack the sheet in position. Tap on the sheet directly over the outlet to transfer the chalked edges to the back of the sheet. Remove the sheet. Cut the opening for the outlet. Openings for wall outlets, such as electrical boxes, must be cut fairly close to the location and

size. The cover plate may not cover if the cutout is not made accurately. This could require replacement of the sheet. A saber saw may be used to cut these openings. When using the saber saw, cut from the back of the panel to avoid splintering the face (Fig. 73-11).

Ending the Application

The final sheet in the wall need not fit snugly in the corner if the adjacent wall is to be paneled or if interior corner molding is to be used. Take measurements at the top, center, and bottom. Cut the sheet to width, and install.

If the last sheet butts against a finished wall and no corner molding is used, the sheet must be cut to fit snugly in the corner. To mark the sheet accurately, first measure the remaining space at the

STEP BY STEP
P R O C E D U R E S

Fig. 73-10 Method of laying out for wall outlets.

Fig. 73-11 A saber saw being used to cut an opening for an electrical outlet box.

top, bottom, and about center. Rip the panel about 1/2 inch wider than the greatest distance. Place the sheet with the cut edge in the corner and the other edge overlapping the last sheet installed. Tack the sheet in position. The amount of overlap should be exactly the same from top to bottom. Set the dividers for the amount of overlap. Scribe this amount on the edge in the corner (Fig. 73-12). Instead of dividers, it is sometimes more exact to use a small block of wood for scribing. The width is cut the same as the amount of overlap. Care must be taken to keep from turning the dividers while scribing along a surface (Fig. 73-13).

Cut to the scribed line. If the line is followed carefully, the sheet should fit snugly between the

STEP BY STEP
P R O C E D U R E S

BACKER BOARD

1. CUT SHEET ABOUT 1/2" WIDER THAN WIDEST DISTANCE.

EDGE OF NEXT TO LAST SHEET

PLAN

OVERLAP MUST BE EQUAL TOP AND BOTTOM

EDGE OF LAST SHEET INSTALLED

IRREGULAR CORNER

LAST SHEET TO BE INSTALLED

LAST SHEET TO BE INSTALLED

SET DIVIDERS FOR AMOUNT OF OVERLAP AND SCRIBE

2. LAP LAST SHEET OVER NEXT TO LAST SHEET. SCRIBE THIS AMOUNT ON EDGE OF LAST SHEET IN CORNER.

3. CUT LAST SHEET TO SCRIBED LINE AND FASTEN IN POSITION

Fig. 73-12 Laying out the last sheet to fit against a finished corner without a molding.

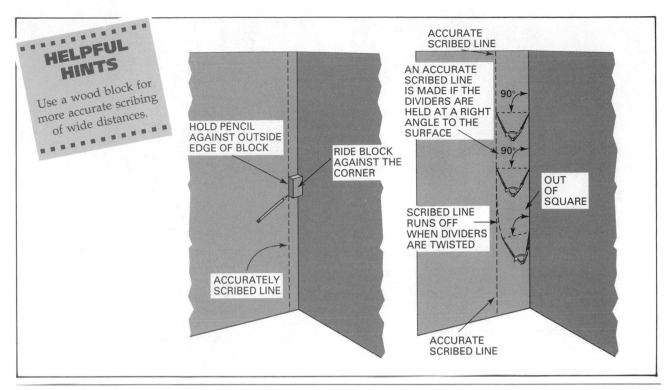

HELPFUL HINTS

Use a wood block for more accurate scribing of wide distances.

HOLD PENCIL AGAINST OUTSIDE EDGE OF BLOCK

RIDE BLOCK AGAINST THE CORNER

ACCURATELY SCRIBED LINE

ACCURATE SCRIBED LINE

AN ACCURATE SCRIBED LINE IS MADE IF THE DIVIDERS ARE HELD AT A RIGHT ANGLE TO THE SURFACE

90°

90°

OUT OF SQUARE

SCRIBED LINE RUNS OFF WHEN DIVIDERS ARE TWISTED

ACCURATE SCRIBED LINE

Fig. 73-13

last sheet installed and the corner, regardless of any irregularities. On exterior corners, a quarter-round molding is sometimes installed against the edges of the sheets. Or, the joint may be covered with a wood, metal, or vinyl corner molding (Fig. 73-14).

Installing Solid Wood Board Paneling

Horizontal board paneling may be fastened to studs in new and existing walls (Fig. 73-15). For vertical application of board paneling in a frame wall, blocking must be provided between studs (Fig. 73-16). On existing and masonry walls, horizontal furring strips must be installed. Blocking or furring must be provided in appropriate locations for diagonal or pattern applications of board paneling.

Allow the boards to adjust to room temperature and humidity by standing them against the walls around the room. At the same time, put them in the order of application. Match them for grain and color (Fig. 73-17). If tongue-and-grooved boards are to be eventually stained or painted, apply the same finish to the tongues so that an unfinished surface is not exposed if the boards shrink after installation.

Starting the Application

Select a straight board to start with. And cut it to length, about 1/4 inch less than the height of the wall. If tongue-and-grooved stock is used, tack it in a plumb position with the grooved edge in the corner. If a tight fit is desired, adjust the dividers to

WOOD OUTSIDE CORNER

GYPSUM BACKER BOARD

SHEET PANELING

METAL OUTSIDE CORNER

WOOD QUARTER ROUND

Fig. 73-14 Exterior corners of sheet paneling may be finished in several ways.

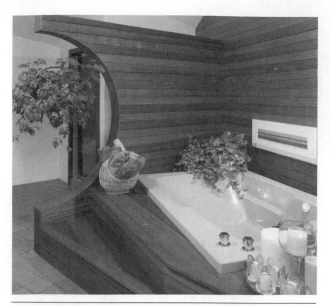

Fig. 73-15 No wall blocking is required for horizontal application of board paneling.

scribe an amount a little more than the depth of the groove. Rip to the scribed line. Face nail along the cut edge into the corner with finish nails about 16 inches apart. Blind nail the other edge through the tongue.

Continuing the Application

Apply succeeding boards by blind nailing into the tongue only (Fig. 73-18). Make sure the joints between boards come up tightly. See Figure 64-15 for methods of bringing edge joints of matched boards

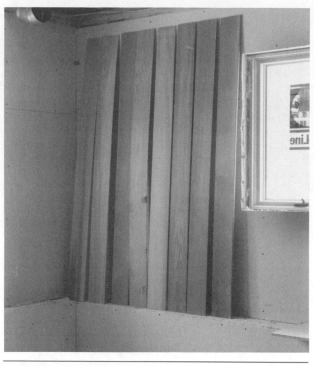

Fig. 73-17 Place boards around the room to adjust to temperature and humidity and to match color and grain. (*Courtesy of California Redwood Association*)

up tightly if they are slightly crooked. Severely warped boards should not be used. As installation progresses, check the paneling for plumb. If out of plumb, gradually bring back by driving one end of several boards a little tighter than the other end. Cut

EXTERIOR WALL

VAPOR BARRIER

CEILING MOLDING (OPTIONAL)

24" O.C. MAXIMUM
FOR 1/2" PANELING

48" O.C. MAXIMUM
FOR 3/4" PANELING

WOOD
PANELING

BLOCKING
BETWEEN
STUDS

SHEATHING

BASE

BLIND NAIL

INSULATION

Fig. 73-16 Blocking must be provided between studs for vertical board paneling.

Fig. 73-18 Tongue-and-grooved paneling is blind nailed.

out openings in the same manner as described for sheet paneling.

Applying the Last Board

If the last board in the installation must fit snugly in the corner without a molding, the layout should be planned so that the last board will be as wide as possible. If boards are a uniform width, the width of the starting board must be planned to avoid ending with a narrow strip. If random widths are used, they can be arranged when nearing the end.

Cut and fit the next to the last board. Then remove it. Tack the last board in the place of the next to the last board. Cut a scrap block about 6 inches long and equal in width to the finished face of the next to the last board. Use this block to scribe the last board by running one edge along the corner and holding a pencil against the other edge (Fig. 73-19). Remove the board from the wall. Cut it to the scribed line. Fasten the next to the last board in position. Fasten the last board in position with the cut edge in the corner. Face nail the edge nearest the corner.

Horizontal application of wood paneling is done in a similar manner. However, blocking between studs on open walls or furring strips on existing walls are not necessary. On existing walls, locate and snap lines to indicate the position of stud centerlines. The thickness of wood paneling should be at least 3/8 inch for 16-inch spacing of frame members and 5/8 inch for 24-inch spacing. Diagonal and pattern application of board paneling is similar to vertical and horizontal applications. If wainscoting

THE LAST BOARD TO BE INSTALLED IS PLACED IN THE POSITION OF THE NEXT TO THE LAST BOARD

MARK INSIDE EDGE OF BLOCK

SCRIBING BLOCK OF SAME WIDTH AS NEXT TO THE LAST BOARD

IRREGULAR CORNER

RANDOM WIDTH VERTICAL BOARD WALL PANELING

Fig. 73-19 Laying out the last board to fit against a finished corner.

THICK WAINSCOT THIN WAINSCOT FLUSH WAINSCOT

Fig. 73-20 Methods of finishing the joint at the top of wainscoting.

is applied to a wall, the joint between the different materials may treated in a several ways (Fig. 73-20).

Applying Plastic Laminates

Plastic laminates are not usually applied to walls unless first prefabricated to sheets of plywood or similar material. They are then installed on walls in the same manner as sheet paneling. Special matching molding is used between sheets and on interior and exterior corners. (The application of plastic laminates is described in greater detail in a later unit on kitchen cabinets and countertops.)

Estimating Paneling

Sheet Paneling

To determine the number of sheets of paneling needed, measure the perimeter of the room. Divide the perimeter by the width of the panels to be used. Deduct from this number any large openings such as doors, windows, or fireplaces. Deduct 2/3 of a panel for a door and 1/2 for a window or fireplace. Round off any remainder to the next highest number.

Estimating Board Paneling

Determine the square foot area to be covered by multiplying the perimeter by the height of each room. Deduct the area of any large openings. An additional percentage of the total area to be covered is needed because of the difference in the nominal size of lumber and its actual size.

Multiply the area to be covered by the area factor shown in Figure 73-21. Add 5 percent for waste in cutting. For example, the total area to be covered is 850 square feet, and 1×8 T&G board paneling is to be used. Multiply 850 by 1.21, the sum of the coverage factor of 1.16 found in the table and .05 for waste in cutting. Round the answer of 1028.50 to 1029 for the number of board feet of paneling needed. To reduce waste in cutting, order suitable lengths.

Nominal Size	WIDTH		Area Factor*
	Dress	Face	
SHIPLAP			
1x6	5 1/2	5 1/8	1.17
1x8	7 1/4	6 7/8	1.16
1x10	9 1/4	8 7/8	1.13
1x12	11 1/4	10 7/8	1.10
TONGUE AND GROOVE			
1x4	3 3/8	3 1/8	1.28
1x6	5 3/8	5 1/8	1.17
1x8	7 1/8	6 7/8	1.16
1x10	9 1/8	8 7/8	1.13
1x12	11 1/8	10 7/8	1.10
S4S			
1x4	3 1/2	3 1/2	1.14
1x6	5 1/2	5 1/2	1.09
1x8	7 1/4	7 1/4	1.10
1x10	9 1/4	9 1/4	1.08
1x12	11 1/4	11 1/4	1.07
PANELING AND SIDING PATTERNS			
1x6	5 7/16	5 1/16	1.19
1x8	7 1/8	6 3/4	1.19
1x10	9 1/8	8 3/4	1.14
1x12	11 1/8	10 3/4	1.12

Fig. 73-21 Factors used to estimate amounts of board paneling.

CHAPTER 74 CERAMIC WALL TILE

Ceramic tile is used to cover floors and walls in rest rooms, baths, showers, and other high-moisture areas that need to be cleaned easily and frequently. On large jobs, the tile is usually applied by specialists. On small jobs, especially in rural or suburban communities, it is sometimes more expe-dient for the general carpenter to install tile. Ceramic tile was once set in *mortar*. Today *adhesive application* to a water-resistant backer board is the method commonly used (Fig. 74-1). This chapter describes the adhesive method of tile application to walls.

Fig. 74-1 Adhesive application of ceramic tile to shower walls. (*Courtesy of U.S. Gypsum Company*)

Description of Ceramic Wall Tile

Ceramic tiles are usually rectangular or square, but many geometric shapes, such as hexagons and octagons, are manufactured. Many solid colors, patterns, designs, and sizes give a wide choice to achieve the desired wall effect.

The most commonly used wall tile are nominal 4- and 6-inch squares, in about about 1/4-inch thickness. Special pieces such as *base, caps, inside corners,* and *outside corners* are used to trim the installation (Fig. 74-2).

Wall Preparation for Tile Application

Water-resistant gypsum board or cement board should be applied as a backing for ceramic wall tile (Fig. 74-3). (Application instructions for these products are given in Chapter 69.)

Minimum Backer and Tile Area in Baths and Showers

Tiles should overlap the lip and be applied down to the shower floor or top edge of the bathtub. On tubs

Fig. 74-2 Special pieces are used to trim a ceramic tile installation.

Fig. 74-3 Applying cement board on bathroom walls. (*Courtesy of U.S. Gypsum Company*)

without a showerhead, they should be installed to extend to a minimum of 6 inches above the rim. Around bathtubs with showerheads, tiles should extend a minimum of 5 feet above the rim or 6 inches above the showerhead, whichever is higher.

In shower stalls, tiles should be a minimum of 6 feet above the shower floor or 6 inches above the showerhead, whichever is higher. A 4-inch minimum extension of the full height is recommended beyond the outside face of the tub or shower (Fig. 74-4).

Calculating Border Tiles

Before beginning the application of ceramic wall tile, the width of the *border tiles* must be calculated. Border tiles are those that fit into the corner of the wall. The installation has a professional appearance if the border tiles are close to the same width and also as wide as possible on both sides of a wall (Fig. 74-5).

Fig. 74-5 Border tiles should be as wide as possible and close to the same width on both sides of the wall.

Fig. 74-4 Minimum areas recommended for the installation of ceramic tile around bathtubs and shower stalls.

Measure the width of the wall from corner to corner. Change the wall measurement to inches. Divide it by the width of a tile. Measure the tile accurately. Sometimes, the actual size of a tile is different than its nominal size. Add the width of a tile to the remainder. Divide by 2 to find the width of border tiles.

For example: a wall measures 8'-6" or 102 inches from corner to corner. If 4-inch, actual width, tiles are used, 102" divided by 4" equals 25 with a remainder of 2". Add 4 to the remainder of 2, which equals 6. Divide 6 by 2 to get 3, the width of border tiles in inches.

Tile Layout

Measure out from the corner near the center of the wall. Mark a point that will be the edge of a full tile. Lay out a plumb line on the wall through this point from the floor to ceiling.

If the floor is level, tiles may be applied by placing the bottom edge of the first row on the floor using plain tile or base tile. If the floor is not level, then base tile cannot be used. The bottom edge of the tiles must be cut to fit the floor while keeping the top edges level.

In this case, place a level on the floor and find the low point of the tile installation. From this point, measure up and mark on the wall the height of a tile. Draw a level line on the wall through the mark. Tack a straightedge on the wall with its top edge to the line. Tiles are then laid to the straightedge. When tiling is completed above, the straightedge is removed. Tiles are then cut and fitted to the floor.

Tile Application

Apply the recommended adhesive to the wall with a trowel. Use a flat trowel with grooved edges that allow the recommended amount of adhesive to remain on the surface. Too heavy a coat results in adhesive being squeezed out between the joints of applied tile, causing a mess. Too little adhesive results in failure of tiles to adhere to the wall. Follow the manufacturer's directions for the type of trowel to use and the amount of adhesive to be spread at any one time. Be careful not to spread adhesive beyond the area to be covered (Fig. 74-6).

Apply the first tile with its bottom edge to the straightedge or floor, as the case may be, and its vertical edge on the plumb line. Press the tile firmly into the adhesive. Apply other tiles in the same manner. Start from the center plumb line and pyramid upward and outward. As tiles are applied, slight adjustments may need to be made to keep them lined up. Keep fingers clean and adhesive off the face of the tiles. Clean tiles with a cloth dampened with the solvent recommended by the manu-

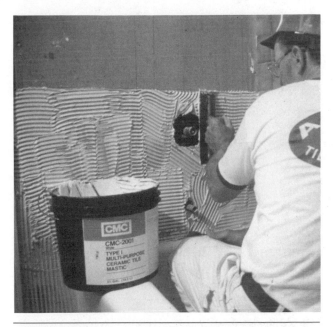

Fig. 74-6 Use a trowel that is properly grooved to allow the correct amount of adhesive to remain on the wall. (*Courtesy of U.S. Gypsum Company*)

facturer to remove any adhesive from the faces of the tile as installation progresses.

Cutting Border Tiles

After all *field* tiles are applied, it is necessary to cut and apply border tiles. Field tiles are whole tiles that are applied in the center of the wall. Specialists on large jobs use a power tool for cutting ceramic tile (Fig. 74-7). On small jobs, the general carpenter may use a hand-operated ceramic tile cutter (Fig. 74-8). The hand cutter scores and snaps the tile in a way similar to that used in cutting glass. However, this tool cannot cut a small amount from the edge of

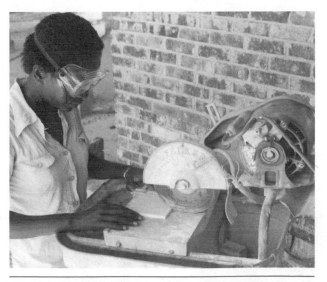

Fig. 74-7 Power tool for cutting ceramic tile.

Fig. 74-8 Hand cutter for ceramic tile.

Fig. 74-9 Nibblers are used to cut a small amount from the tile.

a tile. For this purpose, a *nibbler*, shown in Figure 74-9, is used.

To finish the edges and ends of the installation, *caps* are used. Caps may be 2×6 or 4×4 pieces with one rounded finished edge. Special trim pieces are used to finish interior and exterior corners.

Grouting Tile Joints

After all tile has been applied, the joints are filled with tile *grout*. Grout comes in a powder form. It is mixed with water to form a paste of the desired consistency. It is spread over the face of the tile with a *rubber trowel* to fill the joints. The grout is worked into the joints. Then, the surface is wiped as clean as possible with the trowel.

The grout is allowed to set up, but not harden. The joints are then *pointed*. Removing excess grout and smoothing joints is called pointing. A small hardwood stick with a rounded end can be used as a pointing tool. The entire surface is wiped clean with a dry cloth after the grout has dried.

Estimating Tile

Determine the square foot area to be covered for the amount of tile to order. Divide the area to be covered by the area covered by one tile to find the number of tiles needed. For instance, if the area to be covered is 120 square feet and 4×4 tiles are being used, divide 120 by 9/16 to find that 1,080 tiles are needed.

Note: 9/16 is that part of a square foot that one 4×4 tile will cover. Since there are 144 square inches in one square foot, and there are 16 square inches in one 4×4 tile, the tile covers 16/144ths, or when reduced to its lowest term, 9/16 of a square foot.

The number of straight pieces of cap is found by determining the total linear feet to be covered and dividing by the length of the cap. The number of interior and exterior corners is determined by counting from a layout of the installation.

Review Questions

Select the most appropriate answer.

1. On prefinished plywood paneling that is scored to simulate boards, some scores are always placed in from the edge
   a. 12 and 16 inches.   c. 12 and 24 inches.
   b. 16 and 20 inches.   d. 16 and 32 inches.

2. Tileboard is a type of
   a. tile.   c. fiberboard.
   b. hardboard.   d. plywood.

3. A wainscoting is a wall finish
   a. applied diagonally.
   b. applied partway up the wall.
   c. used as a coating on prefinished wall panels.
   d. used around tubs and showers.

4. For most parts of the country, wood used for interior finish should be dried to a moisture content of about
   a. 8 percent.
   b. 12 percent.
   c. 15 percent.
   d. 20 percent.

5. The thickest plastic laminate described in this unit is called
   a. vertical type.
   b. regular type.
   c. backer type.
   d. all-purpose type.

6. The tool recommended to cut thin plywood paneling is the
   a. portable electric circular saw.
   b. fine-tooth hand crosscut saw.
   c. table saw.
   d. saber saw.

7. If 3/4 inch thick wood paneling is to be applied vertically over open studs, wood blocking must be provided between studs for nailing at intervals of
   a. 16 inches.
   b. 24 inches.
   c. 32 inches.
   d. 48 inches.

8. When applying board paneling vertically to an existing wall
   a. nail paneling to existing studs.
   b. apply horizontal furring strips.
   c. remove the wall covering and install blocking between studs.
   d. use adhesives.

9. The number of 4×4 ceramic tiles needed to cover 150 square feet of wall area is:
   a. 1,152.
   b. 1,350.
   c. 1,500.
   d. 1,674.

10. Bathroom tile should extend over the tops of showerheads a minimum of
   a. 4 inches.
   b. 6 inches.
   c. 8 inches.
   d. 12 inches.

■ BUILDING FOR SUCCESS ■

Planning Is Critical

The old cliché "If you fail to plan, you plan to fail" is important in the construction world. Many builders, remodelers, and other skilled trade representatives will attest to the truth of this statement. The planning process must be at the forefront of all successful construction projects.

The construction industry has structured some ideal planning procedures through critical path processes and other scheduling plans. Why has so much emphasis been placed on the planning process? The following story is humorous, but conveys a lot of truth.

Poor Planning

(The following letter was written to clarify a health insurance claim made for injuries sustained on the job by a bricklayer. Author unknown.)

In response to your request for additional information in block number 3 of this accident report, I put "poor planning" as the cause of my accident. In your letter, you requested that I explain myself more fully. I trust the following details will be sufficient.

I am a bricklayer by trade. On the day of the accident, I was working alone on the roof of a new six story building. When I completed my work, I discovered that I had about 500 pounds of brick left over. Rather than carry the bricks down by hand, I decided to lower them in a barrel using a pulley, which fortunately was attached to the side of the building, at the sixth floor.

Securing the rope at ground level, I went up to the roof, swung the barrel out, and loaded the brick into it. Then I went back to the ground and untied the rope, holding it tightly to ensure a slow descent of the 500 pounds of bricks. You will note in block number 11 of the accident report that I weigh 135 pounds.

Due to my surprise at being jerked off the ground so suddenly, I lost my presence of mind and forgot to let go of the rope. Needless to say, I proceeded at a rather rapid rate up the side of the building.

In the vicinity of the third floor, I met

the barrel coming down. That explains the fractured skull and collarbone.

Slowed only slightly, I continued my rapid ascent, not stopping until the fingers of my right hand were two knuckles deep into the pulley.

Fortunately, by this time I had regained my presence of mind and was able to hold tightly to the rope despite my pain.

At approximately the same time, however, the barrel of bricks hit the ground and the bottom fell out of the barrel. Devoid of the weight of the bricks, the barrel now weighed approximately 50 pounds.

Again, I refer you to my weight in block number 11. As you might imagine I began a rapid descent down the side of the building.

In the vicinity of the third floor, I met the barrel coming up. This accounts for the two fractured ankles and the lacerations of my legs and lower body.

The encounter with the barrel slowed me enough to lessen my injuries when I fell into the pile of bricks. Fortunately, only three vertebrae were cracked.

I'm sorry to report, however, that as I lay there on the bricks—in pain, unable to stand, and watching the empty barrel six stories above me—again I lost my presence of mind. I let go of the rope!

As far fetched as this account may seem, there is something to be learned in planning a project. Arriving at the job site in the morning with no real advance thought about the project usually proves disastrous.

The purpose of planning is manifold. In order to begin and end a project on time, a plan must be formulated and carried out. Economics alone underscores the need to plan well. There are few builders, if any, who can afford to carry out their daily work with no plan to guide them.

In order to structure a realistic work flow, materials and job schedules have to become part of the overall plan. Experience becomes the best gauge when developing a project completion plan. For the beginning carpenter or other skilled craftsperson, planning skills can be learned daily on the job. Good record keeping is a good place to begin. Time cards can be used to accumulate information that will assist each person in defining how much work of each type can be accomplished in a day, week, or month. With this valuable information documented, a person can estimate the amount of time that will be necessary to complete the job under normal conditions. From this, cost estimates can be made.

Another criterion for good planning is to have an adequate understanding of the task at hand. The more that is understood about the building process, the more accurate the plan to build. Education can be interwoven with work experience to provide a big picture of the entire construction process involving all of the trades. With a broad knowledge of the various trade responsibilities and when all construction tasks are to occur, the plan can serve everyone well. Undoubtedly there will be interruptions, illnesses, weather delays, equipment breakdowns, and a myriad of other problems, but a good plan will include problem solving. "What if" situations should be discussed even if they are only remotely possible. Playing the odds and not looking for solutions to the "what if" problems can be a costly gamble.

Good communication among workers, crews, trades, and suppliers is essential for success. A major part of good planning is to openly communicate before, during, and after the construction process. With the complexity of the building industry today, it is essential to include all pertinent people in the planning process. General contractors should work consistently to improve communications within and outside their companies. Each person within the company should be a part of the planning process. Work incentives will inspire each employee to contribute valuable information to aid the planning process.

As the plans for a construction project are being formulated, the project coordinator needs to initiate the planning process well in advance of the actual construction. This allows for time to alter or rearrange any potential problem areas. One of the greatest planning problems involves extremes in weather conditions. Excessively hot or cold weather may necessitate altering a "typical" planning schedule. Characteristically, these elements will affect job and material scheduling. The best way to schedule a job under any condition is to recall all similar past experiences, and then use them as you plan the next project. Good records will help in this process.

Project planning and scheduling will be successful when the process is given full attention. As the bricklayer surely noticed when the brick removal process was completed, more planning could have saved time, money, and heartache.

Remember to plan your work and work your plan. It is critical.

FOCUS QUESTIONS: For individual or group discussion.

1. Review any planning process that you have been exposed to in the classroom or on the job. List the main consid-erations in outline form. Create a detailed plan, from the earliest phases of a construction project through completion.

2. Explain how good planning can affect the profit/loss margin of a construction project. Include how planning is reflected in overhead, scheduling, time management, and cost concerns for time overruns.

3. Select a specific part of a construction project, such as framing. Create a list of the various considerations necessary for good planning by the crew leader.

Ceiling Finish

OBJECTIVES

After completing this unit, the student will be able to:

- describe the sizes, kinds, and shapes of ceiling tile and lay in suspended ceiling panels.

- lay out and install suspended ceilings.

- lay out and install ceiling tile.

- estimate quantities of ceiling finish materials.

CONTENTS

Inexpensive and highly attractive ceilings may be created by installing suspended ceilings or ceiling tiles. They may be installed in new construction beneath exposed joists or in remodeling below existing ceilings.

Suspended ceilings are widely used in commercial and residential construction as a ceiling finish. They also provide space for recessed lighting, ducts, pipes, and other necessary conduits (Fig. 75-1). Besides improving the appearance of a room, a suspended ceiling conserves energy by increasing the insulating value of the ceiling. It also aids in controlling sound transmission. In remodeling work, a suspended system can be easily installed beneath the existing ceiling. In basements, where overhead pipes and ducts may make other types of ceiling application difficult, a suspended type is easily installed. In addition, removable panels make pipes, ducts, and wiring accessible.

Suspended Ceiling Components

A *suspended ceiling* system consists of *panels* that are laid into a metal *grid*. The grid consists of *main runners, cross tees,* and *wall angles.* It is constructed in a 2×4 rectangular or 2×2 square pattern (Fig. 75-2). Grid members come prefinished in white, black,

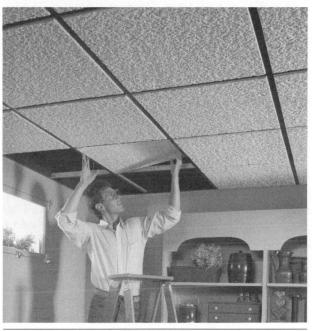

Fig. 75-1 Installing panels in a suspended ceiling grid. (*Courtesy of Armstrong World Industries*)

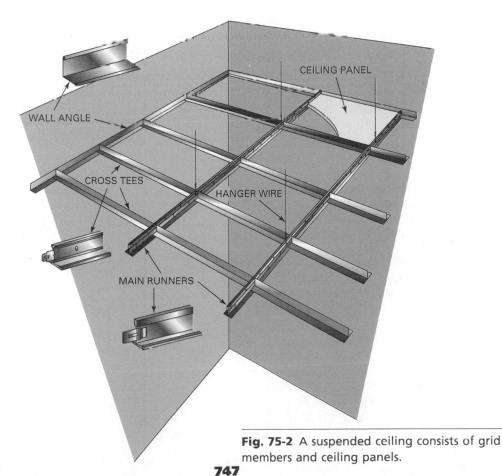

Fig. 75-2 A suspended ceiling consists of grid members and ceiling panels.

brass, chrome, and wood grain patterns, among others.

Wall Angles

Wall angles are L-shaped pieces that are fastened to the wall to support the ends of main runners and cross tees. They come in 10- and 12-foot lengths. They provide a continuous finished edge around the perimeter of the ceiling, where it meets the wall.

Main Runners

Main runners or *tees* are shaped in the form of an upside-down T. They come in 12-foot lengths. End splices make it possible to join lengths of main runners together. Slots are punched in the side of the runner at 12-inch intervals to receive cross tees. Along the top edge, punched holes are spaced at intervals for suspending main runners with *hanger wire*. Main runners extend from wall to wall. They are the primary support of a ceiling's weight.

Cross Tees

Cross tees come in 2- and 4-foot lengths. A slot, of similar shape and size as those in main runners, is centered on 4-foot cross tees for use when making a 2×2 grid pattern. They come with connecting tabs on each end. These tabs are inserted and locked into main runners and other cross tees.

Ceiling Panels

Ceiling panels are manufactured of many different kinds of material, such as gypsum, glass fibers, mineral fibers, and wood fibers. Panel selection is based on considerations such as fire resistance, sound control, thermal insulation, light reflectance, moisture resistance, maintenance, appearance, and cost. Panels are given a variety of surface textures, designs, and finishes. They are available in 2×2 and 2×4 sizes with square or rabbeted edges (Fig. 75-3).

Suspended Ceiling Layout

Before the actual installation of a suspended ceiling, a scaled sketch of the ceiling grid should be made. The sketch should indicate the direction and location of the main runners, cross tees, light panels, and border panels.

Main runners are spaced 4 feet apart. They usually run parallel with the long dimension of the room. For a standard 2×4 pattern, 4-foot cross tees are then spaced 2 feet apart between main runners. If a 2×2 pattern is used, 2-foot cross tees are installed between the midpoints of the 4-foot cross tees. Main runners and cross tees should be located in such a way that *border panels* on both sides of the room are equal and as large as possible (Fig. 75-4).

SQUARE EDGE

RABBETED EDGE

Fig. 75-3 Suspended ceiling panels may have square or rabbeted edges and ends.

Sketching the ceiling layout also helps when estimating materials.

Sketching the Layout

Sketch a grid plan by first drawing the overall size of the ceiling to a convenient scale. Use special care in measuring around irregular walls.

Locating Main Runners. To locate main runners, change the width of room to inches and divide by 48. Add 48 inches to any remainder. Divide the sum by 2 to find the distance from the wall to the first main runner. This distance is also the length of border panels.

For example, if the width of the room is 15′-8″, changing to inches equals 188. Dividing 188 by 48 equals 3, with a remainder of 44. Adding 48 to 44 equals 92. Dividing 92 by 2 equals 46 inches, the distance from the wall to the first main runner.

Draw a main runner the calculated distance from, and parallel to, the long dimension of the ceiling. Draw the rest of the main runners parallel to the first, and at 4-foot intervals. The distance between the last main runner and the wall should be the same as the distance between the first main runner and the opposite wall (Fig. 75-5).

Locating Cross Tees. To locate 4-foot cross tees between main runners, first change the long dimension of the ceiling to inches. Divide by 24. Add 24 to the remainder. Divide the sum by 2 to find the width of the border panels on the other walls.

For example, if the long dimension of the room is 27′-10″, changing it to inches equals 334.

MAIN RUNNERS
4-FOOT CROSS TEES
2-FOOT CROSS TEES

Fig. 75-4 A grid system with 4-foot cross tees spaced 2 feet apart along main runners is the recommended method of constructing a 2×4 foot grid. A 2×2 foot grid is made by installing 2-foot cross tees between the midpoints of the 4-foot cross tees.

Dividing 334 by 24 equals 13, with a remainder of 22. Adding 24 to 22 equals 46. Dividing 46 by 2 equals 23, the distance from the wall to the first row of cross tees.

Draw the first row of cross tees the calculated distance from, and parallel to the short wall. Draw the remaining rows of cross tees parallel to each other at 2-foot intervals. The distance from the last row of cross tees to the wall should be the same as the distance from the first row of cross tees to the opposite wall (Fig. 75-6).

Constructing the Ceiling Grid

The ceiling grid is constructed by first installing *wall angles*, then installing *screw eyes* and *hanger wires,*

suspending the *main runners,* inserting full-length *cross tees,* and, finally, cutting and inserting *border* cross tees.

Installing Wall Angles

A suspended ceiling must be installed with at least 3 inches for clearance below the lowest air duct, pipe, or beam. This clearance provides enough room to insert ceiling panels in the grid. If recessed lighting is to be used, allow a minimum of 6 inches for clearance. The height of the ceiling may be located by measuring up from the floor. If the floor is rough or out of level, the ceiling line may be located with various leveling devices previously described. A combination of a hand level and straightedge, a

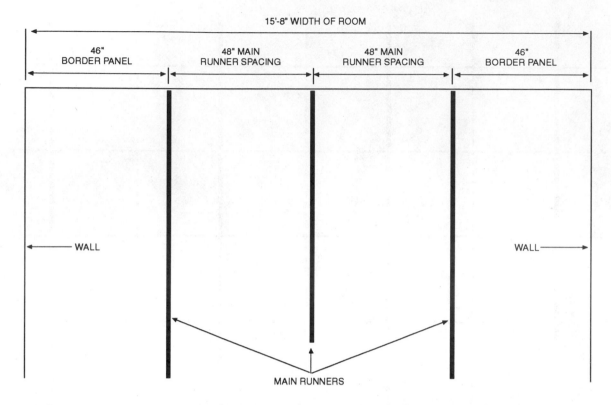

1. CHANGE ROOM WIDTH DIMENSION TO INCHES - 188
2. DIVIDE BY 48 = 3 WITH 44 REMAINDER
3. ADD 48 TO REMAINDER = 92
4. DIVIDE BY 2 = 46, LENGTH OF BORDER PANELS

Fig. 75-5 Method of determining the location of main runners.

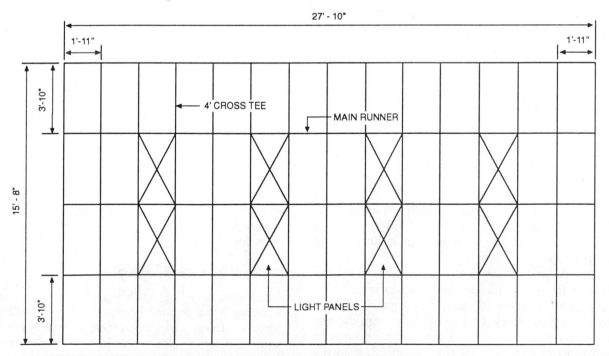

Fig. 75-6 Completed sketch of a suspended ceiling layout.

water level, builders' level, transit-level, or laser level can be used. Snap chalk lines on all walls around the room to the height of the top edge of the wall angle (Fig. 75-7).

Fasten wall angles around the room with their top edge lined up with the chalk line. It may be easier to fasten the wall angle by prepunching holes with a center punch or spike. Fasten into framing

Fig. 75-7 Snap chalk lines on the walls for the location of wall angles.

Fig. 75-8 Installing wall angle. (*Courtesy of Armstrong World Industries*)

wherever possible, not more than 24 inches apart (Fig. 75-8).

To fasten wall angles to concrete walls, short masonry nails are sometimes used. However, they are difficult to hold and drive. Use a small strip of cardboard to hold the nail while driving it with the

hammer (Fig. 75-9). Lead or plastic inserts and screws may also be used to fasten the wall angles. Their use does require more time. If available, power nailers can be used for efficient fastening of wall angles to masonry walls.

Make miter joints on outside corners. Make butt joints in interior corners and between straight lengths of wall angle (Fig. 75-10). Use tin snips or a

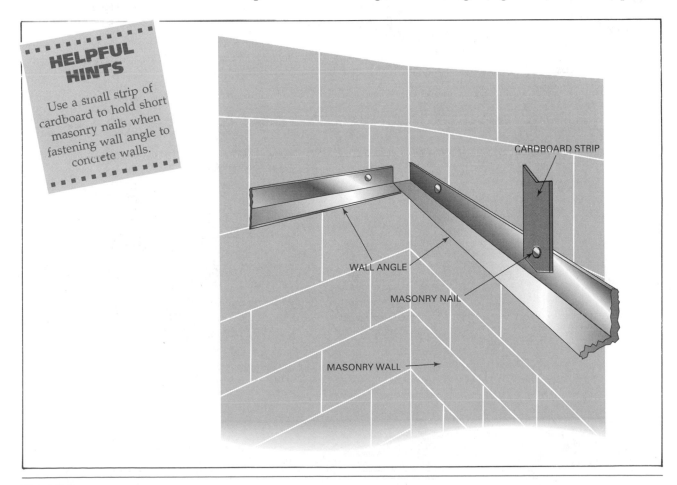

HELPFUL HINTS

Use a small strip of cardboard to hold short masonry nails when fastening wall angle to concrete walls.

CARDBOARD STRIP

WALL ANGLE

MASONRY NAIL

MASONRY WALL

Fig. 75-9

BUTT JOINT

INSIDE CORNER

MITER JOINT

OUTSIDE CORNER

Fig. 75-10 Methods of joining wall angle.

hacksaw with a fine-toothed blade. If a hacksaw is used, a temporary wooden miter box can be made to steady the wall angle and to guide the hacksaw. If tin snips are used, first lay out the cut with a square. Then follow the line carefully.

CAUTION Use care in handling the cut ends of wall angle and other grid members. Cut metal ends are very sharp and can cause serious injury.

Installing Screw Eyes

From the ceiling sketch determine the position of the first main runner. Stretch a line at this location across the room from the top edges of the wall angle. Stretch the line tightly on nails inserted between the wall and wall angle (Fig. 75-11). The line serves as a guide for installing *screw eyes* and *hanger wires* from which main runners are suspended.

HELPFUL HINTS

Stretch lines for main runners on nails inserted between the wall and wall angle.

NAIL

STRETCHED LINE

WALL ANGLE

Fig. 75-11

Install screw eyes, not over 4 feet apart, directly over the stretched line. Screw eyes should be of the type commonly used for suspended ceilings. They must be long enough to penetrate wood joists a minimum of one inch to provide strong support. *Eye pins* are driven into concrete with a *powder-actuated* fastening tool. Hanger wires may also be attached directly around the lower chord of bar joists or trusses.

Installing Hanger Wire

Cut a number of hanger wires using wire cutters. The wires should be about 12 inches longer than the distance between the overhead construction and the stretched line. For residential work, 16-gauge wire is usually used. For commercial work, 12-gauge and heavier wire is used.

Attach the hanger wires to the screw eyes. Insert about 6 inches of the wire through the screw eye. Securely wrap the wire around itself three times. Pull on each wire to remove any kinks. Then make a 90 degree bend where it crosses the stretched line (Fig. 75-12). Stretch lines, install screw eyes, attach and bend hanger wires in the same manner at each main runner location. Leave the last line stretched tightly in position.

Installing Main Runners

The ends of the main runners rest on the wall angle. They must be cut so that a cross tee slot in the web of the runner lines up with the first row of cross tees. A cross tee line must be stretched, at wall angle height, across the short dimension of the room to line up the slots in the main runners. The line must run exactly at right angles to the main runner line and at a distance from the wall equal to the width of

the border panels. If the walls are at right angles to each other, the location of the cross tee line can be determined by measuring out from both ends of the wall.

When walls are not at right angles, a 3, 4, 5, method of squaring lines is used. Otherwise, the entire grid system will be out of square. Measure out, from the short wall, along the stretched main runner line, a distance equal to the width of the border panel. Mark the line. Stretch the cross tee line through this mark and at right angles to the main runner line (Fig. 75-13). Use the 3, 4, 5 method of squaring layout lines that was described in Chapter 26.

At each main runner location, measure from the short wall to the stretched cross tee line. Transfer the measurement to the main runner. Measure from the first cross tee slot beyond the measurement, so as to cut as little as possible from the end of the main runner (Fig. 75-14). Cut the main runners about 1/8 inch less to allow for the thickness of the wall angle. Backcut the web slightly for easier installation at the wall. Measure and cut main runners individually. Do not use the first one as a pattern to cut the rest.

Hang the main runners by resting the cut end on the wall angle and inserting suspension wires in the appropriate holes in the top of the main runner. Bring the runners up to the bend in the wires. Twist the wires with at least three turns to hold the main runners securely (Fig. 75-15).

More than one length of main runner may be needed to reach the opposite wall. Connect lengths of main runners together by inserting tabs into matching ends. Make sure end joints come up tight. The length of the last section is measured from the end of the last one installed to the opposite wall, allowing about 1/8 inch less for the thickness of the wall angle.

JOIST

SCREWEYE

TWIST HANGER WIRE AROUND ITSELF THREE TIMES

ALLOW 6" & BEND AT 90 DEGREES AT MAIN RUNNER LINE

STRETCHED MAIN RUNNER LINE

WHEN CUTTING HANGER WIRE, ALLOW 6" AT EACH END FOR TWISTING

Fig. 75-12 Cut and fasten hanger wire to screw eyes. Bend the lower end at a 90 degree angle.

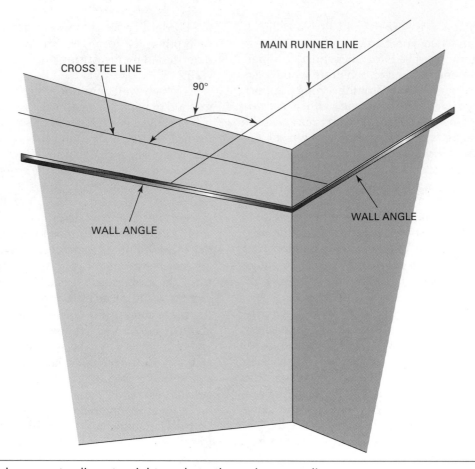

MAIN RUNNER LINE

CROSS TEE LINE

90°

WALL ANGLE

WALL ANGLE

Fig. 75-13 Stretch a cross tee line at a right angle to the main runner line.

STEP BY STEP
P R O C E D U R E S

1. MEASURE FROM FIRST CROSS TEE SLOT BEYOND BORDER PANEL WIDTH

2. MEASURE BACK WIDTH OF BORDER PANEL

3. CUT MAIN RUNNER HERE

END OF MAIN RUNNER

CROSS TEE SLOT

23"

12"

12"

12"

6"

CROSS TEE SLOTS ARE LOCATED 6" FROM THE ENDS OF MAIN RUNNERS AND SPACED 12" O.C. ALONG ITS LENGTH

Fig. 75-14 Method of locating the end cut on main runners.

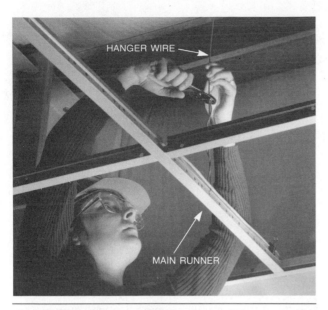

Fig. 75-15 Hanging the main runner. (*Courtesy of Armstrong World Industries*)

Installing Cross Tees

Cross tees are installed by inserting the tabs on the ends into the slots in the main runners. These fit into position easily, although the method of attaching varies from one manufacturer to another.

Install all full-length cross tees between main runners first. Lay in a few full-size ceiling panels. This stabilizes the grid while installing cross tees for border panels. Cut and install cross tees along the border. Insert the connecting tab of one end in the main runner and rest the cut end on the wall angle (Fig. 75-16). It may be necessary to measure and cut cross tees for border panels individually, if walls are not straight or square. For 2×2 panels, install 2-foot cross tees at the midpoints of the 4-foot cross tees. After the grid is complete, sight sections by eye. Straighten where necessary by making adjustments to border cross tees or hanger wires.

Installing Ceiling Panels

Ceiling panels are placed in position by tilting them slightly, lifting them above the grid, and letting them fall into place (Fig. 75-17). Be careful when handling panels to avoid marring the finished surface. Install all full-sized field panels first. Then cut and install border panels. Measure each border panel individually, if necessary. Cut them slightly smaller than measured so they can drop into place easily. Cut the panels with a sharp utility knife using a straightedge as a guide. A scrap piece of cross tee material can be used as a straightedge. Always cut with the finished side of the panel up.

Cutting Ceiling Panels Around Columns

When a column is near the center of a ceiling panel, cut the panel at the midpoint of the column. Cut semicircles from the cut edge to the size required for the panel pieces to fit snugly around the column. After the two pieces are rejoined around the column, glue scrap pieces of panel material to the back of the installed panel.

If the column is close to the edge or end of a panel, cut the panel from the nearest edge or end to fit around the column. The small piece is also fitted

Fig. 75-16 Inserting a cross tee in the main runner. (*Courtesy of Armstrong World Industries*)

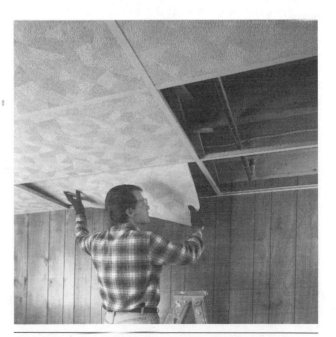

Fig. 75-17 Installing ceiling panels. (*Courtesy of Armstrong World Industries*)

around the column and joined to the panel by gluing scrap pieces to its back side (Fig. 75-18).

Estimating Suspended Ceiling Materials

■ Divide the perimeter of the room by 10, the length of a wall angle, to find the number needed.

■ Find the number of main runners needed from the sketch. No more than two pieces can be cut from one 12-foot length.

■ Count the number of 2- and 4-foot cross tees from the sketch. Border cross tees must be cut from full-length tees.

■ From the sketch, count the number of hanger wires and screw eyes needed. Multiply the number needed by the length of each hanger wire to find the total linear feet of hanger wire needed.

■ Count the number of ceiling panels from the sketch. Each border panel requires a full-size

Fig. 75-18 Fitting ceiling panels around columns. (*Courtesy of Armstrong World Industries*)

ceiling panel. There will be one less for each light fixture.

CHAPTER 76   CEILING TILE

*C*eiling tile is usually stapled to furring strips that are fastened to exposed joists. They may also be cemented to existing ceilings provided the ceilings are solid and level. If the existing ceiling is not sound, furring strips should be installed and fastened through the ceiling into the joists above (Fig. 76-1).

Fig. 76-1 Installing ceiling tile to wood furring strips. (*Courtesy of Armstrong World Industries*)

Description of Ceiling Tiles

Most ceiling tiles are made of *wood fiber* or *mineral fiber*. Wood fiber tiles are lowest in cost and are adequate for many installations. Mineral fiber tiles are used when a more fire-resistant type is required.

Manufacture

Wood fibers are pressed into large sheets that are 7/16 to 3/4 inch thick. Mineral fiber tiles are made of rock that is heated to a molten state. The fibers are then sprayed into a sheet form. The surfaces of some sheets are *fissured* or *perforated* for sound absorption. The surfaces of other sheets are embossed with different designs or left smooth. Then, they are given a factory finish and cut into individual tiles. Most tiles are cut with *chamfered, tongue-and-grooved* edges with two adjacent *stapling flanges* for concealed fastening (Fig. 76-2).

Sizes

The most popular sizes of ceiling tile are a 12-inch square and a 12×24 inch rectangle in thicknesses of 1/2 inch. Tiles are also manufactured in squares of 16 inches and in rectangles of 16×32 inches.

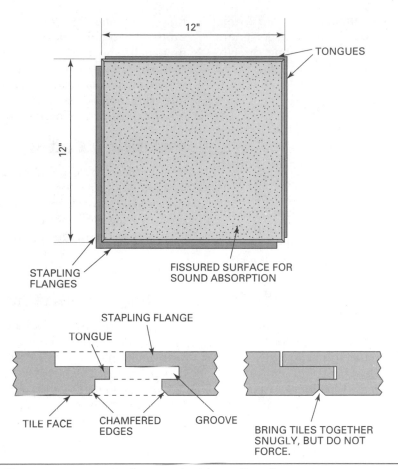

Fig. 76-2 A typical ceiling tile is tongue-and-grooved with stapling flanges.

Ceiling Tile Layout

Before installation begins, it is necessary to calculate the size of *border tiles* that run along the walls. It is desirable that border tiles be as wide as possible and of equal widths on opposite walls.

Calculating Border Tile Sizes

To find the width of the border tiles along the long walls of a room, first determine the dimension of the short wall. In most cases, the measurement will be a number of full feet, plus a few inches. Not counting the foot measurement, add 12 more inches to the remaining inch measurement. Divide the sum by 2 to find the width of border tiles for the long wall. The following example is shown when the distance between the long walls is 10'-6".

EXAMPLE

Room width	10'-6"
Add to inches	12"
Width of border tiles	18" ÷ 2 = 9"

The width of border tiles along the short walls is calculated in the same manner. The following ex-

ample is used when the distance between the short walls is 19'-8".

EXAMPLE

Room length	19'-8"
Add to inches	12"
Width of border tiles	20" ÷ 2 = 10"

Preparation for Ceiling Tile Application

Unless an adhesive application to an existing ceiling is to be made, *furring strips* must be provided on which to fasten ceiling tiles. Furring strips are usually fastened directly to exposed joists. They are sometimes applied to an existing ceiling and fastened into the concealed joists above.

Locating Concealed Joists

If the joists are hidden by an existing ceiling, tap on the ceiling with a hammer. Drive a nail into the spot where a dull thud is heard, to find a concealed joist. Locate other joists by the same method or by measuring from the first location. Usually, ceiling joists are spaced 16 inches OC and run parallel to the short dimension of the room. When all joists are located, snap lines on the existing ceiling directly below and in line with the concealed joists.

Laying Out and Applying Furring Strips

For fastening 12-inch tiles, furring strips must be must be installed 12 inches OC. From the corner, measure out the width of the border tiles. This measurement is the center of the first furring strip away from the wall. To mark the edge, measure from the center, in either direction, half the width of the furring strip. Mark an X on the side of the mark toward the center of the furring strip. From the edge of the first furring strip, measure and mark, every 12 inches, across the room. Place Xs on the same side of the mark as the first one (Fig. 76-3).

Lay out the other end of the room in the same manner. Snap lines between the marks for the location of the furring strips. The strips are fastened by keeping one edge to the chalk line with the strip on the side of the line indicated by the X. Fasteners must penetrate at least one inch into the joist. Starting and ending furring strips are also installed against both walls. (The procedure for applying furring strips to ceiling joists is described in Chapter 36.)

Squaring the Room

First, snap a chalk line on a furring strip as a guide for installing border tiles against the long wall. The line is snapped parallel to, and the width of the border tiles, away from the wall.

A second chalk line must be snapped to guide the application of the short wall border tiles. The line must be snapped at exactly at 90 degrees to the first chalk line. Otherwise tiles will not line up properly. From the short wall, measure in along the first chalk line, the width of the short wall border tiles. From this point, use the 3, 4, 5 squaring method and snap another line at a right angle to the first line. This method of squaring lines has been previously described in Chapter 26.

Installing Ceiling Tile

Ceiling tiles should be allowed to adjust to normal interior conditions for twenty-four hours before installation. Some carpenters sprinkle talcum powder on their hands to keep them dry. This prevents fingerprints and smudges on the finished ceiling. Cut tiles face up with a sharp utility knife guided by a straightedge. All cut edges should be against a wall.

Starting the Installation

To start the installation, cut a tile to fit in the corner. The outside edges of the tile should line up exactly with both chalk lines. Because this tile fits in the corner, it must be cut to the size of border tiles on both long and short sides of the room. For example, if the border tiles on the long wall are 9 inches and those on the short wall are 10 inches, the corner tile should be cut twice to make it 9×10. Staple the tile in position. Be careful to line up the outside edges with both chalk lines (Fig. 76-4).

Completing the Installation

After the corner tile is in place, work across the ceiling. Install two or three border tiles at a time. Then fill in with full-sized field tiles (Fig. 76-5). Tiles are applied so they are snug to each other, but not jammed tightly. Fasten each tile with four 1/2- or 9/16-inch staples. Place two in each flange, using a

1. MEASURE FROM WALL THE WIDTH OF BORDER TO CENTERLINE OF FURRING STRIPS

2. MEASURE BACK 1/2 WIDTH OF FURRING STRIP PLACE X ON SIDE TOWARD CENTERLINE

4. PLACE X ON SAME SIDE OF MARK AS FIRST ONE

WALL

12″

3. MEASURE AND MARK EVERY 12″ ACROSS ROOM

WALL

←FURRING STRIPS→

PLAN OF CEILING

Fig. 76-3 Furring strip layout for ceiling tile.

Fig. 76-4 Install the first tile in the corner with its outside edges to the layout lines.

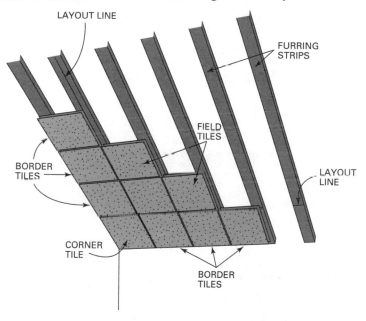

Fig. 76-5 Install a few border tiles. Then fill in with full-sized field tiles.

hand stapler. Use six staples in 12×24 tiles. Continue applying tiles in this manner until the last row is reached.

When the last row is reached, measure and cut each border tile individually. Cut the tiles slightly less than measured for easy installation. Do not force the tiles in place. Face nail the last tile in the corner near the wall where the nailhead will be covered by the wall molding. After all tiles are in place, the ceiling is finished by applying molding between the wall and ceiling. (The application of wall molding is discussed in a later unit.)

Adhesive Application of Ceiling Tile

Ceiling tile is sometimes cemented to existing plaster and drywall ceilings. These ceilings must be completely dry, solid, level, and free of dust and dirt. If the existing ceiling is in poor condition or has loose paint, adhesive application is not recommended.

Special tile adhesive is used to cement tiles to ceilings. Four daubs of cement are used on 12×12 tile (Fig. 76-6). Six daubs are used on 12×24 tile.

Fig. 76-6 Applying adhesive to the back of ceiling tiles.

Before applying the daubs, prime each spot by using the putty knife to force a thin layer into the back surface of the tile. Apply the daubs, about the size of a walnut. Press the tile into position. Keep the adhesive away from the edges to allow for spreading when the tile is pressed into position. No staples are required to hold the tile in place.

Easy Up® Installation System

Ceiling tile may also be installed using a patented system called *Easy Up®*. This system consists of spe-

Fig. 76-7 Installing ceiling tile using the Easy Up® system. (*Courtesy of Armstrong World Industries*)

cially shaped metal tracks and clips that are attached to the tracks to hold the ceiling tile (Fig. 76-7).

Installing Tracks

Border tile sizes are determined in the same manner as previously described. If ceiling joists are hidden, they must be located. The tracks must be screwed or nailed at right angles to the joists. The procedure for installing the tracks is shown in Figure 76-8.

Installing Tile

Two lines must be stretched or snapped at right angles to each other to locate the inner edges of the

STEP BY STEP
P R O C E D U R E S

1. INSTALL 1ST ROW OF TRACK 1" FROM STARTING WALL. UP TO 1/4" GAP MAY BE LEFT BETWEEN ENDS OF TRACK LENGTHS. SCREW OR NAIL THROUGH PRE-PUNCHED HOLES IN TRACK.

2. INSTALL 2ND ROW OF TRACK A DISTANCE FROM STARTING WALL THAT IS 2" LESS THAN THE WIDTH OF BORDER TILES. INSTALL 3RD AND SUBSEQUENT ROWS 12" O.C. STARTING FROM 2ND ROW.

3. INSTALL LAST ROW OF TRACK 1" AWAY FROM FINISHING WALL. SHIM TRACKS AS NEEDED TO STRAIGHTEN.

Fig. 76-8 Procedure for installing Easy Up® metal tracks. (*Courtesy of Armstrong World Industries*)

border tiles. The installation is started in the corner of the room where the layout lines intersect. The procedure for installing tile with the Easy Up® system is described in Figure 76-9.

Estimating Materials for Tile Installation

To estimate ceiling tile, measure the width and length of the room to the next whole foot measurement. Multiply these figures together to find the area of the ceiling in square feet. Divide the ceiling area by the number of square feet contained in one of the ceiling tiles being used to find the number of ceiling tiles needed.

To estimate furring strips, measure to the next whole foot the length of the room in the direction that the furring strips are to run. Multiply this by the number of rows of furring strips to find the total number of linear feet of furring strip stock needed. To find the number of rows, divide the width of the room by the furring strip spacing and add one.

STEP BY STEP
P R O C E D U R E S

Fig. 76-9 Procedure for installing ceiling tile using the Easy Up® system. (*Courtesy of Armstrong World Industries*)

Review Questions

Select the most appropriate answer.

1. The most common sizes in inches of ceiling tile are
   a. 8×12 and 12×12.
   b. 12×12 and 16×16.
   c. 12×12 and 12×24.
   d. 14×24 and 24×48.

2. The most common sizes in inches of lay-in suspended ceiling panels are
   a. 8×12 and 12×12.
   b. 12×12 and 16×16.
   c. 12×12 and 12×24.
   d. 24×24 and 24×48.

3. The width in inches of border tiles for a room 14'-9"×22'-3", when 12×12 inch tiles are used, is
a. 9 and 3.
c. 10 1/2 and 7 1/2.
b. 8 1/2 and 6 1/2.
d. 11 and 8.

4. The width in inches of border panel for a room 12'-6"×18'-8", when 24"×24" suspended ceiling panels are used, is
a. 6 and 8.
c. 14 1/2 and 18 1/2.
b. 10 and 12.
d. 15 and 16.

5. When squaring a room, use measurements of
a. 2, 3, and 4 feet.
c. 4, 5, and 6 feet.
b. 3, 4, and 5 feet.
d. 5, 6, and 7 feet.

6. To locate joists above an existing ceiling
a. determine the direction of the finished floor above.
b. tap on the ceiling with a hammer until a dull thud is heard.
c. look for lines of discoloration on the existing ceiling.
d. determine the location of studs in the wall below.

7. Nails used to fasten furring to ceiling joists must penetrate into the joists at least
a. 3/4 inch.
c. 1 1/4 inches.
b. one inch.
d. 1 1/2 inches.

8. Fastening 12"×12" tiles requires
a. two staples.
c. four staples.
b. three staples.
d. six staples.

9. In a suspended ceiling, hanger wire is used to suspend
a. cross tees.
c. wall angle.
b. main runners.
d. furring strips.

10. The first step in installing a suspended ceiling is to
a. square the room.
b. make a sketch of the planned ceiling.
c. calculate border tiles.
d. install the wall angle.

■ BUILDING FOR SUCCESS ■

The Builder and Community Involvement

The professional builder has established an active, quality business in a community. This reputation stems from a desire to help people realize their dream of home ownership and continues on throughout an entire career. During his or her career, the builder becomes a very important part of his or her community, as people benefit from that builder's voluntary services.

The builder, like many others, is responsible for assuring that various populations are assisted and supported in their quest to improve their living standards. Builders represent a source of expertise and strength that can help solve both existing and potential roadblocks to community development. All construction-related businesses in each community should view their position as one of great potential. By helping those who lack the expertise, time, and financial resources to upgrade their living conditions, building trade representatives can make a valuable contribution toward community development.

With the acknowledged responsibility to assist in dealing with housing problems, health concerns, inadequate community facilities, needy children, and homeless families, the builder has a great opportunity to become involved in community concerns. One of the most accessible avenues to assist with community needs is through involvement with the local home builders' association.

Housing associations at the national, state, and local levels are but one of many organizations that assist people in solving housing-related problems. The builder, subcontractor, retiree, self-employed person, businessperson, and citizen who have the desire to offer assistance to the community may unite to tackle local problems that are obstacles to community improvement.

The person who has a construction-related business must be service-oriented. The desire to provide housing, remodeling, and products to a community must not stop with the issue of payment for each service rendered. The competitive builder will show great vision for the future by looking ahead to how the community will grow during the next one or two decades. This vision will involve participating in the work that attacks immediate and future constructed-related problems. With this investment of time and resources will come visibility, credibility, and integrity for those who participate.

Most local builders' associations have identified community needs that they can participate in and help provide solutions to. Habitat for Humanity projects and Christmas toy projects for local homeless or handicapped children are popular. Teaming up with other local organizations such as police and fire departments has assisted many diverse groups with health or housing needs. As these groups undertake united efforts to assist others, they identify community needs, form common goals, and work toward reasonable solutions in solving problems.

For example, the Home Builders' Association of Lincoln, Nebraska, annually works with Habitat for Humanity. This project assembles a wide variety of builders, materials suppliers, volunteer workers (including college National Association of Home Builders student chapters), businesses, lending agencies, and community organizations to assist families that cannot afford housing by traditional means. Builders and business representatives take time from their busy schedules to head up efforts of this type in order to help less fortunate people gain access to better housing and higher living standards.

Development projects of this magnitude are a high priority for each community. Local elected leaders such as the mayor, city council members, or other administrators should be visibly involved in all ongoing efforts to improve living conditions. There will always be a need to involve as many people as possible in each community for goals to be reached. The success of any national, state, or local development project can be evaluated by the involvement of those who assist others. There will be a continual need for the builder to rec-

ognize his or her place in the community as quality living conditions are established and health hazards are eliminated. The element of charity has not lost its appeal or its place in a community. People will always need to assist others in the constant effort to raise living conditions. This brings into play the often neglected realm of responsibility to do all we can to help each other.

The access to better living conditions in the future will be based on the actions we take in our communities today. What initially appears to be an expense in time or finances up front usually turns out to be an intelligent investment. We must remember that potential clients make contact on the basis of what they know about each builder. The builder who invests time and finances in his or her community may be the recipient of unexpected business opportunities. Each builder and tradesperson has an opportunity to contribute something to his or her community. The decision to get involved can be made at any time.

FOCUS QUESTIONS: For individual or group discussion.

1. What considerations must a person address before volunteering a service to the community? Are there advantages and disadvantages? Evaluate any.

2. As you prepare for a career in the construction industry, how do you see your future participation in volunteer community service? What is appealing and what is not?

3. What are some benefits to members of the community who are recipients of volunteer service by builders and others in the construction trades?

4. What are some local community housing problems that you as a construction student could help solve? How would you do this?

UNIT
27

Interior Doors and Door Frames

OBJECTIVES

After completing this unit, the student will be able to:

- describe the sizes and kinds of interior doors.
- make and set interior door frames.
- hang an interior swinging door.
- install locksets on interior swinging doors.
- set a prehung door and frame.
- install bypass, bifold, pocket, and folding doors.

Interior doors used in residential and light commercial buildings are less dense than exterior doors. They are not ordinarily subjected to as much use and are not exposed to the weather. In commercial buildings, such as hospitals and schools, heavier and larger interior doors are specified that meet special conditions.

CONTENTS

CHAPTER 77

DESCRIPTION OF INTERIOR DOORS

Interior doors are classified by style as *flush, panel, French, louver,* and *cafe* doors. Interior flush doors have a smooth surface, are usually less expensive and are widely used when a plain appearance is desired. Some of the other styles have special uses. Doors are also classified by the way they operate, such as *swinging, sliding,* or *folding.*

Interior Door Sizes and Styles

For residential and light commercial use, most interior doors are manufactured in 1 3/8-inch thickness. Some, like cafe and bifold doors, are also made in 1 1/4- and 1 1/8-inch thicknesses. Most doors are manufactured in 6'-8" heights. Some types may be obtained in heights of 6'-0" and 6'-6". Door widths range from 1'-0" to 3'-0" in increments of 2 inches. However, not all sizes are available in every style.

Flush Doors

Flush doors are made with *solid* or *hollow* cores. Solid core doors are generally used as entrance doors. (They have been previously described in Unit 21, Exterior Doors.) Hollow core doors are commonly used in the interior except when fire resistance or sound transmission is critical.

A hollow core door consists of a light perimeter frame. This frame encloses a mesh of thin wood or composition material supporting the faces of the door. Solid wood blocks are appropriately placed in the core for the installation of locks. The frame and mesh are covered with a thin plywood called a *skin.* *Lauan* plywood is used extensively for flush door skins. Flush doors are also available with veneer faces of *birch, gum, oak,* and *mahogany,* among others (Fig. 77-1). When flush doors are to be painted, an overlay plywood or tempered hardboard is used for the skin.

Panel Doors

Interior *panel doors* consist of a frame with usually one to eight wood panels in various designs (Fig. 77-2). They are similar in style to some exterior panel doors. (The construction of panel doors has been previously described in Unit 21, Exterior Doors.)

French Doors

French doors may contain from one to fifteen lights of glass. They are made in a 1 3/4-inch thickness for exterior doors and 1 3/8-inch thickness for interior doors (Fig. 77-3).

MESH OR CELLULAR CORE
7 PLY CONSTRUCTION ILLUSTRATED

TOP AND BOTTOM RAILS
MINIMUM 2-1/4 INCHES (57.2 mm)

WIDTH OF STILES
MINIMUM 1 INCH (925.4 mm)

WOOD OR WOOD
DERIVATIVE STRIPS

LOCK BLOCKS

FACE VENEER

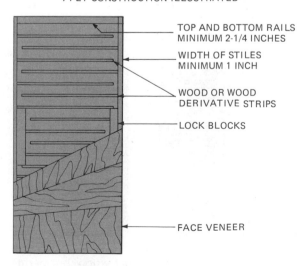

LADDER CORE
7 PLY CONSTRUCTION ILLUSTRATED

TOP AND BOTTOM RAILS
MINIMUM 2-1/4 INCHES

WIDTH OF STILES
MINIMUM 1 INCH

WOOD OR WOOD
DERIVATIVE STRIPS

LOCK BLOCKS

FACE VENEER

Fig. 77-1 The construction of hollow core flush doors.

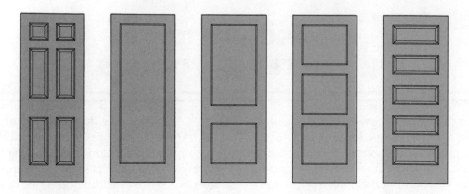

Fig. 77-2 Styles of commonly used interior panel doors. (*Courtesy of Morgan Manufacturing*)

Fig. 77-3 French doors are used in the interior as well as for entrances. (*Courtesy of Morgan Manufacturing*)

Louver Doors

Louver doors are made with spaced horizontal slats called louvers, used in place of panels. The louvers are installed at an angle to obstruct vision but permit the flow of air through the door. Louvered doors are widely used on clothes closets (Fig. 77-4).

Cafe Doors

Cafe doors are short panel or louver doors. They are hung in pairs that swing in both directions. They are used to partially screen an area, yet allow easy

and safe passage through the opening. The tops and bottoms of the doors are usually shaped in a pleasing design (Fig. 77-5).

Methods of Interior Door Operation

Doors are also identified by their method of operation. Doors either swing on hinges or slide on

Fig. 77-4 Louver doors obstruct vision, but permit the circulation of air. (*Courtesy of Morgan Manufacturing*)

Fig. 77-5 Cafe doors are usually used between kitchens and dining areas. (*Courtesy of Morgan Manufacturing*)

tracks. The choice of door operation depends on such factors as convenience, cost, safety, and space.

Swinging Doors

Swinging doors are hinged on one edge. They swing out of the opening. When closed, they cover the total opening. Swinging doors that swing in one direction are called *single-acting*. With special hinges, they can swing in both directions. They are then called *double-acting* doors (Fig. 77-6). Swinging doors are the most commonly used type of door. They have the disadvantage of requiring space for the swing.

Bypass Doors

Bypass doors are commonly used on wide clothes closet openings. A double track is mounted on the header jamb of the door frame. Rollers that ride in the track are attached to the doors so that they slide by each other. A floor guide keeps the doors in alignment at the bottom (Fig. 77-7). Usually two doors are used in a single opening. Three or more doors may used, depending on the situation.

The disadvantage of bypass doors is that although they do not project out into the room, access to the complete width of the opening is not possible. They are easy to install, but are not practical in openings less than 6 feet wide.

Pocket Doors

The *pocket door* is opened by sliding it sideways into the interior of the partition. When opened, only the lockedge of the door is visible (Fig. 77-8). Pocket doors may be installed as a single unit, sliding in one direction, or as a double unit sliding in opposite directions. When opened, the total width of the opening is obtained, and the door does not project out into the room. Pocket doors are used when these advantages are desired.

The installation of pocket doors requires more time and material than other methods of door operation. A special pocket door frame unit and track must be installed during the rough framing stage (Fig. 77-9). The rough opening in the partition must be large enough for the door opening and the pocket.

Bifold Doors

Bifold doors are made in flush, panel, louver, or combination panel and louver styles. They are made in narrower widths than other doors. This allows them to operate in a folding fashion on closet and similar type openings (Fig. 77-10).

Bifold doors consist of pairs of doors hinged at their edges. The doors on the jamb side swing on pivots installed at the top and bottom. Other doors

Fig. 77-6 A single-acting swinging door is the most widely used type of interior door.

Fig. 77-7 Bypass doors are used on wide closet openings.

fold up against the jamb door as it is swung open. The end door has a guide pin installed at the top. The pin rides in a track to guide the set when opening or closing (Fig. 77-11). On very wide openings

Fig. 77-8 The pocket door slides into the interior of the partition.

the guide pin is replaced by a combination guide and support to keep the doors from sagging.

Bifold doors may be installed in double sets, opening and closing from each side of the opening. They have the advantage of providing access to al-

Fig. 77-9 A pocket door frame comes preassembled from the factory. It is installed when the interior partitions are framed.

Fig. 77-10 Bifold doors are manufactured in many styles. (*Courtesy of Morgan Manufacturing*)

most the total width of the opening. Yet they do not project out much into the room.

Folding Doors

Folding doors are similar to bifold doors except they consist of many narrow panels. Each panel is about the same width as the door jamb. The entire door assembly is supported by rollers that slide in an overhead track. When opened, the panels stack against each other on side jambs. A folding door requires only a small amount of space and does not project out into the room (Fig. 77-12). Folding doors are frequently used for wardrobe closets, laundry rooms, and storage spaces.

Fig. 77-11 Bifold doors provide access to almost the total width of the opening.

Fig. 77-12 Folding doors have many narrow panels that stack against each other when opened.

CHAPTER **78**

INSTALLATION OF INTERIOR DOORS AND DOOR FRAMES

Many interior doors come *prehung* in their frames for easier and faster installation on the job. However, it is often necessary to build and set the door frame, hang the door, and install the locksets.

Interior Door Frames

Special *rabbeted* jamb stock or nominal one-inch square-edge lumber is used to make interior *door frames*. If square-edge lumber is used, a separate *stop*, if needed, is applied to the inside faces of the door frame.

Checking Rough Openings

The first step in making an interior door frame is to measure the door opening to make sure it is the correct width and height. The rough opening width for single-acting swinging doors should be the width of the door plus twice the thickness of the side jamb, plus 1/2 inch for shimming between the door frame and the opening. For example, if the thickness of the side jamb beyond the door is 3/4 inch, the rough opening width is 2 inches more than the door width.

The rough opening height should be the height of the door, plus the thickness of the header jamb, plus 1/4 inch for clearance at the top, plus the thickness of the finished floor, plus a desired clearance under the door. An allowance of 1/2 inch is usually made for clearance between the finished floor and the door.

For example, if the header jamb and finished floor thickness are both 3/4 inch, the rough opening height should be 2 1/4 inches over the door height, if 1/2 inch is allowed for clearance under the door (Fig. 78-1).

The rough opening size for other than single-acting swinging doors, such as bypass and bifold doors, should be checked against the manufacturer's directions. The sizes of the doors and allowances for hardware may differ with the manufacturer.

Making an Interior Door Frame

Interior door frames are constructed like exterior door frames except they have no sill. Interior door frames are usually installed after the interior wall

Fig. 78-1 Check the size of rough openings for swinging doors.

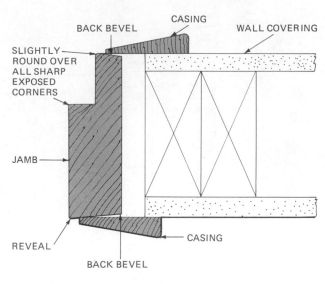

Fig. 78-2 Back-bevel jamb edges slightly to permit casings to fit snugly against them.

covering has been applied. Measure the total thickness of the wall, including the wall covering, to find the jamb width.

Cutting Jambs to Width

If necessary, rip the door jamb stock so its width is the same as the wall thickness. If rabbeted jamb stock is used, cut the edge opposite the rabbet. Plane and smooth both edges to a slight *back bevel*. The back bevel permits the door casings, when later applied, to fit tightly against the edges of the door frame in case there are irregularities in the wall (Fig. 78-2). *Ease* all sharp exposed corners.

Cutting Jambs to Length

On interior door frames, the *header jamb* is usually cut to fit between, and is dadoed into, the *side jambs*. The side jambs run the total height of the rough opening. Cut both side jambs to a length equal to the height of the opening.

Header Jambs of Door Frames for Hinged Doors.
If rabbeted jambs are used, the length of the header jamb is the width of the door plus 3/16 inch. The extra 3/16 inch is for joints of 3/32 inch on each side, between the edges of the door and the

side jambs. If square-edge lumber is used, its length is the same as a rabbeted header jamb. However, 1/2 inch is added, for dadoing 1/4 inch deep into each side jamb (Fig. 78-3). Cut the header jamb to length with both ends square.

Side Jambs of Door Frames for Hinged Doors.
Measure up from the bottom ends. Square lines across the side jambs to mark the location of the bottom side of the header jamb. This dimension is the sum of:

- the thickness of the finish floor, if the door frame rests on the subfloor.
- an allowance of 1/2 inch between the door and the finish floor.
- the height of the door.
- 3/32 inch for a joint between the door and the header jamb.
- on rabbeted jambs, subtract 1/2 inch for the depth of the rabbet.

Hold a scrap piece of jamb stock to the squared lines. Mark its other side to lay out the width of the dado. Mark the depth of the dadoes on both edges of the side jambs. Cut the dadoes to receive the header jamb. On rabbeted jambs, dado depth is to the face of the rabbet. A dado depth of 1/4 inch is sufficient on plain jambs (Fig. 78-4).

Jamb Lengths for Other Types of Doors.
The length of header and side jambs for other types of doors, such as bypass and bifold, must be determined from instructions provided by the manufacturer of the hardware and the door. Door hardware and door sizes differ with the manufacturer. This affects the length of the door jambs.

Fig. 78-3 Length of plain and rabbeted header jambs of door frames for swinging doors.

Plain, square-edge lumber jambs are used to make door frames for doors other than single-acting swinging doors. The rest of the procedures, such as checking rough openings, cutting jambs to width, assembling, and setting, is the same for all door frames.

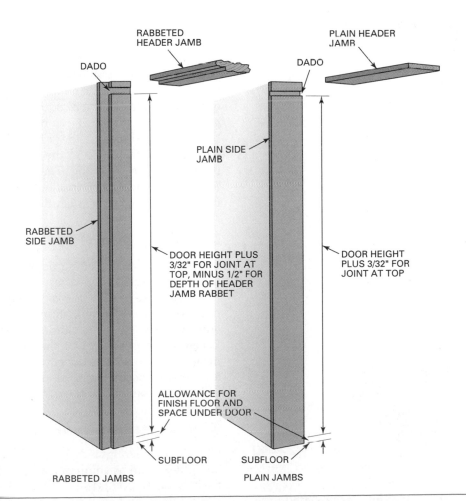

Fig. 78-4 Laying out plain and rabbeted side jambs of door frames for swinging doors.

Assembling the Door Frame

Fasten the side jambs to the header jamb, keeping the edges flush. If there is play in the dado, first wedge the header jamb with a chisel so the face side comes up tight against the dado shoulders before fastening.

Cut a narrow strip of wood. Tack it to the side jambs a few inches up from the bottom so that the frame width is the same at the bottom as it is at the top. This strip is commonly called a *spreader* (Fig. 78-5).

Setting the Door Frame

Door frames must be set so that the jambs are straight, level, and plumb. They are usually set before the finish floor is laid. If a rabbeted frame is used, determine the swing of the door so that the rabbet is facing toward the correct side.

Cut any *horns* from the top ends of side jambs. Place the frame in the opening. The horns are cut off

in case the side jambs need to be shimmed to level the header jamb.

Install shims directly opposite the ends of the header jamb between the opening and the side jambs. Shim an equal amount on both sides so that the frame is close to being centered at the top of the opening. Drive shims up snugly.

Leveling the Header Jamb

Keep the edges of the frame flush with the wall. With the bottom ends of both side jambs resting on the subfloor, check the header jamb for levelness. Level the header jamb, if necessary, by placing shims between the bottom end of the appropriate side jamb and the subfloor. When the header jamb is level, tack the frame in place on both sides, close to the top. Drive fasteners through side jambs and shims into the studs.

If the door frame rests on a finish floor, then a tight joint must be made between the bottom ends of the side jambs and the finish floor. If the floor is level, side jambs will fit the floor if their lengths are exactly the same and their bottom ends have been cut square. When the frame is set, the header jamb should be level when the ends of the side jamb are resting on the floor.

If the floor is not quite level, or if the side jambs are of unequal length, level the header jamb by shimming under the bottom end of the side jamb on the low side. Set the dividers for the amount the jamb has been shimmed. Scribe that amount on the bottom of the opposite side jamb. Remove the frame from the opening. Cut to the scribed line. Replace the frame in the opening. The header jamb should be level. The bottom ends of both side jambs should fit snugly against the finish floor (Fig. 78-6).

Plumbing the Side Jambs

Several ways of plumbing door frames have been previously described. An accurate and fast method is with the use of a *plumb bob* (see Fig. 60–18). When one side jamb is plumb, shim and tack its bottom end in place. Only one side needs to be plumbed. Locate the bottom end of the other side jamb by measuring across. The door frame width at the bottom should be the same as on the top. Shim and tack the bottom end of the other side jamb in place.

Straightening Jambs

Use a 6-foot straightedge against the side jambs. Straighten them by shimming at intermediate points. Besides other points, shims should be placed opposite hinge and lock locations. Fasten the jambs by nailing through the shims. Header jambs on

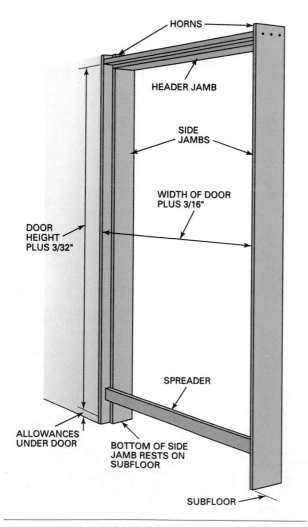

HORNS

HEADER JAMB

SIDE JAMBS

WIDTH OF DOOR PLUS 3/16"

DOOR HEIGHT PLUS 3/32"

SPREADER

ALLOWANCES UNDER DOOR

BOTTOM OF SIDE JAMB RESTS ON SUBFLOOR

SUBFLOOR

Fig. 78-5 An assembled door interior rabbeted door frame.

HELPFUL HINTS

How to fit side jambs to an out of level finish floor.

SHIMS

SHIMS

LEVEL HEADER JAMB

1. SHIM BOTTOM END OF SIDE JAMB UNTIL HEADER JAMB IS LEVEL

3. SCRIBE AMOUNT ON OPPOSITE SIDE JAMB

2. ADJUST DIVIDERS TO AMOUNT JAMB IS SHIMMED

OUT OF LEVEL FINISH FLOOR

Fig. 78-6

wide door frames are straightened, shimmed, and fastened in a similar way.

Sighting the Door Frame for a Wind

Before any nails are driven home, sight the door frame to see if it has a *wind*. The frame must be sighted by eye to make sure that side jambs line up vertically with each other and that the frame is not twisted. This is important when installing rabbeted jambs. The method of checking for a wind in door frames has been previously described (see Fig. 60-20).

If the frame has a wind, move the top or bottom ends of the side jambs slightly until they line up with each other. Fasten top and bottom ends of the side jambs securely. Set the nails (Fig. 78-7).

Applying Door Stops

At this time, *door stops* may be applied to plain jambs. The stops are not permanently fastened, in case they have to be adjusted when locksets are installed. A *back miter* joint is usually made between molded side and header stops. A *butt* joint is made between square-edge stops (Fig. 78-8).

Hanging Interior Doors

The method of fitting and hanging single-acting, hinged, interior doors is the same for exterior doors. See Chapter 61, Door Fitting and Hanging.

Double-Acting Doors

Double-acting doors are installed with either special pivoting hardware installed on the floor and the header jamb or with spring-loaded double-acting hinges. Both types return the door to a closed position after being opened. When opened wide, the doors can be held in the open position. A different type of light-duty, double-acting hardware is used on *cafe* doors. To install double-acting door hardware, follow the manufacturer's directions.

STEP BY STEP
P R O C E D U R E S

1. CUT HORNS FROM TOP OF SIDE JAMBS

ELEVATION

2. SET FRAME IN OPENING
   SHIM ON BOTH SIDES OPPOSITE HEADER
   LEVEL HEADER AND FASTEN AT TOP

ELEVATION

3. PLUMB SIDE JAMBS
   SHIM AND TACK AT BOTTOM

ELEVATION

4. STRAIGHTEN SIDE JAMBS
   INSTALL INTERMEDIATE SHIMS
   TACK IN PLACE

SIGHT FROM ONE SIDE
OF THE OPENING TO
LINE UP THE EDGE ON
ONE SIDE WITH THE
OPPOSITE EDGE ON
THE OTHER SIDE

5. SIGHT THROUGH THE FRAME
   TO CHECK FOR A WIND.
   DRIVE AND SET ALL NAILS
   WHEN EDGES LINE UP.

Fig. 78-7 Procedure for setting an interior door frame.

Fig. 78-8 On molded stops, a back miter joint is used at the header jamb. A butt joint is usually made on square-edge stops.

Installing Bypass Doors

Bypass doors are installed so they overlap each other by about one inch when closed. Cut the track to length. Install it on the header jamb according to the manufacturer's directions.

Installing Rollers. Install pairs of *roller hangers* on each door. The roller hangers may be offset a different amount for the door on the outside than the door on the inside. They are also offset differently for doors of various thicknesses. Make sure that rollers with the same and correct offset are used on each door (Fig. 78-9). The location of the rollers from the edge of the door is usually specified in the manufacturer's instruction sheet.

Installing Door Pulls. Mark the location and bore holes for *door pulls*. Flush pulls must be used so that bypassing is not obstructed (Fig. 78-10). The proper size hole is bored partway into the door. The pull is tapped into place with a hammer and wood block. The press fit holds the pull in place. Rectangular flush pulls, also used on bypass doors, are held in place with small recessed screws.

Hanging Doors. Hang the doors by holding the bottom outward. Insert the rollers in the overhead track. Then gently let the door come to a vertical position. Install the inside door first, then the outside door (Fig. 78-11).

Fitting Doors. Test the door operation and the fit against side jambs. Door edges must fit against side jambs evenly from top to bottom. If the top or bottom portion of the edge strikes the side jamb first, it may cause the door to jump from the track. The door rollers have adjustments for raising and lowering. Adjust one or the other to make the door edges fit against side jambs.

Installing Floor Guides. A *floor guide* is included with bypass door hardware to keep the doors in alignment. The guide is centered on the lap to steady the doors at the bottom. Mark the location of the guide. Remove the doors. Install the inside section of the guide. Replace the inside door. Replace the outside door. Install the rest of the guide (Fig. 78-12).

Installing Bifold Doors

Before installing *bifold* doors, make sure the opening size is as specified by the hardware or door manufacturer. Usually bifold doors come hinged to-

Fig. 78-9 Bypass door rollers are offset different distances for use on doors of various thicknesses.

Fig. 78-10 Bypass doors must have flush pulls.

gether in pairs. The hardware consists of the track, pivot sockets, pivot pins and guides, door aligners, door pulls, and necessary fasteners (Fig. 78-13).

Installing the Track. Cut the track to length. Fasten it to the header jamb with screws provided in the kit. The track contains adjustable *sockets* for the door *pivot pins.* Make sure these are inserted before fastening the track in position. The position

Fig. 78-11 Bypass doors are hung on the overhead track by holding the bottom of the door outward.

OUTSIDE
BYPASS DOOR

FLOOR GUIDE

INSIDE
BYPASS DOOR

FLOOR GUIDE IS ADJUSTABLE FOR VARIOUS DOOR THICKNESSES

Fig. 78-12 A floor guide is installed to keep bypass doors aligned.

UPPER SOCKET
IN TRACK

TRACK

UPPER SOCKET
IN TRACK

GUIDE
PINS

TOP PIVOT
PIN

TOP PIVOT
PIN

BOTTOM PIVOT
PIN IN DOOR

BOTTOM PIVOT
PIN IN DOOR

BOTTOM SOCKET
IN FLOOR BRACKET

DOOR
ALIGNERS

BOTTOM SOCKET
IN FLOOR BRACKET

Fig. 78-13 Installation of the bifold door requires several kinds of special hardware.

of the track on the header jamb is not critical. It may be positioned as desired (Fig. 78-14).

Installing Bottom Pivot Sockets. Locate the bottom pivot sockets. Fasten one on each side, at the bottom of the opening. The pivot socket bracket is

L-shaped. It rests on the floor against the side jamb. It is centered on a plumb line from the center of the pivot sockets in the track on the header jamb above.

Installing Pivot and Guide Pins. In most cases, bifold doors come with prebored holes for

CASING   TRACK   HEADER JAMB

SECTIONS THROUGH HEADER

Fig. 78-14 The bifold door track may be located in any position on the header jamb in several ways. Trim conceals the track from view.

pivot and *guide pins.* If not, it is necessary to bore them. Follow the manufacturer's directions as to size and location. Install pivot pins at the top and bottom ends of the door in the prebored holes closest to the jamb. Sometimes the top pivot pin is spring loaded. It can then be depressed for easier installation of the door. The bottom pivot pin is threaded and can be adjusted for height. The guide pin rides in the track. It is installed in the hole provided at the top end of the door farthest away from the jamb.

Hanging the Doors. After all the necessary hardware has been applied, the doors are ready for installation. Loosen the set screw in the top pivot socket. Slide it along the track toward the center of the opening about one foot away from the side jamb. Place the door in position by inserting the bottom pivot pin in the bottom pivot socket. Tilt the doors to an upright position. At the same time insert the top pivot pin in the top socket, and the guide pin in the track, while sliding the socket toward the jamb.

Adjusting the Doors. Adjust top and bottom pivot sockets in or out so the desired joint is obtained between the door and the jamb. Lock top and bottom pivot sockets in position. Adjust the bottom pivot pin to raise or lower the doors, if necessary.

If more than one set of bifold doors are to be installed in an opening, install the other set on the opposite side in the same manner. Install knobs in the manner and location recommended by the manufacturer.

Where sets of bifold doors meet at the middle of an opening, door aligners are installed, near the bottom, on the inside of each of the meeting doors. The door aligners keep the faces of the center doors lined up when closed (Fig. 78-15).

Installing Pocket Doors

The pocket door frame, complete with track, is in-

JOINT BETWEEN SETS OF DOORS

DOOR ALIGNERS

INSIDE OF BIFOLD DOOR

Fig. 78-15 Door aligners are used near the bottom where sets of bifold doors meet.

stalled when interior partitions are framed. The pocket consists of two ladder-like frames between which the door slides. A steel channel is fastened to the floor. The channel keeps the pocket opening spread the proper distance apart.

The frame, which is usually preassembled at the factory, is made of nominal one-inch stock. The pocket is covered by the interior wall finish. Care must be taken when covering the pocket frame not to use fasteners that are so long that they penetrate the frame. If fasteners penetrate through the pocket door frame, they will probably scratch the side of the door as it is operated or stop its complete entrance into the pocket.

Installing Door Hardware. Attach rollers to the top of the door in the location specified by the manufacturer. Install pulls on the door. On pocket doors an edge pull is necessary, in addition to recessed pulls on the sides of the door. A special pocket door pull contains edge and side pulls. It is mortised in the edge of the door. In most cases, all the necessary hardware is supplied when the pocket door frame is purchased.

Hanging the Door. Engage the rollers in the track by holding the bottom of the door outward in a way similar to that used with bypass doors. Test the operation of the door to make sure it slides easily and butts against the side jamb evenly. Make adjustments to the rollers, if necessary. Stops are later applied to the jambs on both sides of the door. The stops serve as guides for the door. When the door is closed, the stops prevent it from being pushed out of the opening (Fig. 78-16).

Fig. 78-16 Plan of a pocket door.

Hanging Folding Doors

Because of many styles, manufacturer's instructions supplied with the *folding door* unit should be closely followed. Make sure the door frame is the size specified by the manufacturer.

The folding door track is fastened with screws, usually along the centerline, to the header jamb. The door is hung on the track with the *jamb panel* toward the side jamb on which the panels will be stacked. Fasten the jamb panel to the side jamb. Fasten the *stop*, against which the door closes, on the opposite side jamb. Usually, trim strips are fastened on both sides of track to conceal it (Fig. 78-17). Check the operation of the door and make any necessary adjustments.

Setting a Prehung Door

A *prehung* single-acting, hinged door unit consists of a door frame with the door hinged and casings installed. Holes are provided, if locksets have not already been installed. Small cardboard shims are stapled to the lock edge and top end of the door to maintain proper clearance between the door and frame.

Prehung units are available in several jamb widths to accommodate various wall thicknesses. Some prehung units have split jambs that are adjustable for varying wall thicknesses (Fig. 78-18).

A prehung door unit can be set in a matter of minutes. Remove the casings carefully from one side of solid jamb units. Center the unit in the opening, so the door will swing in the desired direction. Be sure the door is closed and spacer shims are in place between the jamb and door. Plumb the door unit. Tack it to the wall through the casing.

SECTION THROUGH HEADER

DETAIL OF FIXED PANEL

PLAN AT LOCK EDGE

Fig. 78-17 Installation details for folding doors.

JAMB

SOLID RABBETED JAMB

SHIMS

GYPSUM
BOARD

STOP

**SQUARE EDGE LUMBER JAMB
WITH APPLIED STOP**

SHIMS

STUD

SPLIT RABBETED JAMB

STOP

**SPLIT JAMB
WITH APPLIED STOP**

Fig. 78-18 Prehung door units come with solid or split jambs.

Open the door and move to the other side. Install shims between the side jambs and the rough opening at intermediate points, keeping side jambs straight. Nail through the side jambs and shims. Remove spacers. Check the operation of the door. Make any necessary adjustments. Replace the previously removed casings. Drive and set all nails (Fig. 78-19).

Prehung door units with split jambs are set in a similar manner. However, there is no need to remove the casings. One section is installed as de-

scribed above. The remaining section is inserted into the one already in place.

Installing Locksets

Locksets are installed on interior doors in the same manner as for exterior doors and as described in Chapter 62, Door Lock Installation. Although their installation is basically the same, some locks are used exclusively on interior doors.

STEP BY STEP
P R O C E D U R E S

1. REMOVE CASINGS FROM ONE SIDE OF UNIT
2. PLACE UNIT IN OPENING AND PLUMB SIDE CASING
3. FASTEN THROUGH CASINGS INTO WALL

4. MOVE TO OTHER SIDE
5. INSTALL SHIMS BETWEEN JAMB AND WALL
6. FASTEN THROUGH JAMB AND SHIMS

7. REPLACE CASINGS THAT WERE PREVIOUSLY REMOVED
8. FASTEN THROUGH CASINGS INTO WALL.

Fig. 78-19 Procedure for installing a prehung door unit.

The *privacy* lock is often used on bathroom and bedroom doors. It is locked by pushing or turning a button on the room side (Fig. 78-20). On most privacy locks, a turn of the knob on the room side unlocks the door. On the opposite side, the door can be unlocked by a pin or key inserted into a hole in the knob. The unlocking device should be kept close by, in a prominent location, in case the door needs to be opened quickly in an emergency.

The *passage* lockset has knobs on both sides that are turned to unlatch the door. This lockset is are used when it is not desirable to lock the door.

Fig. 78-20 A bathroom door is locked by pushing on the button in the center of the knob of a privacy lockset.

Review Questions

Select the most appropriate answer.

1. Most interior doors are manufactured in a thickness of
   a. one inch.
   b. 1 3/8 inches.
   c. 1 1/2 inches.
   d. 1 3/4 inches.

2. The height of most interior doors is
   a. 6'-0".
   b. 6'-6".
   c. 6'-8".
   d. 7'-0".

3. Interior door widths usually range from
   a. 1'-6" to 2'-6".
   b. 2'-2" to 2'-8".
   c. 2'-6" to 2'-8".
   d. 1'-0" to 3'-0".

4. Used extensively for flush door skins is
   a. fir plywood.
   b. lauan plywood.
   c. metal.
   d. plastic laminate.

5. The usual distance allowed between the finish floor and the bottom of swinging doors for clearance is
   a. 1/4 inch.
   b. 1/2 inch.
   c. 3/4 inch.
   d. one inch.

6. A disadvantage of bypass doors is that they
   a. project out into the room.
   b. cost more and require more time to install.
   c. are difficult to operate.
   d. do not provide total access to the opening.

7. If the jamb stock is 3/4 inch thick, the rough opening width for a swinging door should be the door width plus
   a. 3/4 inches.
   b. 1 1/2 inches.
   c. 2 inches.
   d. 2 1/2 inches.

8. If the jamb stock and the finished floor are both 3/4-inch thick, the rough opening height for a 6'-8" swinging door should be
   a. 7'-0".
   b. 6'-11 1/2".
   c. 6'-10 1/4".
   d. 6'-9 1/4".

9. When a rabbeted door frame is made for a swinging door, the header jamb length is equal to the width of the door plus
   a. 3/16 inches.
   b. 3/32 inches.
   c. 3/8 inches.
   d. 1/2 inches.

10. An accurate and fast method of plumbing side jambs of a door frame is by the use of a
    a. builder's level.
    b. carpenter's 26-inch hand level.
    c. plumb bob.
    d. straightedge.

■ BUILDING FOR SUCCESS ■

Environmental Issues in Construction

Each year brings new environmental concerns to nations throughout the world. In certain areas of the United States, the availability of selected natural resources such as water, timber, and land suitable for building has become a front-page issue. Other issues, such as waste disposal and wetlands, will continue to affect the quality of land available for development. With larger populations to feed, clothe, house, and employ, every nation must conservatively utilize its natural resources. At the same time, it must maintain a strong economy in a well thought out fashion.

The United States has experienced confrontations involving construction material producers, environmentalists, and conservationists. The issues vary, but generally center on the consumption of natural resources or preservation of those resources and the creating or maintaining of natural habitat for wildlife. Directly and indirectly, many of these issues affect housing and construction. How should we, as people associated with the construction industry, look at these and other controversial issues? Do we consider ourselves only consumers of natural resources or only environmentalists? Most builders, producers of construction products, and material suppliers understand how critical both the preservation of the natural environment and economic growth are to our economy. How should members of the construction industry become informed and work with others as environmental, ecological, and economic disputes are resolved? How can they stay informed about the issues that have a direct bearing on construction-related industries?

Initially, all those earning their living in the construction industry should be as informed as possible about current environmental, ecological, and economic issues. They should participate in discussions and information exchanges that deal with local and national environmental concerns. Most national and local associations representing the construction industry will be prepared to take a stand on issues that affect their membership. The best decisions to be made about these issues will be made with the most information and concern for who and what are involved. The research process should involve data from all perspectives.

One large group with a great deal of concern for the preservation of our natural resources and habitats is elementary through postsecondary school students. Many young people are responding to the major concerns that will affect the environment of tomorrow. Many school-age children and young adults participate in awareness and problem resolution activities. They are willing to become involved in attempts to help others understand the urgency of responding. They realize that cooperation between opposing factions must be established and maintained. They are also involved with education and economic concerns that affect future job opportunities. In many ways, these concerned and motivated young people demonstrate how people can work together to find solutions. It is encouraging to know that other organizations and special interest groups have taken the same approach to solving controversial disputes.

Almost everyone has heard of the spotted owl issue and the desire of many to preserve the owl's natural habitat in the Pacific Northwest. Some opposition to this cause has come from the logging industry and lumber product producers. Industries' concerns to continue utilizing the forests in the production of building materials are also very important and require study. This dispute has involved people representing a large variety of associations, special interest groups, federal and state governments, private industry, and even the president of the United States.

This issue is representative of others that involve the future preservation and/or consumption of natural resources. Attempting to place blame in many cases should not be the goal. Realistically, the goal should be to look beyond selfish interests and focus on possible solutions that will preserve the environment and stimulate the economy at the same time. Left unaddressed, many controversial issues may prove to be devastating to our environment, our economy, and our ecological well-being. Few of these concerns, if any, can be solved on short notice. They will take time, understanding, flexibility, commitment, and money to resolve.

Historically, many solutions to issues of national concern have come about by compromise resulting from discussion. The goals on both sides of a controversial issue should be to bring solutions to the problems that benefit both people and the environment. The greatest challenge may be in arriving at a compromise that satisfies all opponents.

Economically, there is a great need to maintain growth in jobs affected by environmental issues. Stimulating the economy of a community that has been hit by a drastic reduction in producing construction materials and products is a tremendous challenge. The solutions need to be immediate, sound, and valuable long-term. Finding alternative industrial production and business potential must be an integral part of any discussion.

The construction industry as a whole must continue to be informed and desire to work with others in the process of working to save our natural environment and maintain economic stability. Each student of construction must understand the need to solve the ongoing environmental and economic problems that put our nation's workforce at risk. A continual desire to be a part of the solution and not the problem will be required. We have a good history of addressing concerns of all types in this country and finding workable solutions. That should continue to be a goal for everyone.

FOCUS QUESTIONS: For individual or group discussion.

1. Identify one important environmental, ecological, or economic issue that directly affects a construction-related industry. Provide an overview of the issue.

2. How have issues such as the spotted owl controversy been addressed by the construction-related industries in your specific geographic area?

3. What solutions do you feel are workable in addressing the spotted owl issue where environmentalist, ecological, and industrial groups have not reached agreements? How can opposing parties begin to work toward "reasonable" solutions?

4. As a future participant in the nation's construction workforce, what controversial issues do you think will affect construction-related industries? What do you see as your personal obligation in following these types of issues?

Interior Trim

Interior trim, also called *interior finish*, involves the application of *molding* around windows and doors; at the intersection of walls, floors, and ceilings; and to other inside surfaces. Moldings are strips of material, shaped in numerous patterns, for use in specific locations. Wood is used to make most moldings, but some are made of plastic or metal.

OBJECTIVES

After completing this unit, the student will be able to:

- identify standard moldings and describe their use.
- apply ceiling and wall molding.
- apply interior door casings, baseboard, base cap, and base shoe.
- install window trim, including stools, aprons, jamb extensions, casings, and stop beads.
- install closet shelves and closet pole.
- install mantels.

CONTENTS

CHAPTER 79 DESCRIPTION AND APPLICATION OF MOLDING

Moldings are available in many *standard* types. Each type is manufactured in several sizes and patterns. Standard patterns are usually made only from softwood. When other kinds of wood, or special patterns, are desired, mills make *custom* moldings to order. All moldings must be applied with a tight fitting joints to present a suitable appearance.

Standard Molding Patterns

Standard moldings are designated as *bed, crown, cove, full round, half round, quarter round, base, base shoe, base cap, casing, chair rail, back band, apron, stool, stop,* and others (Fig. 79-1).

Molding usually comes in lengths of 8, 10, 12, 14, and 16 feet. Some moldings are available in odd lengths. Door casings, in particular, are available in lengths of 7 feet to reduce waste.

Finger-jointed lengths are made of short pieces joined together. These are used only when a paint finish is to be applied. The joints show through a stained or natural finish.

Molding Shape and Use

Some moldings are classified by the way they are shaped. Others are designated by location. For example, *beds, crowns, and coves* are terms related to shape. Although they may be placed in other locations, they are usually used at the intersection of walls and ceilings (Fig. 79-2). Also classified by their shape are *full rounds, half rounds,* and *quarter rounds.* They are used in many locations. Full rounds are used for such things as closet poles. Half rounds may be used to conceal joints between panels or to trim shelf edges. Quarter rounds may be used to trim outside corners of wall paneling and for many other purposes (Fig. 79-3).

Designated by location, *base, base shoe,* and *base cap* are moldings applied at the bottom of walls where they meet the floor. When square-edge base is used, a base cap is usually applied to its top edge. Base shoe is normally used to conceal the joint between the bottom of the base and the finish floor (Fig. 79-4).

Casings are used to trim around windows, doors, and other openings. They cover the space between the frame and the wall. *Back bands* are applied to the outside edges of casings for a more decorative appearance.

Aprons, stools, and *stops* are parts of window trim. Stops are also applied to door frames. On the same window, aprons should have the same molded shape as casings. Aprons, however, are not *backed out.* They have straight, smooth backs and sharp, square, top edges that butt against the bottom of the stool (Fig. 79-5).

Corner guards are also called *outside corners.* They are used to finish exterior corners of interior wall finish. *Caps* and *chair rail* trim the top edge of wainscoting. (These moldings have been previously described in Chapter 23, Application of Wall Paneling.) Others, such as *astragals, battens, panel,* and *picture* moldings are used for various purposes.

Making Joints on Molding

End joints between lengths of ceiling molding may be made square or at a 45 degree angle. Many carpenters prefer to make square joints between moldings because less joint line is shown. Also, the square end acts as a stop when bowing and snapping the last length of molding into place at a corner.

Usually, the last piece of molding along a wall is cut slightly longer. It is bowed outward in the center, then pressed into place when the ends are in position. This makes the joints come up tight. After the molding has been fastened, joints between lengths should be sanded flush, except on prefinished moldings. Failure to sand butted ends flush with each other results in a shadow being cast at the joint line. This gives the appearance of an open joint.

Joints on exterior corners are *mitered.* (Miter joints are defined in Chapter 12, Boring and Cutting Tools.) Joints on interior corners are usually *coped,* especially on large moldings. A coped joint is made by fitting the piece on one wall with a square end into the corner. The end of the molding on the other wall is cut to fit against the shaped face of the molding on the first wall (Fig. 79-6).

Methods of Mitering Using Miter Boxes

Moldings of all types may be mitered by using either hand or power *miter boxes.* A metal miter box

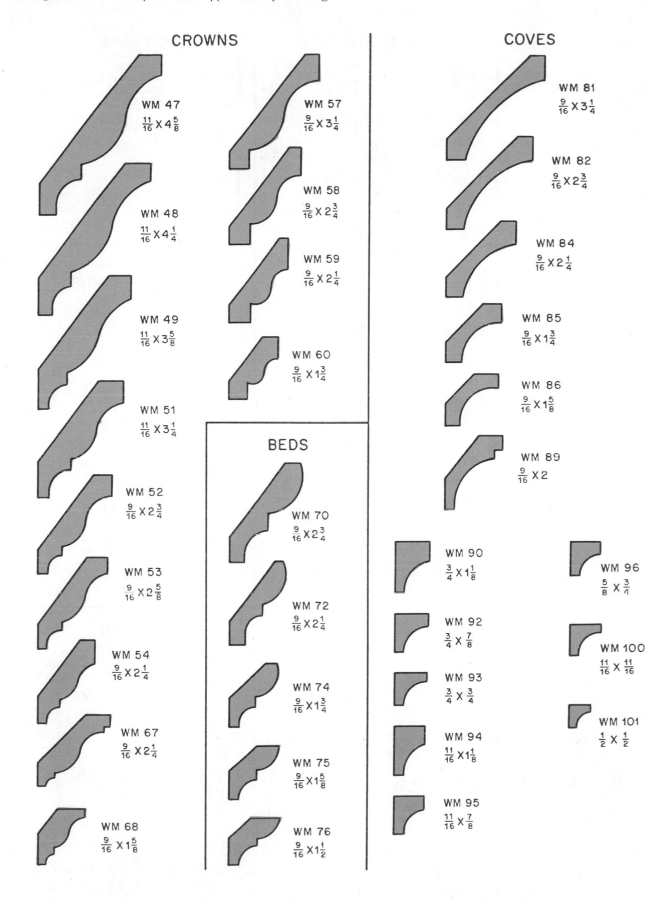

Fig. 79-1 Standard molding patterns (*Courtesy of Wood Molding and Millwork Producers, Inc., P.O. Box 25278, Portland, Oregon 97225*)

QUARTER ROUNDS

WM 103
$1\frac{1}{16} \times 1\frac{1}{16}$

WM 104
$\frac{11}{16} \times 1\frac{3}{8}$

WM 105
$\frac{3}{4} \times \frac{3}{4}$

WM 106
$\frac{11}{16} \times \frac{11}{16}$

WM 107
$\frac{5}{8} \times \frac{5}{8}$

WM 108
$\frac{1}{2} \times \frac{1}{2}$

WM 109
$\frac{3}{8} \times \frac{3}{8}$

WM 110
$\frac{1}{4} \times \frac{1}{4}$

HALF ROUNDS

WM 120
$\frac{1}{2} \times 1$

WM 122
$\frac{3}{8} \times \frac{11}{16}$

WM 123
$\frac{5}{16} \times \frac{5}{8}$

WM 124
$\frac{1}{4} \times \frac{1}{2}$

FLAT ASTRAGALS

WM 133
$\frac{11}{16} \times 1\frac{3}{4}$

WM 134
$\frac{11}{16} \times 1\frac{3}{8}$

WM 135
$\frac{7}{16} \times \frac{3}{4}$

BASE SHOES

WM 126
$\frac{1}{2} \times \frac{3}{4}$

WM 129
$\frac{7}{16} \times \frac{11}{16}$

WM 127
$\frac{7}{16} \times \frac{3}{4}$

WM 131
$\frac{1}{2} \times \frac{3}{4}$

SHELF EDGE / SCREEN MOULD

WM 137
$\frac{3}{8} \times \frac{3}{4}$

WM 141
$\frac{1}{4} \times \frac{5}{8}$

WM 138
$\frac{5}{16} \times \frac{5}{8}$

WM 142
$\frac{1}{4} \times \frac{3}{4}$

WM 140
$\frac{1}{4} \times \frac{3}{4}$

WM 144
$\frac{1}{4} \times \frac{3}{4}$

GLASS BEADS

WM 147
$\frac{1}{2} \times \frac{9}{16}$

WM 148
$\frac{3}{8} \times \frac{3}{8}$

BASE CAPS

WM 163
$\frac{11}{16} \times 1\frac{3}{8}$

WM 167
$\frac{11}{16} \times 1\frac{1}{8}$

WM 164
$\frac{11}{16} \times 1\frac{1}{8}$

WM 172
$\frac{5}{8} \times \frac{3}{4}$

WM 166
$\frac{11}{16} \times 1\frac{1}{4}$

BRICK MOULD

WM 175
$1\frac{1}{16} \times 2$

WM 176
$1\frac{1}{16} \times 1\frac{3}{4}$

WM 180
$1\frac{1}{4} \times 2$

DRIP CAPS

WM 187
$1\frac{11}{16} \times 2$

WM 196
$\frac{11}{16} \times 1\frac{3}{4}$

WM 188
$1\frac{1}{16} \times 1\frac{5}{8}$

WM 197
$\frac{11}{16} \times 1\frac{5}{8}$

CORNER GUARDS

WM 199
1×1

WM 200
$\frac{3}{4} \times \frac{3}{4}$

WM 201
$1\frac{5}{16} \times 1\frac{5}{16}$

WM 204
$1\frac{5}{16} \times 1\frac{5}{16}$

WM 202
$1\frac{1}{8} \times 1\frac{1}{8}$

WM 205
$1\frac{1}{8} \times 1\frac{1}{8}$

WM 203
$\frac{3}{4} \times \frac{3}{4}$

WM 206
$\frac{3}{4} \times \frac{3}{4}$

BATTENS

WM 224
$\frac{9}{16} \times 2\frac{1}{4}$

WM 229
$\frac{11}{16} \times 1\frac{5}{8}$

ROUNDS

WM 232 $1\frac{5}{8}$

WM 233 $1\frac{5}{16}$

WM 234 $1\frac{1}{16}$

SQUARES

WM 236 $1\frac{5}{8} \times 1\frac{5}{8}$

WM 237 $1\frac{5}{16} \times 1\frac{5}{16}$

WM 238 $1\frac{1}{16} \times 1\frac{1}{16}$

WM 239 $\frac{3}{4} \times \frac{3}{4}$

Fig. 79-1 *(continued)*

SHINGLE PANEL MOULDINGS

WM 207 $\frac{11}{16} \times 2\frac{1}{2}$

WM 209 $\frac{11}{16} \times 2$

WM 210 $\frac{11}{16} \times 1\frac{5}{8}$

WM 212 $\frac{11}{16} \times 2\frac{1}{2}$

WM 213 $\frac{9}{16} \times 2$

WM 217 $\frac{11}{16} \times 1\frac{3}{4}$

WM 218 $\frac{11}{16} \times 1\frac{1}{2}$

HAND RAIL

WM 230 $1\frac{1}{2} \times 1\frac{11}{16}$

WM 231 $1\frac{1}{2} \times 1\frac{11}{16}$

WM 240 $1\frac{1}{4} \times 2\frac{1}{4}$

PICTURE MOULDINGS

WM 273 $\frac{11}{16} \times 1\frac{3}{4}$

WM 276 $\frac{11}{16} \times 1\frac{3}{4}$

SCREEN/S4S STOCK

WM 241	$1\frac{1}{16} \times 2\frac{3}{4}$
WM 243	$1\frac{1}{16} \times 1\frac{3}{4}$
WM 246	$\frac{3}{4} \times 2\frac{3}{4}$
WM 247	$\frac{3}{4} \times 2$
WM 248	$\frac{3}{4} \times 1\frac{3}{4}$
WM 249	$\frac{3}{4} \times 1\frac{5}{8}$
WM 250	$\frac{3}{4} \times 1\frac{1}{2}$
WM 251	$\frac{3}{4} \times 1\frac{3}{8}$
WM 252	$\frac{3}{4} \times 1\frac{1}{4}$
WM 254	$\frac{1}{2} \times 1\frac{3}{4}$

LATTICE

WM 265	$\frac{9}{32} \times 1\frac{3}{4}$
WM 266	$\frac{9}{32} \times 1\frac{5}{8}$
WM 267	$\frac{9}{32} \times 1\frac{3}{8}$
WM 268	$\frac{9}{32} \times 1\frac{1}{8}$

BACK BANDS

WM 280 $\frac{11}{16} \times 1\frac{1}{16}$

WM 281 $\frac{11}{16} \times 1\frac{1}{8}$

WAINSCOT/PLY CAP MOULDINGS

WM 290 $\frac{11}{16} \times 1\frac{3}{8}$

WM 292 $\frac{9}{16} \times 1\frac{1}{8}$

WM 294 $\frac{11}{16} \times 1\frac{1}{8}$

WM 295 $\frac{1}{2} \times 1\frac{1}{4}$

WM 296 $\frac{3}{4} \times \frac{3}{4}$

CHAIR RAILS

WM 297 $\frac{11}{16} \times 3$

WM 298 $\frac{11}{16} \times 2\frac{1}{2}$

WM 300 $1\frac{1}{16} \times 3$

WM 303 $\frac{9}{16} \times 2\frac{1}{2}$

WM 304 $\frac{1}{2} \times 2\frac{1}{4}$

WM 390 $\frac{11}{16} \times 2\frac{5}{8}$

FLAT STOOLS

WM 1021 $\frac{11}{16} \times$ WIDTH SPECIFIED

T-ASTRAGALS

WM 1300 $1\frac{1}{4} \times 2\frac{1}{4}$

WM 1305 $1\frac{1}{4} \times 2$

WM 1310 $1\frac{1}{4} \times 2\frac{1}{4}$

WM 1315 $1\frac{1}{4} \times 2$

Fig. 79-1 (continued)

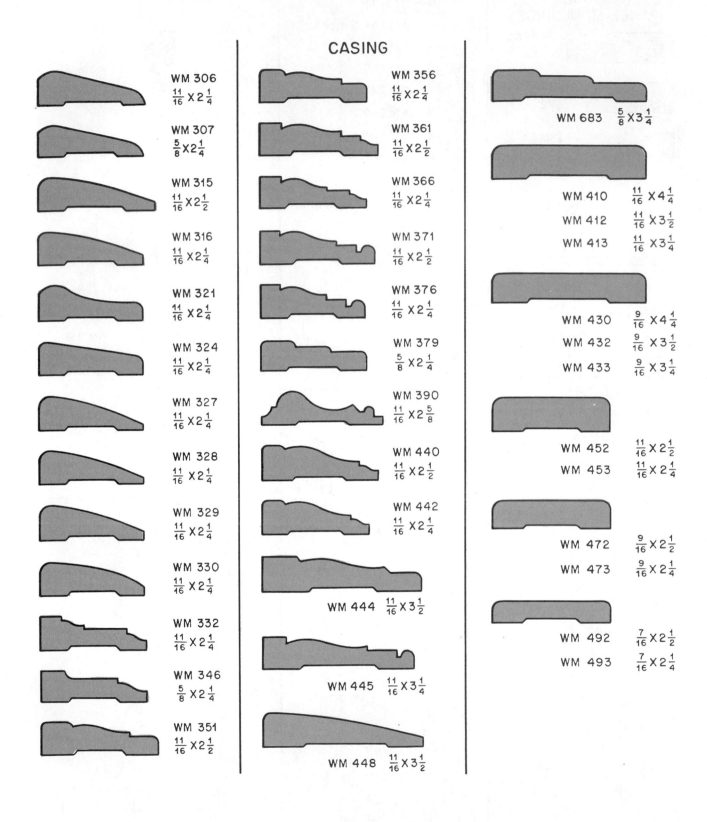

CASING

WM 306 $\frac{11}{16} \times 2\frac{1}{4}$

WM 307 $\frac{5}{8} \times 2\frac{1}{4}$

WM 315 $\frac{11}{16} \times 2\frac{1}{2}$

WM 316 $\frac{11}{16} \times 2\frac{1}{4}$

WM 321 $\frac{11}{16} \times 2\frac{1}{4}$

WM 324 $\frac{11}{16} \times 2\frac{1}{4}$

WM 327 $\frac{11}{16} \times 2\frac{1}{4}$

WM 328 $\frac{11}{16} \times 2\frac{1}{4}$

WM 329 $\frac{11}{16} \times 2\frac{1}{4}$

WM 330 $\frac{11}{16} \times 2\frac{1}{4}$

WM 332 $\frac{11}{16} \times 2\frac{1}{4}$

WM 346 $\frac{5}{8} \times 2\frac{1}{4}$

WM 351 $\frac{11}{16} \times 2\frac{1}{2}$

WM 356 $\frac{11}{16} \times 2\frac{1}{4}$

WM 361 $\frac{11}{16} \times 2\frac{1}{2}$

WM 366 $\frac{11}{16} \times 2\frac{1}{4}$

WM 371 $\frac{11}{16} \times 2\frac{1}{2}$

WM 376 $\frac{11}{16} \times 2\frac{1}{4}$

WM 379 $\frac{5}{8} \times 2\frac{1}{4}$

WM 390 $\frac{11}{16} \times 2\frac{5}{8}$

WM 440 $\frac{11}{16} \times 2\frac{1}{2}$

WM 442 $\frac{11}{16} \times 2\frac{1}{4}$

WM 444 $\frac{11}{16} \times 3\frac{1}{2}$

WM 445 $\frac{11}{16} \times 3\frac{1}{4}$

WM 448 $\frac{11}{16} \times 3\frac{1}{2}$

WM 683 $\frac{5}{8} \times 3\frac{1}{4}$

WM 410 $\frac{11}{16} \times 4\frac{1}{4}$

WM 412 $\frac{11}{16} \times 3\frac{1}{2}$

WM 413 $\frac{11}{16} \times 3\frac{1}{4}$

WM 430 $\frac{9}{16} \times 4\frac{1}{4}$

WM 432 $\frac{9}{16} \times 3\frac{1}{2}$

WM 433 $\frac{9}{16} \times 3\frac{1}{4}$

WM 452 $\frac{11}{16} \times 2\frac{1}{2}$

WM 453 $\frac{11}{16} \times 2\frac{1}{4}$

WM 472 $\frac{9}{16} \times 2\frac{1}{2}$

WM 473 $\frac{9}{16} \times 2\frac{1}{4}$

WM 492 $\frac{7}{16} \times 2\frac{1}{2}$

WM 493 $\frac{7}{16} \times 2\frac{1}{4}$

Fig. 79-1 (continued)

BASE MOULDINGS

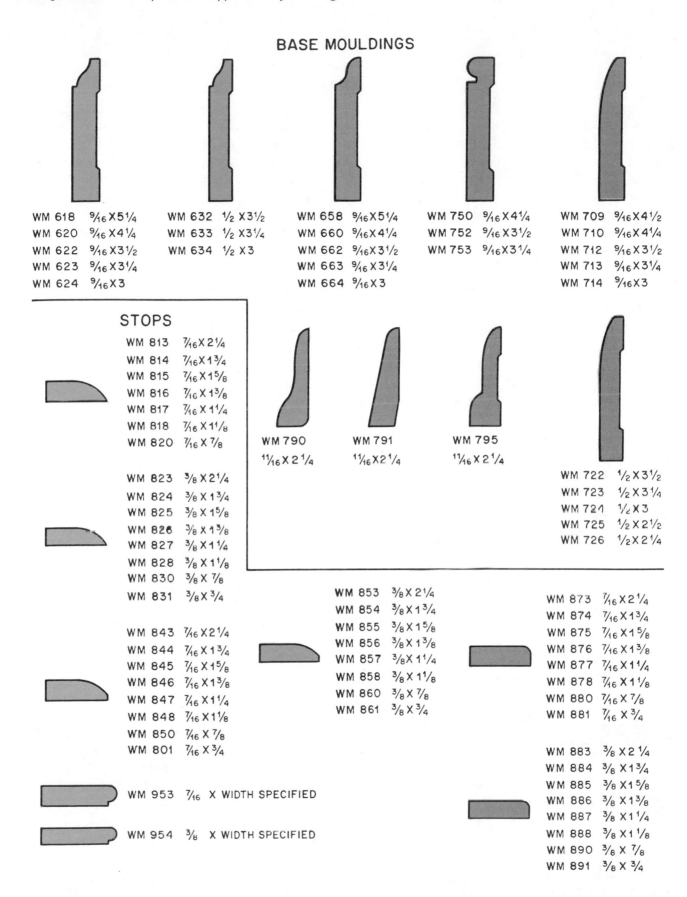

WM 618 $^9/_{16}$X5$^1/_4$	WM 632 $^1/_2$X3$^1/_2$	WM 658 $^9/_{16}$X5$^1/_4$	WM 750 $^9/_{16}$X4$^1/_4$	WM 709 $^9/_{16}$X4$^1/_2$
WM 620 $^9/_{16}$X4$^1/_4$	WM 633 $^1/_2$X3$^1/_4$	WM 660 $^9/_{16}$X4$^1/_4$	WM 752 $^9/_{16}$X3$^1/_2$	WM 710 $^9/_{16}$X4$^1/_4$
WM 622 $^9/_{16}$X3$^1/_2$	WM 634 $^1/_2$X3	WM 662 $^9/_{16}$X3$^1/_2$	WM 753 $^9/_{16}$X3$^1/_4$	WM 712 $^9/_{16}$X3$^1/_2$
WM 623 $^9/_{16}$X3$^1/_4$		WM 663 $^9/_{16}$X3$^1/_4$		WM 713 $^9/_{16}$X3$^1/_4$
WM 624 $^9/_{16}$X3		WM 664 $^9/_{16}$X3		WM 714 $^9/_{16}$X3

STOPS

WM 813 $^7/_{16}$X2$^1/_4$
WM 814 $^7/_{16}$X1$^3/_4$
WM 815 $^7/_{16}$X1$^5/_8$
WM 816 $^7/_{16}$X1$^3/_8$
WM 817 $^7/_{16}$X1$^1/_4$
WM 818 $^7/_{16}$X1$^1/_8$
WM 820 $^7/_{16}$X$^7/_8$

WM 823 $^3/_8$X2$^1/_4$
WM 824 $^3/_8$X1$^3/_4$
WM 825 $^3/_8$X1$^5/_8$
WM 826 $^3/_8$X1$^3/_8$
WM 827 $^3/_8$X1$^1/_4$
WM 828 $^3/_8$X1$^1/_8$
WM 830 $^3/_8$X$^7/_8$
WM 831 $^3/_8$X$^3/_4$

WM 843 $^7/_{16}$X2$^1/_4$
WM 844 $^7/_{16}$X1$^3/_4$
WM 845 $^7/_{16}$X1$^5/_8$
WM 846 $^7/_{16}$X1$^3/_8$
WM 847 $^7/_{16}$X1$^1/_4$
WM 848 $^7/_{16}$X1$^1/_8$
WM 850 $^7/_{16}$X$^7/_8$
WM 801 $^7/_{16}$X$^3/_4$

WM 953 $^7/_{16}$ X WIDTH SPECIFIED

WM 954 $^3/_8$ X WIDTH SPECIFIED

WM 790
$^{11}/_{16}$X2$^1/_4$

WM 791
$^{11}/_{16}$X2$^1/_4$

WM 795
$^{11}/_{16}$X2$^1/_4$

WM 722 $^1/_2$X3$^1/_2$
WM 723 $^1/_2$X3$^1/_4$
WM 724 $^1/_2$X3
WM 725 $^1/_2$X2$^1/_2$
WM 726 $^1/_2$X2$^1/_4$

WM 853 $^3/_8$X2$^1/_4$
WM 854 $^3/_8$X1$^3/_4$
WM 855 $^3/_8$X1$^5/_8$
WM 856 $^3/_8$X1$^3/_8$
WM 857 $^3/_8$X1$^1/_4$
WM 858 $^3/_8$X1$^1/_8$
WM 860 $^3/_8$X$^7/_8$
WM 861 $^3/_8$X$^3/_4$

WM 873 $^7/_{16}$X2$^1/_4$
WM 874 $^7/_{16}$X1$^3/_4$
WM 875 $^7/_{16}$X1$^5/_8$
WM 876 $^7/_{16}$X1$^3/_8$
WM 877 $^7/_{16}$X1$^1/_4$
WM 878 $^7/_{16}$X1$^1/_8$
WM 880 $^7/_{16}$X$^7/_8$
WM 881 $^7/_{16}$ X $^3/_4$

WM 883 $^3/_8$X2$^1/_4$
WM 884 $^3/_8$X1$^3/_4$
WM 885 $^3/_8$X1$^5/_8$
WM 886 $^3/_8$X1$^3/_8$
WM 887 $^3/_8$X1$^1/_4$
WM 888 $^3/_8$X1$^1/_8$
WM 890 $^3/_8$ X $^7/_8$
WM 891 $^3/_8$ X $^3/_4$

Fig. 79-1 (continued)

STOPS

WM 903 $7/16 \times 2\frac{1}{4}$	WM 913 $3/8 \times 2\frac{1}{4}$	WM 933 $7/16 \times 2\frac{1}{4}$	WM 943 $3/8 \times 2\frac{1}{4}$
WM 904 $7/16 \times 1\frac{3}{4}$	WM 914 $3/8 \times 1\frac{3}{4}$	WM 934 $7/16 \times 1\frac{3}{4}$	WM 944 $3/8 \times 1\frac{3}{4}$
WM 905 $7/16 \times 1\frac{5}{8}$	WM 915 $3/8 \times 1\frac{5}{8}$	WM 935 $7/16 \times 1\frac{5}{8}$	WM 945 $3/8 \times 1\frac{5}{8}$
WM 906 $7/16 \times 1\frac{3}{8}$	WM 916 $3/8 \times 1\frac{3}{8}$	WM 936 $7/16 \times 1\frac{3}{8}$	WM 946 $3/8 \times 1\frac{3}{8}$
WM 907 $7/16 \times 1\frac{1}{4}$	WM 917 $3/8 \times 1\frac{1}{4}$	WM 937 $7/16 \times 1\frac{1}{4}$	WM 947 $3/8 \times 1\frac{1}{4}$
WM 908 $7/16 \times 1\frac{1}{8}$	WM 918 $3/8 \times 1\frac{1}{8}$	WM 938 $7/16 \times 1\frac{1}{8}$	WM 948 $3/8 \times 1\frac{1}{8}$
WM 910 $7/16 \times 7/8$	WM 920 $3/8 \times 7/8$	WM 940 $7/16 \times 7/8$	WM 950 $3/8 \times 7/8$
WM 911 $7/16 \times 3/4$	WM 921 $3/8 \times 3/4$	WM 941 $7/16 \times 3/4$	WM 951 $3/8 \times 3/4$

PANEL STRIPS MULLION CASINGS

WM 955 $3/8 \times 2\frac{1}{4}$
WM 956 $3/8 \times 2$
WM 957 $3/8 \times 1\frac{3}{4}$

WM 960 $3/8 \times 2\frac{1}{4}$
WM 962 $3/8 \times 2$
WM 963 $3/8 \times 1\frac{3}{4}$

WM 965 $3/8 \times 2\frac{1}{4}$
WM 967 $3/8 \times 2$
WM 968 $3/8 \times 1\frac{3}{4}$

WM 970 $3/8 \times 2\frac{1}{4}$
WM 972 $3/8 \times 2$
WM 973 $3/8 \times 1\frac{3}{4}$

WM 975 $3/8 \times 2\frac{1}{4}$
WM 977 $3/8 \times 2$
WM 978 $3/8 \times 1\frac{3}{4}$

WM 980 $3/8 \times 2\frac{1}{4}$
WM 982 $3/8 \times 2$
WM 983 $3/8 \times 1\frac{3}{4}$

WM 985 $3/8 \times 2\frac{1}{4}$
WM 987 $3/8 \times 2$
WM 988 $3/8 \times 1\frac{3}{4}$

RABBETED STOOLS

SPECIFY WIDTH OF RABBET
AND DEGREE OF BEVEL

WM 1131 $1\frac{1}{16} \times 3\frac{5}{8}$
WM 1133 $1\frac{1}{16} \times 3\frac{1}{4}$
WM 1134 $1\frac{1}{16} \times 2\frac{3}{4}$

WM 1153 $1\frac{1}{16} \times 3\frac{1}{4}$
WM 1154 $1\frac{1}{16} \times 2\frac{3}{4}$
WM 1155 $1\frac{1}{16} \times 2\frac{1}{2}$
WM 1156 $1\frac{1}{16} \times 2\frac{1}{4}$

WM 1161 $1\frac{1}{16} \times 3\frac{5}{8}$
WM 1163 $1\frac{1}{16} \times 3\frac{1}{4}$
WM 1164 $1\frac{1}{16} \times 2\frac{3}{4}$

WM 1193 $1\frac{1}{16} \times 3\frac{1}{4}$
WM 1194 $1\frac{1}{16} \times 2\frac{3}{4}$
WM 1195 $1\frac{1}{16} \times 2\frac{1}{2}$
WM 1196 $1\frac{1}{16} \times 2\frac{1}{4}$

Fig. 79-1 (*continued*)

Fig. 79-2 Bed, crown, and cove moldings are often used at the intersection of walls and ceilings.

Fig. 79-3 Half round and quarter round moldings are used for many purposes.

Fig. 79-4 Base, base shoe, and base cap are used to trim the bottom of the wall.

Fig. 79-5 Casing, back bands, and stops are used for window and door trim. Stools and aprons are part of window trim.

is the most commonly used hand tool for making miters (Fig. 79-7).

A wood miter box may be used to guide a fine-tooth hand crosscut saw to cut miters. The box can be easily made on the job with three pieces of scrap lumber (Fig. 79-8). The saw kerfs used to guide the saw eventually widen with use. However, it takes only a little time to make another wood miter box. A wood miter box may be constructed to any size that is needed. When extrawide crown moldings need to be mitered, a specially constructed wood miter box may be the only tool large enough to hold the moldings and guide the handsaw.

Probably the most widely used tool for making miters and other end cuts on interior trim is a *power miter box*. It is commonly called a *chop saw* (Fig. 79-9). (The operation of the power miter box is described in Chapter 18, Radial Arm and Miter Saw.)

Positioning Molding in the Miter Box

Placing molding in the correct position in the miter box is essential for accurate *mitering*. Cut all moldings with their face sides or edges up or toward the

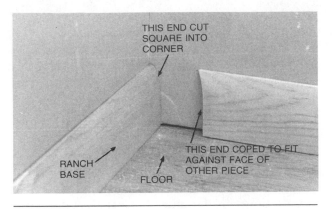

Fig. 79-6 A coped joint is made by fitting the end of one piece of molding against the shaped face of the other piece.

Fig. 79-7 Using a metal miter box.

STEP BY STEP
P R O C E D U R E S

1. CUT 3 PEICES OF 1"×4" STOCK 16" LONG. FASTEN SIDES TO THE EDGES OF THE BOTTOM AS SHOWN

NOTE: MITER BOX CAN BE MADE TO ACCOMODATE ANY SIZE MOULDING.

2. LAY OUT MITER BOX AS SHOWN ABOVE. KEEP DIMENSION "X" CLOSE TO EQUAL TO EACH OTHER TO HELP PREVENT SECTIONS BETWEEN SAW KERFS FROM BREAKING OUT.

3. SQUARE LINES DOWN ON BOTH SIDES. USE A FINE-TOOTHED HAND CROSS-CUT (10 TO 12 PT) SAW AND MAKE CUTS. CUT FROM BOTH SIDES IN ORDER TO FOLLOW LINES. USE THE SAME SAW TO CUT MITERS.

Fig. 79-8 How to make a wood miter box.

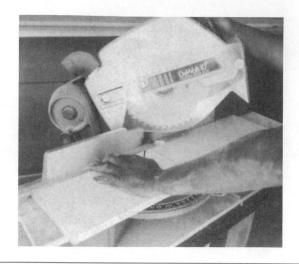

Fig. 79-9 Using a power miter box. It is commonly called a chop saw by workers in the trade.

operator so that the saw splinters out the back side, not the face side. Position the molding with one back side or edge against the bottom of the miter box and the other against the side of the miter box. On a wood miter box, which ordinarily has two

sides, hold the molding against the side farthest away from the worker.

Flat miters are cut by holding the molding with its face side up and its thicker edge against the side of the miter box. Some moldings, such as base, base cap and shoe, and chair rail, are held right side up. Their bottom edge should be against the bottom of the miter box and their back against the side of the miter box (Fig. 79-10).

Mitering Bed, Crown, and Cove molding.

Bed, crown, and *cove* molding must be positioned upside down in the miter box. Their top edge is placed against the bottom of the miter box and their bottom edge against the far side (Fig. 79-11). This is the only way these types of moldings can be positioned accurately and held steady while being mitered. This position is also convenient because the bottom edge of this type of molding is the edge that is usually marked for cutting to length. With the bottom edge up, the mark, on which to start the cut, can be easily seen.

A thin, narrow strip of wood fastened to the bottom of the miter box and against the molding helps prevent the molding from moving when be-

MITER BOX

CASING

THICK EDGE AGAINST
SIDE OF MITER BOX

FACE SIDE UP

BACK SIDE
DOWN

POSITION OF CASING IN MITER BOX TO CUT
FLAT MITERS.

BACK SIDE
AGAINST MITER
BOX

BOTTOM EDGE
DOWN

FACE SIDE OUT

POSITION OF BASE CAP WHEN
CUTTING EDGE MITERS.

Fig. 79-10 Position of molding in a miter box to cut flat and edge miters.

CROWN MOULDING

MITER BOX

BOTTOM EDGE UP

BACK SIDE AGAINST MITER BOX

FACE SIDE OUT

TOP EDGE DOWN

Fig. 79-11 To miter bed, crown, and cove moldings, they must be positioned upside down in the miter box.

ing mitered. The strip also assures that subsequent pieces of the same type of molding will be positioned at the same angle. Therefore, they will be mitered the same as the first piece (Fig. 79-12).

To position *corner guards* and other rabbeted molding, such as *back bands* and *caps,* for mitering, temporarily fasten a small strip of wood into the far corner of the miter box. The width and thickness of the strip should be slightly more than the rabbet of the molding. The molding is positioned on the strip and held steady while being mitered (Fig. 79-13).

Using a Miter Trimmer

Miters may also be made with a hand-operated *miter trimmer* (Fig. 79-14). All cuts on the trimmer must be made at or near the end of the molding. All softwood molding and trim are usually rough mitered with a chop cut. They are trimmed with one or two shaving slices. Hardwoods must be sawed at the approximate miter angle and then trimmed. The harder the wood, the thinner the cut must be made.

When miter joints need fitting, paper-thin corrective cuts are made quickly and easily with the miter trimmer.

Mitering with the Table and Radial Arm Saw

Miters may also be made by using the table saw or the radial arm saw. The use of mitering jigs is helpful when making flat miters on window and door casings. The jigs allows both right- and left-hand miters to be cut quickly and easily without any changes in the setup (Fig. 79-15). (The construction and use of mitering jigs for table and radial arm saws are more fully described in Unit 7, Stationary Power Tools.)

Making a Coped Joint

To cope the end of molding, first make a *back miter* on the end. A back miter starts from the end and is cut back on the face of the molding (Fig. 79-16. The

MITER BOX

BED MOULDING

HELPFUL HINTS

Fasten a thin, narrow strip of wood to the bottom of the miter box. This assures accurate positioning of bed, crown, cove, and similar type moldings.

THIN, NARROW STRIP OF WOOD

SMALL FINISH NAILS PLACE CLEAR OF SAW BLADE

Fig. 79-12

HELPFUL HINTS

Fasten a strip of wood in the corner of the miter box on which to position rabbeted type moldings, such as outside corners, back bands, and caps.

MITER BOX

OUTSIDE CORNER MOULDING

WOOD STRIP

Fig. 79-13

Fig. 79-14 The miter trimmer is widely used for fast and accurate mitering. (*Courtesy of Pootatuck Corporation*)

Fig. 79-15 With a mitering jig, left- and right-hand miters can be made quickly and easily without changing the setup (Guard has been removed for clarity).

MITER BOX SAW

MITER BOX

Fig. 79-16 Making a back miter on a piece of crown molding.

edge of the cut along the face forms the profile of the cope. Rub the side of a pencil point lightly along the profile to outline it more clearly.

Use a coping saw with a fine-tooth blade. Cut along the outlined profile with a slight undercut. Cut with the handle of the coping saw above the work and the teeth of the blade pointing away from the handle (Fig. 79-17). Hold the molding so it is over the end of a sawhorse. The side of the molding that will butt the wall should be lying flat on the top of the sawhorse. It is important that the molding be

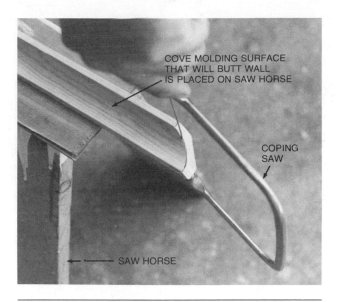

Fig. 79-17 Coping the end of a cove molding.

MARK THIS DISTANCE FROM THE CEILING ON BOTH ENDS OF WALL AND SNAP A LINE BETWEEN MARKS

CEILING

CROWN MOULDING

CHALK LINE

WALL

Fig. 79-18 Hold a scrap piece of molding against the ceiling and wall to determine the distance from the ceiling to its bottom edge.

held in this position. Holding it any other way makes it difficult to cut the cope with an undercut so that it will fit properly.

Coped joints are used on interior wall corners. They will not open up when the molding is nailed in place, especially if the backing is not solid. Miter joints may open up in interior corners when the ends are fastened.

Applying Molding

To apply chair rail, caps, or some other type molding located on the wall, chalk lines should be snapped. This makes sure molding is applied in a straight line. No lines need to be snapped for base moldings or for small-size moldings applied at the intersection of walls and ceiling.

For large-size ceiling moldings, such as beds, crowns, and coves, a chalk line should be snapped. This assures straight application of the molding and easier joining of the pieces. Without a straight line to guide application, the molding may be forced at different angles along its length. This results in a noticeably crooked bottom edge and difficulty making tight-fitting miters and copes.

Hold a short scrap piece of the large-size molding at the proper angle at the wall and ceiling intersection (Fig. 79-18). Lightly mark the wall along the bottom edge of the molding. Measure the distance from the ceiling down to the mark. Measure and mark this same distance down from the ceiling on each end of each wall to which the molding is to be applied. Snap lines between the marks. Apply the molding so its bottom edge is to the chalk line.

Apply the molding to the first wall with square ends in both corners. If more than one piece

is required to go from corner to corner, install the first piece with both ends square. If desired to make mitered joints between lengths, cut a square end into the corner and a back miter on the other end. On subsequent lengths, make matching end joints until the other corner is reached.

On some moldings, such as quarter rounds and small cove moldings, the straight, back surfaces should, but may not always, be of equal width. One of the back surfaces of these moldings should be marked with a pencil to assure positioning them in the miter box the same way each time. Mitering the molding with the same side down each time helps make fitting more accurate, faster, and easier (Fig. 79-19).

If a small-size molding is used, fasten it with finish nails in the center. Use nails of sufficient length to penetrate into solid wood at least one inch. If large-size molding is used, fastening is required along both edges (Fig. 79-20).

Press the molding in against the wall or intersection with one hand while driving the nail almost home. Then set the nail below the surface. Nail at about 16-inch intervals and in other locations as necessary to bring the molding tight against the surface. End nails should be placed 2 to 3 inches from the end to keep from splitting the molding. If it is likely that the molding may split, blunt the pointed end of the nail. Or, drill a hole slightly smaller than the nail diameter.

Install the last piece on the first wall by first squaring one end. Place the square end in the corner. Let the other end overlap the first piece. Mark and cut it at the overlap. This method is more accurate than measuring and then transferring the measurement to the piece. Mark all pieces of interior

COVE MOULDING

MARK ONE OF THE BACK SIDES

QUARTER ROUND

Fig. 79-19

FURRING STRIP   TOP PLATE

CEILING

SMALL MOLDING
MAY REQUIRE ONE
NAIL AT EACH
LOCATION

LARGER MOLDING
REQUIRES FASTENING
IN CEILING AND WALL

WALL COVERING

STUD

Fig. 79-20 Methods of fastening molding.

trim for length in this manner whenever possible. Cut and fasten the last piece with its square end into the corner.

Cope the starting end of the first piece on each succeeding wall against the face of the last piece installed on the previous wall. Work around the room in one direction, either clockwise or counterclockwise. The end of the last piece installed must be coped to fit against the face of the first piece.

CHAPTER **80** APPLICATION OF DOOR CASINGS, BASE, AND WINDOW TRIM

In addition to wall and ceiling molding, the application of *door casings*, *base*, *base cap*, *base shoe*, and *window trim* is a major part of interior finish work. Care must be taken to avoid marring the work and to make neat, tight-fitting joints.

Door Casings

Door casings are moldings applied around the door opening. They trim and cover the space between the door frame and the wall. Casings must be applied before any base moldings because the base butts against the edge of door casing (Fig. 80-1). Door casings extend to the floor.

Design of Door Casings

Moldings or *S4S* stock may be used for door casings. S4S is the abbreviation for *surfaced four sides*. It is used to describe smooth, square-edge lumber. When molded casings are used, the joint at the head must be mitered unless butted against *plinth blocks*. Plinth blocks are small decorative blocks. They are thicker and wider than the door casing. They are used as part of the door trim at the base and at the head (Fig. 80-2).

When using S4S lumber, the joint may be mitered or butted. If a butt joint is used, the head casing overlaps the side casing. The appearance of

DOOR CASING

DOOR JAMB

DOOR STOP

BASE

CASING EXTENDS TO THE FLOOR

BASE BUTTS AGAINST CASING

Fig. 80-1 Door casings are applied before the base is installed.

HEAD PLINTH BLOCKS

MOULDED CASING

DOOR JAMB

DOOR STOP

BASE

BASE PLINTH BLOCKS

Fig. 80-2 Molded casings are mitered at the head unless plinth blocks are used.

S4S casings and some molded casings may be enhanced with the application of *back bands* (Fig. 80-3).

Molded casings usually have their back sides *backed out*. In cases where the jamb edges and the wall surfaces may not be exactly flush with each other, the backed out surfaces allow the casing to come up tight on both wall and jamb (Fig. 80-4). If S4S casings are used, they must be backed out on the job. (A method of backing out S4S lumber is described in Chapter 19, Table Saw, and is illustrated in Fig. 19–6.)

Applying Door Casings

Door casings are set back from the inside face of the door frame a distance of about 5/16 inch. This allows room for the door hinges and the striker plate of the door lock. This setback is called a *reveal* (Fig. 80-5). The reveal also improves the appearance of the door trim.

Set the blade of the combination square so that it extends 5/16 inch beyond the body of the square. Gauge lines at intervals along the side and head jamb edges by riding the square against the inside face of the jamb. Let the lines intersect where side and head jambs meet.

The following procedure applies to molded door casings mitered at the head. If several door openings are to be cased, cut the necessary number of *casings* to rough lengths with a miter cut on one end of each piece. Rough lengths are a few inches longer than actually needed. For each interior door opening, four *side casings* and two *header casings* are required. Cut side casings in pairs with right- and left-hand miters for use on both sides of the opening.

Applying the Header Casing

Miter one end of the header casing. Hold it against the header jamb of the door frame so that the miter is on the intersection of the gauged lines. Mark the length of the header casing at the intersection of the gauged lines on the opposite side of the door frame. Miter the casing to length at the mark.

Fasten the header casing in position. Its inside edge should be to the gauged lines on the header jamb. The mitered ends should be in line with the gauged lines on the side jambs. Use finish nails along the inside edge of the casing into the header jamb. If the casing edge is thin, use 3d or 4d finish nails spaced about 12 inches apart. Keep the edge of

Fig. 80-3 Back bands may be applied to improve the appearance of S4S door casings. They may also be used on some types of molded casings.

INTERIOR WALL SURFACE

CASING

BACKED OUT SIDE

EDGE OF JAMB AND WALL SURFACE ARE IN LINE

DOOR STOP

DOOR JAMB

EDGE OF JAMB

NOT IN LINE

WALL SURFACE

INTERIOR WALL SURFACE

BACKING OUT CASING PROVIDES ROOM IN CASE JAMB AND WALL DO NOT LINE UP

PLAN OF SIDE JAMB

Fig. 80-4 Backing out door casings allows for a tight fit on wall and jamb.

SHIMS

DOOR STOP

BASE

CASING

JAMB

THIS DISTANCE IS CALLED A REVEAL

Fig. 80-5 The setback of the door casing on the jamb is called a reveal.

the casing to the gauged lines on the jamb. Straighten the casing as necessary as nailing progresses. Drive nails at the proper angle to keep them from coming through the face or back side of the jamb. Pneumatic finish nailers speed up the job of fastening interior trim.

Fasten the top edge of the casing into the framing. The outside edge is thicker, so longer nails are used, usually 6d or 8d finish. They are spaced farther apart, about 16 inches OC (Fig. 80-6). Do not drive end nails at this time. It may be necessary to move the ends slightly to fit the mitered joint between head and side casings.

Applying the Side Casings

Mark one of the previously mitered side casings by turning it upside down with the point of the miter touching the floor. If the finish floor has not been laid, hold the point of the miter on a scrap block of wood that is equal in thickness to the finish floor.

Mark the side casing in line with the top edge of the header casing (Fig. 80-7). Make a square cut on the casing at the mark.

Place the side casing in position. Try the fit at the mitered joint. If the joint needs fitting, trim the mitered end of the side casing by planing thin shavings with a sharp block plane. The joint may also be fitted by shimming the casing away from the side of the chop saw and making a thin corrective cut. Shim either near or far from the saw blade as needed to hold the casing at the desired angle (Fig. 80-8). When fitted, apply a little glue to the joint. Nail the side casing in the same manner as the head casing.

Avoid sanding the joint to bring the casing faces flush. It is difficult to keep from sanding across the grain on one or the other of the pieces. Cross-grain scratches will be very noticeable, especially if the trim is to have a stained finish. Bring the faces flush, if necessary, by shimming between the back of the casing and the wall. Usually, only very thin shims are needed. Any small space between

Fig. 80-6 The header casing is mitered on both ends and fastened in place.

HELPFUL HINTS

How to lay out the length of side casings.

HEADER CASING INSTALLED

MARK SIDE CASING IN LINE WITH TOP EDGE OF HEADER CASING

EDGE OF OPENING

DOOR JAMB

CUT SQUARE END AND PLACE ON FLOOR TO INSTALL SIDE CASING

MARK AND INSTALL OTHER SIDE IN LIKE MANNER

DOOR JAMB

MITERED END OF SIDE CASING AGAINST FLOOR

ELEVATION OF DOOR OPENING

Fig. 80-7

the casing and the wall can be filled later with joint filling compound. Most carpenters prefer to do this rather than try to sand the joint.

Drive a 4d finish nail into the edge of the casing and through the mitered joint. Drive end nails. Then set all fasteners. Keep nails 2 or 3 inches from the end to avoid splitting the casing. If there is danger of splitting, blunt the pointed end or drill a hole slightly smaller in diameter than the nail.

Applying Base Moldings

Molded or *S4S* stock may be used for base. If S4S base is used, it should be backed out. A *base cap* should be applied to its top edge. The base cap conforms easily to the wall surface, resulting in a tight fit against the wall. The base trim should be thinner than the door casings against which it butts. This makes a more attractive appearance.

HELPFUL HINTS

Shim casings in the power miter box to make corrective cuts for fitting miter joints.

SHIM

CASING

SHIM

POWER MITER BOX

PLAN VIEW OF POWER MITER BOX

Fig. 80-8

The base is applied in a manner similar to wall and ceiling molding. However, copes are laid out for joints in interior corners. Instead of back-mitering to outline the cope, it is usually more accurate to lay out the cope by *scribing* (Fig. 80-9). When placed against the wall, the face of the base may not always be square with the floor. Therefore, if the base is tilted slightly, back-mitering to obtain the outline of the cope will result in a poor fit against it.

Apply the base to the first wall with square ends in each corner. Drive and set two finishing nails, of sufficient length, at each stud location. Nailing blocks previously installed during framing provide solid wood for fastening the ends of the base in interior corners.

Cut the base to go on the next wall about an inch longer than required. Lay the base against the wall by bending it so the end to be scribed lies flat against the wall and against the first base. Set the dividers to scribe about 1/2 inch. Lay out the cope by riding the dividers along the face of the base on the first wall.

Hold the dividers while scribing so that a line between the two points is parallel to the floor. Twisting the dividers while making the scribe results in an inaccurate layout. Cut the end to the scribed line with a slight undercut. Bend the base back in position and try the fit. If scribed and cut accurately, no adjustments should be necessary. Its overall length must now be determined.

Cutting the Base to Length

The length of a baseboard that fits between two walls may be determined by measuring from corner to corner. Then, transfer the measurement to the baseboard. Another method of determining its length eliminates using the rule. It may be faster and more accurate.

With the base in the last position described above, place marks, near the center, on the top edge of the base and the wall so they line up with each other. Place the other end in the opposite corner. Press the base against the wall at the mark. The difference between the mark on the wall and the mark on the base is the amount to scribe off the end in the corner. Set the dividers to this distance. Scribe the end. Cut to the scribed line (Fig. 80-10). If a tighter fit is desired, set the dividers slightly less than the distance between the marks. This method of fitting long lengths between walls may be applied to other kinds of trim. However, this works especially well with the base.

Place one end in the corner, and bow out the center. Place the other end in the opposite corner, and press the center against the wall. Fasten in place. Continue in this manner around the room in a previously planned order. Make regular miter joints on outside corners.

If both ends of a single piece are to have regular miters for outside corners, it is imperative that it be fastened in the same position as it was marked. Tack the rough length in position with one finish nail in the center. Mark both ends. Remove, and cut the miters. Installing the piece by first fastening into the original nail hole assures that the piece is fastened in the same position as marked (Fig. 80-11).

Applying the Base Cap and Base Shoe

The *base cap* is applied in the same manner as most wall or ceiling molding. Cope interior corners and miter exterior corners. The *base shoe* is also applied in a similar manner as other molding. However, it is ordinarily nailed into the floor and not into the baseboard. This prevents the joint under the shoe

Fig. 80-9 Laying out a coped joint on base by scribing.

HELPFUL HINTS

How to lay out the length of a base that fits between walls.

WALL

2. MARK WALL AND TOP EDGE OF BASE NEAR CENTER

BASE

WALL

1. BEND BASE AGAINST WALL WITH ONE END IN CORNER

WALL

PLAN

WALL

MARK ON WALL → ← MARK ON BASE

3. PLACE OTHER END OF BASE IN OPPOSITE CORNER
4. SCRIBE THE DIFFERENCE BETWEEN MARKS OFF THE END IN THE CORNER

WALL →

← WALL

PLAN

Fig. 80-10

from opening should the shrinkage take place in the baseboard (Fig. 80-12).

Because the base shoe is a small-size molding and has solid backing, both interior and exterior corners are mitered. When the base shoe must be stopped at a door opening or other location, with nothing to butt against, its exposed end is generally *back-mitered* and sanded smooth (Fig. 80-13). No base shoe is required if carpeting is to be used as a finish floor.

Installing Window Trim

Interior window trim, in order of installation, consists of the *stool*, also called *stool cap, apron, jamb extensions, casings*, and *stops* or *stop bead*. Although

HELPFUL HINTS

How to lay out and position base or other molding that has both ends mitered on outside corners.

OUTSIDE CORNER

OUTSIDE CORNER

END OVERLAPS CORNER

1. TACK BASE IN PLACE WITH ONE NAIL.

END OVERLAPS CORNER

BASE

FINISH NAIL

2. LAY OUT MITERS ON BOTH ENDS

OUTSIDE CORNER

OUTSIDE CORNER

3. REMOVE FROM WALL AND CUT MITERS

4. REPLACE BASE IN SAME POSITION BY DRIVING FIRST NAIL IN SAME HOLE.

MITERED END

BASE

FINISH NAIL

MITERED END

Fig. 80-11

Fig. 80-12 The base shoe is fastened into the floor, not into the baseboard. This prevents the joint at the floor from opening due to any movement of the baseboard.

Fig. 80-13 The exposed ends of base shoe are usually back-mitered and sanded smooth.

the kind and amount of trim may differ, depending on the style of window, the application is basically the same. The procedure described in this chapter applies to most double hung windows.

Installing the Stool

The bottom side of the *stool* is rabbeted at an angle to fit on the sill of the window frame so its top side will be level. Its final position has the outside edge against the sash. The inside edge should extend beyond the window casing by the casing thickness. Both ends are notched around the side jambs of the window frame. Each end projects beyond the casings by an amount equal to the casing thickness.

On both sides of the window, just above the sill, hold a scrap piece of casing stock on the wall. Its inside edge should be flush with the inside face

of the side jamb of the window frame. Draw a light line on the wall along the outside edge of the casing stock. Lay out a distance outward from each line equal to the thickness of the window casing. Cut a piece of stool stock to length equal to the distance between the outermost marks.

Raise the lower sash slightly. Place a short, thin strip of wood under it, on each side, which projects inward to support the stool while it is being laid out (Fig. 80-14). Place the stool on the strips. Raise or lower the sash slightly so the top of the stool is level. Position the stool with its outside edge against the wall. Its ends should be in line with the marks previously made on the wall.

Square lines, across the face of the stool, even with the inside face of each side jamb of the window frame. Set the pencil dividers or scribers so that, on both sides, an amount equal to twice the

RAISE LOWER SASH SLIGHTLY TO INSERT SUPPORT FOR STOOL

STOOL

WINDOW SILL

THIN STRIP OF WOOD OR WOOD SHINGLE TO SUPPORT STOOL

Fig. 80-14

casing thickness will be left on the stool. Scribe the stool by riding the dividers along the wall on both sides and along the bottom rail of the window sash (Fig. 80-15).

Cut to the lines, using a handsaw. Smooth the sawed edge that will be against the sash. Shape and smooth both ends of the stool the same as the inside edge. Apply a small amount of caulking compound to the bottom of the stool along its outside edge. Fasten the stool in position by driving finish nails along its outside edge into the sill. Set the nails.

Applying the Apron

The apron covers the joint between the sill and the wall. It is applied with its ends in line with the outside edges of the window casing. Cut a length of apron stock equal to the distance between the outer edges of the window casings.

Each end of the apron is then *returned upon itself.* This means that the ends are shaped the same as its face. To return an end upon itself, hold a scrap piece on the apron. Draw its profile flush with the end. Cut to the line with a coping saw. Sand the cut end smooth (Fig. 80-16). Return the other end upon itself in the same manner.

Place the apron in position with its upper edge against the bottom of the stool. Be careful not to force the stool upward. Keep the top side of the stool level by holding a square between it and the edge of the side jamb. Fasten the apron along its bottom edge into the wall. Then drive nails through the stool into the top edge of the apron. When nailing through the stool, wedge a short length of 1×4

SIDE JAMB OF WINDOW

SIDE JAMB OF WINDOW

CASING WIDTH PLUS CASING THICKNESS

TWICE CASING THICKNESS

OVERALL LENGTH = DISTANCE BETWEEN JAMBS PLUS TWICE THE CASING WIDTH PLUS TWICE THE CASING THICKNESS

Fig. 80-15 Method of laying out a stool.

USE SCRAP PIECE TO LAY OUT PROFILE

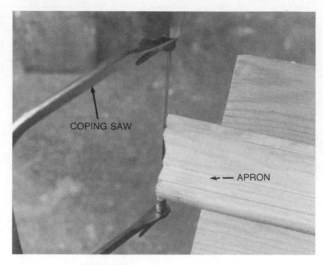

CUT PROFILE WITH COPING SAW

Fig. 80-16 Returning the end of an apron upon itself.

stock between the apron and the floor at each nail location. This supports the apron while nails are being driven. Failure to support the apron results in an open joint between it and the stool. Take care not to damage the bottom edge of the apron with the supporting piece (Fig. 80-17).

Installing Jamb Extensions

Windows are often installed with jambs that are narrower than the wall thickness. Strips must be fastened to these narrow jambs to bring the inside edges flush with inside wall surface. These strips are called *jamb extensions.*

Some manufacturers provide jamb extensions with the window unit. However, they are not applied when the window is installed, but when the window is trimmed. Therefore when windows are set, these pieces should be carefully stored and then retrieved when it is time to apply the trim. They are usually precut to length and need only to be cut to width.

Measure the distance from the inside edge of the jamb to the finished wall. Rip the jamb extensions to this width with a slight back-bevel on the inside edge. Cut the pieces to length, if necessary, and apply them to the header and side jambs. Drive finish nails through the edges into the edge of the jambs (Fig. 80-18).

HELPFUL HINTS

Support the apron when fastening the stool to it.

NAIL THROUGH STOOL INTO TOP EDGE OF APRON

STOOL

APRON

SUPPORT APRON AT EACH NAIL LOCATION WITH A LENGTH OF 1X4 STOCK

FLOOR

Fig. 80-17

Fig. 80-18 Jamb extensions are used to widen the window jamb.

Applying the Casings

Casings are applied with their inside edges flush with the inside face of the jambs. The bottom ends of the side casings rest on the stool. The window casing pattern is usually the same as the door casings. Window casings are applied in the same manner as door casings.

Cut the number of window casings needed to a rough length with a miter on one end. Cut side casings with left- and right-hand miters. Install the header casing first and then the side casings. Find the length of side casings by turning them upside down with the point of the miter on the stool in the same manner as door casings. Fasten casings with their inside edges flush with the inside face of the jamb. Make neat, tight-fitting joints at the stool and at the head.

Installing the Stops

Stops are applied to the head and side jambs. Usually the stop applied to the head jamb is fastened with small finish nails. Stops applied to side jambs are usually fastened with screws. This allows for easy removal in case the sash needs to be taken out of the frame. Joints at the top between head and side stops are usually made with a combination miter and butt joint. The molded edge of the stop is mitered. The flat sides are butted (Fig. 80-19). The bottom ends are cut square to fit against the top surface of the stool.

Cut the stops to width, if necessary. Install the stop at the head first and then the side stops. Fasten the stops so that the sash slides smoothly. The stop covers the joint between the jamb and casing and also between the jamb and the jamb extension.

Installing Closet Trim

A simple clothes closet is normally furnished with a *shelf* and a *rod* for hanging clothes. Usually a piece of 1×5 stock is installed around the walls of the closet to support the shelf and the rod. This piece is called a *cleat*. The shelf is installed on top of it. The closet pole is installed in the center of it. Shelves are not fastened to the cleat. Rods are installed for easy removal in case the closet walls need refinishing.

Shelves are usually 1×12 boards. Rods may be 3/4-inch steel pipe, 1 5/16-inch full round wood poles, or chrome plated rods manufactured for this purpose. On long spans, the rod may be supported in its center by special metal closet pole supports. On each end, the closet pole is supported by plastic or metal closet pole *sockets*. In place of sockets, holes and notches are made in the cleat to support the ends of the closet pole.

For ordinary clothes closets, the height from the floor to the top edge of the cleat is 66 inches (Fig. 80-20). Measure up from the floor this distance. Draw a level line on the back wall and two end walls of the closet. Ease the bottom outside corner and install the cleat so its top edges are to the line.

Fig. 80-19 The joint between head and side window stops is usually a combination miter and butt joint. The miter is made only the width of the molded edge.

The cleat is installed in the same manner as baseboard. Butt the interior corners. Fasten with two finish nails at each stud.

Fig. 80-20 Specifications for an ordinary clothes closet.

Install the closet pole sockets on the end cleats. The center of the socket should be at least 12 inches from the back wall and centered on the width of the cleat. Fasten the sockets through the predrilled holes with the screws provided.

Installing Closet Shelves

For a professional job, fit the ends and back edge of the shelf to the wall. Cut the shelf about 1/2 inch longer than the distance between end walls.

Place the shelf in position by laying one end on the cleat and tilting the other end up and resting against the wall. Scribe about 1/4 inch off the end resting on the cleat. Remove the shelf. Cut to the scribed line. Measure the distance between corners along the back wall. Transfer this measurement to the shelf, measuring from the scribed cut along the back edge of the shelf.

Place the shelf in position, tilted in the opposite direction. Set the dividers to scribe the distance from the wall to the mark on the shelf. Scribe and cut the other end of the shelf. Place the shelf into position, resting it on the cleats. Scribe the back edge to the wall to take off as little as possible. Cut to the scribed line. Ease the corners on the front edge of the shelf with a hand plane. Sand and place the shelf in position (Fig. 80-21).

STEP BY STEP

P R O C E D U R E S

PENCIL DIVIDERS
OR SCRIBERS

SHELF

CLEAT

1. CUT SHELF ABOUT 1/2 " LONGER THAN WIDTH OF CLOSET.
2. TILT SHELF IN POSITION WITH ONE END ON CLEAT. SCRIBE ABOUT 1/4 ON THIS END.
3. REMOVE SHELF AND CUT TO SCRIBED LINE.
4. FROM SCRIBED END, LAY OUT WIDTH OF CLOSET ON BACK EDGE.

SCRIBED END

WIDTH OF CLOSET

5. REPLACE SHELF IN TILTED POSITION WITH OPPOSITE END ON CLEAT.
6. SET DIVIDERS FOR DISTANCE FROM WALL TO MARK INDICATING SHELF LENGTH AT BACK EDGE.
7. SCRIBE ALONG END OF SHELF.
8. REMOVE SHELF AND CUT END TO SCRIBED LINE.

9. REPLACE SHELF WITH BOTH ENDS ON CLEAT.
10. SCRIBE BACK EDGE TO BACK WALL. SCRIBE ONLY ENOUGH TO FIT SHELF.
11. REMOVE SHELF AND CUT TO SCRIBED LINE. EASE CORNERS ON FRONT EDGE. REPLACE SHELF.

Fig. 80-21 Procedure for fitting a closet shelf.

Installing the Closet Pole

Measure the distance between pole sockets. Cut the pole to length. Install the pole on the sockets. One socket is closed. The opposite socket has an open top. Place one end of the pole in the closed socket. Then rest the other end on the opposite socket.

Linen Closets

Linen closets usually consist of a series of shelves spaced 12 to 16 inches apart. Cleats used to support shelves are 3/4×1 stock, chamfered on the bottom outside corner. A *chamfer* is a bevel that extends only partway through the thickness of the stock.

Lay out level lines for the top edges of each set of cleats. Install the cleats and shelves in the same manner as described for clothes closets.

Mantels

Mantels are used to decorate fireplaces and to cover the joint between the fireplace and the wall. Most mantels come preassembled from the factory. They are available in a number of sizes and styles (Fig. 80-22).

Study the manufacturer's directions carefully. Place the mantel against the wall. Center it on the fireplace. Scribe it to the floor or wall as necessary. Carefully fasten the mantel in place and set all nails.

Conclusion

All pieces of interior trim should be sanded smooth after they have been cut and fitted, and before they are fastened. The sanding of interior finish provides a smooth base for the application of stains, paints, and clear coatings. Always sand with the grain, never across the grain.

All sharp, exposed corners of trim should be rounded over slightly. Use a block plane to make a slight chamfer. Then round over with sandpaper.

If the trim is to be stained, make sure every trace of glue is removed. Excess glue, allowed to dry, seals the surface. It does not allow the stain to penetrate, resulting in a blotchy finish.

Be careful not to make hammer marks in the finish. Occasionally rubbing the face of the hammer with sandpaper to clean it helps prevent it from glancing off the head of a nail.

Make sure any pencil lines left along the edge of a cut are removed before fastening the pieces. Pencil marks in interior corners are difficult to re-

Fig. 80-22 Mantels come preassembled in a number of styles and sizes.

move after the pieces are fastened in position. Pencil marks show through a stained or clear finish and make the joint appear open. When marking interior trim make light, fine pencil marks.

NOTE Layout lines in the illustrations are purposely made dark and heavy only for the sake of clarity.

Make sure all joints are tight-fitting. Measure, mark, and cut carefully. Do not leave a poor fit. Do it over, if necessary!

Review Questions

Select the most appropriate answer.

1. Bed, crown, and cove moldings are used frequently as
   a. window trim.   c. part of the base.
   b. ceiling molding.   d. door casings.

2. Back bands are applied to
   a. wainscoting.   c. casings.
   b. exterior corners.   d. interior corners.

3. A stool is part of the
   a. soffit.   c. base.
   b. door trim.   d. window trim.

4. The joint between moldings in interior corners is usually
   a. coped.   c. butted.
   b. mitered.   d. bisected.

5. The setback of door casings from the face of the jamb is often referred to as a
   a. gain.   c. reveal.
   b. backset.   d. quirk.

6. Find the length of side casings by
   a. measuring the distance from floor to the header casing.
   b. marking the length on a scrap strip and transferring it to the side casing.
   c. turning the side casing upside down with the point of the miter against the floor.
   d. holding the side casing with the right end up and marking the miter.

7. If the joint between side and head casings needs fitting
   a. sand it.   c. plane it.
   b. fill it.   d. nail it.

8. The cope on baseboard is laid out more accurately by
   a. back mitering.
   b. returning it.
   c. a combination square.
   d. scribing.

9. The base shoe is fastened
   a. to the baseboard only.
   b. to both the base and the floor.
   c. to the floor only.
   d. directly to the wall.

10. When the end of a molding has no material to butt against, its end is
   a. back-mitered.   c. returned upon itself.
   b. mitered.   d. coped.

■ BUILDING FOR SUCCESS ■

The Role of Subcontractors in Construction

Over the past four decades subcontractors representing various construction-related trades have served a vital role in successful building projects. This role has been altered occasionally by influences from various sectors. Today subcontracting is affected by economic factors, federal and state laws, management trends, and the financial world, to name a few. Therefore, reliance on subcontracting in new home construction and remodeling has fluctuated.

The need to streamline a business and cut unnecessary operational costs has caused many contractors to reduce their number of employees. In turn, they utilize more subs as a way to hold costs down, reduce overhead, and optimize on any potential profit. This need is a variable with each contractor in every section of the country. Once the decision is made to utilize more independent subcontractors, a basic plan must be set in motion to secure and work along with com-

patible subs who reflect the builder's concern for quality.

In selecting subs, a builder must be careful not to look at the low estimate first. The quality builder must be certain that the sub will also be concerned about quality customer service. The chances for a successful project are greatly increased by creating a workable plan with each sub in which contracts are understood, communications are established, priorities are set, schedules are agreed on, and expected outcomes are known.

Many times the best sources for locating a quality subcontractor may be other subs, contractors, clients, or material suppliers. It is advisable to have a firm knowledge of the sub's financial stability. This will help avoid dissatisfaction over when and how payments are to be made during construction. By providing all potential subs with a complete bid package, the builder will receive the most accurate and detailed bids. The subs are to be included in any necessary preconstruction meetings in which important information about the planned project is disseminated. At these meetings the builder's quality standards can be presented, written performances can be explained, and tolerances can be described. Anticipated construction schedules must be shared with all subs, material suppliers, and the client. During the basic information exchange among the builder, client, and any potential subs, communications can help assure that problems are solved as soon as possible. A preliminary "walk-thru" of the project can occur. Once the subs are selected and scheduled into the construction process, they should be introduced to the client. This begins the process of building positive relationships based on trust.

Flexibility in scheduling will help assure that the project is completed as planned. As both known and unknown obstacles are encountered during the planning and construction phases of the project, the builder and subs will better adapt and find solutions to problems.

By establishing a policy of being firm but fair to all subcontractors, the builder will increase the potential for success. Before the bottom line of generating a profit from a job can be realized, many other goals must be met. The goal of finding and utilizing the services of quality subcontractors begins with good planning. By keeping his or her word and establishing a good business relationship with all subs, the builder will not become frustrated in the continual hunt for quality subs.

FOCUS QUESTIONS: For individual or group discussion.

1. Explain the need to find good subcontractors who demonstrate quality work.

2. As a builder, how would you locate and utilize the services of a subcontractor?

3. What are some of the greatest challenges to be overcome in working with subcontractors?

4. As a subcontractor, how would you work with a general contractor or home builder? List at least six considerations for building successful relationships with the builder and the client.

Stair Finish

The staircase is usually one of the most outstanding features of a building's interior. All stair finish work must be done in a first-class manner. Joints between stair finish members must be tight-fitting for appearance. They must also be strong enough to provide support and prevent accidents.

OBJECTIVES

After completing this unit, the student will be able to:

- name various stair finish parts and describe their location and function.

- lay out, dado, and assemble a housed-stringer staircase.

- apply finish to the stair body of open and closed staircases.

- lay out treads for winding steps.

- install a post-to-post balustrade, without fittings, from floor to balcony on the open end of a staircase.

- install an over-the-post balustrade, with fittings, on an open staircase that runs from a starting step to an intermediate landing and, then, to a balcony.

CONTENTS

Many kinds of stair finish parts are manufactured in a wide variety of wood species, such as oak, beech, cherry, poplar, pine, and hemlock. It is important to identify each of the parts, know their location, and understand their function when learning to apply stair finish.

Types of Staircases

The stair finish may be separated in two parts: the *stair body* and the *balustrade*. Important components of the stair body finish are *treads, risers,* and *finish stringers.* The stair body may be constructed as an *open* or *closed* staircase. In an open staircase, the ends of the treads are exposed to view. In a closed staircase, they butt against the wall. Staircases may be open or closed on one or both sides.

Major parts of the balustrade include *handrails, newel posts,* and *balusters.* Balustrades are constructed in either a *post-to-post* or *over-the-post* method. In the post-to-post method, the handrail is fitted between the newel posts (Fig. 81-1). In the over-the-post

method, the handrail runs continuously from top to bottom. It requires special curved sections, called *fittings,* where the handrail changes height or direction (Fig. 81-2).

Stair Body Parts

Many kinds of stair parts are required to finish the stair body.

Risers

Risers are vertical members that enclose the space between treads. They are manufactured in a thickness of 3/4 inch and in widths of 7 1/2 and 8 inches.

Treads

Treads are horizontal finish members upon which the feet are placed when ascending or descending stairs. High-quality treads are made from oak. Others are made from poplar or hard pine. They nor-

Fig. 81-1 A closed staircase with a post-to-post balustrade on a kneewall. (*Courtesy of L. J. Smith*)

818

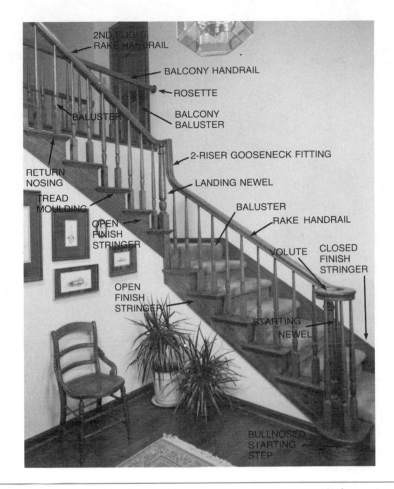

Fig. 81-2 This staircase is open on one side. It is finished with an over-the-post balustrade. (*Courtesy of L. J. Smith*)

mally come in 3/4 and 1 1/32 inch thicknesses, in 10 1/2 and 11 1/2 inch widths, and in lengths from 36 up to 72 inches.

Nosings. The outside edge of the tread beyond the riser has a half round shape. It is called the *nosing.*

A *return nosing* is a separate piece mitered to the open end of a tread to conceal its end grain. Return nosings are available in the same thickness as treads and in 1 1/4 inch widths. Treads are available with the return nosing already applied to one end (Fig. 81-3).

Landing Treads. *Landing treads* are used at the edge of landings and balconies. They match the tread thickness at the nosing. However, they are rabbeted to match the finish floor thickness on the landing. They come in 3 1/2- and 5 1/2-inch widths. The wider landing tread is used when newel posts are more than 3 1/2 inches wide (Fig. 81-4).

Tread Molding. The *tread molding* is a small cove molding used to finish the joint under stair and landing treads. The molding should be the same kind of wood as the treads. Its usual size is 5/8× 13/16.

Fig. 81-3 Treads and risers are often fastened to supporting stair carriages.

Fig. 81-4 Landing treads are rabbeted to match the thickness of the finish floor. (*Courtesy of L. J. Smith*)

Finish Stringers

Finish stringers, are sometimes called *skirt boards*. They are members of the stair body used to trim the intersection of risers and treads with the wall. They are called *closed finish stringers* when they are located above treads that butt the wall. They are termed *open finish stringers* when they are placed on the open side of a stairway below the treads (Fig. 81-5. Finish stringer lineal stock is available in a 3/4-inch thickness and in widths of 9 1/4 and 11 1/4 inches.

Starting Steps

A *starting step* is the first tread-and-riser unit sometimes used at the bottom of a stairway. The starting step is used when the staircase is open on one or both sides and the handrail curves outward at the bottom. They are available in a number of styles with *bullnosed* ends, preassembled, and ready for installation (Fig. 81-6).

Balustrade Members

Finish members of the balustrade are available in many designs that are combined to complement each other. Various types of fittings are sometimes joined to straight lengths of handrail when turns in direction are required.

Newel Posts

Newel posts are anchored securely to the staircase to support the handrail. In post-to-post balustrades, the newel posts have flat square surfaces near the top, against which the handrails are fitted, and also at the bottom for fitting and securing the post to the staircase. In between the flat surfaces, the posts may be *turned* in a variety of designs (Fig. 81-7).

In over-the-post systems, a round pin at the top of newel posts fits into the underside of handrail fittings. The posts are tapered toward the top end in a number of turned designs (Fig. 81-8).

Fig. 81-5 Open and closed finish stringers trim the intersection of the stair body and the wall.

OVER THE POST STARTING NEWELS

Fig. 81-8 Newels, in over-the-post balustrades, are made with a pin at the top. (*Courtesy of L. J. Smith*)

Fig. 81-6 Starting steps are available in a number of styles and sizes. Two commonly used types are shown. (*Courtesy of L. J. Smith*)

POST TO POST STARTING NEWELS

Fig. 81-7 Newels, in post-to-post balustrades, must have flat surfaces against which handrails are fitted. (*Courtesy of L. J. Smith*)

Three types of newel posts are used in a post-to-post balustrade. *Starting newels* are used at the bottom of a staircase. They are fitted against the first or second riser. If fitted against the second riser, the flat, square surface at the bottom must be longer. At the top of the staircase, *second floor newels* are used. *Intermediate landing newels* are also available. Because part of the bottom end of these newels

are exposed, turned *buttons* are available to finish the end. The same design is used in the same staircase for each of the three types of posts. They differ only in their overall length and in the length of the flat surfaces (Fig. 81-9).

Four types of newel posts are used in an over-the-post balustrade. There are three types of *starting newels* depending on the type of handrail fitting used. If a *volute* or *turnout* is used, a newel post with a dowel at the bottom is installed on top of a required *starting step*. The fourth type is a longer newel for landings, where a gooseneck handrail fitting is used (Fig. 81-10).

When the balustrade ends against a wall, a *half newel* is sometimes fastened to the wall. The handrail is then butted to it. In place of a half newel, the handrail may butt against an oval or round *rosette* (Fig. 81-11).

Handrails

The *handrail* is the sloping finish member grasped by the hand of the person ascending or descending the stairs. It is installed horizontally when it runs along the edge of a balcony. Several styles of handrails come in lineal lengths that are cut to fit on the job. Some handrails are *plowed* with a wide groove on the bottom side to hold square top balusters in place (Fig. 81-12).

On closed staircases, a balustrade may be installed on top of a *kneewall*. In relation to stairs, a kneewall is a short wall that projects a short distance above and on the same rake as the stair body. A *shoe rail*, which is plowed on the top side, is usually applied to the top of the kneewall on which the bottom end of balusters are fastened (Fig. 81-13). Narrow strips, called *fillets*, are used between balusters to fill the plowed groove on handrails and shoe rails.

Fig. 81-9 Three types of newel posts are used in a post-to-post balustrade. (*Courtesy of L. J. Smith*)

Fig. 81-10 Some newels for over-the-post balustrades have pins at the bottom for installation on a starting step. Others have square bottom ends for use in many locations. (*Courtesy of L. J. Smith*)

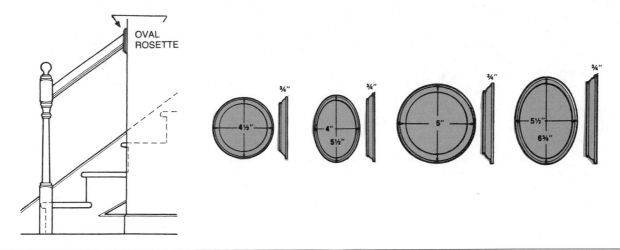

Fig. 81-11 Rosettes come in round and oval shapes. They are fastened to the wall to provide a surface on which to butt and fasten the end of a handrail. (*Courtesy of L. J. Smith*)

HANDRAILS FOR OVER THE POST AND POST TO POST

HANDRAILS FOR POST TO POST ONLY

Fig. 81-12 Straight lengths of handrail are manufactured in many styles. (*Courtesy of L. J. Smith*)

FILLET

SHOE RAIL

Fig. 81-13 A shoe rail is often used at the bottom of a balustrade that is constructed on a kneewall. (*Courtesy of L. J. Smith*)

Handrail Fittings

Short sections of specially curved handrail are called *fittings*. They are used at various locations, joined to straight sections, to change the direction of the handrail. They are classified as *starting, gooseneck,* and *miscellaneous* fittings.

VOLUTE TURNOUT STARTING EASING

THREE WAYS TO START AN OVER-THE-POST HANDRAIL

HANDRAIL

STARTING NEWEL UPEASING

AN UPEASING CAN BE USED TO START A POST-TO-POST HANDRAIL

Fig. 81-14 Handrail fittings, called volutes, turnouts, and starting easings, are used to start an over-the-post handrail. An upeasing is sometimes used to start a post-to-post handrail. (*Courtesy of L. J. Smith*)

Starting Fittings. To start an over-the-post handrail, starting fittings called *volutes, turnouts* or *starting easings* may be used. In a post-to-post system, a straight length of handrail may be used at the bottom. To start with a soft, graceful curve, an *upeasing* is used (Fig. 81-14).

Gooseneck Fittings. Over-the-post systems, in which the handrail is continuous, fittings called *goosenecks* are required at intermediate landings and at the top. This is because of changes in the handrail

height or direction. In post-to-post systems, goosenecks are not required. However, their use makes the balustrade more attractive (Fig. 81-15).

Goosenecks are available for handrails that continue level or sloping or that turn 90 or 180 degrees right or left. They are made with or without caps for both types of handrail systems.

Miscellaneous Fittings. Among the miscellaneous handrail fittings are *easings* of various kinds, *coped* and *returned ends, quarterturns,* and *caps* (Fig.

BALCONY GOOSENECK WITH CAP

LANDING

2-RISE GOOSENECK WITH CAP

R.H. 1-RISER WITH CAP

2ND FLOOR NEWEL UP

STARTING VOLUTE

R.H. 2-RISER WITH CAP

LANDING NEWEL UP

Fig. 81-15 Gooseneck fittings are used at landings when handrails change direction or height. Many shapes, with or without caps, are manufactured. (*Courtesy of L. J. Smith*)

Fig. 81-16 Many miscellaneous fittings are available to meet the need of any type of handrail installation. (*Courtesy of L. J. Smith*)

81-16). They are used where necessary to meet the specifications for the staircase. All handrail fittings are ordered to match the straight lengths of handrail being used. When the handrail being used is plowed, a matching plow is specified for the handrail fittings. A special fillet, shaped to fit the plow, comes with the handrail fitting.

Balusters

Balusters are vertical, usually decorative pieces between newel posts. They are spaced close together and support the handrail. On a kneewall, they run from the handrail to the shoe rail. On an open stair-

case, they run from the handrail to the treads (see Figs. 81-1 and 81-2).

Balusters are manufacturered in many styles. They should be selected to complement the newel posts being used (Fig. 81-17). For example, in a post-to-post staircase, balusters with square tops are usually used. In an over-the-post staircase, balusters that are tapered at the top complement the newel posts.

Most balusters are made in lengths of 31, 34, 36, 39, and 42 inches for use in any part of the stairway. Several lengths of the same style baluster are needed for each tread of the staircase because of the rake of the handrail.

Fig. 81-17 Balusters are made in designs that match newel post design. (*Courtesy of L. J. Smith*)

CHAPTER 82

FINISHING THE STAIR BODY OF OPEN AND CLOSED STAIRCASES

Review the previous chapter for the description, location, and function of the finish members of the stair body. This chapter describes their installation.

Making a Housed Stringer

Because all the major components of a housed-stringer staircase are finish members, its construction is described in this chapter. The method of determining the riser height and tread width for a housed-stringer staircase is the same as for any staircase. (It is explained in Unit 16, Stair Framing.) After these dimensions have been determined, the layout is made on the stringer stock.

Using a Pitch Board

A *pitch board* is often used instead of a framing square for laying out a housed stringer. A pitch board is a piece of stock, usually 3/4 inch thick. It is cut to the rise and run of the stairs. A strip of wood, usually a 1×2, is fastened to the *rake* edge of the pitch board. This is used to hold the pitch board against the edge while laying out the stringer (Fig. 82-1).

On the face side of the stringer stock, draw a line parallel to and about 2 inches down from the top edge. This distance may vary, depending on the width of the stringer stock and the desired height of the top edge of the stringer above the stair treads. The line is the intersection of the tread and riser faces. Using the pitch board, lay out the risers and

treads for each step of the staircase. These lines show the location of the face side of each riser and tread and are the outside edges of the housing (Fig. 82-2).

Cut and fit the bottom end of the stringer to the floor and the top end to the landing. Make end cuts that will join with the baseboard properly. The housed stringer and the baseboard should be joined in a professional manner to provide a continuous line of finish from one floor to the next. Since the stringer is usually S4S stock, the base should also be S4S. A base cap may then be used that continues from floor to floor on the raked top edge of the stringer (Fig. 82-3).

Housing the Stringer

After the stringer has been laid out, it is ready to be housed. One method of housing the stringer is to use a *template* to guide a router. The router must be equipped with a straight bit and a template guide of the correct size. Stair routing templates are manufactured that are adjustable for different rises and runs. Templates may also be made by cutting out thin plywood or hardboard to the shape of the housing. The template is shaped so the dadoes will be the exact width at the nosing and wider toward the inside. Then the treads and risers can be wedged tightly against the shoulders of the dadoes (Fig. 82-4). Clamp the template to the stringer. Rout the stringer, about 1/4 inch deep, for all treads and risers.

Fig. 82-1 A pitch board is often used for laying out a housed stringer.

Fig. 82-2 Using a pitch board to lay out a housed finish stringer.

Fig. 82-3 The baseboard is joined to the housed stringer at the bottom and top of the staircase.

STAIR ROUTING TEMPLATE

Fig. 82-4 Stair routing templates are used when routing finish stringers. Stringers are routed to allow treads and risers to be wedged and glued in place. (*Part 1, Courtesy of Porter-Cable*)

Laying Out an Open Stringer

The layout of an open (or *mitered*) stringer is similar to that of a housed stringer. However, riser and tread layout lines intersect at the top edge of the stringer, instead of against a line in from the edge. The riser layout line is the outside face of the riser. This layout line is mitered to fit the mitered end of the riser. The tread layout is to the face side of the tread. The risers and treads are marked lightly with a sharp pencil (Fig. 82-5).

To lay out the *miter cut* for the risers, measure in at right angles from the riser layout line a distance equal to the thickness of the riser stock. Draw another plumb line at this point. Square both lines across the top edge of the stringer stock. Draw a diagonal line on the top edge to mark the miter angle (Fig. 82-6).

To mark the tread cut on the stringer, measure down from the tread layout line a distance equal to the thickness of the tread stock. Draw a level line at

this point for the tread cut. The tread cut is square through the thickness of the stringer. Fit the bottom end to the floor. Fit the top end against the landing. Make the mitered plumb cuts for the risers and the square level cuts for the treads.

Fig. 82-5 Laying out the open finish stringer of a housed staircase.

Fig. 82-6 Laying out the miter angle on an open finish stringer.

Installing Risers and Treads

Cut the required number of risers to a rough length. Determine the face side of each piece. Then cut a 3/8×3/8 groove near the bottom edge on the face side of all but the starting riser. The groove is located so its top is a distance from the bottom edge equal to the tread thickness. The groove is made for the rabbeted inner edge of the tread to fit into it (Fig. 82-7). Rip the risers to width by cutting the edge opposite the groove. Rip the treads to width. Rabbet their back edges to fit in the riser grooves. Cut the risers and tread to exact lengths.

On a closed staircase, the risers are installed with wedges, glue, and screws between housed stringers. On the open side of a staircase, where the riser and open stringer meet, a miter joint is made

so no end grain is exposed (Fig. 82-8). The treads are then installed with wedges, glue, and screws on the closed side and with screws through screw blocks on the open side. Screw blocks reinforce interior corners on the underside at appropriate locations and intervals.

Applying Return Nosings and Tread Molding

If the staircase is open, a *return nosing* is mitered to the end of the tread. The back end of the return nosing projects past the riser the same amount as the tread overhangs the riser. The end is returned upon itself.

The tread molding is then applied under the overhang of the tread. If the staircase is closed on both sides, the molding is cut to fit between finish stringers. On the open end of a staircase, the molding is mitered around and under the return nosing. It is stopped and returned on itself at a point so the end assembly appears the same as at the edge (Fig. 82-9).

After the housed-stringer staircase is assembled, it is installed in position. The balustrade is then constructed in a manner similar to that described later in this chapter for framed staircases.

Finishing the Body of a Closed Staircase Supported by Stair Framing

The following section describes the installation of finish to a *closed* staircase in which the supporting

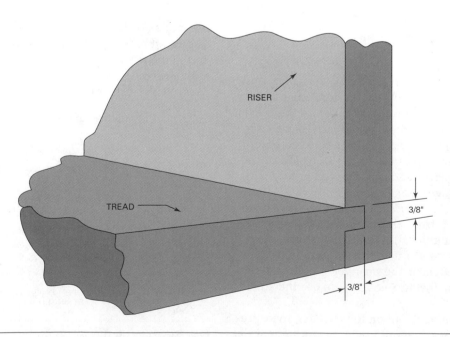

Fig. 82-7 A groove is made in the riser to receive the rabbeted inner edge of the tread.

RISERS

STAIR CARRIAGE

MITER JOINT

OPEN FINISH STRINGER

Fig. 82-8 A mitered joint is made between the risers and open finish stringer so no end grain is exposed.

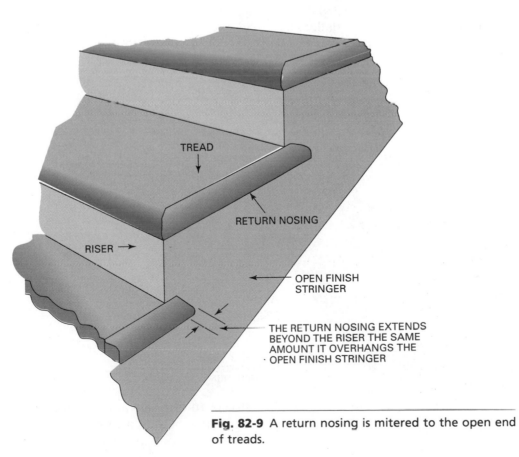

TREAD

RETURN NOSING

RISER →

OPEN FINISH STRINGER

THE RETURN NOSING EXTENDS BEYOND THE RISER THE SAME AMOUNT IT OVERHANGS THE OPEN FINISH STRINGER

Fig. 82-9 A return nosing is mitered to the open end of treads.

stair carriages have already been installed between two walls that extend from floor to ceiling.

Applying Risers

The first trim applied to the stair carriages are the *risers*. Rip the riser stock to the proper width. Cut grooves as described previously for housed-stringer construction. Cut the risers to a length, with square ends, about 1/4 inch less than the distance between walls.

Fasten the risers in position with three 8d finish nails into each stair carriage. Start at the top and work down. Remove the temporary treads installed previously as work progresses downward (Fig. 82-10).

> **CAUTION** Put up positive barriers at the top and bottom of the stairs so that the stairs cannot be used while the finish is applied. A serious accident can happen if a person who does not realize that the temporary treads have been removed uses the stairs.

Layout and Installation of the Closed Finish Stringer

After the risers have been installed, the *closed finish stringer* is cut around the previously installed risers. Usually 1×10 lumber is used. When installed, its top edge will be about 3 inches above the tread

nosing. A 1×12 may be used if a wider finish stringer is desired.

Tack a length of stringer stock to the wall. Its bottom edge should rest on the top edges of the previously installed risers. Its bottom end should rest on the floor. The top end should extend about 6 inches beyond the landing.

Lay out plumb lines, from the face of each riser, across the face of the finish stringer. If the riser itself is out of plumb, then plumb upward from that part of the riser that projects farthest outward. Then, lay out level lines on the stringer, from each tread cut of the stair carriage and also from the floor of the landing above (Fig. 82-11).

Remove the stringer from the wall. Cut to the layout lines. Use a fine-toothed crosscut handsaw. Follow the plumb lines carefully with a slight undercut. Plumb cuts will butt against the face of the risers, so a careful cut needs to be made. Not as much care needs to be taken with level cuts because treads will later butt against and cover them.

After the cutouts are made in the finish stringer, tack it back in position. Fit it to the floor. Then, lay out top and bottom ends to join the base that will later be installed on the walls. Remove the stringer. Make the end cuts. Sand the board, and place it back in position. Fasten the stringer securely to the wall with finishing nails. Do not nail too low to avoid splitting the lower end of the stringer. Install the finish stringer on the other wall in the same manner.

Drive shims, at each step, between the back

WALL

STAIR CARRIAGE

RISER

SECTION OF WALL REMOVED TO SHOW CONSTRUCTION

Fig. 82-10 Risers are ordinarily the first finish members applied to the stair carriage in a closed staircase.

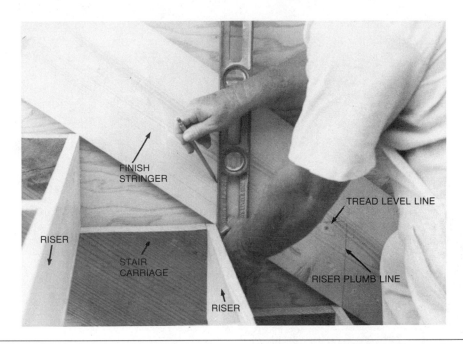

Fig. 82-11 The closed finish stringer is laid out using a level to extend plumb and level lines from the stair carriage.

side of the risers and the stair carriage. The shims force the risers tightly against the plumb cut of the finish stringer. Shim at intermediate stair carriages to straighten the risers, from end to end, between walls.

Installing Treads

Treads are cut on both ends to fit snugly between the finish stringers. The nosed edge of the tread projects beyond the face of the riser by 1 1/8 inches (Fig. 82-12).

Along the top edge of the riser, measure carefully the distance between finish stringers. Transfer the measurement and square lines across the tread. Cut in from the nosed edge. Square through the thickness for a short distance. Then undercut slightly. Smooth the cut ends with a block plane. Rub one

end with wax. Place the other end in position. Press on the waxed end until the tread lays flat on the stair carriages.

Place a short block on the nosed edge. Tap it until the inner rabbeted edge is firmly seated in the groove of the riser.

If it is possible to work from the underside, the tread may be fastened by the use of screw blocks at each stair carriage and at intermediate locations.

If it is not possible to work from the underside, the treads must be face nailed. Fasten each tread in place with three 8d finish nails into each stair carriage. It may be necessary to drill holes in hardwood treads to prevent splitting the tread or bending the nail. A little wax applied to the nail makes driving easier and helps keep the nail from bending.

Start from the bottom and work up, installing the treads in a similar manner. At the top of the stairs, install a landing tread. If 1 1/32-inch thick treads are used on the staircase, use a landing tread that is rabbeted to match the thickness of the finish floor (Fig. 82-13).

Tread Molding

The *tread molding* is installed under the overhang of the tread and against the riser. Cut the molding to the same length as the treads, using a miter box. Predrill holes. Fasten the molding in place with 4d finish nails spaced about 12 inches apart. Nails are driven at about a 45 degree angle through the center of the molding.

Fig. 82-12 Tread and riser detail.

Fig. 82-13 A rabbeted landing tread is used at the top of the stairway.

Finishing the Body of an Open Staircase Supported by Stair Framing

The following section describes the installation of finish to the stair body of a staircase, supported by stair carriages, which is closed on one side and open on the other side.

Installing the Finish Stringers

The *open finish stringer* must be installed before the *risers* and the *closed finish stringer*. To lay out the *open finish stringer*, cut a length of finish stringer stock. Fit it to the floor and against the landing. Its top edge should be flush with the top edge of the stair carriage. Tack it in this position to keep it from moving while it is being laid out.

First, lay out level lines on the face of the stringer in line with the tread cut on the stair carriage. Next, plumb lines must be laid out on the face of the finish stringer for making miter joints with risers.

Using a Preacher to Lay Out Plumb Lines.

Use a *preacher* to lay out the plumb lines on the open finish stringer. A preacher is made from a piece of nominal 4-inch stock about 12 inches long. Its thickness must be the same as the riser stock. The preacher is notched in the center. It should be wide enough to fit over the finish stringer. It should be long enough to allow the preacher to rest on the tread cut of the stair carriage when held against the rise cut.

Place the preacher over the stringer and against the rise cut of the stair carriage. Plumb the preacher with a hand level. Lay out the plumb cut on the stringer by marking along the side of the preacher that faces the bottom of the staircase (Fig. 82-14).

Mark the top edge of the stringer along the side of the preacher that faces the top of the staircase. Draw a diagonal line across the top edge of the stringer for the miter cut. Lay out all plumb lines on the stringer in this manner.

Remove the stringer. Cut to the layout lines. Make miter cuts along the plumb lines. Cut square through the thickness along the level lines. Sand the piece. Fasten it in position. To assure getting the piece in the same position as it was when laid out, fasten it first in same holes where the piece was originally tacked.

Installing Risers

Cut *risers* to length by making a square cut on the end that goes against the wall. Make miters on the other end to fit the mitered plumb cuts of the open finish stringer. Sand all pieces before installation. Apply a small amount of glue to the miters. Fasten

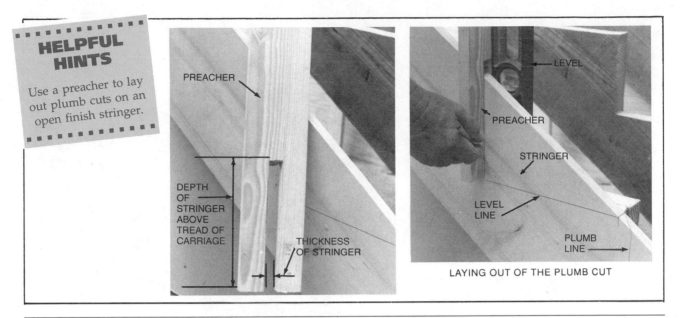

PREACHER

DEPTH OF STRINGER ABOVE TREAD OF CARRIAGE

THICKNESS OF STRINGER

LEVEL

PREACHER

STRINGER

LEVEL LINE

PLUMB LINE

LAYING OUT OF THE PLUMB CUT

Fig. 82-14

RISERS

STAIR CARRIAGE

STRINGER

MITERED JOINT

Fig. 82-15 The ends of the risers are mitered to fit against the miters of the open finish stringer.

them in position to each stair carriage. Drive finish nails both ways through the miter to hold the joint tight (Fig. 82-15). Wipe off any excess glue. Set all nails. Lay out and install the *closed finish stringer* in the same manner as described previously.

Installing the Treads

Rip the treads to width. Rabbet the back edges. Make allowance for the rabbet when ripping treads to width. Cut one end to fit against the closed finish stringer. Make a cut on the other end to receive the return nosing. This is a combination square and miter cut. The square cut is made flush with the outside face of the open finish stringer. The miter

starts a distance equal to the width of the return nosing beyond the square cut as shown in Figure 82-9.

Applying the Return Nosings

The return nosings are applied to the open ends of the treads. Miter one end of the return nosing to fit against the miter on the tread. Cut the back end square. Return the end on itself. The end of the return nosing extends beyond the face of the riser, the same amount as its width.

Predrill pilot holes in the return nosing for nails. Locate the holes so they are not in line with any balusters that will later be installed on the treads. Holes must be bored in the treads to receive the balusters. Any nails in line with the holes will damage the boring tool (Fig. 82-16).

Apply glue to the joint. Fasten the return nosing to the end of the tread with three 8d finishing nails. Set all nails. Sand the joint flush. Apply all other return nosings in the same manner. Treads may be purchased with the return nosing applied in the factory. If used, the closed end of the tread is cut so the nosed end overhangs the finish stringer by the proper amount.

Applying the Tread Molding

The *tread molding* is applied in the same manner as for closed staircases. However, it is mitered on the open end and returned back onto the open stringer. The back end of the return molding is cut and returned upon itself at a point so the end assembly shows the same as at the edge (Fig. 82-17). Predrill pilot holes in the molding. Fasten it in place. Mold-

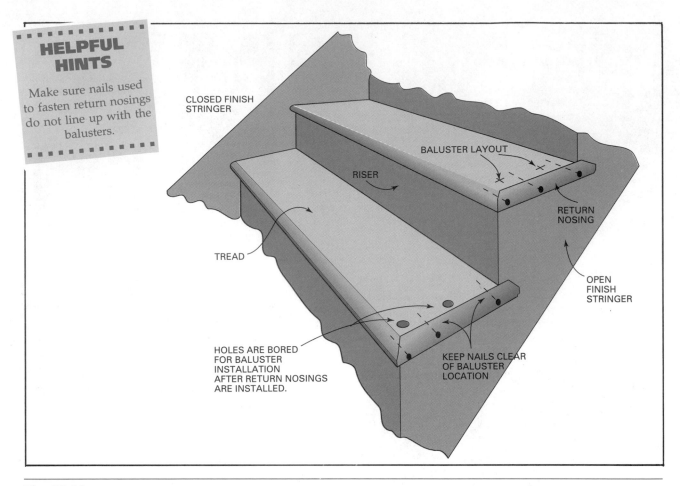

CLOSED FINISH STRINGER

BALUSTER LAYOUT

RISER

RETURN NOSING

TREAD

OPEN FINISH STRINGER

HOLES ARE BORED FOR BALUSTER INSTALLATION AFTER RETURN NOSINGS ARE INSTALLED.

KEEP NAILS CLEAR OF BALUSTER LOCATION

Fig. 82-16

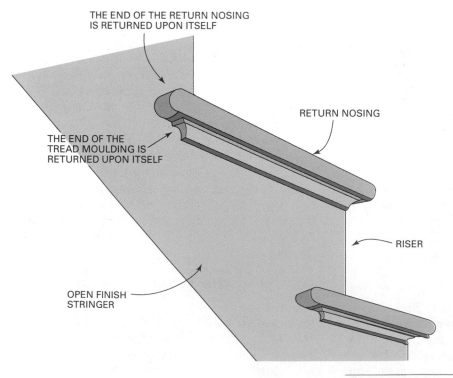

THE END OF THE RETURN NOSING IS RETURNED UPON ITSELF

RETURN NOSING

THE END OF THE TREAD MOULDING IS RETURNED UPON ITSELF

RISER

OPEN FINISH STRINGER

Fig. 82-17 The back ends of the return nosing and molding are returned upon themselves.

ing on starting and landing treads should only be tacked in case it needs to be removed for fitting after newel posts have been installed.

Installing Treads on Winders

Treads on winding steps are especially difficult to fit because of the angles on both ends. A method used by many carpenters involves the use of a pattern and a scribing block. Cut a thin piece of plywood so it fits in the tread space within 1/2 inch on the ends and back edge. The outside edge should be straight and in line with the nosed edge of the tread when installed.

Tack the plywood pattern in position. Use a 3/4-inch block, rabbeted on one side by the thickness of the pattern, to scribe the ends and back edge. Scribe by riding the block against stringers, riser, and post while marking its inside edge on the pattern. Remove the pattern. Tack it on the tread stock. Place the block with its rabbeted side down and inside edge to the scribed lines on the pattern. Mark the tread stock on the outside edge of the block (Fig. 82-18).

Protecting the Finished Stairs

Protect the risers and treads by applying a width of building paper to them. Unroll a length down the stairway. Hold the paper in position by tacking thin strips of wood to the risers.

Fig. 82-18

CHAPTER 83 — BALUSTRADE INSTALLATION

Balustrades are the most visible and complex component of a staircase (see Fig. I-9). Installing them is one of the most intricate kinds of interior finish work. This chapter describes installation of post-to-post and over-the-post balustrades. Mastering the techniques described in this chapter will enable the student to install balustrades for practically any situation.

Laying Out Balustrade Centerlines

For the installation of any balustrade, its centerline is first laid out. On an open staircase, the centerline should be located a distance inward, from the face of the finish stringer, that is equal to half the baluster width. It is laid out on top of the treads. If the balustrade is constructed on a *kneewall,* it is centered and laid out on the top of the wall (Fig. 83-1).

Lay Out Baluster Centers

The next step is to lay out the baluster centers. Code requirements for maximum baluster spacing may vary. Check the local building code for allowable spacing. Most codes require that balusters be spaced so that no object 5 inches in diameter can pass through.

On open treads, the center of the front baluster is located, a distance equal to half its thickness, back from the face of the riser. If two balusters can be used on each tread, the spacing is half the run. If codes require three balusters per tread, the spacing is one-third the run (Fig. 83-2).

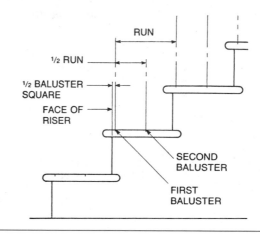

Fig. 83-2 Layout of baluster centers on open treads. (*Courtesy of L. J. Smith*)

Fig. 83-1 The centerline of the balustrade is laid out on a kneewall or open treads. (*Courtesy of L. J. Smith*)

Installing a Post-to-Post Balustrade

The following procedure applies to a post-to-post balustrade running, without interruption, from floor to floor.

Laying Out the Handrail

Clamp the handrail to the tread nosings. Use a short bar clamp from the bottom of the finish stringer to the top of the handrail. Clamp opposite a nosing to avoid bowing the handrail. Use only enough pressure to keep the handrail from moving. Protect the edges of the handrail and finish stringer with blocks to avoid marring the pieces. Use a framing square to mark the handrail where it will fit between starting and balcony newel posts (Fig. 83-3).

While the handrail is clamped in this position, use a framing square at the landing nosing to measure the *vertical thickness* of the rake handrail. Also, at the bottom, measure the height from the first tread to the top of the handrail where it butts the

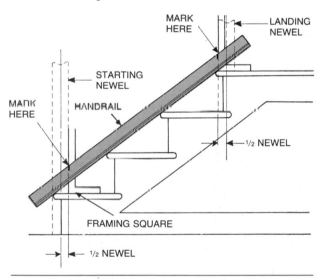

Fig. 83-3 The handrail is laid out to fit between starting and balcony newel posts. (*Courtesy of L. J. Smith*)

newel post. Record and save the measurements for later use (Fig. 83-4).

Determining the Height of the Starting Newel

Most building codes state that the rake handrail height shall be no less than 30 inches or more than 34 inches. However, some codes require heights of no less than 34 inches or more than 38 inches. Verify the height requirement with local building codes.

The height of the stair handrail is taken from the top of the tread along a plumb line flush with the face of the riser (Fig. 83-5). Handrails are required only on one side in stairways of less than 44 inches in width. Stairways wider than 44 inches require a handrail on both sides. A center handrail must be provided in stairways more than 88 inches wide.

If a *turned* starting newel post is used, add the *difference* between the two previously recorded measurements above to the required rake handrail height. Add one inch for a *block reveal*. The block reveal is the distance from the top of the handrail to the top of the square section of the post. This sum is the distance from the top of the first tread to the top

ALWAYS CHECK LOCAL BUILDING CODES TO DETERMINE HANDRAIL HEIGHT REQUIREMENTS.

Fig. 83-5 The rake handrail height is the vertical distance from the tread nosing to the top of the handrail. (*Courtesy of L. J. Smith*)

DETAIL "A"

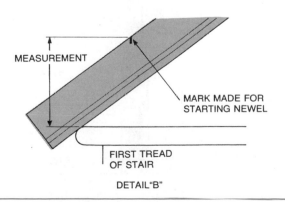

DETAIL "B"

Fig. 83-4 Determine the two measurements shown above and record for future use. (*Courtesy of L. J. Smith*)

of the upper block. To this measurement, add the height of the turned top and the distance the newel extends below the top of the first tread to the floor (Fig. 83-6). Cut the starting newel to its total length.

Fig. 83-6 Determining the height of the starting newel. (*Courtesy of L. J. Smith*)

Installing the Starting Newel

The starting newel is notched over the outside corner of the first step. One-half of its bottom thickness is left on from the front face of the post to the face of the riser. In the other direction, it is notched so its centerline will be aligned with the handrail centerline (Fig. 83-7). The post is then fastened to the first step with lag screws. The lag screws are counterbored and later concealed with wood plugs.

Newel posts must be set plumb. They must be strong enough to resist lateral force applied by persons using the staircase. The post may be slightly out of plumb after it is fastened. If so, loosen the lag screws slightly. Install thin shims, between the post and riser or finish stringer, to plumb the post. On one or both sides, install the shims, near the bottom or top of the notch as necessary, to plumb the post. When plumb, retighten the lag screws.

Installing the Balcony Newel

Generally, codes require that *balcony rails* for homes be no less than 36 inches. For commercial or public structures, the rails are required to be no less than 42 inches. Check local codes for requirements.

Fig. 83-7 The starting newel is notched to fit over the first step. It is then fastened in place. (*Courtesy of L. J. Smith*)

The height of the *balcony newel* is determined by finding the sum of the required balcony handrail height, a block reveal of one inch, the height of the turned top, and the distance the newel extends below the balcony floor.

Trim the balcony newel to the calculated height. Notch and fit it over the top riser with its centerlines aligned with both the rake and balcony handrail centerlines. Plumb it in both directions. Fasten it in place with counterbored lag bolts (Fig. 83-8).

Preparing Treads and Handrail for Baluster Installation

Bore holes in the treads at the center of each baluster. The diameter of the hole should be equal to the diameter of the pin at the bottom end of the baluster. The depth of the hole should be slightly more than the length of the pin (Fig. 83-9).

Cut the handrail to fit between starting and balcony newels. Lay it back on the tread nosings. The handrail can be cut with a handsaw, a radial arm saw, or a compound miter saw with the blade tilted. If using a power saw, make a practice cut to be sure the setup is correct, before cutting the handrail. Transfer the baluster centerlines from the treads to the handrail (Fig. 83-10).

Turn the handrail upside down. Set it back on the tread nosings with the starting newel end facing up the stairs. Bore holes at baluster centers at least 3/4 inch deep, if balusters with *round tops* are to be used (Fig. 83-11).

Fig. 83-8 The height of the balcony newel post is calculated, notched at the bottom, and fastened in place. (*Courtesy of L. J. Smith*)

Fig. 83-9 Holes are bored in the top of the treads at each baluster center point. (*Courtesy of L. J. Smith*)

Installing Handrail and Balusters

Prepare the posts for fastening the handrail by counterboring and drilling shank holes for lag bolts through the posts. Place the handrail at the correct height between newel posts. Drill pilot holes. Temporarily fasten the handrail to the posts (Fig. 83-12).

Cut the balusters to length. Allow 3/4 inch for insertion in the hole in the bottom of the handrail. The handrail may have to be removed for baluster installation and then fastened permanently. The bottom pin is inserted in the holes in the treads. The top of the baluster is inserted in the holes in the handrail bottom.

Fig. 83-10 The handrail is fitted between newel posts. The baluster centers are transferred to it. (*Courtesy of L. J. Smith*)

Fig. 83-11 Holes are bored in the bottom of the rake handrail to receive round-top balusters. (*Courtesy of L. J. Smith*)

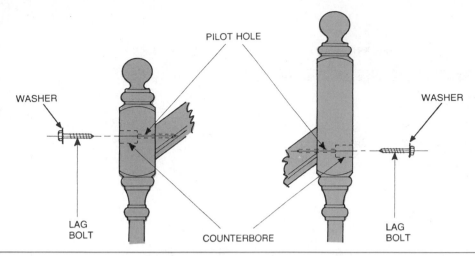

Fig. 83-12 The handrail is temporarily fastened to newel posts with lag bolts. The use of nails when constructing balustrades is discouraged. (*Courtesy of L. J. Smith*)

If *square top balusters* are used, they are trimmed to length at the rake angle. They are inserted into a *plowed handrail.* The balusters are then fastened to

the handrail with finish nails and glue (Fig. 83-13). Care must be taken to keep the handrail in a straight line from top to bottom when fastening square top

Fig. 83-13 Balusters are cut to length and installed between handrail and treads. (*Courtesy of L. J. Smith*)

Fig. 83-14 The balcony rail is fitted between the landing newel and a half newel placed against the wall. (*Courtesy of L. J. Smith*)

balusters. Care must also be taken to keep each baluster in a plumb line. Install *fillets* in the plow of the handrail, between the balusters.

Installing the Balcony Balustrade

Cut a *half newel* to the same height as the balcony newel. Temporarily place it against the wall. Mark the length of the balcony handrail (Fig. 83-14). Cut the handrail to length. Fasten the half newel to one end of it. Temporarily fasten the half newel to the wall and the other end of the handrail to the landing newel, if they must be removed to install the balcony balusters.

If the balcony handrail ends at the wall against a *rosette*, first fasten the rosette to the end of the handrail. Hold the rosette against the wall. Mark the length of the handrail at the landing new-

el. Cut the handrail to length. Temporarily fasten it in place (Fig. 83-15).

Spacing and Installing Balcony Balusters

The balcony balusters are spaced by adding the thickness of one baluster to the distance between the balcony newel and the half newel. The overall distance is then divided into spaces that equal, as close as possible, the spacing of the rake balusters (Fig. 83-16). The balcony balusters are then installed in a manner similar to the rake balusters.

Installing an Over-the-Post Balustrade

The following procedures apply to an over-the-post

Fig. 83-15 A rosette is sometimes used to end the balcony handrail instead of a half newel. (*Courtesy of L. J. Smith*)

Fig. 83-16 Balcony balusters are installed as close to possible to the same spacing as the rake balusters. (*Courtesy of L. J. Smith*)

balustrade running from floor to floor with an intermediate landing (see Fig. 81–2). An over-the-post balustrade is more complicated. *Handrail fittings* are required to be joined to straight sections to construct a continuous handrail from start to end. The procedure consists of constructing the entire handrail first, then setting newel posts and installing balusters.

Constructing the Handrail

The first step is to lay out balustrade and baluster centerlines on the stair treads as described previously. If a *starting step* is used, lay out the baluster and starting newel centers using a template provided with the starting fitting (Fig. 83-17).

Laying Out the Starting Fitting

Make a *pitch block* by cutting a piece of wood in the shape of a right triangle whose sides are equal in length to the rise and tread run of the stairs. Hold the cap of the *starting fitting* and the run side of the pitch block on a flat surface. The rake edge of the pitch block should be against the bottom side of the fitting. Mark the fitting at the *tangent point,* where its curved surface touches the pitch block. A straight line, *tangent* to a curved line, touches the curved line at one point only.

Then turn the pitch block so its rise side is on the flat surface. Mark the fitting along the rake edge of the pitch block. Place the fitting in a power miter box, supported by the pitch block. Cut it to the layout line (Fig. 83-18).

Fig. 83-17 Baluster and newel post centers are laid out on a starting step with a template. (*Courtesy of L. J. Smith*)

Fig. 83-18 Procedure for laying out and cutting the starting fitting. (*Courtesy of L. J. Smith*)

CAUTION When cutting handrail fittings, supported by a pitch block in a power miter box, clamp the pitch block and fitting securely to make sure they will not move when cut. If either one moves, during cutting, a serious injury could result. Even if no injury occurs, the fitting would probably be ruined.

Joining the Starting Fitting to the Handrail

The starting fitting is joined to a straight section of handrail by means of a special *handrail bolt*. The bolt has threads on one end designed for fastening into wood. The other end is threaded for a nut. Holes must be drilled in the end of both the fitting and the handrail in a manner that assures their alignment when joined together.

To mark the hole locations, make a template by cutting about a 1/8-inch piece of handrail. Drill a 1/16-inch hole centered on the template width and 15/16 inch from the bottom side. Mark one side *rail* and the other side *fitting*. This will ensure that the template is facing in the right direction when making the layout. If the hole in the template is drilled slightly off center and the template turned when making the layout, the handrail and fitting will not be in alignment when joined. Use the template to mark the location of the rail bolt on the end of the fitting and the handrail.

Drill all holes to the depth and diameter shown in Figure 83-19. Double nut the rail bolt and turn it into the fitting. Remove the nuts, place the handrail on the bolt. Install a washer and nut, and tighten to join the sections together. Clamp the assembly to the tread nosings. The newel center points on the fitting and the starting tread should be in a plumb line (Fig. 83-20).

Installing the Landing Fitting

The second flight of stairs turns at the landing, so a right hand, two-riser *gooseneck* fitting is used. It is laid out and joined to the bottom end of the handrail of the second flight in a similar manner as the starting fitting (Fig. 83-21). The assembled fitting and rail are then clamped to the nosings of the treads in the second flight. Position the rail so that

Fig. 83-19 A template is made to mark the ends of handrails and fittings for joining with rail bolts. The ends are drilled to specific depths and diameters.

Fig. 83-20 The starting fitting and handrail assembly is clamped to the tread nosings of the first flight of stairs. (*Courtesy of L. J. Smith*)

the *newel cap* of the fitting is in a plumb line with the baluster centerlines of both flights (Fig. 83-22).

Fig. 83-21 Procedure for laying out and cutting the landing fitting for joining to the second flight handrail. (*Courtesy of L. J. Smith*)

Joining First and Second Flight Handrails

In preparation for laying out the *easement* used to join the first and second flight handrails, tack a piece of plywood about 5 inches wide to the bottom side of the gooseneck fitting and the handrail of the first flight. These pieces are used to rest the connecting easement against when laying out the joint.

Clamp the handrails back on the treads. Make sure the newel cap centers are plumb with starting and landing newel post centers. Rest the connecting easement on the plywood blocks. Level its upper end. Mark the gooseneck in line with the upper end. Mark the lower end of the easement at a point *tangent* with the block under the handrail (Fig. 83-23).

Fig. 83-22 The second flight handrail is clamped to the tread nosings after the landing fitting is joined to it. (*Courtesy of L. J. Smith*)

Fig. 83-23 The connecting easement is positioned on small blocks while it and the landing fitting are marked. (*Courtesy of L. J. Smith*)

Make a square cut at the mark laid out on the lower end of the gooseneck. Join the gooseneck and easement together with a rail bolt. Lay out the lower end of the easement with the pitch block. Cut it using the pitch block for support (Fig. 83-24).

Place the assembled landing fitting and handrail back on the stair nosings. Mark the handrail of the lower flight where it meets the end of the fitting.

Cut the handrail square at the mark. Join it to the landing fitting with a rail bolt. Clamp the entire handrail assembly to the tread nosings with newel cap centers plumb with newel post centerlines (Fig. 83-25).

Fig. 83-24 The connecting easement and landing fitting are joined. Then the lower end is then trimmed to form a two-riser gooseneck. (*Courtesy of L. J. Smith*)

Fig. 83-25 Lower and upper handrail assemblies are joined. They are then clamped to the tread nosings. (*Courtesy of L. J. Smith*)

Installing the Balcony Gooseneck Fitting

A one-riser *balcony gooseneck* fitting is used when balcony rails are 36 inches high. Two-riser fittings are used for rails 42 inches high. In this case, a one-riser fitting is used at the balcony. Laying out and fitting a two-riser fitting at the landing has been previously described.

Hold the fitting so the center of its cap is directly above the balcony newel post centerline. Hold it against the handrail of the upper flight. Mark it and the handrail at the point of tangent (Fig. 83-26). Lay out and make the cut on the gooseneck with the use of a pitch block (Fig. 83-27). Cut the handrail square at the mark. Join the gooseneck fitting and the handrail with a rail bolt.

Clamp the entire rail assembly back on the nosing of the treads in line with the balustrade centerlines and the three newel post centers. Use a framing square to transfer the baluster centers from the treads to the side of the handrails. Remove the handrail assembly out of the way of newel post installation.

Installing Newel Posts

In this stair case, the starting newel is installed on a starting step. The height of the *rake handrail* is calculated, from the height of the starting newel, to make sure the handrail will conform to the height required by the building code.

From the height of the starting newel to be

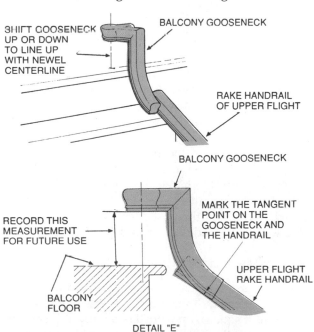

Fig. 83-26 Method used to mark the balcony gooseneck fitting and handrail of the upper flight. (*Courtesy of L. J. Smith*)

Fig. 83-27 Procedure for laying out and cutting the balcony gooseneck fitting. (*Courtesy of L. J. Smith*)

used, subtract the previously recorded distance between the starting fitting and the starting tread as shown in Detail "C" of Figure 83-20. Then, add the vertical thickness of the rake handrail shown in Figure 83-4, Detail "A." The result is the rake handrail height (Fig. 83-28). If the height does not conform to the building code, the starting newel post height must be changed.

Installing the Starting Newel Post

Before installing the starting step, measure the diameter of the *dowel* at the bottom of the starting newel post. At the centerpoint of the newel post, bore a hole for the dowel through the tread and floor. Install the post. Wedge it under the floor. Wedges are driven in a *through mortise* cut in the doweled end of the post.

An alternate method is used when there is no access under the floor. Bore holes only through the tread and upper riser block of the starting step. Cut the dowel to fit against the lower riser block so the newel post rests snugly on the tread. Fasten the end of the dowel with a lag screw through the lower riser block (Fig. 83-29). Fasten the assembled starting newel and starting step in position.

Installing the Landing Newel Post

The height of *landing newel* above the landing is found by subtracting the previously recorded distance between the starting fitting and the starting tread (Fig. 83-20, Detail "C") from the height of the starting newel. Then add the distance between the

PIN

A

DIMENSION
RECORDED

DOWEL

① FROM DIMENSION A
  SUBTRACT ──────→

② DIMENSION RECORDED.
  (SEE FIG. 83-20, DETAIL "C")

RAKE
HANDRAIL

VERTICAL
THICKNESS

TREAD
NOSING

③ THEN ADD VERTICAL
  THICKNESS OF HANDRAIL
  (SEE FIG. 83-4, DETAIL "A")

RAKE
HANDRAIL
HEIGHT

④ RESULT EQUALS HEIGHT
  OF HANDRAIL FOR STARTING
  NEWEL SELECTED

Fig. 83-28 The rake handrail height is calculated from the height of the starting newel post. (*Courtesy of L. J. Smith*)

NEWELL POST

LAG SCREW
AND WASHER

RISER
BLOCKS

ALTERNATE METHOD

NEWEL POST

STARTING
STEP

DOWEL END
OF NEWEL
POST

WEDGES

SECTIONS THROUGH
STARTING STEP

Fig. 83-29 Methods of installing the starting newel on a starting step.

landing fitting and the landing as previously recorded and shown in Figure 83-23, Detail "D." To this length, add the distance that the landing newel extends below the landing.

Notch the landing newel to fit over the landing and the first step of the upper flight. Fasten the post in position with lag bolts in counterbored holes (Fig. 83-30).

Installing the Balcony Newel Post

The height of the *balcony handrail* must be calculated before the height of the *balcony newel* can be determined. The height of the balcony handrail is found by subtracting the previously recorded distance between the starting fitting and the starting tread (Fig. 83-20, Detail "C",) from the height of the starting newel. Then add the previously recorded distance between the balcony gooseneck fitting and the landing (Fig. 83-27, Detail "E"). Then, add the thick-

ness of the balcony handrail. The balcony handrail height must conform to the building code. If not, substitute a two-riser gooseneck fitting instead of a one-riser fitting.

The height of the balcony newel, above the balcony floor, is found by subtracting the handrail thickness from the handrail height. To this length, add the distance the post extends below the floor. Notch and install the post over the balcony riser in line with balustrade centerlines (Fig. 83-31).

Installing Balusters

Bore holes for balusters in the tread and bottom edge of the handrail. No holes are bored in the handrail if square top balusters are used. Install the handrail on the posts. Cut the balusters to length. Install them in the manner described previously for post-to-post balustrades.

TO FIND THE NEWEL HEIGHT ABOVE
THE LEVEL OF THE LANDING:

① FROM THE STARTING NEWEL HEIGHT,
SUBTRACT PREVIOUSLY RECORDED MEASUREMENT
SEE FIG. 83-20, DETAIL "C"

② THEN, ADD PREVIOUSLY RECORDED
MEASUREMENT. SEE FIG. 83-23 DETAIL "D"

Fig. 83-30 The landing newel is notched to fit over the landing and first step of the upper flight. (*Courtesy of L. J. Smith*)

Installing the Balcony Balustrade

Cut a *half newel* to the same height as the balcony newel extends above the floor. Install it against the wall on the balustrade centerline. Cut an *opening cap*

Fig. 83-31 The height of the balcony newel is determined. The post is then fastened in place. (*Courtesy of L. J. Smith*)

so it fits on top of the half newel. Join the cap to the end of the balcony handrail with a rail bolt. Place the cap on the half newel. Mark the length of the handrail at the balcony newel. Cut the handrail and join it, on one end, to the balcony newel post and, on the other end, to the half newel (Fig. 83-32).

A *rosette* may be used against the wall instead of the half newel and opening cap. The procedure for installing a rosette has been previously described (see Fig. 83–15).

Installing Balcony Balusters

Balcony balusters are laid out and installed in the same manner as described previously for post-to-post balustrades. Balconies with a span of 10 feet or more should have intermediate balcony newels installed every 5 or 6 feet.

Fig. 83-32 Procedure for installing a half newel and the balcony handrail. (*Courtesy of L. J. Smith*)

Review Questions

Select the most appropriate answer.

1. The rounded outside edge of a tread that extends beyond the riser is called a
   a. housing.   c. coving.
   b. turnout.   d. nosing.

2. Finish boards between the stairway and the wall are called
   a. returns.   c. stringers.
   b. balusters.   d. casings.

3. Treads are rabbeted on their back edge to fit into
   a. risers.   c. return nosings.
   b. housed stringers.   d. newel posts.

4. An open stringer is
   a. housed to receive risers.
   b. mitered to receive risers.
   c. housed to receive treads.
   d. mitered to receive treads.

5. A volute is part of a
   a. tread.   c. newel post.
   b. baluster.   d. handrail.

6. The entire rail assembly on the open side of a stairway is called a
   a. baluster.
   b. balustrade.
   c. guardrail.
   d. finish stringer assembly.

7. In a framed staircase, the treads and risers are supported by
   a. stair carriages.   c. each other.
   b. housed stringers.   d. blocking.

8. One of the first things to do when trimming a staircase is
   a. check the rough framing for rise and run.
   b. block the staircase so no one can use it.
   c. straighten the stair carriages.
   d. install all the risers.

9. Treads usually project beyond the face of the riser
   a. 3/4 inch.   c. 1 1/4 inches.
   b. 1 1/8 inches.   d. 1 3/8 inches.

10. Newel posts are notched around the stairs so that their centerline lines up with the
    a. centerline of the stair carriage.
    b. centerline of the balustrade.
    c. outside face of the open stringer.
    d. outside face of the stair carriage.

■ BUILDING FOR SUCCESS ■

Selling Your Product

If you produce a quality product, does this mean that you will have no difficulty selling it? Not necessarily. The marketing of a product is not automatically assured. There are many challenges in convincing someone to purchase your product instead of a competitor's. How can a person be certain the market is out there? How can you know that your product is better than the others, or that it will sell?

The free enterprise system demands that quality be a part of every structure that is built. Assuming that quality is there, how does a builder persuade the public to purchase his or her product? The builder can possess excellent technical skills and a good understanding of the construction industry, but fail to convince the public to buy. There are other pieces of the puzzle that must fall into place in order for a sale to be made. Let's see what criteria are necessary to convince someone to buy.

What do people want when they consider which builder or craftsperson to hire? Up front, each potential buyer is searching for a product that will meet his or her needs, is affordable, and is one with which he or she will feel satisfied. Many buyers who are considering a remodel or a new home are very consumer-smart. The single person, couple, or family that is researching the feasibility of a construction project probably will ask good questions of any builder, remodeler, or other construction technician. The builder must be knowledgeable and informed about all possible construction products and processes. He or she must be prepared to communicate with an informed public. With a large sum of money to be invested in a new home or home improvement, many buyers do their homework thoroughly.

The builder has an obligation to demonstrate that he or she is capable of successfully producing the desired product, by having the expertise, workforce, financial stability, and experience necessary to do the job.

Assuming that the need is there, the economy is good, the interest rates are low enough to encourage borrowing, and construction materials are affordable and available, many people will entertain the thought of building. They will be expecting the most for their money. The builder must thoroughly understand the potential client's needs and wants in a new home or remodel. In most cases, the money available may not be enough to purchase all of the amenities the buyer desires. The builder has to be able to discern what can be bought for the money at hand. Above all, the buyer must be satisfied that the proposed structure is a good value and investment. The builder can help assure this through good builder/client preconstruction discussions.

Potential clients may have looked at many homes or remodels in the community. Over a period of time, they will acquire many ideas that are incorporated into their desired structure. They will offer a multitude of pictures, examples, and suggestions for the builder to sort through in the design process. The challenge for the builder is how to put the puzzle pieces together and satisfy the customer within the prescribed price range.

On the surface, the buyer wants quality. He or she also wants the largest possible amount of amenities, convenience, and comfort for the least amount of money. But there are equally important considerations that the builder must explain to the customer. Today many people may desire all of the above. But they must understand that the lowest bid may not be the best in all cases. What else is there? Part of the answer lies in the area of service, project coordination, and assistance with the client's needs during and after the construction process.

Explaining to the client how the remodel construction process will not disturb the family living situation will be of great value. Assuring the customer that all of the scheduling, coordination, and problem solving will be the responsibility of the builder will establish trust. People must understand that they are paying for things other than lumber, concrete, shingles, and paint when they engage a builder. This new home or remodel could be the largest single investment they will make in their life. They need to be sure it is the right thing to do and that they have employed the right person to do it.

In order to effectively sell a quality product, a builder must demonstrate the ability to provide more than tangibles to the buyer. The builder has to legitimately convince the potential client that his or her company can provide more quality, service, and trustworthiness than competitors. Each builder or craftsperson will be challenged to do this on a regular basis. In order to provide this type of assurance and service to the customer, the builder will be looking for construction technicians that can meet these same standards. Each carpenter and other skilled technician must understand this and buy into the concept.

The construction business today has many marketing and selling demands placed on it that become everyday responsibilities for builders. Competition will dictate to each person in the industry a need to strive for excellence. Excellence will be visible not only in the durable product, but in the construction process that is used to build and market that product.

Before, during, and after the project is completed, clients will communicate to others how satisfied they were with the project and the builder. If they feel that the builder met their needs, provided quality, gained their trust, and satisfied their desires, the report will be positive. Satisfied customers will be one of the most effective marketing tools for the builder. That resource will prove to be invaluable in the quest to build and maintain a reputable business. By providing and maintaining customer satisfaction, many challenges in the selling process are lessened. It is a great advantage to any skilled worker to know what people want in a construction product and process, then meet the customer's desires. The selling process will be easier and the competition will be met.

FOCUS QUESTIONS: For individual or group discussion.

1. As you begin to address the challenges of selling your construction projects on a regular basis, what do you see as considerations affecting a successful sale?

2. How should the builder structure his or her sales strategy to develop trust with the client?

3. How could a good understanding of the competition's marketing/sales plan help your strategy as a builder? Is it possible to learn from others in the same profession?

Finish Floors

A number of materials are used for finish floors. Wall to wall *carpet*, applied over an underlayment, is widely used. The carpenter usually applies the underlayment, but not the carpet. *Terrazo floors* are the responsibility of masons. Terrazo floors are made by grinding and polishing concrete in which small marble chips or colored stone are embedded. *Resilient sheet* and *tile* floors are widely used in bathrooms and kitchens. They are usually installed by specialists. On occasion, however, the carpenter may be required to install a resilient tile floor. Carpenters install *wood finish floors*. These floors are long-time favorites, because of their durability, beauty, and warmth.

OBJECTIVES

After completing this unit, the student will be able to:

■ describe the kinds, sizes, and grades of hardwood finish flooring.

■ apply strip, plank, and parquet finish flooring.

■ estimate quantities of wood finish flooring required for various installations.

■ apply underlayment and resilient tile flooring.

■ estimate required amounts of underlayment and resilient tile for various installations.

CONTENTS

CHAPTER 84

DESCRIPTION OF WOOD FINISH FLOORS

Most hardwood finish flooring is made from white or red oak. Beech, birch, hard maple, and pecan finish flooring are also manufactured. For less expensive finish floors, some softwoods such as Douglas fir, hemlock, and southern yellow pine are used.

Kinds of Hardwood Flooring

The four basic types of solid wood finish flooring are *strip, plank, parquet strip,* and *parquet block. Laminated strip* wood flooring is a relatively new type that is gaining in popularity. *Laminated parquet blocks* are also manufactured.

Solid Wood Strip Flooring

Solid wood *strip* flooring is probably the most widely used type. Most strips are tongue-and-grooved on edges and ends to fit precisely together.

Unfinished strip flooring is milled with square, sharp corners at the intersections of the face and edges. After the floor is laid, any unevenness in the faces of adjoining pieces is removed by sanding the surface so strips are flush with each other.

Prefinished strips are sanded, finished, and waxed at the factory. They cannot be sanded after installation. A *chamfer* is machined between the face side and edges of the flooring prior to prefinishing. When installed, these chamfered edges form small V-grooves between adjoining pieces. This obscures any unevenness in the surface.

The most popular size of hardwood strip flooring is 3/4 inch thick with a *face width* of 2 1/4 inches. The face width is the width of the exposed surface between adjoining strips. It does not include the tongue. Other thicknesses and widths are manufactured (Fig. 84-1).

FACE WIDTH

OPPOSITE END TONGUED

TONGUE

HOLLOW BACK

END GROOVE

GROOVE

UNFINISHED FLOORING
SANDED SMOOTH AFTER INSTALLATION

A V-GROOVE IS PRODUCED WHEN CHAMFERED EDGES ARE JOINED.

PREFINISHED FLOORING
CHAMFERED EDGES ARE NECESSARY
TO OBSCURE UNEVENESS OF SURFACE.

Fig. 84-1 Hardwood strip flooring is edge and end matched. The edges of prefinished flooring are chamfered.

Laminated Wood Strip Flooring

Laminated strip flooring is a five-ply prefinished wood assembly. Each board is 9/16 inch thick, 7 1/2 inches wide, and 7 feet, 11 1/2 inches long. The board consists of a bottom veneer, a three-ply cross-laminated core, and a face layer. The face layer consists of three rows of hardwood strips joined snugly edge to edge (Fig. 84-2).

The uniqueness of this flooring is in the exact milling of edge and end tongue and grooves. This precision allows the boards to be joined with no noticeable unevenness of the prefinished surface. This eliminates the need to chamfer the edges. Without the chamfers, a smooth continuous surface without V-grooves results when the floor is laid (Fig. 84-3).

Plank Flooring

Solid wood *plank* flooring is similar to strip flooring. However, it comes in various mixed combinations ranging from 3 to 8 inches in width. For instance, plank flooring may be laid with alternating widths of 3 and 4 inches, 3, 4, and 6 inches, 3, 5, and 7 inches, or any random width combination.

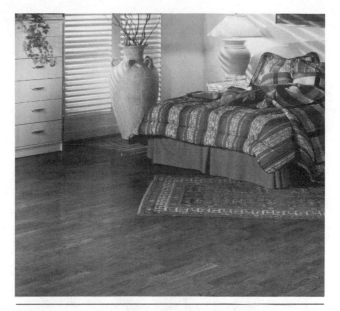

Fig. 84-3 Completed installation of a laminated strip floor. (*Courtesy of Harris-Tarkett*)

Like strips, planks are available unfinished or prefinished. The edges of some prefinished planks have deeper chamfers to accentuate the plank widths. The surface of some prefinished plank flooring may have plugs of contrasting color already installed to simulate screw fastening. One or more plugs, depending on the width of the plank, are used across the face at each end (Fig. 84-4).

Unfinished plank flooring comes with either square or chamfered edges and with or without plugs. The planks may be bored for plugs on the job, if desired.

Parquet Strips

Parquet strip flooring is a type in which short strips are laid to form various mosaic designs. The original parquet floors were laid by using short strips. Some, at the present time, are laid in the same manner. This type is manufactured in precise, short lengths, which are multiples of its width. For instance, 2 1/4 inch parquet strips come in lengths of 9, 11 1/4, 13 1/2, and 15 3/4 inches. Each piece is tongue-and-grooved on the edges and ends. Herringbone, basketweave, and other interesting patterns can be made using parquet strips (Fig. 84-5).

Parquet Blocks

Parquet block flooring consists of square or rectangular blocks, sometimes installed in combination with strips, to form mosaic designs. The three basic types are the *unit*, *laminated*, and *slat* block (Fig. 84-6).

Unit Blocks. The highest-quality parquet block is made with 3/4-inch thick, tongue-and-groove solid hardwood, usually oak. The widely used 9×9 *unit*

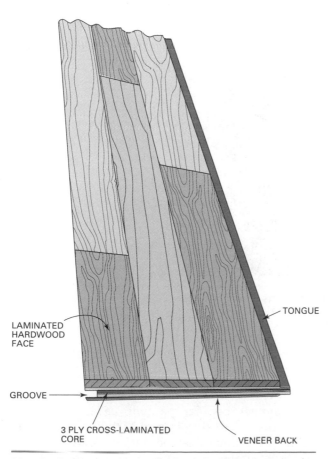

LAMINATED HARDWOOD FACE

TONGUE

GROOVE

3 PLY CROSS-LAMINATED CORE

VENEER BACK

Fig. 84-2 Cross-section of a board of laminated strip flooring.

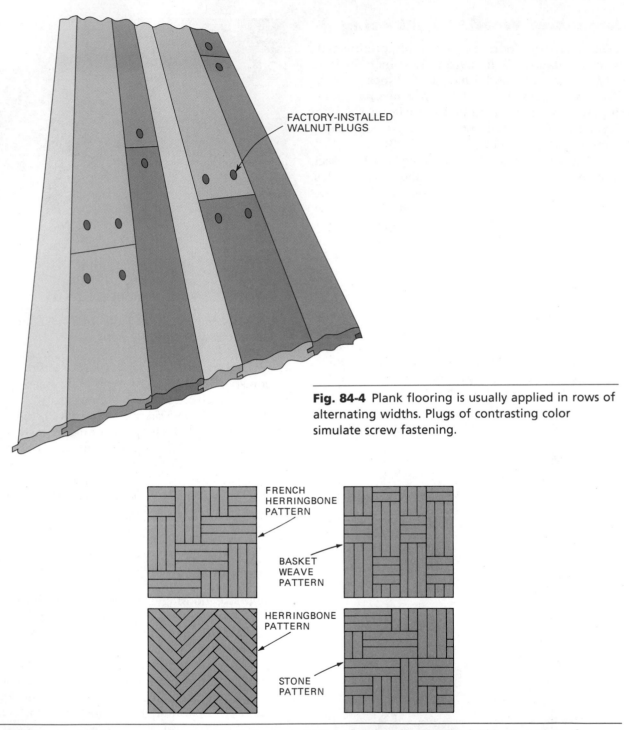

Fig. 84-4 Plank flooring is usually applied in rows of alternating widths. Plugs of contrasting color simulate screw fastening.

FACTORY-INSTALLED WALNUT PLUGS

FRENCH HERRINGBONE PATTERN

BASKET WEAVE PATTERN

HERRINGBONE PATTERN

STONE PATTERN

Fig. 84-5 Parquet strips are made in lengths that are multiples of its width. They are laid in a number of interesting patterns.

block is made with six strips, 1 1/2 inches wide, or with four strips 2 1/4 inches wide. Unit blocks are laid with the direction of the strips at right angles to adjacent blocks (Fig. 84-7). Unit blocks are made in other sizes and used in combination with parquet strips. Several patterns have gained popularity.

Monticello is the name of a parquet originally designed by Thomas Jefferson, the third president of the United States. The pattern consists of a 6×6

center unit block surrounded by 2 1/4-inch wide pointed *pickets*. Each center unit block is made of four 1 1/2-inch wide strips (Fig. 84-8). Each block comes with three pickets joined to it at the factory.

Another popular parquet, called the *Marie Antoinette*, is copied from part of the Versailles Palace in France. Square center unit blocks are enclosed by strips applied in a basketweave design (Fig. 84-9).

GROOVE

UNIT BLOCK

USUAL SIZE 3/4"×9"×9"

TONGUE

TONGUE

LAMINATED BLOCK

USUAL SIZE 3/8"×9"×9"

GROOVE

USUALLY 3 PLY.
SOME 5 PLY.

TONGUE

SLAT BLOCK

USUAL SIZE 5/16"×12"×12"

GROOVE

GROOVE

SLATS ARE
JOINED IN MANY
DIFFERENT PATTERNS.

TONGUE

Fig. 84-6 Three basic types of parquet block.

Fig. 84-7 Unit blocks are widely used in an alternating pattern.

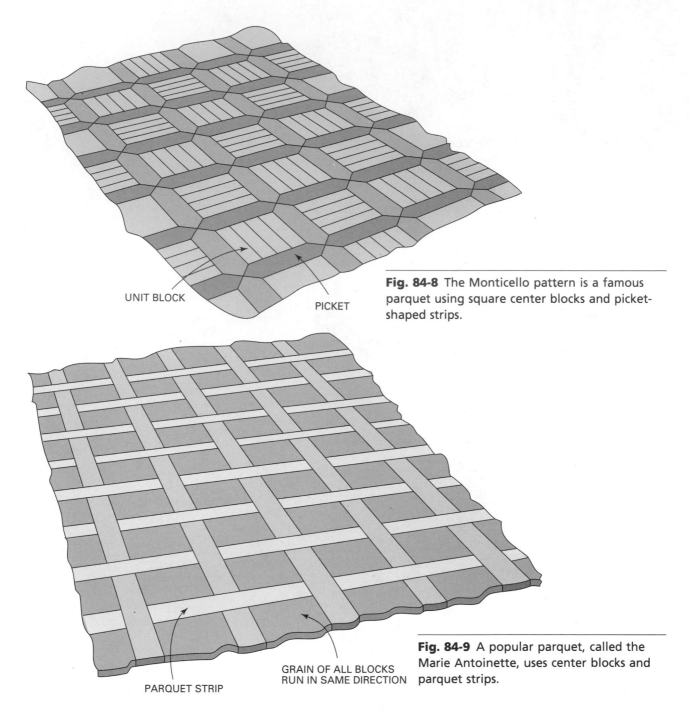

UNIT BLOCK

PICKET

Fig. 84-8 The Monticello pattern is a famous parquet using square center blocks and picket-shaped strips.

PARQUET STRIP

GRAIN OF ALL BLOCKS
RUN IN SAME DIRECTION

Fig. 84-9 A popular parquet, called the Marie Antoinette, uses center blocks and parquet strips.

Rectangular parquet blocks are often used in a *herringbone* pattern. One commonly used block is 4 1/2×9 and made of three strips each 1 1/2 inches wide and 9 inches long (Fig. 84-10).

Laminated Blocks. *Laminated blocks* are generally made of three-ply laminated oak in a 3/8-inch thickness. Most blocks come in 8×8 or 9×9 sizes. A 2×12 laminated strip is manufactured for use in herringbone and similar patterns.

Slat Blocks. *Slat blocks* are also called *finger* blocks. They are made by joining many short, narrow strips together in small squares of various patterns. Some strips may be as narrow as 5/8 inch

and as short as 2 inches or less. Several squares are assembled to make the block (Fig. 84-11). The squares are held together with a mesh backing or with a paper on the face side that is removed after the block is laid.

Presanded Flooring

Some strip, plank, and parquet finish flooring can be obtained *presanded* at the factory, but without the finish. The presanded surface eliminates the necessity for all but a touch-up sanding after installation to remove surface marks before the finish is applied.

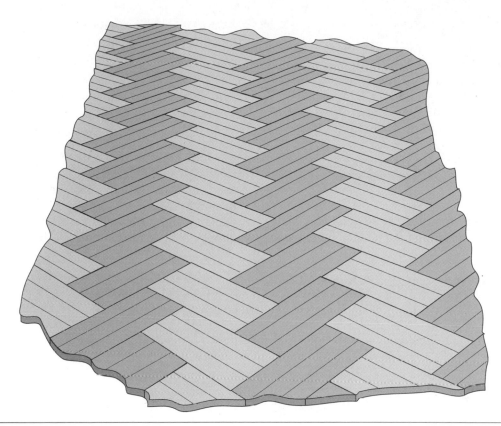

Fig. 84-10 Rectangular parquet blocks are often used to make herringbone and similar patterns.

17½ x 17½
EACH PIECE IN THIS BLOCK
MEASURES ⅝" x 4⅜"

12" x 12"

12" x 12"

12" x 12"

Fig. 84-11 Slat blocks are made of short, narrow strips joined together in various patterns. (*Courtesy of Bruce Hardwood Floors*)

Grades of Hardwood Flooring

Uniform grading rules have been established for strip and plank solid hardwood flooring by the National Oak Flooring Manufacturers Association. The association's trademark on flooring assures consumers that the flooring is manufactured and graded in compliance with established quality standards. Other types of wood finish flooring, such as parquet and laminated flooring, have no official grade rules.

Unfinished Flooring

Oak flooring is available *quarter-sawed* and *plain-sawed*. The grades for unfinished oak flooring, in declining order, are *clear, select, no. 1 common, no. 2 common,* and *1 1/4-foot shorts.* Quarter-sawed flooring is available in clear and select grades only.

Birch, beech, and hard maple flooring are graded in declining order as *first grade, second grade, third grade,* and *special grade.* Grades of pecan flooring are *first grade, first grade red, first grade white, second grade, second grade red,* and *third grade.* Red grades contain all heartwood. White grades are all bright sapwood.

In addition to appearance, grades are based on length. For instance, bundles of 1 1/4-foot shorts contain pieces from 9 to 18 inches long. The average length of clear bundles is 3 3/4 feet. The flooring comes in bundles in lengths of 1 1/4 feet and up. Pieces in each bundle are not of equal lengths. A bundle may include pieces from 6 inches under to 6 inches over the nominal length of the bundle. No pieces shorter than 9 inches are allowed.

Prefinished Flooring

Grades of *prefinished* flooring are determined after it has been sanded and finished. In declining order, they are *prime, standard and better, standard,* and *tavern*. Prefinished beech and pecan are furnished only in a combination grade called *tavern or better.* A guide to hardwood flooring grades is shown in Figure 84-12.

Estimating Hardwood Flooring

To estimate the amount of hardwood flooring material needed, first determine the area to be covered.

Add to this a percentage of the area depending on the width of the flooring to be used. The percentages include an additional 5 percent for end matching and normal waste:

   55 percent for flooring 1 1/2 inches wide
   42.5 percent for flooring 2 inches wide
   38.33 percent for flooring 2 1/4 inches wide.

For example, the area of a room 16 feet by 24 feet is 384 square feet. If 2 1/4-inch flooring is to be used, multiply 384 by .3833 to get 147.18. Round this off to 147 square feet. Add 147 to 384 to get 531, which is the number of board feet of flooring required. A calculator simplifies the mathematics.

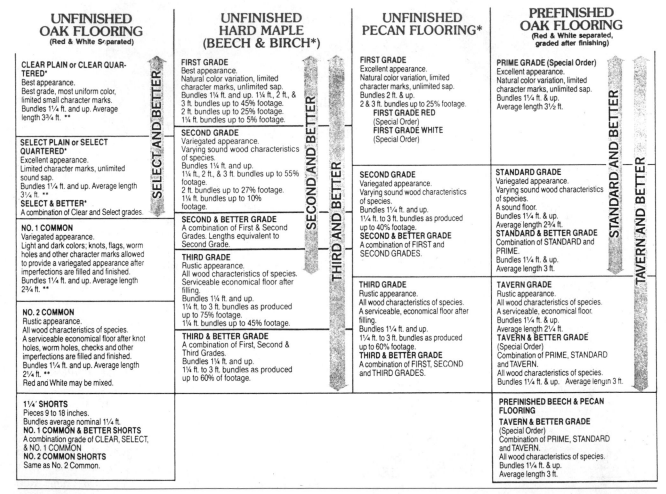

Fig. 84-12 Guide to hardwood flooring grades. (*Courtesy of National Oak Flooring Manufacturers Association*)

CHAPTER 85 LAYING WOOD FINISH FLOOR

Lumber used in the manufacture of hardwood flooring has been air-dried, kiln-dried, cooled, and then accurately machined to exacting standards. It is a fine product that should receive proper care during handling and installation.

Handling and Storage of Wood Finish Floor

Maintain moisture content of the flooring by observing recommended procedures. Flooring should not be delivered to the job site until the building has been closed in. Outside windows and doors should be in place. Cement work, plastering, and other materials must be thoroughly dry. In warm seasons, the building should be well ventilated. During cold months, the building should be heated, not exceeding 72 degrees Fahrenheit, for at least five days before delivery and until flooring is installed and finished.

Do not unload flooring in the rain. Stack the bundles in small lots in the rooms where the flooring is to be laid. Leave adequate space around the bundles for good air circulation. Let the flooring become acclimated to the atmosphere of the building for four or five days or more before installation.

Concrete Slab Preparation

Wood finish floors can be installed on an on-grade or above-grade *concrete slab*. Floors should not be installed on below-grade slabs. New slabs should be at least ninety days old. Flooring should not be installed when tests indicate excessive moisture in the slab.

Testing for Moisture

A test can be made by laying a smooth *rubber mat* on the slab. Put weight on it to prevent moisture from escaping. Allow the mat to remain in place for twenty-four hours. If moisture shows when the mat is removed, the slab is too wet. Another method of testing is by taping and sealing the edges of about a one-foot square of 6 mil clear *polyethylene film* to the slab. If moisture is present, it can be easily seen on the inside of the film after twenty-four hours.

If no moisture is present, prepare the slab. Grind off any high spots. Fill low spots with level-ing compound. The slab must be free of grease, oil, or dust.

Applying a Moisture Barrier

A *moisture barrier* must be installed over all concrete slabs. This insures a trouble-free finish floor installation. Spread a skim coat of *mastic* with a straight trowel over the entire slab. Allow to dry at least two to three hours. Then, cover the slab with poly-ethylene film. Lap the edges of the film 4 to 6 inches and extending up all walls enough to be covered by the baseboard, when installed.

When the film is in place, *walk it in*. Step on the film, over every square inch of the floor, to make sure it is completely adhered to the cement. Small bubbles of entrapped air may appear. The film may be punctured, without concern, to let the air escape.

Applying Plywood Subfloor

A *plywood subfloor* may be installed over the moisture barrier on which to fasten the finish floor. Exterior grade sheathing plywood of at least 3/4-inch thickness is used. The plywood is laid with staggered joints. Leave a 3/4-inch space at walls and 1/4- to 1/2-inch space between panel edges and ends. Fasten the plywood to the concrete with at least nine nails per panel (Fig. 85-1).

Instead of driving fasteners, the plywood may be cemented to the moisture barrier. Cut the plywood in 4×4 squares. Use a portable electric circular saw. Make scores 3/8 inch deep on the back of each panel that form a 12×12 grid. Lay the panel in asphalt cement spread with a 1/4-inch notched trowel.

Applying Sleepers

Finish flooring may also be fastened to *sleepers* installed on the slab. Sleepers are short lengths of lumber cemented to the slab. They must be pressure-treated and dried to a suitable moisture content. Usually, 2×4 lumber, from 18 to 48 inches long, is used.

Sleepers are laid on their side and cemented to the slab with mastic. They are staggered, with end laps of at least 4 inches, in rows 12 inches on center and at right angles to the direction of the finish floor. A polyethylene vapor barrier is then placed

Fig. 85-1 Installation details of a plywood subfloor over a concrete slab.

Fig. 85-2 Sleepers are cemented to a concrete slab. They provide fastening for strip or plank finish flooring.

over the sleepers. The edges are lapped over the rows (Fig. 85-2). With end-matched flooring end joints need not meet over the sleepers.

Subfloors on Joists

Exterior plywood or boards are recommended for use as subfloors on joists when wood finish floors are installed on them. If plywood is used, a full 1/2-inch thickness is required. Thickness of 5/8 inch or more is preferred for 3/4-inch strip finish flooring. Use 3/4-inch thick subfloor for 1/2-inch strip flooring. The National Oak Flooring Manufacturers Association does not recommend fastening finish flooring to subfloors of nonveneered panels.

Laying Wood Finish Floor

In new construction, the base or door casings are not usually applied yet for easier application of the finish floor. In remodeling, the base and base shoe must be removed. Use a scrap piece of finish flooring as a guide on which to lay a handsaw. Cut the ends of any door casings that are extending below the finish floor surface.

Before laying any type of wood finish flooring, nail any loose areas. Sweep the subfloor clean. Scraping may be necessary to remove all plaster, taping compound, or other materials.

Laying Strip Flooring

Strip flooring laid in the direction of the longest dimension of the room gives the best appearance. The flooring may be laid in either direction on a plywood or diagonal board subfloor. It must be laid perpendicular to subfloor boards that run at right angles to the joists (Fig. 85-3).

When the subfloor is clean, cover it with building paper. Lap it 4 inches at the seams, and at right angles to the direction of the finish floor. The paper helps keep out dust, prevents squeaks in dry seasons, and retards moisture from below that could cause warping of the floor.

Snap chalk lines on the paper showing the centerline of floor joists so flooring can be nailed into them. For better holding power, fasten flooring through the subfloor and into the floor joists whenever possible. On 1/2-inch plywood subfloors, flooring fasteners must penetrate into the joists.

Starting Strip. The location and straight alignment of the first course is important. Place a strip of flooring 3/4 inch from the starter wall with the groove side toward the wall. On each end of the room, mark along the edge of the flooring tongue. Snap a chalk line between the two points. The gap between the flooring and the wall is needed for expansion. It will eventually be covered by the base.

STRIP FINISH FLOOR

SUBFLOOR

FINISH FLOOR
MUST RUN AT A
RIGHT ANGLE TO
SUBFLOOR BOARDS

FLOOR
JOIST

BOARD SUBFLOOR PERPENDICULAR TO JOISTS

FINISH FLOOR MAY
RUN IN EITHER
DIRECTION, BUT
PERPENDICULAR TO
FLOOR JOIST IS
PREFERRED.

PREFERRED
DIRECTION

SUBFLOOR

DIAGONAL BOARD SUBFLOOR

FINISH FLOOR MAY
RUN IN EITHER
DIRECTION, BUT
PERPENDICULAR TO
JOISTS IS
PREFERRED

PREFERRED
DIRECTION

SUBFLOOR

PANEL SUBFLOOR

Fig. 85-3 Several factors determine the direction in which strip flooring is laid.

Hold the strip with its tongue edge to the chalk line. Face nail it with 8d finish nails, alternating from one edge to the other, not less than 8 inches apart. Work from left to right with the grooved end of the first piece toward the wall. Left is determined by having the back of the person laying the floor to the wall where the starting strip is laid. Make sure end joints between strips are driven up tight (Fig. 85-4).

When necessary to cut a strip to fit to the right wall, use a strip long enough so that the cut-off piece is 8 inches or longer. Start the next course on the left wall with this piece.

Blind Nailing

Flooring is *blind nailed* by driving nails at about a 45 degree angle through the flooring. Start the nail in the corner at the top edge of the tongue. Usually 7d hardened cut or spiral screw nails are used. Recommendations for fastening are shown in Figure 85-5.

For the first two or three courses of flooring, a hammer must be used to drive the fasteners. For floor laying, a heavier than usual hammer, from 20 to 28 ounces, is generally used for extra driving power. Care must be taken not to let the hammer glance off the nail. This may damage the edge of the flooring. Care also must be taken that, on the final

Fig. 85-4 A chalk line is snapped on the floor for alignment of the starting row of strip flooring. (*Courtesy of Chickasaw Hardwood Floors*)

STRIP T & G		
SIZE FLOORING	**SIZE NAIL TO BE USED**	**BLIND NAIL SPACING ALONG THE LENGTH OF STRIPS. MINIMUM 2 NAILS PER PIECE NEAR THE ENDS. (1"-3")**
3/4X1 1/2", 2 1/4", & 3 1/4"	2" SERRATED EDGE BARBED FASTENER, 7D OR 8D SCREW OR CUT NAIL, 2" 15 GAUGE STAPLES WITH 1/2" CROWN. ON SLAB WITH 3/4" PLYWOOD SUBFLOOR USE 1 1/2" BARBED FASTENER. 1/2" PLYWOOD SUBFLOOR WITH JOISTS A MAXIMUM 16" O.C., FASTEN INTO EACH JOIST WITH ADDITIONAL FASTENING BETWEEN, OR 8" APART.	IN ADDITION-10-12" APART-8-10" PREFERRED
	MUST INSTALL ON A SUBFLOOR	
1/2X1 1/2" & 2"	1 1/2" SERRATED EDGE BARBED FASTENER, 5D SCREW, CUT STEEL, OR WIRE CASING NAIL.	10" APART 1/2" FLOORING MUST BE INSTALLED OVER A MINIMUM 5/8" THICK PLYWOOD SUBFLOOR.
3/8X1 1/4" & 2"	1 1/4" SERRATED EDGE BARBED FASTENER, 4D BRIGHT WIRE CASING NAIL.	8" APART
SQUARE-EDGE FLOORING		
5/16X1 1/2" & 2"	1" 15 GAUGE FULLY BARBED FLOORING BRAD.	2 NAILS EVERY 7"
5/16X1 1/3"	1" 15 GAUGE FULLY BARBED FLOORING BRAD.	1 NAIL EVERY 5" ON ALTERNATE SIDES OF STRIP
PLANK 3/4X3" TO 8"	2" SERRATED EDGE BARBED FASTENER, 7D OR 8D SCREW, OR CUT NAIL, USE 1 1/2" LENGTH WITH 3/4" PLY-WOOD SUBFLOOR ON SLAB.	8" APART
FOLLOW MANUFACTURER'S INSTRUCTIONS FOR INSTALLING PLANK FLOORING		
WIDTHS 4" AND OVER MUST BE INSTALLED ON A SUBFLOOR OF 5/8" OR THICKER PLYWOOD OR 3/4" BOARDS. ON SLAB USE 3/4" OR THICKER PLYWOOD.		

Fig. 85-5 Nailing guide for strip and plank finish flooring. (*Courtesy of National Oak Flooring Manufacturers Association*)

HELPFUL HINTS

Method of driving a blind nail without damaging the edge of the flooring.

DRIVE NAIL TO THIS POSITION HOLDING HAMMER AT CONVENIENT ANGLE

ON FINAL BLOW, RAISE HAMMER HANDLE AND STRIKE NAIL HEAD AND TONGUE EDGE AT THE SAME TIME TO DRIVE FLOORING UP TIGHT

DO NOT LET HAMMER HIT CORNER

Fig. 85-6

blows, the hammer head does not hit the top corner of the flooring. To prevent this, raise the hammer handle slightly on the final blow so that the hammer head hits the nail head and the tongue, but not the corner of the flooring (Fig. 85-6).

CAUTION Eye protection should be worn when driving hardened steel nails with a hammer. A small piece of steel may break off the hammer or the nail, and fly out in any direction. This could cause serious injury to an unprotected eye.

After the nail is driven home, its head must be set slightly. This allows adjoining strips to come up tightly against each other. Floor layers use the head of the next nail to be driven to set the nail just driven.

When fastening flooring, the floor layer holds a hammer in one hand and a number of nails in the other. While driving one nail, the floor layer *fingers* the next nail to be driven into position to be used as a set. When the nail being driven is home, the fingered nail is laid on edge with its head on the nail to be set. With one sharp blow, the nail is set (Fig. 85-7). The setting nail is then the next nail to be driven. In this manner, the floor layer maintains a smooth, continuous motion when fastening flooring.

NOTE A nail set should not be used to set hardened flooring nails. If used, the tip of the nail set will be flattened, thus rendering the nail set useless. Do not lay the nail set flat along the tongue, on top of the nail head, and then set the nail by hitting the side of the nail set with a hammer. Not only is this method is slower, but it invariably breaks the nail set, possibly causing an injury.

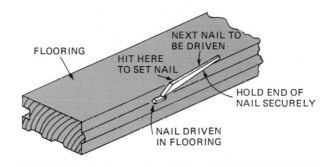

FLOORING

NEXT NAIL TO BE DRIVEN

HIT HERE TO SET NAIL

HOLD END OF NAIL SECURELY

NAIL DRIVEN IN FLOORING

Fig. 85-7 Method used by floor layers to set nails driven by hand.

Racking the Floor

After the second course of flooring is fastened, lay out seven or eight loose rows of flooring, end to end. Lay out in a staggered pattern. End joints should be at least 6 inches apart. Find or cut pieces to fit within 1/2 inch of the end wall. Distribute long and short pieces evenly for the best appearance. Avoid clusters of short strips. Laying out loose flooring in this manner is called *racking the floor*. Racking is done to save time and material (Fig. 85-8).

Using the Power Nailer

At least two courses of flooring must be laid by hand to provide clearance from the wall before a *power nailer* can be used. The power nailer holds strips of special barbed fasteners. The fasteners are driven and set through the tongue of the flooring at the proper angle. Although it is called a power nailer, a heavy hammer is swung by the operator against a plunger to drive the fastener (Fig. 85-9).

The hammer is double-ended. One end is rubber and the other end is steel. The flooring strip is placed in position. The rubber end of the hammer is used to drive the edges and ends of the strips up

FASTEN FIRST 2 OR 3
COURSES IN PLACE

GPYSUM BD. WALL

LAY OUT SEVERAL COURSES
LOOSELY FROM WALL TO WALL
WITH END JOINTS STAGGERED
AT LEAST 6 INCHES.

Fig. 85-8 Racking the floor places the strips in a handy position for efficient installation.

tight. The steel end is used against the plunger of the power nailer to drive the fasteners. Slide the power nailer along the tongue edge. Drive fasteners about 8 to 10 inches apart or as needed to bring the strip up tight against previously laid strips.

NOTE When using the power nailer, one and only one heavy blow must be used to drive the fastener. After the first blow, another fastener drops into place ready to be driven. Make sure the first blow is powerful enough to drive the fastener.

Whether laying floor with a power nailer or driving nails with a hammer, the floor layer stands with heels on strips already fastened, and toes on the loose strip to be fastened. With weight applied to the joint, easier alignment of the tongue and

groove is possible (Fig. 85-10). The weight of the worker also prevents the loose strip from bouncing when it is driven to make the edge joint tight.

Ending the Flooring

Continue across the room. Rack seven or eight courses as work progresses. The last three or four courses from the opposite wall must be nailed by hand. This is because of limited room to place the power nailer and swing the hammer. The next-to-the-last row can be blind nailed if care is taken. However, the flooring must be brought up tightly by prying between the flooring and the wall. Use a bar to pry the pieces tight at each nail location (Fig. 85-11).

The last course is installed in a similar manner. However, it must be face nailed. It may need to be ripped to the proper width. If it appears that the installation will end with an undesirable, difficult to

Fig. 85-9 A power nailer is widely used to fasten strip flooring. (*Courtesy of National Oak Flooring Manufacturers Association*)

RACKED FLOORING

STEP ON STRIP BEING FASTENED

Fig. 85-10

apply, narrow strip, lay wider strips in the last row (Fig. 85-12).

Framing Around Obstructions

A much more professional and finished look is given to a strip flooring installation if *hearths* and other floor obstructions are framed. Use flooring, with mitered joints at the corners, as framework around the obstructions (Fig. 85-13).

Changing Direction of Flooring

Sometimes it is necessary to change direction of flooring when it extends from a room into another room, hallway, or closet. To do this, face nail the extended piece to a chalk line. Change directions by joining groove edge to groove edge and inserting a *spline*, ordinarily supplied with the flooring

WALL

PRY IN THIS DIRECTION

BAR

FLOORING

WOOD BLOCK

Fig. 85-11 The last two courses of strip flooring must be brought tight with a pry bar.

HELPFUL HINTS

How to overcome the difficulty of ending with a narrow strip.

GYPSUM BD.

BASE

2 1/4" FLOORING

BASE SHOE

1. A NARROW ENDING STRIP IS DIFFICULT TO INSTALL

SUBFLOOR

ENDING STRIP

NEXT TO LAST STRIP

2. TO OVERCOME THE DIFFICULTY, FASTEN THE NARROW ENDING STRIP TO THE NEXT TO LAST STRIP BEFORE INSTALLING.

RIP THE ENDING ROW FROM 3 1/4"-WIDE FLOORING

3. ANOTHER WAY TO OVERCOME THE DIFFICULTY IS TO USE WIDER FLOORING FOR THE ENDING ROW.

Fig. 85-12

HEARTH

FRAME

FINISH FLOOR

Fig. 85-13 Frame around floor obstructions, such as hearths, with strips that are mitered at the corners.

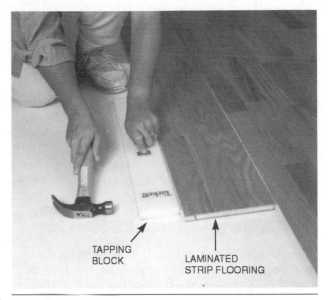

Fig. 85-14 The direction of strip flooring is changed by the use of a spline.

(Fig. 85-14). For best appearance, avoid bunching short or long strips. Open extra bundles, if necessary, to get the right selection of lengths.

Laying Laminated Strip Flooring

Before laying *laminated strip* flooring, a 1/8-inch foam underlayment, supplied or approved by the manufacturer, is applied to the subfloor or slab. The flooring is not fastened or cemented to the floor. However, the boards must be glued to each other along the edges and ends. Apply glue on edges in 8-inch long beads with 12-inch spaces between them. Apply a full bead across the ends.

The first row is laid in a straight line with end joints glued. The groove edge is placed toward the wall, leaving about a 1/2-inch expansion space. Subsequent courses are installed. Edges and ends are glued. Each piece is brought tight against the other with a hammer and *tapping block*. The tapping block should be used only against the tongue. It should never be used against the grooved edge (Fig. 85-15). Tapping the grooved edge will damage it. Stagger end joints at least 2 feet from those in adjacent courses.

Fig. 85-15 A tapping block is used to drive laminated strip flooring boards tight against those already installed. (*Courtesy of Harris-Tarkett*)

Usually the last course must be cut to width. To lay out the last course to fit, lay a complete row of boards, unglued, tongue toward wall, directly on top of the already installed next to last course. Cut a short piece of flooring for use as a *scribing* block. Hold the tongue edge against the wall. Move the block along the wall while holding a pencil against the other edge to lay out the flooring. The width of the tongue of the scribing block provides the necessary expansion space (Fig. 85-16). The last row, when cut, can be glued and wedged tightly in place with a pry bar.

HELPFUL HINTS

How to lay out the width of the last row of laminated strip flooring.

TONGUE EDGE

TONGUE EDGE

MARK HERE WHILE MOVING SCRIBING BLOCK ALONG WALL.

LAST ROW PLACED DIRECTLY ON TOP OF NEXT TO LAST ROW.

Fig. 85-16

Installing Plank Flooring

Plank flooring is installed like strip flooring. Alternate the courses by widths. Start with the narrowest pieces. Then use increasingly wider courses, and repeat the pattern. Stagger the joints in adjacent courses. Use lengths so they present the best appearance.

Manufacturers' instructions for fastening the flooring vary and should be followed. Generally, the flooring is blind nailed through the tongue of the plank and at intervals along the plank in a manner similar to strip flooring.

Installing Parquet Flooring

Procedures for the application of parquet flooring vary with the style and the manufacturer. Detailed installation directions are usually provided with the flooring. Generally, both parquet blocks and strips are laid in *mastic*. Use the recommended type. Apply with a notched trowel. The depth and spacing of the notches are important to leave the correct amount of mastic on the floor. Parquet may be installed either square with the walls or diagonally.

Square Pattern

When laying *unit blocks* in a square pattern, two layout lines are snapped, at right angles to each other, and parallel to the walls. Blocks are laid with their edges to the lines. Lines are usually laid out so that rows of blocks are either centered on the floor or half the width of a block off center. This depends on which layout produces border blocks of equal and maximum widths against opposite walls.

To determine the location of the layout lines, measure the distance to the center of the room's width. Divide this distance by the width of a block. If the remainder is half or more, snap the layout line in the center. If the remainder is less than half, snap the layout line off center by half the width of the block. Find the location of the other layout line in the same way. It is possible that one of the layout lines will be centered, while the other must be snapped off center.

Other factors may determine the location of layout lines, such as ending with full blocks under a door, or where they meet another type of floor. Regardless of the location, two lines, at right angles to each other, must be snapped.

Place one unit at the intersection of the lines. Position the grooved edges exactly on the lines. Lay the next units ahead and to one side of the first one and along the lines. Install blocks in a pyramid. Work from the center outward toward the walls in all directions. Make adjustments as installation progresses to prevent misalignment (Fig. 85-17).

Fig. 85-17 Parquet blocks are laid to center or off-center lines. This depends on which produces the best size border blocks.

Diagonal Pattern

To lay unit blocks in a diagonal pattern, measure an equal distance from one corner of the room along both walls. Snap a starting line between the two marks. At the center of the starting line, snap another line at right angles to it. The location of both lines may need to be changed in order to end with border blocks of equal and largest possible size against opposite walls. The diagonal pattern is then laid in a manner similar to the square pattern (Fig. 85-18).

Special Patterns

Many parquet patterns can be laid out with square and diagonal layout lines. The *herringbone* pattern requires two layout lines. One will be the 90 degree line used for a square pattern. The other line crosses the intersection at a 45 degree angle. Align the first block or strip with its edge on the diagonal line. The corner of the piece should be lined up at the intersection. Continue the pattern in rows of three units wide, aligning the units with the layout lines (Fig. 85-19).

The *Monticello* pattern can be laid square or diagonally in the same way as unit blocks. For best

Fig. 85-18 Layout of a diagonal pattern for unit blocks. Make adjustments for satisfactory borders.

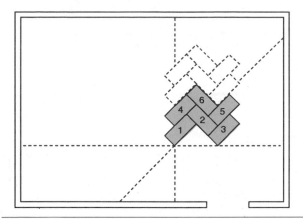

Fig. 85-19 The herringbone pattern requires 90 and 45 degree layout lines. (*Courtesy of Chickasaw Hardwood Floors*)

results, lay the parquet in a pyramid pattern. Alternate the grain of the center blocks. Keep *picket* points in precise alignment (Fig. 85-20).

The *Marie Antoinette* pattern may also be laid square or diagonally. It is started by laying a *band* with its grooved edge and end aligned with the layout lines with the tongued edge to the right. Continue laying the pattern by placing center blocks and bands in a sequence so that bands appear woven (Fig. 85-21). Tongues of all members face toward the right or ahead. In this pattern, the grain of center blocks runs in the same direction.

Many other parquet patterns are manufactured. With the use of parquet blocks and strips, the design possibilities are almost endless. Competence in laying the popular patterns described in this chapter will enable the student to apply the principles for professional installations of many different parquet floors.

SQUARE PATTERN

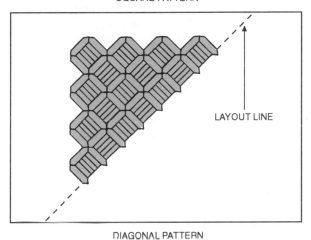

DIAGONAL PATTERN

Fig. 85-20 Layout of the Monticello parquet pattern. (*Courtesy of Chickasaw Hardwood Floors*)

Fig. 85-21 Layout of the Marie Antoinette parquet pattern. (*Courtesy of Chickasaw Hardwood Floors*)

CHAPTER 86

UNDERLAYMENT AND RESILIENT TILE

Resilient flooring is widely used in residential and commercial buildings. It is a thin, flexible material that comes in *sheet* or *tile* form. It is applied on a smooth concrete slab or an *underlayment*.

Underlayment

Plywood, strandboard, hardboard, or particleboard may be used for *underlayment*. It is installed on top of the subfloor. This provides a base for the application of resilient sheet or tile flooring. The number of underlayment panels required is found by dividing the area to be covered by the area of the panel.

Underlayment thickness may range from 1/4 to 3/4 inch, depending on the material and job requirements. In many cases, where a finish wood floor meets a tile floor, the underlayment thickness is determined by the difference in the thickness of the two types of finish floor. Both floor surfaces should come flush with each other.

Installing Underlayment

Sweep the subfloor as clean as possible. Cover with asphalt felt lapped 4 inches at the edges. Stagger joints between the subfloor and underlayment. If installation is started in the same corner as the subfloor, the thickness of the walls will be enough to offset the joints. Leave about 1/32 inch between underlayment panels to allow for expansion.

If the underlayment is to go over a board subfloor, install the underlayment with its face grain across the boards. In all other cases, a stiffer and stronger floor is obtained if the face grain is across the floor joists (Fig. 86-1).

Fasten the first row in place with staples or nails. Underlayment requires more nails than subfloor to provide a squeak-free and stiff floor. Nail spacing and size are shown in Figure 86-2 for various thicknesses of plywood underlayment. Similar

Fig. 86-1 The joints between underlayment and subfloor are offset. (*Courtesy of American Plywood Association*)

Plywood Thickness (inch)	Fastener Size (approx.) and Type	Fastener Spacing (inches)	
		Panel Edges	Panel Interior
1/4	3d ring-shank nails	3	6 each way
	18 gauge staples	3	6 each way
3/8 or 1/2	3d ring-shank nails	6	8 each way
	16 gauge staples	3	6 each way
5/8 or 3/4	4d ring-shank nails	6	8 each way
	16 gauge staples	3	6 each way
1/4	3d ring-shank nails	3	6 each way
	18 gauge staples	3	6 each way

Fig. 86-2 Nailing specifications for plywood underlayment.

nail spacing and size are appropriate for other types of underlayment.

Install remaining rows of underlayment with end joints staggered from the previous courses. The last row of panels is ripped to width to fit the remaining space.

If APA Sturd-I-Floor is used, it can double as a subfloor and underlayment. If square-edged panels are used, blocking must be provided under the joints to support the edges. If tongue-and-groove panels are used, blocking is not required (Fig. 86-3).

Resilient Tile Flooring

Most *resilient floor tiles* are made from asphalt, vinyl, rubber, or cork. Many different colors, textures, and patterns are available. Generally, tiles come in 9×9 and 12×12 squares. They are applied to the floor in a manner similar to that used for applying parquet blocks. The most common tile thicknesses are 1/16 inch and 1/8 inch. Thicker tiles are used in commercial applications subjected to considerable use.

Long strips of the same material are called

RESILIENT TILE FLOORING

STURD-I-FLOOR 16",
20" OR 24" O.C.

STAGGER END JOINTS

BLOCKING WITH SQUARE EDGE PANELS

LEAVE 1/16" SPACING AT ALL EDGE AND END JOINTS (3/32" AT T & G EDGES)

TONGUE & GROOVE EDGES (OR PROVIDE EDGE BLOCKING)

Fig. 86-3 APA Sturd-I-Floor requires no underlayment for the installation of resilient floors. (*Courtesy of American Plywood Association*) *feature strips.* They are available from the manufacturer. The strips vary in width from 1/4 inch to 2 inches. They are used between tiles to create unique floor patterns.

Installing Resilient Tile

CAUTION If the application involves the removal of existing resilient floor covering, be aware that it, and the adhesive used, may contain *asbestos.* The presence of asbestos in the material is not easily determined. If there is any doubt, always assume that the existing flooring and adhesive do contain asbestos. Practices for removal of existing flooring or any other building material containing asbestos should comply with standards set by the U.S. Occupational Safety and Health Administration (OSHA).

If the application is over an existing resilient floor covering, do not sand the existing surface unless absolutely sure it does not contain asbestos. Inhalation of asbestos dust can cause serious harm.

Before installing resilient tile, make sure underlayment fasteners are not projecting above the surface. Fill any open areas, such as splits, with a floor leveling compound. Use a portable disc sander to bring underlayment joints flush. Skim the entire surface with the sander to make sure the surface is smooth. Sweep the floor clean.

Layout

Snap layout lines across the floor at right angles to each other. The lines are centered if border tiles are a half width or more. If border tiles are less than a half width, layout lines are snapped half the width of a tile off center.

Applying Adhesive

Some floor tiles are manufactured with the adhesive applied at the factory. They are commonly known as *peel and stick* tiles. The adhesive is protected by a *release paper.* The paper is peeled off when ready to apply the tile.

CAUTION The release paper is very smooth and slippery. When peeled from the tile, dispose of it in a container. If left on the floor, a serious accident could occur if a person stepped on it, slipped, and fell.

If adhesive must be applied to the underlayment, apply it to one-quarter of the floor with a properly notched trowel, up to the layout lines (Fig. 86-4). Use an adhesive and notched trowel recommended by the manufacturer of the tile. If the notches in the trowel are too shallow, not enough adhesive will be applied. If the notches are too deep, more adhesive than necessary will be applied. This will result in the adhesive squeezing up through the joints onto the face of the tile. It will probably require removal of the application.

Laying Tiles

Start by applying a tile to intersection of the layout lines with the two adjacent edges on the lines. Lay tiles with edges tight. Work toward the walls (Fig. 86-5). Watch the grain pattern. It may be desired to alternate the run of the patterns or to lay the patterns all in one direction. Some tiles are stamped with an arrow on the back. They are placed so the arrows on all tiles point in the same direction.

Lay tiles in place instead of sliding them into position. Sliding the tile pushes the adhesive through the joint. With some types of adhesives, it

Fig. 86-4 Applying adhesive to the underlayment for the application of resilient floor tile.

Fig. 86-5 Laying resilient floor tile.

Fig. 86-7 Fitting a border tile in place

may be difficult or impossible to slide the tile. Apply all tiles in the quarter section except border tiles. Apply tiles to all other quarter sections in the same manner.

Applying Border Tiles

Border tiles may be cut by scoring with a sharp utility knife and bending or by the use of a special tile cutter. To lay out and score a border tile to fit snugly, first place the tile to be cut directly on top of the last tile installed. Make sure all edges are in line. Place a full tile with its edge against the wall and on top of the one to be fitted. Score the border tile along the outside edge of the top tile (Fig. 86-6). Bend the tile. Break it along the scored line, and fit it into place (Fig. 86-7). If a base has not yet been installed, then not so much care need be taken with the fit of border tiles. Simply measure and cut to an approximate width so the space can be covered by the base.

Applying a Vinyl Cove Base

Many times a vinyl cove base is used to trim a tile floor. A special vinyl base cement is applied to its back. The base is pressed into place (Fig. 86-8).

Estimating Resilient Tile

To estimate the amount of tile flooring needed, find the area of the room in square feet. To do this, measure the length and width of the room to the next whole foot. Multiply these figures together to find the area. For 12×12 tiles, the result is the number of pieces needed. Multiply the area by 1 1/3 to find the number of 9×9 tiles needed.

Fig. 86-8 Applying cement to the back of a vinyl cove base and pressing it into place.

BORDER TILE

FULL TILE

Fig. 86-6 Scoring a border tile to fit against the wall.

Review Questions

Select the most appropriate answer.

1. If hardwood flooring is stored in a heated building, the temperature should not exceed
   a. 72 degrees.   c. 85 degrees.
   b. 78 degrees.   d. 90 degrees.

2. Most hardwood finish flooring is made from
   a. Douglas fir.   c. southern pine.
   b. hemlock.   d. oak.

3. Bundles of strip flooring may contain pieces over and under the nominal length of the bundle by
   a. 4 inches.   c. 8 inches.
   b. 6 inches.   d. 9 inches.

4. No pieces are allowed in bundles of hardwood strip flooring shorter than
   a. 4 inches.   c. 8 inches.
   b. 6 inches.   d. 9 inches.

5. The edges of prefinished strip flooring are chamfered to
   a. prevent splitting.
   b. apply the finish.
   c. simulate cracks between adjoining pieces.
   d. obscure any unevenness in the floor surface.

6. The best grade of unfinished oak strip flooring is
   a. prime.   c. select.
   b. clear.   d. quarter-sawed.

7. To estimate the amount of 2 1/4-inch face hardwood flooring, add to the area to be covered a percentage of the area of
   a. 42.5.   c. 29.
   b. 38.33.   d. 33.33.

8. When it is necessary to cut the last strip in a course of flooring, the waste is used to start the next course and should be at least
   a. 8 inches long.   c. 12 inches long.
   b. 10 inches long.   d. 16 inches long.

9. For floor laying, the hammer weight is generally
   a. 13 to 16 ounces.   c. 20 to 28 ounces.
   b. 16 to 20 ounces.   d. 25 to 30 ounces.

10. To change direction of strip flooring,
    a. face nail both strips.
    b. turn the extended strip around.
    c. blind nail both strips.
    d. use a spline.

■ BUILDING FOR SUCCESS ■

Communication Infrastructure

In any building project, communication must be a priority. The flow of pertinent information has to be continual if success is to be obtained. Each builder must create a personal communication infrastructure within the company and community that will leave no weak link for potential problems.

The successful skilled technician, subcontractor or builder will always work to improve his or her communication network within the industry. Experience usually proves to be the best teacher and will be the most valuable resource. By combining actual business communication experiences with observations in the field and continued edu-

cation, each person can create the most meaningful communication system for his or her company.

The following is a sample of a communication infrastructure that has proven to be successful for a home builder. Sam and Mike Manzitto, owners of Manzitto Brothers Custom Homes and Remodeling in Lincoln, Nebraska, have graciously provided this example. This planned communication infrastructure has the capacity for ongoing modification and improvement. Its flexibility makes it attractive for others who are addressing their own communications structure.

For seventeen years, goals have been met, volume in production has peaked, profit margins are high, and a good company team has been established. We have made many changes to keep up with the economy, competition, and technology.

Then comes a new problem: a breakdown in the communications infrastructure (inner office, supervisor to lead carpenters, production manager to client, production manager to job supervisor, lead carpenter to production manager, lead carpenter to client, and so on.)

To correct and counteract these problems, we built a system to accomplish the following goals:

1. Keep communication lines open with the customer throughout its construction process.

2. Improve communication between lead carpenters and subcontractors.

3. Improve communication between supervisor and lead carpenters.

4. Improve the scheduling and organization of the job.

5. Improve the quality of the finished product.

6. Improve our followup with a punch list procedure.

This system consists of a series of ongoing weekly meetings, involving reports on jobs under construction, jobs to be completed, current clients, and marketing strategies by our sales and management team. Separate weekly meetings are held with supervisors and lead carpenters for job updates. Daily meetings begin with job supervisors. These activities involve completing written forms and procedures to be carried on throughout construction, from start to finish.

The following documents have been created and are used on every job in the following order.

PROJECT CHECKLIST: This list becomes part of each job file. These procedures must be met before the project can completed.

LEADPERSON CHECKLIST: This document is reviewed by the lead carpenter at the time the project is being estimated or at the signing of the contract.

PRECONSTRUCTION CONFERENCE: This document is used at the first meeting of the lead carpenter or supervisor, the client, and the salesperson.

SUBCONTRACTOR LETTER: This letter is sent out to all subcontractors working on the project.

GENERAL PROJECT PROCEDURES: This document is posted in poster form on the project before the work is to begin.

NEIGHBORHOOD LETTER: This document is sent to neighbors in the area of the project.

WALK-THRU PROCEDURES: This document helps the lead carpenter or supervisor with the final walk-thru at the completion of the job.

PUNCHLIST PROJECT: This list helps the final walk-thru of the project to run more smoothly. Any unfinished items can be agreed on at that point. The amounts of those items only are withheld from the final settlement of the project.

The following exhibit is of the *General Project Procedures* established by Manzitto Brothers Builders.

1. INTRODUCE YOURSELF TO HOME OWNER BY NAME AND COMPANY.

2. PLEASE AVOID USING PROJECT PROPERTY FOR BATHROOM FACILITIES.

3. PUNCTUALITY—MAKE A QUICK CALL IF YOU ARE RUNNING LATE.

4. USE DROP CLOTHS TO PROTECT FINISHED AREAS AT ALL TIMES.

5. PROPERLY DISPOSE OF ALL TRASH, INCLUDING PERSONAL TRASH.

6. DEMONSTRATE PROFESSIONAL ATTITUDE TOWARD FELLOW EMPLOYEES AND SUBCONTRACTORS.

7. BE JOB ACCESSIBLE AT ALL TIMES FOR HOME OWNER AND SUBCONTRACTORS.

8. NO PARKING IN DRIVEWAY. AVOID PARKING ON LOT.

9. WEAR APPROPRIATE ATTIRE (wear shirts in the house at all times).

10. NO PROFANITY.

11. NO SMOKING.

12. NO LOUD MUSIC.

CUSTOMER SERVICE DEPARTMENT

A customer service department has been created so that our followup or call-back procedures run smoothly. We feel that following up on our product is one of the most important parts of our business.

CUSTOMER SERVICE DEPARTMENT: Explains how our customer service department is structured, and shows the steps that are taken throughout the year.

CUSTOMER EVALUATION: Each customer completes this after the project is completed so that our services can be evaluated.

SERVICE REQUEST LETTER: Each customer responds three months after the project is completed, requesting items that may need our attention.

WARRANTY SERVICE REQUEST FORM: This is sent out with the service request letter in reference to the new home warranty.

SERVICE REQUEST LETTER #2: This is sent out after the client has been in the home for approximately nine months.

FOCUS QUESTIONS: For individual or group discussion.

1. As you review this communication infrastructure concept, what may be the most productive uses for a plan of this type? Why?

2. What appears to be the best way this plan can be implemented? Who are the key people that need to participate in this plan for it to work?

3. As expectations of performance, scheduling, and quality are set forth in this communication process, what can it tell the prospective client, subcontractors, and public about the builder? Does this appear to be an advantage over a builder that does not communicate these important elements to others? Why?

4. How can the following "checklists" and service considerations be of benefit to the company's management employees and clients?
    a. Project checklist
    b. Leadman checklist
    c. Preconstruction conference
    d. General project procedures list
    e. Walk-thru procedures
    f. Punch list

5. What function does the customer service department serve?

6. How can the company followup process benefit the company?

Cabinets and Countertops

C abinets and countertops may be purchased in preassembled units and installed. Or, they may be custom-made, piece by piece, on the job. More manufactured cabinets are installed because of the great variety and shorter installation time. Cabinets are custom-built on the job because the specifications cannot be met by manufactured cabinets, or because of certain job conditions.

OBJECTIVES

After completing this unit, the student will be able to:

- state the sizes and describe the construction of typical base and wall kitchen cabinet units.

- plan, order, and install manufactured kitchen cabinets.

- build kitchen cabinets on the job site.

- make, laminate, and install a countertop.

CONTENTS

CHAPTER 87

DESCRIPTION AND INSTALLATION OF MANUFACTURED CABINETS

Manufactured kitchen and bath cabinets come in a wide variety of styles, materials, and finishes (Fig. 87-1). The carpenter must be familiar with the various kinds, sizes, uses, and construction of the cabinets to know how to plan, order, and install them.

Description of Manufactured Cabinets

For commercial buildings, many kinds of specialty cabinets are manufactured. They are designed for specific uses in offices, hospitals, laboratories, schools, libraries, and other buildings. Most cabinets used in residential construction are manufactured for the kitchen or bathroom. All cabinets, whether for commercial or residential use, consist of a *case* which is fitted with shelves, doors, and/or drawers. Cabinets are manufactured and installed in essentially the same way. Designs vary considerably with the manufacturer, but sizes are close to the same.

Kinds and Sizes

One method of cabinet construction utilizes a *face frame*. This frame provides openings for doors and drawers. Another method, called *European* or *frameless,* eliminates the face frame (Fig. 87-2). Face-framed cabinets usually give a traditional look. Frameless cabinets are used when a contemporary appearance is desired.

The two basic kinds of kitchen cabinets are the *wall* unit and the *base* unit. The surface of the countertop is usually about 36 inches from the floor. Wall units are installed about 18 inches above the countertop. This distance is enough to accommodate such articles as coffee makers, toasters, blenders, and mixers. Yet it keeps the top shelf within reach, not over 6 feet from the floor. The usual overall height of a kitchen cabinet installation is 7'-0" (Fig. 87-3).

Wall Cabinets. Standard *wall* cabinets are 12 inches deep. They normally come in heights of 42, 30, 24, 18, 15, and 12 inches. The standard height is

Fig. 87-1 Manufactured kitchen cabinets are available in a wide variety of styles and sizes. (*Courtesy of Merillat Industries*)

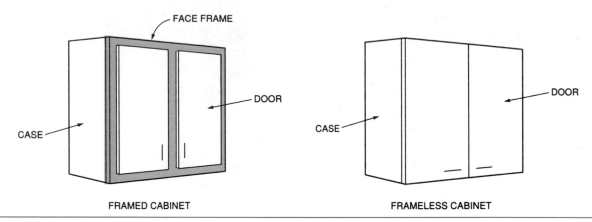

Fig. 87-2 Two basic methods of cabinet construction are with a face frame or frameless.

30 inches. Shorter cabinets are used above sinks, refrigerators, and ranges. The 42-inch cabinets are for use in kitchens without soffits where more storage space is desired. A standard height wall unit usually contains two adjustable shelves.

Usual wall cabinet widths range from 9 to 48 inches in 3-inch increments. They come with single or double doors depending on their width. Single-door cabinets can be hung so doors can swing in either direction.

Wall *corner* cabinets make access into corners easier. *Double-faced* cabinets have doors on both sides for use above island and peninsular bases. Some wall cabinets are made 24 inches deep for installation above refrigerators. A microwave oven case, with a 30-inch wide shelf, is available (Fig. 87-4).

Base Cabinets. Most *base* cabinets are manufactured 34 1/2 inches high and 24 inches deep. By

Fig. 87-3 Common kitchen cabinet heights and dimensions. (*Courtesy of Merillat Industries*)

WALL CABINETS

Fig. 87-4 Kinds and sizes of manufactured wall cabinets. (*Courtesy of Merillat Industries*)

adding the usual countertop thickness of 1 1/2 inches, the work surface is at the standard height of 36 inches from the floor. Base cabinets come in widths to match wall cabinets. Single-door cabinets are manufactured in widths from 9 to 24 inches. Double-door cabinets come in widths from 27 to 48 inches. A recess called a *toe space* is provided at the bottom of the cabinet.

The standard base cabinet contains one drawer, one door, and an adjustable shelf. Some base units have no drawers; others contain all drawers. *Double-faced* cabinets provide access from both sides. *Corner* units, with round revolving shelves, make corner storage easily accessible (Fig. 87-5).

Tall Cabinets. *Tall* cabinets are usually manufactured 24 inches deep, the same depth as base cabinets. Some *utility* cabinets are 12 inches deep. They are made 66 inches high and in widths of 27, 30, and 33 inches for use as *oven* cabinets. Single-door *utility* cabinets are made 18 and 24 inches wide. Double-door *pantry* cabinets are made 36 inches wide (Fig. 87-6). Wall cabinets with a 24-inch depth are usually installed above tall cabinets.

Vanity Cabinets. Most *vanity* base cabinets are made 31 1/2 inches high and 21 inches deep. Some are made in depths of 16 and 18 inches. Usual widths range from 24 to 36 inches in increments of 3 inches, then 42, 48, and 60 inches. They are available with several combinations of doors and drawers, depending on their width. Various sizes and styles of vanity wall cabinets are also manufactured (Fig. 87-7).

Accessories. *Accessories* are essential to or enhance a cabinet installation. *Filler* pieces fill small gaps in width between wall and base units when no combination of sizes can fill the existing space. They are cut to necessary widths on the job. Other accessories include cabinet end panels, face panels for dishwashers and refrigerators, open shelves for cabinet ends, and spice racks.

Laying Out Manufactured Kitchen Cabinets

The blueprints for a building contain plans, elevations, and details that show the cabinet layout. Ar-

Available in 9" width. Left or right hinged. **Tray Divider**

Available in 36" width. **Single Drawer**

Available in 30", 36" width. **Three Drawer**

Available in 12", 15", 18", 21", 24" width. **Four Drawer**

Available in 12", 15", 18", 21", 24" width. Left or right hinged. **Single Door**

Available in 27", 30", 33", 36", 39", 42", 45", 48" width. **Double Door**

Requires 36" or 42" of wall space. **Sink Front Diagonal**

Available in 24", 30", 33", 36" width. **Sink Front Straight**

Available in 24", 27", 30", 33", 36", 39", 42", 48"* width. **Sink Base**

Available in 24" width. **Double Faced Starter**

Available in 36", 39", 42", 45", 48" width. **Blind Corner**

Available in 18", 21", 24" width. **Double Faced Single Door**

Available in 30", 36" width. **Double Faced Double Door**

Lazy Susan

Base End

Left or right hinge. **Base Corner Revolving**

BASE CABINETS

Fig. 87-5 Most base cabinets are manufactured to match wall units. (*Courtesy of Merillat Industries*)

Available in 27", 30", 33" width. Drawers optional. **Oven Cabinet — Universal**

Available in 18", 24" width. **Utility Cabinet 12" Deep**

Available in 18", 24" width. **Utility Cabinet**

Available in 36" width. **Pantry Cabinet**

Fig. 87-6 Tall cabinets are manufactured as oven, utility, and pantry units. (*Courtesy of Merillat Industries*)

chitects may draw the layout. But they may not specify the size or the manufacturer's identification for each individual unit of the installation. In residential construction, particularly in remodeling, no plans are usually available to show the cabinet arrangement. In addition to installation, it becomes the responsibility of the carpentry contractor to plan, lay out, and order the cabinets, in accordance with the customer's specifications.

The first step is to measure carefully and accurately the length of the walls on which the cabinets are to be installed. A plan is then drawn to scale. It must show the location of all appliances, sinks, windows, and other necessary items (Fig. 87-8).

Next, draw elevations of the base cabinets, referring to the manufacturer's catalog for sizes. Always use the largest size cabinets available instead of two or three smaller ones. This reduces the cost and makes installation easier.

Match up the wall cabinets with the base cabinets, where feasible. If filler strips are necessary, place them between a wall and a cabinet or between

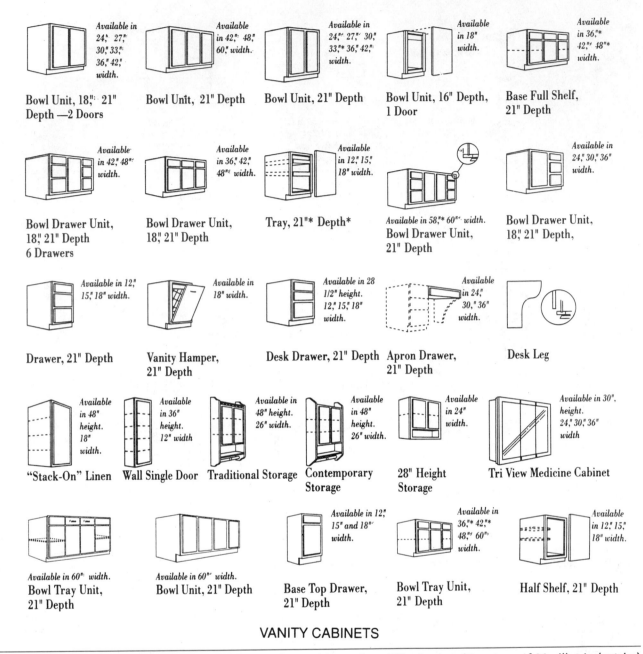

Bowl Unit, 18," 21"
Depth —2 Doors
Available in 24," 27," 30," 33," 36," 42," width.

Bowl Unit, 21" Depth
Available in 42," 48," 60," width.

Bowl Unit, 21" Depth
Available in 24," 27," 30," 33," 36," 42," width.

Bowl Unit, 16" Depth, 1 Door
Available in 18" width.

Base Full Shelf, 21" Depth
Available in 36," 42,"* 48"* width.*

Bowl Drawer Unit, 18," 21" Depth 6 Drawers
Available in 42," 48" width.*

Bowl Drawer Unit, 18," 21" Depth
Available in 36," 42," 48" width.*

Tray, 21"* Depth*
Available in 12," 15," 18" width.

Bowl Drawer Unit, 21" Depth
Available in 58," 60"** width.*

Bowl Drawer Unit, 18," 21" Depth,
Available in 24," 30," 36" width.

Drawer, 21" Depth
Available in 12," 15," 18" width.

Vanity Hamper, 21" Depth
Available in 18" width.

Desk Drawer, 21" Depth
Available in 28 1/2" height. 12," 15," 18" width.

Apron Drawer, 21" Depth
Available in 24," 30," 36" width.

Desk Leg

"Stack-On" Linen
Available in 48" height. 18" width.

Wall Single Door
Available in 36" height. 12" width

Traditional Storage
Available in 48" height. 26" width.

Contemporary Storage
Available in 48" height. 26" width.

28" Height Storage
Available in 24" width.

Tri View Medicine Cabinet
Available in 30", height. 24," 30," 36" width

Bowl Tray Unit, 21" Depth
Available in 60" width.*

Bowl Unit, 21" Depth
Available in 60" width.*

Base Top Drawer, 21" Depth
Available in 12," 15" and 18" width.

Bowl Tray Unit, 21" Depth
Available in 36," 42,"* 48,"* 60" width.*

Half Shelf, 21" Depth
Available in 12," 15," 18" width.

VANITY CABINETS

Fig. 87-7 Vanity cabinets are made similar to kitchen cabinets, but differ in size. (*Courtesy of Merillat Industries*)

cabinets in the corner. Identify each unit on the elevations with the manufacturer's identification (Fig. 87-9). Make a list of the units in the layout. Order from the distributor.

Computer Layouts

Computer programs are available to help in laying out manufactured kitchen cabinets. When the required information is fed into the computer, a number of different layouts can be quickly made. When a acceptable layout is made, it can be printed with each of the cabinets in the layout identified and priced. Most large kitchen cabinet distributors will supply computerized layouts on request.

Installing Manufactured Cabinets

Cabinets must be installed level and plumb even though floors are not always level and walls not always plumb. Level lines are first drawn on the wall for base and wall cabinets. In order to level base cabinets that set on an unlevel floor, either shim the cabinets from the high point of the floor or scribe and fit the cabinets to the floor from lowest point on the floor. Shimming the base cabinets leaves a space that must be later covered by a molding. Scribing and fitting the cabinets to the floor eliminate the need for a molding. The method used

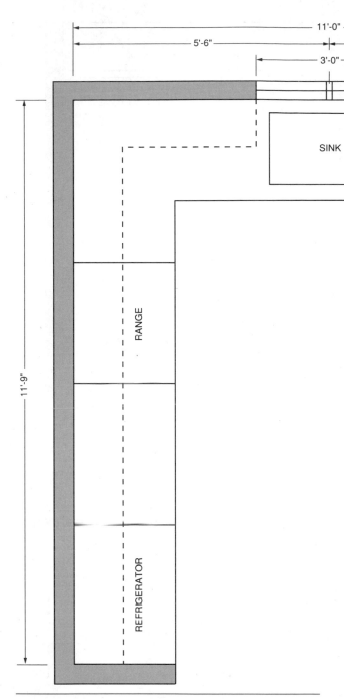

Fig. 87-8 Typical plan of a kitchen cabinet layout showing location of walls, windows, and appliances.

depends on various conditions of the job. If shimming base cabinets, lay out the level lines on the wall from the highest point on the floor where cabinets are to be installed. If fitting cabinets to the floor, measure up from the low point.

Laying Out the Wall

Measure 34 1/2 inches up the wall. Draw a level line to indicate the tops of the base cabinets. Use the most accurate method of leveling available (described in Chapter 25, Leveling and Layout Instruments). Another level line must be made on the wall 54 inches from the floor. The bottoms of the wall units are installed to this line. It is more accurate to measure 19 1/2 inches up from the first level line and snap lines parallel to it than to level another line.

The next step is to mark the stud locations in a framed wall. (Cabinet mounting screws will be driven into the studs.) Lightly tap on and across a short distance of the wall with a hammer. Drive a finish nail in at the point where a solid sound is heard. Drive the nail where holes are later covered by a cabinet. If a stud is found, mark the location with a pencil. If no stud is found, try a little over to one side or the other.

Measure at 16-inch intervals in both directions from the first stud to locate other studs. Drive a finish nail to test for solid wood. Mark each stud location. If studs are not found at 16-inch centers, try 24-inch centers. At each stud location, draw plumb lines on the wall. Mark the outlines of all cabinets on the wall to visualize and check the cabinet locations against the layout (Fig. 87-10.

Installing Wall Units

Many installers prefer to mount the wall units first so that work does not have to be done over the base units. First, remove all doors and adjustable shelves. This makes the cabinets lighter and easier to clamp together. If possible, screw a 1×3 strip of lumber so its top edge is on the level line for the bottom of the wall cabinets. This is used to support the wall units while they are being fastened. If it is not possible to screw to the wall, build a stand on which to support the unit near the line of installation (Fig. 87-11).

Start the installation of wall cabinets in a corner. On the wall, measure from the line representing the outside of the cabinet to the stud centers. Transfer the measurements to the cabinets. Drill

SINK WALL ELEVATION

RANGE WALL ELEVATION

87-9 Elevations of the installation are drawn and the cabinets identified.

Fig. 87-10 The wall is laid out with outlines of the cabinets.

Fig. 87-11 Support the wall unit with a job-built stand. (*Courtesy of Merillat Industries*)

shank holes for mounting screws through mounting rails usually installed at the top and bottom of the cabinet. Place the cabinet on the supporting strip or stand so its bottom is on the level layout line. Fasten the cabinet in place with mounting screws of sufficient length to hold the cabinet securely. Do not fully tighten the screws (Fig. 87-12).

On masonry walls, first drill holes through the mounting strips. Place the cabinet in position, and mark the location of the drilled holes on the wall. Remove the cabinet. Drill holes into the masonry wall for lead inserts. Replace the cabinet, and screw in place.

The next cabinet is installed in the same manner. Align the adjoining *stiles* so their faces are flush with each other. Clamp them together with C-

Fig. 87-12 Installation of wall cabinets is started in the corner.

TOP RAILS

C-CLAMP

STILES

BOTTOM RAILS

Fig. 87-13 The stiles of adjoining cabinets are joined together with screws.

clamps. Screw the stiles tightly together (Fig. 87-13). Continue this procedure around the room. Tighten all mounting screws.

If a *filler* needs to be used, it is better to add it next to a *blind corner* cabinet or at the end of a run. It may be necessary to scribe the filler to the wall. This procedure is shown in Figure 87-14.

Installing Base Cabinets

Start the installation of base cabinets in a corner. Shim the bottom until the cabinet top is on the layout line. Then level and shim the cabinet from back to front.

If cabinets are to be fitted to the floor, shim until their tops are level across width and depth. This will bring the tops above the layout line that was measured from the low point of the floor. Adjust the pencil dividers so the distance between the points is equal to the amount the top of the unit is above the layout line. Scribe this amount on the bottom end of the cabinets by running the dividers along the floor (Fig. 87-15).

Cut both ends and toeboard to the scribed lines. There is no need to cut the cabinet backs because they do not, ordinarily, extend to the floor.

Place the cabinet in position. The top ends should be on the layout line. Fasten it loosely to the wall.

The remaining base cabinets are installed in the same manner. Align and clamp the stiles of adjoining cabinets. Fasten them together. Finally, fasten all units securely to the wall (Fig. 87-16).

Installing Manufactured Countertops

Countertops are manufactured in various standard lengths. They can be cut to fit any installation against walls. They are also available with one end precut at a 45 degree angle for joining with a similar one at corners. Special hardware is used to join the sections. The countertops are covered with a thin, tough *high-pressure plastic laminate.* This is generally known as *mica.* It is available in many colors and patterns. The countertops are called *postformed countertops.* This term comes from the method of forming the mica to the rounded edges and corners of the countertop (Fig. 87-17). Postforming is bending the mica with heat to a radius of 3/4 inch or less. This can only be done with special equipment.

After the base units are fastened in position, the countertop is cut to length. It is fastened on top

SCRIBING BLOCK SAME WIDTH AS STILE

2. RIDE SCRIBING BLOCK AGAINST WALL AND MARK FILLER

STILE

1. CLAMP FILLER TO STILE WITH INSIDE EDGES FLUSH WITH EACH OTHER

FILLER

Fig. 87-14

of the base units and against the wall. The back-splash can be scribed, limited by the thickness of its scribing strip, to an irregular wall surface. Use pencil dividers to scribe a line on the top edge of the backsplash. Then plane or belt sand to the scribed line.

Fasten the countertop to the base cabinets with screws up through triangular blocks usually installed in the top corners of base units. Use a stop on the drill bit. This prevents drilling through the countertop. Use screws of sufficient length, but not so long that they penetrate the countertop.

Exposed cut ends of postformed countertops are covered by specially shaped pieces of plastic laminate. (The application of plastic laminates is described in the next chapter.)

Sink cutouts are made by carefully outlining the cutout and cutting with a saber saw. The cutout pattern usually comes with the sink. Use a fine tooth blade to prevent chipping out the face of the mica beyond the sink. Some duct tape applied to the base of the saber saw will prevent scratching of the countertop when making the cutout.

SHIM BETWEEN
CABINET AND FLOOR
UNTIL LEVEL IN BOTH
DIRECTIONS

LAYOUT

AMOUNT TO
SCRIBE OFF
BOTTOM

34 1/2" FROM
LOW POINT

SCRIBED LINE

SHIMS

Fig. 87-15 Method of scribing base cabinets to the floor.

Fig. 87-16 Base cabinets are secured with screws to wall studs. (*Courtesy of Merillat Industries*)

Fig. 87-17 A section of a manufactured postformed countertop. The edges and interior corner are rounded.

CHAPTER **88**   CUSTOM-MADE CABINETS AND COUNTERTOPS

For various reasons, cabinets are sometimes custom-made on the job. There are several advantages to job-built cabinets. Material is conserved because the wall acts as the cabinet backs. Also, there is no need to double cabinet ends as in the case of individual units installed side by side. Cabinet bottoms, ends, and shelves are scribed and fitted to walls and floors. In addition, the cabinets are built to fit the space. There is no need for filler blocks, resulting in a neater and more pleasing appearance. The cost, in many cases, may be even less than factory-built cabinets. A section through typical job-built wall and base cabinets is shown in Figure 88-1.

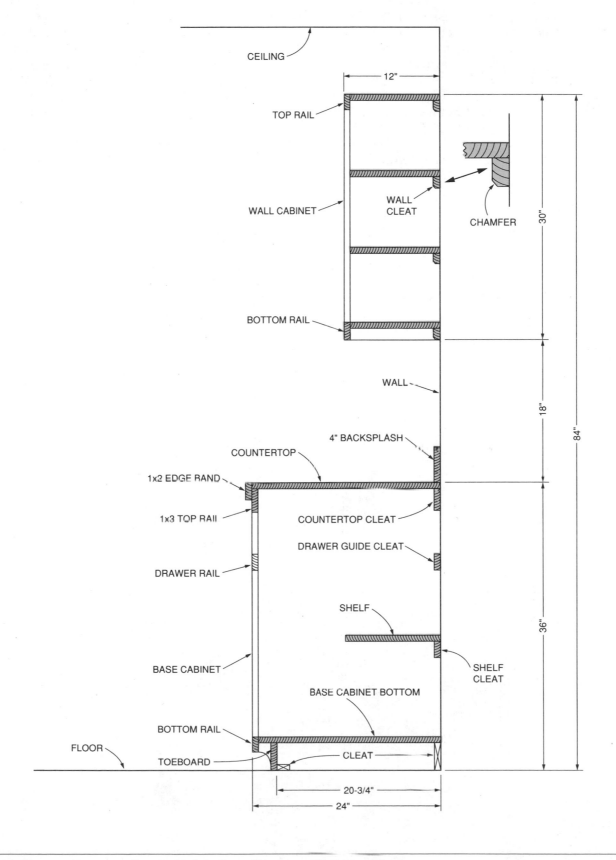

Fig. 88-1 Cross-section of typical job-built wall and base kitchen cabinets.

Building Wall Cabinets

Lay out the wall with level lines, 35 1/4 inches from the floor, for the top of the base cabinets, and 54 inches for the bottom of the wall cabinets. Measure from any point unless the floor is extremely out of level. In that case, measure from an average height. Find and mark the centerlines of all studs in the wall. Snap lines, 11 1/4 inches from the wall, on the ceiling or soffit, above the line of wall cabinets. This distance is equal to the depth of the wall cabinet minus the face frame (Fig. 88-2). The wall cabinets are usually constructed first to avoid working over the base units.

Applying Wall Cleats

Draw level lines for the bottom of each shelf and the cabinet top. Draw plumb lines for the inside face of each cabinet end. Nail 3/4×1 *cleats* with their top edges to the level lines and between the plumb lines. For a better appearance, *chamfer* the bottom outside corner of each cleat. Fasten the bottom cleat with its bottom edge to the line (Fig. 88-3).

Installing Cabinet Ends

Cut the cabinet ends to length with both ends square. Place a cabinet end in position against the ends of the cleats. Its bottom end should be flush with the bottom edge of the bottom cleat. Plumb the outside edge of the cabinet end by placing a shim, at the top or bottom, between it and the wall. Determine the amount the outside edge projects beyond the chalk line on the ceiling or soffit. Scribe this amount off the back edge of the cabinet end (Fig. 88-4). Cut the end to the scribed line. Fasten it to the ends of the cleats. Install the other end in the same manner.

Installing Cabinet Shelves

Draw level lines on the inside faces of the cabinet ends from the top of the cleats for the position of the shelves. Cut the cabinet bottom, top, and shelves the same length as the cleats. Place the cabinet bottom in position, against the wall, on top of the cleat and between the cabinet ends. Move one end or the other so that both ends project the same distance

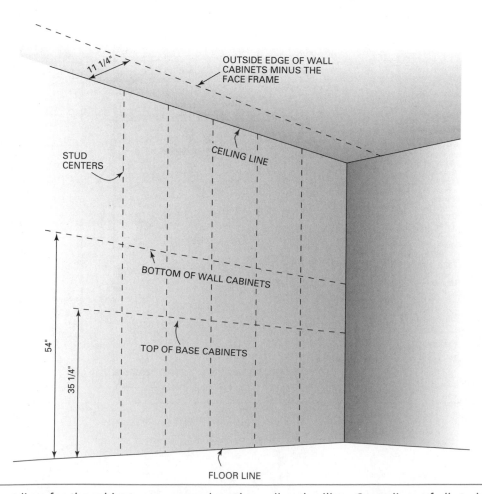

Fig. 88-2 Layout lines for the cabinets are snapped on the wall and ceiling. Centerlines of all studs are marked.

11 1/4"

TOP OF CABINET

CEILING LINE

30"

EQUAL
EQUAL
EQUAL
EQUAL

54" FROM FLOOR

BOTTOM OF WALL CABINET

3/4" × 1" WALL CLEAT

3/4"

1"

CHAMFER

CROSS-SECTION

OTHER
CABINET
OUTLINES

Fig. 88-3 Cleats are fastened to the wall to support the wall cabinet top, bottom, and shelves.

STEP BY STEP
P R O C E D U R E S

2. ADJUST PENCIL DIVIDERS
TO THIS DISTANCE

11 1/4"

LEVEL

SHIM, IF NECESSARY
TO PLUMB CABINET
END.

CEILING LINE

WALL CLEATS

3. SCRIBE BACK
EDGE OF CABINET
END TO THE WALL

1. PLUMB CABINET
END

CABINET END

NOTE: WIDTH OF SCRIBE
IS EXAGGERATED FOR
CLARITY PURPOSES

Fig. 88-4 Procedure for fitting wall cabinet ends.

Fig. 88-5 Method of fitting and fastening the wall cabinet bottoms, shelves, and top.

beyond the outside edge of the cabinet ends. Scribe the amount of projection on the back edge of the cabinet bottom. Cut the bottom to the scribed line. Fasten it in position. Besides fastening through the cabinet ends, toenail through the back edge into each stud (Fig. 88-5). Set all nails. Mark and fasten the shelves and cabinet top in the same manner.

In the same manner, continue installing cleats, cabinet ends, bottoms, shelves, and tops on the walls, where required. In wide cabinets, over about 4 feet, install *partitions* for support between the cabinet top and bottom (Fig. 88-6).

Making the Face Frame

Members of the face frame are usually made of 1×2 nominal lumber. In corners, the exposed width of the cabinet stiles should be at least 2 1/2 inches to allow for the swing of doors.

Cut the stiles the same length as the cabinet ends. Apply them with their ends flush with the cabinet ends. Horizontal members of the face frame are called *rails*. They are fitted and fastened between the stiles.

Fit and fasten the bottom rails so their top

Fig. 88-6 Cabinet construction is continued along the wall. Partitions are installed in wide cabinets for support.

edge is flush with the bottom shelf. Fasten the top rail with its top edge flush with the top of the cabinet. Fasten the rails into the outside edge of the shelves. Toenail their ends into the stiles. Apply a little glue to the joint when fastening face frame members.

Install *mullions* to divide the total cabinet width into door openings of equal width. The openings should not, normally, exceed 24 inches in width. Mullions are intermediate vertical members of the face frame installed between top and bottom rails. When fastening mullions to the shelf edges, straighten the shelves by sighting them from one end to the other (Fig. 88-7). Set all nails, sand all joints flush. Ease all exposed sharp corners.

Building Base Cabinets

The level line for the top of the *base* cabinets has already been laid out. Begin by plumbing lines to mark the inside faces of the cabinet at each end of

Fig. 88-7 The face frame is fastened to the outside edges of the cabinet case. Mullions are installed to divide the width into equal-sized door openings.

the run. Then, draw a level line, between the plumb lines, 3 1/4 inches above the floor. Install a cleat, to support the cabinet bottom, against the wall (Fig. 88-8). Its top edge should be to the line, its bottom edge against the floor, and its ends to the plumb lines previously laid out. At each end of the wall cleat, mark lines on the floor about 2 feet out and square with the wall. Snap a line on the floor between the squared lines, 20 3/4 inches away from the wall. This line marks the inside of the *toeboard*. Nail a 1×2 cleat on the floor. Its outside edge is to the chalk line. This provides a stop for the toeboard. The cleat ends at each squared line.

Installing the Toeboard

The toeboard must be fastened against the floor cleat so it fits the floor and its top edge is level with the cleat previously installed on the wall. Temporarily tack the toeboard to the floor cleat so its top edge is 3/8 inch above the level of the wall cleat. Scribe 3/8 inch off the bottom of the toeboard. Remove and cut the toeboard to the scribed line. Fasten it against the cleat by toenailing into the floor (Fig. 88-9). The ends of the toeboard are installed flush with the ends of the cleat. The top edges of the toeboard should then fit the floor with its top edge level with the wall cleat.

Installing the Cabinet Bottom

The *cabinet bottom* must fit the wall with its outside edge overhanging the toeboard by 1 3/4 inches from one end of the cabinet run to the other. Cut 3/4-inch or 5/8-inch plywood or other suitable panel to a rough width of 23 1/2 inches. To con-

Fig. 88-8 To start the base cabinet installation, cleats are fastened to the wall and floor.

Fig. 88-9 Method of laying out and fastening the toeboard.

serve material, cut one width from each sheet. Save the other width for making the countertop. Place the pieces on top of the wall cleat and toeboard. Tack the pieces with one edge against the wall. The outside edge should overhang the toeboard the same amount for the entire length.

Adjust the pencil dividers or scribers for the difference between the amount of overhang and 1 3/4 inches. Scribe this amount on the wall edge of the cabinet bottom. Cut the bottoms to the scribed lines. Install *backing* under joints between lengths of cabinet bottoms. Then fasten them in position. Keep the ends flush with the ends of the toeboard and the wall cleat (Fig. 88-10). Mark plumb lines on the wall in line with each end of the cabinet bottom.

Installing Other Wall Cleats

Other horizontal *wall cleats* are needed to support the *countertop, drawer slides,* and *shelves.* The same level line that was previously made for the base cabinet top is used to position the cleat to support

the countertop. Fasten a 1×3 from one end of the cabinet to the other. Its top edge sgiykd be to this line. The height of cleats used to support the inner end of drawer slides is determined by the depth of

Fig. 88-10 Procedure for scribing the cabinet bottom to the wall. Cutting the back edge leaves a straight outside edge.

the drawer and the type of drawer guides. Some sections of the cabinet may have stacks of drawers. Determine the height. Install cleats for fastening inner ends of drawer slide hardware, where required. In most cases, one shelf is installed in the base cabinet. Snap lines and install cleats, where necessary, to support the inner edges of the cabinet shelves (Fig. 88-11).

Installing the Cabinet Ends

The back edge of the *cabinet end* must fit the wall. Its front edge must be flush with the front edge of the cabinet bottom and be plumb. Its bottom end must fit the floor. Its top end must be flush with the top edge of the countertop cleat and be level.

Cut the cabinet ends to a rough length and width with square ends. The length should be about 36 inches and the width about 23 1/2 inches. Place one end in position. Plumb the outside edge. Use shims between it and the wall and floor, if necessary, to hold it steady.

Adjust the pencil dividers to the same distance the top end projects above the top cleat. Scribe this amount on the bottom. Cut the bottom end to the scribed line to fit to the floor. Place the end back in position. Its bottom end should fit the floor. Top

end should be flush and level with the top edge of the top cleat, and its outside edge plumb.

Adjust the pencil dividers to the same distance the outside edge projects beyond the cabinet bottom. Scribe this amount on the edge against the wall. Cut to the scribed line.

Before fastening the end in position, mark and cut a circular arc out of the bottom outside corner for the toespace. The top of the cut is started 1 1/2 inches below the cabinet bottom. The cut is flush with the face of the toeboard at the floor line. Smooth the cut edge. Fasten the cabinet end in position. Nail into the end of the cabinet bottom and into the ends of the wall cleats (Fig. 88-12). Cut and install the other end in the same manner.

Making the Face Frame

The face frame for the base unit is made in the same manner as for wall units. However, the stiles project 1 1/2 inches below the face of the cabinet bottom. Extra rails need to be installed for drawer openings. In this installation, the top rails are made of 1×3 stock because of the method to be used for countertop construction.

Install stiles on each cabinet end. Install the bottom rail between the stiles. Keep the top edge flush with the top side of the cabinet bottom. Cut the top rail and drawer rail to the same length as the bottom rail. The top rails are fastened by clamping and gluing them to short blocks of 1×2 stock installed in back of the joint (Fig. 88-13). Sand all joints flush. Ease all sharp exposed corners.

Fig. 88-11 Location of wall cleats for countertop, drawer guides, and shelves of the base cabinet.

STEP BY STEP
P R O C E D U R E S

③ SCRIBE AMOUNT CABINET END PROJECTS ABOVE TOP FROM BOTTOM END

CABINET END

LINE OF CABINET TOP

CABINET BOTTOM

② SCRIBE AMOUNT CABINET END PROJECTS BEYOND CABINET BOTTOM FROM BACK EDGE

① PLUMB EDGE OF CABINET END

Fig. 88-12 Method of fitting the base cabinet end panels.

Fig. 88-13 Construction of the base cabinet face frame.

Making the Countertop

Use the pieces of 3/4- or 5/8-inch panel material left from making the cabinet bottoms to make the *countertop.* If more than one length is required, join them with glue and screws to a short piece of backing plywood. The width of the pieces should be about 24 1/2 inches.

Fitting the Countertop

Place the countertop on the base cabinets, against the wall. Its outside edge should overhang the face frame the same amount along the entire length. Open the pencil dividers or scribers to the amount of overhang. Scribe the back edge of the countertop to the wall. Cut the countertop to the scribed line. Place it back on top of the base cabinets. The ends should be flush with the ends of the base cabinets. The front edge should be flush with the face of the face frame (Fig. 88-14). Install a 1×2 on the front edge and at the ends if an end overhang is desired. Keep the top edge flush with the top side of the countertop.

Applying the Backsplash

If a *backsplash* is used, rip a 4-inch wide length of 3/4-inch stock the same length as the countertop. Use lumber for the backsplash, if lengths over 8 feet are required, to eliminate joints. Fasten the backsplash on top of and flush with the back edge of the countertop by driving screws up through the countertop and into the bottom edge of the backsplash

Fig. 88-14 Scribing the countertop to fit the wall with its outside edge flush with the face of the cabinet.

(Fig. 88-15). In corners, fasten the ends of the backsplash together with screws.

Laminating a Countertop

Most countertops are covered with *plastic laminate.* Before laminating a countertop, make sure all surfaces are flush. Check for protruding nailheads. Fill in all holes and open joints. Lightly hand or power sand the entire surface, making sure joints are sanded flush.

Laminate Trimming Tools and Methods

Pieces of laminate are first cut to a *rough size,* about 1/4 to 1/2 inch wider and longer than the surface

Fig. 88-15 Drive screws into the bottom edge of the backsplash to fasten it to the countertop.

to be covered. A strip is then cemented to the edge of the countertop. Its edges are *flush-trimmed* even with the top and bottom surfaces. Laminate is then cemented to the top surface, overhanging the edge strip. The overhang is then *bevel-trimmed* even with the laminated edge. A *laminate trimmer* or a small *router* fitted with *laminate trimming bits* is used for rough cutting and flush and bevel trimming of the laminate (Fig. 88-16).

Cutting Laminate to Rough Sizes

Sheets of laminate are large, thin, and flexible. This makes them difficult to cut on a table saw. One method of cutting laminates to rough sizes is by

FLUSH TRIM

LAMINATE TRIMMER

clamping a straightedge to the sheet. Cut it by guiding a laminate trimmer with a flush trimming bit along the straightedge (Fig. 88-17). It is easier to run the trimmer across the sheet than to run the sheet across the table saw. Also, the router bit leaves a smooth, clean cut edge. Use a *solid carbide* trimming bit, which is smaller in diameter than one with ball bearings. It makes a narrower cut. It is is easier to control and creates less waste. With this method, cut all the pieces of laminate needed to a rough width and length. Cut the narrow edge strips from the sheet first.

Using Contact Cement

Contact cement is used for bonding plastic laminates and other thin, flexible material to surfaces. A coat of cement is applied to the back side of the laminate and to the countertop surface. The cement must be dry before the laminate is bonded to the core. The bond is made on contact without the need of clamps. There are several reasons why a contact cement bond may fail.

■ Not enough cement is applied. If the material is porous, like the edge of particleboard or plywood, a second coat is required after the first coat dries. When enough cement has been applied, a glossy film appears over the entire surface when dry.

■ Too little time is allowed for the cement to dry. Both surfaces must be dry before contact is made. To test for dryness, lightly press your finger on the surface. Although it may feel sticky, the cement is dry if no cement remains on the finger.

■ The cement is allowed to dry too long. If contact cement dries too long (more than about 2

BEVEL TRIM

ADJUST BEVEL TRIMMING BIT TO CUT FLUSH
WITH, BUT NOT INTO EDGE LAMINATE.
THE BEVEL KEEPS THE CUTTING EDGE FROM
GRAZING THE FIRST LAYER OF LAMINATE.

Fig. 88-16 The laminate trimmer is used with flush and bevel bits to trim overhanging edges of laminate. (*Part 1, Courtesy of Porter-Cable*)

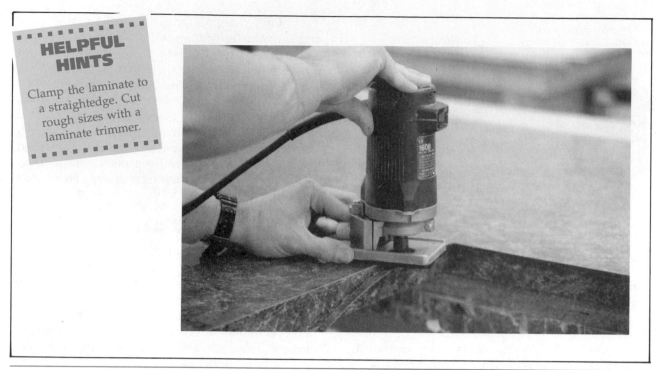

Fig. 88-17

hours, depending on the humidity), it will not bond properly. To correct this condition, merely apply another coat of cement and let it dry.

■ The surface is not rolled out or tapped after the bond is made. Pressure must be applied to the entire surface using a 3-inch *J-roller* or by tapping with a hammer on a small block of wood.

CAUTION Some contact cements are flammable. Apply only in a well-ventilated area around no open flame. Avoid inhaling the fumes.

Laminating the Countertop Edges

Remove the backsplash from the countertop. Apply coats of cement to the countertop edges and the back of the edge laminate with a narrow brush or small paint roller. After the cement is dry, apply the laminate to the front edge of the countertop (Fig. 88-18). Position it so the bottom edge, top edge, and ends overhang. A permanent bond is made when the two surfaces make contact. A mistake in positioning means removing the bonded piece—a time-consuming, frustrating, and difficult job. Roll out or tap the surface.

Apply the laminate to the ends in the same manner as to the front edge piece. Make sure that the square ends butt up firmly against the back side of the overhanging ends of the front edge piece to make a tight joint.

Fig. 88-18 Applying laminate to the edge of a countertop.

Trimming Laminated Edges

The overhanging ends of the edge laminate must be trimmed before the top and bottom edges. If the laminate has been applied to the ends, a bevel trimming bit must be used to trim the overhanging ends.

Using a Bevel Trim Bit. When using a bevel trimming bit, the router base is gradually adjusted to expose the bit so that the laminate is trimmed flush with the first piece but not cutting into it. The bevel of the cutting edge allows the laminate to be trimmed without cutting into the adjacent piece (see Fig. 88–16). A flush trimming bit cannot be used when the pilot rides against another piece of laminate because the cutting edge may damage it.

Ball bearing trimming bits have *live pilots.* Solid carbide bits have *dead pilots* that turn with the bit. When using a trimming bit with a dead pilot, the laminate must be lubricated where the pilot will ride. Rub a short piece of white candle or some solid shortening on the laminate to prevent marring the laminate by the bit.

Using the bevel trimming bit, trim the overhanging ends of the edge laminate. Then, using the flush trimming bit, trim off the bottom and top edges of both front and end edge pieces (Fig. 88-19). To save the time required to change and adjust trimming bits, some installers use two laminate trimmers, one with a flush bit and the other with a bevel bit.

Use a belt sander or a file to smooth the top edge flush with the surface. Sand or file *flat* on the countertop core so a sharp square edge is made. This assures a tight joint with the countertop laminate. Sand or file *toward* the core to prevent chipping the laminate. Smooth the bottom edge. Ease the sharp outside corner with a sanding block.

Laminating the Countertop Surface

Apply contact bond cement to the countertop and the back side of the laminate. Let dry. To position large pieces of countertop laminate, first place thin strips of wood or metal venetian blind slats about a foot apart on the surface. Lay the laminate to be bonded on the strips or slats. Then position the laminate correctly (Fig. 88-20).

Make contact on one end. Gradually remove the slats one by one until all are removed. The laminate should then be positioned correctly with no costly errors. Roll out the laminate (Fig. 88-21). Trim the overhanging back edge with a flush trimming bit. Trim the ends and front edge with a bevel trimming bit (Fig. 88-22). Use a flat file to smooth the trimmed edge. Slightly ease the sharp corner.

Fig. 88-20 Position the laminate on the countertop using venetian blind slats.

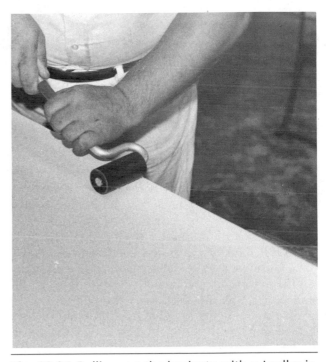

Fig. 88-21 Rolling out the laminate with a J-roller is required to assure a proper bond.

Fig. 88-19 Flush trimming the countertop edge laminate.

Fig. 88-22 The outside edge of the countertop laminate is bevel-trimmed.

HELPFUL HINTS

How to make a tight joint between the ends of two lengths of laminate.

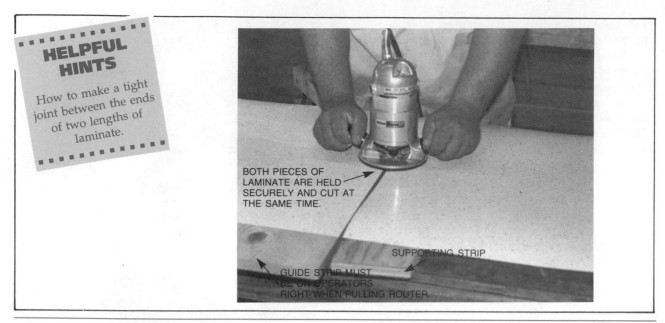

BOTH PIECES OF LAMINATE ARE HELD SECURELY AND CUT AT THE SAME TIME.

SUPPORTING STRIP

GUIDE STRIP MUST BE ON OPERATORS RIGHT WHEN PULLING ROUTER.

Fig. 88-23

Laminating a Countertop with Two or More Pieces

When the countertop is laminated with two or more lengths, tight joints must be made between them. Tight joints can be made by clamping the two pieces of laminate in a straight line on some strips of 3/4-inch stock. Butt the ends together or leave a space less than 1/4 inch between them.

Using one of the strips as a guide, run the laminate trimmer, with a flush trimming bit installed, through the joint. Keep the pilot of the bit against the straightedge. Cut the ends of both pieces at the same time to assure making a tight joint (Fig. 88-23). Bond the sheets as previously described. Apply *seam-filling compound,* especially made for laminates, to make a practically invisible joint. Wipe off excess compound with the recommended solvent.

Laminating Backsplashes

Backsplashes are laminated in the same manner as countertops. Laminate the backsplash. Then reattach it to the countertop with the same screws. Use a little caulking compound between the backsplash and countertop. This prevents any water from seeping through the joint (Fig. 88-24).

Laminating Rounded Corners

If the edge of a countertop has a rounded corner, the laminate can be bent. Strips of laminate can be cold bent to a minimum radius of about 6 inches.

Fig. 88-24 Apply the laminate to the backsplash. Then fasten it to the laminated countertop by driving screws through the countertop into its bottom edge.

Heating the laminate to 325 degrees Farenheit uniformly over the entire bend will facilitate bending to a minimum radius of about 2 1/2 inches. Heat the laminate carefully with a *heat gun.* Bend it until the desired radius is obtained (Fig. 88-25). Experimentation may be necessary until success in bending is achieved.

CAUTION Keep fingers away from the heated area of the laminate. Remember that the laminate retains heat for some time.

Fig. 88-25 Heating and bending laminate with a heat gun.

Review Questions

Select the most appropriate answer.

1. The vertical distance between the base unit and the wall unit is usually
a. 12 inches.   c. 18 inches.
b. 15 inches.   d. 20 inches.

2. The maximum distance, in feet, that the top shelf of a wall unit should be from the floor is
a. 5.   c. 7.
b. 6.   d. 8.

3. The distance from the floor to the surface of the countertop is
a. 30 inches.   c. 36 inches.
b. 32 inches.   d. 42 inches.

4. In order to accommodate sinks and provide adequate working space, the width of the countertop is
a. 25 inches.   c. 30 inches.
b. 28 inches.   d. 32 inches.

5. Standard wall unit height is
a. 24 inches.   c. 32 inches.
b. 30 inches.   d. 36 inches.

6. The depth of standard wall units is
a. 10 inches.   c. 14 inches.
b. 12 inches.   d. 16 inches.

7. The height of most manufactured base kitchen cabinets is
a. 30 3/4 inches.   c. 34 1/2 inches.
b. 32 1/2 inches.   d. 35 1/4 inches.

8. To find the line of installation for the bottom edge of wall units, measure up from the low point on the floor
a. 38 inches.   c. 54 inches.
b. 50 inches.   d. 84 inches.

9. The amount scribed off the wall edge of job-built base unit end panels is
a. one inch.
b. the amount the outside edge projects beyond the cabinet bottom.
c. the amount the top end extends above the top cleat.
d. an amount equal to the cabinet end thickness.

10. The base cabinet bottom is fitted to the wall by scribing an amount
a. of 1 3/4 inches.
b. of the difference between the overhang at the toeboard and 1 3/4 inches.
c. of the difference between the overhang at the toeboard and 3 1/2 inches.
d. which is the farthest distance the back edge is away from the wall.

■ BUILDING FOR SUCCESS ■

Demonstrating Quality Workmanship

The demonstration of quality workmanship begins with two basic premises: one involving need and one involving desire. First, each skilled worker must accept and understand the need to do quality work. Skill competencies cannot be reached without a commitment to quality. Second, each person has to want to do quality work. Without meeting these two requirements, quality workmanship cannot be accomplished.

Since customers expect quality materials and workmanship and are willing to pay for them, they will not settle for anything less. By looking at successful building and remodeling companies, quality work will surface as a trademark. With those companies quality will become a product. Therefore, all workers associated with those companies are expected to deliver quality. The mind-set for each person has to be a total buy-in to the quality concept.

How is quality to be defined? What are the characteristics of quality? In some cases, definitions of quality work will differ. Its definition is often based on expectations. At the beginning of each construction project, written statements of quality in materials, features, and workmanship must relate to expectations held by all parties. This in turn relates to satisfaction. Therefore it becomes imperative to agree on quality materials and workmanship before the construction process begins.

There are quality standard guidelines available for both new construction and remodeling. For example, the *Quality Standards for the Professional Remodeler*, published by the NAHB Remodeler's Council, has established good descriptions of industry quality standards. Other ideal publications by the NAHB Home Builder Bookstore are: *Production Checklist for Builders and Superintendents, Customer Service for Home Builders, The Positive Walk-Through: Your Blueprint for Success*, and *Warranty Service for Builders and Remodelers*.

Typically, these standards manuals and guides present criteria, written specifications, and suggestions that help each builder or remodeler provide quality products, workmanship, and service. The quality standards presented in these publications define measureable criteria to be used throughout the construction process. These manuals serve well in resolving complaints and disputes between the builder or remodeler and the client. They are written in layperson's terms that are easily understood by everyone.

Written materials that explain quality workmanship and construction procedures set the stage for credibility in relationships in a construction project. Written standards that are presented at the beginning of the construction discussions reassure clients that they will receive the quality they are paying for. With that assurance, clients will be more trusting and willing to cooperate fully with the builder or remodeler.

Quality standards, along with construction building codes, can be used by the builder or remodeler as the very minimum in quality to be provided throughout the building process. The goal should be to meet or exceed these minimums during the entire construction process.

The minimum quality standards adopted by a company also become the basic guidelines for training employees in company expectations. Each prospective employee should be exposed to these quality standards before actual employment. This will give everyone a chance to see what is expected on the job. At that time the potential employee can see the company philosophy, the commitment to quality workmanship, and how it is provided to the customer. This preview of expectations will undoubtedly increase each employee's chances for success.

By involving all employees in developing and using quality standards, a company can better promote and defend its commitment to excellence. Each skilled craftsperson should desire to protect his or her image of doing quality work. Communicating in writing what quality workmanship is and what each client can expect from the skilled workers should be presented in the early stages of the project. The commitment by the builder or remodeler to meet or exceed those standards will be satisfying to the client.

Competition will never allow the skilled worker to abandon the need to demonstrate quality workmanship. Each worker must develop good technical skills and gain a good understanding of the construction processes used in building. Obtaining and maintaining quality tools will help install quality materials. People are willing to pay for these things and more if they are satisfied the results will be the best possible. Customer satisfaction will remain one of the primary gauges for determining the level of quality workmanship and service being provided by a construction company. The margin for error is small and the tolerance level for poor workmanship will continue to decrease. Each person entering the construction industry today will have to buy into the quality workmanship concept to be successful.

FOCUS QUESTIONS: For individual or group discussion.

1. What criteria is necessary in order for a carpenter to demonstrate quality workmanship? Explain how this criteria will be observable to the client.

2. Is there a correlation between a carpenter's attitude toward quality workmanship and customer satisfaction? Explain your answer in detail.

3. What would be some advantages of starting the degree of quality materials and workmanship before a contract is agreed upon?

4. How could meeting or exceeding the quality standards and workmanship stated in the agreement prove to be a positive long-term strategy for the builder? Explain how this might affect customer satisfaction.

UNIT

32

Cabinet Doors and Drawers

OBJECTIVES

After completing this unit, the student will be able to:

- identify cabinet doors and drawers according to the type of construction and method of installation.

- make solid and paneled cabinet doors.

- hang overlay, lipped, and flush cabinet doors.

- make overlay, lipped, and flush cabinet drawers.

- install drawer guides.

- install overlay, lipped, and flush drawers in a cabinet.

- apply door and drawer hardware.

Cabinet doors and drawers may be purchased in a wide variety of styles and sizes for installation on job-built cabinets. If the style does not require special equipment for its construction, the carpenter may design, build, fit, and install cabinet doors and drawers on the job.

CONTENTS

CHAPTER 89

DOOR TYPES, CONSTRUCTION, AND INSTALLATION

Doors are classified by their construction and also by the method of installation. Sliding doors are occasionally installed, but most cabinets are fitted with hinged doors that swing.

Kinds of Doors

Hinged cabinet doors are classified as *overlay, lipped,* and *flush,* based on the method of installation (Fig. 89-1). The overlay method of hanging cabinet doors is the most widely used.

Overlay Doors

The *overlay* type of door laps over the opening, usually 3/8 inch on all sides. However, it may overlay any amount. In many cases, it may cover the entire face frame. The overlay door is easy to install. It does not require fitting in the opening and the face frame of the cabinet acts as a stop for the door. *European-style* cabinets omit the face frame. Doors completely overlay the front edges of the cabinet (Fig. 89-2).

Lipped Doors

The *lipped* door has rabbeted edges that overlap the opening by about 3/8 inch on all sides. Usually the ends and edges are rounded over to give a more pleasing appearance. Lipped doors and drawers are easy to install. No fitting is required and the rabbeted edges stop against the face frame of the cabi-

Fig. 89-2 Overlay doors lap the face frame by varying amounts. European style doors are hinged to and completely overlay the case.

net. However, a little more time is required to shape the rabbeted edges.

Flush Type

The *flush type* door fits into and flush with the face of the opening. They are a little more difficult to hang because they must be fitted in the opening. A fine joint, about the thickness of a dime, must be made between the opening and the door. Stops must be provided in the cabinet against which to close the door.

Door Construction

Doors are also classified, by their construction, as *solid* or *paneled.* Solid doors are made of plywood, particleboard, or solid lumber. Particleboard doors are ordinarily covered with plastic laminate. Matched boards with V-grooves and other designs, such as used for wall paneling, are often used to make solid

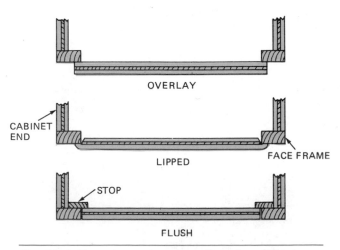

Fig. 89-1 Plan views of overlay, lipped, and flush doors.

Fig. 89-3 Solid doors are usually made of plywood or particleboard.

doors (Fig. 89-3). Designs may be grooved into the face of the door with a router. Small moldings may be applied for a more attractive appearance.

Paneled doors have an exterior framework of solid wood with panels of solid wood, plywood, hardboard, plastic, glass, or other panel material. Many complicated designs are manufactured by millworkers with specialized equipment. With the equipment available, carpenters can make paneled doors of simple design only (Fig. 89-4). Both solid doors and paneled doors may be hinged in overlay, lipped, or flush fashion.

Fig. 89-4 Panel doors of simple design can be made on the job.

Making Solid Doors
Overlay Doors

To make a *solid overlay* door from plywood, particleboard, or other panel material, simply cut the stock to size. The size of an overlay door is the size of the door opening plus the overlay. For instance, cut the door 3/4 inch larger than the width and height of the opening for a door that overlays 3/8 inch.

Tilt the table saw blade 30 degrees to cut reverse bevel edges, if desired. Overlay doors are, ordinarily, the only type that are laminated. They may also be finished in other ways.

Lipped Doors

Particleboard is not normally used to make *solid lipped* doors. This is because the thin edge tends to break off. Lipped doors are usually not laminated because of the difficulty of laminating the rabbeted edges.

For a lipped door, cut the stock 5/8 inch larger than the opening. Then rabbet the stock 3/8×3/8 on the edges and ends. This gives 1/16-inch clearance between the frame and rabbeted edges and ends of the door when installed. Cutting the rabbet on a slight bevel gives added clearance to swing the door (Fig. 89-5).

Use a block plane to round over the edges and ends of the door. Then sand them smooth. A router with a 3/8-inch-radius rounding-over bit may also be used. However, round over the edges before making the rabbets so that the pilot of the bit will have enough wood to ride on. Instead of rounded edges, a reverse bevel may also be made on lipped doors.

Flush Doors

For a flush door, cut the stock the same size as the opening. Then, joint the door in the opening with a hand plane. A joint about 1/16 inch should be

Fig. 89-5 Edge designs on overlay and lipped doors.

Fig. 89-6 Solid matched board door.

made all around the door. The edge opposite the hinges should be back-beveled slightly for clearance when the door is swung open.

Solid Matched Board Door

To make a *solid matched board* door, enough pieces are cut and joined edge to edge. To hold the pieces together, cleats are screwed to the back side near the top and the bottom of the door with a diagonal brace between them (Fig. 89-6).

Cut the number of pieces needed a little longer than the height of the opening. Plan the width of the pieces to end up with a piece as wide as possible on both edges of the door. Rip square edges on both outside pieces.

Bar clamps for edge-gluing boards are not usually available to the carpenter. Instead of using bar clamps, tack a straightedge on the workbench or subfloor. Place the first piece, face down, with its square edge against the straightedge. Then, glue and place the rest of the pieces, edge to edge, with faces down, against the first piece. Toenail another straightedge to the bench against the edge of the last piece. Drive the edges of the pieces tightly against each other (Fig. 89-7).

Cut two 1×3 cleats about one inch shorter than the width of the door. Fasten them to the back side of the door with screws staggered into each piece. Keep the cleats about 2 inches from the top and bottom edges of the door.

Cut a diagonal brace of the same size material between the cleats. The bottom end of the brace should be toward the hinge edge of the door. Fasten

Fig. 89-7 Method of driving matched board edges tightly together.

with screws staggered into each piece of the door. Chamfer all exposed edges and ends of the cleats and brace. Shape the edges for an overlay, lipped, or flush installation as desired.

Covering Solid Door Edges

The edges of solid plywood or particleboard doors should be covered for a more attractive appearance. The edges of particleboard doors are laminated with the same pattern as their faces. Plywood doors may also be laminated in the same manner. When laminating doors, the back side should be covered with a thin laminate called a *backer* laminate to achieve a *balanced core*. The core is balanced when both sides are sealed to the same degree. Warpage, due to moisture absorbed more by one side than the other, is thus prevented. It is important that both sides of a door, regardless of the type of finish, be sealed equally to prevent warpage.

Veneer Strips

Another method used to band door edges is to apply a thin wood veneer of the same material as the door to all the edges (Fig. 89-8). These veneer strips may be purchased in rolls of different kinds of wood. Veneers may also be made on the job by slicing thin strips with the table saw. Make the strips slightly wider than the thickness of the door and from the same kind of wood.

The veneers may be applied to door edges with contact cement. Coat both the underside of the strips and the door edges with the cement. Let them dry. Apply the strips to the edges of the door, making tight joints at the corners. Trim the excess with a sharp knife. Sand smooth. Some wood veneer strips come with a backing of *heat-sensitive* adhesive. They are applied by simply using a heating iron pressed on the strip.

Fig. 89-8 Applying wood veneer strips to the edges of a plywood door.

Fig. 89-9 Covering door edges with plastic tape.

Fig. 89-10 Straight line designs can be routed on door faces by guiding the router against a straightedge.

Plastic Tapes

Rolls of plastic tape are available to cover edges of plywood and other core material. These tapes come in a variety of colors and wood grains. They are backed with a heat-sensitive adhesive. They are applied with a heating iron and trimmed flush (Fig. 89-9). To save time when laminating doors, tape of the same color and pattern as the face laminate is used frequently to cover the edges.

Wood Strips

Strips of wood are often used to band door edges. The strips may be thin, from 1/4 to 1/2 inch. They are applied with glue and brads. The corners are mitered to avoid exposing any end grain. After application the edges of the strips are sanded flush with the faces of the door. The sharp corners should be eased. The strips should be of the same wood as the door, if a natural finish is used.

Treating Solid Door Faces

The appearance of solid doors may be improved by routing designs in or by applying moulding to the door faces.

Routing Door Faces

To improve the appearance of some solid doors, a number of designs may be routed in the door faces with *straight*, *V-groove*, or *core box* bits. Straight line cuts may be made in various ways by guiding the router base against a straightedge while the door is held securely (Fig. 89-10).

A *French Provincial* design is often routed on the face of laminated solid doors. The design is made by using *veining bit* adjusted to cut about 1/16 inch deep. The design may be made on the job by removing the router base and replacing it with an elongated base made with rounded ends. The base can be made on the job from 1/4-inch tempered hardboard or plastic. Equal radii must be used to curve both ends. The centers of both curved

ends must be located a distance equal to the radii from the center of the router. The size of the base is determined by the desired distance from the edge of the door to the groove and its radius in the corners.

Lay the door on a bench. Tack some strips a little thicker than the door around it. Start routing along one side. Stop the router into each corner. Rotate it until the base touches the other side. Continue in this manner until the design is complete (Fig. 89-11).

Applying Molding to Door Faces

Small molding that has one flat side, such as *astragal*, *screen*, or *half-round*, may be applied to door faces in a number of interesting designs (Fig. 89-12). Sand the door faces before the moulding is applied. Lightly mark the outer edge of the molding on the door face. Cut the molding to length, mitering the corners, and apply it with a little glue and brads. Apply the molding so the outside edge just covers the line. Wipe off any excess glue with a damp rag. Set all brads below the surface.

Making Paneled Doors

Paneled doors of simple design may be made by the carpenter on the job site. The frame is usually made of 1×2 or 1×3 solid lumber. Flat panels of plywood, hardboard, tempered glass, or decorative plastic may be used.

Some joints strong enough for the frame of the door can be made with the tools usually available to the carpenter. A *through mortise-and-tenon* joint is a strong joint that can be made with hand tools or by using a table saw (Fig. 89-13).

A strong joint can also be made with a portable electric *plate joiner*, also called a *biscuit joiner*, if available. The plate joiner drives a horizontal circular saw blade that cuts a slot with a curved bottom into each of the mitered ends of the frame's stiles and rails. Three sizes of oval-shaped, compressed, plate joining wood *biscuits* are available, depending on the depth of cut, to be used as *blind splines* in the joint. When glue is applied to the joint, the compressed biscuit swells. However, it does not shrink

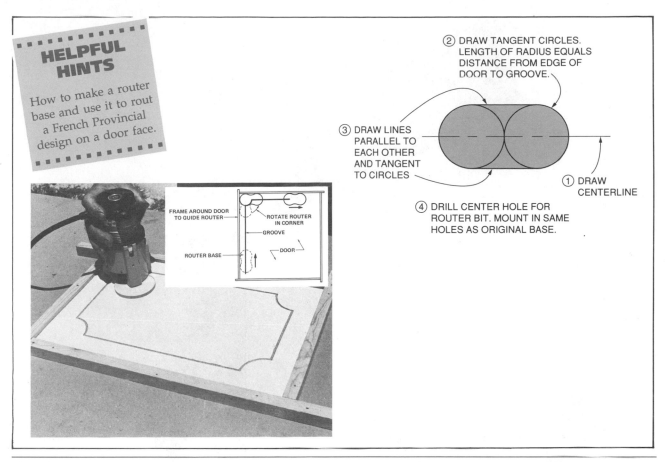

FRAME AROUND DOOR TO GUIDE ROUTER

ROTATE ROUTER IN CORNER

GROOVE

ROUTER BASE

DOOR

② DRAW TANGENT CIRCLES. LENGTH OF RADIUS EQUALS DISTANCE FROM EDGE OF DOOR TO GROOVE.

③ DRAW LINES PARALLEL TO EACH OTHER AND TANGENT TO CIRCLES

① DRAW CENTERLINE

④ DRILL CENTER HOLE FOR ROUTER BIT. MOUNT IN SAME HOLES AS ORIGINAL BASE.

Fig. 89-11

HALF ROUND   SCREEN MOULD   SMALL ASTRAGAL

Fig. 89-12 Small molding may be applied to door faces in a number of interesting patterns.

back when dry. It makes an extremely strong joint, well-suited for the frame of a panel door (Fig. 89-14).

Making the Frame

Through mortise-and-tenon joints are used in the procedure that follows for making the frame of a panel door. Cut the stiles to the overall height of the door, the rails to the overall width of the door. The bottom rail is usually made wider than the rest of the frame members.

Cut a groove, centered on one edge of each piece, about 1/4 inch deep and as wide as the thickness of the panel stock. Cut a through mortise on each end of the stiles. This may be done by marking out the joints and cutting them with a handsaw. The cuts may also be made by using the table saw.

Raise the blade of the table saw so its height is equal to the width of the rails minus the depth of the groove. Make the mortises by using a *tenoning jig.* One-quarter inch is removed from the center by standing the piece on end, making a cut, turning the piece around, and making another cut with a single saw blade. Cut tenons on each end of all the rails in the same manner, but leave 1/4 inch in the center between saw cuts. Then, clamp a small block to the rip fence on the operator's side of the blade. Set the rip fence so the distance between the block and the outside of the saw blade is equal to the width of the stiles. Lower the blade to remove the waste on each side of the tenon. Use the miter gauge as a guide, and the block on the rip fence as a stop, and cut the waste from the sides of the tenon.

Assembling the Door

Cut the panels about 1/16 inch undersize to make sure the frame joints will come up tight. Assemble

Fig. 89-13 The through mortise-and-tenon is a strong joint for doors. It can be made with carpentry tools.

the frame and panel (Fig. 89-15). Use glue in the joints, but not on the panel edges. Hold each joint tight and drive a screw in the center of the joint from the back side of the door. The screw should be

Fig. 89-14 A plate jointer, if available, may be used to reinforce miter joints of the door frame. (*Courtesy of Porter-Cable Professional Tools*)

Fig. 89-15 Assembly of the frame and panel door.

as long as possible, but not so long as to penetrate the door frame. Countersink the screw slightly so the head can be filled. Measure the door from corner to corner to make sure it is square. If the door is not square, rack it slightly. This must be done before the glue sets. Sand the joints and faces smooth, ease all sharp corners. Shape the door edges and ends as desired.

Types of Hinges

Several types of cabinet hinges are *surface, offset, overlay, pivot,* and *butt.* For each type there are many styles and finishes (Fig. 89-16). Some types are *self-closing* hinges that hold the door closed and eliminate the need for door catches.

Surface Hinges

Surface hinges are applied to the exterior surface of the door and frame. The back side of the hinge

leaves may lie in a straight line for flush doors. One leaf may be *offset* for lipped doors (Fig. 89-17). The surface type is used when it is desired to expose the hardware, as in the case of wrought iron and other decorative hinges.

Offset Hinges

Offset hinges are used on lipped doors. They are called *offset surface* hinges when both leaves are fas-

Fig. 89-17 Surface hinges.

Fig. 89-16 Cabinet door hinges come in many styles and finishes. (*Courtesy of Amerock Corporation*)

Fig. 89-18 Offset hinges.

tened, to outside surfaces. The *semiconcealed offset* hinge has one leaf bent to a 3/8-inch offset that is screwed to the back of the door. The other leaf screws to the exterior surface of the face frame. A *concealed offset* type is designed in which only the pin is exposed when the door is closed (Fig. 89-18).

Overlay Hinges

Overlay hinges are available in *semiconcealed* and *concealed* types. With semiconcealed types, the amount of overlay is variable. Certain concealed overlay hinges are made for a specific amount of overlay, such as 1/4, 5/16, 3/8, and 1/2 inch. European-style hinges are completely concealed. They are not usually installed by the carpenter because of the equipment needed to bore the holes to receive the hinge. Some overlay hinges, with one leaf bent at a 30 degrees angle, are used on doors with reverse beveled edges (Fig. 89-19).

Fig. 89-20 Pivot hinges for an overlay door.

Pivot Hinges

Pivot hinges are usually used on overlay doors. They are fastened to the top and bottom of the door and to the inside of the case. They are frequently used when there is no face frame and the door completely covers the face of the case (Fig. 89-20).

Butt Hinges

Butt hinges are used on flush doors. Butt hinges for cabinet doors are a smaller version of those used on entrance doors. The leaves of the hinge are set into *gains* in the edges of the frame and the door, in the same manner as for entrance doors. Butt hinges are used on flush doors when it is desired to conceal most of the hardware. They are not often used on cabinets because they take more time to install than other types (Fig. 89-21).

Hanging Cabinet Doors
Surface Hinges

To hang cabinet doors with surface hinges, first apply the hinges to the door. Then shim the door in

Fig. 89-19 Overlay hinges. (Note: European style hinge line-art is Courtesy of Hettich America L.P.)

135° AUTOMATIC STOP

DOOR FRAME

FOR FLUSH DOORS

Fig. 89-21 Butt hinges.

the opening so an even joint is obtained all around. Screw the hinges to the face frame.

Semiconcealed Hinges

For semiconcealed hinges, screw the hinges to the back of the door. Then center the door in the opening. Fasten the hinges to the face frame. When more than one door is to be installed side by side, clamp a straightedge to the face frame along the bottom of the openings for the full length of the cabinet. Rest the doors on the straightedge to keep them in line (Fig. 89-22).

Concealed Hinges

When installing concealed hinges, first screw the hinges on the door. Center the door in the opening. Press or tap on the hinge opposite the face frame. Small projections on the hinge make indentations to mark its location on the face frame. Open the door. Place the projections of the hinges into the indentations. Screw the hinges to the face frame.

Butt Hinges

Hanging flush cabinet doors with butt hinges is done in the same manner as hanging entrance doors. Drill pilot holes for all screws so they are centered on the holes in the hinge leaf. Drilling the holes off center throws the hinge to one side when the screws are driven. This usually causes the door to be out of alignment when hung. Many carpenters use a a self-centering tool, called a *VIX bit,* when drilling pilot holes for screw fastening of cabinet door hinges of all types (Fig. 89-23). The tool centers a twist drill on the hinge leaf screw hole. It also stops at a set depth to prevent drilling through the door or face frame.

Installing Pulls and Knobs

Cabinet *pulls* or *knobs* are used on cabinet doors and drawers. They come in many styles and designs. They are made of metal, plastic, wood, porcelain, or other material (Fig. 89-24).

Pulls and knobs are installed by drilling holes through the door. Then fasten them with machine screws from the inside. When two screws are used to fasten a pull, the holes are drilled slightly oversize in case they are a little off center. This allows the pulls to be fastened easily without cross-threading the screws. Usually 3/16-inch diameter holes are drilled for 1/8 inch machine screws. To drill holes quickly and accurately, make a *template* from scrap wood that fits over the door. The template can be made so that holes can be drilled for doors that swing in either direction (Fig. 89-25).

Installing Door Catches

Doors without self-closing hinges need *catches* to hold them closed. There are many kinds of catches available (Fig. 89-26). Catches should be placed where they are not in the way, such as on the bottom of shelves, instead of the top.

CABINET   DOORS

STRAIGHTEDGE

Fig. 89-22 When installing doors, use a straightedge to keep them in line.

Fig. 89-23 The VIX bit is a self-centering drill stop used for drilling holes for cabinet hinges.

Magnetic catches are widely used. They are available with single or double magnets of varying holding power. An adjustable magnet is attached to the inside of the case. A metal plate is attached to the door. First attach the magnet. Then place the plate on the magnet. Close the door and tap it opposite the plate. Projections on the plate mark its location on the door. Attach the plate to the door where marked. Try the door. Adjust the magnet, if necessary.

Friction catches are installed in a similar manner to that used for magnetic catches. Fasten the adjustable section to the case and the other section to the door.

Elbow catches are used to hold one door of a double set. They are released by reaching to the back side of the door. These catches are usually used when one of the doors is locked against the other.

Bullet catches are spring loaded. They fit into the edge of the door. When the door is closed, the catch fits into a recessed plate mounted on the frame.

Fig. 89-24 A few of the many styles of pulls and knobs used on cabinet doors and drawers. (*Courtesy of Amerock Corporation*)

HELPFUL HINTS

Use a template when drilling holes for cabinet door pulls.

TEMPLATE CAN BE USED RIGHT OR LEFT ON TOP

TEMPLATE

DOOR

TEMPLATE CAN BE USED RIGHT OR LEFT ON BOTTOM

EQUAL

3/4" × 3/4" STRIP

CENTER TO CENTER DISTANCE OF PULL

3/16" HOLE

1/4" PLY OR HDBD.

DOOR SIDE OF TEMPLATE

Fig. 89-25

ROLLER-TYPE FRICTION CATCH

BULLET CATCH

DOUBLE MAGNETIC CATCH

ELBOW CATCH

FRICTION CATCH

DECORATIVE CATCH FOR FLUSH DOORS

Fig. 89-26 Several types of catches are available for use on cabinet doors. (*Courtesy of Amerock Corporation*)

CHAPTER 90 DRAWER CONSTRUCTION AND INSTALLATION

Drawers are classified as overlay, lipped, and flush in the same way as doors. In a cabinet installation, the drawer type should match the door type.

Drawer Construction

Drawer fronts are generally made from the same material as the cabinet doors. Drawer sides and backs are generally 1/2 inch thick. They may be made of solid lumber, plywood, or particleboard. Medium-density fiberboard with a printed wood grain is also manufactured for use as drawer sides and backs. The drawer bottom is usually made of 1/4-inch plywood or hardboard. Small drawers may have 1/8-inch hardboard bottoms.

Drawer Joints

Typical joints between the front and sides of drawers are the *dovetail, lock,* and *rabbet* joints. The dovetail joint is used in higher-quality drawer construction. It takes a longer time to make, but is the strongest. Dovetail drawer joints may be made using a router and a dovetail template (Fig. 90-1). The lock joint is simpler. It can be easily made using a table saw. The rabbet joint is the easiest to make. However, it must be strengthened with fasteners in addition to glue (Fig. 90-2).

Joints normally used between the sides and back are the *dovetail, dado and rabbet, dado,* and *butt* joints. With the exception of the dovetail joint, the drawer back is usually set in at least 1/2 inch from the back ends of the sides to provide added strength.

DOVETAIL JOINT

LOCK JOINT

RABBETED JOINT

Fig. 90-2 Typical joints between drawer front and side.

Fig. 90-1 Dovetail joints can be made with a router and a dovetail template.

This helps prevents the drawer back from being pulled off if the contents get stuck while opening the drawer (Fig. 90-3).

Drawer Bottom Joints

The drawer bottom is fitted into a groove on all four sides of the drawer (Fig. 90-4). In some cases, the drawer back is made narrower, the four sides assembled, the bottom slipped in the groove, and its back edge fastened to the bottom edge of the drawer back (Fig. 90-5).

Fig. 90-3 Typical joints between drawer back and side.

Fig. 90-4 Drawer bottom fitted in groove at drawer back.

Fig. 90-5 Drawer bottom fastened to bottom edge of drawer back.

Making Drawers

Making drawers involves the use of simple joints that can be made using a table saw. To make more complex joints, such as dovetail joints, requires special equipment that may not always be available. When such equipment is available and it is desired

to use dovetail joinery in the making of drawers, follow the dovetail template manufacturer's directions.

Overlay Drawers

The overlay drawer consists of the bottom, two sides, a back, a false front, and an overlay front. The overlay front is fastened to the false front with screws from the inside (Fig. 90-6).

Cut the drawer sides to height and length. The height should be about 1/8 inch less than the height of the drawer opening. If metal drawer guides are used, the height of the drawer sides depends on the type and manufacturer of the drawer guide. Care must be taken not to make the drawer sides too long. They are usually made 22 inches long for a standard base cabinet.

Dado the sides for the back and front of the drawer, 1/4 inch deep as shown in Figure 90-7. The

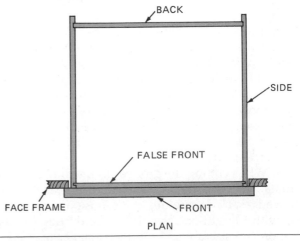

Fig. 90-6 Plan view of the overlay drawer.

Fig. 90-7 Overlay drawer sides are dadoed on the front and back ends.

dadoes are cut to the same depth so the front and back can be cut to the same length.

Cut the false fronts and backs to width and length. The width is the same as the drawer sides. The length determines the overall width of the drawer. The length should be such to allow the proper clearance between the drawer sides and the opening. This depends on the type of drawer guides used. A clearance of 1/2 inch on each side is required for most metal side guides. If the drawers are to slide on wood guides, generally about 1/8 inch clearance is required on each side.

Rabbet both ends of the false front to fit into the dado in the drawer sides. Round off and smooth the top edges of all sides, fronts, and back.

Make a groove on all parts for the drawer bottom. The width of the groove should be such that the drawer bottom will slide easily into it without being forced. The depth of the groove should be about 1/4 inch, if 1/2-inch drawer sides are used. If the groove is too deep, it will weaken the sides. If the groove is too shallow, the bottom may fall out after assembly. The distance from the bottom edge should be about 3/8 inch to the bottom of the groove (Fig. 90-8).

When cutting the grooves in the drawer sides, fronts, and back, be careful to groove near the bottom edges. If not carefully done, it is possible to end up with all right-hand or all left-hand sides. Cut the bottoms to size about 1/16 inch smaller than measured.

Fig. 90-8 Location of the groove for the drawer bottom.

Glue and fasten the sides to the back. Slip in the bottom. Do not apply glue to the bottom. Fasten the false front to the sides. Cut and fasten the overlay front to the drawer with screws driven from the inside. Care must be taken to position the drawer front correctly.

Making a Lipped Drawer

The sides, back, and bottom of a lipped drawer are machined in the same way as for an overlay drawer. However, the lipped drawer does not have a false front. Both ends of the drawer front are rabbeted to receive the drawer sides (Fig. 90-9). The sides of the drawer are dadoed only on the back end to receive the drawer back.

Cut the front 5/8 inch over the opening size on both the width and length. Rabbet the top and

BACK

SIDE

FACE FRAME

ONE-PIECE RABBETED FRONT

PLAN

Fig. 90-9 Plan view of the lipped drawer.

bottom edges 3/8×3/8. On each end make, rabbets 3/8×7/8 to allow for the overlap, the thickness of the drawer sides, and clearance. The depth of the rabbet may be greater if more clearance is needed (depending on the type of drawer guide used). Shape the edges and ends of the drawer front as required. The drawer is then assembled in a manner similar to that used for the overlay drawer. However, the sides are fastened to the rabbeted ends of the drawer front.

Making a Flush Drawer

Cut out the drawer front to the overall height and width of the drawer opening. Rabbet the two ends of the drawer front to receive the drawer sides plus an allowance for side clearance. The amount of clearance depends on the type of drawer guide used. Cut out the other drawer parts. Assemble in a manner similar to that used for other type drawers

(Fig. 90-10). After the drawer is assembled, try it in the opening. Fit the drawer front to the opening by hand planing, if necessary.

Drawer Guides

There are many ways of guiding drawers (Fig. 90-11). The type of drawer guide selected affects the size of the drawer. The drawer must be supported level and guided sideways. It must also be kept from tilting down when opened.

Wood Guides

Probably the simplest wood guide is the center strip. It is installed in the bottom center of the opening from front to back (Fig. 90-12). The strip projects above the bottom of the opening about 1/4 inch. The bottom edge of the drawer back is notched to ride in the guide. A *kicker* is installed. It is centered above the drawer to keep it from tilting downward when opened.

Another type of wood guide is the grooved center strip (Fig. 90-13). The strip is placed in the center of the opening from front to back. A matching strip is fastened to the drawer bottom. In addition to guiding the drawer, this system keeps it from tilting when opened, eliminating the need for drawer kickers.

Another type of wood guide is a rabbeted strip. Strips are used on each side of the drawer opening (Fig. 90-14). The drawer sides fit into and slide along the rabbeted pieces. Sometimes these guides are made up of two pieces instead of rabbeting one piece. A kicker above the drawer is necessary with this type guide.

DRAWER BACK

A STOP MUST BE INSTALLED ON THE BACK END

DRAWER SIDE

FACE FRAME

DRAWER FRONT

PLAN

Fig. 90-10 Plan view of the flush drawer.

Fig. 90-11 Wood drawer guides are installed in several ways.

Fig. 90-12 Simple center wood drawer guide. The back of the drawer is notched to run on the guide.

Fig. 90-13 The grooved center wood drawer guide eliminates the need of a kicker.

Fig. 90-14 Rabbeted wood guides are installed on each side of the drawer.

Metal Drawer Guides

There are many different types of metal drawer guides. Some have a single track mounted on the bottom center of the opening. Others may be centered above or on each side of the drawer. Nylon rollers mounted on the drawer ride in the track of the guide (Fig. 90-15).

Instructions for installation differ with each type and manufacturer. When using commercially made drawer guides, read the instructions first, before making the drawer, so proper allowances for the drawer guide can be made.

Installing Pulls and Knobs

The same style pulls or knobs that are used on the doors are used on the drawers. They are installed in the same manner as on doors. However, a differ-

Fig. 90-15 Installing metal drawer guides. (Note: Line-art is Courtesy of Knape and Vogt Mfg. Co.)

ent template is made to locate the holes. The template center is lined up with the center of the drawer. By using the template, all that is required to locate holes is to find the center of any width drawer. The template can be made for use in drawers of different heights (Fig. 90-16).

HELPFUL HINTS

Make a template for quick and accurate installation of drawer pulls.

MACHINE SCREWS

DRAWER PULL

FACE OF DRAWER FRONT

METHOD OF INSTALLING DRAWER PULLS

② REST TEMPLATE ON TOP OF DRAWER

CENTER TO CENTER DISTANCE OF PULL

TEMPLATE

LINE UP TEMPLATE CENTER WITH DRAWER CENTER

① LIGHTLY MARK CENTER OF DRAWER

HOLES FOR DEEPER DRAWERS

DRAWER FRONT

METHOD OF LAYING OUT HOLES FOR DRAWER PULLS

Fig. 90-16

Review Questions

Select the most appropriate answer.

1. Overlay and lipped doors and drawers are easier to install than flush types because
   a. no catches are necessary.
   b. no jointing is necessary.
   c. they take less material.
   d. simpler joints can be used.

2. A drawer front or door with its edges and ends rabbeted to fit over the opening is called
   a. an overlay type.   c. a lipped type.
   b. a flush type.   d. a rabbeted type.

3. In contrast to other types of doors and drawers, the flush type requires
   a. more time to construct.
   b. no fitting.
   c. extra clearance.
   d. stops.

4. A lipped door is cut larger than the openings by
   a. 1/4 inch.   c. 1/2 inch.
   b. 3/8 inch.   d. 5/8 inch.

5. Door panels and drawer bottoms are usually cut
   a. undersize.
   b. exactly to size.
   c. oversize.
   d. with rabbeted edges.

6. The offset hinge is used on
   a. paneled doors.   c. lipped doors.
   b. flush doors.   d. overlay doors.

7. Butt hinges are used on
   a. flush doors.   c. overlay doors.
   b. lipped doors.   d. solid doors.

8. The thickness of drawer sides and backs is usually
   a. 1/4 inch.   c. 5/8 inch.
   b. 1/2 inch.   d. 3/4 inch.

9. The joint used on high-quality drawers is the
   a. dado joint.
   b. dado and rabbet joint.
   c. dovetail joint.
   d. rabbeted joint.

10. A wood strip installed in the cabinet to prevent a drawer from tilting downward when opened is called a
   a. top guide.   c. sleeper.
   b. kicker.   d. tilt strip.

■ BUILDING FOR SUCCESS ■

Developing Your Individual Abilities

As people seek employment in the construction industry, technical and personal skills and abilities become the area of focus.

Assuming that the person looking for employment has the necessary technical skills needed for gainful employment, concentration can be centered on other personal skills and abilities that help assure success in work. Individual abilities to communicate and function well on the job will be of primary concern to the employer. Each person participating in a construction education program should ask what personal skills and abilities an employer will be looking for.

Often the young person seeking construction employment for the first time will have doubts about what he or she can really do. Students must remember that they have proven themselves competent many times during their studies in school. What many people call "life skills" are not all learned in formal educational experiences, but that is one of the places where many life skills are started. Developing life skills and abilities to function well in the world of work are learned and strengthened as each person lives and works. It is a process in which everyone grows and builds present skills and abilities. Successes and mistakes are the building blocks that create abilities to perform, mature, and be effective citizens.

Many things are important in the process of developing individual abilities. Self-esteem is a major contributor to individual success. Self-esteem is self-image, or how each person views himself or herself. It is formed in the early years through experiences with environments, activities, other people, and personal perceptions. People who exhibit low self-esteem may have a poor opinion of themselves and almost everything in life. What causes this to happen is difficult to define at times. Through involvement with positive environments, self-esteem can be increased.

One of the many goals of construction education program should be to assist each student in developing personal skills and abilities that supplement technical competencies. Each student has the ability to do many good and worthwhile things; all he or she may need is the opportunity to experience success. Each program of study that has the student at heart will center the curriculum on developing all of a student's skills and abilities. These curriculums should culminate in helping people become quality workers and responsible citizens.

The world of construction will call for personal abilities in each individual that help the industry meet established goals. Some of those abilities will center on solving problems, helping people, analyzing, organizing, influencing others, working with others, learning from mistakes, accepting responsibility, developing flexibility, motivating self and others, competing, and maintaining a positive outlook. A common goal for all construction workers should be that of job satisfaction. Each person's perspective will have a great effect on his or her job enjoyment. This will call for developing many of the previously stated abilities.

Developing self-confidence should continue to be a personal goal. Many times this can be enhanced by gaining a good understanding of how people learn. Personal learning comes through reading, listening, doing, or a combination of these. Each person should be aware of how he or she learns the most.

Knowing how you are motivated will enable you to excel in the pursuit of knowledge and skills. With knowledge, skills, abilities, and motivation, a person should feel confident to apply that knowledge.

There was a young man who was seeking employment with a builder. During their conversation, the man asked the employer what wages he paid for the position. The builder responded that he would "pay you what you are worth." In response, the young man said "I can't work for that." This young man did not appear to have a high opinion of himself, nor of what he could do for the employer. This story points out the need to feel good about your personal skills and abilities, and to have an attitude of self-confidence that can be seen by others.

Goal setting will continue to be a driving force in the construction industry. Each laborer, skilled worker, supervisor, and self-employed person will have to set personal goals to be successful. It will be a task that involves education, training, other people, work, successes, and hard knocks. Without goals people tend to wander around in life, not knowing where they have been or where they are going. The construction industry will continue to require that all workers desiring a successful construction career be goal-oriented. The challenges will not diminish. Success and satisfaction will come through hard work, time spent in study, and adaptability. The industry will experience new ideas, new materials, and new methods. This has proven to be the avenue through which higher-quality products are produced. The process calls for flexibility, determination, and a desire to learn and succeed. This can only be accomplished by dedication to developing effective skills and abilities that promote learning.

The process of developing individual skills and abilities that lead to success can be thought of as a journey. This journey will expose people to experiences through which they can learn, work, and adapt to new things. Anyone who tells another person that he or she has learned it all has actually closed the door to learning somewhere in the past. Set goals, subject yourself to a positive learning environment, and spend time with successful people, and your potential for success will increase substantially.

FOCUS QUESTIONS: For individual or group discussions.

1. What individual abilities do you possess that will be necessary for a successful career?

2. How do you anticipate continuing to develop these individual abilities? Where would be a good place to begin that continuation?

3. Who are some experienced people that may be willing to share their experiences in developing self-esteem and confidence? Ask them to respond to specific questions you have in this area.

4. By combining talents and technical skills with good individual or personal skills, the outcome should produce a well-adjusted worker. Describe what that person would be like. Define any of the special abilities that would enhance that person's success.

STUDENT ACTIVITIES for

Section 4: Interior Finish

Unit 24 Drywall Construction

1. Explain the drywall construction process. Include the product types and descriptions, tools used, cutting, fasteners, mastics, taping procedures, trims, moldings, and accessories.

2. Demonstrate the estimating, measuring, cutting, hanging (nailing and screwing) process, and taping and sanding of drywall products.

3. Demonstrate the wall preparation and application of textures to drywall surfaces.

4. Using a set of blueprints, estimate the quantity and cost of the drywall needed for the interior walls and ceilings of the house.

Unit 25 Wall Paneling and Wall Tile

1. Create a display board that illustrates a variety of interior wall paneling, moldings, mastics, and fasteners.

2. Using a set of blueprints, estimate the necessary quantity of paneling needed to cover the walls of three rooms.

3. Explain and demonstrate the procedure for installing solid lumber paneling. Include furring strips, scribing, measuring, matching, cutting, and adhering with nails and mastic.

4. Explain and demonstrate the process of installing wainscoting on walls (less than full height). Include the use of moldings and trim.

5. Explain and demonstrate the installation of plastic laminate products. Include the measuring, cutting, fitting, and adhering processes.

6. Explain and demonstrate the process of installing ceramic tile and accessories on both walls and floors. Include the measuring, cutting, adhering, and grouting processes.

7. Using a blueprint, estimate the ceramic tile needed for the bathroom and kitchen.

Unit 26 Ceiling Finish

1. Explain the variety of ceiling tile and panel products used to finish interior ceilings. Include the product names, structure, accessories, sizes, fastening methods, and materials cost estimating.
2. Demonstrate the ceiling preparation, room squaring, applying furring strips, and the installation of acoustical ceiling tile.

3. Explain the process of installing a suspended ceiling in a room. Include the room preparation, squaring, leveling, installing wall angles, runners, cross tees, ceiling panels, and materials cost estimating.
4. Repeat the process with the concealed grid suspended ceilings.

Unit 27 Interior Doors and Door Frames

1. Describe the types of interior doors and jambs, their structure, operating functions, and the method of determining the hand of door.

2. Demonstrate installing a jamb, trimming and hanging a swinging door, and installing the hardware. Repeat the process for a bifold and bypass door. Use the hinge butt template, portable electric plane, and router.

Unit 28 Interior Trim

1. Identify the types of standard molding patterns and describe their use.
2. Demonstrate the installation of as many of these as possible, including the process of coping and mitering a joint. Use the hand miter box, the power miter box, the table

saw, a coping saw, and basic hand tools. Include these moldings: beds, crowns, full rounds, half rounds, quarter rounds, base, base shoes, base caps, corner molds, back bands, aprons, stools, paneling caps, casings, and stops.

Unit 29 Stair Finish

1. Name and show the parts of stair finish. Describe their location and function.
2. Using a 4-foot piece of stringer material, lay out a housed stringer and a job-built (rough) stringer with given dimensions from the instructor.

3. Apply the following finish to a job-built staircase: risers, newel posts, open and closed finish stringers, treads, return nosings, tread moldings, handrail, and balusters.
4. Using a stair manufacturer's catalog, assemble a price list of the previously listed finish parts.

Unit 30 Finish Floors

1. Prepare and present a display of the various kinds and sizes of hardwood finish flooring. Include strip, plank, and parquet products. Explain the installation process for hardwood floors.

2. Prepare a wood or concrete floor to receive either strip, plank, or parquet flooring. Install a hardwood finish floor. Use both a hammer power nailer and mastics if possible. Install base trim.
3. Prepare a floor to receive resilient tiles or sheet products, and install resilient floor tile.

Unit 31 Kitchen Cabinets

1. Describe in detail the construction process for the shop-built and the site-built kitchen cabinet. Illustrate and use examples of the cabinet components, materials, and plans.
2. Using a residential blueprint, assemble a list of factory-built kitchen cabinets. Estimate the cost of the cabinets.

3. Using a cabinet plan, build and install a base and wall cabinet in the shop or on the job. Make and laminate a countertop. Include a material list with costs. The cabinet is to have doors and drawers.

Unit 32 Cabinet Doors and Drawers

1. Make and install overlay, lipped, and flush cabinet doors (solid or panel) and drawers. Include any necessary hardwares. Use any quality drawer construction.

2. Make a routed design cabinet door face using the router and any appropriate jigs.

GLOSSARY

Acoustical board—material used to control or deaden sound

Actual size—the size of lumber after it has been surfaced

Admixture—material used in concrete or mortar to produce special qualities

Aggregate—materials, such as sand, rock, and gravel, used to make concrete

Air-dried lumber—lumber that has been seasoned by drying in the air

Anchor—a device used to fasten structural members in place

Annular ring—the rings seen when viewing a cross-section of a tree trunk; each ring constitutes one year of tree growth

Apprentice—a beginner who serves for a stated period of time to learn a trade

Apron—a piece of the window trim used under the stool

Areaway—below-grade, walled area around basement windows

Arbor—a shaft on which circular saw blades are inserted

Asphalt felt—a building paper saturated with asphalt for waterproofing

Astragal—a semicircular molding often used to cover a joint between doors

Awning window—a type of window in which the sash are hinged at the top and swing outward

Back band—molding applied around the sides and tops of windows and doors

Back-bevel—a bevel on the edge or end of stock toward the back side

Backing—strips or blocks installed in walls or ceilings for the purpose of fastening or supporting trim or fixtures

Backing the hip—beveling the top edge of a hip rafter to line it up with adjacent roof surfaces

Back miter—an angle cut starting from the end and coming back on the face of the stock

Backset—the distance an object is set back from an edge, side, or end of stock, such as the distance a hinge is set back from the edge of a door

Backsplash—a raised portion on the back edge of a countertop to protect the wall

Balloon frame—a type of frame in which the studs are continuous from foundation to roof

Baluster—vertical members of a stair rail, usually decorative and spaced closely together

Balustrade—the entire stair rail assembly, including handrail, balusters, and posts

Baseboard—finish board used to cover the joint at the intersection of wall and floor

Base cap—a molding applied to the top edge of the baseboard

Base shoe—a molding applied between the baseboard and floor

Batten—a thin, narrow strip usually used to cover joints between vertical boards

Batter board—a temporary framework erected to hold the stretched lines of a building layout

Bay window—a window, usually three-sided, which projects out from the wall line

Bearer—horizontal members of a wood scaffold that support scaffold plank

Bearing partition—an interior wall that supports the floor above

Bedding—filling of mortar, putty, or similar substance used to secure a firm bearing, as in a bed of putty for glass panes

Benchmark—a reference point for determining elevations during the construction of a building

Bevel—the sloping edge or side of a piece with the angle between not a right angle

Bird's mouth—a notch cut in the underside of a rafter to fit on top of the wall plate

Blind joint—a type of joint in which the cuts do not go all the way through

Blind nail—a method of fastening that conceals the nails

Blind stop—part of a window finish applied just inside the exterior casing

Blocking—pieces installed between studs in a wall usually used to provide fastening

Board—lumber usually 8 inches or more in width and less than 2 inches thick

Board foot—a board that measures one foot square and one inch thick or any equivalent lumber volume

Bow—a type of warp in which the side of lumber is curved from end to end

Box nail—a thin nail with a head, usually coated with a material to increase its holding power

Brad—a thin, short, finishing nail

Bridging—diagonal braces or solid wood blocks between floor joists used to distribute the load imposed on the floor

Break joints—to stagger joints in adjacent rows of sheathing, siding, roofing, flooring, and similar materials

Buck—a rough frame used to form openings in concrete walls

Bullnose—a starting step that has one or both ends rounded

Butt—the joint formed when one square cut piece is placed against another; also, a type of hinge

Cambium layer—a layer just inside the bark of a tree where new cells are formed

Cant strip—a thin strip of wood placed under a piece to tilt the piece at a slant

Carbide-tipped—in reference to cutting tools that have small, extremely hard pieces of carbide steel welded to the tips

Casement window—a type of window in which the sash are hinged at the edge and usually swing outward

Casing—Molding used to trim around doors, windows, and other openings

Caulking—putty-like mastic used to seal cracks and crevices

Center punch—a tool used to make an indentation at the centerline of holes

Chair rail—molding applied horizontally along the wall to prevent chair backs from marring the wall

Chamfer—an edge or end bevel that does not go all the way across the edge or end

Chase—a channel formed in buildings to run electrical, plumbing, or mechanical lines

Check—lengthwise split in the end of lumber usually resulting from more rapid drying of the end than the rest of the piece

Cheek cut—a beveled cut on the end of certain roof rafters

Cleat—a small strip applied to support a shelf or similar piece

Closed grain—wood in which the pores are small and closely spaced

Closed valley—a roof valley in which the roof covering meets in the center of the valley, completely covering the valley

Collar tie—a horizontal member placed close to the ridge at right angles to the plate

Compound miter—a bevel cut across the width and also through the thickness of a piece

Concrete—a building material made from portland cement, sand, gravel, and water

Condensation—when water, in a vapor form, changes to a liquid due to cooling of the air; the resulting drops of water that accumulate on the cool surface

Conductor—a vertical member used to carry water from the gutter downward to the ground; also called *downspout* or *leader*

Coniferous—cone-bearing tree; also known as *evergreen* tree

Contact cement—an adhesive used to bond plastic laminates or other thin material; so called because the bond is made on contact, eliminating the need for clamps

Contour line—lines on a drawing representing a certain elevation of the land

Coped joint—a type of joint between moldings in which the end of one piece is cut to fit the molded surface of the other

Corner bead—metal trim used on exterior corners of walls to trim and reinforce them

Corner boards—boards used to trim corners on the exterior walls of a building

Corner brace—diagonal member of the wall frame used at the corners to stiffen and strengthen the wall

Cornerite—metal lath, cut into strips and bent at right angles, used in interior corners of walls and ceilings, on top of lath to prevent cracks in plaster

Corner post—built-up stud used in the corner of a wall frame

Cornice—the entire finished assembly where the walls of a structure meet the roof; sometimes called the *eaves*

Counterbore—boring a larger hole partway through the stock so that the head of a fastener can be recessed

Countersink—making a flared depression around the top of a hole to receive the head of a flathead screw; also, the tool used to make the depression

Course—a continous row of building material, such as brick, siding, roofing, or flooring

Cove—a concave-shaped molding

Crawl space—a shallow space below the living quarters of a structure without a basement

Cricket—a small, false roof built behind a chimney or other roof obstacle for the purpose of shedding water

Crook—a type of warp in which the edge of lumber is not straight

Crosscut—a cut made across the grain of lumber

Crown—usually referred to as the high point of the crooked edge of joists, rafters, and other framing members

Cup—a type of warp in which the side of a board is curved from edge to edge

Dado—a cut, partway through, and across the grain of lumber

Deadbolt—door-locking bolt operated by a key from the outside and by a handle or key from the inside

Deadman—a T-shaped wood device used to support ceiling drywall panels and other objects

Deciduous—trees that shed leaves each year

Dew point—temperature at which moisture in the air condenses into drops

Diagonal—at an angle, usually from corner to corner in a straight line

Door stop—molding fastened to door jambs for the door to stop against

Dormer—a structure that projects out from a sloping roof to form another roofed area to provide a surface for the installation of windows

Double acting—doors that swing in both directions or the hinges used on these doors

Double-hung window—a window in which two sash slide vertically by each other

Dovetail—a type of interlocking joint resembling the shape of a dove's tail

Dowel—hardwood rods of various diameters

Downspout-—*see* Conductor

Drip—that part of a cornice or a course of horizontal siding that projects below another part; also, a channel cut in the underside of a windowsill that causes water to drop off instead of running back and down the wall

Drip cap—a molding placed on the top of exterior door and window casings for the purpose of shedding water away from the units

Drip edge—metal edging strips placed on roof edges to provide a support for the overhang of the roofing material

Dropping the hip—increasing the depth of the hip rafter seat cut so that the centerline of its top edge will lie in the plane of adjacent roof surfaces

Dry rot—dry, powdery residue of wood left after fungus destruction of wood due to excessive moisture

Drywall—a type of construction usually referred to as the installation of gypsum board

Dry well—gravel or stone-filled excavation for catching water so it can be absorbed into the earth

Duplex nail—a double-headed nail used for temporary fastening such as in the construction of wood scaffolds

Dutch hip—a roof consisting of a partial hip and partial gable roof

Dutchman—an odd-shaped piece usually used to fill or cover an opening

Eased edge—an edge of lumber whose sharp corners have been rounded

Easement—a curved member of a stair handrail. Also, an area of land that cannot be built upon because it provides access to a structure or utilities

Eaves—that part of a roof that extends beyond the sidewall

Edge—the narrow surface of lumber running with the grain

Edge grain—boards in which the annular rings are at or near perpendicular to the face; sometimes called *vertical grain*

Electrolysis—the decomposition of the softer of two unlike metals in contact with each other in the presence of water

Elevation—a drawing in which the height of the structure or object is shown; also, the height of a specific point in relation to another point

End—the extremities of a piece of lumber

Engineered Lumber Products (ELP)—manufactured lumber substitutes, such as wood I-beams, glue-laminated beams, laminated veneer lumber, parallel strand lumber, and laminated strand lumber

Equilibrium moisture content—the point at which the moisture content of wood is equal to the moisture content of the surrounding air

Escutcheon—protective plate covering the knob or key hole in doors

Exposure—the amount that courses of siding or roofing are exposed to the weather

Face—the best-appearing side of a piece of wood or the side that is exposed when installed, such as finish flooring

Face frame—a framework of narrow pieces on the face of a cabinet containing door and drawer openings

Fascia—a vertical member of the cornice finish installed on the tail end of rafters

Feather edge—the edge of material brought down in a long taper to a very thin edge, such as a wood shingle tip

Fence—a guide for ripping lumber on a table saw

Fiber saturation point—the moisture content of wood when the cell cavities are empty but the cell walls are still saturated

Fillet—small strips used to fill a space, such as between balusters in a plowed handrail

Finish carpentry—that part of the carpentry trade involved with the application of exterior and interior finish

Finish nail—a thin nail with a small head designed for setting below the surface of finish material

Finish stringer—the finish board running with the slope of the stairs and covering the joint between the stairs and the wall; also called a *skirt board*

Finger joint—joints made in a mill used to join short lengths together to make long lengths

Firecut—an angle cut made on the ends of floor joists bearing on a masonry wall designed to prevent the masonry wall from toppling in case the joists are burned through and collapse

Firestop—material used to fill air passages in a frame to prevent the spread of fire; in a wood frame, might consist of 2×4 blocking between studs

First & Seconds—the best grade of hardwood lumber

Fissured—irregular shaped grooves made in material, such as ceiling tile, for acoustical purposes

Flashing—material used at intersections such as roof valleys, dormers, and above windows and doors to prevent the entrance of water

Flat grain—grain in which the annular rings of lumber lie close to parallel to the sides; opposite of edge grain

Flute—concave groove in lumber; usually a number are used closely spaced for decorative purposes as in a column, post, or pilaster

Footing—a foundation for a column, wall, or chimney made wider than the object it supports, to distribute the weight over a greater area

Foundation—that part of a wall on which the building is erected

Frieze—a part of the exterior finish applied at the intersection of a overhanging cornice and the wall

Frostline—the depth to which frost penetrates into the ground in a particular area; footings must be placed below this depth

Furring strip—strips of lumber spaced at desired intervals for the attachment of wall or ceiling covering

Gable end—the triangular-shaped section on the end of a building formed by the rafters in a common or gable roof and the top plate line

Gable roof—a type of roof that pitches in two directions

Gain—a cutout made in a piece to receive another piece, such as a cutout for a butt hinge

Galvanized—protected from rusting by a coating of zinc

Gambrel roof—a type of roof that has two slopes of different pitches on each side of center

Girder—a heavy timber or beam used to support the ends of floor joists

Glass bead—small molding used to hold lights of glass in place

Glaze—to install glass in a frame

Glazier—a person who installs glass in a frame

Glazing—the act of installing glass in a frame

Glazing compound—a soft, plastic-type material, similar to putty, used for sealing lights of glass in a frame

Glazing points—small, triangular, or diamond-shaped pieces of metal used to secure and hold lights of glass in a frame

Glue-Laminated Lumber (Glulam)—large beams or columns made by gluing smaller-dimension lumber together side to side

Gooseneck—a curved section of handrail used when approaching a landing; also, an outlet in a roof gutter

Grade—the level of the ground; also identifies the quality of lumber

Grain—in wood, the design on the surfaces caused by the contrast, spacing, and direction of the annular rings

Graphite—a mineral used as pencil lead and also as a lubricant for the working parts of locks and certain tools

Green lumber—lumber that has not been dried to a suitable moisture content

Groove—a cut, partway through, and running with the grain of lumber

Ground—strips of wood placed at the base of walls and around openings and used as a guide for the application of an even thickness of plaster; also, a system used for electrical safety

Ground Fault Circuit Interrupter (GFCI)—device used in electrical circuits for protection against electrical shock; it detects a short circuit instantly and shuts off the power automatically

Grout—a mixture of cement, sand, and water used to fill joints in masonry and tile

Gusset—a pad of wood or metal used over a joint to stiffen and strengthen it

Gutter—a wood or metal trough used at the roof edge to carry off rain water and water from melting snow

Gypsum—a chalky type rock that is the basic ingredient of plaster

Gypsum board—a sheet product made by encasing gypsum in a heavy paper wrapping

Half round—a molding with its end section in the shape of a semicircle

Handrail—a railing on a stairway intended to be grasped by the hand to serve as a support and guard

Hardboard—a building product made by compressing wood fibers into sheet form

Header—pieces placed at right angles to joists, studs, and rafters to form and support openings in a wood frame

Hearth—a section of the floor in front of a fireplace; usually covered with some type of fireproof material

Heartwood—the wood in the inner part of a tree, usually darker and containing inactive cells

Heel—the back end of objects, such as a handsaw or hand plane

Herringbone—a pattern used in parquet floors

Hexagon—a plane figure having six sides

Hip—the intersection of two surfaces of a hip roof

Hip rafter—extends diagonally from the corner of the plate to the ridge at the intersection of two surfaces of a hip roof

Hip-valley cripple jack rafter—a short rafter, running parallel to common rafters, cut between hip and valley rafters

Hopper window—a type of window in which the sash is hinged at the bottom and swings inward

Horn—an extension of the stiles of doors or the side jambs of window and door frames

Housed stringer—a finished stringer that is dadoed to receive treads and risers of a stairway

Housewrap—type of building paper with which the entire sidewalls of a building are covered

Insulated glass—double panes of glass fused together

Insulation—material used to restrict the passage of heat or sound

Intersecting roof—the roof of irregular shaped buildings; valleys are formed at the intersection of the roofs

Isometric—a drawing in which three surfaces of an object are seen in one view, with the base of each surface drawn at a 30° angle

J-roller—a 3-inch wide rubber roller used to apply pressure over the surface of contact cement-bonded plastic laminates

Jalousie window—a type of window containing movable, horizontal slats of glass

Jamb—the sides and top of window and door frames

Jamb extension—narrow strips of wood fastened to the edge of window jambs to increase their width

Jig—any type of fixture designed to hold pieces or guide tools while work is being performed

Joint—as a verb, denotes straightening the edge of lumber; as a noun, means the place where parts meet and unite

Joist—horizontal framing members used in a spaced pattern that provide support for the floor or ceiling system

Joist hanger—metal stirrups used to support the ends of joists

Journeyman—a tradesman who has completed an apprenticeship or who has gained enough experience to perform work without instruction

Kerf—the width of a cut made with a saw

Keyway—a groove made in concrete footings for tying in the concrete foundation wall

Kicker—a strip of wood placed above a drawer to keep it from tilting down when opened

Kiln-dried—lumber dried by placing it in huge ovens called kilns

Kneewall—a wall of less than full height

Knot—a defect in lumber caused by cutting through a branch or limb embedded in the log

Laminate—to build up with thin layers of wood or other material

Laminated Strand Lumber (LSL)—lumber manufactured by bonding thin strands of wood, up to 12 inches long, with adhesive and pressure

Laminated Veneer Lumber (LVL)—lumber manufactured by laminating many veneers of plywood with the grain of all running in the same direction

Landing—the top of a flight of stairs or an intermediate-level platform

Laser—a concentrated, narrow beam of light; laser-equipped devices are used in building construction

Lath—a base for plaster; usually gypsum board or expanded metal sheets

Lattice—thin strips of wood, spaced apart and applied in two layers at angles to each layer resulting in a kind of grillwork

Lazy Susan—a set of revolving circular shelves; used in kitchen cabinets and other places

Ledger—a horizontal member of a wood scaffold that ties the scaffold posts together and supports the bearers; a temporary or permanent supporting member for joists or other members running at right angles

Level—horizontal or perpendicular to the force of gravity

Light—a pane of glass or an opening for a pane of glass

Line length—the length of a rafter along a measuring line without consideration to the width or thickness of the rafter

Linear measure—a measurement of length

Lintel—horizontal support member over an opening; also called a *header*

Lipped door—a cabinet door with rabbeted edges

Lookout—horizontal framing pieces in a cornice, installed to provide fastening for the soffit

Louver—an opening for ventilation consisting of horizontal slats installed at an angle to exclude rain, light, and vision, but to allow the passage of air

Low emissivity glass (LoE)—a coating on double-glazed windows designed to raise the insulating value by reflecting heat

Lumber—wood that is cut from the log to form boards, planks, and timbers

Magazine—a container in power nailers and staplers in which the fasteners to be ejected are placed

Mansard—a type of roof that has two different pitches on all sides of the building, with the lower slopes steeper than the upper

Mantel—the ornamental finish around a fireplace, including the shelf above the opening

Masonry—any construction of stone, brick, tile, concrete plaster, and similar materials

Mastic—a thick adhesive

Matched boards—boards that have been finished with tongue-and-grooved edges

Medullary ray—bands of cells radiating from the cambium layer to the pith of a tree to transport nourishment toward the center

Metes and Bounds—boundaries established by distances and compass directions

Millwork—any wood products that have been manufactured, such as moldings, doors, windows, and stairs for use in building construction; sometimes called *joinery*

Miter—the cutting of the end of a piece at any angle other than a right angle

Miter gauge—a guide used on the table saw for making miters and square ends

Miter joint—the joining of two pieces by cutting the end of each piece by bisecting the angle at which they are joined

Modular construction—a method of construction in which parts are preassembled in convenient-sized units

Moisture content—the amount of moisture in wood expressed as a percentage of the dry weight

Moisture meter—a device used to determine the moisture content of wood

Mortar—a mixture of portland cement, lime, sand, and water

Mortise—a rectangular cavity cut in a piece of wood to receive a tongue or tenon projecting from another piece

Molding—decorative strips of wood used for finishing purposes

Mullion—a vertical division between windows or panels in a door

Muntin—slender strips of wood between lights of glass in windows or doors

Newel post—an upright post supporting the handrail in a flight of stairs

No. 1 common—a lower grade of hardwood lumber

Nominal size—the stated size of the thickness and width of lumber even though it differs from its actual size; the approximate size of rough lumber before it is surfaced

Nosing—the rounded edge of a stair tread projecting over the riser

OC ("on center")—the distance from the center of one structural member to the center of the next one

Octagon—a plane figure with eight sides

Ogee—a molding with an S-shaped curve

Open valley—a roof valley in which the roof covering is kept back from the centerline of the valley

Panel—a large sheet of building material

Panel door—a door in which panels are enclosed by a frame

Parallel Strand Lumber (PSL)—lumber manufactured by bonding strands of structural lumber, up to 8 feet long, with adhesive, heat, and pressure

Parquet—a floor made with strips or blocks to form intricate designs

Particleboard—a building product made by compressing wood chips and sawdust with adhesives to form sheets

Parting strip—a small strip of wood separating the upper and lower sash of a double-hung window

Partition—an interior wall separating one portion of a building from another

Penny (d)—a term used in designating nail sizes

Perforated—material that has closely spaced holes in a regular or irregular pattern

Perm—a measure of water vapor movement through a material

Phillips head—a type of screw head with a cross-slot

Pier—a column of masonry, usually rectangular in horizontal cross-section, used to support other structural members

Pilaster—column built within and usually projecting from a wall to reinforce the wall

Pile—concrete, metal, or wood pillar forced into the earth or cast in place as a foundation support

Pilot—a guide on the end of edge-forming router bits used to control the amount of cut

Pilot hole—a small hole drilled to receive the threaded portion of a wood screw

Pitch—the amount of slope to a roof expressed as ratio of the total rise to the span

Pitch board—a piece of wood cut in the shape of a right triangle; used as a pattern for laying out stair stringers, rafters, and handrail fittings

Pitch pocket—an opening in lumber between annular rings containing pitch in either liquid or solid form

Pith—the small, soft core at the center of a tree

Pivot—revolving around a point

Plan—in an architectural drawing, a object drawn as viewed from above

Plain-sawed—a method of sawing lumber that produces flat-grain

Planchir (or plancier)—the finish member on the underside of a box cornice; also called *soffit*

Plank—lumber that is 6 or more inches in width and from 1 ½ to 6 inches in thickness

Plaster—a mixture of portland cement, sand, and water used for covering walls and ceilings of a building

Plastic laminate—a very tough, thin material in sheet form used to cover countertops; available in a wide variety of colors and designs

Plate—top or bottom horizontal member of a wall frame

Platform frame—method of wood frame construction in which walls are erected on a previously constructed floor deck or platform

Plinth block—a small, decorative block, thicker and wider than a door casing, used as part of the door trim at the base and at the head

Plot plan—a drawing showing a bird's-eye view of the lot, the position of the building, and other pertinent information; also called *site plan*

Plow—a wide groove cut with the grain of lumber

Plumb—vertical; at right angles to level

Plumb bob—a pointed weight attached to a line for testing plumbness

Plunge cut—an interior cut made with a portable saw by a method that does not require boring holes before making the cut

Ply—one thickness of several layers of built-up material, such as one of the layers of plywood

Plywood—a building material in which thin sheets of wood are glued together with the grain of adjacent layers at right angles to each other

Pneumatic—powered by compressed air

Pocket—a recess in a wall to receive a piece, such as a recess in a concrete foundation wall to receive the end of a girder

Pocket door—a type of door that when opened slides into a recess in the wall

Polyethylene film—a thin plastic sheet used as a vapor barrier

Postforming—method used to bend plastic laminate to small radii

Preacher—a small piece of wood of the same thickness as the stair risers; it is notched in the center to fit over the finish stringer and rest on the tread cut of the stair carriage; it is used to lay out open finished stringers in a staircase

Preservative—a substance applied to wood to prevent decay

Pressure-treated—treatment given to lumber that applies preservative under pressure to penetrate the total piece

Primer—the first coat of paint applied to the surface or the paint used to prime a surface

Purlin—horizontal roof member used to support rafters between the plate and ridge

Quarter round—a type of molding, an end section of which is in the form of a quarter-circle

Quartersawed—a method of sawing lumber parallel to the medullary rays to produce edge-grain lumber. *See* edge grain

Rabbet—a cutout along the edge or end of lumber

Rafter—a sloping structural member of a roof frame that supports the roof sheathing and covering

Rail—the horizontal member of a frame

Rake—the sloping portion of the gable ends of a building

Rebar—steel reinforcing rod used in concrete

Reciprocating—a back-and-forth action, as in certain power tools

Return—a turn and continuation for a short distance of a molding, cornice, or other kind of finish

Return nosing—a separate piece mitered to the open end of a stair tread for the purpose of returning the tread nosing

Reveal—the amount of setback of the casing from the face side of window and door jambs or similar pieces

Ribbon—a narrow board let into studs of a balloon frame to support floor joists

Ridge—the highest point of a roof that has sloping sides

Ridge board—a horizontal member of a roof frame that is placed on edge at the ridge and into which the upper ends of rafters are fastened

Rip—sawing lumber in the direction of the grain

Rise—in stairs, the vertical distance of the flight; in roofs, the vertical distance from plate to ridge; may also be the vertical distance through which anything rises

Riser—the finish member in a stairway covering the space between treads

Rosette—a round or oval decorative wood piece used to end a handrail against a wall

Rough carpentry—that part of the trade involved with construction of the building frame or other work that will be dismantled or covered by the finish

Rough opening—opening in a wall in which windows or doors are placed

Rough stringer—cutout supports for the treads and risers in a staircase; also called *stair carriage* and *stair horse*

Run—the horizontal distance over which rafters, stairs, and other like members travel

R-value—a number given to a material to indicate its resistance to the passage of heat

S2S—surfaced two sides

S4S—surfaced four sides

Saddle—same as cricket

Sapwood—the outer part of a tree just beneath the bark containing active cells

Sash—that part of a window into which the glass is set

Sash balance—a device, usually operated by a spring or tensioned weatherstripping, designed to counterbalance double-hung window sash

S-beam—an I-shaped steel beam

Scab—a length of lumber applied over a joint to stiffen and strengthen it

Scaffold—an elevated, temporary working platform

Scratch coat—the first coat of plaster applied to metal lath

Screed—strips of wood, metal, or pipe secured in position and used as guides to level the top surface of concrete

Scribe—laying out woodwork to fit against an irregular surface

Scuttle—attic access

Seasoned lumber—lumber that has been dried to a suitable moisture content

Seat cut—*see* Bird's mouth

Section—drawing showing a vertical cut through an object or part of an object

Selvage—the unexposed part of roll roofing covered by the course above

Set—alternate bending of saw teeth to provide clearance in the saw cut

Shake—a defect in lumber caused by a separation of the annular rings; also, a type of wood shingle

Shank hole—a hole drilled for the thicker portion of a wood screw

Sheathing—boards or sheet material that are fastened to roofs and exterior walls and on which the roof covering and siding are applied

Shed roof—a type of roof that slopes in one direction only

Shim—a thin, wedge-shaped piece of material used behind pieces for the purpose of straightening them, or for bringing their surfaces flush at a joint

Shingle tip—the thin end of a wood shingle

Shortened valley rafter—a valley rafter that runs from the plate to the supporting valley rafter

Side—the wide surfaces of a board, plank, or sheet

Sidelight—a framework containing small lights of glass placed on one or both sides of the entrance door

Siding—exterior sidewall finish covering

Sill—horizontal timbers resting on the foundation supporting the framework of a building; also, the lowest horizontal member in a window or door frame

Skirt board—another name for a finished stringer in a staircase

Skylight—a type of roof window

Sleeper—strips of wood laid over a concrete floor to which finish flooring is fastened

Slump test—a test given to concrete to determine its wetness

Snap tie—a metal device to hold concrete wall forms the desired distance apart

Soffit—the underside trim member of a cornice or any such overhanging assembly

Softwood—wood from coniferous (cone-bearing) trees

Soil stack—part of the plumbing; a vertical pipe extending up through the roof to vent the system

Soleplate—the bottom horizontal member of a wall frame

Specifications—written or printed directions of construction details for a building

Spike—a nail, 16d or over

Spline—a thin, flat strip of wood inserted into the grooved edges of adjoining pieces

Spreader—a strip of wood used to keep other pieces a desired distance apart

Square—the amount of roof covering that will cover 100 square feet of roof area

Staging—*see* Scaffold

Stair carriage—*see* Rough stringer

Stair horse—*see* Rough stringer

Stairwell—an opening in the floor to provide headroom for climbing or descending stairs

Standing cut—a cut made through the thickness of stock at more than a 90° angle between the side and edge or end

Staple—a U-shaped fastener

Starter course—usually used in reference to the first row of shingles applied to a roof or wall

Starting step—the first step in a flight of stairs

Sticker—machine that makes moldings

Sticking—the molded inside edge of the frame of a panel door; also, strips of wood used to separate layers of lumber to allow air to circulate to dry the lumber

Stile—the outside vertical members of a frame, such as in a paneled door

Stool—the bottom horizontal member of a door or window frame; also called *sill*

Stool cap—a horizontal finish piece covering the stool or sill of a window frame on the interior; also called *stool*

Stop bead—a vertical member of the interior finish of a window against which the sash butts or sides

Storm sash—an additional sash placed on the outside of a window to create a dead air space to prevent the loss of heat from the interior in cold weather

Story—the distance between the upper surface of any floor and the upper surface of the floor above

Story pole—a narrow strip of wood used to lay out the heights of members of a wall frame or courses of siding

Straightedge—a length of wood or metal having at least one straight edge to be used for testing straight surfaces

Strapping—application of furring strips at specified spacings for the purpose of attaching wall or ceiling finish; called *stripping* in some locations

Striated—finish material with random and finely spaced grooves running with the grain

Strongback—a member placed on edge and fastened to others to help support them

Stud—vertical framing member in a wall running between plates

Tack—to fasten temporarily in place; also, a small headed nail

Tail cut—a cut on the extreme lower end of a rafter

Tail joist—short joist running from an opening to a bearing

Tangent—straight line touching the circumference of a circle at one point

Taper—becoming thinner from one end to the other

Taper ground—the inner part of a saw blade ground thinner than the outside edge

Tempered—treated in a special way to be harder and stronger

Template—a pattern or a guide for cutting or drilling

Tenon—a tongue cut on the end of a piece usually to fit into a mortise

Termite shield—metal flashing plate over the foundation to protect wood members from termites

Threshold—a piece with chamfered edges placed on the floor under a door

Tile—square or rectangular blocks placed side by side to cover an area

Timber—large pieces of lumber over 5 inches in thickness and width

Toe—the forward end of tools, such as a hand saw and hand plane

Toeboard—a recessed space at the bottom of a base cabinet that provides room for the toes of a person standing close to the cabinet

Toenail—nail driven diagonally to fasten the end of framing

To the weather—a term used to indicate the exposure of roofing or siding

Transom—small sash above a door

Tread—horizontal finish members in a staircase upon which the feet of a person ascending or descending the stairs are placed

Trestle—similar to a sawhorse used to support scaffold plank

Trimmer—members of a frame placed between headers in an opening

Truss—an engineered assembly of wood or wood and metal members used to support roofs or floors

Turnout—a type of handrail fitting

Undercut—a cut made through the thickness of stock at less than 90 degrees between the side and edges or ends

Underlayment—material placed on the subfloor to provide a smooth, even surface for the application of resilient finish floors

Unit of run—horizontal distance over which rafters travel for each foot of span

Valley—the intersection of two roof slopes at interior corners

Valley cripple jack rafter—a rafter running between two valley rafters

Valley jack rafter—a rafter running between a valley rafter and the ridge

Valley rafter—the rafter placed at the intersection of two roof slopes in interior corners

Vapor barrier—a material used to prevent the passage of vapor

Veneer—a very thin sheet or layer of wood

Vermiculite—a mineral closely related to mica with the ability to expand on heating to form a lightweight material with insulating qualities

Volute—a spiral fitting at beginning of a handrail

Wainscoting—a wall finish applied partway up the wall from the floor

Waler—horizontal or vertical members of a concrete form used to brace and stiffen the form and to which ties are fastened

Wane—bark, or lack of wood, on the edge of lumber

Warp—any deviation from straightness in a piece of lumber

Water table—finish work applied just above the foundation that projects beyond it and sheds water away from it

W-beam—a wide-flanged, I-shaped steel beam

Weatherstripping—narrow strips of thin metal or other material applied to windows and doors to prevent the infiltration of air and moisture

Web—wood or metal members connecting top and bottom chords in trusses; also, the center section of a wood or steel I-beam

Whet—the sharpening of a tool on a sharpening stone by rubbing the tool on the stone

Wind—a defect in lumber caused by a twist in the stock from one end to the other

Winder—a tread in a stairway, wider on one end than the other, which changes the direction of travel

Index